# Design and Application of Agricultural Equipment in Tillage System

# Design and Application of Agricultural Equipment in Tillage System

Editors

**Mustafa Ucgul**
**Chung-Liang Chang**

MDPI • Basel • Beijing • Wuhan • Barcelona • Belgrade • Manchester • Tokyo • Cluj • Tianjin

*Editors*
Mustafa Ucgul
Southern Cross University
Australia

Chung-Liang Chang
National Pingtung University
of Science and Technology
Taiwan

*Editorial Office*
MDPI
St. Alban-Anlage 66
4052 Basel, Switzerland

This is a reprint of articles from the Special Issue published online in the open access journal *Agriculture* (ISSN 2077-0472) (available at: https://www.mdpi.com/journal/agriculture/special_issues/Agricultural_Equipment_Tillage_System).

For citation purposes, cite each article independently as indicated on the article page online and as indicated below:

LastName, A.A.; LastName, B.B.; LastName, C.C. Article Title. *Journal Name* **Year**, *Volume Number*, Page Range.

**ISBN 978-3-0365-7294-9 (Hbk)**
**ISBN 978-3-0365-7295-6 (PDF)**

# Contents

# About the Editors

**Mustafa Ucgul**

Dr. Mustafa Ucgul has a Doctor of Philosophy degree (Mechanical Engineering) from the University of South Australia. Before joining Southern Cross University, he worked at the University of South Australia and had a long career as an agricultural engineering researcher. Mustafa has a strong interest in computational mechanics and its applications in machinery design (particularly in agricultural machinery design). With 17 years of research experience in Turkish and Australian contexts, a major component of his research has focused on the discrete element simulations of soil-engaging tools and the design of mechanical systems. His doctorate specialization was in discrete element modelling of soil–sweep tool interactions, and since then, he has undertaken research activities in research and other industry-linked projects.

**Chung-Liang Chang**

Chung-Liang Chang received the B.S. degree in Industrial Education from National Taiwan Normal University, Taipei, Taiwan, in 1998. He gained M.S. and Ph.D. degrees in Electrical Engineering from the National Cheng Kung University, Tainan, Taiwan, in 2004 and 2008, respectively. In 2009, he joined the Department of Biomechatronics Engineering at the National Pingtung University of Science and Technology, Pingtung county, Taiwan, where he is now a distinguished Professor. He has been involved in many research projects financed by the National Science and Technology Council (NSC), Council of Agriculture (CIA), etc. He served as the head of Intelligent Agriculture Production Technology Alliance at the Pingtung University of Science and Technology, creating business opportunities for intelligent agricultural technology. Recently, his work has focused on the autonomy of farming systems with agricultural robots and their impact on the growth of native crops or poultry. His main research interests include smart farming machine, autonomous bio-robot designs for agriculture, the greenhouse system, and artificial intelligence of things.

*agriculture*

MDPI

*Editorial*

# Design and Application of Agricultural Equipment in Tillage Systems

**Mustafa Ucgul [1,*] and Chung-Liang Chang [2,*]**

1 Faculty of Science and Engineering, Southern Cross University, Lismore, NSW 2084, Australia
2 Department of Biomechatronics Engineering, National Pingtung University of Science and Technology, Neipu, Pingtung 91201, Taiwan
* Correspondence: mustafa.ucgul@scu.edu.au (M.U.); chungliang@mail.npust.edu.tw (C.-L.C.)

Due to the rapid increase in world population, the demand for food has increased dramatically. Climate change, labour shortage, and aging problems cause insufficient food supply. Agricultural technologies and farming systems need to be improved to meet the increased demand. Tillage systems are essential for food production and significantly affect crop production. A significant amount of the energy (from fuels) used in crop production is also expended in tillage systems. This energy use increases greenhouse gas emissions. The energy used in tillage systems must be reduced, and the efficiency of tillage tools must be improved to achieve sustainable farming practices. This can help farmers manage their farms better, reach high productivity, save resources, and achieve high-quality food production value. Therefore, designing new tillage systems or optimising existing systems is essential. This Special Issue involves 29 papers—27 research and 2 review articles—from different countries on 6 continents and has eminent research focusing on improving, analysing, and designing tillage systems.

Tractors or related power sources are the main components of tillage systems. Increasing the efficiency of tractors or developing tractor systems can significantly help to reduce power consumption and increase the efficiency of using tillage systems. Therefore, the contributions of [1–6] in this Special Issue focused on developing tractors or their relevant components. The authors of [1,3] investigated the control features of unmanned electric tractor platforms and unmanned self-powered robot trailers, respectively, while in [2,4,5], the authors mainly focused on tractor transmissions.

No-till or minimum-till seeders are vital parts of tillage systems. The authors of [7–16] investigated several seeding systems and their components. No-till planters were researched in [7,8,12,14], while the authors of [9–13] focused on the components of tillage systems either using experimental studies or a computer modelling approach, particularly with a focus on discrete element method (DEM) modelling. Another contribution [15] is considered the only study on pot seedling transplantation in this Special Issue. A detailed review of the DEM modelling of furrow openers was also conducted in [16].

Soil-tillage equipment has a vital role in crop production. There are different tillage tools available for various purposes. In this Special Issue, the authors of Cao et al. [17] designed a residual film-picking device to collect residual films (mulch) from the soil, while an arc-shaped nail-tooth roller-type machine was tested and optimised in [18]. With a similar approach, a cotton recovery device was also designed and tested by the authors of [19]. In [20], the authors designed the spiral soil separation mechanism of compound planters using DEM, while in another contribution, the authors designed a well-cellar cavitating device [21]. The analysis of disc ploughs using DEM–MBD coupling was conducted in [22]. In the study of [23], tillage depth was verified using a machine learning approach. A general review of the application of computational intelligence methods in agricultural soil–machine interaction was also presented in another contribution [24].

**Citation:** Ucgul, M.; Chang, C.-L. Design and Application of Agricultural Equipment in Tillage Systems. *Agriculture* **2023**, *13*, 790. https://doi.org/10.3390/agriculture13040790

Received: 21 March 2023
Accepted: 28 March 2023
Published: 30 March 2023

Weed control is important for improving crop production. A mechanical control method for precise weeding based on deep learning was studied in [25]. Soil compaction significantly reduces crop yield. In this Special Issue, the effects of tyre configuration and axel load of a four-wheel tractor on soil compaction under no-till systems were investigated by the authors of [26]. Soil compaction from wheel traffic under three tillage systems was studied in another contribution [27].

Due to working conditions, tillage tools are prone to wear. In their article, the authors of Gulyarenko and Bembenek [28] proposed a method to calculate the ploughshare's durability for increasing the service life of tillage parts. In Wang et al [29], the authors constructed a harvester's dynamic model of straw throwing and analysed the factors affecting the straw throwing width to create conditions for the smooth implementation of straw returning to the field, which is particularly important for straw management during seeding.

In summary, the studies amassed in this Special Issue reflect the current status and recent development in tillage systems. The guest editors hope that this Special Issue is helpful for researchers and industry experts in this field.

**Funding:** This research received no external funding.

**Acknowledgments:** The guest editors would like to thank all authors who contributed to the Special Issue of *Agriculture* entitled "Design and Application of Agricultural Equipment in Tillage Systems" and reviewers who reviewed these papers.

**Conflicts of Interest:** The authors declare no conflict of interest.

## References

1. Chen, Y.C.; Chen, L.W.; Chang, M.Y. A design of an unmanned electric tractor platform. *Agriculture* **2022**, *1*, 112. [CrossRef]
2. Cheng, Z.; Zhou, H.; Lu, Z. A novel 10-parameter motor efficiency model based on I-SA and its comparative application of energy utilisation efficiency in different driving modes for electric tractor. *Agriculture* **2022**, *3*, 362. [CrossRef]
3. Chang, C.L.; Chen, H.W.; Chen, Y.H.; Yu, C.C. Drip-tape-following approach based on machine vision for a two-wheeled robot trailer in strip farming. *Agriculture* **2022**, *3*, 428. [CrossRef]
4. Cheng, Z.; Lu, Z. Regression-Based Correction and I-PSO-Based Optimisation of HMCVT's Speed Regulating Characteristics for Agricultural Machinery. *Agriculture* **2022**, *5*, 580. [CrossRef]
5. Cheng, Z.; Chen, Y.; Li, W.; Zhou, P.; Liu, J.; Li, L.; Chang, W.; Qian, Y. Optimization design based on I-GA and simulation test verification of 5-stage hydraulic mechanical continuously variable transmission used for tractor. *Agriculture* **2022**, *6*, 807. [CrossRef]
6. Chen, Y.; Cheng, Z.; Qian, Y. Research on Wet Clutch Switching Quality in the Shifting Stage of an Agricultural Tractor Transmission System. *Agriculture* **2022**, *8*, 1174. [CrossRef]
7. Li, Y.; Lu, C.; Li, H.; He, J.; Wang, Q.; Huang, S.; Gao, Z.; Yuan, P.; Wei, X.; Zhan, H. Design and experiment of spiral discharge anti-blocking and row-sorting device of wheat no-till planter. *Agriculture* **2022**, *4*, 68. [CrossRef]
8. Geng, Y.; Wang, X.; Zhong, X.; Zhang, X.; Chen, K.; Wei, Z.; Lu, Q.; Cheng, X.; Wei, M. Design and Optimization of a Soil-Covering Device for a Corn No-Till Planter. *Agriculture* **2022**, *8*, 1218. [CrossRef]
9. Yan, D.; Xu, T.; Yu, J.; Wang, Y.; Guan, W.; Tian, Y.; Zhang, N. Test and Simulation Analysis of the Working Process of Soybean Seeding Monomer. *Agriculture* **2022**, *9*, 1464. [CrossRef]
10. Khosravani, A.; Desbiolles, J.M.; Saunders, C.; Ucgul, M.; Fielke, J.M. Prediction of single disc seeding system forces, using a semi-analytical and discrete element method (DEM) considering Rotation Effects. *Agriculture* **2023**, *1*, 202. [CrossRef]
11. Khosravani, A.; Desbiolles, J.M.; Fielke, J.M.; Ucgul, M.; Saunders, C. Prediction of single disc seeding system forces, using a semi-analytical and discrete element method (DEM). *Agriculture* **2023**, *1*, 206. [CrossRef]
12. Li, H.; Fang, L.; Yuan, P.; Lu, W.; Yang, W. A Seedbed Clearing and Shaping Device for Dry Direct-Seeded Rice. *Agriculture* **2022**, *10*, 1740. [CrossRef]
13. Gao, Z.; Lu, C.; Li, H.; He, J.; Wang, Q.; Huang, S.; Li, Y.; Zhan, H. Measurement Method of Collision Restitution Coefficient between Corn Seed and Soil Based on the Collision Dynamics Theory of Mass Point and Fixed Surface. *Agriculture* **2022**, *10*, 1611. [CrossRef]
14. Bilgili, M.E.; Vurarak, Y.; Aybek, A. Determination of Performance of No-Till Seeder and Stubble Cutting Prototype. *Agriculture* **2023**, *2*, 289. [CrossRef]
15. Quan, W.; Wu, M.; Dai, Z.; Luo, H.; Shi, F. Design and Testing of Reverse-Rotating Soil-Taking-Type Hole-Forming Device of Pot Seedling Transplanting Machine for Rapeseed. *Agriculture* **2022**, *3*, 319. [CrossRef]

16. Aikins, K.A.; Ucgul, M.; Barr, J.B.; Awuah, E.; Antille, D.L.; Jensen, T.A.; Desbiolles, J.M. Review of Discrete Element Method Simulations of Soil Tillage and Furrow Opening. *Agriculture* **2023**, *3*, 541. [CrossRef]
17. Cao, S.; Xie, J.; Wang, H.; Yang, Y.; Zhang, Y.; Zhou, J.; Wu, S. Design and Operating Parameters Optimization of the Hook-and-Tooth Chain Rail Type Residual Film Picking Device. *Agriculture* **2022**, *10*, 1717. [CrossRef]
18. Zhang, Z.; Li, J.; Wang, X.; Zhao, Y.; Xue, S.; Su, Z. Parameters Optimization and Test of an Arc-Shaped Nail-Tooth Roller-Type Recovery Machine for Sowing Layer Residual Film. *Agriculture* **2022**, *5*, 660. [CrossRef]
19. Wang, H.; Cao, S.; Liu, Y.; Yang, Y.; Meng, X.; Ji, P. Design of Cotton Recovery Device and Operation Parameters Optimization. *Agriculture* **2022**, *9*, 1296. [CrossRef]
20. Han, L.; Yuan, W.; Yu, J.; Jin, J.; Xie, D.; Xi, X.; Zhang, Y.; Zhang, R. Simulation and experiment of spiral soil separation mechanism of compound planter based on discrete element method (DEM). *Agriculture* **2022**, *4*, 511. [CrossRef]
21. Xu, G.; Fang, H.; Song, Y.; Du, W. Optimal Design and Analysis of Cavitating Law for Well-Cellar Cavitating Mechanism Based on MBD-DEM Bidirectional Coupling Model. *Agriculture* **2023**, *1*, 142. [CrossRef]
22. Ucgul, M. Simulating Soil–Disc Plough Interaction Using Discrete Element Method–Multi-Body Dynamic Coupling. *Agriculture* **2023**, *2*, 305. [CrossRef]
23. Pang, J.; Zhang, X.; Lin, X.; Liu, J.; Du, X.; Han, J. Tillage-Depth Verification Based on Machine Learning Algorithms. *Agriculture* **2023**, *1*, 130. [CrossRef]
24. Badgujar, C.; Das, S.; Figueroa, D.M.; Flippo, D. Application of computational intelligence methods in agricultural soil–machine interaction: A review. *Agriculture* **2023**, *2*, 357. [CrossRef]
25. Chang, C.L.; Xie, B.X.; Chung, S.C. Mechanical control with a deep learning method for precise weeding on a farm. *Agriculture* **2021**, *11*, 1049. [CrossRef]
26. Rivero, D.; Botta, G.F.; Antille, D.L.; Ezquerra-Canalejo, A.; Bienvenido, F.; Ucgul, M. Tyre Configuration and Axle Load of Front-Wheel Assist and Four-Wheel Drive Tractors Effects on Soil Compaction and Rolling Resistance under No-Tillage. *Agriculture* **2022**, *11*, 1961. [CrossRef]
27. Acquah, K.; Chen, Y. Soil compaction from wheel traffic under three tillage systems. *Agriculture* **2022**, *2*, 219. [CrossRef]
28. Gulyarenko, A.; Bembenek, M. The Method of Calculating Ploughshares Durability in Agricultural Machines Verified on Plasma-Hardened Parts. *Agriculture* **2022**, *6*, 841. [CrossRef]
29. Wang, J.; Wang, X.; Li, H.; Lu, C.; He, J.; Wang, Q.; Liu, D.; Deng, B.; Zhang, M. Improvement of Straw Throwing Performance of Harvester Based on Matching Header Width. *Agriculture* **2022**, *9*, 1291. [CrossRef]

*agriculture*

MDPI

*Article*

# Mechanical Control with a Deep Learning Method for Precise Weeding on a Farm

**Chung-Liang Chang *, Bo-Xuan Xie and Sheng-Cheng Chung**

Department of Biomechatronics Engineering, National Pingtung University of Science and Technology, Neipu 91201, Taiwan; m10944002@mail.npust.edu.tw (B.-X.X.); m10844005@mail.npust.edu.tw (S.-C.C.)
* Correspondence: chungliang@mail.npust.edu.tw; Tel.: +886-8-7703202 (ext. 7586)

**Abstract:** This paper presents a mechanical control method for precise weeding based on deep learning. Deep convolutional neural network was used to identify and locate weeds. A special modular weeder was designed, which can be installed on the rear of a mobile platform. An inverted pyramid-shaped weeding tool equipped in the modular weeder can shovel out weeds without being contaminated by soil. The weed detection and control method was implemented on an embedded system with a high-speed graphics processing unit and integrated with the weeder. The experimental results showed that even if the speed of the mobile platform reaches 20 cm/s, the weeds can still be accurately detected and the position of the weeds can be located by the system. Moreover, the weeding mechanism can successfully shovel out the roots of the weeds. The proposed weeder has been tested in the field, and its performance and weed coverage have been verified to be precise for weeding.

**Keywords:** deep learning; machine vision; weeder; smart agriculture; mechanical control

**Citation:** Chang, C.-L.; Xie, B.-X.; Chung, S.-C. Mechanical Control with a Deep Learning Method for Precise Weeding on a Farm. *Agriculture* **2021**, *11*, 1049. https://doi.org/10.3390/agriculture11111049

Academic Editor: Yanbo Huang

Received: 25 September 2021
Accepted: 23 October 2021
Published: 26 October 2021



## 1. Introduction

The type of crop production and management has been toward knowledge- and automation-intensive practices, which use automated machine, information communication technology, and biotechnology for large-scale production, which can be combined with precision agriculture technique to increase productivity, reduce resource waste and production costs, and improve environmental quality [1–3]. Among them, weed management is regarded as one of the most challenging tasks in crop production. Effective weed control can increase the productivity per unit area to meet the growing demand for crop production [4]. Improper weed management can lead to a potential loss of approximately 32%, which is increasing every year [5]. If weeds are not effectively controlled, most of the fertilizer nutrients applied to the crop are absorbed by the weeds, resulting in 60% reduction in crop yield in organic farming [6].

Since weeds exhibit uneven spatial distribution [7], however, the traditional weed management method is that herbicides are usually applied uniformly across the field. Most herbicides are released into the environment through runoff and drift, which have an impact on the ecological environment and human health [8]. Hand-weeding is a common weed management practice, but it is time-consuming, high cost, labor-intensive, and more difficult due to labor shortage in the agriculture. This practice may also expose farmers to the risk of infected weeds. Some countries have even abandoned this practice [9,10]. Fortunately, some smart agricultural machines have been investigated recently, which use physical or chemical methods to solve the issue of weed management [11–17]. It can be expected that machines will replace humans or assist operators to achieve the purpose of smart production management [18].

The type of weeding machine can be divided into passive and active based on whether there is a power source [19]. Among them, active weeding can realize the behavior of avoiding seedlings and simultaneously weeding. Its weeding behavior can be divided into swing, rotation, hybrid, etc. [18,20–23]. Among them, swing behaviors are mainly powered by ground-driven system to drive the hoe to reciprocate. The rotary type is divided into vertical axis rotation and horizontal axis rotation according to the position of the rotation axis. There are notched hoe knives, claw tooth cycloidal hoe knives, etc., which rotate around the vertical axis of the transmission mechanism of the machine. Hybrid is a combination of swing and rotating, and its motion behavior has a high degree of spatial freedom. However, the design challenge is how to optimize the transmission mechanism and reduce the number of components. The weeding machines are usually mounted behind the tractor. As the tractor moves, the weeding machine will continue to shovel the soil to remove weeds. In fact, the implementation of full-cover mechanized shoveling operations will affect the organic matter content in the soil, which in turn affects the nutrient absorption effect of the crop roots.

Generally, a detector is installed on an automated weeding machine, which is expected to be used to detect whether there is crop in the interrow. At the same time, the actuator can control the knives or hoe knives under the soil. Based on the detection results, the actuator can move the knives into or out of the rows to fork over the soil, so as to remove weeds [9,24]. Since the performance of the end effector (actuator) of weeding machinery directly affects the efficiency of weeding. This kind of variable rate technology is rarely used in actual operation and the cost is also a key factor that needs to be considered [25]. In order to achieve the purpose of precise weeding, some weeding machine combine computer vision technology with a mobile robot capable of autonomous navigation [26]. The mobile manipulator must be able to accurately locate the weeds in real time. At the same time, the weeding tools must cooperate with the actuator to operate the weeding tools at the right time to remove the weed.

In previous study, a machine vision-based smart weeder is peoposed that uses image processing methods to identify crops and weeds, and uses an inference-based control method to drive three direct current (DC) motors, which are driven by gears and chains [27]. The three-claw harrows weeding tool on the connecting rod is inserted into the soil, and then the soil is moved backward to remove the weeds. However, due to the type of claw harrow and the torque limitation of the actuator, this machine is only suitable for soft soil and small weed removal. McCool et al. [28] described mechanical methods as an alternative to weed management. They proposed different types of weeding tools, including arrow- and tine-shaped, which can be mounted on a guided vehicle to perform weeding operations. Statistical analysis proves the effectiveness of these tools and emphasizes the importance of early intervention. Other types of weeding tools, such as intrarow plowshares, comb harrows, spring harrows, and specific plowshares for in-row weeding, are also used for weeding operations [29]. Fennimore and Cutulle [30] developed and implemented machine vision technology in an autonomous weeder. Two robotic arms cooperate with weed actuators to spray herbicides directly on each weed. Raja et al. [31] proposed a weeding system based on a 3D geometry detection algorithm of robot vision. A corresponding mechanical weeding device was also designed, used for automated weeding in tomato and lettuce fields, which can efficiently perform weed removal in a high-density environment. Kumar et al. [32] proposed an mechatronics prototype for interrow weeding and crop damage control, which initiates weeding operations through plant sensing, soil, and plantation parameters. The developed method combines the different conditions of soil, forward speed, and plant spacing to calculate the dynamic lateral movement speed. However, it is still easy to be affected by vibration or other uncontrolled movements during image processing in practical applications, resulting in blurry images, which impact the recognition and positioning performance. Meanwhile, this mechanism is complicated and lacks modular design.

The implementation of machine vision technology for weeding tasks first needs to use image processing methods to extract features such as the color, texture, and shape of the image, and then combine them with machine learning algorithms such as clustering or classification to detect and classify weeds [33–36]. Among them, the shape or feature extraction based on the support vector machine is the most commonly used to distinguish crops and weeds [37–39]. After this, it is necessary to determine the feature of the target object and use some morphology or color space conversion methods to extract the feature and position of the weeds [40,41]. Due to the use of a machine vision system to detect and locate weeds, its system performance is limited by the uncertainty of the environment, including light conditions and color variance of leaves or soil, which also results in a decrease in the performance of weed control. There are currently some weed detection technologies that integrate images taken from multiple perspective sources and multiple feature marks to improve the accuracy of weed recognition and location [42,43]. Because of its complex system design, time-consuming and maintenance costs need to be considered. Other methods include the use of controlled light emitting diode (LED) lighting equipment in the dark box and the use of camera-lighting module to record the reflection spectrum of the object. The system combines the size information of the desired object to distinguish crops, weeds and soil in horticultural crops, which can locate weeds [44]. Currently, this method has not integrated weeding equipment to implement precise weeding operations.

With the improvement of computer computing performance and the increase in the number of available images, deep learning has been able to provide enhanced data expression capabilities for target objects in images. These methods can be used to extract multiscale and multidimensional spatial semantic feature information of objects [5,45–47].

In many cases, the detection and classification results obtained using convolutional neural network method are better than the classification results produced by using machine learning commonly in the early stages [48–55]. However, deep learning needs to rely on a large number of data sets for training, it is not easy to collect crop and weed images [56]. Redmon et al. [57] proposed a fast target detection algorithm called YOLO, which can quickly implement real-time applications. This method is based on the Darknet-53 network architecture and has been modified many times to greatly improve the accuracy of target identification with only a small amount of data samples.

This study proposes a weed identification technology and weeding tool control method based on the YOLOv3 model [58], and implements it in an innovative weeding mechanism. In the early study, an artificial intelligent-enabled shovel weeder is designed and implemented [59]. Nevertheless, the weeder was only tested in a simulated field and its weeding performance is limited by the torque of actuator and unstable transmission mechanism, which requires further design and testing. The earlier designed mechanism was modified and re-made and assembled. The modular weeding tool is attached to an unpowered machine. The motion behavior of the weeding tool is a combination of swing and rotating. The design concept of the transmission mechanism of the weeding machine is derived from the power transmission of a bicycle. An inverted triangle weeding knife is designed. The weeding machine is equipped with a camera module, which can be used to obtain top-in-view images in real time. This weeding tool is used to test and evaluate the effectiveness of deep learning methods. After that, the weeding machine was used in the field to actually test the weeding performance of the method in the presence and absence of crops. The results of different types of knives for weed removal will also be analyzed and compared.When the trailer is moving, the proposed weeding machine can automatically remove weeds in the farmland.

The purpose of this study is as follows: First, a weeder is implemented and can be used to replace manual weeding. Second, the use of deep learning methods to achieve precise removal of individual weeds to improve the existing mechanized weeding. Third, modularize the weeder. Multiple modules can be attached to the back of the vehicle to solve the problem of difficult disassembly and spacing adjustment of large weeders. Fourth, the proposed weeder simultaneously weeds and shovels soil, which can reduce the probability of weed growth. The proposed weeder is particularly suitable for homeworkers and farmers who want to carry out organic cultivation for weeding operations in small fields.

The chapters of this paper are organized as follows: The design method for the weeding machine and the mobile platform, including the design of the weeding mechanism and transmission mode, the software and hardware construction of the weeding system, and the performance evaluation matrices are described in Section 2. The flow of the weed detection program is also explained in this chapter. Section 3 explains how to test the performance of weeding machine, including evaluating the performance of weed detection and testing the weeding efficiency. The last chapter summarizes the characteristics and applications of the weeding methods proposed in this paper and explains future work.

## 2. Materials and Methods

The weeding machine developed in this research will can be attached to a simple four-wheeled trailer with no power source to perform weeding operations. The battery supplies power to the machine. The appearance of the entire mechanism is shown in Figure 1. At most, two sets of weeding machines are attached to the vehicle, which are respectively mounted on the left and right sides of the vehicle. On the right is the advanced intelligent weeding machine (Weeder #1) equipped with an inverted triangle weeding tool. On the left is the first-generation weeding machine (Weeder #2), which is equipped with a claw rake-type weeding tool [59].

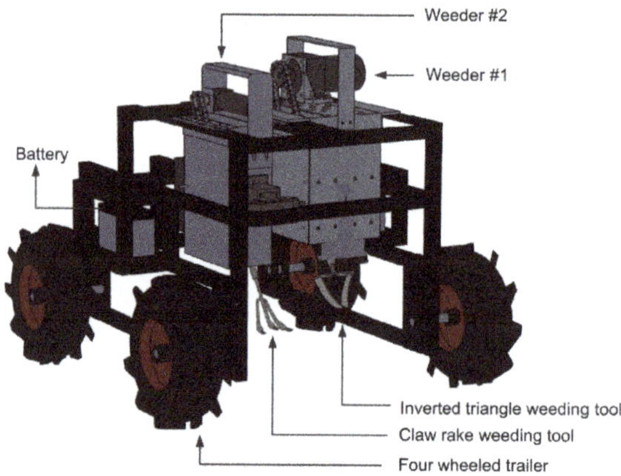


**Figure 1.** Smart weeding machine and its four-wheel trailer.

Based on the YOLOv3 network, the deep learning model is used as weed detection, the network model is trained by multiple feature objects, and the trained network model and the weeding tool control algorithm are integrated and implemented in the embedded system. Through the execution of the program, the weeding tool can swing up and down and back and forth for weeding operations. The following describes the design of the drive mechanism of the weeding machine and the weed detection and control system, including the weed recognition algorithm and the hardware construction, and the software program flow is also described in detail in this chapter.

### 2.1. Mechanism Design

The design and development of weeding equipment must take into account the various agronomical requirements of crops, soil conditions, and weed characteristics for field management operations. For example, the appearance of the field includes different field heights, widths, and densities of cultivated crops. In addition, the height of crops, root length, leaf branch and soil type, water content, bulk density, and strength of the soil also need to be considered. The mechanical design of weeding tools needs to be simplified, so that farmers or craftsmen can repair them quickly and have low maintenance costs. Therefore, based on the above ideas, a DC-driven weeding machine was developed. Its components include a DC motor (model: SWG-24-1800, Xajong), a transmission mechanism, a height-adjustable weeding handle, and a protective case (see Figure 2).

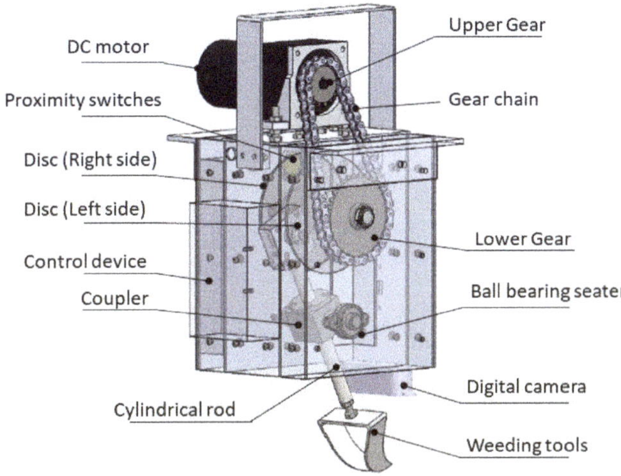


**Figure 2.** A drawing prototype of the weeding mechanism.

The transmission mechanism consists of an upper sprocket (Model: RS35-B-16, New Sheylee CO., Ltd., Taichung City, Taiwan), a lower sprocket (Model: RS35-B-32, New Sheylee CO., Ltd., Taichung City, Taiwan), a drive chain (Model: RS35, Prelead Industrial CO., Ltd., Taiwan), left and right discs, a coupler, a ball bearing seat, and a cylindrical rod (16 mm × 200 mm (diameter [D] × length [L])). The size of the case is 216 mm × 180 mm × 278 mm (L × width (W) × height (H)), and the weight of the whole machine is 6 kg. In terms of tool design, the appearance of traditional weeding tools is mostly designed to imitate the blade geometry. Different types of soil require the use of different shaped cutters to shovel the soil [60]. This type of tool set is installed on a rotating mechanism, which can make the vertical cutting surface of the blade move downward through the rotating torsion force to achieve the purpose of shoveling the soil. However, this tool is suitable for use in fields with a low cultivation density. In contrast, weeding tools, such as the disc, round head, and sawtooth types, are more suitable for use in fields with higher planting density and can effectively treat weeds on the surface of the soil. In addition, the rake-type cutter can be used to dig out weeds with shallow roots [27,59], but the material of this cutter is more likely to stick to the soil.

Therefore, a new type of tool was designed, the material of which was aluminum alloy. The shape of the weeding tool is an inverted triangle (90 mm × 47 mm × 80 mm (L × W × H)) with a sharp end, which is suitable for hard soil. In addition, the bottom of the cutter is wider, which can cover the size of a single weed and shovel out the roots of the weeds. A combination of multiple iron plates is used as the mechanism case. The upper part of the front and rear sides is locked with a pull handle, and a proximity switch is installed inside the upper part of the iron plate (Model: TG1-X3010E1, Prosensor Phototech Co., Ltd., Taoyuan city, Taiwan), which is used to stop the motor. The digital lens (Model: Logitech BRIO, Logitech International S.A., Lausanne, Switzerland) is installed under the case.

The control box is installed on the back side of the case, and it contains an embedded control board (Model: Jeston Nano, NVIDIA Company, Santa Clara, CA, USA) and peripheral circuit boards. A DC 24V lead-acid battery (Model: GP1272 F2, CSB Energy Technology Co., Ltd., Taipei city, Taiwan) is the power source for the entire system. The specifications of the weeding system are shown in Table 1.

**Table 1.** The specifications of the weeding machine.

| Description | Value or Other Details |
|---|---|
| Mechanism body | |
| Size (L × W × H) | 216 mm × 180 mm × 278 mm |
| Weight | 6 kg |
| Weeding body | |
| Upper sprocket (Number of teeth (T) × outer diameter (Ø)) | 16 mm × 54 mm |
| Lower sprocket (T × Ø) | 32 mm × 102 mm |
| Roller chain (length; tension) | RS35-1; 1150 kgf |
| Cylindrical rod (D × L) | 16 × 200 mm |
| Weeding tools (L × W × H) | 90 mm × 47 mm × 80 mm |
| Disc (D × W) | 140 mm × 3 mm |
| Electronics components | |
| Main control board (speed; memory) | 1.43 GHz; 4 GB 64-bit LPDDR4 |
| DC motor (voltage; gear ratio; torque; speed) | 24 V; 1:15; 26.7 kg/cm; 120 rpm |
| Proximity switches (voltage; distance; output) | 24 V; 10 mm; normal open |
| Digital camera (resolution; focus type) | 4 K Ultra HD; auto focus |
| Battery (voltage, capacity, weight) | DC 24 V, 7.2 Ah, 2.4 Kg |

Considering the lowest transmission loss, the double-gear chain transmission mechanism was designed. This design concept was derived from the mechanical transmission principle of the bicycle. The transmission component adopts a sprocket, which is made of medium carbon steel.

First, the DC motor rotates to drive the upper gear, and the chain of the upper gear drives the lower gear. The lower gear is fixed in the case on the left side, and is connected to the left and right disks by a coaxial connector. Close to the center point of the two discs, a square seat is locked, and a cylindrical rod is installed in it, which is inserted into the square coupler and is connected to a ball bearing seat inside the casing. There are holes in different positions on the end of the cylindrical rod, and the user can select a suitable hole position and lock the weeding tool on the cylindrical rod to adjust the distance between the weeding tool and the ground. When the motor rotates, the weeding handle has a reciprocating swinging behavior (Figure 3). This operation mode is like a farmer holding a hoe for weeding. The sequence of this motion involves extending the weeding tools, digging down, turning up the roots of the weeds, throwing away the weeds, and retracting the weeding tools.

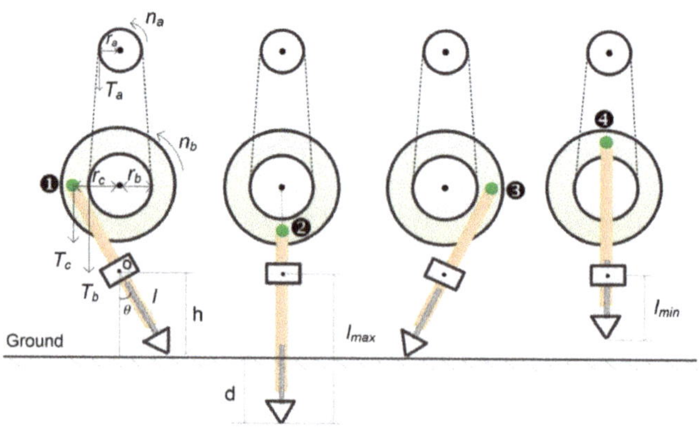


**Figure 3.** Transmission of the weeding mechanism.

Assume that the torque, speed, and radius of the upper sprocket are $T_a$, $n_a$, and $r_a$, respectively. The chain connects the upper and lower gears. Without considering the transmission and mechanical friction, the torque $T_b$ and speed $n_b$ of the lower gear are:

$$T_b = \frac{G_b}{G_a} \times T_a \tag{1}$$

$$n_b = \frac{G_a}{G_b} \times n_a \tag{2}$$

Among them, $G_a$ and $G_b$ represent the ratio of the number of teeth of the upper gear to the lower gear, respectively. Since the lower gear and the two discs are on the same axis, the disc rotation speed $n_c = n_b$, the tangential torque of the fixed point of the cylindrical rod on the disc $T_c$ is:

$$T_c = \frac{r_b}{r_c} \times T_b \tag{3}$$

where $r_b$ represents the radius of the lower gear and $r_c$ is the distance between the center of the disc and the center of the square seat. Assuming that point o is a fixed point, the distance from point $o$ to the ground is defined as h and the depth of weeding as D. When the center point of the square seat is at the positions ❶, ❷, ❸, and ❹ in Figure 3, the length $l$ from point $p$ to the end of the cylindrical rod can be defined as:

$$l = \text{hcos}^{-1}\theta \tag{4}$$

where $\theta$ depicts the angle of weeding. When $\theta = 0^{\circ}$ (position ❷), the length reaches the maximum value $l_{max}$:

$$l_{max} = \text{h} + \text{d} \tag{5}$$

When the center point of the square seat is at position 4 (origin position), $l$ has a minimum length $l_{min}$.

During the weeding process, a digital camera takes an image of the planted area, and a YOLOv3-based deep learning method is used to detect and locate weeds (see Figure 4a). Suppose v depicts the moving speed of the vehicle and s represents the operation range of weeder between the center point p ($x_p$, $y_p$) of the weed detection frame and the point q ($x_q$, $y_q$) below the weed cutter, as shown in Figure 4b. The orange frame represents the detection results. The green arrow indicates the heading of the trailer and the dashed box indicates ground truth. The light gray area represents the weeding range, w is the width of the weeding, and the white color line represents the upper and lower boundary of the weeding range. Once two weeds are detected and appear in the gray area, the object with the largest frame area is selected. In addition, the size of weeds that are too small are ignored because they have little effect on the growth of the crop. When the trailer moves for t = s/v seconds, once the weeding system detects weeds, the system must activate the weeding tool within t seconds to remove the weeds.

### 2.2. *System Description*

#### 2.2.1. Hardware

The sensing and control circuit components in the weeding system include a main control board, relays (JQC-3FF-S-Z, Tongling), DC motors, digital cameras, DC/DC conversion modules (model: XL4005, XLSEMI company, Shanghai, China), proximity switches, and automatic voltage regulators (AVRs). The circuit system architecture is depicted in Figure 5a. The function of the main control unit is to execute weed detection algorithms and motor drive and control decisions. The main control board can receive the images taken by the digital camera via the Universal serial bus (USB) port and store them in the memory. Two sets of relays are connected to the general-purpose input/output (GPIO) port of the main control board, which can receive the driving signal output from the main control board to start and stop the motor.

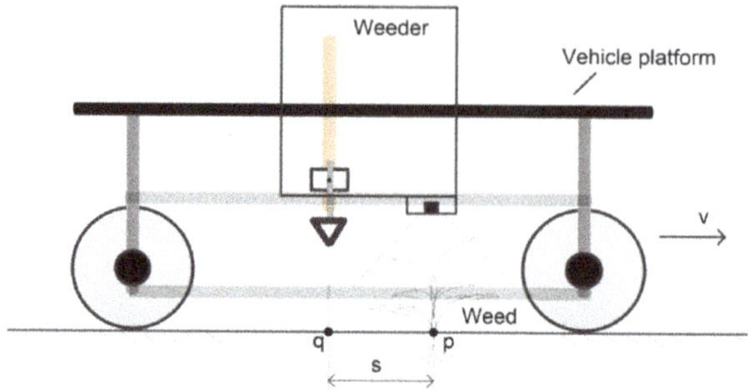


(a)

(b)

**Figure 4.** Operation concept of the weeding process of the weeder. (**a**) The relationship between the two-dimensional coordinate points of the weeding tool, the camera and the weeds; (**b**) Frame selection of weed objects in the snapshot image and description of the range of weeding.

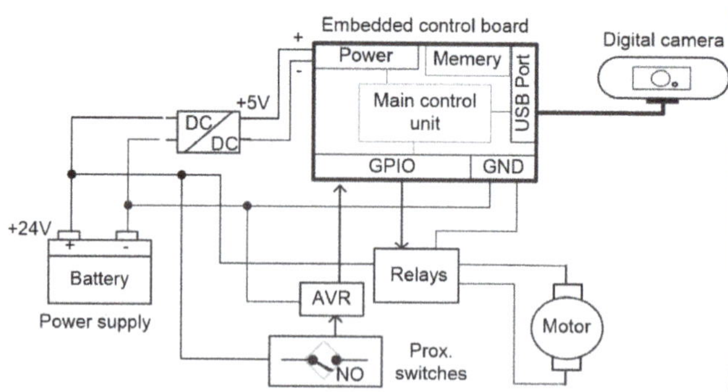


(a)

(b)

**Figure 5.** Sensing and circuit architecture: (**a**) Block diagram of electronic circuit and (**b**) peripheral electronic component board (upper layer) and main control board (lower layer) in the control box.

The proximity switch (type: normal open (NO)) is used to detect whether the square seat in the weeding mechanism has returned to the original position, and the detection signal is then input into the main control unit through the GPIO interface. The 24 V battery provides power for circuit components, including motors and proximity switches. The negative output terminal "−" of the battery is connected to the ground (GND) terminal of the circuit board. The DC/DC module is used to convert 24 V to 5 V for the embedded control board; these components and the control board are integrated in a waterproof control box, as shown in Figure 5b. The upper layer is a circuit board, which mainly integrates DC/DC conversion modules, relays, and other electronic components, and the lower layer is for placing an embedded control board.

#### 2.2.2. Software

The YOLOv3 tool [57] is a common deep learning model used to quickly detect objects. It is executed in the Darknet environment. Residual neural network (RestNet) [61] and feature pyramid networks (FPN) are its main architectures, which can improve the prediction ability of small objects. This network tool is used to detect weed objects. A desktop computer with a high-speed computing processor (Model: Intel i5-8400, Intel Co., Santa Clara, CA, USA) is paired with a high-speed graphics processing unit (GPU) (Model: GTX 1070, Nvidia Co., Santa Clara, CA, USA) to train the YOLOv3 network model. The training model of YOLOv3 is configured as follows: Batch size set to 64, image size resized to 416 $\times$ 426 pixels, subdivision of 32, momentum of 0.9, decay of 0.0005, learning rate of 0.001, batch size of 64, etc. After that, image preprocessing is performed, including image cropping, white balance, and noise filtering processing, which is then marked by trained technicians and used for model training and evaluation. Among them, 80% of the images are used for training and 20% are used for testing. The bounding box of the region of interest is drawn and exported to YOLO format for model development.

During training, the training loss of each epoch is recorded to evaluate the performance of the visualization model in real time. Once the loss is stable and there is no significant change, the training process stops, and the corresponding weights of the model are saved for further evaluation of the weed detection performance. The trained YOLOv3 model integrates the weeding control program and is embedded in the weeding system. Figure 6 shows the program execution flow, which is written in python language. First, the function library is imported, including the external function (ctype.cdll), multi-threading module, and open source computer vision library (cv2). Then the GPIO pins, data type, class, structure, and subfunctions are defined. The next step is to set, import, and load the environmental variables of Darknet; it also includes defining the frame selection parameters and their storage file paths.

The program is executed to perform a while loop, the image is read and converted from the blue (B)–green (G)–red (R) color layer to the RGB color layer, and then weed detection operation is performed. Once the weed object is detected, the value "1" is written to the text file. Otherwise, the value "0" is written to the text file. The detection results, including bounding box and labels, are displayed in the image (see Figure 6a). In the process of program execution, the multi-threaded module is activated and the motor control program is executed synchronously (Figure 6b). In the while loop, the text file value is open and read. When the value is 1, the system outputs a signal to start the motor, otherwise it stops the motor. A function Delay() with a delay time is inserted into the program for starting and stopping the motor.

### *2.3. Performance Evaluation Metrics*

The performance indicators for detecting weeds will be defined in this section, including the precision, recall, and F1-score, as well as descriptions of the efficiency of weeding and the rate of plant damage.

2.3.1. Weed Detection

The detection performance metrics used to evaluate YOLOv3 include the precision, recall, and F1 score [61]. The accuracy index is as per Equation (6):

$$\delta_P = \frac{TP}{TP + FP} \tag{6}$$

*TP* (true positive solution) represents a true positive test result is one that detects the condition when the condition is presented; in contrast, *FP* (false positive solution) is the opposite result.

Ideally, the *FP* should be as small as possible in order to ensure the accuracy of the network in identifying each object. The intersection-over-union (IoU) is a method to define whether the detected object is a positive solution, as shown in Equation (7):

$$u = \frac{U_d \cap U_y}{U_d \cup U_y} \tag{7}$$

where $U_d$ and $U_y$ indicate the ground truth and predict boxes of the deep neural network, respectively, and the symbols "∩" and "∪" depict the intersection and union operator, respectively. If $u$ is larger than the threshold $u_T$, the prediction result is regarded as a *TP*; otherwise, it is regarded as an *FP*.

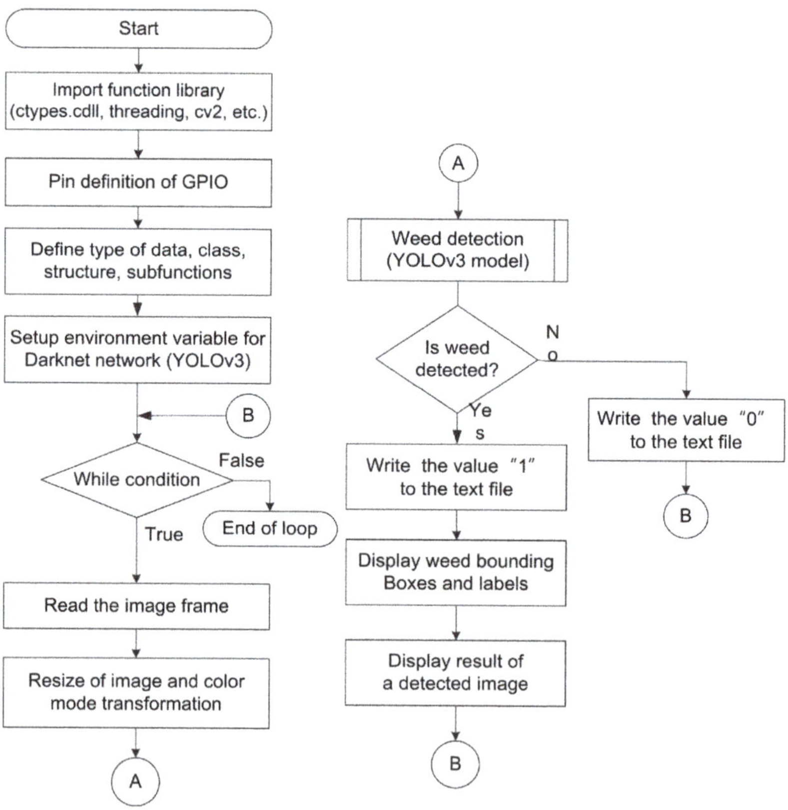


(a)

**Figure 6.** *Cont.*

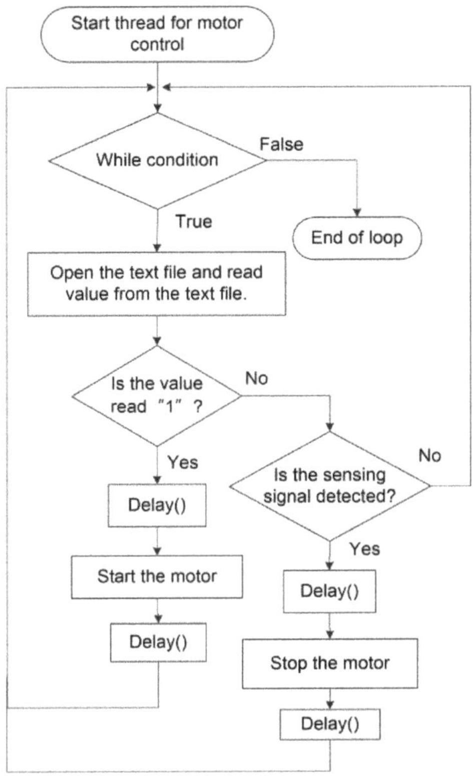


(**b**)

**Figure 6.** Software program flow for weeding system. (**a**) program flow for weed detection; (**b**) program flow for weeding operation.

The recall rate is a metric that quantifies the number of correct positive predictions made from all possible positive predictions, and its definition is shown in Equation (8).

$$\delta_R = \frac{TP}{TP + FN} \tag{8}$$

where *FN* depicts the false negative test result. The sum of *TP* and *FN* in Equation (8) is just the number of ground-truths, so there is no need to compute the number of *FN*. The F1-score ($\delta_f$) is a weighted average of the precision and recall which is performed as a trade-off between $\delta_R$ and $\delta_P$ to demonstrate the comprehensive performance of the trained models.

$$\delta_f = \frac{2\delta_P\delta_R}{\delta_P + \delta_R} \tag{9}$$

The values of $\delta_f$ range from 0 to 1, where 1 means the highest accuracy. Through the $u_T$ setting for the confidence score at various recall levels, different pairs of precision and recall are generated with recall on the *x*-axis and precision on the *y*-axis, which can be drawn as a precision–recall (PR) curve, indicating their association and can be employed to measure the performance of the weed detection.

2.3.2. Weeding Efficiency

We conducted field tests in the field to evaluate the performance of the weeding machine for weeding operations. The evaluation metrics include weeding efficiency and plant damage, which are shown in Equqtions (10) and (11):

$$\eta = (W - \overline{W})/W \tag{10}$$

$$D = \overline{d}/d \tag{11}$$

Among them, $W$ and $\overline{W}$ represent the number of weeds before and after weeding, respectively, and $\overline{d}$ and $d$ represent the damaged crop and the total amount of crops, respectively.

## 3. Experimental Results

This section explains the data collection and model training methods. In addition, two test scenarios were used to evaluate weed detection performance and weeding efficiency

### *3.1. Data Collection and Model Training*

Images were collected in the field under different climates and time periods. A digital camera was used to take a total of 140 images of weeds in the experimental field. Image processing technology, including geometric transformation (resize, crop, rotate, horizontal flip, etc.) and intensity transformation (such as contrast and brightness enhancement, color and noise adjustment), was used to modify the original image, thereby increasing the number of image samples, which totaled 60.

Then, the image size was adjusted from 1920 × 1080 to 416 × 416 pixels to fit the YOLOv3 model network, and then, each weed in each image was marked with an object box for model training. There were 160 images in the training set, 30 images in the validation set, and 10 images in the test set. When the number of iterations reached 20,000 times and the loss function approached 0.135, the training was stopped and the weight value of the network was obtained. Finally, the trained model was used to evaluate the performance of weed detection.

### *3.2. Experimental Test*

The experiment site is located in front of the Department of Biomechanical Engineering of National Pingtung University of Science and Technology (longitude: 120.6059°; latitude: 22.6467°). The experiment period was from 5 August to 15 September 2021. Vegetable crops were grown for 20 days on the cropland ridges. The length of each cropland ridge in the field was approximately 20 m and the width was 25 cm. The spacing between each plant was 50 cm. The number and location of the weeds within the cropland were recorded in advance. These data were used for a comparison with the experimental results. In addition, we set up a hoist machine at the end of the field, and hooked the trailer with a steel shackle. The user was able to adjust the speed of the hoist machine to maintain the forward speed of the trailer.

Two experiments were used to verify the performance of the weeding system. Experiment 1 was mainly to test the weed removal performance of the weeding machine on both sides of the crop. Two weeding machines were used. Among them, the weeder machine (Weeder #1) was mounted on the right side of the vehicle, and the first-generation weeder machine (Weeder #2) was mounted on the other side. An inverted triangle-shaped weeding tool was installed on the right machine, and a claw-shaped weeding tool was installed on the left machine. Experiment 2 was mainly to test the weeding performance of the weeder (Weeder #1) proposed in this study in the intrarow of crops. Weeder #2 was mounted at the center of the rear of the trailer.

The test scenarios of Experiments 1 and 2 are shown in Figure 7. The dashed border represents the area of weed detection. The mechanical design parameters and specifications of the modified weeder (Weeder #1) based on previous research results [60] are demonstrated in Table 2. When the weeding tool was at the origin of the mechanism, the distance between the coupler in the mechanism and the surface of the ground was h = 16.9 cm. When the weeding tool was activated, the excavation depth for the weeding tool was d = 3 cm. The maximum and minimum lengths of the cylindrical rod were $lmax$ = 26 cm and $lmin$ = 15 cm, respectively.

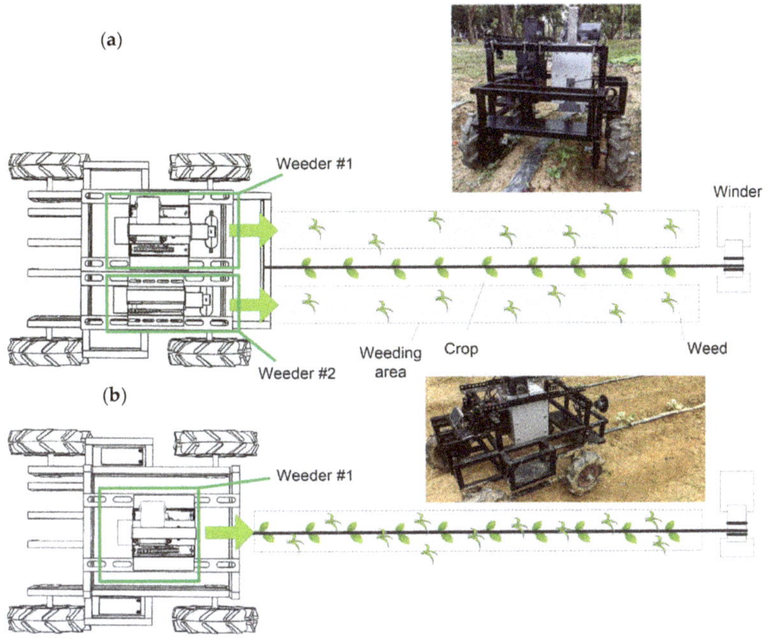


**Figure 7.** Illustration of two scenarios for testing the performance of weeding: (**a**) Using two weeders (Weeder #1 and Weeder #2) to weed the areas on both sides of the cropland ridges (gray areas); (**b**) using a weeding machine (Weeder #1) for intrarow weeding (the area within the dashed frame).

**Table 2.** The parameters and specifications of the modified weeder (Weeder #1).

| Parameters | Value | Parameters | Value |
|---|---|---|---|
| $n_a$ | 120 rpm | $l$ | 185 mm |
| $n_b$ | 60 rpm | $l_{min}$ | 150 mm |
| $G_a$ | 16 | $l_{max}$ | 260 mm |
| $G_b$ | 32 | h | 169 mm |
| $T_a$ | 27 Kg-cm | d | 30 mm |
| $T_b$ | 54 Kg-cm | $\theta$ | 24 degree |
| $T_c$ | 50 Kg-cm | $r_b$ | 51 mm |
| $r_a$ | 27 mm | $r_c$ | 55 mm |

When the weeding operation was completed, manually the number of weeds that had not been removed and the number of damaged crops on the cropland ridges were recorded. Weeds that are too small are ignored. When the roots of the weeds were exposed to the soil surface, it was considered that the weeds had been successfully removed.

### 3.3. Results and Discussion

#### 3.3.1. Performance of Weed Detection Using the YOLOv3 Model

The trained YOLOv3 model was verified to detect weeds in different climatic conditions. During the experiment, the climatic conditions were cloudy in the morning and at noon, cloudy in the afternoon, and cloudy in the afternoon. When the vehicle was moving, the weeding tool was not activated. Only the digital camera under the weeding tool was used to shoot the image on the cropland, and the image samples were taken by a digital camera every 2 h. The image samples were stored in the memory card. The number of weed objects is counted in each image that were framed (or unframed), and Equations (6), (8) and (9) were finally used to evaluate detection performance of the model.

Table 3 shows the results of weed detection using the YOLOv3 model in different time periods. The results show that the F1 score was between 74.3% and 92.8%, especially during the period from 10:00 to 13:00, where the accuracy was up to 95.6% and the F1-score value was also the highest. It is worth noting that due to the low light intensity during 18:00–19:00, the accuracy rate and recall rate are reduced.

**Table 3.** Using deep learning models to detect weeds during the daytime.

| Description | | Evaluation Metrics | | |
|---|---|---|---|---|
| Weather | Time | Precision | Recall | F1-score |
| Cloudy and sunny | 08:00–09:00 | 0.902 | 0.829 | 0.864 |
| | 10:00–11:00 | 0.956 | 0.901 | 0.928 |
| | 12:00–13:00 | 0.936 | 0.885 | 0.910 |
| | 14:00–15:00 | 0.918 | 0.854 | 0.885 |
| Cloudy | 16:00–17:00 | 0.903 | 0.833 | 0.867 |
| | 18:00–19:00 | 0.832 | 0.701 | 0.761 |

Figure 8 shows the weed detection results of each time interval, where the green frame represents the area where weeds are detected. It can be seen from these figures that most of the weed objects were framed, and only a few weeds were not framed between 18:00 and 19:00.

Then, weed detection experiments were carried out on different days, and the climatic conditions during the detection process were variable, including cloudy, sunny, and rainy. Figure 9 shows average detection performance results obtained at different time intervals in the same field using the YOLOv3 network model. The evaluation metrics at different time intervals include precision, recall, and F1-score, each representing a ten day average value.

#### 3.3.2. Performance of Weeder

The experimental weeder tests was conducted from 10:00 to 12:00, and the weather conditions were sunny. Due to the limited area of the site, two experiments were carried out in a single day and repeated on three different days. Finally, the data obtained from the three times were averaged. Figure 10 shows the actuation behavior of the weeding tool. In Figure 10a, "❶" and "❷" in the white frame indicate the visible range of the camera on the left and right weeding tools. The orange line indicates the position of the weeding tool, which is the origin of the mechanism. When the vehicle was moving, once the weeds had been detected, the weeder was activated (the weeding tool on the right side of Figure 10b). In contrast, Weeder #1 was maintained at the origin of the mechanism when the weeds could not be detected (as shown in Figure 10b, the left weeder—Weeder #2).

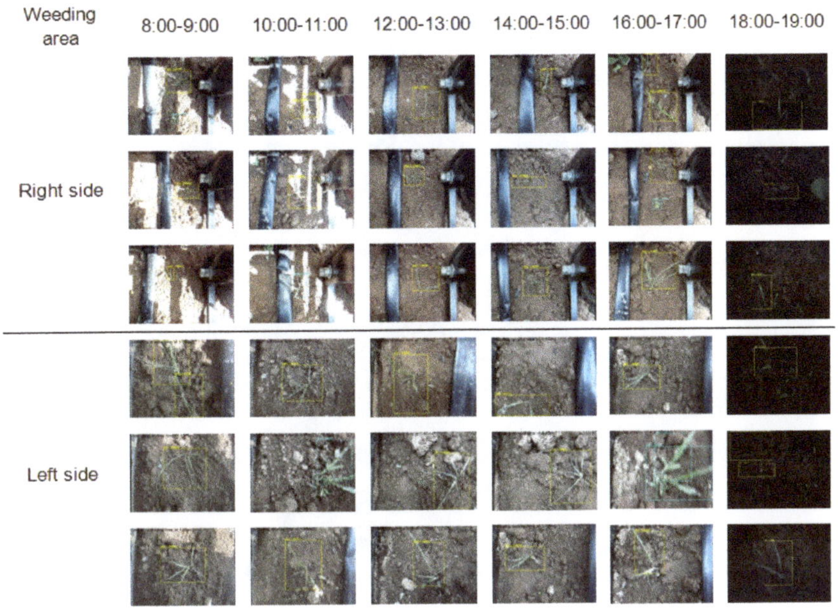


**Figure 8.** Weed identification results in different time intervals.

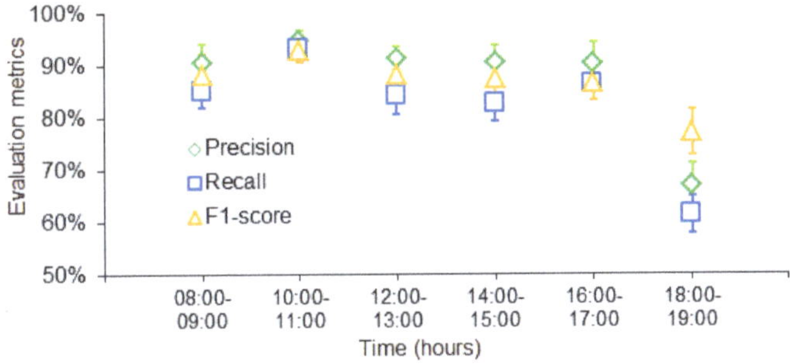


**Figure 9.** Ten day average detection results at different time intervals.

The effective cutting width of the two weeding machines is 20 cm. The data given in Table 4 show that in scenario 1, when the vehicle speed was 10 and 15 cm/s, the weeding efficiency was between 84% and 90.9%, which is equivalent to an hourly working area of up to 72 and 108 $m^2$. The average F1-score values of the deep learning networks in the left and right weeders were between 0.841 and 0.901. When the trailer speed increased to 20 cm/s, its weeding efficiency was significantly reduced, and the F1-score value was able to still reach approximately 0.867.

In scenario 2, when the vehicle moving speed was 10 and 15 cm/s, the weeding efficiency when using Weeder #1 was 92.3% and 82.6%, respectively, the crop damage rate was 5.5% and 11.1%, and the F1-score was at least 0.890. The weeding efficiency of using Weeder #2 was 87.0% (10 cm/s) and 78.6% (15 cm/s), respectively, the crop damage rate was 8.33% and 13.8%, and the F1 score value was above 0.878. Once the vehicle speed increased to 20 cm/s, the weeding efficiency of using Weeder #1 and Weeder #2 dropped to 64% and 56%, respectively, and the crop damage rate increased to 44.4% and 52.7%. However, the F1-score values were still 0.833 and 0.848, respectively.

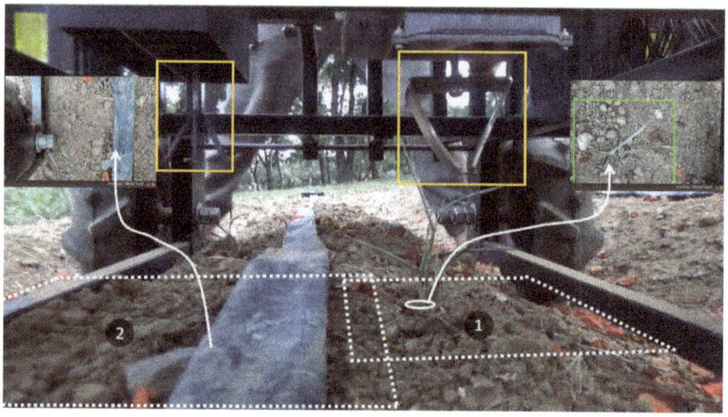


**(a)**

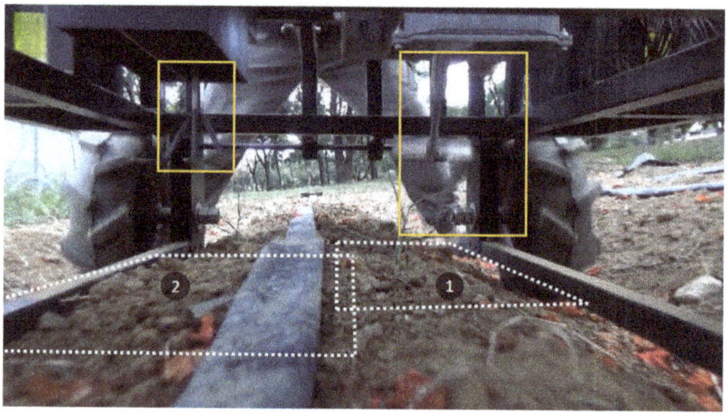


**(b)**

**Figure 10.** The operation of the weeder. The orange lines indicate the claw rake (left) and the inverted triangle (right) weeding tools. The white dotted line indicates the area of view taken by the two cameras on the left and right weeders. (**a**) The weed object is framed (the detection result of area ❶ (upper right corner)) and no weed is detected (the detection result of area ❷ (upper left corner)); (**b**) the weeding tool on the right is activated, and the left weeding cutter is maintained at the origin of the mechanism.

**Table 4.** Performance evaluation results of the weeding system.

| **Experiments** | **Type of Weeder** | **v (cm/s)** | **Number of Weeds** | | $\eta$ | **Damaged Crop** | $D$ **(%)** | **F1-Score** |
|---|---|---|---|---|---|---|---|---|
| | | | $W$ | $\overline{W}$ | | $\overline{d}$ | | |
| Scenario 1 | Weeder #1/ Weeder #2 | 10 | 25 */22 ** | 4/2 | 84.0/90.9 | - | - | 0.852/0.901 |
| | | 15 | 26/27 | 4/3 | 84.6/88.8 | - | - | 0.841/0.889 |
| | | 20 | 24/21 | 11/8 | 54.2/61.9 | - | - | 0.851/0.867 |
| Scenario 2 | Weeder #1 | 10 | 26 | 2 | 92.3 | 2 | 5.5 | 0.910 |
| | | 15 | 23 | 4 | 82.6 | 4 | 11.1 | 0.890 |
| | | 20 | 25 | 9 | 64.0 | 16 | 44.4 | 0.833 |
| | Weeder #2 | 10 | 23 | 3 | 87.0 | 3 | 8.33 | 0.903 |
| | | 15 | 28 | 6 | 78.6 | 5 | 13.8 | 0.878 |
| | | 20 | 25 | 11 | 56.0 | 19 | 52.7 | 0.848 |

*, **: Number of weeds on the left* and right** sides of the cropland.

Figure 11 shows an image of the weeds being removed by two weeding tools and the damage of the crops. Most of the roots of the weeds were turned up to the soil surface (Figure 11a,d), and some of the weeds on the edge of the weeding tool's coverage area were also turned up (Figure 11b,e). Some crops were slightly shifted or damaged from their original position due to the activation of the weeding tools (Figure 11c,f).

**Figure 11.** Snapshot of the soil on the field after weeding by the weeding machine. (**a**) weeds are completely removed by weeder #2, partially removed (**b**) and damaged crops (**c**); (**d**) weeds are completely removed by weeding tool #1, partially removed (**e**) and damaged crop (**f**). The red circle and orange arrow indicates the position of the crop roots and the root of the weed, respectively.

3.3.3. Discussion

There were three types of weeds in the experimental field, namely gramineous weeds, cyperaceae and broadleaf grasses, of which sedges and broadleaf grasses accounted for a higher proportion. At the end of each weeding experiment, we recorded the number of weeds remaining in the field, and most of these weed objects were detected. Part of the weeds did not actually turn up and the roots of some weeds were not removed due to the position of the weeds on both sides of the cutting width of the weeding tools. In addition, different shapes of weeding tools have different effects on different types of weeds. The claw rake-type weeding tool is suitable for shallow-rooted weeds. In contrast, the weeding tools used in this study are more suitable for removing weeds with deep roots, such as the tuber roots of Cyperaceae.

Second, the speed of the vehicle needs to match the weeding time. When the speed is greater than 20 cm/s, the weeding tool cannot accurately turn up the weeds. Especially under high weed density, some weeds cannot be removed immediately. The experimental results showed that the vehicle has a 92.6% success rate of weeding when the moving speed is lower than 15 cm/s. The cutter can shovel 3 cm below the ground. The, the height of the camera and the ground, and the distance between the camera and the weeding tool are 10 cm and 20 cm respectively. However, when the vehicle moves at a speed of 20 cm/sec, the highest success rate is only 64%, and there is a 44.4% crop damage rate. The loop speed of the weeding machine is set to one circle per second. If the moving speed of the trailer exceeds 20 cm/s, it increases the probability of crop damage and reduce the efficiency of weeding. A relatively slow speed is required to achieve a higher weeding success rate without damaging the crop. It is worth noting that when multiple weeds appear in the image simultanously, select the weed object with the largest area to maximize the weeding efficiency. In addition, before using this weeding tools, make sure that there are no large stones or bricks in the soil to avoid damage to the weeder. Because the steel cable is used by the hoist to pull the vehicle, when the vehicle is moving, the ground is relatively uneven, and there are several short speed changes during the movement of the vehicle, resulting in a time deviation. However, the deviation of weeding is still within the acceptable range.

The frame rate of YOLOv3 is set to 5 frame per second (fps), which can meet the requirements of real-time detection. A small number of weed samples were provided to the YOLOv3 model for training. Its network model was able to effectively detect weed objects with an accuracy rate of up to 95.6%. As far as we know, there are no relevant studies that use the YOLOv3 model to detect individual weeds in the field and use weeding tools to weed them. Since the number of image samples has an impact on the model detection performance, too few samples will reduce the model recognition performance [62].

In this study, the images were taken by mobile phones and some of the images were obtained using data augmentation technology. With a limited number of images, the weed detection model will still have different detection performance due to the difference in the brightness of the image background. In Scenario 1, the brightness of the images captured by the cameras on both sides is different due to the mask of the body frame and the asymmetry of the position of the weeding equipment, resulting in different model detection results (F1-score) of the two modules. The F1-score of the deep learning model designed in this research can reach above 0.83. Although the use of image processing technology can achieve a recognition rate of more than 90% in the identification of individual weeds and crops [27]. However, due to the influence of unstable light, the recognition rate fluctuates greatly. Using YOLOv3 model to detect weeds in low light conditions, the accuracy rate dropped slightly, but it remained at 83.2%. On the other hand, when the deep learning model detects eggs, its detection results are not affected by light [63], which is slightly different from the results of this study. The reason may be that the characterization of the detected object is more complicated. In weak light intensity environments, the performance of the model is still affected. This result still needs to be further studied.

The advantage of using the YOLOv3 model based on the Darknet-53 architecture is that it can quickly obtain the main characteristics of a weed or crop, and even features outside of human visual perception [55]. It can be observed from Fig. 11 that tiny weeds still remain on the soil surface. This result is acceptable. The dynamic balance of farmland agroecosystems will be improved when the composition of the weed community is changed, and the biodiversity of farmland will be improved [64].

The weeder is equipped with only one camera, and its weeding system can detect all weeds in the image. The proposed system does not involve the construction of multiple cameras and complex detection systems that require lighting control [42]. Meanwhile, the YOLOv3 model can also solve the identification limit of the same size of crops and weeds [44]. This study proposes an alternative strategy for single weed removal, replacing the traditional all-in-one weeding (chemical or physical) method. Small weeds on the field are neglected, which can improve the dynamic balance of the farmland ecosystem and increase the biodiversity of the farmland [64].

Finally, the use of a new-generation YOLOv4 network can shorten the time for object recognition [65]. If there are multiple different types of objects in the image or there are complex backgrounds, this method should be explored and studied.

## 4. Conclusions

The proposed weeder uses deep learning technology to detect weeds in the field and can use a special weeding tool to remove the weeds. The experimental results herein confirmed the effectiveness of the machine for weeding. At travel speeds of vehicle below

15 cm/s, the weeding system can detect the weed signal with a detection speed 5 fps of YOLOv3 and the average weeding efficiency is 88.6%. With an F1-score of 89.5% and a recall rate of 90.1%, the average detection accuracy rate is 90.7%. These results were from field trials of vegetable under different climate condition, which also included various densities of weeds. Since most of the deep learning model is only used to detect objects in the image; and the operating conditions of the weeder depend on the detection results of the contact or non-contact sensors on the machine. In this study, a smart farming method combining deep learning and weeding control was proposed. Its advantage lies in reducing the number of sensors used and the cost of maintenance. In addition, the powerful deep learning method can also identify different types of crops and weeds, with high flexibility.

The proposed weeder can be installed on the pylon behind the tractor, and multiple units can be made to be used on farmland of different scales and areas. The weeder is suitable for low-density weeds, early germination of weeds, or farming environments with deep roots of weeds, such as rice in wetlands or weeding in fields that have been prepared. The use of the proposed weeder can indeed destroy the growth conditions of weeds while reducing environmental medication. In addition, the weeder adopts DC power supply, which has a low production cost (approximately 1000 US dollars) and power consumption (approximately 500 W/h), which is of great significance for energy saving and environmental protection.

Future work will focus on the improvement of the performance of the weeder, including reducing the weight of the weeder and adjusting the rotation speed of the weeding tool in real time to adapt to different speed of vehicle. This deep learning method will also be tested to distinguish crops or weeds of the same size but different colors. Finally, install this weeder on a large tractor for tillage farming verification.

**Author Contributions:** Conceptualization, C.-L.C. and B.-X.X.; Methodology, C.-L.C.; Software, B.-X.X. and S.-C.C.; Verification, C.-L.C., B.-X.X. and S.-C.C.; Data management, B.-X.X.; Writing-manuscript preparation, C.-L.C. and B.-X.X.; writing—Review and edit, C.-L.C.; visualization, C.-L.C. and B.-X.X.; supervision, C.-L.C.; project management, C.-L.C.; fund acquisition, C.-L.C. All authors have read and agreed to the published version of the manuscript.

**Funding:** This research was funded by Ministry of Science and Technology (MOST), Taiwan, grant number MOST 109-2321-B-020-004; MOST 110-2221-E-020-019.

**Data Availability Statement:** The datasets presented in this study are available from the corresponding author on reasonable request.

**Acknowledgments:** Many thanks to all anonymous reviewers for their constructive comments on this manuscript. Meanwhile, we sincerely thank Wen-Chung Li, Director of the Department of Biomechanical Engineering, National Pingtung University of Science and Technology, for providing administrative support and Wei-Cheng Chen for assisting in the maintenance of the experimental site.

**Conflicts of Interest:** The authors declare no conflict of interest.

## References

1. Zambon, I.; Cecchini, M.; Egidi, G.; Saporito, M.G.; Colantoni, A. Revolution 4.0: Industry vs. agriculture in a future development for SMEs. *Processes* **2019**, *7*, 36. [CrossRef]
2. Pierce, F.J.; Nowak, P. Aspects of Precision Agriculture. *Adv. Agron.* **1999**, *67*, 1–85.
3. McBratney, A.; Whelan, B.; Ancev, T.; Bouma, J. Future Directions of Precision Agriculture. *Precis. Agric.* **2005**, *6*, 7–23. [CrossRef]
4. Oerke, E.C.; Dehne, H.W. Safeguarding production—Losses in major crops and the role of crop protection. *Crop Prot.* **2004**, *23*, 275–285. [CrossRef]
5. Wang, A.; Zhang, W.; Wei, X. A review on weed detection using ground-based machine vision and image processing techniques. *Comput. Electron. Agric.* **2019**, *158*, 226–240. [CrossRef]
6. Singh, G. Development and fabrication techniques of improved grubber. *AMA-Agric. Mech. Asia Afr.* **1988**, *19*, 42–46.
7. Bah, M.D.; Hafiane, A.; Canals, R. Deep Learning with Unsupervised Data Labeling for Weed Detection in Line Crops in UAV Images. *Remote Sens.* **2018**, *10*, 1690. [CrossRef]
8. Rani, L.; Thapa, K.; Kanojia, N.; Sharma, N.; Singh, S.; Grewal, A.S.; Srivastav, A.L.; Kaushal, J. An extensive review on the consequences of chemical pesticides on human health and environment. *J. Clean. Prod.* **2021**, *283*, 124657. [CrossRef]

9. Van Der Weide, R.Y.; Bleeker, P.O.; Achten, V.T.J.M.; Lotz, L.A.P.; Fogelberg, F.; Melander, B. Innovation in mechanical weed control in crop rows. *Weed Res.* **2008**, *48*, 215–224. [CrossRef]
10. Chandel, A.K.; Tewari, V.K.; Kumar, S.P.; Nare, B.; Agarwal, A. On-the-go position sensing and controller predicated contact-type weed eradicator. *Curr. Sci.* **2018**, *114*, 1485–1494. [CrossRef]
11. Steward, B.L.; Tian, L.F.; Tang, L. Distance-based control system for machine vision-based selective spraying. *Trans. ASAE* **2002**, *45*, 1255–1262. [CrossRef]
12. Zaman, Q.U.; Esau, T.J.; Schumann, A.W.; Percival, D.C.; Chang, Y.K.; Read, S.M.; Farooque, A.A. Development of prototype automated variable rate sprayer for real-time spot-application of agrochemicals in wild blueberry fields. *Comput. Electron. Agric.* **2011**, *76*, 175–182. [CrossRef]
13. Ahmad, M.T. Development of an Automated Mechanical Intra-Row Weeder for Vegetable Crops. Master's Dissertation, Iowa State University, Ames, IA, USA, 2012.
14. Bawden, O.; Ball, D.; Kulk, J.; Perez, T.; Russell, R. A lightweight, modular robotic vehicle for the sustainable intensification of agriculture. In Proceedings of the Australasian Conference on Robotics and Automation, Melbourne, Australia, 2–4 December 2014; pp. 1–9.
15. Cordill, C.; Grift, T.E. Design and testing of an intra-row mechanical weeding machine for corn. *Biosyst. Eng.* **2011**, *110*, 247–252. [CrossRef]
16. Perez-Ruiz, M.; Slaughter, D.C.; Fathallah, F.A.; Gliever, C.J.; Miller, B.J. Co-robotic intra-row weed control system. *Biosyst. Eng.* **2014**, *126*, 45–55. [CrossRef]
17. Utstumo, T.; Urdal, F.; Brevik, A.; Dørum, J.; Netland, J.; Overskeid, Ø.; Berge, T.W.; Gravdahl, J.T. Robotic in-row weed control in vegetables. *Comput. Electron. Agric.* **2018**, *154*, 36–45. [CrossRef]
18. Gobor, Z. Mechatronic system for mechanical weed control of the intra-row area in row crops. *KI—Künstliche Intell.* **2013**, *27*, 379–383. [CrossRef]
19. Rask, A.M. A review of non-chemical weed control on hard surfaces. *Weed Res.* **2007**, *47*, 370–380. [CrossRef]
20. Astrand, B.; Baerveldt, A.J. An agricultural mobile robot with vision-based perception for mechanical weed control. *Auton. Robot.* **2002**, *13*, 21–35. [CrossRef]
21. Griepentrog, H.; Nørremark, M.; Nielsen, J. Autonomous intra-row rotor weeding based on GPS. In Proceedings of the CIGR World Congress Agricultural Engineering for a Better World, Bonn, Germany, 3–7 September 2006.
22. Tillett, N.D.; Hague, T.; Grundy, A.C.; Dedousis, A.P. Mechanical within row weed control for transplanted crops using computer vision. *Biosyst. Eng.* **2008**, *99*, 171–178. [CrossRef]
23. Nørremark, M.; Griepentrog, H.W.; Nielsen, J.; Søgaard, H.T. Evaluation of an autonomous GPS-based system for intra-row weed control by assessing the tilled area. *Precis. Agric* **2012**, *13*, 149–162. [CrossRef]
24. Peruzzi, A.; Martelloni, L.; Frasconi, C.; Fontanelli, M.; Pirchio, M.; Raffaelli, M. Machines for non-chemical intra-row weed control in narrow and wide-row crops: A review. *J. Agric. Eng.* **2017**, *48*, 57–70. [CrossRef]
25. Schimmelpfennig, D. *Farm Profits and Adoption of Precision Agriculture*; USDA: Washington, DC, USA, 2016; Volume 217, pp. 1–46.
26. Michaels, A.; Haug, S.; Albert, A. Vision-based high-speed manipulation for robotic ultra-precise weed control. In Proceedings of the 2015 IEEE/RSJ International Conference on Intelligent Robots and Systems (IROS), Hamburg, Germany, 28 September–2 October 2015; pp. 5498–5505.
27. Chang, C.L.; Lin, K.M. Smart agricultural machine with a computer vision-based weeding and variable-rate irrigation scheme. *Robotics* **2018**, *7*, 38. [CrossRef]
28. McCool, C.; Beattie, J.; Firn, J.; Lehnert, C.; Kulk, J.; Bawden, O.; Russell, R. Efficacy of mechanical weeding tools: A study into alternative weed management strategies enabled by robotics. *IEEE Robot. Autom.* **2018**, *3*, 1184–1190. [CrossRef]
29. Naïo Technologies. Autonomous Vegetable Weeding Robot—Dino. Available online: https://www.naio-technologies.com/en/dino/ (accessed on 3 August 2021).
30. Fennimore, S.A.; Cutulle, M. Robotic weeders can improve weed control options for specialty crops. *Pest Manag. Sci.* **2019**, *75*, 1767–1774. [CrossRef] [PubMed]
31. Raja, R.; Nguyen, T.T.; Slaughter, D.; Fennimore, S. Real-time robotic weed knife control system for tomato and lettuce based on geometric appearance of plant labels. *Biosyst. Eng.* **2020**, *194*, 152–164. [CrossRef]
32. Kumar, S.P.; Tewari, V.K.; Chandel, A.K.; Mehta, C.R.; Nare, B.; Chethan, C.R.; Mundhada, K.; Shrivastava, P.; Gupta, C.; Hota, S. A fuzzy logic algorithm derived mechatronic concept prototype for crop damage avoidance during eco-friendly eradication of intra-row weeds. *Artif. Intell. Agric.* **2020**, *4*, 116–126. [CrossRef]
33. Sujaritha, M.; Annadurai, S.; Satheeshkumar, J.; Sharan, S.K.; Mahesh, L. Weed detecting robot in sugarcane fields using fuzzy real time classifier. *Comput. Electron. Agric.* **2017**, *134*, 160–171. [CrossRef]
34. Wu, Z.; Chen, Y.; Zhao, B.; Kang, X.; Ding, Y. Review of weed detection methods based on computer vision. *Sensors* **2021**, *21*, 3647. [CrossRef] [PubMed]
35. Dadashzadeh, M.; Abbaspour-Gilandeh, Y.; Mesri-Gundoshmian, T.; Sabzi, S.; Hernández-Hernández, J.L.; Hernández-Hernández, M.; Arribas, J.I. Weed classification for site-specific weed management using an automated stereo computer-vision machine-learning system in rice fields. *Plants* **2020**, *9*, 559. [CrossRef] [PubMed]

36. Liakos, K.G.; Busato, P.; Moshou, D.; Pearson, S.; Bochtis, D. Machine learning in agriculture: A review. *Sensors* **2018**, *18*, 2674. [CrossRef] [PubMed]
37. Tellaeche, A.; Pajares, G.; Burgos-Artizzu, X.P.; Ribeiro, A. A computer vision approach for weeds identification through Support Vector Machines. *Appl. Soft Comput. J.* **2011**, *11*, 908–915. [CrossRef]
38. Bakhshipour, A.; Jafari, A. Evaluation of support vector machine and artificial neural networks in weed detection using shape features. *Comput. Electron. Agric.* **2018**, *145*, 153–160. [CrossRef]
39. Ahmed, F.; Al-Mamun, H.A.; Bari, A.S.; Hossain, E.; Kwan, P. Classification of crops and weeds from digital images: A support vector machine approach. *Crop Prot.* **2012**, *40*, 98–104. [CrossRef]
40. Tang, J.; Chen, X.; Miao, R.; Wang, D. Weed detection using image processing under different illumination for site-specific areas spraying. *Comput. Electron. Agric.* **2016**, *122*, 103–111. [CrossRef]
41. Mahajan, S.; Raina, A.; Gao, X.Z.; Pandit, A.K. Plant recognition using morphological feature extraction and transfer learning over SVM and AdaBoost. *Symmetry* **2021**, *13*, 356. [CrossRef]
42. Raja, R.; Nguyen, T.T.; Vuong, V.L.; Slaughter, D.; Fennimore, S.A. RTD-SEPs: Real-time detection of stem emerging points and classification of crop-weed for robotic weed control in producing tomato. *Biosyst. Eng.* **2020**, *195*, 152–171. [CrossRef]
43. Chen, Y.; Wu, Z.; Zhao, B.; Fan, C.; Shi, S. Weed and corn seedling detection in field based on multi feature fusion and support vector machine. *Sensors* **2021**, *21*, 212. [CrossRef]
44. Elstone, L.; How, K.Y.; Brodie, S.; Ghazali, M.Z.; Heath, W.P.; Grieve, B. High Speed Crop and Weed Identification in Lettuce Fields for Precision Weeding. *Sensors* **2020**, *20*, 455. [CrossRef]
45. McCool, C.; Perez, T.; Upcroft, B. Mixtures of lightweight deep convolutional neural networks: Applied to agricultural robotics. *IEEE Robot. Autom. Lett.* **2017**, *2*, 1344–1351. [CrossRef]
46. Yu, J.; Sharpe, S.; Schumann, A.; Boyd, N. Deep learning for image-based weed detection in turfgrass. *Eur. J. Agron.* **2019**, *104*, 78–84. [CrossRef]
47. Jiang, S.; Li, X.; Xing, Y. Repair method of data loss in weld surface defect detection based on light intensity and 3D geometry. *IEEE Access* **2020**, *8*, 205814–205820. [CrossRef]
48. Dyrmann, M.; Karstoft, H.; Midtiby, H.S. Plant species classification using deep convolutional neural network. *Biosyst. Eng.* **2016**, *151*, 72–80. [CrossRef]
49. Kamilaris, A.; Prenafeta-Boldú, F.X. Deep learning in agriculture: A survey. *Comput. Electron. Agric.* **2018**, *147*, 70–90. [CrossRef]
50. Kamilaris, A.; Prenafeta-Boldu, F.X. A review of the use of convolutional neural networks in agriculture. *J. Agric. Sci.* **2019**, *156*, 312–322. [CrossRef]
51. Koirala, A.; Walsh, B.; Wang, Z.; McCarthy, C. Deep learning for real-time fruit detection and orchard fruit load estimation: Benchmarking of 'mangoyolo'. *Precis. Agric.* **2019**, *20*, 1107–1135. [CrossRef]
52. Hasan, A.S.M.M.; Sohel, F.; Diepeveen, D.; Laga, H.; Jones, M.G.K. A survey of deep learning techniques for weed detection from images. *Comput. Electron. Agric.* **2021**, *184*, 106067. [CrossRef]
53. Kounalakis, T.; Triantafyllidis, G.; Nalpantidis, L. Deep learning-based visual recognition of rumex for robotic precision farming. *Comput. Electron. Agric.* **2019**, *165*, 104973. [CrossRef]
54. Sun, J.; He, X.; Ge, X.; Wu, X.; Shen, J.; Song, Y. Detection of key organs in tomato based on deep migration learning in a complex background. *Agriculture* **2018**, *8*, 196. [CrossRef]
55. Osorio, K.; Puerto, A.; Pedraza, C.; Jamaica, D.; Rodríguez, L. A deep learning approach for weed detection in lettuce crops using multispectral images. *AgriEngineering* **2020**, *2*, 471–488. [CrossRef]
56. Partel, V.; Kakarla, S.C.; Ampatzidis, Y. Development and evaluation of a low-cost and smart technology for precision weed management utilizing artificial intelligence. *Comput. Electron. Agric.* **2019**, *157*, 339–350. [CrossRef]
57. Redmon, J.; Divvala, S.; Girshick, R. You only look once: Unified, real-time object detection. In Proceedings of the IEEE Conference on Computer Vision and Pattern Recognition (CVPR), Las Vegas, NV, USA, 27–30 June 2016; pp. 779–788.
58. Redmon, J.; Farhadi, A. YOLOv3: An incremental improvement. *arXiv Preprint* **2018**, arXiv:1804.02767.
59. Xie, B.X.; Chung, S.C.; Chang, C.L. Design and implementation of a modular AI-enabled shovel weeder. In Proceedings of the IEEE International Symposium on Computer, Consumer and Control (IS3C2020), Taichung, Taiwan, 13–16 November 2020. [CrossRef]
60. Bernacki, H.; Haman, J.; Kanafojski, C. *Agricultural Machines, Theory and Construction*; U.S. Department of Agriculture and the National Science Foundation: Washington, DC, USA, 1972; Volume 1.
61. He, K.; Zhang, X.; Ren, S.; Sun, J. Deep residual learning for image recognition. *arXiv Preprint* **2015**, arXiv:1512.03385.
62. Ahmad, A.; Saraswat, D.; Aggarwal, V.; Etienne, A.; Hancock, B. Performance of deep learning models for classifying and detecting common weeds in corn and soybean production systems. *Comput. Electron. Agric.* **2021**, *184*, 106081. [CrossRef]
63. Li, G.; Xu, Y.; Zhao, Y.; Du, Q.; Huang, Y. Evaluating convolutional neural networks for cage-free floor egg detection. *Sensors* **2020**, *20*, 332. [CrossRef] [PubMed]
64. MacLaren, C.; Storkey, J.; Menegat, A.; Helen Metcalfe, H.; Dehnen-Schmutz, K. An ecological future for weed science to sustain crop production and the environment. A review. *Agron. Sustain. Dev.* **2020**, *40*, 24. [CrossRef]
65. Bochkovskiy, A.; Wang, C.Y.; Liao, H.Y.M. YOLOv4: Optimal speed and accuracy of object detection. *arXiv Preprint* **2020**, arXiv:2004.10934.

agriculture

Article

# A Design of an Unmanned Electric Tractor Platform

**Yung-Chuan Chen, Li-Wen Chen and Ming-Yen Chang ***

Department of Vehicle Engineering, National Pingtung University of Science and Technology, Pingtung 912, Taiwan; chuan@mail.npust.edu.tw (Y.-C.C.); Liwen@mail.npust.edu.tw (L.-W.C.)
* Correspondence: chang.mingyen@mail.npust.edu.tw

**Abstract:** The tractor is a vehicle often used in agriculture. It is mainly used to tow other unpowered agricultural machinery for farming, harvesting, and seeding. They consume a lot of fuel with emissions that often contain a large amount of toxic gases, which seriously jeopardize human health and the ecological environment. Therefore, the electrical tractor is bound to become a future trend. The objective of this study is to design and implement a lightweight, energy-saving, and less polluting electric tractor, which meets the requirements of existing smallholder farmers, equipped with unmanned technology and multi-functions to assist labor and to provide the potential for unmanned operation. We reduced the weight of the tractor body structure to 101 kg, and the bending rigidity and torsional rigidity reached 11,579 N/mm and 4923 Nm/deg, respectively. Two 7.5 kW induction motors driven by lithium batteries were applied, which allows at least 3.5 h of working time.

**Keywords:** agricultural; unmanned; electrical tractor

**Citation:** Chen, Y.-C.; Chen, L.-W.; Chang, M.-Y. A Design of an Unmanned Electric Tractor Platform. *Agriculture* **2022**, *12*, 112. https://doi.org/10.3390/agriculture12010112

Academic Editor: Mustafa Ucgul

Received: 24 November 2021
Accepted: 10 January 2022
Published: 14 January 2022



## 1. Introduction

In recent years, with the rapid development of industrialization, agricultural machinery has gradually replaced traditional labor-intensive farming methods, improved work efficiency, and reduced manpower requirements. This is a major change in agricultural history. Today, agricultural machinery has developed into different forms, for example, agricultural machinery for arable land, for planting and fertilizing, for field management, etc. However, the wide usage of agricultural machinery increases the use of internal combustion engine vehicles, causing air pollution, environmental damage, and rapid consumption of land resources. According to statistics, global natural gas, oil, and coal resources can be supplied for another 30 years, 55 years, and 152 years, respectively [1]. Global environmental awareness is gradually rising. In order to reduce the harm to the environment, many countries have begun to promote electric vehicle-related industries vigorously, and have achieved good results in batteries, hybrid vehicles, and electric vehicles.

The tractor is a vehicle often used in agriculture. It is mainly used to tow other unpowered agricultural machinery for farming, harvesting, and seeding. For sedans, the key performance criteria are speed, loading force, and traction, but for tractors, high-speed performance and strength for traction are not important. The tractor needs to pull agricultural machinery and implement farmland farming. The tractor can have different operation modes depending on the agricultural tools with which it is equipped. If a Western plow, harrow, or raking machine, etc., is mounted behind the tractor, a plowing operation can be carried out. If the tractor is equipped with a rotary plow, the power transmission device of the tractor can be used for rotary tillage. Adding a flat soil board or a rice transplanter to the tractor can achieve soil leveling or planting operations, so the tractor is essential in agriculture. The agricultural tractors used in Taiwan are less driven by motors. However, as a trend for small-scale farmers, greenhouse planting emphasizes the development of technological agriculture and is environmentally controlled. The development of agricultural machinery needs to fulfill users' requirements and provide a safe and environmental-friendly configuration to reduce labor demand and improve work comfort for better agricultural production.

Electric tractors have been studied since the 19th century. The first electric tractor appeared in the USA [2,3], and subsequent developments were mainly powered by batteries. A 36.8 kW electric tractor manufactured by German Siemens in 1912 was the first electric tractor [4], which was mainly used for rotary tillage operations. The German company, Bungartz, developed an electric tractor called Töpfer in 1945. It was equipped with a gearbox and had a speed control function. Its key characteristic was to move both forward and backward without turning [5]. Later, General Electric (USA) introduced the Elec-Trak series of electric tractors. This electric tractor used lead-acid batteries to drive a permanent magnet brushless motor. The motor power was between 5.9 kW and 11 kW. In addition, the tractor was equipped with a rotary converter, which could tow other agricultural implements [6]. Since the 1990s, the control and battery technology have developed rapidly, and the performance of electric tractors has gradually improved, and there is more research devoted to the development of electric tractors and their electromechanical related design [7–9].

Furthermore, many articles are devoted to studying the performance and stability of tractors [10–13]. Improving the effectiveness of tillage is closely related to the characteristics of soil, and it is one of the areas that cannot be ignored [14–16]. There is a lot of research on autonomous driving [17–21], but there is very little on intelligent electric vehicles for unmanned driving in agriculture. We will enable these devices for autonomous driving on the field either to reach fixed points or to follow planned routes in farming.

## 2. Materials and Methods

The vehicle design was divided into four parts: the vehicle body design, which is lightweight and contains safety considerations; the power and vehicle control, which provides vehicle power and covers a series of system integration and unmanned controls; the mechanism design, which improves the mechanical functions and analyzes the state of the vehicle driving; and the field tests and the implementation of the whole vehicle. The divisions are shown in Figure 1.

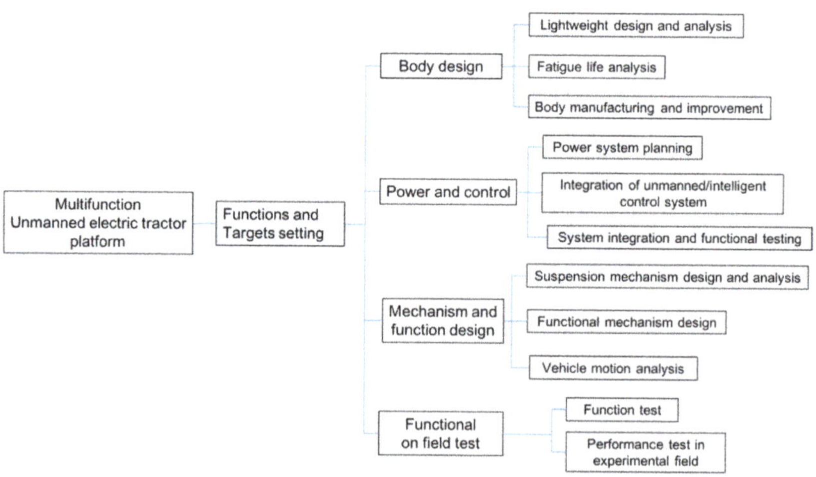


**Figure 1.** Planning of the division for vehicle design.

The tractor type and the related load affect the power consumption and the design of the vehicle. For a small electric tractor, the size, mass, and motor power are smaller than those of a high-load tractor. The designed tractor will be mainly used for rotary tillage and plowing operations, so the resistances are calculated based on the rotary tillage operation. Since the designed electric tractor is mainly operated in greenhouses, which have rather flat terrain, the slope resistance and air resistance of the tractor are not taken into consideration. The configuration and parameters are shown in Figure 2 and Table 1.

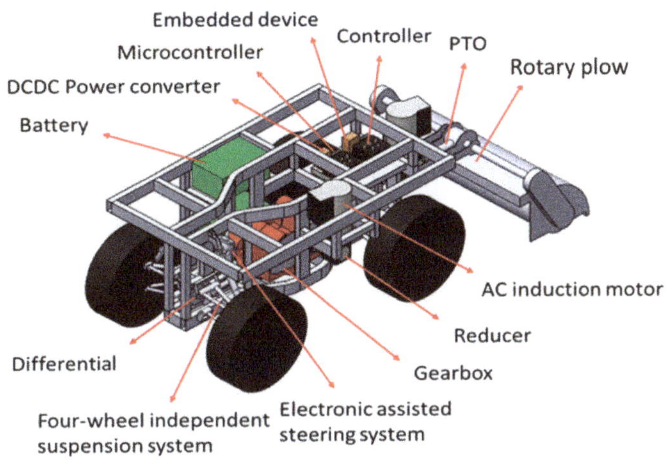


**Figure 2.** System configuration of the electric tractor.

**Table 1.** Vehicle parameters of the electric tractor.

| Vehicle Parameters | | |
|---|---|---|
| Length | Width | Height |
| 1720 mm | 1100 mm | 660 mm |
| Tire size | | Total vehicle weight |
| Front 26 × 8-14 | Rear 26 × 10-14 | 650 kg |

In addition, the tractors require a wide range of force changes, especially when working under heavy loads, which requires larger torque output. Therefore, the reducer must be used to decelerate and increase the torque to respond to different conditions. The drive motor can be adjusted and is equipped forward and reverse rotation to achieve reverse gear requirements. When transporting in the field, it can be switched to a high gear to increase the speed. In addition, when working in the field, the wheels may have insufficient grip due to the road or terrain, so a four-wheel drive system is required. Based on the above analysis, the power system configuration of the electric tractor in this study is shown in Figure 3, which is equipped with a motor, a reducer, a differential, and a controller.

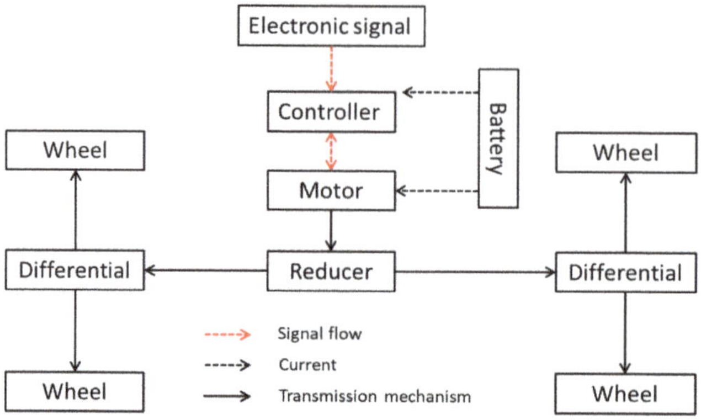


**Figure 3.** Power system configuration of the small electric tractor.

## 3. Body Design

The body of the small tractor is equipped with an on-board motor and a battery system. The tractor must meet the requirements of rigidity, safety, strength, and fatigue durability. The analysis process first established the prototype of the car body, then used TOSCA topology optimization analysis for lightweight analysis and ABAQUS finite element analysis for strength check, and finally used fe-safe fatigue life analysis to calculate the fatigue life of the vehicle body on different road grades.

### 3.1. Lightweight Design and Analysis of Car Body

During the space planning, the load position of each system on the vehicle was considered, such as the vehicle power system, steering system, suspension system, transmission system, electronic control system, battery, counterweight, rotary plow power system and slewing plow, etc., as shown in Figure 4. The required bearing weight of the vehicle included the vehicle power system, steering system, transmission system, electronic control system, battery, counterweight, rotary plow power system, rotary plow, and other weights. The target weight of the vehicle body was 120 kg. The total weight was estimated to be 650 kg. In order to reduce manufacturing costs, a commercially available transmission system and suspension system with a wheelbase of 1297 mm were selected. The power system was placed on the left side of the vehicle body for the power transmission. To balance the center of gravity, two lithium batteries were placed on the right side of the vehicle body to reduce the possibility of overturns. The power system was planned to be 350 mm in length, 338 mm in width, and 285 mm in height, and the battery is 338 mm in width, and 660 mm in height, and 480 mm in length. During transportation, the rotary plow at the rear of the vehicle body will be raised and the center of gravity will be moved backwards. Therefore, a counterweight of about 100 kg was installed in the front of the vehicle body to maintain balance. Based on the above-mentioned configuration, the preliminary frame size was 1720 mm in length, 1100 mm in width, and 660 mm in height.

In order to complete the required setting under the existing conditions, the vehicle types similar to this study were evaluated, space planning of the whole vehicle was carried out, including transmission, suspension type, wheelbase, type and quantity of battery, etc. The prototype of an electric tractor was established, and a topology optimization analysis of this structure was conducted. According to the topological optimization analysis, an overall structural material distribution was obtained for reference, and a preliminary conceptual design of an electric tractor was proposed.

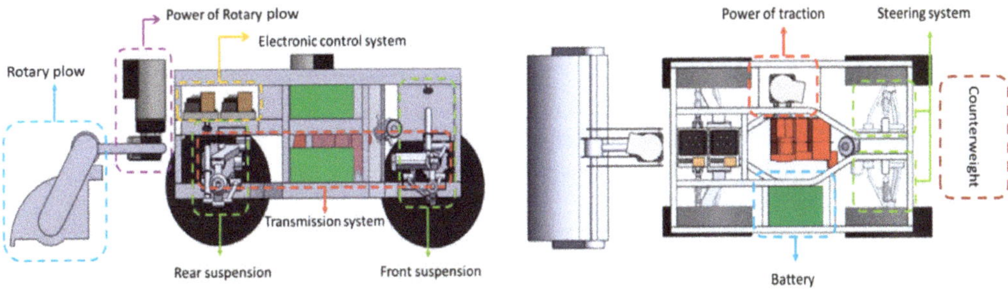


(**a**) (**b**)

**Figure 4.** Space planning of car body, (**a**) Side view; (**b**) Top view.

It was necessary to confirm whether the preliminary conceptual design met the design goals. Once the design goals were met, manufacturing feasibility was considered for welding deformation and component interferences. The body of the preliminary conceptual design was modified to obtain the final conceptual design. The completed vehicle body was imported into ABAQUS to create a finite element model and to conduct rigidity and strength analysis to confirm whether the rigidity and strength of the vehicle body met the design goals. If the design goals were not met, the structure was modified. Once the final vehicle body was obtained, the body was imported into the fe-safe for fatigue analysis. The fatigue life theory and rain flow cycle counting method were used to calculate the fatigue life. Figure 5 shows the analysis process of the lightweight design of the small electric tractor for this study.

First, the prototype of the vehicle body that was planned during the space layout was imported into ABAQUS to build a finite element model, as shown in Figure 6. Then the area to be made lightweight in the TOSCA topology optimization analysis software was defined as the design zone, shown as the green area in Figure 7. Considering the compatibility of the suspension system and the transmission system, the space and hardpoints were reserved for the shock absorbers, upper and lower control arms, power system, steering system, and transmission system, etc. The outer frame of the vehicle was defined as the non-design zone, the red area shown in Figure 7. Next, the material parameters of the vehicle body in the design area and the non-design area were defined. This study mainly used high-strength steel SPFH 590 and STKM 11A for materials. The properties of the materials are shown in Table 2. The topology optimization analysis, which were load condition analyses, including bending, torsion, and full load braking, was mainly static. The load conditions of these three types were all the same, and the weight of all the load-bearing objects was applied to the vehicle body as shown in Figure 8. Then the boundary conditions were set separately according to the bending strength analysis, torsion strength analysis, and load braking strength analysis.

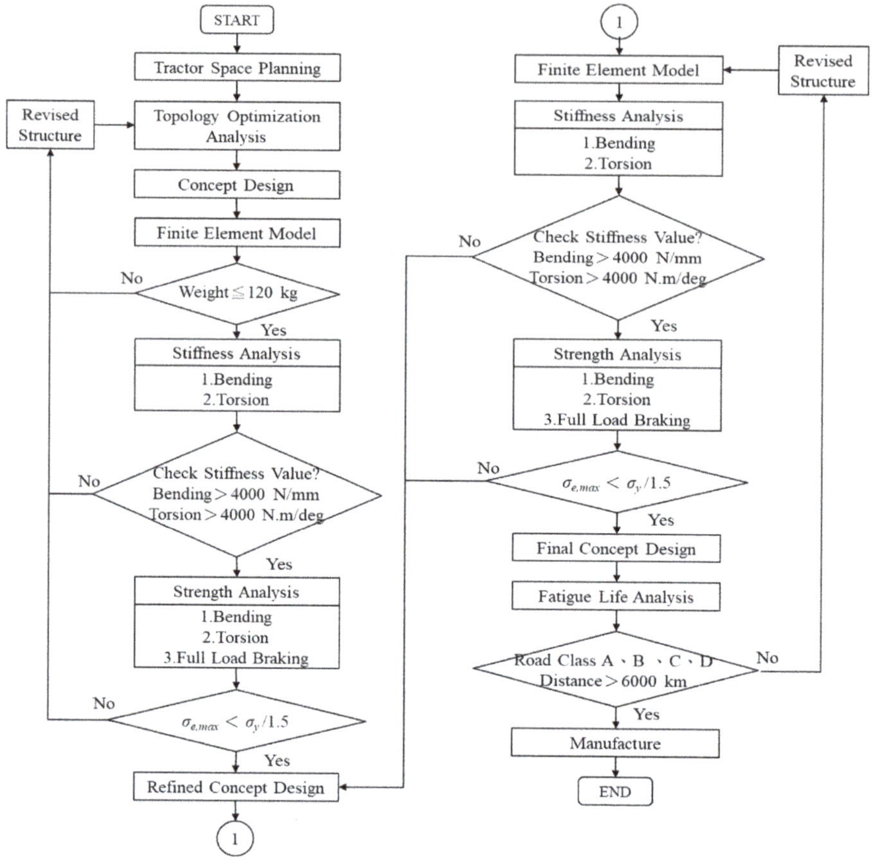


**Figure 5.** Analysis process of the lightweight design of the small electric tractor.

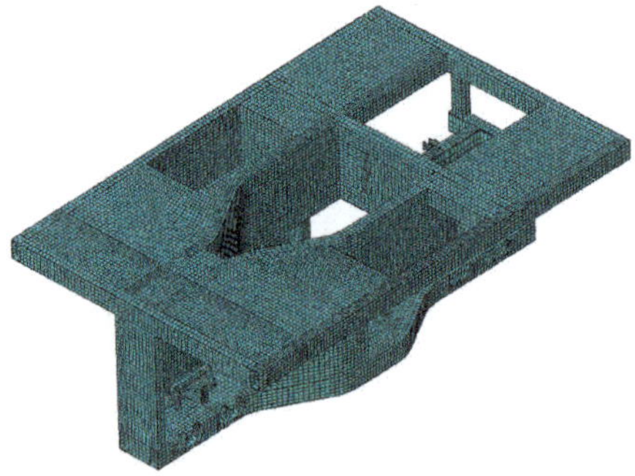

**Figure 6.** The finite element model of the vehicle body prototype.

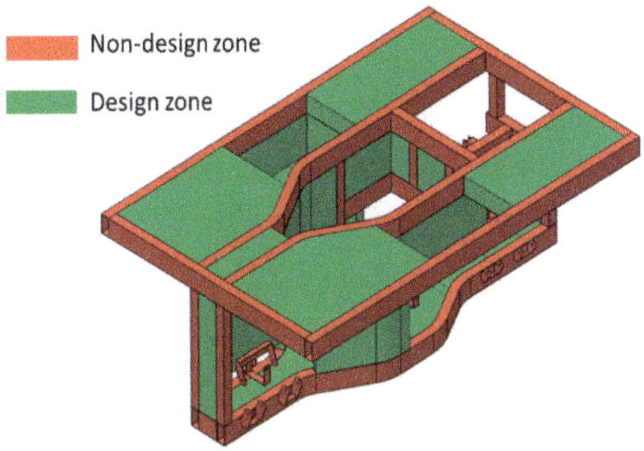


**Figure 7.** The design area of the vehicle body prototype.

**Table 2.** Material properties.

| Type Parameter | SPFH 590 | STKM 11A |
|---|---|---|
| Density (kg/m$^3$) | 7850 | 7850 |
| Young's Modulus (MPa) | 210 | 210 |
| Yielding Stress (MPa) | 420 | 175 |
| Ultimate Stress (MPa) | 590 | 290 |
| Poisson's Ratio | 0.3 | 0.3 |

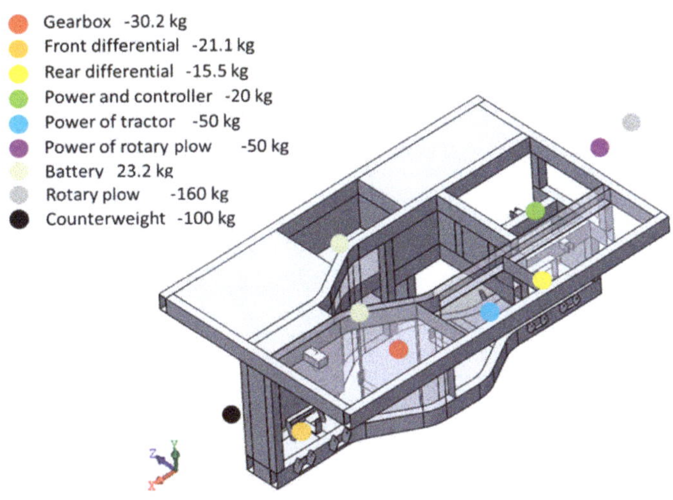


**Figure 8.** Load conditions on the prototype of the vehicle body.

*3.2. Fatigue Life Analysis*

In order to confirm the fatigue life of the frame, the geometric model of the final conceptual design of the vehicle body was imported into the ABAQUS to establish the finite element model. In the fatigue life analysis of the vehicle body, the suspension system did not undergo fatigue life analysis; the suspension model was described as simplified beam elements. In Figure 9, $K_f$ is the stiffness value of the front shock absorber spring, $K_r$ is the stiffness value of the rear shock absorber spring, $C_f$ is the damping value of the front shock absorber, and $C_r$ is the damping value of the rear shock absorber. The simplified suspension model was simulated by a three-dimensional two-node beam element and defined as a rigid body. In the fatigue life analysis, a three-dimensional dynamic elastoplastic finite element model was used, and the element type was eight-node hexahedral.

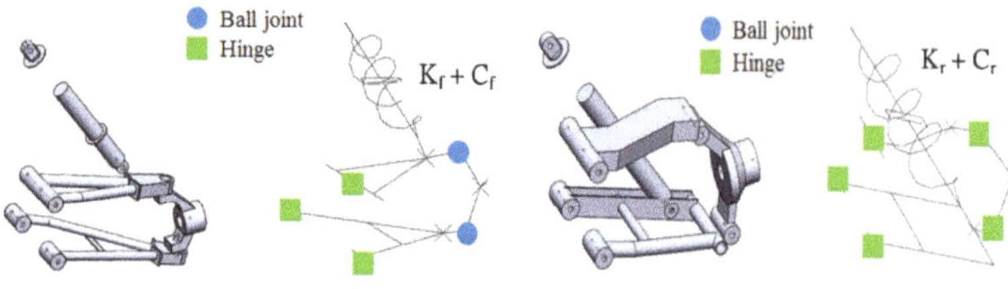


**Figure 9.** Finite element model of the suspension system. (**a**) Front suspension; (**b**) Rear suspension.

A. Material parameters

When the vehicle is running, the structure of each part bears different stresses from different load. Since the tractor often works in the field, and the working environment is relatively harsh compared with ordinary vehicles, SPFH 590 high-strength steel was used as the main material of the vehicle structure. Due to the small load between the upper and lower layers of the vehicle, steel STKM 11A was used between the 2 layers. The suspension hardpoints and the reinforced plate were used to meet the assembly requirements with SPFH 590, which were cut by laser and formed by bending. Other parts of the vehicle body used different materials, as shown in Figure 10.

B. Loading conditions

In the fatigue life analysis, different load conditions were applied on the vehicle body based on the working conditions and operations. For example, the rotary plow was placed on the ground and counted as unsprung mass. There was no load from the rotary plow in the analysis. In addition, traction resistance will be generated during rotary tillage operations, so resistance was applied to the rear of the vehicle body.

C. Boundary conditions

The boundary conditions of the fatigue life analysis are shown in Figure 11. In the figure, $K_f$ and $K_r$ are the stiffness of the front and rear suspension springs, respectively, which both equaled 27.5 N/mm. $C_f$ and $C_r$ are the damping values of the front and rear suspension shock absorbers whose values were set as 0.96 N.s/mm and 2.16 N.s/mm, respectively. During analysis, road signals of different levels in y direction were applied to the wheel center. Table 3 lists the parameters of the spring stiffness and shock absorber damping coefficient of the front and rear suspension systems during the fatigue life analysis.

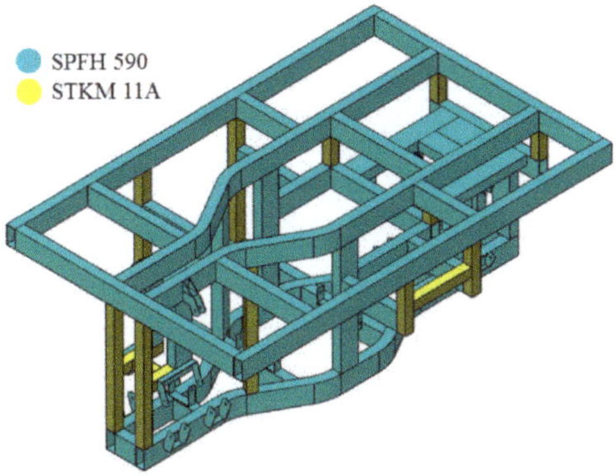


**Figure 10.** Material distribution of car body.

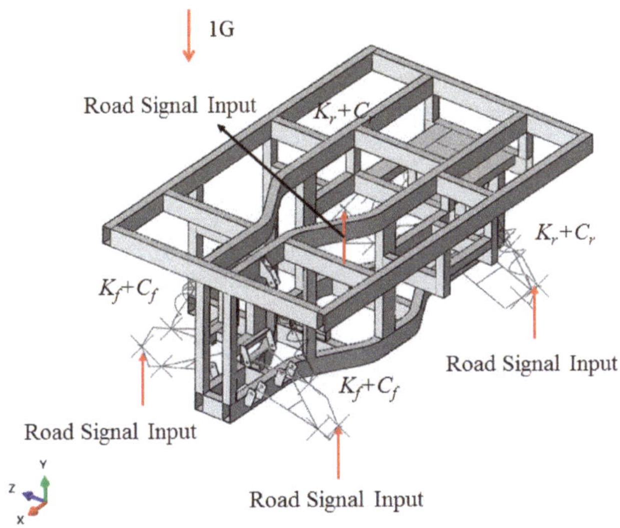


**Figure 11.** Boundary conditions for fatigue life analysis.

**Table 3.** Spring stiffness and damping coefficient value of suspension system.

| | Front Suspension | Rear Suspension |
|---|---|---|
| Spring Constant (N/mm) | 27.5 (Kf) | 27.5 (Kr) |
| Damping Coefficient (N.s/mm) | 0.96 (Cf) | 2.16 (Cr) |

Fe-safe was used to predict fatigue life. It uses the rain flow method to count the number of stress amplitude occurrences and the average stress. Here, the rotary tillage with a larger load was selected for illustration. When the vehicle body is used for rotary tillage, Figures 12a and 13a show the von-Mises stress distribution diagrams obtained from the simulation of the vehicle body driving on the C- and D-level road surfaces at a speed of 2 km/h. It is observed that both $\sigma_{e,max}$ occurred at the hardpoint of the upper control arm of the left front suspension. The stress histories of this location, as shown in Figures 12b and 13b, were respectively imported into the fe-safe software for calculation. The rain flow counting method was used to calculate the stress history of the C-and D-level road surface. The stress amplitude $\sigma_a$ and the number of occurrences of average stress $\sigma_m$ are shown in Figures 12c and 13c. The fatigue life of the vehicle body obtained from the fe-safe simulation was $N = 2.4 \times 10^6$ and $N = 6.7 \times 10^5$, which also means the vehicle can travel 40,724 and 11,190 km, respectively, as shown in Figures 12d and 13d.

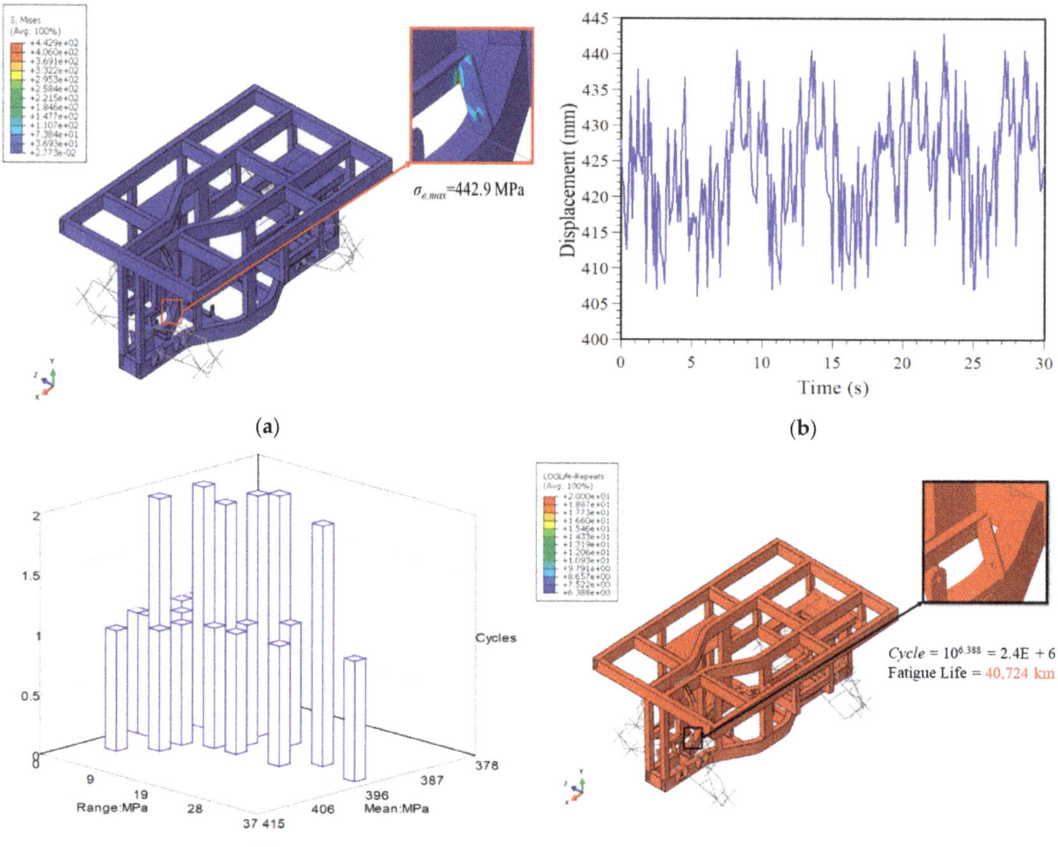


**Figure 12.** Analysis of driving on a C-level road at = 2 km/h during rotary tillage operation. (**a**) Von-Mises stress distribution; (**b**) the von-Mises stress of the car body where $\sigma_{e,max}$ occurs; (**c**) stress amplitude and number of average stress; and (**d**) fe-safe life simulation.

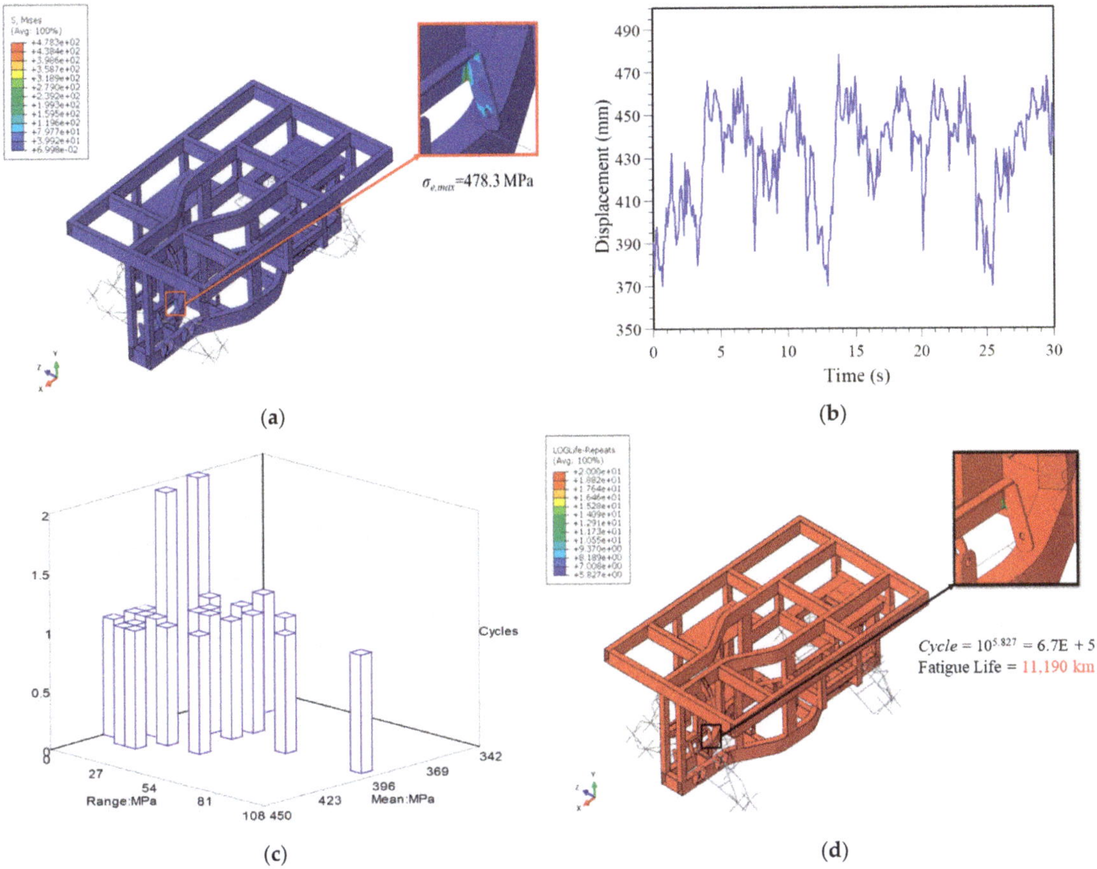


**Figure 13.** Analysis of driving on D-class road surface at = 2 km/h during rotary tillage operation. (**a**) Von-Mises stress distribution; (**b**) the von-Mises stress of the car body where $\sigma_{e,max}$ occurs; (**c**) stress amplitude and number of average stress; and (**d**) fe-safe life simulation.

## 4. Power and System Integration

Due to different operating conditions and load distributions of the tractor, there are also different requirements for the performance of the tractor. The power system of the electric tractor must be designed according to the load distribution in different operating conditions. It was calculated according to the requirements of the vehicle under various operating conditions, with a suitable drive motor and reducer.

### *4.1. Force Estimate and Power System Planning*

The resistance of the tractor is related to the selection of the specifications of the power system, so the total resistance of the tractor during transportation operations and rotary tillage operations can be calculated. In addition, the electric tractor uses two drive motors (induction motors) for rotary tillage operations, one for driving the tractor, and one for driving the rotary plow operation, etc. When the tractor is transporting, the motor driving the rotary plow will not work. At this time, the driving force of the whole vehicle is relatively small.

4.1.1. Force Estimate

In the transportation operation, the required driving force can be obtained [22]:

$$F_{q1} = f\,G = 0.12 \times 6474.6 = 776.95\ \text{N} \quad (1)$$

where f = 0.12 is the rolling resistance coefficient, and G = 6474.6 N is the vehicle weight.

Before the driving force $F_{q2}$ required for the rotary tillage operation can be calculated, it is necessary to know the rotary tillage speed ratio λ and the soil cutting pitch S. After calculating these two values, the rotary tillage specific resistance $K_\lambda$ and the soil resistance $F_L$ can be obtained, and then the rotary tillage can be calculated. The driving force $F_{q2}$ is required for the operation. The rotary tillage speed ratio λ can be obtained by formula (2):

$$\lambda = \frac{2\pi r n_r}{60,000 v_m} = \frac{2\pi \times 205 \times 236}{60,000 \times 0.55} = 9.2 \quad (2)$$

where r = 205 mm is the turning radius of the scimitar, $n_r$ = 236 r/min is the rotary knife shaft speed, and $v_m$ = 0.55 m/s is the tractor speed. After calculating the rotary tillage speed ratio λ, we obtain:

$$S = \frac{\pi r}{5Z\lambda} = \frac{\pi \times 205}{5 \times 2 \times 9.2} = 7\ \text{cm} \quad (3)$$

where Z = 2 is the number of scimitars in the same vertical plane.

Before calculating the soil resistance, it is necessary to find the standard rotary tillage specific resistance $K_g$ corresponding to the soil cutting pitch S and then find the correction coefficient that meets the working conditions to obtain the rotary tillage specific resistance $K_\lambda$. The specific resistance $K_\lambda$ of the rotary tillage is:

$$K_\lambda = K_g K_1 K_2 K_3 K_4 = 15 \times 0.8 \times 0.95 \times 0.8 \times 0.66 = 6.019\ \text{N/cm}^2 \quad (4)$$

where $K_g$ = 15 N/cm$^2$ is the standard rotary tillage specific resistance, $K_1$ = 0.8 is the tillage depth correction coefficient, $K_2$ = 0.95 is the soil moisture content correction coefficient, $K_4$ = 0.66 is the stubble vegetation correction coefficient, and $K_3$ = 0.8 is the operation mode correction coefficient. Knowing the specific resistance of rotary tillage, we can obtain the soil resistance $F_L$:

$$F_L = 100\ K_\lambda\ BH = 100 \times 6.019 \times 1 \times 12 = 7222.8\ \text{N} \quad (5)$$

where B = 1 m is the width of the rotary tillage, and H = 12 cm is the depth of the rotary tillage. After calculating Formulas (2)–(5), we know the rotary tillage speed ratio λ, soil cutting pitch S, rotary tillage specific resistance $K_\lambda$ and soil resistance $F_L$, and then the rotary tillage operation time can be calculated by Formula (6) The required driving force $F_{q2}$ is:

$$\begin{aligned} F_{q2} &= k_1 F_L + f \times (G + k_2 F_L) \\ &= 0.68 \times 7222.8 + 0.12 \times (6474.6 + 0.74 \times 7222.8) \\ &= 6329.8\ \text{N} \end{aligned} \quad (6)$$

where $k_1$ = 0.68 is the horizontal component coefficient, and $k_2$ = 0.74 is the vertical component coefficient.

4.1.2. Power System Planning

The specifications of the electric vehicle power system are critical to the performance of the vehicle. The motor and the reduction ratio of the reducer of the vehicle power system are calculated by the rotary tillage operation that requires more power.

(1) Transportation operations

$$P_1 = (\frac{1}{3600 \times \eta}) \times fG \times V1_{max} = (\frac{1}{3600 \times 0.81}) \times 0.12 \times 6474.6 \times 18 = 4.8\ kW \quad (7)$$

where $\eta$ = 0.81 is the efficiency of the power transmission of the whole vehicle, and $V_{1max}$ = 18 km/h is the highest vehicle speed during transportation.

(2) Rotary tillage operations

$$\begin{aligned} P_2 &= \left(\frac{1}{3600 \times \eta}\right) \times [k_1 F_L + f \times (G + k_2 F_L)] \times V2_{max} \\ &= \left(\frac{1}{3600 \times 0.81}\right) \times [0.68 \times 7222.8 + 0.12 \times (6474.6 + 0.74 \times 7222.8)] \times 2 \\ &= 4.3\ kW \end{aligned} \quad (8)$$

where the highest vehicle speed during rotary tillage operation is $V_{2max}$ = 2 km/h. Generally, the tractor used in the greenhouse requires 20 PS (15 kW) of horsepower, so two motors with a rated power of 7.5 kW were finally selected as the driving motors in this study, one for driving the tractor and the other for driving the rotary plow. Table 4 presents the specifications of the selected motor.

**Table 4.** Selected motor specifications.

| Rated Voltage | DC72V |
|---|---|
| Rated power | 7.5 kW |
| Instantaneous peak | 17.8 kW |
| Maximum speed | 5800 rpm |
| Maximum torque | 23.2 N-m |

4.1.3. Reducer Selection

In order to confirm whether there is sufficient torque, the upper and lower limits of the reduction ratio of the reducer can be calculated.

(1) Transportation operations

$$\begin{gathered} \frac{fGR}{\eta T_m i_H} \le i \le \frac{0.377 n_m R}{V_{1max} i_H} \\ \frac{0.12 \times 6474.6 \times 0.3125}{0.81 \times 23.2 \times 12.5} \le i \le \frac{0.377 \times 5800 \times 0.3125}{18 \times 12.5} \\ 1.03 \le i \le 3.04 \end{gathered} \quad (9)$$

where R = 0.3125 m is the wheel radius, $T_m$ = 23.2 N·m is the motor output maximum torque, $i_H$ = 12.5 is the high gear reduction ratio, $n_m$ = 5800 rpm is the motor maximum speed, and $V_{1max}$ = 18 km/h is the maximum vehicle speed.

(2) Rotary tillage operations

$$\begin{gathered} \frac{f \times (G + k_2 \times F_L) R}{\eta T_m i_L} \le i \le \frac{0.377 n_m R}{V_{2max} i_L} \\ \frac{0.12 \times (6474.6 + 0.74 \times 7222.8) \times 0.3125}{0.81 \times 23.2 \times 23.3} \le i \le \frac{0.377 \times 5800 \times 0.3125}{2 \times 23.3} \\ 1.01 \le i \le 14.7 \end{gathered} \quad (10)$$

where $i_L$ = 23.3 is the low gear reduction ratio, and $V_{2max}$ = 2 km/h is the maximum vehicle speed.

### 4.2. Unmanned/Intelligent Control System Integration

There is a lot of research on autonomous driving [12,13], but there are very few developments on intelligent electric vehicles for unmanned driving in agriculture. The unmanned driving system uses cameras, optical radars, ultrasonic sensors, and other equipment on the vehicle to enable autonomous driving on the field to reach fixed points or to follow planned routes for farming and other tasks.

The unmanned driving system can be divided into three parts, information collection, electronic control unit (ECU), and execution unit. The information collection part refers to the sensing components. Different sensors collect information for different systems. The information collected by the sensors will be transmitted to the ECU for analysis and processing, and commands are given based on the results calculated by the ECU for the Execution unit. Figure 14 is a diagram of the unmanned driving system.

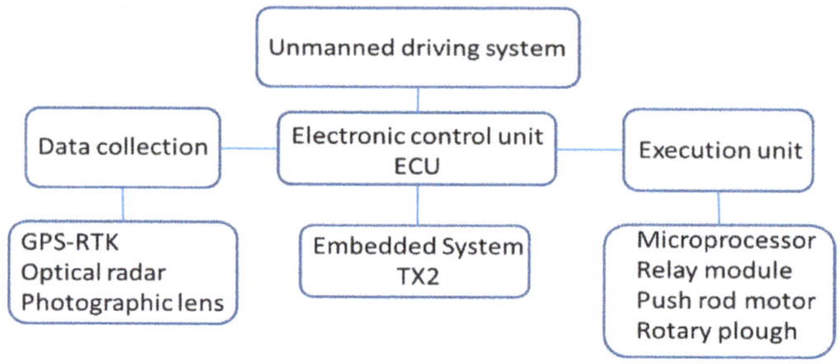


**Figure 14.** Unmanned driving system.

The positioning system used two GPS-RTK. The two were aligned in parallel, one was set at the center of gravity of the vehicle and the other was set at the back of the car. The two GPS-RTK collect the difference between the current heading angle and the heading angle toward the target point. A PID controller receives the angle error from the GPS-RTK, determines whether the vehicle should drive, spin, or stop, and tracks the target point with a GPS-RTK located at the center of gravity to achieve autonomous driving. The tractor uses LiDAR for obstacle avoidance and navigation assistance. The rotary plow carried by the tractor is dangerous. The LiDAR scans 360 degrees around the tractor to stop the tractor when an object enters the hazard zone, which was redefined, and to prevent something from getting caught in the rotary plow.

To assist the autonomous vehicle to touch the ground in excessive ups and downs, three optical radars were added in addition to the existing GPS-RTK. Two of the three optical radars were placed on the front left and right sides of the vehicle to detect both sides of the vehicle and to calibrate the lateral accuracy of the vehicle. The remaining optical radar was mounted in the front of the vehicle to detect obstacles. When an obstacle is detected within a defined distance in front of the vehicle, the operation of the vehicle will be stopped. The vehicle was also equipped with an image processing unit to keep the vehicle stable, as shown in Figure 15. After the camera captures the image, the desired targets in the region of interest (ROI) are processed. The image is then converted to grayscale and binarized. Then the Canny tracks the edges of the objects and performs dilate processing. Then, Hough transfers the edges to calculate the centerline of the road as the driving trajectory.

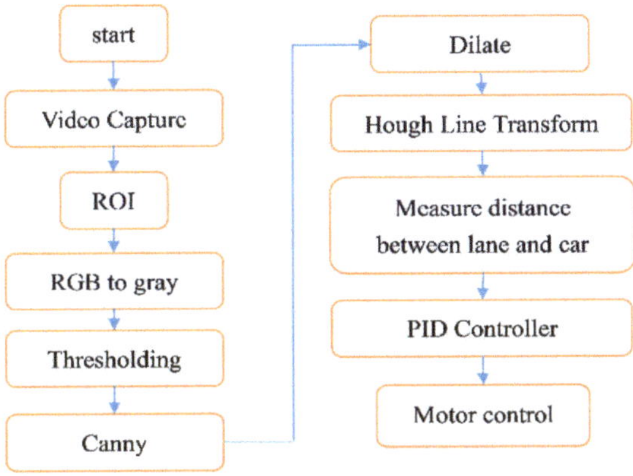


**Figure 15.** Image processing and control flowchart.

The vehicle control uses a NVIDIA TX2 to collect all the information from sensors, calculate the vehicle offset, the current position, and actuation function, and to send signals to the corresponding MCUs through the CAN bus to the control motors. The control system flow chart is shown in Figure 16.

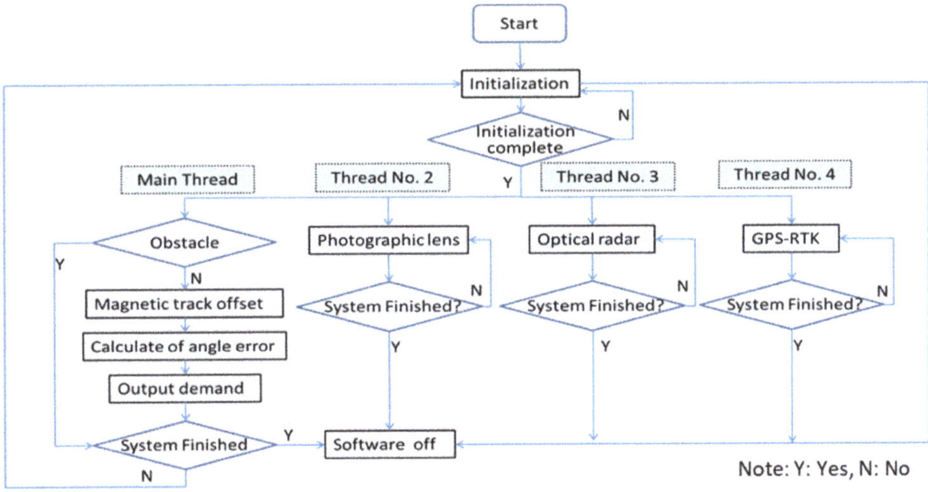


**Figure 16.** Control system flowchart.

### 4.3. Vehicle System Integration and Function Testing

The Smart Farm of Pingtung University of Science and Technology was selected for the functional test of the tractor, as shown in Figure 17. The field is a 10 hectares smart agricultural production demonstration base. The crops in the field can be planted and cultivated according to research needs. In this study, both dry and wet arable land was used for field testing. The greenhouse was also one of the indispensable test areas. Figure 18 is the configuration diagram of the vehicle control system. The software architecture was divided into manual mode and automatic mode. In the manual mode, the operator gave the accelerator signal to the motor controller to achieve the driving requirement, and the steering signal was transmitted to the electronic assisted steering system to steer the whole vehicle. The automatic driving mode processed and captured the current position data at the first second, calculated the driving route with the current position and the set driving route, and then informed the controller and the electronic assisted steering system to make the tractor follow the route. Figure 19 is the completed diagram of the small electric tractor of the study.

**Figure 17.** Smart Farm of Pingtung University of Science and Technology.

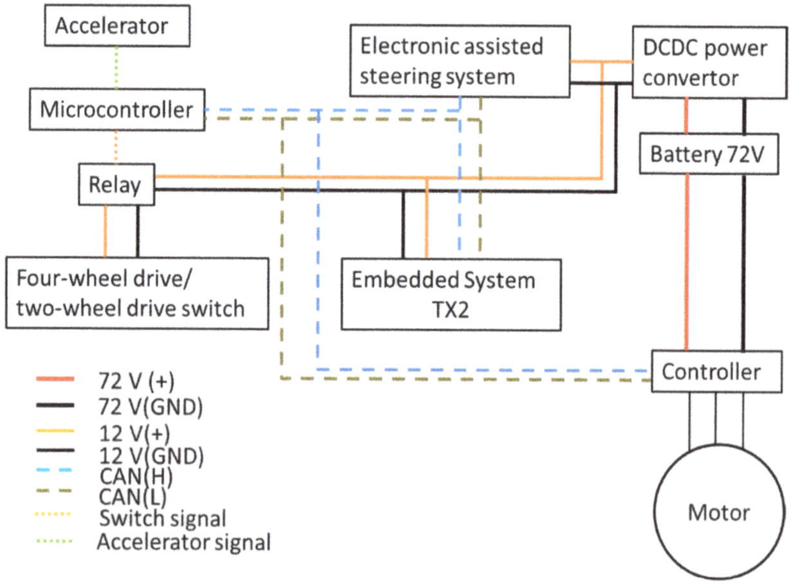


**Figure 18.** Configuration of vehicle control system.

**Figure 19.** The small electric tractor.

## 5. Conclusions

This study focused on the development of a small electric tractor with complete design and fabrication of the whole vehicle system. The conceptual design of a lightweight vehicle body was obtained through topological optimization analysis and the finite element model analysis, which was used to obtain the rigidity, strength, and fatigue life analysis of the vehicle body. Finally, the vehicle body structure, chassis, and electrical system were completed with the integration of the vehicle control system. Based on the results of the above analysis, the following conclusions can be summarized:

1. According to the lightweight analysis process, the weight of the proposed vehicle body was 101 kg, and the bending and torsional rigidity were 11,579 N/mm and 4923 N·m/deg, respectively.
2. In the analysis of the bending, torsion, and full load braking strength of the vehicle body, the maximum von-Mises stress was lower than the material yield strength by 2/3, which met the design requirements.
3. The fatigue life analysis showed that the fatigue life of the designed vehicle body reached $6.5 \times 10^8$ km when driven on a general asphalt road at a speed of 18 km/h. When rotating or plowing at a speed of 2 km/h, the fatigue life reached 11,190 km and 23,166 km, respectively.
4. This research completed the development and fabrication of a small electric tractor, which met the requirements of manual driving and automatic driving. In addition, the tractor was equipped with two 7.5 kW induction motors, driven by lithium batteries, which can achieve at least 3.5 h of working time, and the rotary tillage operations can reach a depth of about 15 cm. The result of field tests on the prototype electric tractor are shown in Tables 5–7.

**Table 5.** Field tests on different kinds of road.

| Field Test | | | |
|---|---|---|---|
| Velocity (km/h) | Road Type | Current (A) | Power (W) |
| 3.05 | Asphalt | 2.38 | 185.64 |
| | Hard | 7.3 | 569.4 |
| | Soft | 3.83 | 298.74 |
| | Grass | 2.5 | 195 |
| 1.02 | Asphalt | 14.16 | 1104.48 |
| | Hard | 26.0 | 2028 |
| | Soft | 10.85 | 846.3 |
| | Grass | 16.7 | 1302.6 |

The current of the drive motor test on the soft soil in the field was the largest, with a low speed of 7.3 A and a high speed of 26 A.

**Table 6.** Influence of driving speed and tillage on motor power.

| Velocity (km/h) | Depth (cm) | Driving Current | | Tillage Current | |
|---|---|---|---|---|---|
| | | Current (A) | Power (W) | Current (A) | Power (W) |
| 1.02 | 5 | 10.9 | 850.2 | 28.1 | 2191.8 |
| | 10 | 7.9 | 616.2 | 32.8 | 2558.4 |
| | 15 | 12.9 | 1006.2 | 37.4 | 2917.2 |
| 3.05 | 5 | 25.7 | 2004.6 | 44.3 | 3455.4 |
| | 10 | 28.2 | 2199.6 | 49.1 | 3829.8 |
| | 15 | 37.2 | 2901.6 | 52.2 | 4071.6 |

**Table 7.** Influence of motor power consumption on battery discharge time.

| Drive Current and Tillage Current (A) | |
|---|---|
| Low speed 1.02 (km/h) | 50.3 A |
| High speed 3.05 (km/h) | 126.6 A |

Ploughing field 15 cm, for low-speed, the current was 50.3 A, for high-speed, the current was 126.6 A.

**Author Contributions:** Y.-C.C., M.-Y.C. and L.-W.C. conceived the idea. We developed the theory and Y.-C.C. performed the CAE. L.-W.C. developed the control. M.-Y.C. and Y.-C.C. verified the analytical methods and supervised the findings of this work. All authors discussed the results and contributed to the final manuscript. All authors have read and agreed to the published version of the manuscript.

**Funding:** This research was funded by the Pingtung University of Science and Technology (NPUST).

**Institutional Review Board Statement:** Not applicable.

**Informed Consent Statement:** Not applicable.

**Acknowledgments:** We thank Pingtung University of Science and Technology (NPUST) for funding implementation and providing the venue for the experiment.

**Conflicts of Interest:** The authors declare no conflict of interest.

## References

1. Ji, W. Research on the Application of Automotive New Energy and Energy-saving Technology. *Sci. Technol. Wind.* **2013**, *8*, 104.
2. Carlini, M.; Abenavoli, R.I.; Kormanski, H.; Rudzinska, K. A Hybrid Electric Propulsion System for a Forest Vehicle. In Proceedings of the Energy Conversion Engineering Conference, Honolulu, HI, USA, 27 July–1 August 1997; p. 5926956.
3. Florentsev, S.; Izosimov, D.; Makarov, L.; Baida, S.; Belousov, A. Complete Traction Electric Equipment Eets of Electro-Mechanical Drive Trains for Tractors. In Proceedings of the IEEE Region 8 International Conference on Computational Technologies in Electrical and Electronics Engineering, Irkutsk, Russia, 11–15 July 2010; pp. 611–616.
4. Xiao, S. Development Status and Trends of Light Electric Agricultural Transport Machinery. *Agric. Dev. Equip.* **2012**, *3*, 49–50.

5. Yu, Z. Research on Human-Machine System Design of Greenhouse Electric Walking Tractor. Master's Thesis, Institute of Mechanical Engineering, Nanjing Agricultural University, Nanjing, China, 2013.
6. Guan, C.S.; Hu, H.; Chen, Y.S.; Yang, Y.T.; Cao, S.M. Study on Small Electric Tractors for Greenhouses. *China J. Agric. Mech.* **2015**, *36*, 67–69.
7. Troncon, D.; Alberti, L. Case of Study of the Electrification of a Tractor: Electric Motor Performance Requirements and Design. *Energies* **2020**, *13*, 2197. [CrossRef]
8. Gao, Y.; Wang, X.; Li, J. Optimization Design of Electric Tractor for Small Seeding Machinery. In Proceedings of the 3rd International Conference on Automation, Mechanical and Electrical Engineering, Shanghai, China, 22–23 July 2018; pp. 59–63. [CrossRef]
9. Sitompul, J.; Zhang, H.; Noguchi, T.; Ahamed, T. Optimization study on the design of utility tractor powered by electric battery. *IOP Conf. Ser. Earth Environ. Sci.* **2019**, *355*, 012058. [CrossRef]
10. Franceschetti, B.; Rondelli, V.; Capacci, E. Lateral Stability Performance of Articulated Narrow-Track Tractors. *Agronomy* **2021**, *11*, 2512. [CrossRef]
11. Xu, J.; Li, R.; Li, Y.; Zhang, Y.; Sun, H.; Ding, X.; Ma, Y. Research on Variable-Universe Fuzzy Control Technology of an Electro-Hydraulic Hitch System. *Processes* **2021**, *9*, 1920. [CrossRef]
12. Askari, M.; Abbaspour-Gilandeh, Y.; Taghinezhad, E.; El Shal, A.M.; Hegazy, R.; Okasha, M. Applying the Response Surface Methodology (RSM) Approach to Predict the Tractive Performance of an Agricultural Tractor during Semi-Deep Tillage. *Agriculture* **2021**, *11*, 1043. [CrossRef]
13. Moinfar, A.; Shahgholi, G.; Abbaspour-Gilandeh, Y.; Herrera-Miranda, I.; Hernández-Hernández, J.L.; Herrera-Miranda, M.A. Investigating the Effect of the Tractor Drive System Type on Soil Behavior under Tractor Tires. *Agronomy* **2021**, *11*, 696. [CrossRef]
14. Du, J.; Heng, Y.; Zheng, K.; Zhang, W.; Zhang, J.; Xia, J. Evaluation of the Performance of a Combined Tillage Implement with Plough and Rotary Tiller by Experiment and DEM Simulation. *Processes* **2021**, *9*, 1174. [CrossRef]
15. Cheng, J.; Zheng, K.; Xia, J.; Liu, G.; Jiang, L.; Li, D. Analysis of Adhesion between Wet Clay Soil and Rotary Tillage Part in Paddy Field Based on Discrete Element Method. *Processes* **2021**, *9*, 845. [CrossRef]
16. Li, X.; Wei, B.; Xu, X.; Zhou, J. Effect of Deep Vertical Rotary Tillage on Soil Properties and Sugarcane Biomass in Rainfed Dry-Land Regions of Southern China. *Sustainability* **2020**, *12*, 10199. [CrossRef]
17. Lygouras, E.; Gasteratos, A. A novel moving-base RTK-GPS-Based wearable apparatus for precise localization of humans in peril. *Microprocess. Microsyst.* **2021**, *82*, 103833. [CrossRef]
18. Yu, W.; Peng, H.; Pan, L.; Dai, W.; Qu, X.; Ren, Z. Performance assessment of high-rate GPS/BDS precise point positioning for vibration monitoring based on shaking table tests. *Adv. Space Res.* **2021**. [CrossRef]
19. Gutmann, E.D.; Larson, K.M.; Williams, M.W. Snow measurement by GPS interferometric reflectometry: An evaluation at Niwot Ridge, Colorado. *Hydrol. Process.* **2012**, *26*, 2951–2961. [CrossRef]
20. Zhao, L.; Qiu, H.; Feng, Y. Analysis of a robust Kalman filter in loosely coupled GPS/INS navigation system. *Measurement* **2016**, *80*, 138–147. [CrossRef]
21. Amini, A.; Vaghefi, R.M.; Jesus, M.; Buehrer, R.M. Improving GPS-based vehicle positioning for intelligent transportation systems. In Proceedings of the 2014 IEEE Intelligent Vehicles Symposium Proceedings, Dearborn, MI, USA, 8–11 June 2014; pp. 1023–1029.
22. Gao, H. Research on Electric Tractor Drive System. Ph.D. Thesis, College of Engineering, Nanjing Agricultural University, Nanjing, China, 2008.

*agriculture*

MDPI

*Article*

# Soil Compaction from Wheel Traffic under Three Tillage Systems

**Kobby Acquah and Ying Chen ***

Department of Biosystems Engineering, University of Manitoba, Winnipeg, MB R3T 2N2, Canada; acquahk@myumanitoba.ca
* Correspondence: ying.chen@umanitoba.ca; Tel.: +1-204-470-6292

**Abstract:** Agricultural fields are usually subjected to high amounts of traffic from field operations. The influence of traffic on sandy loam soil in three tillage systems were investigated in a field experiment. The field was located in a Canadian prairie region. In the experiment, the treatments were three tillage systems: no-tillage, disc tillage, and spring-tine tillage. Following tillage operations, field plots were trafficked with one pass of a sub-compact tractor. Soil properties were measured before and after the traffic to examine the effects of tillage systems and wheel traffic. For the effects of the tillage systems on the soil bulk density, soil shear strength, soil surface resistance, and soil cone index, the no-tillage system had higher values for all the soil properties when compared with the disc and spring-tine tillage systems. The plant (canola) population density ranged from 18.2 plants/m$^2$ to 34.9 plants/m$^2$, with the no-tillage having the lowest plant densities. For the effects of wheel traffic, one pass of the tractor in the disc and spring-tine tillage plots resulted in a 2.7% and 17.4% reduction in soil moisture content, respectively. After wheel traffic, the average soil shear strength for the disc and spring-tine systems were still significantly lower than the no-tilled system. Sinkages of 40 and 50 mm were observed for the spring-tine and disc tillage systems, respectively. The results of this study highlight the importance of preventing the demerits of soil compaction induced by wheel traffic after tillage operations.

**Keywords:** no-tillage; disc; spring-tine; soil; property; traffic

**Citation:** Acquah, K.; Chen, Y. Soil Compaction from Wheel Traffic under Three Tillage Systems. *Agriculture* **2022**, *12*, 219. https://doi.org/10.3390/agriculture12020219

Academic Editors: Mustafa Ucgul and Chung-Liang Chang

Received: 8 December 2021
Accepted: 31 January 2022
Published: 3 February 2022



## 1. Introduction

Tillage operations such as harrowing, ploughing and pulverization of soil crust [1] are usually performed under conventional tillage practises. One of the main purposes of tillage is to loosen the soil for good seedbed conditions. However, a subsequent field operation such as seeding or chemical spraying can compact the loosened soil, erasing the benefit of the tillage. The wheel traffic will alternate the structure of the topsoil layer [2]. Existing studies have been focusing on soil property changes caused solely by different trafficking scenarios, such as by large or small tractors, or different numbers of passes of tractors, in the same field condition. There is a lack of documentation on the soil property changes by trafficking on differently tilled soil conditions, especially in Canadian prairie regions. This study aimed to fill this gap by investigating the effects of trafficking under three different tillage systems.

To avoid soil compaction problems, agricultural producers use different tillage systems such as conventional and conservation tillage systems [3]. No-tillage and minimum tillage systems fall under conservation tillage systems due to reduced reliance on farm machinery for soil tiling purposes. On the other hand, conventional tillage systems rely heavily on farm machinery for seedbed preparations. A conventional tillage system would require more than 20% of the total traffic experienced before seeding operations, as this tillage system often consists of both primary and secondary tillage activities [4]. Disc and spring-tine tillage systems are generally associated with secondary tillage practices. These systems are generally used due to their quick work rates and low draft [5]. There are some demerits associated with both disc and spring-tine tillage systems, such as the ineffective burial of crop residue or weeds and soil smearing [5]. However, tillage involving one pass of such implements (disc or spring-tine harrows) can be classified as a reduced tillage system and, in turn, is beneficial for soil conservation. In addition, the low draft associated with these implements allow agricultural producers to use smaller tractors for tillage operations. Thus, disc and spring-tine tillage systems, together with the no-tillage system were selected in this study. Wheeling or traffic from seeding operations after tillage can affect the soil properties. For example, the increased intensity of tractor movements (traffic) on a field can decrease soil porosity which can lead to improper aeration and drainage, increased soil strength, and impedance of root growth [6,7]. All these may have adverse effects on plant growth and crop yield.

Understanding the influence of traffic or wheeling events under different tillage systems are very important in reducing severe soil compaction. Alaoui and Helbling [8] studied the soil compaction effects on the soil structure, and the results showed that traffic caused severe soil compaction from the soil surface to the 0.10 m depth resulting in the collapse of the soil structure. This stopped or reduced water flow movement from the topsoil layer to the subsoil layer. Moreover, Villanueva et al. [9] observed a 23% decrease in the soil moisture content in the conventional tillage system when compared to the zero tillage and minimum tillage systems. The conventional tillage systems experienced more tractor passes than the other tillage systems investigated.

In terms of the effect of soil compaction on the soil properties, Botta et al. [10] studied the effect of the number of tractor passes on the soil cone index in two tillage systems. That study inferred that irrespective of the mass and size of the tractor used, the cone index was observed to be higher in the conventional tillage system as compared to the direct sowing system (zero tillage). Zhang et al. [11] observed an increase in the soil bulk density and soil penetration resistance with an increase in tractor movement intensity and axle loads when studying the influence of small tractor traffic on soil compaction in northeast China. Lastly, Samuel and Ajav [1] also observed an increase in soil bulk density and soil strength when the number of tractor passes increased. Consequently, the soil compaction from the traffic intensity decreased soil porosity and soil moisture content.

In a nutshell, the aforementioned research studies did not focus on sandy loam soil found in the Canadian prairie regions. Additionally, other research studies made conclusions based on only crop yield and root growth [4,12,13]. However, the studies also involved clayey and Stagnic Luvisol soils and mostly focused on how different tire characteristics and configurations affected soil properties without considering the effect of the different tillage systems. The aim of the research study was to investigate the influence of traffic or wheeling events on sandy loam soil in a Canadian prairie region. The specific objectives of this study were to conduct a field experiment to study (1) the effect of no-tillage, disc tillage, and spring-tine tillage systems on the soil properties and plant population density, and (2) the effect of soil compaction across different tillage systems on soil properties.

## 2. Materials and Methods

### 2.1. Description of Test Equipment and Experimental Field

A John Deere 1023e sub-compact tractor was used for the experiment. The total mass of the tractor was 656 Kg. The tires (Carlisle® multi-trac C/S) were generally elastic and composed of polyurethane rubber material. The front and rear tires had a four-ply and six-ply rating, respectively, and the corresponding recommended inflation pressures were 151.7 and 206.8 KPa. The front tire had a diameter, width, and rim width of 0.47, 0.21, and 0.15 m, respectively. The diameter, width, and rim width of the rear tire were 0.65, 0.29, and 0.22 m, respectively. The tillage equipment included a double-action disc harrow and a spring-tine cultivator. The double-action disc harrow was a type of disc harrow with four gangs arranged in tandem (Figure 1a). It consisted of eight notched front discs and eight smooth rear discs (Figure 1b). Each disc had a diameter of 0.46 m. The implement had a cutting width of 1.52 m. The spring-tine cultivator consisted of nine individual spring-tines (Figure 2a). The tines were arranged in a staggered manner to prevent stepping. The tine consisted of an S-shank and a "spoon" point (Figure 2b). The implement had a working width of 1.5 m. The experiment was performed on a field located in the municipality of Piney, Manitoba, Canada. The soil contained 73.5% sand, 23.6% silt, and 2.9% clay, which was classified as a sandy loam by the USDA soil texture triangle [14]. The previous crop was wheat, and straw was spread in the field at the harvest.

(a) (b)

**Figure 1.** (**a**) Double-action disc harrow; (**b**) notched and smooth discs on disc harrow.

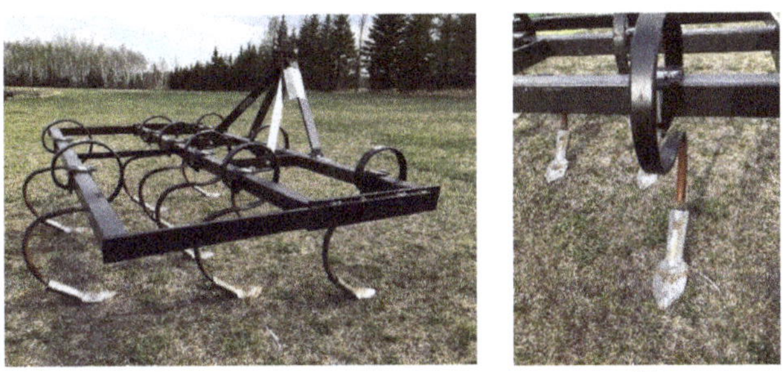

(a) (b)

**Figure 2.** (**a**) Spring-tine cultivator; (**b**) individual spring-tine on the spring-tine cultivator.

*2.2. Experimental Design and Field Layout*

The experimental design was a completely randomized design, which consisted of a factorial treatment layout. The three tillage systems were no-tillage, disc tillage, and spring-tine tillage systems. These tillage systems served as the factors for the experiment. Each treatment was replicated five times in the experimental design, giving a total of 15 field plots.

Each plot was 1.5 m wide to accommodate one pass of the tillage implement and seeder. The plot length was 45.7 m long. The three tillage systems were applied to the plots which had the wheat residue condition, as shown in Figure 3a. After tillage, the soil surface conditions are shown in Figure 3b. A tillage depth of 12.5 mm was used for the disc and spring-tine tillage systems.

(a) (b) (c) (d)

**Figure 3.** (**a**) Initial soil surface condition of field with wheat residue; (**b**) experimental field after tillage; (**c**) presence of a wheel track after passage of the tractor for seeding; (**d**) emerged canola plants in the no-tillage plot.

### 2.3. Experimental Procedure and Measurements

The experiment involved taking measurements under three phases. For the first phase, measurements were taken to characterize the initial soil surface condition of the field before tillage application (Figure 3a). These measurements included soil properties, residue cover, and dry residue mass. In the second phase, the various tillage systems were performed on the field (Figure 3b). Again, soil properties were measured immediately after the tillage operation. Then, the third phase involved tractor passage of the seeding operation on the tilled plots (Figure 3c). Soil properties and compaction variables were measured in the tractor wheel tracks. Finally, the plant population density was measured after canola plants emerged (Figure 3d).

#### 2.3.1. Initial Field Conditions before Field Operations

Soil properties (dry bulk density, soil cone index, soil shear strength, soil moisture content, and soil surface resistance) and residue condition (residue cover and dry residue mass) were measured to determine the initial soil surface conditions of the field. Seven samples were taken for each soil property measured. Soil moisture content and soil dry bulk density of the experimental field were measured using the soil core method. Soil samples were taken with a core sampler at random locations of the field. The core sampler had a diameter and height of 50 and 100 mm, respectively. The soil samples were weighed and oven-dried at 105 °C for 24 h. Soil shear strength was measured using a Geotechnics vane shear meter (Figure 4a). The vane shear meter consisted of four rectangular bladed vanes with a vane height of 29 mm. The vane diameter and area of a single blade were 19 mm and 275.5 $mm^2$, respectively [15]. The soil cone index was measured with a Rimik CP20 cone penetrometer device (Figure 4b). It consisted of a standard cylindrical rod with a cone tip angle of 30°, a load cell, and a chipset. Measurements were taken from the soil surface to a maximum depth of 105 mm at 15 mm intervals. The measurements were taken at a constant penetration velocity of 0.02 m/s [16]. The soil surface resistance was measured with a Humboldt dial pocket penetrometer (Figure 4c). The pocket penetrometer consisted of a 25 mm diameter plunger or foot and a dial with inner and outer scales. The dial had an accuracy of ±1% full scale at 20 °C. Measurements were taken by pushing the plunger into the soil until its foot flushed with the soil surface, then the value on the dial was read. The residue mass on the soil surface was determined by collecting residue within a 1 $m^2$ quadrant placed at seven random locations on the field (Figure 4d). The seven collected samples were oven-dried at 60 °C for 72 h [17] and weighed to determine the dry mass (Kg/ha). The residue cover was measured using the rope method. This method involved counting the number of times marked lines on a rope, spaced at 0.3 m intervals, intersected with a piece of residue on the field (Figure 4e).

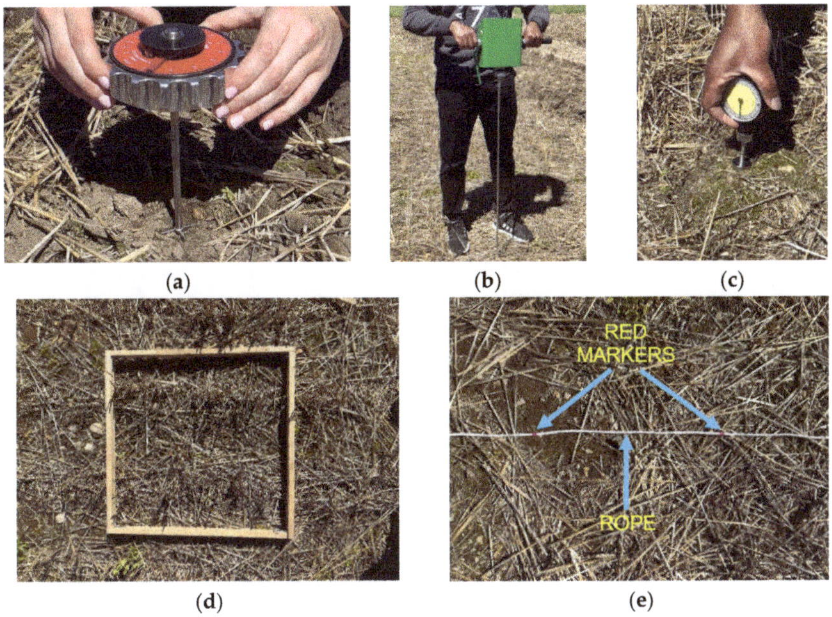


**Figure 4.** (**a**) Vane shear meter; (**b**) Standard penetrometer; (**c**) Pocket penetrometer; (**d**) Quadrant for residue mass measurement; (**e**) Rope method for residue cover measurement.

### 2.3.2. Tillage Operation

The tillage operation was performed based on the plot layout of the experimental design. A tillage depth of 12.5 mm was used for the disc and spring-tine tillage systems. The tractor was driven at a velocity of 1.1 m/s [8]. After the tillage operation, soil dry bulk density, soil cone index, soil shear strength, soil moisture content, and soil surface resistance were measured in each plot. Within each plot, three samples or measurements were taken for each aforementioned soil property.

### 2.3.3. Traffic and Seeding Operations

Afterwards, the tilled plots were trafficked with one pass of the tractor for seeding operations (Figure 3c). Seeding was also performed with a four-row seeder (Plotter Choice, Kasco Manufacturing, 170 W 600 N, Shelbyville, IN 46176, USA). Canola was seeded at a row spacing of 300 mm. The canola seeds (variety: InVigor L340) were certified seeds from a Manitoban seed supplier (Friesen Seeds Ltd, Rosenort, MB, Canada.). The seeds were treated with Prosper Evergol 4.25 M. The travel speed of the tractor for the operation was 1.1 m/s (4 Km/h). The target seeding depth was 12.5 mm, recommended by the canola council of Canada [18].

In the centreline of the wheel track, soil dry bulk density, soil cone index, soil shear strength, soil moisture content, soil surface resistance, imprinted tire width, and soil sinkage were measured in individual plots. Within each individual plot, three samples or measurements were taken for each soil property, imprinted tire width, and soil sinkage. The imprinted tire width and soil deformation (sinkage) were measured using a measuring rule and tape on the wheel track. The plant population density was counted 10 days after seeding. The number of emerged canola seedlings within the central two crop rows were counted at three random locations per plot. At each location, the length of the crop rows used for counting was 2 m. Unfortunately, flea beetles attacked and destroyed the plants. As a result, plant counting was performed only two times, and crop yields could not be obtained.

Field operations were managed based on normal farm practices [19]. A summary of the crop production management information is presented in Table 1.

**Table 1.** Crop production management information.

| Field Activity | Type | Date |
|---|---|---|
| Soil fertility sampling | Soil core | May 3 |
| Herbicide application | Glyphosate Ammonium sulphate (20-0-0-24) | May 3 |
| Fertilizer application | Monoammonium phosphate (11-52-0) and Nitrogen | May 28 |
| Tillage application | No till, Disc, and Spring-tine | May 28 |
| Seeding | Four-row seeder | May 28 |
| 1st Plant count | | June 7 |
| 2nd Plant count | | June 8 |

### 2.4. *Statistical and Data Analysis*

Analysis of variance (ANOVA) was used to determine if the three tillage systems had a significant effect on the measured variables. ANOVA was also used to determine if the soil compaction across the three tillage systems had an effect on the variables. SAS University edition, 2021 statistical software was used to perform the ANOVA and Duncan's multiple range test (a pairwise comparison) at $p < 0.05$.

## 3. Results and Discussion

### 3.1. *Initial Field Conditions*

Residue cover measured using the rope method was 76.4%, which was close to the percentage residue cover range stated by Burgess et al. [20] for classifying no-till plots with high amounts of corn stover residue. The dry residue mass was 1847 Kg/ha, which was lower than that reported in Burgess et al. [20]. The low residue mass was possibly due to the residue decomposition over time. The soil moisture content for the field at the time of the experiment was 18.5% (dry basis). This amount of moisture content level in the topsoil layer was suitable for field operations on sandy loam soil. The values of dry soil bulk density, soil surface resistance, soil shear strength, and soil cone index were typical for the given soil type before tillage. A summary of the initial soil condition and measurements is presented in Table 2.

**Table 2.** Summary of initial soil conditions.

| Measurement | Unit | Value |
|---|---|---|
| Soil moisture content | % | $18.5 \pm 5.4$ |
| Soil dry bulk density | $Kg/m^3$ | $1524.3 \pm 0.2$ |
| Soil surface resistance | $Kg/m^3$ | $1.8 \pm 0.4$ |
| Soil shear strength | Nm | $2.9 \pm 1.1$ |
| Soil cone index | KPa | $2611.4 \pm 511.2$ |
| Crop residue mass | Kg/ha | $1847 \pm 477$ |
| Crop residue cover | % | $76.4 \pm 12.1$ |

### 3.2. *Effect of Tillage Systems on Soil Properties*

#### 3.2.1. Dry Soil Bulk Density

The dry soil bulk density was significantly different between the no-tillage system and the other two tillage systems (Figure 5a). The no-tillage system recorded a dry soil bulk density of 1520 $Kg/m^3$ while the disc and spring-tine tillage systems recorded 23% (1170 $Kg/m^3$) and 27.6% (1100 $Kg/m^3$) reduction, respectively, in dry soil bulk density.

Both conventional tillage systems (disc and spring-tine) aid in reducing the dry soil bulk density by breaking and loosening the soil in the topsoil layer. The difference in dry soil bulk density between the disc and spring-tine tillage systems was not significant. This was due to the similar tilling depth (12.5 mm) used for both tillage systems. In a nutshell, this indicated that both double-action disc and spring-tine harrow implements were suitable for reducing the dry soil bulk density in firm soil.

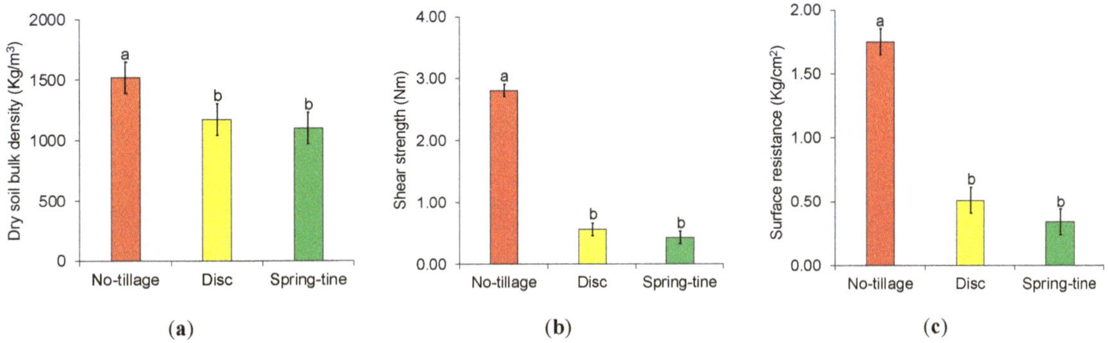


**Figure 5.** Soil properties across the three tillage systems: no-tillage, disc tillage, and spring-tine tillage; (**a**) Dry soil bulk density; (**b**) Soil shear strength; (**c**) Soil surface resistance. Values labelled with different letters are significantly different based on Duncan's multiple-range test at a significance level of 0.05 ($p < 0.05$).

### 3.2.2. Soil Shear Strength

The soil shear strength for the no-tillage system was 2.81 Nm (Figure 5b). This reflected the state of the soil at the no-tillage plots. The soil was typically firm and unresponsive to deformation (compression). The disc and spring-tine tillage systems recorded an 80% (0.56 Nm) and 84% (0.43 Nm) reduction in soil shear strength as compared to the no-tillage systems. Shearing, rotation, and compression are the three forms of soil deformation in soil dynamics [21]. Soil deformation due to shearing was evident when the double-action disc and spring-tine harrow implements were developed to till the soil. When tilling the soil, the internal structure and interlocking soil particles in the topsoil layer experienced friction. This led to the disassociation of bonds between soil particles resulting in shear. This resulted in the observed decrease of the soil shear strength in the conventional tillage systems. The soil shear strength was not significantly different between the disc and spring-tine tillage systems.

### 3.2.3. Soil Surface Resistance

In terms of the surface resistance of the soil surface, the no-tillage system was significantly different from the conventional (disc and spring-tine) tillage systems (Figure 5c). The highest surface resistance (1.75 $Kg/cm^2$) was observed for the no-tillage system, reflecting the firm state of the soil surface in those plots. The disc and spring-tine tillage systems recorded lower surface resistance due to the ability of the double-action disc and spring-tine harrow implements to break up firm soils. However, the surface resistance between the disc and spring-tine tillage systems were not significantly different. This could be due to the primary objective of both harrow implements to achieve soil pulverization in the topsoil layer irrespective of the degree of intensity provided.

### 3.2.4. Soil Cone Index

The relationship between the soil cone index and depth was investigated to understand the effect of both harrow implements on the soil (Figure 6a). As the depth increased, the soil cone index increased, regardless of the tillage system. The soil cone index values for the spring-tine tillage system ranged from 265.5 KPa to 1760 KPa within a specified depth range of 15 mm to 90 mm. In contrast, the soil cone index values for the disc tillage and no-tillage systems varied in a much narrower range within the same depth range. Other researchers observed similar linearly increasing trends for the relationship between soil cone index and the specified depth range of 15 mm to 105 mm in their research studies involving the effect of traffic on different tillage regimes [11,22,23].

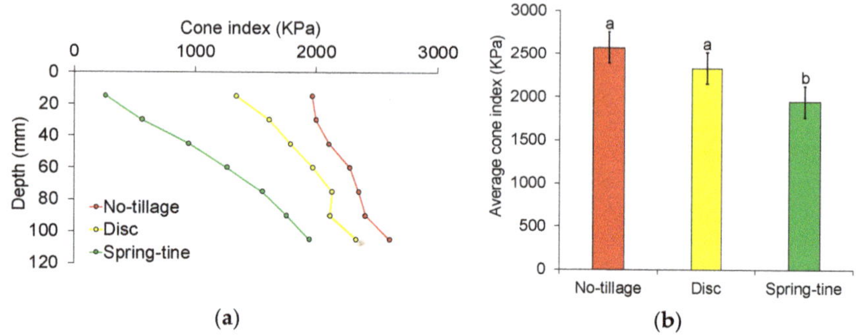


**Figure 6.** Soil cone index across the three tillage systems: no-tillage, disc tillage, and spring-tine tillage; (**a**) index-depth curves; (**b**) average indices over the depth range. Values labelled with different letters are significantly different based on Duncan's multiple-range test at a significance level of 0.05 ($p < 0.05$).

To further demonstrate the effects of the tillage system on the soil cone index, data points over an index curve were averaged over the depth of 105 mm (Figure 6b). The average soil cone index was not significantly different between the no-tillage and disc tillage systems. However, the average soil cone index for the spring-tine tillage system was significantly different when compared to the no-tillage and disc tillage systems. There was an observed reduction in the average soil sone index for the disc and spring-tine tillage systems. However, the spring-tine tillage system recorded a greater reduction in the average soil cone index when compared to the no-tillage and disc tillage systems. This could suggest the efficacy of the spring-tine harrow implement in reducing soil cone index at similar depths in sandy loam soils.

### 3.3. *Plant Population Density*

The first emergence of the canola seedlings began on 4 June. This was seven days after seeding. The spring-tine tillage system was marked with a significantly higher population density of canola plants as compared with the disc tillage and no-tillage systems on both dates when plants were counted (Figure 7). The plant population density for the disc tillage and no-tillage systems were not significantly different. On average, over two dates, the population density per unit area of canola plants for the spring-tine tillage, disc tillage, and no-tillage systems were 33.9, 25.2, and 18.5 plants/$m^2$, respectively.

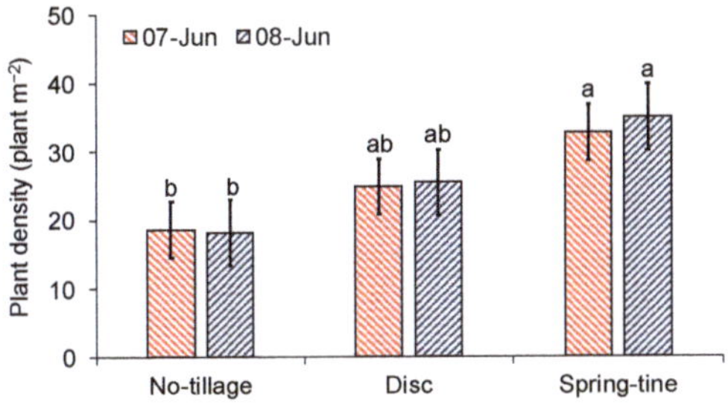


**Figure 7.** Plant population density per unit area in three tillage systems: no-tillage, disc tillage, and spring-tine tillage. Values labelled with different letters are significantly different based on Duncan's multiple-range test at a significance level of 0.05 ($p < 0.05$).

As mentioned in the previous sections, the soil within the no-tillage system was firm and highly resistant to shear and penetration. This clearly affected the seed germination and resulted in the smallest population of canola plants within that tillage system. The population density of canola plants for the disc tillage and no-tillage systems were not significantly different. The observed values of the soil properties in the disc tillage system were similar to the values recorded in the no-tillage system (Figures 8a and 9a–c). In contrast, the spring-tine tillage system observed a significantly higher population density of canola plants as compared to the no-tillage system. This was due to the lower readings for the soil properties (Figures 5a–c and 6a,b). This reduction in the soil properties gave credence to the ability of the spring-tine cultivator to alleviate the heightened effects of soil degradation in order to promote plant growth.

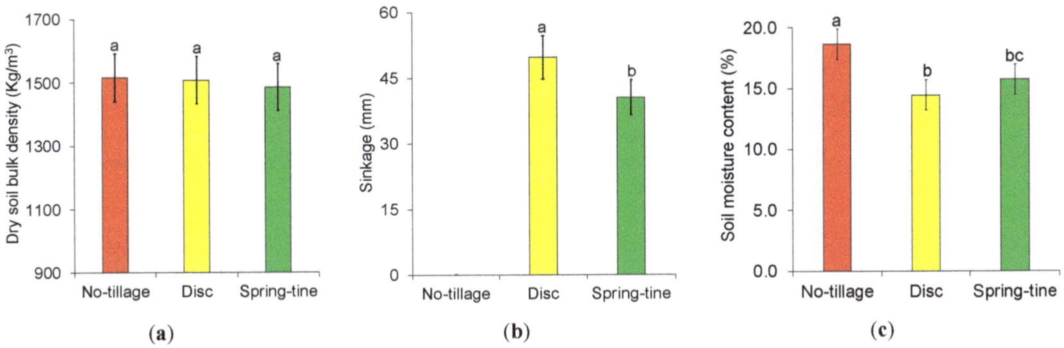


**Figure 8.** Soil properties taken in the wheel tracks under the no-tillage, disc tillage, and spring-tine tillage systems; (**a**) dry soil bulk density, (**b**) soil sinkage and (**c**) soil moisture content. Values labelled with different letters are significantly different based on Duncan's multiple-range test at a significance level of 0.05 ($p < 0.05$).

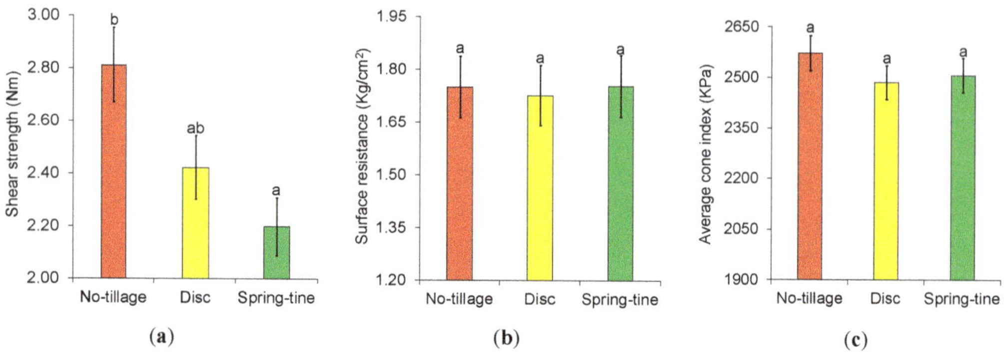


**Figure 9.** Soil properties taken in the wheel tracks under the no-tillage, disc tillage, and spring-tine tillage systems; (**a**) soil shear strength, (**b**) soil surface resistance and (**c**) average indices over the depth range. Values labelled with different letters are significantly different based on Duncan's multiple-range test at a significance level of 0.05 ($p < 0.05$).

### 3.4. *Effect of Wheel Track across Tillage Systems on Soil Properties*

#### 3.4.1. Dry Soil Bulk Density

The dry soil bulk density taken on the wheel tracks was not significantly different between the disc tillage, spring-tine tillage and no-tillage systems (Figure 8a). Considering the before and after effects of traffic, the dry soil bulk density significantly increased after the passage of the tractor in both disc and spring-tine tillage systems. However, the change in dry soil bulk density after tractor passage was not significantly different between the disc tillage, spring-tine tillage, and no-tillage systems. The effect of the passage of the tractor increased the dry soil bulk density towards the initial dry soil bulk density value of the experimental field for the disc and spring-tine tillage systems. The dry soil bulk density value recorded for the no-tillage system remained the same even after the passage of the tractor. Overall, the aim of drastically reducing soil bulk density before seeding would be considered not achievable. Increasing the field traffic by a higher number of tractor passes would lead to higher dry soil bulk density. This would, in turn, affect seed germination and plant growth resulting in lower crop yields.

#### 3.4.2. Soil Sinkage

Due to the firmness of the soil in the no-tillage plots, there was no presence of deformation (compression) on the soil surface after the passage of the tractor. Therefore, the sinkage was recorded as zero. The soil sinkage was significantly different between the disc and spring-tine tillage systems (Figure 8b). Generally, when a load such as a tractor tire is in contact with the soil surface, regional soil shear failure occurs, which results in the deformation of the soil structure beneath the tire [24]. This phenomenon was evident during the passage of the tractor. Additionally, the pore spaces in the soil are greatly reduced for every tractor passage. If pore spaces are greatly reduced or non-existent, further soil compression or deformation cannot occur [25]. The disc tillage system observed higher sinkage than the spring-tine tillage system due to the reduction of the pore spaces after the passage of the tractor. These pore spaces were created during the breaking and loosening of the soil structure by both implements. However, this indicated that the double-action disc harrow increased the porosity in the topsoil layer during tillage.

#### 3.4.3. Soil Moisture Content

The soil moisture content for the no-tillage, disc tillage, and spring-tine tillage systems was not significantly different from each other directly after tillage. After the passage of the tractor (after compaction), the disc and spring-tine tillage systems observed lower soil moisture content readings when compared to its initial readings directly after tillage (Figure 8c). However, the soil moisture content for the disc and spring-tine tillage systems were not significantly different after compaction. The soil moisture content for the disc and spring-tine tillage systems were 14.4% and 15.7%, respectively. The effect of soil compaction was evident on the soil moisture content for the disc and spring-tine tillage systems. The disc tillage system observed a 2.7% (14.4% dry basis) reduction in soil moisture content directly after the passage of the tractor. When comparing the initial (after tillage) and final (after compaction) effects, the soil moisture content was not significantly different. In contrast, the spring-tine tillage system observed a 17.4% (15.7% dry basis) reduction in soil moisture content directly after the passage of the tractor. When comparing the initial (after tillage) and final (after compaction) effects, the soil moisture content was significantly different. Overall, soil compaction as a result of traffic had an observable effect on the soil moisture content. The passage of the tractor influenced the disruption of the soil structure in the topsoil layer resulting in the reduction of macropores (porosity) for water flow [8]. This led to a decrease in soil moisture content after compaction. Samuel and Ajav [1] and Villanueva et al. [9] observed similar findings where soil moisture content decreased after several passes of the tractor in conventional tillage plots.

#### 3.4.4. Soil Shear Strength and Soil Surface Resistance

Shear strength and surface resistance were not significantly different between the wheel tracks under the disc and spring-tine tillage systems (Figure 9a,b). Similar to what was observed with the dry soil bulk density, the shear strength and surface resistance significantly increased after the passage of the tractor in the disc and spring-tine tillage systems. Again, the shear strength and surface resistance values recorded for the no-tillage system remained the same even after the passage of the tractor. The change in shear strength and surface resistance was not significantly different between the disc and spring-tine tillage systems. In the wheel tracks, the shear strength and surface resistance also increased towards the initial shear strength and surface resistance recorded. The influence of the tractor passage on the soil surface led to the collapse of the loosened soil structure. This form of soil compaction combined the soil particles together, thereby creating a dense soil structure in the topsoil layer. As a result, the soil shear strength and surface resistance increased for the topsoil layer as well.

#### 3.4.5. Average Soil Cone Index

The average soil cone index was not significantly different between the wheel tacks under three tillage systems (Figure 9c). There was a significant increase in the average cone index after the passage of the tractor in the disc and spring-tine tillage systems. The change in average cone index was significantly different between the disc and spring-tine tillage systems. In the disc and spring-tine tillage systems, the average cone index increased towards the initial soil cone index value. However, the spring-tine tillage system observed a 28.6% (2507 KPa) increase in the average soil cone index, while the disc tillage system observed a 6.5% (2485 KPa) increase. The average soil cone index for the no-tillage system did not increase or decrease due to its firm nature. As previously mentioned, the spring-tine harrow implement was effective in reducing the soil cone index in the 105 mm depth. However, the spring-tine tillage system was greatly influenced by the passage of the tractor on the soil surface. The resulting soil compaction compressed the topsoil layer, which led to an increase in the soil cone index at the specified depth. The relationship between the depth and soil cone index after tractor passage (Figure 10) was variable as compared to the relationship between the depth and soil cone index after tillage. The disc tillage system observed a higher soil cone index ranging from 2002 KPa to 2217 KPa for the specified depth of 15 mm to 60 mm. In the same specified depth, the spring-tine tillage system observed a lower soil cone index ranging from 1914 KPa to 2069 KPa. Thereafter, the soil cone index in both tillage systems became similar, and the trend increased linearly.

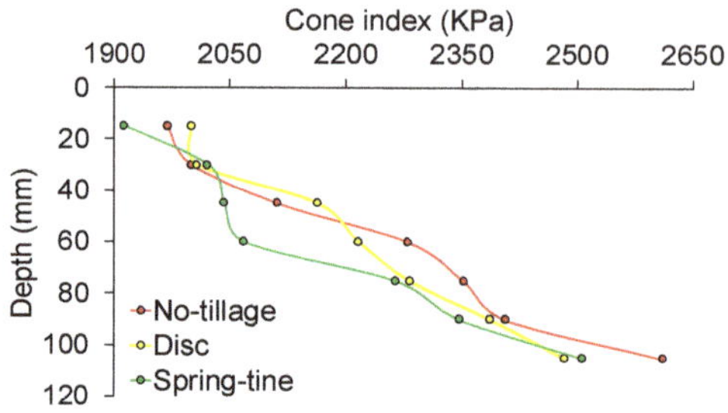


**Figure 10.** Index-depth curves for the no-tillage, disc tillage, and spring-tine tillage systems after passage of the tractor.

## 4. Conclusions

A field experiment was performed to study the influence of tillage and traffic on sandy loam soil in the Canadian prairie region. In terms of the effect of tillage systems on the soil properties, the no-tillage system observed significantly higher dry soil bulk density, soil shear strength, and soil surface resistance when compared with the disc and spring-tine tillage systems. The relationship between the soil cone index and depth was observed to be an increasing trend for all three tillage systems. The spring-tine favoured a higher plant population density when compared to the no-tillage system. After wheel trafficking on the tilled soil, some soil properties on the wheel tracks were significantly different. The soil sinkage was greater in the disc tillage plots than the spring-tine tillage plots. The wheel tracks on the disc and spring-tine tillage plots reduced the soil moisture content compared with the wheel tracks in the no-tillage plots. Overall, soil preparation using tillage was highly advantageous to the establishment of the canola crop. The reduction of the dry soil bulk density and soil strengths in the tilled plots led to the improvement of the plant population density. Specifically, the spring-tine tillage system was the best method of soil preparation for the canola crop considering both soil condition and plant establishment. This study showed that soil properties varied not only with different tillage systems but also with the field traffic from the subsequent field operation. Therefore, both the tillage system and subsequent traffic should be taken into consideration for the management of field operations in supporting crop production.

**Author Contributions:** Conceptualization, methodology, resources, formal analysis, investigation, validation, writing—original draft, data curation, visualization, K.A. and Y.C.; supervision, Y.C. All authors have read and agreed to the published version of the manuscript.

**Funding:** This research was funded by the Natural Sciences and Engineering Research Council of Canada (NSERC). Grant No.: RGPIN-2019-05861.

**Institutional Review Board Statement:** Not applicable.

**Informed Consent Statement:** Not applicable.

**Data Availability Statement:** Not applicable.

**Acknowledgments:** The authors would like to thank Sylvio Tessier, Leno Guzman, Sydni Reimer, and Robyn Birch for their assistance in the field experiments.

**Conflicts of Interest:** We declare that we have no conflict of interest.

## References

1. Samuel, T.M.; Ajav, E.A. Optimum Tillage System for Pepper Production in an Alfisol of South-Western Nigeria. *Afr. J. Agric. Res.* **2010**, *5*, 2647–2652.
2. Badalíková, B. Influence of Soil Tillage on Soil Compaction. *Soil Eng.* **2010**, *20*, 19–30. [CrossRef]
3. Moraru, P.I.; Rusu, T. Effect of Tillage Systems on Soil Moisture, Soil Temperature, Soil Respiration and Production of Wheat, Maize and Soybean Crops. *J. Food Agric. Environ.* **2012**, *10*, 445–448.

4. Botta, G.F.; Rivero, D.; Tourn, M.; Melcon, F.B.; Pozzolo, O.; Nardon, G.; Balbuena, R.; Becerra, A.T.; Rosatto, H.; Stadler, S. Soil Compaction Produced by Tractor with Radial and Cross-Ply Tyres in Two Tillage Regimes. *Soil Tillage Res.* **2008**, *101*, 44–51. [CrossRef]
5. Choudhary, M.A. *Seedbed Preparation Methods and Their Effect on Soil Physical Conditions and Crop Establishment*; In Soil physics: Application under stress environments. Proceedings of the International Symposium on Applied Soil Physics in Stress Environments, Islamabad, Pakistan, 22–26 January 1989; Pakistan Agricultural Research Council (PARC): Islamabad, Pakistan, 1990; pp. 296–300.
6. Hamza, M.A.; Anderson, W.K. Soil Compaction in Cropping Systems: A Review of the Nature, Causes and Possible Solutions. *Soil Tillage Res.* **2005**, *82*, 121–145. [CrossRef]
7. Bayhan, Y.; Kayisoglu, B.; Gonulol, E. Effect of Soil Compaction on Sunflower Growth. *Soil Tillage Res.* **2002**, *68*, 31–38. [CrossRef]
8. Alaoui, A.; Helbling, A. Evaluation of Soil Compaction Using Hydrodynamic Water Content Variation: Comparison between Compacted and Non-Compacted Soil. *Geoderma* **2006**, *134*, 97–108. [CrossRef]
9. Villanueva, J.J.M.; García, V.M.V.; Alberich, M.V.E.; Huerta, A.G.; Arriaga, M.R.; Rodríguez, F.G. Mean Infiltration Speed in a Vertisol under Different Tillage Systems. *TERRA Latinoam.* **2015**, *33*, 95–99.
10. Botta, G.F.; Becerra, A.T.; Tourn, F.B. Effect of the Number of Tractor Passes on Soil Rut Depth and Compaction in Two Tillage Regimes. *Soil Tillage Res.* **2009**, *103*, 381–386. [CrossRef]
11. Zhang, X.Y.; Cruse, R.M.; Sui, Y.Y.; Jhao, Z. Soil Compaction Induced by Small Tractor Traffic in Northeast China. *Soil Sci. Soc. Am. J.* **2006**, *70*, 613–619. [CrossRef]
12. Koch, H.J.; Heuer, H.; Tomanová, O.; Märländer, B. Cumulative Effect of Annually Repeated Passes of Heavy Agricultural Machinery on Soil Structural Properties and Sugar Beet Yield under Two Tillage Systems. *Soil Tillage Res.* **2008**, *101*, 69–77. [CrossRef]
13. Millington, A. The Effect of Low Ground Pressure and Controlled Traffic Farming Systems on Soil Properties and Crop Development for Three Tillage Systems (. Ph.D. Thesis, Harper Adams University, Newport, Shropshire, UK, 2019.
14. USDA. Keys to Soil Taxonomy. *Soil Conserv. Serv.* **2014**, *12*, 410. [CrossRef]
15. ASTM Standards. Standard Test Method for Field Vane Shear Test in Cohesive Soil D2573—01. *Am. Soc. Test. Mater.* **2002**, *4*, 4–11.
16. ASAE Standards. *Soil Cone Penetrometer S 313.3*; ASAE: St. Joseph, MI, USA, 1999; 3–5.
17. ASABE Standards. *S358.3 Moisture Measurement—Forages*; ASAE: St. Joseph, MI, USA, 2012.
18. The Canola Council of Canada. Plant Establishment—Seed Depth. Available online: https://www.canolacouncil.org/canola-encyclopedia/plant-establishment/seed-depth/ (accessed on 8 September 2021).
19. Manitoba Agriculture. Canola Production and Management. Available online: https://www.gov.mb.ca/agriculture/crops/crop-management/canola.html (accessed on 8 September 2021).
20. Burgess, M.S.; Mehuys, G.R.; Madramootoo, C.A. Tillage and Crop Residue Effects on Corn Production in Quebec. *Agron. J.* **1996**, *88*, 792–797. [CrossRef]
21. Pytka, J. Effects of Repeated Rolling of Agricultural Tractors on Soil Stress and Deformation State in Sand and Loess. *Soil Tillage Res.* **2005**, *82*, 77–88. [CrossRef]
22. Kurjenluoma, J.; Alakukku, L.; Ahokas, J. Rolling Resistance and Rut Formation by Implement Tyres on Tilled Clay Soil. *J. Terramechanics* **2009**, *46*, 267–275. [CrossRef]
23. Larney, F.J.; Kladivko, E.J. Soil Strength Properties Under Four Tillage Systems at Three Long-Term Study Sites in Indiana. *Soil Sci. Soc. Am. J.* **1989**, *53*, 1539–1545. [CrossRef]
24. McKyes, E. Soil Cutting and Tillage. *Dev. Agric. Eng.* **1989**, *10*, 192–221. [CrossRef]
25. Taghavifar, H.; Mardani, A. Effect of Velocity, Wheel Load and Multipass on Soil Compaction. *J. Saudi Soc. Agric. Sci.* **2014**, *13*, 57–66. [CrossRef]

*agriculture*

*Article*

# Design and Testing of Reverse-Rotating Soil-Taking-Type Hole-Forming Device of Pot Seedling Transplanting Machine for Rapeseed

**Wei Quan [1,2], Mingliang Wu [1,*], Zhenwei Dai [1], Haifeng Luo [1] and Fanggang Shi [1]**

1 College of Mechanical and Electrical, Hunan Agricultural University, Changsha 410128, China; quanwei@hunau.edu.cn (W.Q.); daizhenwei@stu.hunau.edu.cn (Z.D.); luohaifeng@hunau.edu.cn (H.L.); shifanggang@stu.hunau.edu.cn (F.S.)
2 College of Orient Science & Technology, Hunan Agricultural University, Changsha 410128, China
* Correspondence: mlwu@hunau.edu.cn

**Abstract:** To address the problem whereby the size of the hole formed by the existing hole-forming device of hole-punching transplanters is significantly inconsistent with the theoretical size as it is impacted by the inserting and lifting methods, a scheme for eliminating the forward speed of the whole machine by the horizontal linear velocity of reverse rotation of the hole-forming mechanism is proposed to vertically insert and lift the hole-forming device in accordance with the working characteristics of hole-punching transplanting and the agronomic requirements of rapeseed transplanting. In addition, a novel type of reverse-rotating soil-taking-type hole-forming device for the pot seedling transplanting machine for rapeseed was developed. A test bench for the hole-forming device was set and its effectiveness was verified in the soil bin. It was found, from the test results, that, when the forward speed of the hole-forming device was between 0.25 m/s and 0.45 m/s, the average qualified rates of hole forming of the device were 95.2%, 94.0% and 93.3%, respectively; the average change rates of the hole size were 2.3%, 2.9% and 5.5%, respectively; and the average error between the theoretical value of effective depth and the experimental value was between 2.0% and 5.6%. The average angle between the hole-forming stage trajectory of the hole opener and the horizontal direction at different forward speeds was higher than 88.0°; the coefficient of variation was between 0.16% and 0.64%; the perpendicularity of the hole-forming operation was high; the change rates of soil porosity of the hole wall were between 8.2% and 9.3%; and the average soil heave degrees at the hole mouth after the completion of the hole-forming operation were 3.9%, 4.1% and 4.2%, respectively. The average soil stability rates of the hole wall were 91.9%, 91.2% and 91.0%, respectively. The different performances of the hole-forming device were confirmed to meet the requirements of rapeseed pot seedling transplanting. This study can provide a reference for the structural improvement and optimization of the hole-punching transplanter for rapeseed pot seedlings.

**Keywords:** rapeseed transplanting; hole-forming device; key components; experiment

**Citation:** Quan, W.; Wu, M.; Dai, Z.; Luo, H.; Shi, F. Design and Testing of Reverse-Rotating Soil-Taking-Type Hole-Forming Device of Pot Seedling Transplanting Machine for Rapeseed. *Agriculture* **2022**, *12*, 319. https://doi.org/10.3390/agriculture12030319

Academic Editors: Mustafa Ucgul and Chung-Liang Chang

Received: 5 December 2021
Accepted: 20 February 2022
Published: 22 February 2022



## 1. Introduction

Rapeseed, a vital oil crop worldwide, has been found as the major source of plant edible oil and forage protein; the rapeseed planting area and the rapeseed demand are increasing on a year-to-year basis [1–3]. There have been two main modes of rapeseed planting, including direct seeding and transplanting; conventional manual planting can no longer satisfy the needs of large-scale rapeseed planting and the rational utilization of mechanized production is critical to developing the rapeseed planting industry [4–6]. As indicated by relevant studies, mechanized rapeseed transplanting is capable of alleviating crop stubble contradiction, as well as expanding the planting area, which acts as a vital planting mode for developing the rapeseed industry [7–11].

In accordance with different planting principles, scholars worldwide have separated the mechanized pot seedling transplanter into three categories, including seedling-ditching channel-type transplanters, duckbill-type transplanters and hole-punching-type transplanters [12–15]. The damage rate of seedlings is high and the uprightness of seedlings cannot be ensured when using the seedling-ditching channel-type transplanter. The duckbill-type transplanter raises more rigorous requirements on seedling age and shape; the feeding speed of seedlings should be appropriate and the transplanting speed is limited. Although the hole-punching-type transplanter raises certain requirements on soil texture and moisture content, its operation is less affected by the previous crop straw and weeds and low requirements are raised on transplanting seedlings; the operation mode to form holes first and subsequently release seedlings can ensure the uprightness of pot seedlings and rapid transplanting operation can be achieved under this mode, so the hole-punching transplanter has broad scientific research prospects [16–20].

At present, the essential part of a hole-punching transplanter is the hole-forming device, which primarily falls into two types in accordance with the hole-forming method, including the soil-extruding type and the soil-taking type [18,19,21–26]. The Rain-flo semi-automatic hole-forming transplanter manufactured by Buckeye Tractor Co in the United States uses a hole-forming shovel distributed on the outer edge of the water wheel to form holes in the soil by forcing soil extrusion; then, the pot seedlings are directly put into the planting holes manually. Both bare seedlings and pot seedlings can be transplanted. However, the hole-forming process brings a great disturbance to the soil and the transplanting speed is easily affected by the artificial proficiency. The transplanting mechanisms of hand-held, semi-automatic and automatic onion transplanters were previously summarized and analyzed and the wheel-type, rotary-type and linkage-type planting mechanisms were compared and analyzed. However, planting mechanisms adopt the rotary extrusion method to complete soil inserting, which results in large hole size and reduced soil porosity of the hole wall, thus having an effect on the perpendicularity of seedlings and air permeability in the formed holes [26]. Chen simulated and tested the hole-forming device of a buckwheat planter by applying discrete element software and multi-body dynamics software; the relevant parameters of the vital components (including duckbill and duckbill spring) were obtained and the bench test was performed to calculate the best spring wire diameter and rotational speed of seeding wheel. However, the soil disturbance around the hole and the soil porosity of the hole wall were significantly affected, since the rotary extrusion-type soil-inserting method was applied in the hole-forming device [27]. Han designed a rotary soil-taking hole-digging device for the semi-automatic watermelon pot seedling transplanter and its accessories; the designed device could satisfy the requirements of transplanting, whereas the size of the formed holes was overly large, since the hole-forming device showed a certain inclination angle with the soil level when put into the soil and excavated [28]. Quan conducted an optimization analysis on the vertical soil extruding-type and vertical soil-taking-type hole-forming devices. Based on the comparison of the porosity of the hole wall and the soil return coefficient of the section within the hole, the soil-taking type was confirmed as a relatively optimal hole-forming method and the size of the formed holes was found to be similar to the theoretical size. However, the device had to be initiated and stopped by traction power to vertically take soil and form holes and it had low operation efficiency [29]. In brief, some progress has been made in hole-punching and transplantation technology and equipment [30–34], while the existing hole-forming devices are primarily prone to a significant difference between the size of the formed hole and the theoretical size; due to problems in the inserting and lifting operation modes, the soil porosity of the hole wall decrease greatly, thus affecting the growth environment of pot seedlings and subsequently affecting the yield of crops [29,35–37].

Accordingly, based on the agronomic requirements of rapeseed pot seedling transplanting and combined with the operating characteristics of the hole-punching transplanter, this study designed a type of continuous operation of the reverse-rotating soil-taking-type hole-forming device. Based on the method to eliminate the forward speed of the whole machine by endowing the hole-forming mechanism with the horizontal linear velocity of reverse rotation, the hole-forming device could realize vertical inserting and lifting. Holes meeting the design requirements could be generated by combining the actions of collecting and discharging soil of the hole-forming device. The hole-forming method used in this study slightly affected the soil porosity of the hole wall of the formed hole, while reducing the disturbance to the soil. Moreover, the shape size of the formed hole was generally consistent with the theoretical size, thus facilitating the growth of pot seedlings after transplanting. In this study, the effect of different advancing speeds on the hole-forming effect was investigated and the changes in hole size, soil disturbance and soil porosity of the hole wall before and after the formation of hole were studied. This study can present a novel idea for the design of hole-forming devices for rapeseed pot seedling transplanters and theoretically improve this design.

## 2. Materials and Methods

### 2.1. Agronomic and Technical Requirements

In general, the existing rapeseed seedling pots have a squared cone shape, with a top-side length of 35 or 50 mm and a bottom-side length of 25 mm, as well as a height of 40 or 45 mm. The planting density of rape transplanting is approximately 110,000 plants/hm$^2$ and the planting depth is 30–60 mm [29,38,39]. Based on the agronomic requirements of dry land transplantation of rapeseed pot seedlings and the existing research results [29,40], soil taking was selected as the method of hole forming in this study and the shape of the hole-forming device was designed as a cylindrical table. The main technical parameters of the hole-forming device of the pot seedling transplanting machine for rapeseed were determined as shown in Table 1.

**Table 1.** Main technical parameters of the soil-taking-type hole-forming device.

| Overall Dimensions (mm) | Hole-Forming Method | Transplant Row Number | Hole-Forming Mechanism Number | Distance between Adjacent Hole-Forming Mechanisms (mm) | Maximum Transplant Depth (mm) |
|---|---|---|---|---|---|
| L × W × H: 2130 × 1130 × 1340 | soil picking | 1 | 5 | 300 | 80 |

### 2.2. Overall Structure and Working Principle

The reverse-rotating soil-picking-type hole-forming device consisted of a hole-forming mechanism and a transmission system, the structure of which is shown in Figure 1. The main structure included a frame, a hole-forming mechanism (5 sets), a transmission system, a side plate (transparent in the 3D picture), a hole-forming cam and a putter cam. To be specific, the transmission system was largely composed of a chain and sprockets, a circular guideway, a slider, a motor, a transmission shaft, etc. The side plate was welded on the frame, the circular guideway was symmetrically fixed on the side plate and the slider was installed on the circular guideway. In addition, the hole-forming mechanism was fixed with the slider, capable of rotating around the circular guideway through the slider; the hole-forming cam and the putter cam were fixed in the frame through the transmission shaft. The retaining plate was installed inside the frame and in the same plane as the soil opener to prevent the taken soil from being thrown back into the formed hole.

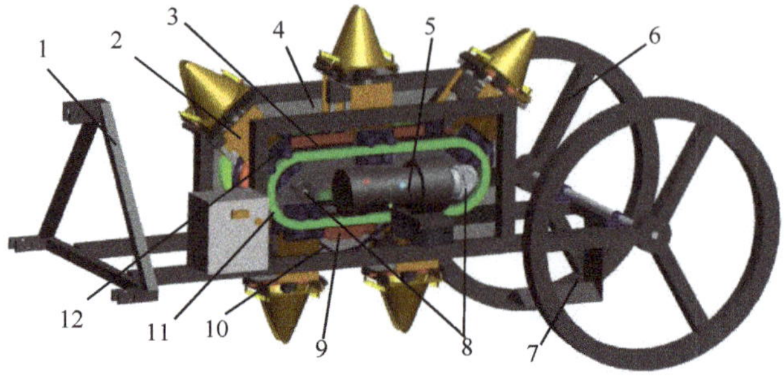


**Figure 1.** Structure diagram of reverse rotary soil-picking-type hole-forming device: 1—frame; 2—hole-forming mechanism; 3—chain and sprockets; 4—side plate; 5—motor; 6—ground wheel; 7—retaining plate; 8—transmission shaft; 9—hole-forming cam; 10—putter cam; 11—circular guideway; 12—slider.

When working, the soil bin tank pulled the hole-forming device at a uniform speed and the power of reverse rotation of the 5 sets of the hole-forming mechanism was inputted by the motor via the transmission shaft, the chains and sprocket and transmitted to the slider installed on the annular guide. Thus, the hole-forming mechanism fixed on the slider was driven to rotate and the linear speed of rotation of the hole-forming mechanism was equated with the forward speed of the hole-forming device. When the hole-forming mechanism entered the hole-forming stage, its speed was opposed to the forward speed of the whole machine and its size was equal. Thus, the hole-forming mechanism did not exhibit a horizontal speed relative to the soil surface at this stage. Then, under the joint action of the hole-forming cam and the putter cam, the soil opener in the hole-forming mechanism carried out the following actions: vertical soil inserting (with an open state of the soil opener); rotary soil taking (with a close state of the soil opener); vertical soil lifting (with a close state of the soil opener); and soil discharge and reset (the soil opener was first opened to the maximum angle to discharge soil and then reverted to the initial state). These operations were conducted according to the sequence as required by design, so as to form holes in the soil that met the design requirements. As shown in Figure 2, to intuitively understand the interaction between the key components of the hole-forming device, a three-dimensional schematic diagram of each working stage was drawn.

### 2.3. Design of the Hole-Forming Mechanism

#### 2.3.1. Overall Structure

The hole-forming mechanism studied in this paper was an important part of the reverse rotary soil-picking-type hole-forming device. It included a small side plate, a linear guide rail, a putter of open and close, a putter of bearing, a putter of hole forming, a putter spring, a support side plate, a return spring, a soil opener and a bracket of the soil opener. The structure is shown in Figure 3.

The small side plate was fixed on the slider and the support side plate was fixed on the small side plate by the support axis. The linear guide rail and the putter of hole forming were installed on four linear bearings, respectively, which could achieve up-and-down reciprocating motions. The upper end of the putter of hole forming was installed with a putter bearing that always moved in the hole-forming chute; the linear bearing was fixed on the small side plate and the support side plate, respectively, while the bottom end of the linear guide rail was welded on the soil-opener bracket. The soil opener was installed on the soil-opener bracket by the bracket axis, the putter of open and close, the return spring and the putter spring, which could rotate around the bracket axis. Among them, the putter of open and close was installed in the soil-opener bracket by linear bearing; its upper end was always in contact with the outer contour of the putter cam under the action of the putter spring and the return spring; the lower end was always in contact with the inner wall of the soil opener through the roller. The putter spring was set on the putter of open and close in a compression state and the return spring was hung on the soil opener in a stretching state. The up-and-down motions of the putter of open and close in the linear bearing could make the soil opener close and open.

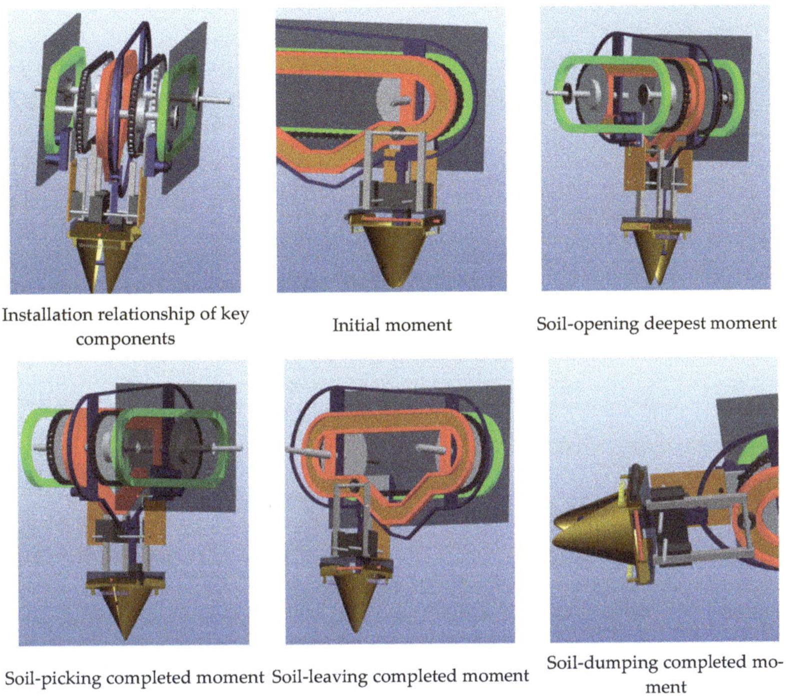


**Figure 2.** Schematic diagram of the working process of the hole-forming mechanism.

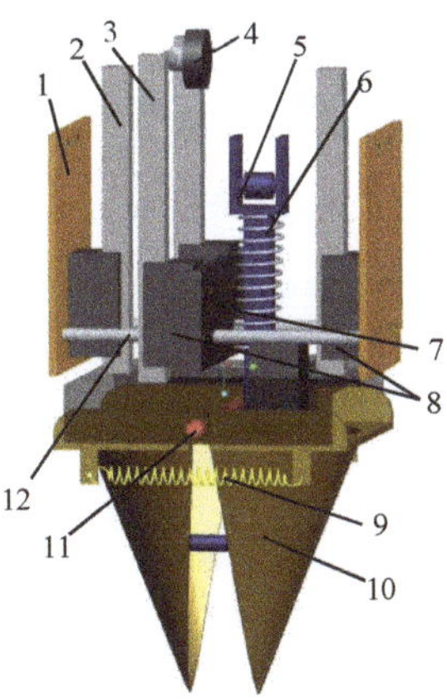


**Figure 3.** Structure diagram of the hole-forming mechanism: 1—small side plate; 2—linear guide rail; 3—putter of hole forming; 4—putter of bearing; 5—putter of open and close; 6—putter spring; 7—support side plate; 8—slider; 9—return spring; 10—soil opener; 11—bracket of axis; 12—support axis.

2.3.2. Design of Soil Opener

The soil opener was a vital part of the hole-forming mechanism and the soil opener formed the planting hole in the soil by inserting the soil, taking the soil and lifting the soil. In addition, the shape of the hole would directly affect the quality of pot seedling transplanting. Figure 4 presents the schematic diagram of the conical table soil-taking soil opener designed in this study. The soil opener acted as the soil-touching part of the hole-forming device, so the design of its parameters significantly affected the quality of the shape of the formed holes; the cutting-edge section thickness was considered a main parameter affecting the soil-cutting resistance. In general, the thinner the cutting-edge section thickness, the smaller the soil cutting resistance was, whereas the soil opener was prone to deform and even collapse. Thus, the cutting-edge section thickness b was selected as 2 mm [28,41].

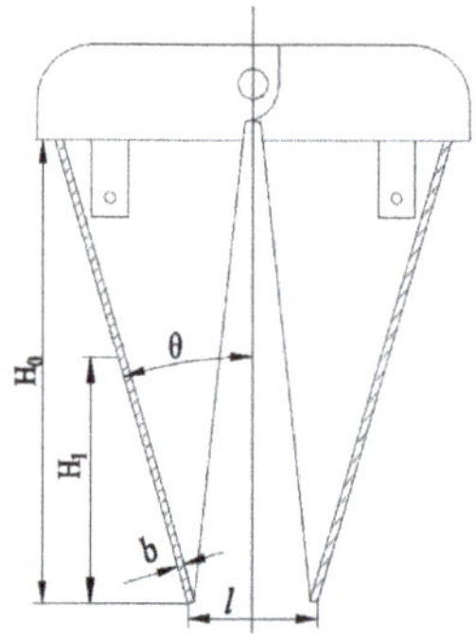


**Figure 4.** Schematic diagram of soil opener. Note: b is cutting edge section thickness (mm); *l* is bottom size of soil opener (mm); $H_0$ is height of conical-table-shaped soil opener (mm); $H_1$ is soil-opening depth (mm); θ is penetration angle (°).

According to the previous research results [40], the bottom size *l* of the soil opener was selected as 35 mm. In accordance with the requirements of planting depth and the structural characteristics of the hole-forming mechanism, the height $H_0$ of the soil opener was selected as 150 mm. If the angle of entry θ was overly large, the resistance of entry increased and the porosity of the soil in the hole wall was reduced. If the angle of entry θ was excessively small, it was not conducive to the stability of the soil in the hole wall. Based on the existing research results, the angle of entry θ was selected as 17° [29].

2.3.3. Movement Trajectory Analysis of Soil Opener

In this study, the movement trajectory of the soil opener comprised two parts, including the vertical inserting, vertical lifting and rotation process of the soil opener relative to the soil surface and the process of soil taking. Three-dimensional drawing software was applied for 3D modeling of the soil-forming device; the motion parameters of the respective component were defined in the mechanism of the application program and the linear speed of rotary motion of the soil opener was equal to the forward speed of the whole machine. The centers of the circle of the two adjacent shafts of the bracket of the soil opener were selected to track the trajectory (Figure 5). The movement track of a complete rotation consisting of $b_0$–$a_0$–$b_0$–$c_0$–$d_0$–$e_0$–$f_0$ relative to the soil surface of the soil opener was determined and $b_1$–$a_1$–$b_1$–$c_1$–$d_1$–$e_1$–$f_1$ was the movement track of the adjacent soil opener. To be specific, the corresponding sections $b_0a_0$ and $a_0b_0$ were found as the movement track of the soil opener in the soil-inserting and soil-lifting stage; the $b_0c_0$ section was found as the movement track of the soil discharging stage. The $c_0d_0e_0$ and $e_0f_0$ sections were the track of the reset and return stage of the soil opener, respectively. The soil-opening depth

$H_1$ was 65 mm and the plant spacing $L_1$ was 300 mm, thus meeting the requirements of rapeseed pot seedling planting.

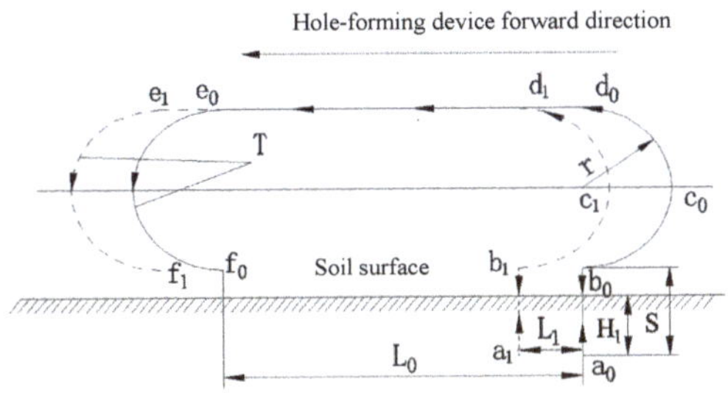


**Figure 5.** Movement trajectory of soil opener. Note: T is movement trajectory of soil opener; $L_0$ represents the displacement of a soil opener in one cycle (mm); $H_1$ is soil-opening depth (mm); $L_1$ denotes the center distance between two adjacent soil opener, namely, the plant distance between formed holes (mm); S expresses the maximum vertical displacement of the soil opener (mm); r is the radius of motion of the soil opener (mm); Point $b_0$ is the initial position of the soil opener.

2.3.4. Hole Contour and Shape Parametric Equations

The hole was formed by the interaction between the contour of the soil opener and the soil, thus confirming that the theoretical contour shape of the hole formed by the soil opener in the soil was the envelope of the movement track of the soil opener. According to Figure 6, the coordinate system of the soil-inserting and soil-taking process of the soil opener was built; the soil opener selected the O point as the center of the circle. After the maximum soil-inserting depth was reached, the soil opener rotated inward by a $\theta - \theta_1$ angle for soil taking; the movement trajectories of points A, B, C and D constituted the theoretical contour of the hole; and the hole was represented by an envelope composed of $D_1$-$A_1$-$O_1$-$B_1$-$C_1$. $D_1A_1$ and $C_1B_1$ were the movement trajectories of endpoints A and B and the movement trajectories of resistance surfaces DA and CB at the soil-inserting stage, while $A_1O_1$ and $B_1O_1$ were the movement trajectories formed by endpoints A and B at the soil-taking stage.

(1) The parameter equations of segment $D_1A_1$ and segment $C_1B_1$ of the hole side formed by the external contour of the soil opener contacting the soil at the vertical soil-inserting stage are written as

$$\begin{cases} y_{D_1A_1} = -\frac{1}{2\tan\theta}(2x + l) & -H_1\tan\theta - \frac{l}{2} \le x < -\frac{l}{2} \\ y_{C_1B_1} = \frac{1}{2\tan\theta}(2x - l) & \frac{l}{2} \le x < H_1\tan\theta + \frac{l}{2} \end{cases} \tag{1}$$

(2) At the soil-taking stage, A and B rotated around point O as the center of the circle till the soil opener was completely closed; the parameter equations of the bottom edges' sections $A_1O_1$ and $B_1O_1$ of the hole are written as

$$\begin{cases} y_{A_1O_1} = -\sqrt{R^2 - x^2} + \sqrt{R^2 - \frac{l^2}{4}} & -\frac{l}{2} \le x < 0 \\ y_{B_1O_1} = -\sqrt{R^2 - x^2} + \sqrt{R^2 - \frac{l^2}{4}} & 0 \le x < \frac{l}{2} \end{cases} \tag{2}$$

Since the formation of a hole is the result of the interaction between the external contour of the soil opener and the soil, the theoretical shape of the hole acts as the envelope of the movement track of the soil opener in the soil [21,22]. Through two sets of Equations (1) and (2), the contour of the hole can be expressed in two parts. In the process of soil inserting and soil taking, the contour of the hole consists of segments DA and CB and the movement track of special points A and B, thus forming the hole with a large upper end and a small lower end, which contributes to the stability of the hole. The coordinates of the respective point on the curve could be directly calculated from the parameter equation of hole contour, by calculating the coordinates of the respective point; the shape of the hole was described and parameterized, helping understand the main factors of the size of the hole. According to Figure 6, the intersection points between the straight line at an arbitrary depth from the soil surface $h_1$ ($h_1 \leq H_1$) on contour lines $D_1A_1$ and $C_1B_1$ and the hole contour line were assumed to be m and n, respectively. Since the beaks on the left- and right-hand sides of the soil opener were symmetrical about axis-y and were vertically inserted into soil, the hole top size at $h_1$ could be defined as

$$l_{mn} = x_n - x_m = 2x_n = 2(H_1 - h_1)\tan\theta + l \quad (3)$$

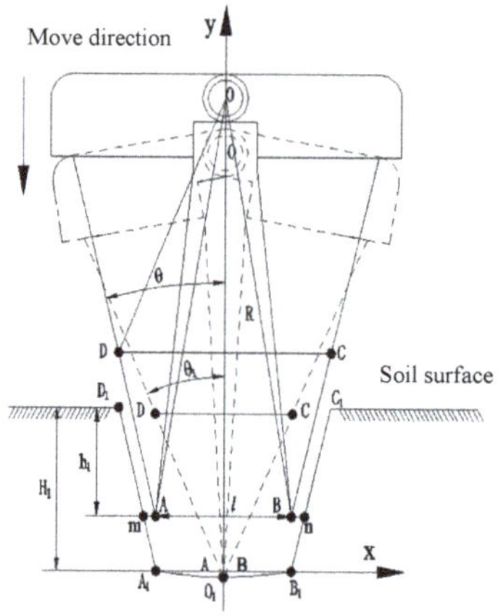


**Figure 6.** Movement trajectory of soil opener in soil-inserting and soil-taking stage. Note: The solid line represents the soil-inserting stage of the soil opener. The dotted line illustrates the soil-taking stage of the soil opener. Point O denotes the center of rotation of the right and left beaks; R is the radius of rotation of the soil opener (mm); θ is the soil entry angle of soil opener (°); $\theta_1$ denotes the angle between the cross-section of the lateral wall and the vertical direction when the soil opener is closed (°); Points D, C, A and B represent the projection points in the direction of the vertical view of the soil opener in the highest position and the lowest position of the external contour surface contacting the soil when the soil opener reaches the lowest position; $h_1$ is actual soil-opening depth of the soil opener (mm); $H_1$ is theoretical soil-opening depth (mm); $l$ is the bottom size of soil opener (mm).

By analyzing the mentioned parameter (Equation (3)), it could be seen that the soil-opening depth $H_1$, soil entry angle θ and bottom size of the soil opener $l$ were the vital factors of the size of hole top and hole bottom.

### 2.4. Design of Putter Cam and Hole-Forming Cam

To keep the bottom size and middle size of soil opener unchanged during the soil-inserting and soil-lifting stages and to ensure that the actions of soil inserting, soil taking, soil lifting, soil discharge and reset of the soil opener were coordinated to the sequential design requirements, the hole-forming cam, the putter cam and its affiliated components were designed. As shown in Figure 7, the putter cam and hole-forming cam were fixed in the side plate of the frame through the transmission shaft. Impacted by the putter spring, the guide roller fixedly connected with the putter of open and close always kept in contact with the outer contour surface of the putter cam (connecting the line of points $A_3$, $B_3$, $C_3$, $D_3$, $E_3$, $F_3$, $G_3$, $H_3$, O and $I_3$); the putter bearing installed on the hole-forming putter always moved in the inner groove of the hole-forming cam (the movement trajectory of the putter bearing center was the line connecting points $A_2$, $B_2$, $C_2$, $D_2$, $E_2$, $F_2$, $G_2$, $H_2$ and $I_2$).

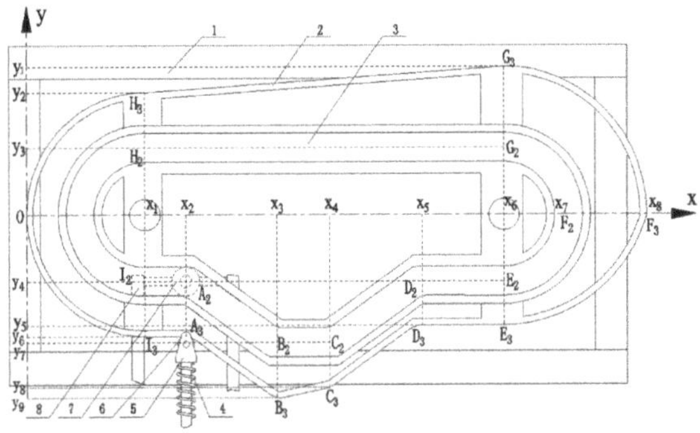


**Figure 7.** Structure diagram of putter cam and hole-forming cam: 1—frame; 2—putter cam; 3—hole-forming cam; 4—putter spring; 5—putter of open and close; 6—guide roller; 7—putter bearing; 8—putter of hole forming. Note: $A_3$, $B_3$, $C_3$, $D_3$, $E_3$, $F_3$, $G_3$, $H_3$, O and $I_3$ represent the limit points of outer contour of follower cam. $(x_2,y_6)$, $(x_3,y_9)$, $(x_4,y_8)$, $(x_5,y_5)$, $(x_6,y_5)$, $(x_7,0)$, $(x_6,y_1)$, $(x_1,y_2)$, $(0,0)$ and $(x_1,y_6)$ represent the coordinate values of points $A_3$, $B_3$, $C_3$, $D_3$, $E_3$, $F_3$, $G_3$, $H_3$, O and $I_3$. $A_2$, $B_2$, $C_2$, $D_2$, $E_2$, $F_2$, $G_2$, $H_2$ and $I_2$ are the special points of the movement trajectory of the center of the putter bearing in the putter of hole forming. $(x_2,y_4)$, $(x_3,y_6)$, $(x_4,y_5)$, $(x_5,y_3)$, $(x_6,y_3)$, $(x_7,0)$, $(x_6,y_3)$, $(x_1,y_3)$ and $(x_1,y_4)$ are the coordinate values of points $A_2$, $B_2$, $C_2$, $D_2$, $E_2$, $F_2$, $G_2$, $H_2$ and $I_2$, respectively. In order to keep the slider horizontal in the hole-forming stage, according to the overall size of the slider, the line segments $I_3A_3$ and $D_3E_3$ were designed as the buffer segments of the guide slider.

Line segments $A_3B_3C_3D_3$ and $A_2B_2C_2D_2$ refer to the hole-forming stage of the soil opener, where point $A_3$ represents the initial position of the hole-forming stage and segment $A_2B_2$ refers to the soil-inserting stage of the soil opener. At this stage, the hole-forming putter moved down at a certain speed to drive the soil opener into the soil vertically. At the same time, the putter of open and close moved down at the same speed in segment $A_3B_3$ to ensure that the bottom end of the soil opener was always open in the process of soil inserting and the size of the bottom end remained unchanged. The segment $B_2C_2$ refers to the soil-collecting stage of the soil opener; relative to the soil surface, the horizontal and vertical velocities of the soil opener were zero at this stage (the soil opener had reached the maximum soil-inserting depth), the putter of open and close moved upward along the contour line of segment $B_3C_3$ and the soil opener rotated inward with the bracket axis as the rotation center to complete soil taking under the joint action of the putter of open and close, the return spring and the putter spring. The segment $C_2D_2$ refers to the soil-lifting stage of the soil opener and the segment $C_3D_3$ represents the movement track of the guide roller on the putter of open and close. At this stage, the hole-forming putter and the putter of open and close moved upward at the same speed to ensure that the soil opener always conducted vertical soil lifting in a closed state. The segment $E_3F_3$ refers to the soil discharge stage of the soil opener. At this stage, the soil opener did not produce soil inserting or soil lifting under the action of the contour of the groove in segment $E_2F_2$. Under the action of the external contour of segment $E_3F_3$, the putter of open and close moved along the direction of soil inserting to the maximum displacement, so that the soil opener rotated outward to the maximum angle with the bracket axis as the rotation center to complete soil discharge. The discharged soil was thrown onto the retaining plate under the action of the forward speed of the machine and the rotary speed of the soil opener, thus sliding to the periphery of the formed hole. Segments $F_3G_3H_3OI_3A_3$ and $F_2G_2H_2I_2A_2$ represent the movement tracks of the guide roller and the putter of hole forming in the return stage of the soil opener.

### 2.4.1. Parametric Equations of Putter Cam

Three-dimensional drawing software was adopted to express the movement track of the guide roller on the upper end of the putter of open and close of the soil opener; given the movement of the respective stage of the soil opener, the hole-forming mechanism was covered around the slewing guide, the center of the guide roller of the respective two putters of open and close was connected with a curve and the curve Ii formed was the central contour curve of the putter cam. Subsequently, the limit contour curve of the putter cam was generated in accordance with the size of the guide roller. The coordinate system was established (Figure 7). Axis-x coincides with the line between Point $F_3$ and O and axis-y passes through the far-left end of the putter cam, tangent to the arc $H_3OA_3$. The abscissas of limit points $A_3$, $B_3$, $C_3$, $D_3$, $E_3$ and $G_3$ are expressed as $x_2$~$x_6$, respectively; that of $F_3$ as $x_8$; that of $H_3$ and $I_3$ as $x_1$; and that of O as 0. The ordinates of limit points $G_3$ and $H_3$ are expressed as $y_1$~$y_2$; those of $D_3$, $E_3$, $I_3$ and $A_3$ as $y_5$~$y_6$; those of $C_3$ and $B_3$ as $y_8$~$y_9$; and those of O and $F_3$ as 0, respectively. The interpolation method was adopted to work out the parameter equations of the respective step of the putter cam [19].

$A_3B_3$ step:

$$y = -\frac{75}{107}x + \frac{386}{107} \qquad 206 \le x < 313 \tag{4}$$

$B_3C_3$ step:

$$y = \frac{7}{34}x - \frac{9773}{68} \qquad 313 \le x < 381 \tag{5}$$

$C_3D_3$ step:

$$y = \frac{75}{107}x - \frac{50938}{107} \qquad 381 \le x < 488 \tag{6}$$

$D_3E_3F_3$ step:

$$\begin{cases} y = -134 & 488 \le x < 600 \\ (y-47)^2 + (x-605)^2 = 180^2 & 600 \le x < 779 \end{cases} \tag{7}$$

$F_3G_3H_3$ step:

$$\begin{cases} y^2 + (x-600)^2 = 179^2 & 600 < x \le 779 \\ y = \frac{31}{452}x + \frac{62308}{452} & 148 < x \le 600 \end{cases} \tag{8}$$

$H_3OI_3A_3$ step:

$$\begin{cases} y^2 + (x-148)^2 = 148^2 & 0 \le x \le 148 \\ y = -148 & 148 < x < 206 \end{cases} \tag{9}$$

### 2.4.2. Parametric Equations of the Hole-Forming Cam

According to the coordinate system shown (Figure 7), the abscissas of limit points $A_2$, $B_2$, $C_2$, $D_2$, $E_2$, $F_2$ and $G_2$ are expressed as $x_2$~$x_7$; and those of $H_2$ and $I_2$ as $x_1$, respectively. The ordinates of limit points $G_2$ and $H_2$ as $y_3$; $I_2$, $A_2$, $D_2$ and $E_2$ are expressed as $y_4$; those of $C_2$ and $B_2$ as $y_7$; and those of O and $F_2$ as 0, respectively. The interpolation method was adopted to work out the parameter equations of the respective step of the putter cam [19].

$A_2B_2$ step:

$$y = -\frac{75}{107}x + \frac{6783}{107} \qquad 206 \le x < 313 \tag{10}$$

$B_2C_2$ step:

$$y = -156 \qquad 313 \le x < 381 \tag{11}$$

$C_2D_2$ step:

$$y = \frac{75}{107}x - \frac{45267}{107} \qquad 381 \le x < 488 \tag{12}$$

$D_2E_2F_2G_2H_2$ step:

$$\begin{cases} y = -81 & 488 \leq x < 600 \\ y^2 + (x-600)^2 = 81^2 & 600 \leq x < 681 \\ y = 81 & 148 \leq x < 600 \end{cases} \tag{13}$$

$H_2I_2A_2$ step:

$$\begin{cases} y^2 + (x-148)^2 = 81^2 & 67 \leq x < 148 \\ y = -81 & 148 \leq x < 206 \end{cases} \tag{14}$$

To ensure the rationality of the external contour curve of the putter cam and the inner groove curve of the hole-forming cam, the pressure angle between the putter of open and close and the putter cam at the hole-forming stage of the soil opener was examined. According to Figure 7, the putter of open and close in section $B_3C_3$ moved upward as it was impacted by the putter spring, so there was no self-locking phenomenon at this stage. According to the obtained contour parameter equation of the putter cam, the included angle between the section $A_3B_3$ of the contour curve of the putter cam and the horizontal direction, and the included angle between the section $C_3D_3$ of the contour curve of the putter cam and the horizontal direction were 35°. In other words, the pressure angle between the inner groove curve of the hole-forming cam and the putter of hole forming was 35° when the putter bearing moved in sections $A_2B_2$ and $C_2D_2$, which was within a reasonable range [42].

### 2.5. *Number of the Hole-Forming Mechanism*

In accordance with the requirements of rapeseed agronomy, the hole distance of pot seedlings was determined as 300 mm; n hole-forming mechanisms had an even distribution on the guide rail of the rotary system. In the hole-forming operation, there might be interference between two adjacent hole-forming mechanisms; thus, when the soil-lifting stage of the previous hole-forming mechanism was completed, the adjacent hole-forming mechanism had just entered the soil-inserting stage. In this design, the sprocket model was selected as 10A; based on the pitch of the corresponding chain, the number of the hole-forming mechanism n should satisfy the following equation:

$$\begin{cases} L_1 = n_1 \cdot p \\ n_2 \cdot L_1 = S_1 \\ S_1 = \pi \cdot D + 2L_2 \\ L_2 \geq L_1 + L_3 \end{cases} \tag{15}$$

where $L_1$ denotes the center distance between the respective hole-forming mechanisms, which is the distance between the formed holes (mm); $n_1$ represents the number of chain pitches between the symmetrical centers of two adjacent hole-forming mechanisms; p expresses the chain pitch (mm); $n_2$ represents the number of hole-forming mechanisms (number); $S_1$ denotes the chain length (mm); D expresses the diameter of the sprocket graduation circle (mm); $L_2$ represents the center distance between the two sprockets (mm); and $L_3$ denotes the length of the slider (mm).

Since the slider should be in a horizontal position at the beginning of the hole-forming stage, the size of the slider was known as $L_3$ = 100 mm, the center distance between the respective hole-forming mechanism was designed to be 300 mm and the diameter of the circular arc section of the annular guide rail and the sprocket indexing circle D was 192 mm. Since the hole-forming mechanism should have a uniform distribution on the annular guide rail, given Formula (15), $n_2$ was 5, sprocket center distance $L_2$ was 452.4 mm, chain length $S_1$ was 1508 mm and $n_1$ was 19.

### 2.6. Soil Bin Test Condition and Equipment

#### 2.6.1. Test Condition and Equipment

To verify the accuracy of the mentioned design method and the reliability of the hole-forming device, a test bench of the hole-forming device was set for soil bin test, which was performed in the digital soil bin of the Agricultural Mechanization Engineering Training Center of Hunan Agricultural University (Figure 8). In general, the soil was clay loam. The testing equipment consisted of a soil trough tractor, a reverse rotary soil-picking-type hole-forming device, a laser pen (it was installed on the hole opener), a stopwatch (used to determine the frequency of hole forming), a firmness tester (used to determine soil firmness), an electronic scale, an aluminum box (used to determine soil water content and soil porosity), a steel ruler, a tape, a tool knife, a soil bulk density tester, a hole shape mapper, coordinate paper, etc.

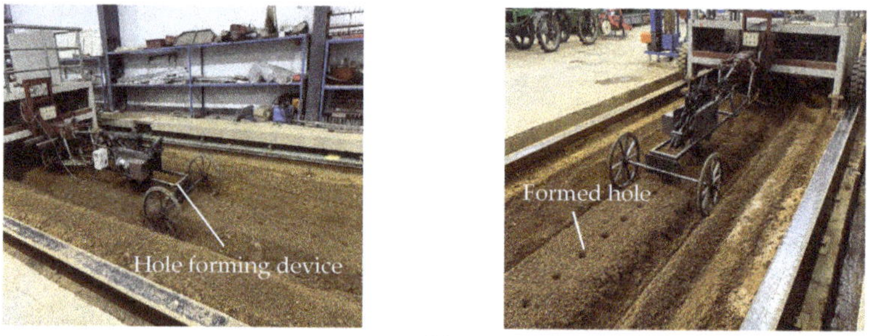


(a)

(b)

**Figure 8.** Hole-forming device and soil bin test: (**a**) soil bin test; (**b**) hole size measurement.

#### 2.6.2. Test Method

Before the test, the soil in the soil tank was compacted by the pressing roller, watered by the sprinkler system and tilled by the rotary cultivator. The soil water content at 0–150 mm reached 19.2% on average and soil firmness was 195 kPa. The hole-forming device moved forward at 0.25, 0.35 and 0.45 m/s under the traction of the soil bin car (the rotary linear velocity of the hole-forming mechanism was regulated to be equal to the forward velocity, so the frequency of hole forming was 50, 70 and 90 hole/min, satisfying the frequency requirements of rapeseed pot seedling transplanting); since the speed of the soil trough tractor was unstable during starting and stopping, the middle 15 m within 20 m of the operation of the soil trough tractor was taken as the test area for each test. The test was repeated for three times and data results were taken as the average values and parameters (e.g., hole-opening diameter, effective depth, hole distance, qualified number of holes, soil porosity of the hole wall and soil disturbance amount of the hole wall), determined for the respective tests.

(1) The shape and size of holes

As shown in Figure 9, the center distance of two adjacent holes was set as the hole distance. During the test, the center distance of all adjacent holes within 15 m was determined ($l_1$, $l_2$, $l_3$) and the mean value was taken as the hole distance of this test. The size of the maximum distance was measured between two points on the contour line of the end face of the hole along the forward direction of the whole machine $X_1$; the size of the maximum distance was measured between two points on the contour line of the hole face perpendicular to the forward direction of the whole machine $Y_1$; $(X_1 + Y_1)/2$ expresses the average value of the two, which was taken as the hole opening diameter of a single hole. During the test, the opening diameter of all holes within 15 m was measured and the average value was adopted as the diameter of formed holes. Its vertical height (h) was measured as the effective depth along the upper-end face of the formed hole until the minimum end-face size within the hole was equal to $\sqrt{2}\,b_2$. During the test, the effective depth of all holes within 15 m was measured and the average value was adopted as the effective depth of the hole.

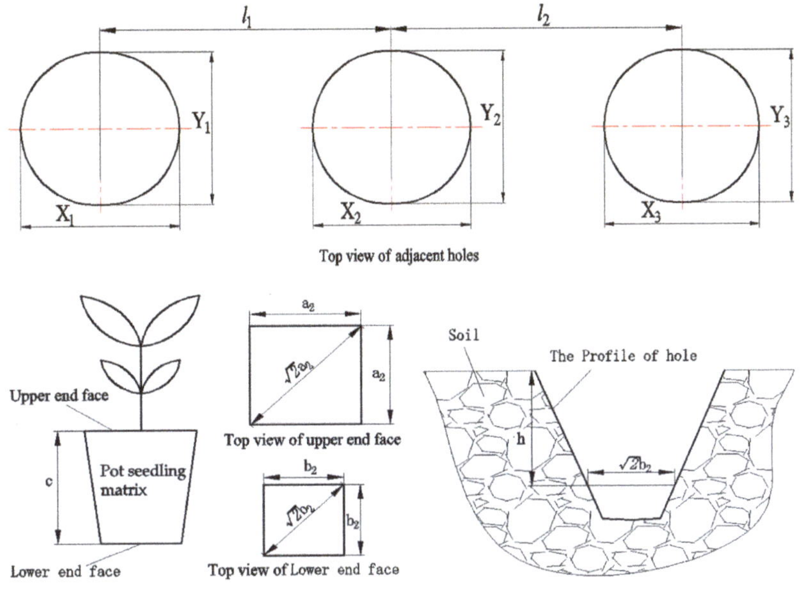


**Figure 9.** Schematic diagram of the determination of hole dimensions.

Qualified holes: If the hole opening diameter exceeded $\sqrt{2}\,a_2$ and the effective depth of the hole h was not less than the height of the pot seedling matrix c, the hole qualified. Based on the analysis and calculation of different parameters measured in the field, the performance parameters of the hole-forming device were obtained as follows: qualified rate of hole forming Q, change rate of hole size k [19].

Qualified rate of hole forming Q (%):

$$Q = \frac{N_{hg}}{N} \times 100\% \tag{16}$$

In the equation, $N_{hg}$ and N represent the number of qualifying holes and the actual number of formed holes within the measured distance, respectively.

Change rate of hole size k (%):

$$k = \frac{|D_0 - d_0|}{d_0} \times 100\% \tag{17}$$

In the equation, $D_0$ represents the hole-opening diameter of actual holes (mm) and $d_0$ represents the hole opening diameter of theoretical holes (mm).

(2) Perpendicularity of hole-forming operation

The glass plate was vertically inserted into the soil bin in accordance with the forward direction of the machine and the coordinate paper was evenly attached to the glass plate, as shown in Figure 10. These two processes had the aim to ensure that the laser pen installed on the respective soil opener frame could be irradiated to the coordinate paper and leave a laser point when the corresponding hole-forming mechanism entered the hole-forming stage. The laser spot followed the inserting and lifting action of the soil opener to leave a movement track on the coordinate paper; after photographing the laser point's movement track using the continuous capture function of SLR camera, we extracted the laser point's contour from the shooting images using Photoshop software and processed its gray scale. Then, the coordinate values of the laser points in each shooting image were read using the coordinate paper and the obtained coordinate values were input into Excel software. The linear regression of each coordinate point was carried out and the fitting straight line of the movement track of the laser points was obtained, corresponding to the movement tracks of the inserting and lifting process of the soil opener; the angle between the two fitting straight lines and the horizontal coordinate axis could reflect the perpendicularity of the soil opener during the hole-forming process. The five sets of soil openers carried out the hole-forming operation at different forward speeds and the hole-forming operation was repeated three times. The perpendicularity of the hole-forming operation of each set of soil openers was measured and the average value was taken.

**(a)** **(b)**

**Figure 10.** Physical experiment on movement trajectory of soil opener: (**a**) the original image of laser point; (**b**) the grayscale image of laser point.

(3) The soil porosity of the hole wall and soil disturbance amount

The soil porosity of the hole wall of the formed hole was measured using a self-made soil bulk density tester [29]. The bulk density tester consisted of a pressing handle and a self-made rectangular ring knife; the length, width and height of the inner cavity of the rectangular ring knife were 60 × 40 × 20 mm, as shown in Figure 11a. After the completion of the hole-forming operation, the surveyor pressed the soil bulk density tester down into the soil (the soil within 20 mm perpendicular to the direction of the hole wall) along the hole wall of the formed hole until the soil overflowed on the upper-end face of the rectangular ring knife; then, the excess soil on the upper and lower ends of the rectangular ring knife was cut with a tool knife and the soil in the ring knife was put into an aluminum box. The soil bulk density was calculated according to $\rho_b = m/v$, where m is the mass of dried soil in the aluminum box (g) and v is the volume of soil in the rectangular ring knife, that is, the volume of the rectangular ring knife inside the cavity ($cm^3$). Then, the soil porosity of the hole wall was calculated by the soil porosity calculation formula.

$$\varepsilon = (1 - \rho_b/\rho_s) \times 100\% \quad (18)$$

**Figure 11.** Soil porosity and hole contour test: (**a**) soil bulk density tester; (**b**) hole shape mapper.

In the equation, $\varepsilon$ is the soil porosity of the hole wall (%); $\rho_b$ is the soil bulk density ($g/cm^3$); and $\rho_s$ is the soil density ($g/cm^3$).

According to the soil disturbance contour measurement method [43,44], the mapping process is shown in Figure 11b. The hole shape mapper was composed of a base, a paper rod and a coordinate paper and its length, width and height were 400 × 100 × 300 mm. The diameter of the paper rod was 4 mm, the length was 250 mm, the paper bars were arranged adjacent to each other and a coordinate paper with a square-edge length of 1 mm was fixed on the hole shape mapper to map the soil disturbance contour after the hole was formed. After the completion of the hole-forming operation, the hole shape mapper was placed right above the hole; the plane of the coordinate paper coincided with the center section of the hole and the paper rod slid downward under the action of gravity until its bottom contacted the soil. At this time, the top end of the paper rod formed the contour curve of the hole section and the soil surface on the coordinate paper and the position of the top end of the paper rod on the coordinate paper was marked by a marker. The coordinates of each marker point were read and the synthetic curve was fitted, which was the contour section curve of the inner wall of the hole.

The soil heave degree of the hole mouth and soil stability rate of the hole wall were taken as the evaluation indexes of the soil disturbance amount and their calculation formulas can be written as follows:

$$P = \frac{A_L - A}{A} \times 100\% \quad (19)$$

$$\delta = \frac{A_h}{A} \times 100\% \quad (20)$$

where P denotes the soil heave degree of the hole mouth (the smaller the *p*-value, the lower the soil heave degree of the hole mouth) (%); A and $A_L$ are the cross-sectional areas from the soil surface to the theoretical hole bottom before and after the hole forming, respectively ($mm^2$); $\delta$ is the soil stability rate of the hole wall (the greater $\delta$, the closer the cross-section area of the actual hole to the theoretical hole, the less the soil returned in the hole) (%); and $A_h$ is the cross-sectional area from the soil surface to the actual hole bottom after the hole forming ($mm^2$). A was determined by the relevant parameters of the soil opener and $A_L$ and $A_h$ could be calculated by MATLAB software.

## 3. Results and Discussion

### *3.1. Analysis of Hole Size and Hole-Forming Quality*

Table 2 shows the hole size and hole performance test results of the reverse-rotating soil-taking-type hole-forming device.

**Table 2.** Test results of hole size and hole-forming performance.

| Test Number | Forward Speed (m/s) | Hole Distance (mm) | Hole-Opening Diameter (mm) | Effective Depth (mm) | Qualified Number of Holes | Qualified Rate of Hole (%) | Change Rate of Hole Size (%) |
|---|---|---|---|---|---|---|---|
| 1 | 0.25 | 299.8 | 76.3 | 48.5 | 47.0 | 94.0 | 3.1 |
| 2 | 0.25 | 302.2 | 75.2 | 49.7 | 47.0 | 94.0 | 1.6 |
| 3 | 0.25 | 298.7 | 75.7 | 48.9 | 49.0 | 98.0 | 2.3 |
| Average value | 0.25 | 300.2 | 75.7 | 49.0 | 47.6 | 95.2 | 2.3 |
| 4 | 0.35 | 296.5 | 75.8 | 49.5 | 47.0 | 94.0 | 2.4 |
| 5 | 0.35 | 302.1 | 77.5 | 47.8 | 46.0 | 92.0 | 4.7 |
| 6 | 0.35 | 298.7 | 75.2 | 48.9 | 48.0 | 96.0 | 1.6 |
| Average value | 0.35 | 299.1 | 76.2 | 48.7 | 47.0 | 94.0 | 2.9 |
| 7 | 0.45 | 299.2 | 77.5 | 47.4 | 48.0 | 96.0 | 4.7 |
| 8 | 0.45 | 297.9 | 78.7 | 46.9 | 46.0 | 92.0 | 6.4 |
| 9 | 0.45 | 297.8 | 78.1 | 47.3 | 46.0 | 92.0 | 5.5 |
| Average value | 0.45 | 298.3 | 78.1 | 47.2 | 46.7 | 93.3 | 5.5 |

According to Table 2, when the forward velocities were 0.25, 0.35 and 0.45 m/s, the average values of the hole spacing of the adjacent two holes were 300.2, 299.1 and 298.3 mm, respectively, which met the design requirements. The change range of hole spacing at different forward speeds was small; the hole spacing was less affected by the forward speed and the slip rate had a slight effect on the hole-forming effect during the whole machine operation. When the forward speed increased from 0.25 m/s to 0.45 m/s, the average hole diameter was 75.7 mm, 76.2 mm and 78.1 mm, respectively, and the average change rates of hole size were 2.3%, 2.9% and 5.5%, respectively, thus indicating that the hole size at different speeds was not significantly different from the theoretical hole size. There was little clay in the closed soil-taking and vertical soil-lifting period of the soil opener and there was no soil scraping phenomenon in the rotation movement. However, with the increase in the forward speed, the change rate of the hole size tended to increase, thus indicating that the faster the speed of the machine (the faster the speed of the soil opener to insert in the soil, take the soil and lift the soil), the more obvious the extrusion and scraping of the soil during the hole-forming operation, which resulted in the collapse of a small part of the soil in the hole wall and the increase in the hole size. The average effective depths of each hole measured in the experiment at different forward speeds were 49.0 mm, 48.7 mm and 47.2 mm, respectively. The average error between the theoretical value of the effective depth and the experimental value was 2.0%, 2.6% and 5.6%, respectively. The above results were achieved because a small part of soil returned to the formed hole during the hole-forming process; the faster the forward speed was, the greater the error between the theoretical value and the experimental value of the effective depth and the more the soil reflux in the hole. Through the measurement and calculation of each test, it was concluded that when the forward speed was between 0.25 m/s and 0.45 m/s, the average qualified rate of forming holes was between 93.3% and 95.2%, indicating that the working stability of the whole machine was high and met the design requirements.

### 3.2. Results and Analysis of Perpendicularity of Hole-Forming Operation

In the process of the hole-forming experiment, the laser point images of one set of soil opener taken by the SLR camera are shown in Figure 10. After the hole-forming operation was completed, all the images were extracted and the gray scale was processed; the coordinate values of the laser points on the coordinate paper in each group of images were read and the linear regression of each point was conducted. The results are shown in Figure 12.

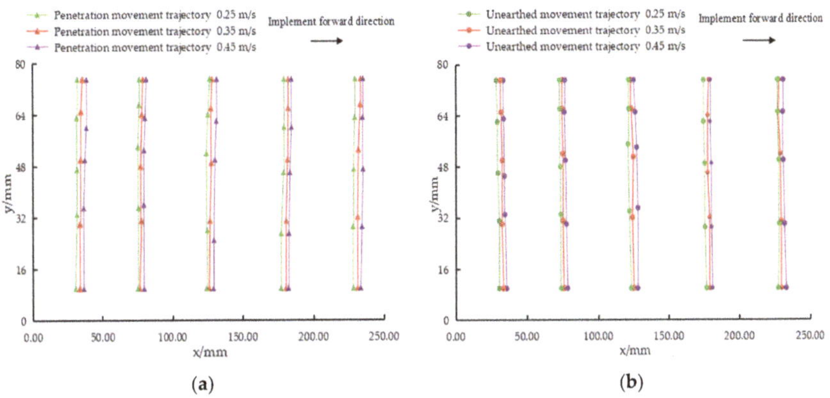


**Figure 12.** The penetration and unearthed movement trajectory of soil opener: (**a**) penetration movement trajectory of soil opener; (**b**) unearthed movement trajectory of soil opener.

Figure 12 shows that the soil-inserting trajectory and the soil-lifting trajectory under different velocity conditions were basically vertical. It can be seen from Figure 12a,b that the angle between the soil-inserting trajectory of the soil opener and the x-axis was slightly less than 90° and the angle between the soil-lifting trajectory of the soil opener and the x-axis was slightly larger than 90°. When the hole-forming operation was carried out, the soil-inserting trajectory and the soil-lifting trajectory showed a "V"-type trend, indicating that, in the operation stage, the actual forward speed of the whole machine was slightly smaller than the rotary linear speed of the hole-forming mechanism. It can be seen from Table 3 that, when the forward speed of the machine was 0.25 m/s, 0.35 m/s and 0.45 m/s, respectively, the average angle between the trajectory of the soil opener during the soil-inserting stage and the horizontal direction was 88.4°, 88.0°and 88.1°, respectively, and the variation coefficients were 0.53, 0.16 and 0.40, respectively. The average angle between the trajectory of the soil opener during the soil-lifting stage and the horizontal direction was 88.7°, 88.5° and 88.1°, respectively, and the variation coefficients were 0.64, 0.54 and 0.47, respectively. The minimum value of the angle between the motion trajectory of the soil opener and the horizontal direction was higher than 88°, indicating that the perpendicularity of the soil opener during soil inserting and soil lifting was high under the actual working conditions [23]. The forward speed of the machine did not affect the perpendicularity of the hole-forming operation and the soil opener met the design requirements of the vertical soil inserting and lifting.

### 3.3. Analysis of the Soil Porosity of the Hole Wall and Soil Disturbance Effect

The contour of the soil disturbance hole shape section after the hole-forming operation at three forward speeds is shown in Figure 13. According to Formula (18), the soil porosity of the hole wall before and after hole-forming operation was calculated. According to the soil disturbance contour (Figure 13) and Equations (19) and (20), the soil heave degree of the hole mouth and the soil disturbance amount were calculated; the results are shown in Table 4.

**Table 3.** Angle between the horizontal direction and the penetration and unearthed movement trajectory of soil opener.

| Forward Speed (m/s) | Test Index | Mean Value (°) | Standard Deviation | Coefficient of Variation (%) |
|---|---|---|---|---|
| 0.25 | Angle of penetration trajectory | 88.4 | 0.47 | 0.53 |
| | Angle of unearthed trajectory | 88.7 | 0.56 | 0.64 |
| 0.35 | Angle of penetration trajectory | 88.0 | 0.14 | 0.16 |
| | Angle of unearthed trajectory | 88.5 | 0.48 | 0.54 |
| 0.45 | Angle of penetration trajectory | 88.1 | 0.34 | 0.40 |
| | Angle of unearthed trajectory | 88.1 | 0.41 | 0.47 |

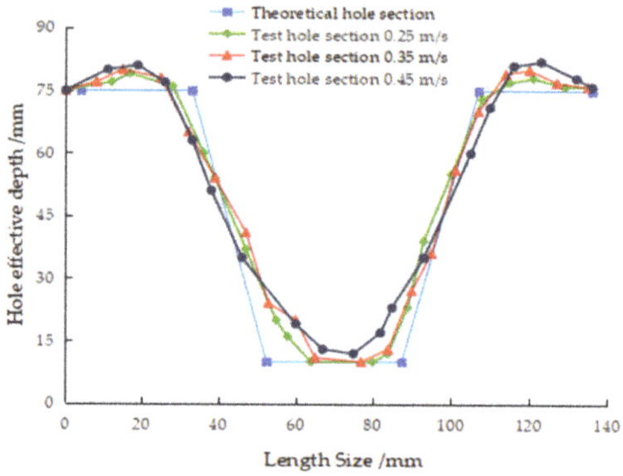


**Figure 13.** The figure of hole section comparison at different speeds.

**Table 4.** Soil porosity of the hole wall comparison results.

| Test Number | Forward Speed (m/s) | Soil Porosity before Test (%) | Soil Porosity of the Hole Wall after Test (%) | The Change Rate of Porosity (%) | Soil Heave Degree of the Hole Mouth (%) | Soil Stability Rate (%) |
|---|---|---|---|---|---|---|
| 1 | 0.25 | 51.8 | 47.8 | 7.7 | 3.9 | 92.5 |
| 2 | 0.25 | 53.5 | 48.9 | 8.6 | 3.7 | 91.5 |
| 3 | 0.25 | 52.9 | 48.6 | 8.5 | 4.0 | 91.6 |
| Average value | 0.25 | 52.7 | 48.4 | 8.2 | 3.9 | 91.9 |
| 4 | 0.35 | 52.7 | 47.9 | 9.1 | 4.3 | 91.6 |
| 5 | 0.35 | 53.2 | 48.7 | 8.4 | 3.8 | 90.8 |
| 6 | 0.35 | 53.9 | 48.3 | 10.4 | 4.1 | 91.3 |
| Average value | 0.35 | 53.3 | 48.3 | 9.3 | 4.1 | 91.2 |
| 7 | 0.45 | 52.9 | 47.5 | 10.2 | 4.3 | 91.3 |
| 8 | 0.45 | 51.8 | 48.3 | 6.8 | 4.3 | 90.7 |
| 9 | 0.45 | 52.7 | 47.6 | 9.7 | 4.1 | 91.1 |
| Average value | 0.45 | 52.5 | 47.8 | 8.9 | 4.2 | 91.0 |

Table 4 shows that, when the forward speeds were 0.25 m/s, 0.35 m/s and 0.45 m/s, the average soil porosity of the hole wall after the completion of the hole-forming operation was 48.4%, 48.3% and 47.8%, respectively; compared with before the hole-forming operation, the average change rates of soil porosity were 8.2%, 9.3% and 8.9%, respectively. With the increase in the forward speed, the soil porosity of the hole wall showed a downward trend. Although the soil porosity decreased, the change rate was not large and the soil porosity of the hole wall was within the range of agronomic requirements, which would not affect the growth of crops [45]. After the hole-forming operation, the mean soil heave degree of the hole mouth was 3.9%, 4.1% and 4.2% and the average soil stability rate of the hole wall was 91.9%, 91.2% and 91.0%, respectively. At different forward speeds, the soil around the formed hole was of low heave degree under the extrusion effect of the soil opener and the contour section of the actual formed hole was slightly different from the theoretical one. At the same time, it was also found that there was an insignificant difference in the soil porosity of the hole wall of the hole formed by the hole-forming device designed in this paper before and after the hole-forming operation and the hole-forming device had slight disturbance to the soil during the hole-forming operation.

## 4. Conclusions

1. A type of reverse-rotating soil-taking-type hole-forming device of pot seedling transplanting machine for rapeseed was designed based on the working characteristics of the hole-punching transplanter and the agronomy requirements of rapeseed transplanting. In this study, the design scheme of a reverse-rotating hole-forming mechanism is proposed. At the hole-forming stage, the vertical soil-inserting and -lifting actions of the soil opener were achieved by providing the speed equal to the forward speed of the whole machine and in the opposite direction; then, a series of holes were formed in the soil by combining the soil-taking and soil-discharging actions of the soil opener.

2. A hole-forming device test bench was built for the soil bin tests at different forward speeds. It was found from the test results that, when the hole-forming device advanced at the speeds of 0.25 m/s, 0.35 m/s and 0.45 m/s, the average qualified rates of hole forming of the device were 95.2%, 94.0% and 93.3% and the average change rates of the hole size were 2.3%, 2.9% and 5.5%, respectively. The average error between the theoretical value of effective depth and the experimental value was 2.0%, 2.6% and 5.6%. The average value of the angle between the trajectory of the soil opener at the hole-forming stage and the horizontal direction was between 88.0° and 88.7° and the variation coefficient was between 0.16% and 0.64%; the perpendicularity of the hole-forming operation was high. At different forward speeds, the average soil porosity of the hole wall after the completion of the hole-forming operation was 48.4%, 48.3% and 47.8%, respectively, while the change rates of soil porosity were 8.2%, 9.3% and 8.9%, respectively, compared with those before the hole-forming operation. The average soil heave degree at the hole mouth after the completion of the hole-forming operation was 3.9%, 4.1% and 4.2%, respectively, and the average soil stability rate of the hole wall was 91.9%, 91.2% and 91.0%, respectively.

3. In this paper, rapeseed seedlings could still have better verticality after being transplanted, so the soil porosity around the pot seedling was still high after operation, thus leading to high air permeability and water permeability for the growth of the pot seedlings. In subsequent research, the power transmission mode should be changed, the power of the rotary movement of the hole-forming mechanism should be inputted by the power output shaft of the tractor via the gearbox. In the future, a supporting mechanical seedling feeding device should be introduced to solve the problem whereby the machine can only achieve mechanical hole forming and then seedlings need to be fed manually. During the operation of hole forming, the change in the working parameters of the soil opener (e.g., the bottom size, soil entry angle, soil entry depth and soil entry speed of the soil opener) have effects on soil disturbance. At a subsequent stage, the above factors resulting in soil disturbance will be investigated in depth. Furthermore, the working parameters of the soil opener will be optimized to improve the qualified rate and quality of the hole.

## 5. Patents

Three patents have been applied for in China for Hole-Forming Device of Pot Seedling Transplanting Machine for Rapeseed in this manuscript (Patent No. ZL201910984983.X and Application Nos. CN202110794519.1, CN202110794523.8).

**Author Contributions:** Conceptualization, W.Q., M.W. and H.L.; methodology, W.Q.; software, W.Q. and Z.D.; validation, W.Q., M.W., Z.D., H.L. and F.S.; formal analysis, W.Q.; investigation, W.Q., Z.D. and F.S.; resources, W.Q.; data curation, W.Q.; writing—original draft preparation, W.Q.; writing—review and editing, W.Q. and M.W.; visualization, W.Q.; supervision, M.W.; project administration, M.W.; funding acquisition, M.W. and H.L. All authors have read and agreed to the published version of the manuscript.

**Funding:** This research study was funded by Natural Science Foundation of Hunan Province, China (No. 2021JJ30344 and 2020JJ5254) and Key Projects of Hunan Provincial Science and Technology Department (No. 2019NK2151).

**Institutional Review Board Statement:** Not applicable.

**Informed Consent Statement:** Not applicable.

**Data Availability Statement:** The datasets used and/or analyzed during the current study are available from the corresponding author on reasonable request.

**Acknowledgments:** The authors would like to thank their schools and colleges, as well as the funding providers of the project. All support and assistance are sincerely appreciated.

**Conflicts of Interest:** The authors declare no conflict of interest.

## Nomenclature

| | |
|---|---|
| b—Cutting-edge section thickness (mm). | $l$—Bottom size of soil opener (mm). |
| $H_0$—Height of soil opener (mm). | θ—Angle of entry (°). |
| $H_1$—Soil-opening depth (mm). | $L_1$—Center distance between two adjacent soil opener (mm). |
| $h_1$—Arbitrary depth from the soil surface (mm). | $l_{mn}$—Hole top size at $h_1$ (mm). |
| $n_1$—Number of chain pitches between the symmetrical centers of two adjacent hole-forming mechanisms. | p—Chain pitch (mm). |
| $n_2$—Number of hole-forming mechanisms. | $S_1$—Chain length (mm). |
| D—Diameter of sprocket graduation circle (mm). | $L_2$—Center distance between the two sprockets (mm). |
| $L_3$—Length of slider (mm). | $a_2$—Side length of upper end face (mm). |
| $l_1/l_2/l_3$—Center distance between adjacent holes (mm). | $b_2$—Side length of lower end face (mm). |
| $X_1$—Points on the contour line of the end face of the hole along the forward direction of the whole machine (mm). | c—Height of the pot seedling matrix (mm). |
| $Y_1$—Two points on the contour line of the hole face perpendicular to the forward direction of the whole machine (mm). | h—Effective depth of hole (mm). |
| Q—Qualified rate of hole forming (%). | $N_{hg}$—Number of qualified holes. |
| N—Actual number of formed holes. | k—Change rate of hole size (%). |
| $D_0$—Hole opening diameter of actual holes (mm); | $d_0$—Hole opening diameter of theoretical holes (mm); |
| ε—Soil porosity of hole wall (%). | $\rho_b$—Soil bulk density (g/cm$^3$). |
| $\rho_s$—Soil density (g/cm$^3$). | P—Soil heave degree of hole mouth. |
| A—Cross-sectional area from the soil surface to the theoretical hole bottom before hole forming (%). | $A_L$—Cross-sectional area from the soil surface to the theoretical hole bottom after the hole forming (%). |
| δ—Soil stability rate of hole wall (%). | $A_h$—Cross-sectional area from the soil surface to the actual hole bottom after the hole forming (mm$^2$). |

## References

1. Jarecki, W. The Reaction of Winter Oilseed Rape to Different Foliar Fertilization with Macro- and Micronutrients. *Agriculture* **2021**, *11*, 515. [CrossRef]
2. Wei, H.; Jun, L.; Xiaomei, W.; Qiao, L.; Xiaowei, Y. Current status of global rapeseed industry and problems, countermeasures of China's rapeseed industry. *China Oils Fats* **2021**, 1–8.
3. Yin, Y.; Yin, L.; Zhang, X.; Guo, J.; Wang, J. Status and Countermeasure of the High-quality Development of Rapeseed Industry in China. *J. Agric. Sci. Technol.-Iran* **2021**, *23*, 1–7.
4. Can, M.; Haifeng, L.; Yi, Z.; Beibei, Z.; Xiao, X. Research status and development trend of seeding transplanting machinery. *J. Chin. Agric. Mech.* **2020**, *41*, 215–222.
5. Wang, X.; Huang, M. Application of rape planting mechanization technology model in rice stubble field. *Hubei Agric. Mech.* **2020**, 3–4. [CrossRef]

6. Mengyuan, C. Problems and Countermeasures of developing rape planting mechanization. *Jiangxi Agric.* **2020**, 104.
7. Yongtao, L.; Yaxiong, L.; Yang, L.; Bing, L.; Tao, W. Status analysis of abroad's transplanting machine and transplanting technology. *Xinjiang Agric. Mech.* **2011**, *3*, 29–32.
8. Qiming, P. Current situation and development trend of domestic and foreign transplanting machine technology. *J. Anhui Agric. Sci.* **2013**, *41*, 12478–12479.
9. Chunyun, G. The development trend of China's rape industry. *Grain Sci. Technol. Econ.* **2011**, *36*, 5–6.
10. Jin, X.; Du, X.W.; Geng, G.S.; Ji, J.T.; Dong, X.; Liu, W.X. Experiment on Planting System of 2ZDJ-2 Transplanter. *Trans. Chin. Soc. Agric. Mach.* **2015**, *46*, 26–31.
11. Xiaoxu, Y.; Yun, Z.; Baocheng, C.; Maile, Z.; Hao, Z.; Zhichao, Z. Current situation and prospect of transplanter. *Trans. Chin. Soc. Agric. Mach.* **2014**, *45*, 44–53.
12. Subo, T.; Lichun, Q.; Naoshi, K.; Ting, Y. Development of Automatic Transplanter for Plug Seedling. *IFAC Proc. Vol.* **2010**, *43*, 79–82.
13. Felezi, M.E.; Vahabi, S.; Nariman-Zadeh, N. Pareto optimal design of reconfigurable rice seedling transplanting mechanisms using multi-objective genetic algorithm. *Neural Comput. Appl.* **2016**, *27*, 1907–1916. [CrossRef]
14. Nandede, B.M.; Raheman, H. Multi-Stage Metering Mechanism for Transplanting of Vegetable Seedlings in Paper Pots. *J. Inst. Eng. Ser. A* **2015**, *96*, 295–300. [CrossRef]
15. Dihingia, P.C.; Prasanna, K.G.V.; Sarma, P.K. Development of a Hopper-Type Planting Device for a Walk-Behind Hand-Tractor-Powered Vegetable Transplanter. *J. Biosyst. Eng.* **2016**, *41*, 21–33. [CrossRef]
16. Chandra, S.; Kumar, S.; Kumar, V. Comparative Performance Evaluation of Self Propelled Paddy Transplanters in Calcareous Soil. *AMA Agric. Mech. Asia Afr. Lat. Am.* **2013**, *44*, 33–38.
17. Selvan, M.M.; Annamalai, S.J.K.; Thavaprakash, N.; Ananathakrishnan, D. Design and development of three-row improved pull-type rice transplanter for small farmers. *Indian J. Agric. Sci.* **2014**, *84*, 1422–1427.
18. Wei, X. Experiment Study on Planting Hole Forming Device of Transplanting Machine for Rape. Master's Thesis, Hunan Agriculture University, Changsha, China, 2014.
19. Xiang, W.; Mingliang, W.; Chunyun, G. Design and experiment of planting hole forming device of crawler transplanter for rape (*Brassica napus*) seedlings. *Trans. Chin. Soc. Agric. Eng.* **2015**, *31*, 12–18.
20. Chaopeng, C.; Wei, Q.; Mingliang, W.; Wentao, Z. Parameter optimization of vertical soil-filling hole-forming parts for rapeseed transplantation based on discrete element method. *J. Hunan Agric. Univ. Nat. Sci.* **2019**, *45*, 433–439.
21. Changjie, H.; Yang, X.; Jia, Y.; Jing, Z.; Panpan, Y. Parameter optimization of opener of semi-automatic transplanter for watermelon seedlings raised on compression substrate. *Trans. Chin. Soc. Agric. Eng.* **2019**, *35*, 48–56.
22. Wei, X.; Mingliang, W.; Chunyun, G.; Yiming, H.; Haifeng, L.; Bo, Y. Design and Test of Transplanting Hole-forming Machine for Rapeseed Potted Seedlings. *Trans. Chin. Soc. Agric. Mach.* **2017**, *48*, 40–48.
23. Zhang, Y.L.H.H. Design and experiment on rear suspended passive aerator in saline-alkali land. *Trans. Chin. Soc. Agric. Eng.* **2016**, 32.
24. Maile, Z.; Xianglei, X.; Mengbo, Q.; Daqing, Y. Optimal Design and Experiment of Potted Vegetable Seedling Transplanting Mechanism with Punching Hole. *Trans. Chin. Soc. Agric. Mach.* **2020**, *51*, 77–83.
25. Wei, S.; Xiaolong, L.; Hucun, W.; Hua, Z.; Jianmin, W.; Xiaoping, Y.; Guanping, W. Design and test of double crank multi-rod hill-drop potato planter on plastic film. *Trans. Chin. Soc. Agric. Eng.* **2018**, *34*, 34–42.
26. Rasool, K.; Islam, N.; Ali, M.; Jang, B.E.; Kwon, H.J. Onion transplanting mechanisms: A review. *Precis. Agric. Sci. Technol.* **2020**, *2*, 196.
27. Chen, Y.; Cheng, Y.; Chen, J.; Zheng, Z.; Hu, C.; Cao, J. Design and Experiment of the Buckwheat Hill-Drop Planter Hole Forming Device. *Agriculture* **2021**, *11*, 1085. [CrossRef]
28. Changjie, H.; Yang, X.; Jing, Z.; Jia, Y.; Hui, G. Design and experiment of semi-automatic transplanter for watermelon seedlings raised on compression substrate. *Trans. Chin. Soc. Agric. Eng.* **2018**, *34*, 54–61.
29. Wei, Q.; Mingliang, W.; Haifeng, L.; Chaopeng, C.; Wei, X. Optimization of soil hole opening method and parameters of pot seedling transplanting machine for rapeseed. *Trans. Chin. Soc. Agric. Eng.* **2020**, *36*, 13–21.
30. Lihua, Y.; Changshou, H.; Zhibo, X.; Fugui, Z.; Delong, F. Design and Test of Spiral Cavitation Device for Transplanting in Large Den. *J. Agric. Mech. Res.* **2020**, *42*, 198–202.
31. Chen, J.W.B.Z. Kinematics modeling and characteristic analysis of multi-linkage transplanting mechanism of pot seeding transplanter with zero speed. *Trans. Chin. Soc. Agric. Eng.* **2011**, *27*, 7–12.
32. Wang, Y.H.Z.W. Experiment on transplanting performance of automatic vegetable pot seedling transplanter for dry land. *Trans. Chin. Soc. Agric. Eng.* **2018**, *34*, 19–25.
33. Xin, J.; Jiangtao, J.; Weixiang, L.; Yakai, H.; Du, X. Structural optimization of duckbilled transplanter based on dynamic model of pot seedling movement. *Trans. Chin. Soc. Agric. Eng.* **2018**, *34*, 58–67.
34. Zhou, Y.; Chaohai, C.; Jieli, D.; Guoqi, Y.; Xuewen, P.; Liangli, Y.; Jialong, L. Performance test of hand-held electric hole-digger for fertilization in orchard. *Trans. Chin. Soc. Agric. Eng.* **2013**, *29*, 25–31.
35. Orzech, K.; Wanic, M.; Załuski, D. The Effects of Soil Compaction and Different Tillage Systems on the Bulk Density and Moisture Content of Soil and the Yields of Winter Oilseed Rape and Cereals. *Agriculture* **2021**, *11*, 666. [CrossRef]

36. Hu, H.L.H.W. Design and experiment of targeted hole-pricking and deep-application fertilizer applicator between corn rows. *Trans. Chin. Soc. Agric. Eng.* **2016**, *32*, 26–35.
37. Jinwu, W.; Wenqi, Z.; Chunfeng, Z.; Zhenwei, P.; Jinfeng, W.; Han, T. Reverse Design and Experiment of Non-circular Gear Planetary System Picking Hole Mechanism. *Trans. Chin. Soc. Agric. Eng.* **2015**, *46*, 70–75.
38. Mingliang, W.; Bo, Y.; Shijie, F.; Wentao, Z. Design of movable tray for bowl seedlings. *J. Hunan Agric. Univ. Nat. Sci.* **2018**, *44*, 666–671.
39. Honglin, C.; Shanghong, C.; Xueshan, S.; Liangcai, J.; Dinghui, L. Effects of Different Density to the Agronomic Traits and Yield of Transplanting Rapeseed in the Hilly Area of Sichuan Basin. *Chin. Agric. Sci. Bull.* **2012**, *28*, 142–145.
40. Wei, Q.; Mingliang, W.; Chunyun, G.; Haifeng, L. Experimental Study on the Shape Optimization of Soil Opener for Rapeseed Pot Seedling Transplanter. *J. Agrc. Sci. Technol.-Iran* **2021**, *23*, 97–106.
41. Xueqing, Z.; Hongwen, L.; Jin, H.; Qingjie, W.; Xirui, Z. Design of multi-segment in situ soil sampler testing bulk density. *Trans. Chin. Soc. Agric. Eng.* **2008**, 127–130. [CrossRef]
42. Zhengxing, L. *Fundamentals of Mechanical Design*; Chongqing University Press: Chongqing, China, 2016.
43. Chengguang, H.; Xijie, G.; Mengchan, Y.; Yuxiang, H.; Ruixiang, Z. Discrete element simulations and experiments of soil disturbance as affected by the tine spacing of subsoiler. *Biosyst. Eng.* **2018**, *168*, 73–82.
44. Yuxiang, H.; Chengguang, H.; Mengchan, Y.; Botao, W.; Ruixiang, Z. Discrete Element Simulation and Experiment on Disturbance Behavior of Subsoiling. *Trans. Chin. Soc. Agric. Eng.* **2016**, *47*, 80–88.
45. Wei, Y.; Liang, G.Z.; Feng, B. Effect of Plants at Different Fertilization Treatments on Soil Bulk Density and Porosity in the Gully Region of Loess Plateau. *Shaanxi For. Sci. Technol.* **2014**, 1–5. [CrossRef]

*agriculture*

MDPI

*Article*

# A Novel 10-Parameter Motor Efficiency Model Based on I-SA and Its Comparative Application of Energy Utilization Efficiency in Different Driving Modes for Electric Tractor

**Zhun Cheng [1], Huadong Zhou [2] and Zhixiong Lu [2,*]**

[1] Department of Vehicle Engineering, College of Automobile and Traffic Engineering, Nanjing Forestry University, Nanjing 210037, China; chengzhun38@163.com

[2] College of Engineering, Nanjing Agricultural University, Nanjing 210031, China; lesprince_3@163.com

* Correspondence: luzx@njau.edu.cn

**Abstract:** To build a more accurate motor efficiency model with a strong generalization ability in order to evaluate and improve the efficiency characteristics of electric vehicles, this paper researches motor efficiency modeling based on the bench tests of two motor efficiencies with differently rated powers. This paper compares and analyzes three motor efficiency modeling methods and finds that, when the measured values in motor efficiency tests are insufficient, the bilinear interpolation method and radial basis kernel function neural networks have poor generalization abilities in full working conditions, and the precision of polynomial regression is limited. On this basis, this paper proposes a new modeling method combining correlation analysis, polynomial regression, and an improved simulated annealing (I-SA) algorithm. Using the mean and the standard deviation of the mean absolute percentage error of the 5-fold Cross Validation (CV) of 100 random tests as the evaluation indices of the precision of the motor efficiency model, and based on the motor efficiency models with verified precision, this paper makes a comparative analysis on the full vehicle efficiency of electric tractors of three types of drive in five working conditions. Research results show that the proposed novel method has a high modeling precision of motor efficiency; tractors with a dual motor coupling drive system have optimal economic performance.

**Keywords:** electric tractor; motor efficiency; dual motor coupling drive; I-SA algorithm; generalization ability; parameter identification

**Citation:** Cheng, Z.; Zhou, H.; Lu, Z. A Novel 10-Parameter Motor Efficiency Model Based on I-SA and Its Comparative Application of Energy Utilization Efficiency in Different Driving Modes for Electric Tractor. *Agriculture* **2022**, *12*, 362. https://doi.org/10.3390/agriculture12030362

Academic Editors: Mustafa Ucgul and Chung-Liang Chang

Received: 28 January 2022
Accepted: 28 February 2022
Published: 3 March 2022



## 1. Introduction

Electric vehicles use a motor as the core power source [1,2] and electric energy as the driving energy [3]. This is not only convenient for recycling energy, but it can also achieve low emissions or even zero emissions [4–6]. In addition, electric energy can be obtained widely and conveniently. Compared with internal combustion engine vehicles, electric vehicles also have the advantages of low noise and low maintenance costs [7]. These characteristics provide great help for the sustainability of the environment and its ecology. A purely electric vehicle can realize zero release, and its energy is widely and easily available. Currently, for all types of vehicles, researchers are continuously studying and developing drive systems with the motor as the power supply unit, including those in vehicles such as the electric truck [8], the refuse collection truck [9], the electric urban delivery truck [10], the electric car [11–14], electric agricultural machinery [15], etc. Especially in the research field of agricultural machinery, agricultural machinery faces complicated working conditions. Agricultural machinery such as tractors generally needs to meet the requirements of operation, including ploughing, rotary tillage, transportation, and normal road driving [16,17]. Therefore, the research and application of electric tractors can help to overcome the problems of the traditional tractor, such as complicated variable transmission structures, frequent operation of the driver, limited energy utilization efficiency, etc. [18–20].

At the current stage, the multi-power-source purely electric vehicle has certain advantages in price, energy consumption, power transmission, system efficiency, etc., compared with the single-motor drive vehicle and the hybrid-power vehicle [21]. The two main types of purely electric vehicles are as follows: distributed independent drive and centralized coupling drive. In addition, there are few studies on electrical agricultural machinery. Zhou et al. [22] proposed a wheel-rim motor independent drive scheme, aiming to give the four-wheel drive electric tractor better tractive performance in the working condition of traction operation. Han et al. [23] made a simulated analysis on the steering characteristics of wheel-rim drive electric tractors in heavy-load low-speed working conditions and in light-load high-speed working conditions. Xie et al. [24] proposed an overall structure scheme of dual-wheel drive electric tractors (the two rear driving wheels are driven by the motor independently) and tested the tractive performance, loaded starting and transportation working con-dition through the bench test. Li et al. [25] designed, compared and analyzed many dual-motor power coupling structure schemes based on the planetary gear train. The research built a dual-motor power coupling unit transmission simulation model using the software Simulation X.

Besides the type of energy used, energy consumption and emission control are also affected by system transmission efficiency. Using the electric tractor as an example, the efficiency characteristics of a power transmission system with the core of a motor plays a decisive role in the full vehicle's economic performance and in its energy utilization performance. Therefore, the correct description and modeling of efficiency characteristics of motors have great significance and necessity. However, currently there are few studies on the modeling of motor efficiency characteristics. Most researchers used polynomial regression, the bilinear interpolation method, and the table look-up method to build efficiency models, and they made subsequent research and applications based on the models. Additionally, most studies on electric vehicles' kinetic analysis, control, and performance improvements require considering motor efficiency. Ma et al. [26] proposed a planetary gear coupling mechanism based on the dual motor structure and used the particle swarm optimization method to optimize the parameters of the transmission system. Their research was made according to the efficiency characteristics of motors. Hu et al. [27] proposed a new dual motor coupling power drive system and combined the efficiency characteristics of motors to analyze and make power distribution strategies for different drive modes. Li et al. [28] used a motor efficiency model for parameter matching, power management strategy development, and power distribution control research of a dual motor coupling drive system. Their research measured and recorded the data of torque, speed, and efficiency through a bench test and then obtained the motor efficiency in any state using the interpolation method. Li et al. [29] designed a dual motor multi-mode power coupling drive system for the purely electric tractor (the system can realize four drive modes) and made a simulation test on the ploughing and transportation working conditions of tractors. Their research used the quasi-static maps of output speed and torque to build the dual motor efficiency model and to obtain the motor efficiency in any working state though the table look-up method. Chen et al. [30] pointed out that the current studies on power system parameter design focused mainly on electric cars but less on electrical agricultural machinery. Their research used a quantified motor efficiency to optimize the system in order to improve the performance of electric tractors.

Therefore, for electric vehicles, the evaluation, analysis, and improvement on their power performances and economic performances should be made according to the efficiency characteristics of motors. In addition, the design of reasonable and effective power transmission systems and parameter matches also depend on the efficiency characteristics of motors. Therefore, building an accurate model of motor efficiency characteristics is an important premise of the design, control, and improvement of electric vehicles. However, different motors have different efficiency characteristics and change laws. Therefore, only the modeling of motor efficiency characteristics with specific test data and scientific verification methods is effective. In addition, building a motor efficiency model with a specific mathematical expression form can further help the kinetic analysis and control of the system. In addition, current studies on electric vehicles such as the electric tractor focus mainly on the comparison of an electrically driven car and a railway motor car and on the comparison of a single motor and multiple motors, but there are few comparative studies on the working efficiency of dual motor coupling drives, of dual motor rear wheel independent drives, and of four-wheel independent drives based on the wheel hub motor.

Based on the above literature, there are some deficiencies in the research of electric vehicles at the present stage: (1) There is more research on electric vehicles but less research on electric agricultural machinery. (2) Comparative studies on distributed independent drives and on centralized coupling drives of tractors with multiple power sources are relatively less abundant. This makes it difficult to develop a drive system with relatively better performance in the conceptual design stage, although the actual characteristics of the motor have been obtained. (3) There are many research papers on the application of motor efficiency but few studies on the establishment of the motor efficiency characteristics model. (4) Polynomial regression, bilinear interpolation, and the table look-up method are often used to establish the motor efficiency characteristics model, which lacks the application and comparison of advanced modeling methods, resulting in limited precision of model estimation. (5) The interpolation method and the table look-up method are difficult to use to form specific mathematical expressions, so the effect of extrapolation prediction is limited. (6) When polynomial regression is applied, only the power function form of speed and torque and the interaction term of each order are considered, and the influence of other data transformation forms on the efficiency characteristics of the motor is not explored.

In order to solve the above problems, combined with the bench test data of tractor motors, this paper aims to put forward a model to correctly describe the efficiency characteristics of motors by comparing four different modeling methods and by exploring the influence of other data transformation forms of physical quantities (speed and torque) on the efficiency characteristics of motors. In addition, the efficiency of tractors' different driving systems is compared and analyzed through the measured data and the new model (based on the typical multiple working conditions of tractors). This provides a direct basis for the further research and development of electric tractor drive systems and for the establishment of a high precision motor model. Specifically, this paper proposes a 10-parameter motor efficiency modeling method combining the correlation coefficient and the improved simulated annealing (I-SA) algorithm. The method mainly adopts the idea of polynomial regression and error compensation. According to the measured data of motor efficiency through the bench test, an initial fitting model made through polynomial regression analysis is built. In the case of the lack of polynomial regression precision, the variable with the maximum correlation is determined according to the calculation of the correlation coefficient, and it is introduced into the initial fitting model as a compensation variable to obtain a new motor efficiency model. Additionally, the proposed method identifies the parameters of a new model of motor efficiency using the I-SA algorithm according to the bench test data. Then, the new model of motor efficiency is built. To improve and verify the effectiveness and precision of the new efficiency model, this paper uses the mean of the 5-fold Cross Validation (CV) accuracy of multiple tests as the evaluation index and comparatively analyzes the novel modeling method with those using polynomial regression (PR), the bilinear interpolation method (BI), and the radial basis kernel function neural network (RBF-NN). This paper uses two motors with different powers of tractors in the research. Finally, this paper researches and comparatively analyzes the differences among three drive types (the dual motor coupling drive, the dual motor rear wheel independent drive, and the four-wheel independent drive based on wheel hub motor) of multi-power-source purely electric vehicles in working efficiency with their motor efficiency models built. This paper uses different reserve power states of electric tractors, including the running speed working condition, the normal running working condition, the transportation working condition, the ploughing working condition, and the rotary tillage working condition, as examples for the comparison of the efficiency characteristics of full vehicles.

## 2. Materials and Methods

### *2.1. Characteristic Analysis of the Distributed Independent Drive System*

Figure 1 gives a structural diagram of a dual motor dual-wheel distributed independent drive system [31]. (The drive system is mainly composed of two independent motors and two independent reducers. The left rear wheel or right rear wheel is driven by a separate "wheel side motor-reducer" system.) Figure 2 gives a structural diagram of a four-motor four-wheel distributed independent drive system [22,32]. (Approximate to the dual motor dual-wheel distributed drive system, four wheels of the vehicle are driven by four separate "wheel motor-reducer" systems.) Different from the traditional "combustion-motor-variable-transmission-unit" and the single-motor-variable-transmission-unit structural forms, the distributed independent drive system removes most mechanical connection mechanisms (mainly referring to complicated mechanical transmission mechanisms such as the gearbox, the differential mechanism, etc.). Using a tractor as an example, the drive system increases the flexibility of the spatial structure arrangement greatly and helps realize the light weight of a full vehicle.

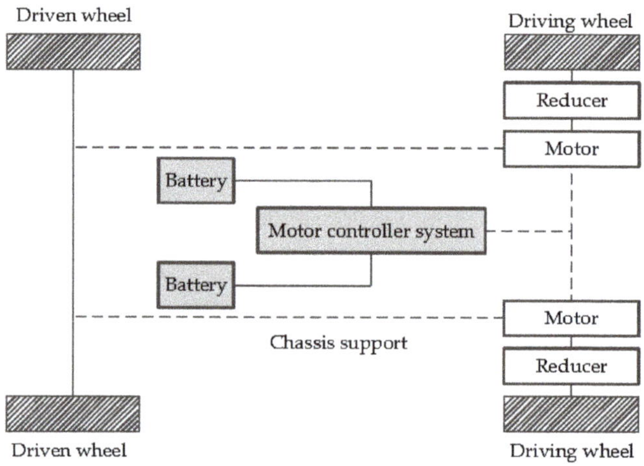


**Figure 1.** Structural diagram of dual motor dual-wheel distributed independent drive system.

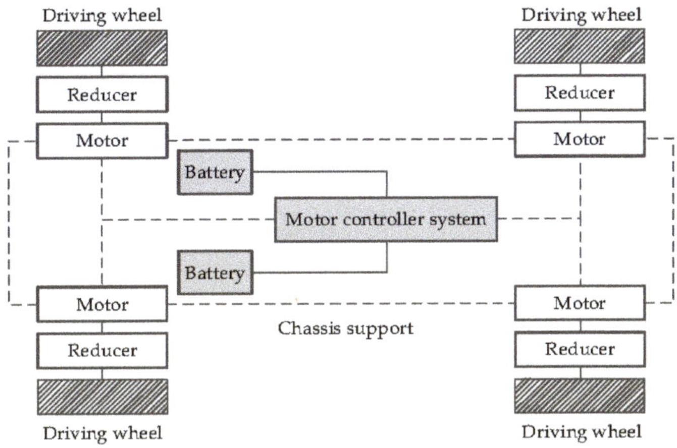


**Figure 2.** Structural diagram of four-motor four-wheel distributed independent drive system.

According to Figures 1 and 2, the calculation formula of speed-regulating characteristics of a distributed independent drive system is as follows:

$$u_a = 3.6 \times 2\pi(n_m r_d)/(60 i_g) \quad (1)$$

where $u_a$ is the running speed of the electric vehicle (km/h), $r_d$ is the rolling radius of the driving wheel (m), $i_g$ is the transmission ratio of the reducer, and $n_m$ is the output speed of the motor (r/min).

The following is the calculation formula of the torque of the distributed independent drive system:

$$n T_m i_g / r_d = F_L \quad (2)$$

where $n$ is the number of the motor of the distributed independent drive system, $T_m$ is the output torque of a single motor (Nm), and $F_L$ is the carrying capacity of the electric vehicle (N).

The following is the calculation formula of efficiency characteristics of the distributed independent drive system:

$$\eta_{sys1} = \eta_m(n_m, T_m)\eta_0 \quad (3)$$

where $\eta_{sys1}$ is the efficiency value of the distributed independent drive system, $\eta_m$ is the efficiency of the motor (a function with motor speed $n_m$ and torque $T_m$ as independent variables), and $\eta_0$ is the overall working efficiency of other systems (mainly composed of the efficiency of the storage battery, the efficiency of the reducer, etc.) and is also a fixed value in this paper.

### 2.2. *Characteristic Analysis of a Dual Motor Centralized Coupling Drive System*

Figure 3 gives a structural diagram of a dual motor centralized coupling drive system used in this paper [33]. The system is a speed coupling drive system. The power output from motor 1 and motor 2 is transmitted into the planetary gear train's gear ring and sun gear. The two parts of power converge through the planetary gear structure and then are output through the planetary carrier. The power output passes through the reducer and the differential mechanism and then forms the power of the driving wheels.

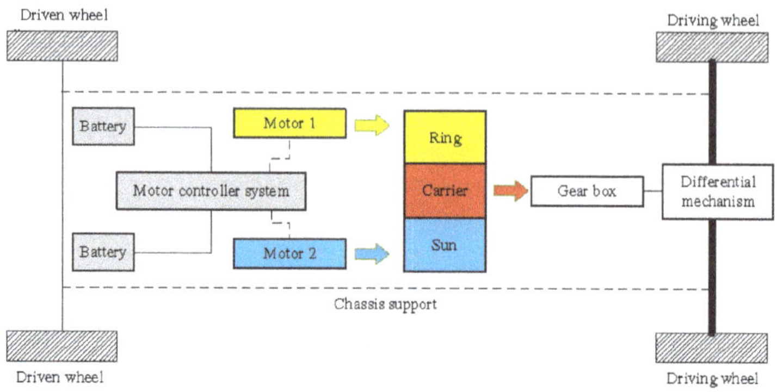


**Figure 3.** Structural diagram of dual motor centralized coupling drive system.

The following is the calculation formula of the speed-regulating characteristics of the dual motor centralized coupling drive system:

$$u_a = 0.377\frac{k_p n_{m1} + n_{m2}}{(1 + k_p)i_g} r_d \tag{4}$$

where $k_p$ is the characteristic parameter of the planetary row and $n_{m1}$ and $n_{m2}$ (r/min) are the output speeds of motor 1 and motor 2, respectively.

The following is the calculation formula of the torque of the dual motor centralized coupling drive system:

$$\frac{T_{m1}}{k_p} = T_{m2} = \frac{F_L r_d}{i_g} \tag{5}$$

where $T_{m1}$ and $T_{m2}$ (Nm) are the output torques of motor 1 and motor 2, respectively.

This paper analyzes the loss of power flow in motor 1, motor 2, and the planetary gear train to deduce the overall efficiency characteristics of the system. By dividing the whole system into a series subsystem (mainly motor 1 to the gear ring; motor 2 to the sun wheel; planetary carrier to the other transmission systems) and a parallel subsystem (mainly confluence mechanisms), the efficiency characteristics model is then obtained by combining the series and parallel subsystems [16,34].

The calculation formula of the efficiency characteristics of the dual motor centralized coupling drive system is as follows:

$$\eta_{sys2} = k_p(\eta_{rc}\eta_{m1}n_{m1} + \eta_{sc}\eta_{m2}n_{m2}/k_p)/(k_p n_{m1} + n_{m2})\eta_0 \tag{6}$$

where $\eta_{sys2}$ is the efficiency of the dual motor centralized coupling drive system, $\eta_{rc}$ is the transmission efficiency of the power flow from the gear ring to the planetary carrier, and $\eta_{sc}$ is the transmission efficiency of power flow from the sun gear to the planetary carrier.

This paper uses the transmission ratio method [16,35] to calculate the transmission efficiency $\eta_{rc}$ of power flow from the gear ring to the planetary carrier and the transmission efficiency $\eta_{sc}$ of the power flow from the sun gear to the planetary carrier. The calculation formula is as follows:

$$\eta_{sc} = \frac{1 + k_p\eta_c^{sign[\frac{k_p}{1+k_p} \times \frac{\partial(1+k_p)}{\partial(k_p)}]}}{1 + k_p} \tag{7}$$

$$\eta_{rc} = \frac{1 + k_p \eta_c^{sign[\frac{k_p^2}{1+k_p} \times \frac{\partial(\frac{1+k_p}{k_p})}{\partial(k_p)}]}}{\eta_c^{sign[\frac{k_p^2}{1+k_p} \times \frac{\partial(\frac{1+k_p}{k_p})}{\partial(k_p)}]}(1+k_p)} \quad (8)$$

where $\eta_c$ is the friction transmission efficiency of the 2K-H planetary gear train.

### 2.3. Obtaining the Measured Data of the Bench Test of Motor Efficiency

This paper uses two motors of electric tractors as examples to obtain related data of the motors through the bench test. The two motors are both brushless direct current motors. Table 1 shows the specific parameters.

**Table 1.** Related parameters of the motor.

| Type | Rated Power/kW | Rated Speed/r/min | Rated Torque/Nm |
|---|---|---|---|
| Brushless DC motor | 5 | 750 | 63 |
| Brushless DC motor | 8 | 850 | 90 |

The test bench uses a lithium battery pack with a total voltage of 211.2 V to offer power (see Table 2 for the main parameters of the power battery) and a magnetic powder brake as the simulation loading unit (Table 3 shows the parameters of the magnetic powder brake). The information of the overall layout of the test bench, the design of the observation and control plan, the equipment installation and debugging, and the calibration of the sensors (mainly including the current sensor, the voltage sensor and the speed torque sensor) can be found in previous studies [22,36].

**Table 2.** Related parameters of power battery.

| Type | Nominal Voltage of Monomer/V | Number of Monomers | Nominal Capacity of Monomer/(A·h) | Total Voltage/V |
|---|---|---|---|---|
| Lithium battery | 3.3 | 64 | 100 | 211.2 |

**Table 3.** Related parameters of magnetic powder brake.

| Model | Rated Torque/Nm | Magnetizing Current/A | Allowable Slip Power/kW | Cooling Mode |
|---|---|---|---|---|
| CZ-20 | 200 | 2 | 10 | Water cooling |

Figure 4 shows the test bench of the motor.

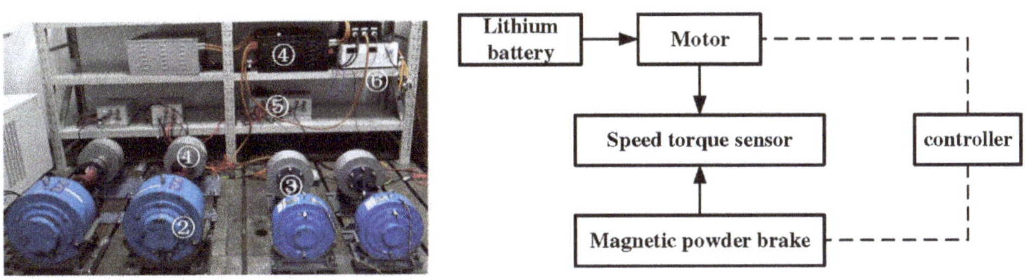


(a) (b)

**Figure 4.** Structural composition of motor test bench. Note: ① Motor used in the research; ② magnetic powder brake; ③ speed torque sensor; ④ lithium battery; ⑤ motor controller; ⑥ power management system. (**a**) The picture of test site; (**b**) Schematic diagram.

The detailed steps of the motor efficiency test can be found in previous studies [22,36]. For the test, load the motor through the magnetic powder brake. Each loading should be performed after the motor's working speed remains stable. The efficiency test results of the 8 kW motor came from a reference document [36]. The following is the calculation formula of the motor efficiency:

$$\eta_m = \frac{P_{out}}{P_{in}} = \frac{T_m n_m}{9.55UI} \tag{9}$$

where $P_{out}$ is the mechanical power output of the motor (W), $P_{in}$ is the input power of the motor (W), $U$ is the input voltage of the motor (V), and $I$ is the current input of the motor (A).

Figure 5 shows the measured results of efficiency of two motors used in the analysis.

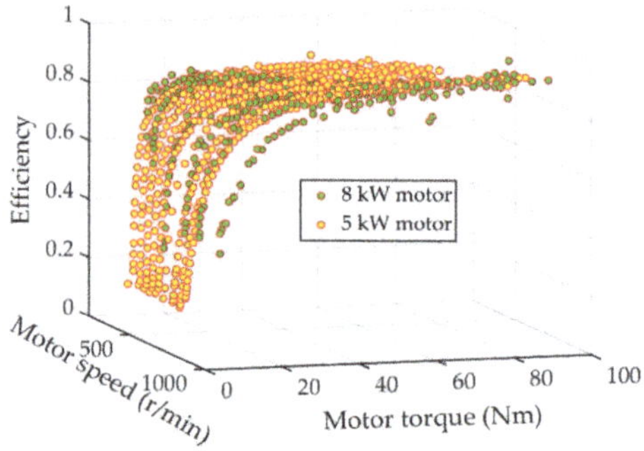


**Figure 5.** Measured results of efficiency of two motors.

### 2.4. Three Modeling Methods of Motor Efficiency

Current studies generally use PR [37] and BI as the modeling methods of motor efficiency. Therefore, to comparatively analyze the differences among the modeling methods of motor efficiency comprehensively, this paper also uses PR and BI as the modeling methods for the comparative analysis. Additionally, this paper chooses the neural network as one of modeling methods for comparative analysis. This paper uses the RBF-NN.

The PR model considers the 1-order term, the 2-order term, and the interaction term of the motor speed $n_m$ and the motor torque $T_m$, and its expression is as follows:

$$\eta_{m\_PR} = a_0 + a_1 n_m + a_2 T_m + a_3 n_m T_m + a_4 n_m^2 + a_5 T_m^2 \tag{10}$$

where $\eta_{m\_PR}$ is the motor efficiency model built using PR and $a_0$~$a_5$ are the coefficients of the order terms of the polynomial.

The BI method [38] requires the information of four data points and makes the linear interpolation in two directions of coordinate axes to finally determine the data value of the points to be interposed.

The RBF-NN was proposed by J. Moody and C. Darken in 1989, and it is a forward local neural network that can approach any nonlinear function with any arbitrary small error [39–41]. Currently, the RBF-NN has a wide range of applications, and many scholars have found in studies that the RBF-NN is superior to the conventional BP neutral network in every respect. Figure 6 shows the topological structure of the RBF-NN.

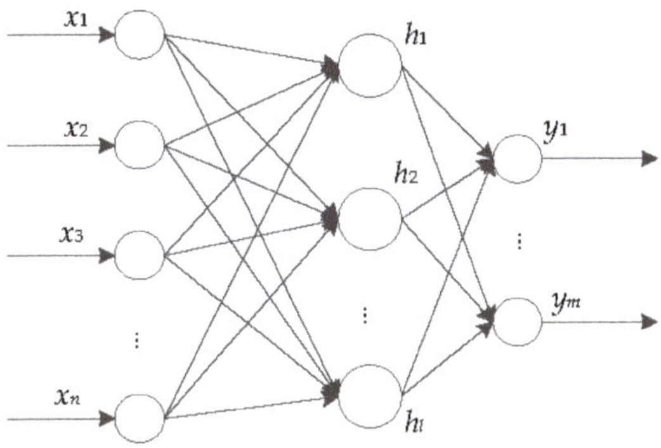


**Figure 6.** Technical route of establishing dynamic load characteristics prediction model. Note: $x = [x_1, x_2, \ldots, x_n]$ represents the input vector; $h = [h_1, h_2, \ldots, h_l]$ represents the middle layer; $y = [y_1, y_2, \ldots, y_m]$ represents output vector.

In Figure 6, $x = [x_1, x_2, \ldots, x_n]$ represents the input vector. The middle layer uses a Gauss function for the nonlinear transformation of the input vector to obtain $h = [h_1, h_2, \ldots, h_l]$ and then a linear weighting combination to obtain $y = [y_1, y_2, \ldots, y_m]$ and considers it as the output. The calculation formula is as follows:

$$y_i = \sum_{j=1}^{l} w_{ij}h_j = \sum_{j=1}^{l} [w_{ij} \exp\{-\frac{(x-c_j)^{\mathrm{T}}(x-c_j)}{\delta_j^2}\}] \tag{11}$$

where $i = 1, \ldots\ldots, m$, of which $m$ is the dimension of the output, $l$ is the number of neurons in the hidden layer, and $w_{ij}$ is the connection weight of the $i$ output and the $j$ neuron in the hidden layer.

To build an efficiency model based on the RBF-NN more comprehensively, this paper uses PR's idea of 1-order terms and 2-order terms using the motor speed $n_m$ and the motor torque $T_m$, and it also uses interaction term modeling, which considers the 1-order term, the 2-order term, and the interaction term of the motor speed $n_m$ and the motor torque $T_m$ as the characteristic variables of neural network learning and training.

Additionally, to further compare the modeling methods of motor efficiency comprehensively, this paper uses the 5-fold CV method [42]. This paper uses the mean and standard deviation of the *MAPE* (mean absolute percentage error) of the 5-fold CV of 100 random tests as the indices to evaluate the precision of the motor efficiency model.

The following is the calculation formula of the *MAPE*:

$$MAPE = \frac{1}{n}\sum_{i=1}^{n}\left|\frac{\eta_{m\_estimated} - \eta_{m\_measured}}{\eta_{m\_measured}}\right| \times 100\% \tag{12}$$

where $\eta_{m\_estimated}$ is the evaluated value based on the motor efficiency model, $\eta_{m\_measured}$ is the measured value of motor efficiency, and $n$ is the total number of data of the training set or the test set.

### *2.5. A Novel 10-Parameter Modeling Method of Motor Efficiency Using the I-SA for Parameter Identification*

Although BI and the RBF-NN can build the motor efficiency model effectively (see Section 3.1), the models built with the two methods (have passed the 5-fold CV, see Section 3.1) have ideal and high precision for both the learning and training samples and the test samples. However, according to research results in this paper (Section 3.1), the precision of the motor efficiency models built with BI and RBF-NN depends on the value range of the learning and training sample data (i.e., the value range of motor speed $n_m$ and motor torque $T_m$ in the learning and training samples). If the set of test data is in the value range of motor speed $n_m$ and motor torque $T_m$ in the learning and training samples, the two methods can offer the estimated value of motor efficiency with high precision. If the set of test data is out of the value range of the motor speed $n_m$ and the motor torque $T_m$ in the learning and training samples, BI cannot be implemented due to the limitations of required conditions. Additionally, the RBF-NN's extrapolation prediction result is poor.

In addition, BI and the RBF-NN have no directly simple relational expression, so they are rather complicated and tedious in practical applications.

To sum it up, the two methods are limited in their generalization abilities and are applicable to the case with a mass of bench test data. Therefore, the two methods are inapplicable to the technical research and development of real-time adjustment full vehicle control strategies according to the current efficiency of intelligent vehicle systems.

However, PR has limited precision (see Section 3.1). According to the problems above, this paper proposes a modeling method of motor efficiency, combining the correlation coefficient and PR. Using the idea of error compensation for reference and considering the limited precision of PR, the method uses the enumeration method and correlation coefficient analysis to determine the variable with the maximum correlation, it introduces the variable into the PR model as a compensation term, and it finally forms a new model. In addition, the motor torque $T_m$'s correlation with efficiency is bigger than that of the motor speed.. with efficiency, so the 3-order term of the motor torque $T_m$ is introduced into the new model. The new model's expression is as follows:

$$\eta_{m\_novel} = a_0 + a_1 n_m + a_2 T_m + a_3 n_m T_m + a_4 n_m^2 + a_5 T_m^2 + a_6 T_m^3 + a_7 f(a_8 x + a_9) \quad (13)$$

where $\eta_{m\_novel}$ is the novel 10-parameter model of motor efficiency, $a_0$~$a_9$ are the coefficients of the order terms of the polynomial, $f(x)$ is the mapping function with $x$ as the independent variable, and $x$ is the motor speed $n_m$ or the motor torque $T_m$.

The novel model has high nonlinearity. This paper uses the I-SA algorithm and combines the bench test data for the parameter identification of the novel model of motor efficiency and then completes the building of the novel model of motor efficiency.

As a result of the sample data involved in this case being large, the optimization process is nonlinear in the pursuit of high modeling precision. The heuristic intelligent optimization algorithm [43–45] is suitable for solving a series of complex engineering problems accurately. This paper adopts the SA algorithm as the method of parameter identification. The SA algorithm used in this paper refers to the flow of the I-SA algorithm that has been proposed and that verified the engineering application effect in previous studies [46,47]. The applied I-SA algorithm is mainly improved in the following aspects: (1) termination conditions of algorithm iteration; (2) disturbance function; and (3) the updating method of decision variables. The specific algorithm flow is shown in Figure 7. This paper considers the mean of the *MAPE* of the 5-fold CV as the objective function of the I-SA algorithm.

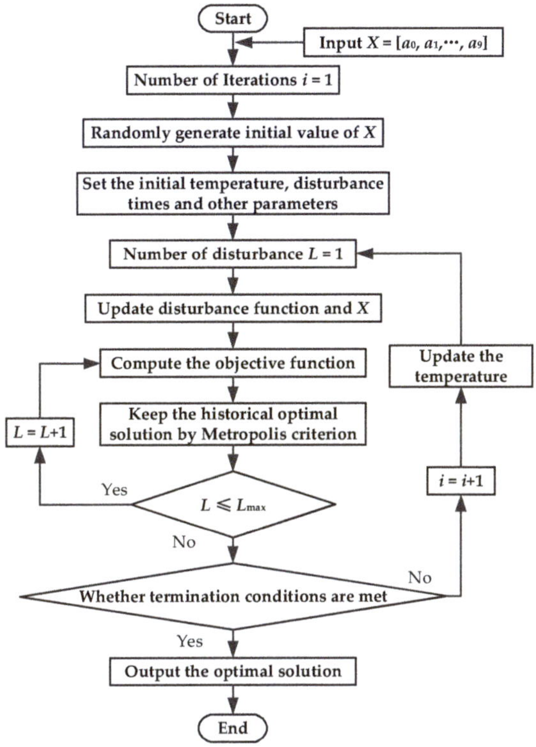


**Figure 7.** Flow diagram of the I-SA algorithm. Note: $L$ is number of disturbances; $X$ is the set of decision variables; $i$ is number of Iterations; $L_{max}$ is maximum number of disturbances.

In this paper, the I-SA algorithm is used to complete the parameter identification of the new model with ten parameters to finally complete the establishment of the motor efficiency characteristics model. Figure 8 shows the building process of the novel model of motor efficiency.

### *2.6. Comparative Analysis Method of the Full Vehicle Efficiency Characteristics of Electric Tractors in Five Working Conditions*

This paper applies the motor efficiency model to the comparative analysis of the work efficiency of multi-power-source electric vehicles of three drive types (i.e., the dual motor coupling drive, the dual motor rear wheel independent drive, and the four-wheel independent drive based on hub motor). This paper uses the electric tractor as an example. See Table 4 for the related parameters of the full vehicle.

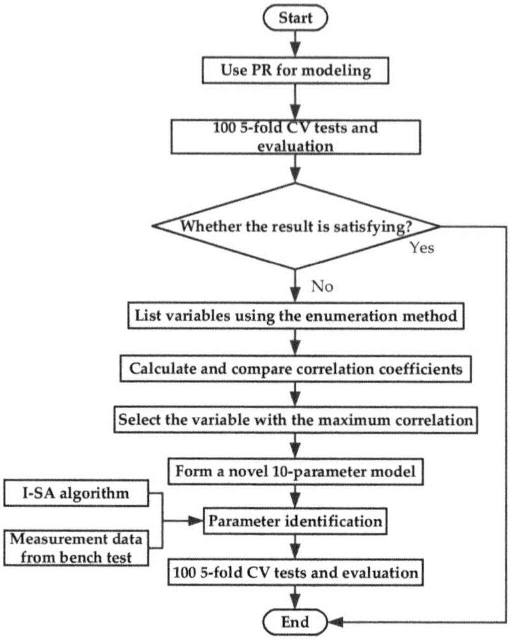


**Figure 8.** Flow diagram of building method of motor efficiency. Note: PR is Polynomial Regression; CV is Cross Validation; I-SA algorithm is Improved-Simulated Annealing algorithm.

**Table 4.** Related parameters of electric tractor.

| Total Weight/kg | Radius of Driving Wheel/m | Coefficient of Air Resistance | Windward Area/m$^2$ | Maximum Trailer Weight/kg |
|---|---|---|---|---|
| 1535 | 0.64 | 0.9 | 3.135 | 1000 |

As for the running speed working condition of the electric tractor, this paper uses scholar Wang's analysis on the research results of Resch and Renius for reference [48]. The results show that, in the overall life cycle, the tractor spends 61–68% of its time working in the speed range of 4–12 km/h.

This paper considers the average full vehicle efficiency of full working ranges in five working conditions of electric tractors as the evaluation index, and it uses it to compare the working efficiency of three drive types. The five working conditions are as follows:

(1) The electric tractors have different reserve power (the residual power eliminating the basic running resistance) and running speeds. The working condition is a general working condition for analysis, used to analyze the full vehicle working efficiency of electric tractors with different loads and running speeds.

The calculation formula of the reserve power is as follows:

$$F_{RF} = \frac{T_{out}}{r_d} - F_f - F_w = \frac{T_{out}}{r_d} - m_1 g f - \frac{C_D A u_a^2}{21.15} \tag{14}$$

where $F_{RF}$ is the reserve power (N), $T_{out}$ is the torque of the system power output to the driving wheel (Nm), $F_f$ is the rolling resistance (N), $F_w$ is the air resistance (N), $m_1$ is the overall weight of the electric tractor (kg), $f$ is the rolling resistance coefficient, $g$ is the acceleration of gravity, $C_D$ is the coefficient of air resistance, and $A$ is the windward area (m$^2$).

(2) The electric tractor is in the normal road running condition with a constant speed. In this working condition, the drive system composed of the motor only needs to overcome the basic running resistance.

(3) The electric tractor is in the transportation working condition with a running speed of 12 km/h. The calculation formula of the loading force in the current working condition is as follows:

$$F_{TC} = (m_1 + m_2)gf + \frac{C_D A u_a^2}{21.15} \tag{15}$$

where $F_{TC}$ is the loading force in the transportation working condition (N) and $m_2$ is the mass of the trailer of the electric tractor (kg).

(4) The electric tractor is in the ploughing working condition with a running speed of 10 km/h. The calculation formula of the loading force in the current working condition is as follows:

$$F_{PC} = F_f + F_w + kn_1 bH \tag{16}$$

where $F_{PC}$ is the loading force in the ploughing working condition (N), $k$ is the soil-specific resistance, $b$ is the width of a single plough, $n_1$ is the number of the plough share, and $H$ is the tillage depth.

(5) The electric tractor is in the rotary tillage working condition with a running speed of 5 km/h. The calculation formula of the loading force in the current working condition is as follows:

$$F_{RTC} = F_f + F_w + k_2 k_\lambda B_n h \tag{17}$$

where $F_{RTC}$ is the loading force in the rotary tillage working condition (N), $k_2$ is the rotary tillage resistance calculation coefficient, $k_\lambda$ is the soil-specific resistance of the rotary tillage, $B_n$ is the tillage width, and $h$ is the tillage depth.

Specifically, the multi-power-source purely electric tractors of three drive types include the 4 wheel independent drive electric tractor composed of four 5 kW motors, the rear wheel independent drive electric tractor composed of two 8 kW motors, and the coupling drive electric tractor composed of two 8 kW motors.

## 3. Results and Discussion

### *3.1. Modeling Results of Motor Efficiency Based on PR, BI, and the RBF-NN and Analysis*

This paper uses the 5 kW motor of the electric tractor in Section 2.3 as an example, and there are a total of 510 measured data samples of the bench.

This paper uses the 5-fold CV method of 100 random tests. Figures 9 and 10, and Table 5 show the modeling results of the motor efficiency based on PR, BI, and the RBF-NN.

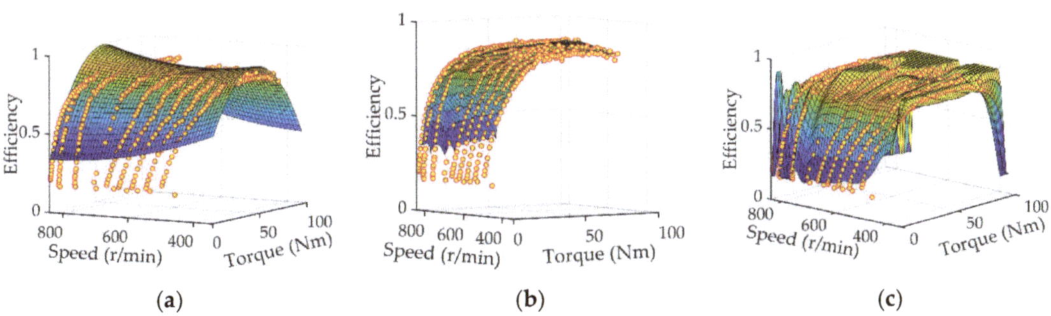


**Figure 9.** The comparison of precision of motor efficiency models in full working conditions. (**a**) Polynomial Regression; (**b**) Bilinear Interpolation method; (**c**) Radial Basis Kernel Function Neural Network.

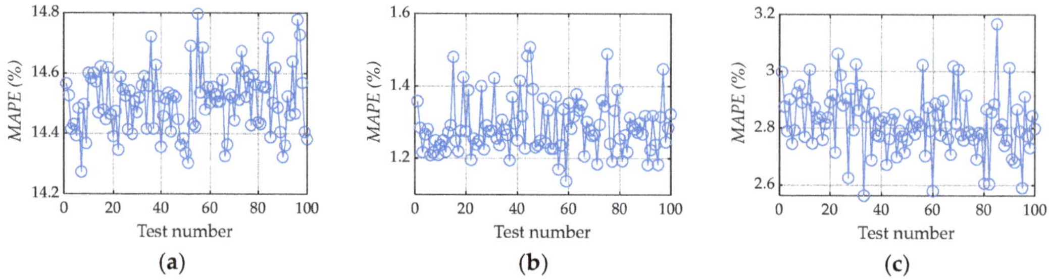


**Figure 10.** Comparison of results of 100 5-fold CV tests. (**a**) Polynomial Regression; (**b**) Bilinear Interpolation method; (**c**) Radial Basis Kernel Function Neural Network.

**Table 5.** Results of 5-fold CV of 100 random tests of four methods.

| | PR | BI | RBF-NN | Novel Method Proposed |
|---|---|---|---|---|
| **Mean (%)** | 14.51 | 1.29 | 2.84 | 2.45 |
| **Standard Deviation (%)** | 0.11 | 0.08 | 0.14 | 0.12 |

According to Figures 9 and 10 and Table 5, BI has the highest modeling precision followed by the RBF-NN, whereas PR has a certain difference compared with the other two methods. The results of the 5-fold CV of 100 random tests (of which the standard deviation is generally small) show that the three modeling methods of motor efficiency all have high reliability.

However, according to the research process in this paper, when using BI, there are invalid data points of motor efficiency generated in the test data set as a result of BI, which is limited by required calculation conditions, not being able to predict the working conditions out of the range of the learning and training set. This paper adds together the number of invalid data points in the 100 5-fold CV tests, and the results are shown in Figure 11.

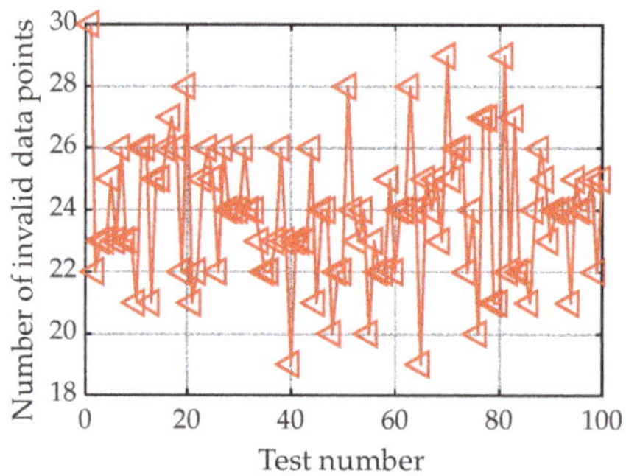


**Figure 11.** Statistical results of the number of invalid data points.

According to Figure 11, when using BI, if the number of measured data samples of motor efficiency is insufficient, there is an interpolation invalidation phenomenon. According to the results of the 100 5-fold CV tests, each model can generate about 24 invalid data points on average. Additionally, if the number of measured data samples of motor efficiency is insufficient, the estimation precision of BI for the nonlinear change section can decrease significantly.

According to Figures 9 and 10, and Table 5, although using the RBF-NN can estimate motor efficiency with high precision, the estimation precision for the change sections of the motor speed $n_m$ and the motor torque $T_m$ out of the change range of the learning and training set decreases significantly.

To sum it up, BI and the RBF-NN depend on learning and training samples greatly. When the change ranges of the motor speed $n_m$ and the motor torque $T_m$ in the learning and training samples cover the whole domain of definition, BI and the RBF-NN have high motor efficiency estimation and modeling precision. Therefore, for the modeling of motor efficiency, BI and the RBF-NN have weak generalization abilities. In addition, neither BI nor the RBF-NN has a directly easy relational expression, causing complicated and tedious applications in practice (compared with PR). In addition, the requirement of many learning and training data for effective modeling also causes the two methods to be inapplicable to the research and development of adaptively adjusting motor efficiency MAP charts for real-time correcting control strategies of intelligent vehicle systems.

The advantages and disadvantages of the three modeling methods are summarized in Table 6.

**Table 6.** Comparison of advantages and disadvantages of three modeling methods.

| Methods | Advantages | Disadvantages |
|---|---|---|
| Polynomial regression | Small standard deviation, high reliability of the model; fewer model parameters and less dependence on the number of data; clear and simple mathematical expression; convenient application | Relatively low precision; limited estimation ability |
| Bilinear interpolation | Small standard deviation, high reliability of the model; high precision (highest precision among three methods) | A certain number of invalid data points can be generated; highly dependent on the learning and training samples; no direct and simple mathematical model; inconvenient practical application |
| Radial Basis Kernel Function Neural Network | Small standard deviation, high reliability of the model; relatively high precision | Error larger than bilinear interpolation; highly dependent on the learning and training samples; no direct and simple mathematical model; inconvenient practical application |

### *3.2. Modeling Results of Motor Efficiency Based on the Novel Method Proposed and Analysis*

According to Section 3.1, the results of modeling purely using PR have certain deficiencies with limited precision. Considering the process of the novel method proposed in Section 2.5, this paper chooses some new variables formed through the mapping transformation of the motor speed $n_m$ and the motor torque $T_m$ and then selects the variable with the maximum correlation coefficient through a correlation analysis. Table 7 shows the research results.

**Table 7.** Correlation coefficients of measured motor efficiency with multiple variables.

| Variable | $\ln(n_m)$ | $\ln(T_m)$ | $\sin(n_m)$ | $\sin(T_m)$ | $\cos(n_m)$ | $\cos(T_m)$ | $\tan(n_m)$ | $\tan(T_m)$ | $e^{T_m}$ |
|---|---|---|---|---|---|---|---|---|---|
| **Correlation Coefficient** | −0.184 | 0.836 | 0.027 | −0.096 | −0.049 | 0.087 | 0.028 | 0.008 | 0.019 |

Therefore, the research chooses $\ln(T_m)$, the variable with the maximum correlation, and introduces it into formula (13) and then forms the 10-parameter model for the efficiency of the motor used. The following is the model's expression:

$$\eta_{m\_novel} = a_0 + a_1 n_m + a_2 T_m + a_3 n_m T_m + a_4 n_m^2 + a_5 T_m^2 + a_6 T_m^3 + a_7 \ln(a_8 T_m + a_9) \quad (18)$$

Use the I-SA algorithm for parameter identification of the 10-parameter motor efficiency model. Figure 12 and Table 5 show the results of the 5-fold CV of 100 random tests.

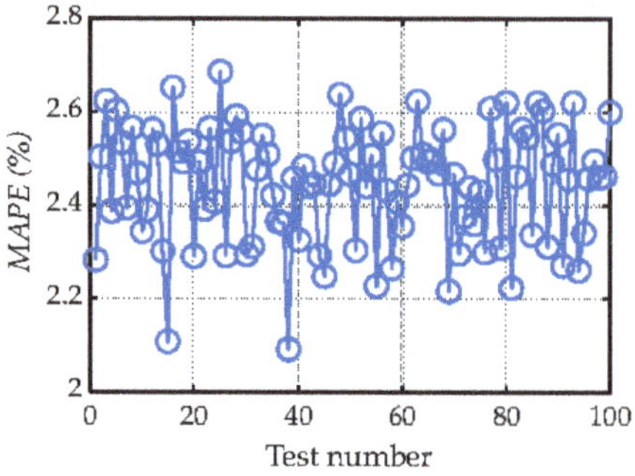


**Figure 12.** Results of 100 5-fold CV tests of novel method proposed. Note: *MAPE* is mean absolute percentage error.

According to Figure 12 and Table 5, the novel method proposed has high precision in motor efficiency modeling. The modeling result of the novel method improves greatly on the basis of PR. Additionally, the 5-fold CV results of 100 random tests (with a small standard deviation) show that the novel method has high reliability.

This paper uses the I-SA algorithm for the parameter identification of the novel model of a 5 kW motor and a 8 kW motor. Figure 13 shows iterative evolution curves.

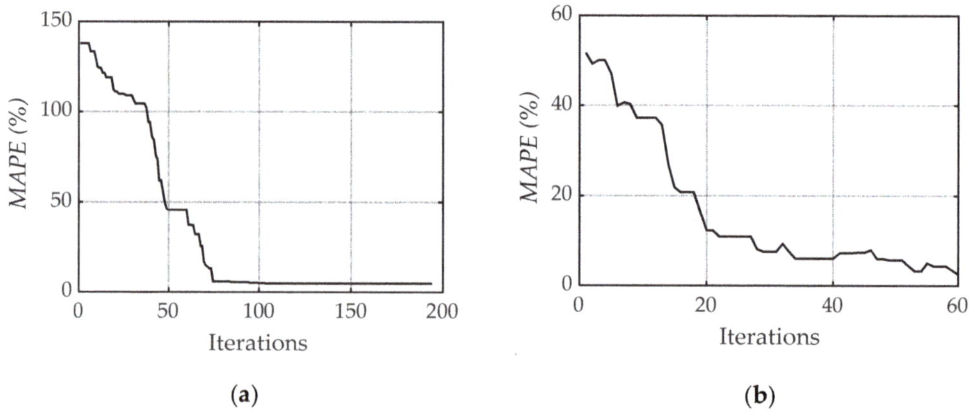


**Figure 13.** Iterative evolution curves of parameter identification of two motor efficiency models. (**a**) 8 kW motor; (**b**) 5 kW motor.

Figure 13 shows that the I-SA algorithm has a good application result for the parameter identification of the motor efficiency mode. The algorithm has a smaller number of iterations (declining to the ideal precision after about 60 iterations). In the implementing process of the algorithm, the value of an objective function is always declining, indicating that the I-SA avoids the problem of prematurity effectively.

Figure 14 shows the motor efficiency models in full working conditions.

Figure 14 shows that the motor efficiency models built with the novel method have ideal estimation results in the full working conditions of the motor speed $n_m$ and the motor torque $T_m$. The MAPE of the 8 kW motor efficiency model's estimated value and all the test data is 4.768%, and the *MAPE* of the 5 kW motor efficiency model's estimated value and all the test data is 2.539%.

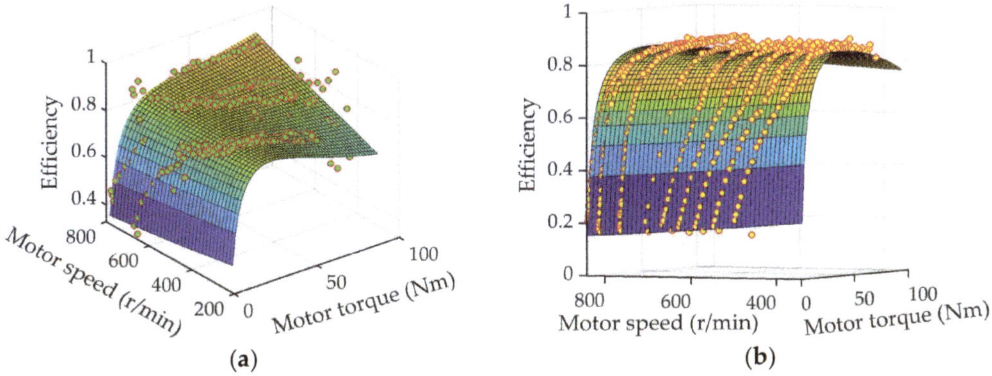


**Figure 14.** Efficiency models of two motors in full working conditions. (**a**) 8 kW motor; (**b**) 5 kW motor.

*3.3. Comparison Results of Full Vehicle Efficiency Characteristics of Electric Tractors in Five Working Conditions and Analysis*

Figure 15 shows the full vehicle working efficiency results of electric tractors of three different drive systems with different loads (showed as different reserve power) and running speeds (shown in the form of statistical histogram for the convenience of evaluation and analysis).

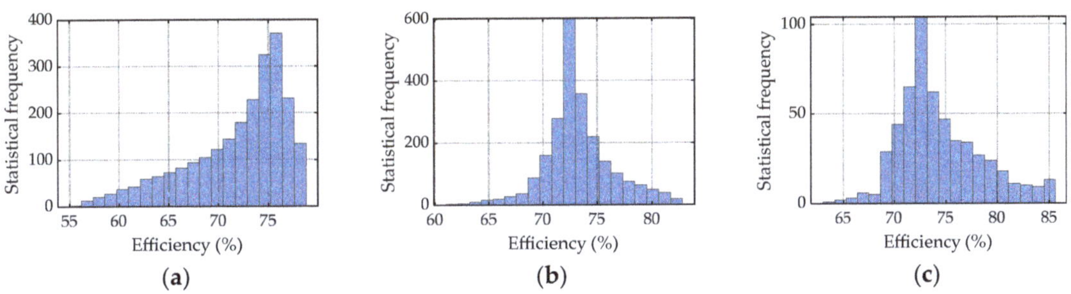


**Figure 15.** Estimation results of the full vehicle working efficiency of electric tractors with different drive systems. (**a**) 4 wheel independent drive system; (**b**) Rear wheel independent drive system; (**c**) Dual motor coupling drive system.

From Figure 15 we can see that the 4 wheel independent drive electric tractor's full vehicle efficiency shows the skewed distribution of which an efficiency lower than 70% accounts for 26.14%. The rear wheel independent drive electric tractor's efficiency distribution is nearly symmetric around the axis of 72%. The dual motor coupling drive electric tractor shows the slightly skewed distribution, but its distribution characteristic is opposite to that of the 4 wheel independent drive electric tractor, of which an efficiency higher than 75% accounts for 32.97%. Considering the range of the full working conditions selected, the dual motor coupling drive tractor shows better work efficiency as a whole.

Figure 16 shows the results of the normal road constant-speed running, transportation, ploughing, and rotary tillage working conditions. Table 8 gives the mean of the full vehicle efficiency of the electric tractors of three drive systems.

According to Figure 16 and Table 8, the dual motor coupling drive mode shows a relatively good full vehicle efficiency in most working conditions and has the highest average working efficiency with different loads and speeds. However, for the ploughing condition and the rotary tillage condition, the dual motor coupling drive electric tractor shows better efficiency characteristics. In the ploughing and rotary tillage conditions, the performance of the efficiency characteristics of the three types of drives are not invariable, and each has its optimal working range.

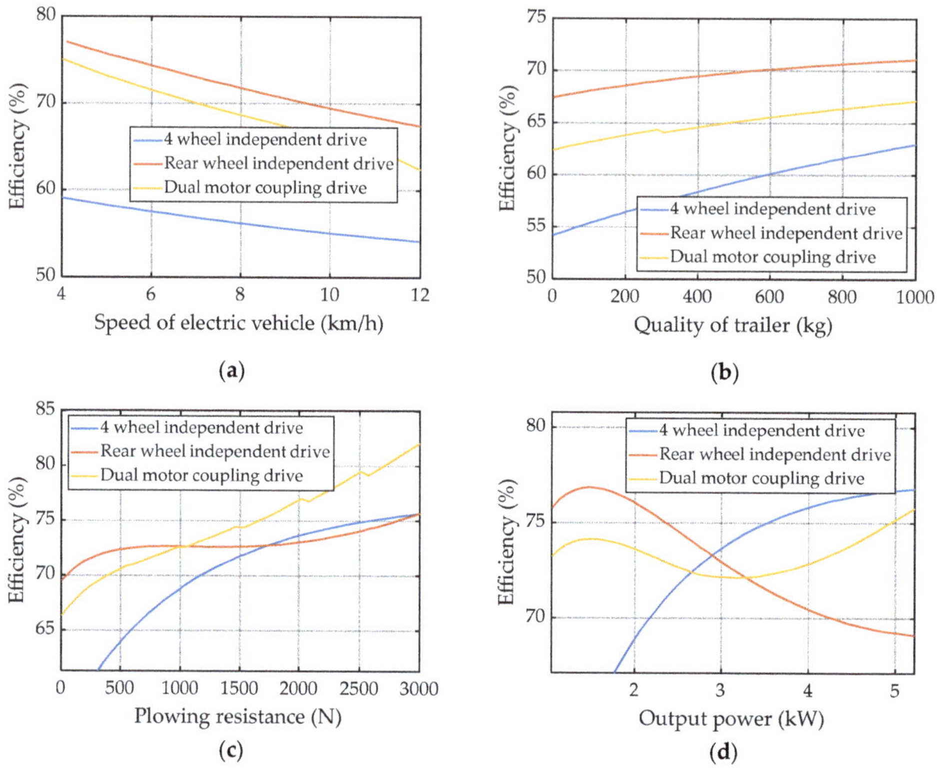


**Figure 16.** Estimation results of full vehicle efficiency of tractor in four working conditions. (**a**) Working condition of constant-speed running; (**b**) Working condition of transportation; (**c**) Working condition of ploughing; (**d**) Working condition of rotary tillage.

**Table 8.** Estimation results of mean of full vehicle efficiency of three drive systems for electric tractors.

| Type of Drive | Mean of Electric Tractor Efficiency/% | | | | |
|---|---|---|---|---|---|
| | Different Loads and Speeds | Constant-Speed Running | Transportation | Ploughing | Rotary Tillage |
| 4 wheel independent | 71.90 | 57.41 | 59.05 | 69.82 | 72.06 |
| Rear wheel independent | 73.26 | 71.86 | 69.65 | 72.96 | 72.86 |
| Dual motor coupling | 74.43 | 68.83 | 65.05 | 74.62 | 73.37 |

## 4. Conclusions

Compared with the three modeling methods (PR, BI, and RBF-NN), the proposed method in this paper not only has high model estimation precision (the mean of the MAPE of the 5-fold-CV of 100 random tests is 2.45 %, and the MAPE of all 853 measured data is 3.65 %) but also has high modeling reliability (the standard deviation of the MAPE of the 5-fold-CV of 100 random tests is 0.12 %). The structure of the new model proposed in this paper is relatively simple. Combined with a heuristic intelligent optimization algorithm, the model can be established only by identifying 10 unknown parameters. This is less dependent on the number of the test data. For the two motors used in this paper, the efficiency characteristics are highly correlated with the variables $\ln(T_m)$.

According to the various working conditions used in this study, this type of electric tractor equipped with a dual motor coupling drive system has the overall optimal economic performance (the highest full vehicle efficiency). From the analysis results of ploughing and rotary tillage conditions, the design of the electric tractor drive system should have the ability to switch between multiple driving modes (four wheel drive, rear wheel drive and coupling drive) to maximize the efficiency of the tractor under different working conditions.

**Author Contributions:** Methodology, Z.C.; software, Z.C. and H.Z.; validation, Z.C.; investigation, Z.C. and H.Z.; resources, Z.L. and H.Z.; writing—original draft preparation, Z.C.; writing—review and editing, Z.C. and Z.L.; supervision, Z.L.; project administration, Z.L. and Z.C. All authors have read and agreed to the published version of the manuscript.

**Funding:** This study was funded by the National Natural Science Foundation of China (grant number: 52105063), the National Key Research and Development Plan (2016YFD0701103), and the Metasequoia teacher research start-up fund of Nanjing Forestry University (163106061).

**Institutional Review Board Statement:** Not applicable.

**Informed Consent Statement:** Not applicable.

**Data Availability Statement:** The data presented in this study are available on demand from the corresponding author or first author at luzx@njau.edu.cn or chengzhun38@163.com.

**Acknowledgments:** The authors thank the National Natural Science Foundation of China (grant number: 52105063), the National Key Research and Development Plan (2016YFD0701103), and the Metasequoia teacher research start-up fund of Nanjing Forestry University (163106061) for funding. We also thank anonymous reviewers for providing critical comments and suggestions that improved the manuscript.

**Conflicts of Interest:** The authors declare no conflict of interest.

## References

1. Tian, J.; Tong, J.; Luo, S. Differential steering control of four-wheel independent-drive electric vehicles. *Energies* **2018**, *11*, 2892. [CrossRef]
2. Tian, J.; Wang, Q.; Ding, J.; Wang, Y.Q.; Ma, Z.S. Integrated control with DYC and DSS for 4WID electric vehicles. *IEEE Access* **2019**, *7*, 124077–124086. [CrossRef]
3. Zhou, W.L.; Zheng, Y.P.; Pan, Z.J.; Lu, Q. Review on the Battery Model and SOC Estimation Method. *Processes* **2021**, *9*, 1685. [CrossRef]
4. Ren, G.Z.; Wang, J.Z.; Chen, C.L.; Wang, H.R. A variable-voltage ultra-capacitor/battery hybrid power source for extended range electric vehicle. *Energy* **2021**, *231*, 120837. [CrossRef]
5. Yu, Y.; Jiang, J.; Min, Z.; Wang, P.; Shen, W. Research on energy management strategies of extended-range electric vehicles based on driving characteristics. *World Electr. Veh. J.* **2020**, *11*, 54. [CrossRef]
6. Chang, C.C.; Zheng, Y.P.; Yu, Y. Estimation for battery state of charge based on temperature effect and fractional extended kalman filter. *Energies* **2020**, *13*, 5947. [CrossRef]
7. Nie, X.W.; Liu, Y.L.; Liu, Z.S. Parameter matching and transmission ratio optimization of pure electric car transmission system. *Mod. Manuf. Eng.* **2020**, *8*, 53–59.
8. Vijayagopal, R.; Rousseau, A. Electric truck economic feasibility analysis. *World Electr. Veh. J.* **2021**, *12*, 75. [CrossRef]
9. Zhao, Y.; Tatari, O. Carbon and energy footprints of refuse collection trucks: A hybrid life cycle evaluation. *Sustain. Prod. Consum.* **2017**, *12*, 180–192. [CrossRef]
10. Lee, D.Y.; Thomas, V.M.; Brown, M.A. Electric urban delivery trucks: Energy use, greenhouse gas emissions, and cost-effectiveness. *Environ. Sci. Technol.* **2013**, *47*, 8022–8030. [CrossRef]
11. Du, X.J.; Tian, Y.M.; Wang, Z.Y.; Wang, Y.Q.; Qiao, K.K. Simulation and optimization of aluminum frame electric car frontal collision. *J. Highw. Transp. Res. Dev.* **2017**, *34*, 136–142.
12. Vervaeke, M.; Calabrese, G. Prospective design in the automotive sector and the trajectory of the Bluecar project: An electric car sharing system. *Int. J. Veh. Des.* **2015**, *68*, 245–264. [CrossRef]
13. Petrovic, D.T.; Pesic, D.R.; Petrovic, M.M.; Mijailovic, R.M. Electric cars are they solution to reduce CO2 emission? *Therm. Sci.* **2020**, *24*, 2879–2889. [CrossRef]

14. Wang, W.W.; Gao, F.L.; Cheng, Y.T.; Lin, C. Multidisciplinary design optimization for front structure of an electric car body-in-white based on improved collaborative optimization method. *Int. J. Automot. Technol.* **2017**, *18*, 1007–1015. [CrossRef]
15. Moreda, G.P.; Munoz-Garcia, M.A.; Barreiro, P. High voltage electrification of tractor and agricultural machinery—A review. *Energy Convers. Manag.* **2016**, *115*, 117–131. [CrossRef]
16. Cheng, Z.; Lu, Z.X.; Qian, J. A new non-geometric transmission parameter optimization design method for HMCVT based on improved GA and maximum transmission efficiency. *Comput. Electron. Agric.* **2019**, *167*, 105034. [CrossRef]
17. Raikwara, S.; Tewaria, V.K.; Mukhopadhyayb, S.; Verma, C.R.B.; Rao, M.S. Simulation of components of a power shuttle transmission system for an agricultural tractor. *Comput. Electron. Agric.* **2015**, *114*, 114–124. [CrossRef]
18. Fang, Q.R.; Shi, A.P.; Chen, A.Y.; Yin, S.Y. Research on driving torque control strategy of electric tractor. *J. Agric. Mech. Res.* **2021**, *43*, 243–248.
19. Chen, Y.N.; Xie, B.; Mao, E.R. Electric Tractor Motor Drive Control Based on FPGA. *IFAC Pap.* **2016**, *49*, 271–276. [CrossRef]
20. Gao, H.S.; Xue, J.L. Modeling and economic assessment of electric transformation of agricultural tractors fueled with diesel. *Sustain. Energy Technol. Assess.* **2020**, *39*, 100697. [CrossRef]
21. Li, T.H.; Xie, B.; Song, Z.H.; Li, J. Transmission characteristics of dual-motor coupling system for electric tractors. *Trans. Chin. Soc. Agric. Mach.* **2019**, *50*, 379–388.
22. Zhou, H.D.; Lu, Z.X.; Deng, X.T.; Zhang, C.; Luo, G.J.; Zhou, R.D. Study on torque distribution of traction operation of four wheel independent driven electric tractor. *J. Nanjing Agric. Univ.* **2018**, *41*, 962–970.
23. Han, B.; Liu, M.N.; Xu, L.Y. Analysis of steering characteristics of wheel-drive electric tractor. *J. Henan Univ. Sci. Technol. (Nat. Sci.)* **2021**, *42*, 38–45.
24. Xie, B.; Zhang, C.; Chen, S.; Mao, E.R.; Du, Y.F. Transmission performance of two-wheel drive electric tractor. *Trans. Chin. Soc. Agric. Mach.* **2015**, *46*, 8–13.
25. Li, S.Y.; Liu, M.N.; Xu, L.Y. Design and selection of power coupling device for electric tractor. *J. Mech. Transm.* **2021**, *45*, 75–82.
26. Ma, C.; Jin, S.W.; Yang, K.; Tan, D.; Gao, J.; Yan, D.C. Particle swarm optimization and real-road/driving-cycle analysis based powertrain system design for dual motor coupling electric vehicle. *World Electr. Veh. J.* **2020**, *11*, 69. [CrossRef]
27. Hu, J.J.; Zheng, L.L.; Jia, M.X.; Zhang, Y.; Pang, T. Optimization and model validation of operation control strategies for a novel dual-motor coupling-propulsion pure electric vehicle. *Energies* **2018**, *11*, 754. [CrossRef]
28. Li, T.H.; Xie, B.; Li, Z.; Li, J.K. Design and optimization of a dual-input coupling powertrain system: A case study for electric tractors. *Appl. Sci.* **2020**, *10*, 1608. [CrossRef]
29. Li, T.H.; Xie, B.; Wang, D.Q.; Zhang, S.L.; Wu, L.P. Real-time adaptive energy management strategy for dual-motor-driven electric tractors. *Trans. Chin. Soc. Agric. Mach.* **2020**, *51* (Suppl. S2), S530–S543.
30. Chen, Y.N.; Xie, B.; Du, Y.F.; Mao, E.R. Powertrain parameter matching and optimal design of dual-motor driven electric tractor. *Int. J. Agric. Biol. Eng.* **2019**, *12*, 33–41. [CrossRef]
31. Zhang, X.; Wang, S.T.; Zhang, X.; Wang, J. Torque coordination control strategy of in-wheel drive electric vehicle. *J. Beijing Jiaotong Univ.* **2017**, *41*, 121–129.
32. Hao, L.; Sun, B.H.; Li, G.; Guo, L.X. The eco-driving considering coordinated control strategy for the intelligent electric vehicles. *IEEE Access* **2021**, *9*, 10686–10698. [CrossRef]
33. Cheng, Z. I-SA algorithm based optimization design and mode-switching strategy for a novel 3-axis-simpson dual-motor coupling drive system of PEV. *World Electr. Veh. J.* **2021**, *12*, 221. [CrossRef]
34. Cheng, Z.; Lu, Z.X.; Dai, F. Research on HMCVT Efficiency Model Based on the Improved SA Algorithm. *Math. Probl. Eng.* **2019**, *2019*, 2856908. [CrossRef]
35. Jiao, W.M.; Ma, F.; Yang, Y.D.; Zhao, X.X. Calculation of transmission ratio and efficiency of planetary gear box. *J. Mech. Transm.* **2012**, *36*, 114–116.
36. Zhou, H.D. *Research on Torque Distribution Strategy of Four-Wheel Independent Drive Electric Tractor*; Nanjing Agricultural University: Nanjing, China, 2018.
37. Qian, Y.; Cheng, Z.; Lu, Z.X. Bench testing and modeling analysis of optimum shifting point of HMCVT. *Complexity* **2021**, *2021*, 6629561. [CrossRef]
38. Ji, X.Y.; Hao, H.M.; Yang, K.; Wang, H.; Li, B.; Huang, J.H. Dead-zone compensation for proportional directional valve based on bilinear interpolation control strategy. *Chin. Hydraul. Pneum.* **2021**, *45*, 56–62.
39. Khan, S.; Naseem, I.; Malik, M.A.; Togneri, R.; Bennamoun, M. A fractional gradient descent-based RBF neural network. *Circuits Syst. Signal Processing* **2018**, *37*, 5311–5332. [CrossRef]
40. Khan, S.; Naseem, I.; Togneri, R.; Bennamoun, M. A novel adaptive kernel for the RBF neural networks. *Circuits Syst. Signal Processing* **2017**, *36*, 1639–1653. [CrossRef]
41. Kilic, E.; Ozcalik, H.R.; Sit, S. Adaptive controller with RBF neural network for induction motor drive. *Int. J. Numer. Model. Electron. Netw. Devices Fields* **2018**, *31*, e2280. [CrossRef]
42. Cheng, Z.; Lu, Z.X. A novel efficient feature dimensionality reduction method and its application in engineering. *Complexity* **2018**, *2018*, 2879640. [CrossRef]
43. Xu, X.M.; Lin, P. Parameter identification of sound absorption model of porous materials based on modified particle swarm optimization algorithm. *PLoS ONE* **2021**, *16*, e0250950. [CrossRef] [PubMed]

44. Wang, H.; Zheng, Y.P.; Yu, Y. Joint estimation of soc of lithium battery based on dual kalman filter. *Processes* **2021**, *9*, 1412. [CrossRef]
45. Li, Y.J.; Ma, Z.S.; Zheng, M.; Li, D.X.; Lu, Z.H.; Xu, B. Performance analysis and optimization of a high-temperature PEMFC vehicle based on particle swarm optimization algorithm. *Membranes* **2021**, *11*, 691. [CrossRef] [PubMed]
46. Cheng, Z.; Lu, Z.X. Research on the PID control of the ESP system of tractor based on improved AFSA and improved SA. *Comput. Electron. Agric.* **2018**, *148*, 142–147. [CrossRef]
47. Cheng, Z.; Lu, Z.X. Semi-empirical model for elastic tyre trafficability and methods for the rapid determination of its related parameters. *Biosyst. Eng.* **2018**, *174*, 204–218. [CrossRef]
48. Wang, G. *Study on Characteristics, Control and Fault Diagnosis of Tractor Hydro-Mechanical CVT*; Nanjing Agricultural University: Nanjing, China, 2014.

agriculture

MDPI

Article

# Drip-Tape-Following Approach Based on Machine Vision for a Two-Wheeled Robot Trailer in Strip Farming

Chung-Liang Chang [1,*], Hung-Wen Chen [1], Yung-Hsiang Chen [2] and Chang-Chen Yu [2]

[1] Department of Biomechatronics Engineering, National Pingtung University of Science and Technology, Neipu 91201, Pingtung, Taiwan; m10944005@mail.npust.edu.tw
[2] Department of Vehicle Engineering, National Pingtung University of Science and Technology, Neipu 91201, Pingtung, Taiwan; m10838016@g4e.npust.edu.tw (Y.-H.C.); chyu@mail.npust.edu.tw (C.-C.Y.)
* Correspondence: chungliang@mail.npust.edu.tw; Tel.: +886-8-770-3202 (ext. 7586)

**Abstract:** Due to the complex environment in the field, using machine vision technology to enable the robot to travel autonomously was a challenging task. This study investigates a method based on mathematical morphology and Hough transformation for drip tape following by a two-wheeled robot trailer. First, an image processing technique was utilized to extract the drip tape in the image, including the selection of the region of interest (ROI), Red-Green-Blue (RGB) to Hue-Saturation-Value (HSV) color space conversion, color channel selection, Otsu's binarization, and morphological operations. The line segments were obtained from the extracted drip tapes image by a Hough line transform operation. Next, the deviation angle between the line segment and the vertical line in the center of the image was estimated through the two-dimensional law of cosines. The steering control system could adjust the rotation speed of the left and right wheels of the robot to reduce the deviation angle, so that the robot could stably travel along the drip tape, including turning. The guiding performance was evaluated on the test path formed by a drip tape in the field. The experimental results show that the proposed method could achieve an average line detection rate of 97.3% and an average lateral error of 2.6 ± 1.1 cm, which was superior to other drip-tape-following methods combined with edge detection, such as Canny and Laplacian.

**Keywords:** machine vision; image processing; two-wheeled robot trailer; steering control; strip farming

**Citation:** Chang, C.-L.; Chen, H.-W.; Chen, Y.-H.; Yu, C.-C. Drip-Tape-Following Approach Based on Machine Vision for a Two-Wheeled Robot Trailer in Strip Farming. *Agriculture* **2022**, *12*, 428. https://doi.org/10.3390/agriculture12030428

Academic Editor: Yanbo Huang

Received: 30 January 2022
Accepted: 16 March 2022
Published: 18 March 2022



## 1. Introduction

Monitoring the growth response of crops in the field could allow farmers to make production decisions in advance, avoid agricultural losses caused by extreme weather and pest invasion, while ensuring crop quality and yield, which is of great significance to sustainable development and food security. Crop management was generally performed manually, including spraying, fertilizing, watering, and weeding. With the advancement of semiconductor microfabrication technology, the development concepts of miniaturization and electrification have been gradually introduced into the design of novel farming tools [1–6]. These smart machines often have to be attached to the end of a tractor to perform farming tasks, such as weeding, spraying, or soil preparation. However, tractors are usually large in size and less convenient to use in small-scale farmland. Meanwhile, the use of large tractors also causes the soil to be too compact during the farming process, which is not conducive to the growth and water absorption of crop roots.

In the past 10 years, many small- and medium-sized autonomous mobile robots have been used to assist farming operations [7–12]. In principle, these mobile robots were equipped with global satellite navigation and positioning system (GNSS) receivers and a path planner, allowing the robots to move autonomously in the field [13,14]. Installing magnetic markers on the ground [15,16] was another localization method, which could serve as a reference point for the robot while traveling. It requires path planning in advance and needs to consider issues such as farmland plowing, land preparation, and crop rotation.

Machine vision technology has been widely used for automatic inspection and sorting of object defects [17,18], and its procedural flow includes image preprocessing, feature extraction, and image postprocessing. This technology could be used to identify the distribution of objects or features in images [19–21]. The purpose of image preprocessing was to remove irrelevant information in the image through color space transformation, filtering, denoising, affine transformation, or morphological operations. Meanwhile, it enhances the detectability of key information and simplifies the data to be processed, thereby improving the reliability of the feature extraction process. Image segmentation, matching, and recognition were procedures for feature extraction, the core concept of which was to divide the input image into several classes with the same properties. This method could be utilized to extract the region of interest (ROI) in the image, including region growing [22], gray-level histogram segmentation [23,24], edge-based segmentation [25–31], and clustering [32–35]. In addition, neural networks, support vector machines, or deep learning, etc., [36–39] are also common image segmentation methods. Other specialized machine vision methods include the use of a variable field of view [40], adaptive thresholds [41], and spectral filters for automatic vehicle guidance and obstacle detection [42]. The image segmentation methods based on the Otsu method [23] are very favorable for use when there is a large variance in the foreground and background of the image, and many rapid image segmentation methods with high real-time performance have also been proposed [29–31]. The edge-based segmentation method has been used to detect the pixel positions of discontinuous grayscale or color regions in a digital image. Canny's algorithm [26] is one of the common methods that could extract useful structural information from images and reduce the amount of data to be processed. Another edge detection method is called the Laplacian method, which is a second derivative operator that produces a steep zero crossing [43] at the edges of objects within the image.

Machine vision has been applied to detect crops or weeds in the field and allow robots to track specific objects for autonomous navigation [44–47]. With the development of high-speed computing, some high-complexity image processing algorithms could be implemented in embedded systems to detect crop rows in real time [48,49]. However, the instability of light intensity and the unevenness of the ground would still cause color distortion in the image, degrading the detection performance. It has been demonstrated that deep learning could overcome the impact of light and shadow on image recognition performance [50,51]; however, the detection model is still slightly affected in low light conditions [6].

In recent years, some automated drip irrigation systems have been used for precise crop irrigation in the field to save water. This system generally lays a long water line or pipeline on the field. Based on this premise, this study proposes a drip-tape-following approach based on machine vision. A digital red (R)-green (G)-blue (B) (RGB) image was converted to the hue (H)-saturation (S)-value (V) (HSV) color space and then binarized using a V-channel image. The mathematical morphology and Hough line transformation were utilized to extract the drip tape line in the binarized image, and the deviation angle between the drip tape line and the vertical line in the center of the image was estimated. The steering control system could adjust the rotation speed of the left and right wheels of the robot so that the robot could follow the drip tape line and perform U-turns to move to the next vegetable strip.

The organization of this study was as follows: The second section presents the mechanism design, system architecture, and hardware and software specifications of the two-wheeled robot trailer; the kinematic model, drip tape line detection, and following methods are also described in this section. The third section describes the test results of the proposed robot following the drip tape in the vegetable strips, including moving straight forward and U-turning. The evaluation results and comparative analysis of different drip tape line detection methods are also illustrated and discussed in this section. The last section summarizes the highlights of this paper and presents future work.

## 2. Materials and Methods

### *2.1. Description of the Two-Wheeled Robot Trailer*

The appearance of a two-wheeled robot trailer (which had a length (L) × width (W) × height (H) of 132 cm × 92 cm × 87.5 cm) is shown in Figure 1. Table 1 represents the specifications of the hardware and mechanism components of the robot trailer. The auxiliary wheel was 6 inches (15 cm) in diameter and was attached to the bottom of the support frame (50 cm in length). The left and right sides of a piece of aluminum alloy plate, whose length could be adjusted, were connected to the left and right brackets of the robot, respectively. A brushless DC motor (Model: 9B200P-DM, TROY Enterprise Co., Ltd., New Taipei City, Taiwan) was selected as the power source for the two-wheeled robot trailer. The rated speed, torque, and maximum torque of the motor were 3000 rpm, 0.8 N-m, and 1.0 N-m, respectively. The gear ratio of the reducer was 36, and the transmission efficiency was 80%. The bearing of the reducer was connected to the wheel through a ball bearing seat.

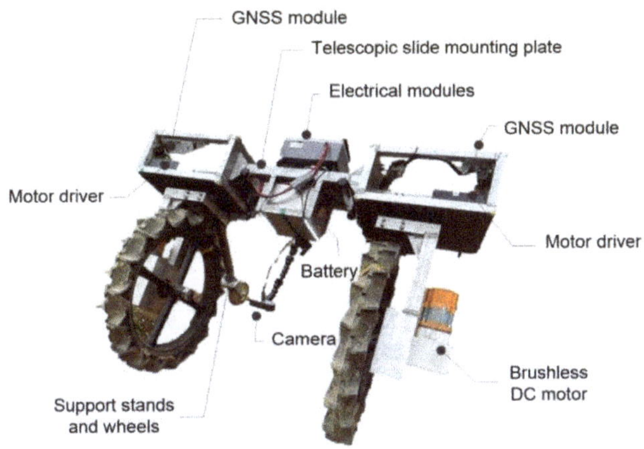


**Figure 1.** Appearance of the two-wheeled robot trailer.

**Table 1.** Component specifications for the robot trailer.

| Description | Value or Feature |
|---|---|
| Mechanism body | |
| Size: L [cm] × W [cm] × H [cm] | 132 × 92 × 87.5 |
| Front wheel: D [cm] × W [cm] | 65 × 8 |
| Rear wheel: D [cm] × W [cm] | 15 × 4 |
| Electronics | |
| Data processing board (speed; memory) | 1 GHz single-core ARMv6 CPU (BCM2835); 512 MB RAM |
| Image processing board (speed; memory) | 6-core NVIDIA Carmel ARM; 8 GB LPDDR4x |
| Guidance control board (speed; memory) | ARM Cortex-M processor; 256 KB single-cycle Flash memory |
| Motor (voltage;velocity; gear ratio; power; torque) | 24 V; 3000 rpm; 1:36; 200 W; 0.8 N-m |
| Driver (voltage; power; communication) | 24 V; 200 W; RS-232/RS-485 |
| Camera (connection; resolution; focus) | USB 2.0/3.0; 4096 × 2160 (30 frame per second (fps)); Auto |
| GNSS board (interface; type; voltage; precision) | USB/UART; Multi constellation; 5 V; <4 cm with state space representation (SSR) corrections (precise point positioning (PPP) in Real-Time Kinematic (RTK)). |
| DC–DC module (input voltage; output voltage) | 28 V; 5 V |

**Table 1.** *Cont.*

| Description | Value or Feature |
|---|---|
| Others | |
| Battery (voltage, capacity) | 24 V; 30 Ah |
| Antennas (type; voltage) | Passive; 3.3 V |

GNSS modules and motor drivers were installed in the left and right frames of the robot, and electrical modules, including the image-processing board and guidance control board, were mounted on the aluminum alloy plate. The rack under the aluminum alloy plate was used to place the battery. A tube was clamped to the rack. The end of the tube was connected to a camera module (Model: Logitech BRIO, Logitech International S.A., Lausanne, Switzerland). The driver could output 0~24 VDC, the power was 200 W, and it supports the communication specifications of RS-232 and RS-485. The size of the battery pack was 22 (L) cm × 16 (W) cm × 11 (H) cm, which could output 24 V/30 Ah. The direct current (DC)-to-DC module could step down the output voltage of the battery pack from 24 V to 5 V, thereby supplying power to electrical modules and others.

The architecture of the guidance and control system is shown in Figure 2. A GNSS module (Model: F9P, u-blox Company, Thalwil, Switzerland) with antennas was utilized to determine the location of the robot. The location data were transmitted from the GNSS module via a universal asynchronous receiver/transmitter (UART) interface to the micro-controller, which also received the sensing data obtained by the optical quantum meter and the environmental sensor. The location and sensing data were transmitted to the data processing unit through UART, and the data processing unit transmitted these data to the local server for storage through WiFi.

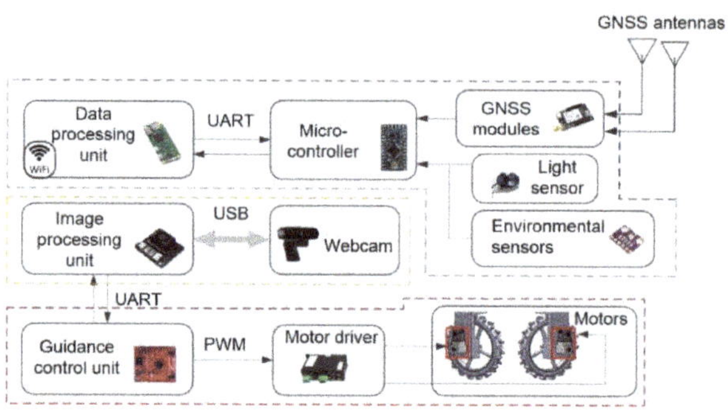


**Figure 2.** The guidance and control system architecture for a two-wheeled robot trailer.

The images captured by the camera module were sent to the image-processing unit (Model: Xarier NX, NVIDIA Company, Santa Clara, CA, USA), which performed a drip tape detection algorithm to estimate the deviation angle. After that, this deviation angle value was transmitted to the steering control unit (Model: EK-TM4C123GXL, TI Company, Dallas, TX, USA) via UART, and then the built-in steering control program of the unit was executed to estimate the drive voltages of left and right wheel motor. Finally, the voltage value was converted into a pulse width modulated (PWM) signal and sent to a motor driver (Model: BLD-350-A, Zi-Sheng Company, Taichung City, Taiwan) to drive the two motors.

### 2.2. *Drip Tape Following Approach*

This section presents the kinematic model of a two-wheeled robot trailer in a Cartesian coordinate frame under nonholonomic constraints, and the drip-tape-following methods are also illustrated in this section.

2.2.1. Kinematic Model

The two wheels of the robot were actuated on a plane by two independent motors, which provide torque to the two wheels. The radius of the two wheels was assumed to be $r$, and the distance between the two wheels was $l$. The pose of the robot in the inertial Cartesian frame {O-X-Y} could be described by the position $P_o(x_o, y_o)$ and the orientation angle $\phi$ measured relative to the X-axis, which is represented in Figure 3. The symbol $P_c$ represented the center of mass of the two-wheeled robot, and its distance from point $P_o$ was $d$. Assuming that the velocities in the X-axis and Y-axis directions of $P_o$ were $\dot{x}_o$ and $\dot{y}_o$, respectively; the angular velocity was $\dot{\phi}$; and the sideslip was ignored, the kinematics model of the robot under constraint conditions could be defined as follows [52,53]:

$$\begin{bmatrix} \dot{x}_o \\ \dot{y}_o \\ \dot{\varphi} \end{bmatrix} = \begin{bmatrix} r\cos\varphi/2 & r\cos\varphi/2 \\ \sin\varphi & \sin\varphi \\ \frac{r}{l} & \frac{-r}{l} \end{bmatrix} \begin{bmatrix} v_R/r \\ v_L/r \end{bmatrix} \tag{1}$$

where $v_L$ and $v_R$ represented the volocities of the left and right wheels of the robot, respectively.

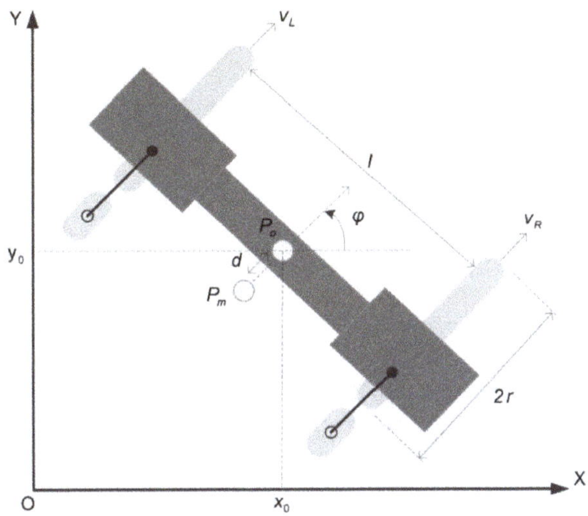


**Figure 3.** The representation of the position and orientation of the two-wheeled robot in the Cartesian coordinate system. Dark gray, light gray, and black color represent the mechanism body, wheels, and support stands, respectively.

2.2.2. Drip Tape Detection

This section presented the detection methods of drip irrigation tapes, including two edge detection methods (Canny and Laplacian) and the proposed approach. Simple examples were also presented.

- Canny

The premise of implementing this method was to first convert the red (R)-green (G)-blue (B) (RGB) color space of the original digital image (Figure 4a) into grayscale (Figure 4b) and then perform a Gaussian blur on the image (Figure 4c). Its main purpose was to reduce the effect of noise components on edge detection. Next, assume the first-order derivatives in the horizontal and vertical directions within the image, which were denoted as $\mathbf{G}_u$ and $\mathbf{G}_v$, respectively, as in Equations (2) and (3):

$$\mathbf{G}_u = \mathbf{S}_u * \mathbf{F}_{img} \tag{2}$$

$$\mathbf{G}_v = \mathbf{S}_v * \mathbf{F}_{img} \tag{3}$$

where Sobel operators $\mathbf{S}_u$ and $\mathbf{S}_v$ were designed to maximize the response to edges running vertically and horizontally with respect to the pixel grid. The operator consists of a pair of $3 \times 3$ convolution masks. The masks could be employed individually to the input image, which produce separate measurements of the gradient components in each direction. $\mathbf{F}_{img}$ indicated the measured pixel grids in the image. The symbol "$*$" depicted the convolution operator.

**Figure 4.** Process flow of edge detection: (**a**) original image; (**b**) grayscale; (**c**) Gaussian blur; (**d**) Canny edge detection; (**e**) Laplacian edge detection.

These two gradient components could then be combined together to learn the absolute magnitude $\mathbf{M}_g$ of the gradient at each point and the direction of its gradient (see Equation (4)). The angle of the directional gradients could also be measured by Equation (5):

$$\mathbf{M}_g = \sqrt{\mathbf{G}_u^2 + \mathbf{G}_v^2} \tag{4}$$

$$\mathbf{M}_\varphi = \arctan^{-1}(\mathbf{G}_v/\mathbf{G}_u) \tag{5}$$

Nonmaximum suppression was used to remove the blurred gradient values of edge detection pixel grids. The main concept of this method was to compare the gradient intensity of the current pixel with the intensity values of two pixels along the positive and negative gradient directions of the point. If the gradient value of the current pixel was greater than the gradient values of the other two pixels, the current pixel was reserved as an edge point; otherwise, the value of the pixel point was discarded.

Next, a double threshold was used to find some potential edge pixels caused by noise and color changes. True weak edge points could be preserved by suppressing weak edge pixels extracted from noise/color variations. The result of edge detection is shown in Figure 4d.

- Laplacian

The Laplacian-based edge detector used only one kernel to compute the second derivative of the image. It was also necessary to convert the RGB color space of the digital image into grayscale. Before performing Laplacian edge detection, the Gaussian blur operation was employed to reduce the influence of non-dominant features on detected objects, such as noise. Then, Equation (6) was utilized to find the edge points in the image:

$$\mathbf{G} = \mathbf{L_p} * \mathbf{F}_{img} \tag{6}$$

where $\mathbf{L_p}$ represented the Laplacian operator. After the Laplacian operation, the median filter was also used to filter out spot noise in the image, so that the edges could be represented very clearly. The edge detection results obtained in Figure 4c using the Laplacian method with a median filter was shown in Figure 4e.

- Proposed-HLT

It was difficult to observe the brightness, saturation, and hue of the object pixels in the digital image based on the RGB model. Therefore, the HSV model was used to replace the original RGB model. Assume an 8-bit digital image was captured by a camera. The image was cropped into an ROI of L× width (Q), which could be regarded as a two-dimensional matrix, and the R, G, and B channels presented in each pixel $U_{rgb}(p,q)$, $p = 1, 2, 3, \ldots, L$, and $q = 1, 2, 3, \ldots, Q$ in the matrix were divided. Each of the color channels represented $U_r(p,q)$, $U_g(p,q)$, and $U_b(p,q)$, and these values were between 0 and 255:

$$\begin{aligned} U_h(p,q) &= \cos^{-1}\left\{ \frac{\left(\left(U_r(p,q)-U_g(p,q)\right)+\left(U_r(p,q)-U_b(p,q)\right)\right)}{2\sqrt{\left(U_r(p,q)-U_g(p,q)\right)^2-\left(U_r(p,q)-U_b(p,q)\right)\left(U_g(p,q)-U_b(p,q)\right)}} \right\} \\ U_s(p,q) &= \frac{\max\left(U_r(p,q),U_g(p,q),U_b(p,q)\right)-\min\left(U_r(p,q),U_g(p,q),U_b(p,q)\right)}{\max\left(U_r(p,q),U_g(p,q),U_b(p,q)\right)} \\ U_v(p,q) &= \frac{\max\left(U_r(p,q),U_g(p,q),U_b(p,q)\right)}{255} \end{aligned} \tag{7}$$

where $U_h \in [0,\ 360)$, $U_s(p,\ q)$, $U_v(p,\ q) \in [0, 1]$. Operators $\max(\cdot)$ andmin$(\cdot)$ represented the selection of the maximum and minimum values, respectively. For example, Figure 5a,b are the original RGB image and the ROI, respectively. The ROI image was converted from the RGB model to the HSV model via Equation (7) (see Figure 5c).

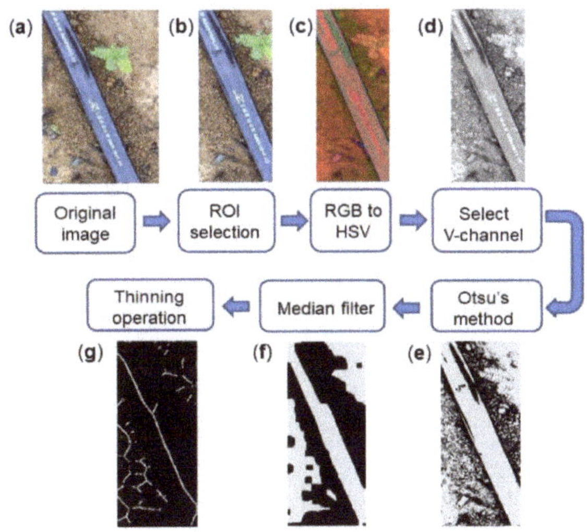


**Figure 5.** Drip tape extraction process: (**a**) Original image; (**b**) ROI image; (**c**) HSV image; (**d**) V-channel; (**e**) Otsu's binarization method; (**f**) median filter; (**g**) thinning operation.

For an image with a uniform gray distribution, the larger the variance value was, the larger the difference was between the two parts in the image. Otsu's method utilized the maximum interclass variance that was relatively common as a measure of classification criteria. This method was used to separate the background and objects in the V-channel of the HSV image (see Figure 5d). Using threshold segmentation to maximize the interclass variance means that the probability of misclassification was minimized. Assume $N$ represents the total number of pixels in an ROI of the image depicted in Equation (8):

$$N = \sum_{i=0}^{I-1} g_i = g_0 + g_1 + \ldots + g_{I-1} \quad (8)$$

where $g_i$ $(i = 0,\ 1,\ \ldots,\ I-1)$ illustrated the number of pixels whose gray values were $i$, and $I$ denoted the maximum gray level of the ROI in the image. The probability of gray level $i$ of pixel occurrence was depicted by $h_i = g_i/N$. The threshold $m$ could be employed to divide the gray level of the image into two groups: $G_0 = \{0,\ 1, \ldots, j\}$ and $G_1 = \{j+1, j+2, \ldots,\ I\}$. The grayscale numbers of $G_0$ and $G_1$ were $j+1$ and $I-j$, respectively. Then, $\alpha_0 = P_{G_0} = \sum_{i=0}^{j} h_i$ and $\alpha_1 = P_{G_1} = \sum_{i=j+1}^{I-1} h_i$ were the probabilities of the two groups. The means of $G_0$ and $G_1$ were depicted as $u_{\alpha_0} = (\sum_{i=0}^{j} ih_i)/\alpha_0$ and $u_{\alpha_1} = (\sum_{i=j+1}^{I-1} ih_i)/\alpha_1$, respectively. The mean value of gray level over the whole image was $u = \sum_{i=0}^{I-1} ih_i$. For any value of $j$, equations $\alpha_0 u_{\alpha_0} + \alpha_1 u_{\alpha_1} = u$ and $\alpha_0 + \alpha_1 = 1$ could be easily verified. The variances of $G_0$ and $G_1$ were $\sigma^2_{\alpha_0} = (\sum_{i=0}^{j} (i - \mu_{\alpha_0})^2 h_i)/\alpha_0$ and $\sigma^2_{\alpha_1} = (\sum_{i=j+1}^{I-1} (i - \mu_{\alpha_1})^2 h_i)/\alpha_1$, respectively. The intraclass and interclass variance are shown in Equations (9) and (10), respectively:

$$\sigma^2_{intra} = \alpha_0 \sigma^2_{\alpha_0} + \alpha_1 \sigma^2_{\alpha_1} \quad (9)$$

$$\begin{aligned} \sigma^2_{inter} &= \alpha_0 (u_{\alpha_0} + u)^2 + \alpha_1 (u_{\alpha_1} + u)^2 \\ &= \alpha_0 \alpha_1 (u_{\alpha_1} - u_{\alpha_0})^2 \end{aligned} \quad (10)$$

The sum of the variance of the intraclass and interclass was $\sigma^2 = \sigma^2_{inter} + \sigma^2_{intra}$. Therefore, maximizing the interclass variation was equivalent to minimizing the intraclass variation so that the optimal critical threshold $T$ could be obtained as follows:

$$T = \underset{0 \le j \le I-1}{\arg} \max\left(\sigma^2_{inter}/\sigma^2\right) \quad (11)$$

Each pixel $P_v(p, q)$ of the V-channel in the HSV image after execution by Otsu's method was shown in Figure 5e, and the image was then denoised (Figure 5f) and thinned (Figure 5g).

### 2.2.3. Hough Transformation

The Hough line transform (HLT) was used to detect the line in the image, which could be presented in the form of $p = bq + c$ in the Hough space (Hough, 1962). $b$ and $c$ denoted the slope and constant, respectively. There would be an infinite number of lines passing through the edge points $(p, q)$ on the edge image, except the vertical line, because $b$ was an undefined value (Leavers, 1992). Therefore, the alternative equation $p \cos \varphi + q \sin \varphi = \rho$ in the polar coordinate frame was used to replace the original equation. The symbol $\rho \in R$ represented the shortest length from the origin to the line, and $\varphi \in [0,\ \pi]$ expressed the angle between the line and horizontal axis. For all values of $\rho$, each pixel in the image was mapped to Hough space. Assuming there were two pixels on the same line, their corresponding cosine curves would intersect on a particular $(\rho, \varphi)$ pair. Each pair represented each line that passes by $(p, q)$. This detection process was carried out in parameter steps or accumulators, which was a voting process. By finding the highest bin in the parameter space, the most likely line and its geometric definition could be extracted. The results (red lines) of Figure 4d,e,g processed by HLT, respectively, are shown in Figure 6a–c.

(**a**) (**b**) (**c**)

**Figure 6.** Results of different drip tape detection methods combined with HLT (red color): (**a**) Canny with HLT; (**b**) Laplacian with HLT; (**c**) proposed HLT.

The white solid and dashed frames represented the field of view captured by the camera when the robot turned (Figure 7a). The upper and lower graphs in Figure 7b–f represented the image-processing results within the solid frame and the dashed frame, respectively. A line (Figure 7g) could be obtained by applying the HLT in Figure 7f.

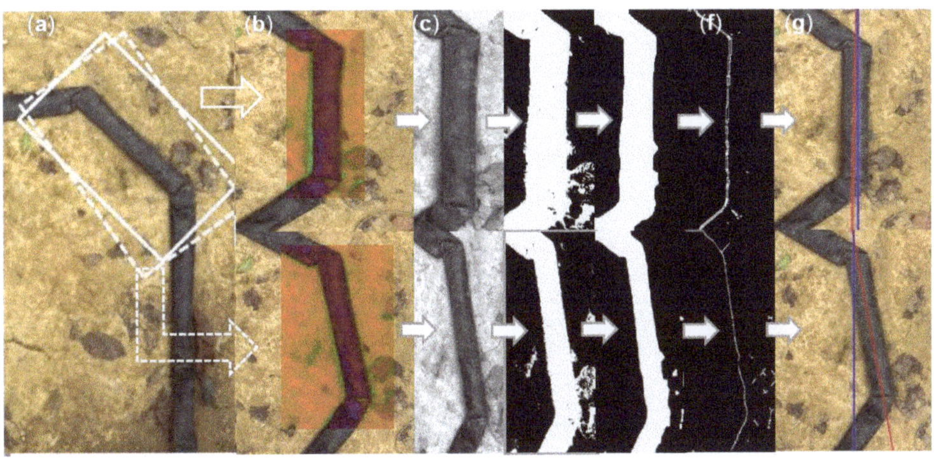


**Figure 7.** Line detection result when the robot turned: (**a**) Two frames within the snapshot that require image processing (solid frame and dashed frame); (**b**) The result of HSV transformation of the ROI in the solid frame (top) and the dashed frame (bottom) in the image (**a**); (**c**) The V channel of the HSV image; (**d**) Binarization; (**e**) Median filtering; (**f**) Thinning; (**g**) Hough line transformation (red line) and the central vertical line (blue) of the image.

### 2.3. Determination of the Deviation Angle

Once the drip tape was detected (as shown in the red line in Figure 8), two points $P_a(x_a, y_a)$ and $P_b(x_b, y_b)$ were selected on the red line. At the same time, a vertical line (blue color) was drawn in the center of the image. The blue line intersected the red line at point $P_c(x_c, y_c)$. Next, taking a point $P_m(x_m, y_m)$ on the line segment $\overline{P_cP_a}$, the point horizontally extended a line segment, and the vertical blue line intersected at point $P_n(x_n, y_n)$, where $y_n = y_m$. After obtaining the length of $\overline{P_mP_n}$, the deviation angle $\theta = \tan^{-1}(\overline{P_mP_n}/\overline{P_nP_c})$ between the drip tape and the centerline of the image could be obtained. When $\theta = 0$, the robot heading was parallel to the drip tape.

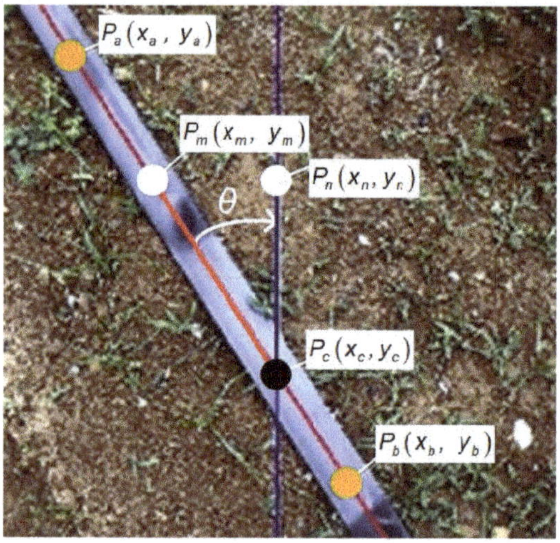


**Figure 8.** The deviation of heading angle estimation (red line: desired line; blue line: vertical line in the center of the image; white dotted line: horizontal line in the center of the image).

### 2.4. Heading Angle Control

The field navigation mode of the robot was divided into straight forward and turning. A flow chart of the path following control was demonstrated in Figure 8. First, the digital RGB image was captured by the camera module, the lines in the image were extracted by drip tape detection algorithms (Section 2.2), and a deviation angle estimation method (described in Section 2.3) was used to estimate the heading angle. Then, set the PWM value $\kappa$, the motor speed control gain $k_p$, the minimum threshold $\varepsilon_{\text{min}}$, and the maximum threshold $\varepsilon_{\text{max}}$ of the heading angle, etc. When $|\theta| \leq \varepsilon_{\text{max}}$, that is, in the area of "❶" or "❷" or "❸" (see the top right of Figure 9), the speed difference control was executed. If $|\theta| \leq \varepsilon_{\text{max}}$, that is, in the "❸" area, then the left and right motor speed control parameters, denoted as $\text{PWM}_\text{L}$ and $\text{PWM}_\text{R}$, respectively, were equal. Finally, the robot stopped when it moved to the target position. Conversely, when $|\theta| > \varepsilon_{\text{min}}$ and $\theta > \varepsilon_{\text{min}}$ (the area of "❶"), then $\text{PWM}_\text{L} = k_p\kappa + \kappa$ and $\text{PWM}_\text{R} = \kappa$; in contrast, when $\theta < \varepsilon_{\text{min}}$ (the area of "❷"), then $\text{PWM}_\text{L} = \kappa$ and $\text{PWM}_\text{R} = k_p\kappa + \kappa$. When $|\theta| > \varepsilon_{\text{max}}$, it means that $\theta$ was in the "❹" area, and the steering control program would be executed. When $\theta < -\varepsilon_{\text{max}}$, set $\text{PWM}_\text{L} = \kappa(-)$ and $\text{PWM}_\text{R} = \kappa(+)$. Conversely, set $\text{PWM}_\text{L} = \kappa(+)$ and $\text{PWM}_\text{R} = \kappa(-)$. Among them, the symbols "+" and "−" depict the forward rotation and the reverse rotation of the motor, respectively. It was worth noting that the robot stopped ($\text{PWM}_\text{L} = \text{PWM}_\text{R} = 0$) once the drip tape was not extracted or an abnormal deviation angle was acquired.

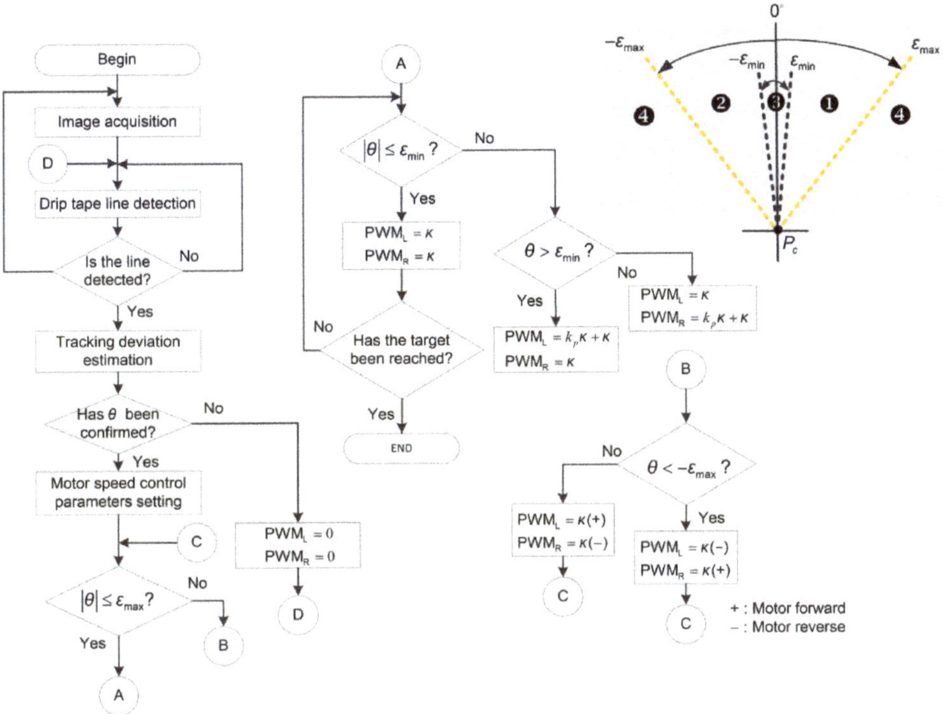


**Figure 9.** Line of drip tape detection and steering control process. The numbers "❶", "❷", "❸" and "❹" respectively demonstrated the areas corresponding to the heading of the robot; the letters A, B, C, D inside the circle represented flowchart connectors.

## 3. Experimental Results

This section illustrates different drip tape detection methods for testing and verification of two-wheeled robot trailers in the field. The performance comparison, analysis, and discussion of different drip tape following methods are also presented in this section.

The experimental site was located in front of the experimental factory of the Department of Biomechatronics Engineering of Pingtung University of Science and Technology (Longitude: 120.60659 Latitude: 22.64606). The field had a size of L × W × H of 10 × 0.25 × 0.2 m (see Figure 10a). The experiment was carried out during the spring, and the weather conditions were mostly cloudy and sunny in the morning and cloudy in the afternoon. According to the climate conditions, butter lettuce (LS-047, Known-You Seed Co., Ltd., Kaohsiung, Taiwan) and red lettuce (HV-067, Known-You Seed Co., Ltd., Kaohsiung, Taiwan) were selected for planting in the field. A black drip tape was laid on the field. The robot would continuously follow the drip tape and move to another tape area (see Figure 10b,c). The control parameters of the motors were set to $\kappa = 1500$ and $k_p = 200$. The image processing speed was 5 fps. $\varepsilon_{min}$ and $\varepsilon_{max}$ were set to $1^\circ$ and $6^\circ$ degrees, respectively.

The drip tape was configured as a polygon in the turning area. As shown in Figure 11, there were four corner points, which were represented as "①", "②", "③", and "④". This figure also showed the drip tape detection results for each segment (red line within a black box).

**Figure 10.** The appearance of experimental field and the two-wheeled robot: (**a**) the appearance of the field; (**b**) the snapshot of the two-wheeled trailer traveling autonomously; (**c**) the U-turning process of the robot trailer.

**Figure 11.** The configuration of the drip tape in the turning area and its detection result (red line). The blue line represented the vertical line in the center of the image. The numbers "①", "②", "③", and "④" represented the corner points, which can divide the drip tape into five line segments.

Then, the guiding performance of the robot was tested. The test duration was from 6:00 in the morning to 10:00 in the evening, which was divided into eight time intervals, and the robot followed the drip tape to rewind the field twice in each time interval. During the experiment, the line detection rate, $LDR_i = N_s^i / N_{total}^i$, was estimated. Among them, $N_{total}^i$ and $N_s^i$ represented the total number of processed images and the number of images that successfully detected the drip tape in the $i-$ th test duration of each interval, respectively. Finally, the average detection rate $LDR_{avg} = \sum_{i=1}^{T} LDR_i / T$ of the line could be obtained, and $T$ denoted the total number of test durations. The processing rate of the image was 20 fps.

Figure 12a presents the movement trajectory of the robot obtained by the GNSS-RTK positioning module at 06:00 in the morning using the proposed-HLT; the heading angle of the robot (the first loop: black-dot color; the second loop: brown hollow-dot color) was also shown in Figure 12b. The gray area represents the variation in heading angle when the robot turns. The movement trajectory of the robot during the turning process was shown in the blue box in Figure 12a, which was enlarged in Figure 12c (green color).

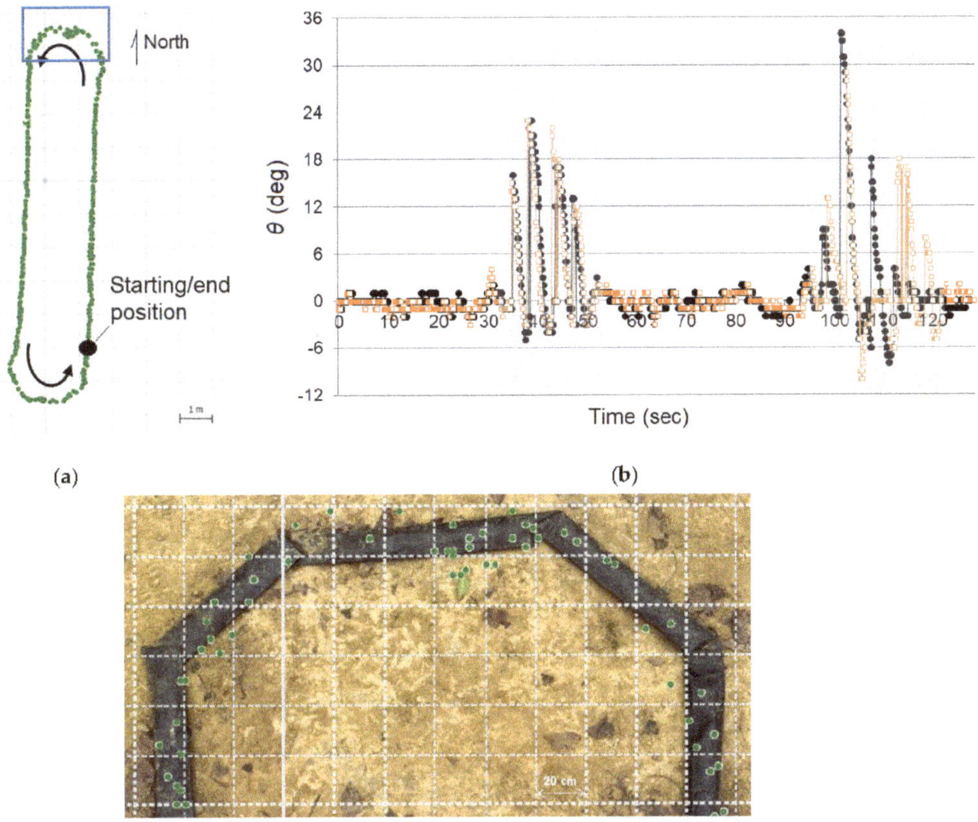


(a) (b) (c)

**Figure 12.** The movement trajectory and deviation angle of the robot (test time interval: 06:00–08:00 a.m.): (**a**) Movement trajectory of the robot, initial location: 22.64657° N, 120.6060° E (the area of grid: 1 (length) × 1 (width) m); (**b**) The variation in heading angle when the robot travels (black-dot color: the first loop; brown hollow-dot color: the second loop); (**c**) The position point (green-dot color) distribution of the moving trajectories of the two loop (the area of grid: 20 (L) × 20 (W) cm).

Similarly, two drip tape detection methods based on Canny and Laplacian combined with HLT (called Canny-HLT and Laplacian-HLT) were used to conduct the above experiments and evaluate the performance of different drip tape detection methods. The performance test was repeated for three days ($T = 6$), and the results are shown in Figure 13. The average line detection rates obtained by the Canny-HLT, Laplacian-HLT, and proposed-HLT methods in different time intervals were 65~84%, 91~93%, and 93~97%, respectively.

During the experiment, the guiding performance for following straight and polyline trajectories was evaluated with a total length of 44 m. Table 2 presents the evaluation results of the robot's guidance performance, including the mean maximum error (MAE), maximum lateral error (ME), and root mean square error (RMSE). The MAE of the proposed-HLT was 2.6 ± 1.1 cm, which was lower than that of the Canny-HLT method (3.2 ± 1.2 cm) and the Laplacian-HLT (2.9 ± 1.6 cm). When using the Canny-HLT method, the robot has the largest ME when moving in a straight line, which reaches 12.3 cm. Using the proposed-HLT results in the smallest RMSE of 2.9 cm in total traveling length. A video of experimental results is demonstrated in Supplementary material.

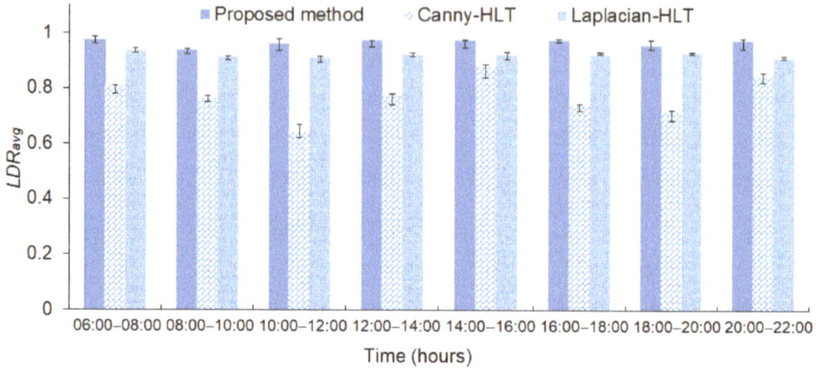


**Figure 13.** Comparison of the drip tape detection performance of the three methods at different time intervals within three days.

**Table 2.** The performance evaluation results of different line detection methods in the guidance system of the robot.

| | Travel Distance (m) | MAE ± SD (cm) | ME (cm) | RMSE (cm) |
|---|---|---|---|---|
| Canny-HLT | 32 (Straight) | 4.2 ± 1.7 | 12.3 | 5.8 |
| | 12 (Polyline) | 2.3 ± 1.4 | 10.7 | 2.7 |
| | 44 (Total) | 3.2 ± 1.2 | 11.2 | 3.6 |
| Laplacian-HLT | 32 (Straight) | 3.9± 1.4 | 11.4 | 5.1 |
| | 12 (Polyline) | 2.1 ± 1.9 | 7.6 | 2.2 |
| | 44 (Total) | 2.9 ± 1.6 | 7.8 | 3.3 |
| Proposed-HLT | 32 (Straight) | 3.6 ± 1.5 | 8.6 | 4.9 |
| | 12 (Polyline) | 1.9 ± 1.4 | 6.4 | 1.7 |
| | 44 (Total) | 2.6 ± 1.1 | 6.9 | 2.9 |

## 4. Discussion

The proposed-HLT has been verified to obtain an average *LDR* of 96.6% under the condition of unstable light intensity outdoors. Select the V-channel of the HSV image for drip tape detection, which could reduce the brightness and contrast of the image to enhance the differences between objects and backgrounds in the image. The experimental results demonstrated that the *LDR* during the daytime was at least 93.8%. It was worth noting that the LED lighting device installed on the robot could be used at night (20:00–22:00), and the results depicted that the *LDR* during this time interval could reach 97.5%. In sunny and cloudless summer conditions, the image would be overexposed due to excessive light and cause a failed recognition result. Therefore, once the drip tape was not detected or an abnormal deviation angle was estimated, the robot would stop. The robot did not start moving until the deviation angle was confirmed and the drip tape line has been extracted. Therefore, once the planting season comes to summer, the safety of the robot during the movement could still be guaranteed under uncertain climatic factors.

The drip tape detection performance using the Canny-HLT was the worst among the three methods. Since the Canny algorithm has its own execution flow, it was difficult to use it with other image-processing algorithms, which also limits its flexibility in use. For images with uniform grayscale variation, it was hard to detect boundaries by only using the first derivative (such as the Canny operation). At this time, the use of a second derivative operation could provide critical information, such as the Laplacian operation. This method was often used to determine whether the edge pixels in the image were bright or dark areas. By smoothing the image to reduce noise, the Laplacian method could also achieve a detection rate of up to 93.4% after combining with Hough line transformation. Compared with the Canny-HLT method, the detection performance of the Laplacian-HLT method was more stable.

When the Canny-HLT or Laplacian-HLT program was executed in the high-speed embedded system, lines on the left or right side of the drip tape were detected (Figure 7a,b) under the condition of an image processing speed of 7 fps. The line segment of drip tape detected first would be used for heading angle estimation.

The configuration of the drip tape as a polygon allowed the robot to perform the U-turning operation smoothly, and a larger ROI could prevent the intersection of the line segment and the vertical line from falling outside the ROI. At this time, it would also cause three line segments to appear at the same time in the ROI of the image (as shown in the black frame in Figure 11), which indirectly increased the probability of misjudgment of the heading angle. Therefore, the estimated angle was checked by monitoring the variation in the heading angle to ensure the stability of the robot's traveling. This configuration of the drip tape could enable the robot trailer to move autonomously and turn to another narrow strip. This study only investigated a drip tape line detection operation based on mathematical morphology and Hough transformation, which was used in complex environments and unstable light intensity conditions. Once the detected target or object characteristics change, the parameters of the proposed approach need to be adjusted specifically. In principle, the color of the drip tape was different from that of most soils (except for black soils). Generally speaking, the material was very suitable for two-wheeled robots to be used as object tracking. Although green crops could also be selected as objects, different types of crops have different characteristics and sizes, and the planting interval of each crop would also be adjusted according to the size of the crop. Due to the many factors or parameters to be considered, the object detection process would be more complicated.

On the other hand, although the color of the soil was similar to that of the drip tape, fortunately, the surface of the drip tape and the soil were still different in brightness under uniform lighting. Therefore, using Otsu's binarization method could still distinguish the foreground object from the background. Using a drip tape that was similar in color to the soil was undesirable when working in the field. Secondly, the recognition performance of the proposed solution was limited by the existence of shadows in the objects in the image. Therefore, it was more suitable to use the proposed approach when there were no shelters (such as trees) around the field.

When using deep learning for object detection, its detection performance was limited by the diversity of images and the number of labeled samples used in model training [54]. Once the experimental site was changed or the climate changes (such as solar radiation), the images to be identified have to be collected again and the detection model has to be rebuilt. In addition, the operation of labeling target objects was also time-consuming.

The advantage of using image processing was that it could only extract the features of the target object in the image, especially the objects with obvious features. However, this detection method is limited by the image quality and the high complexity of the feature extraction would indirectly increase the computational load. These problems have been solved due to the gradual improvement in camera hardware level, the gradual popularization of high-speed computing processors, and reduction in costs.

The proposed robot trailer adopted differential speed steering control, which has a small turning radius and was suitable for fields with narrow planting spacing and turning areas. In addition, deep tread tires could repel mud, which could be used on sticky soils or upland fields. The narrow tire could reduce the friction between the tire and the ground, making the steering control more flexible.

During the experiment, although the drip tape in some areas was blocked by mud or weeds, the proposed approach could successfully detect the line of drip tape. In addition, during drip irrigation, the drip tape would be filled with water, and the appearance would expand slightly, but it has no effect on the line detection for the robot trailer.

## 5. Conclusions

The proposed machine-vision-based approach could provide the two-wheeled robot trailer to move along the drip tape in the field. Three line detection methods were used to evaluate the autonomous guiding performance of the robot. Among them, the proposed image recognition strategy could effectively detect the drip tape on the strip-planting area and estimate the angle of heading deviation to make the robot travel stably between planting areas, especially in the case of unstable light conditions.

The robot trailer was small in size, and it was suitable for autonomous guiding operations in 1 hectare fields, especially for strip, till, or no-till farming applications. In the future, the proposed approach would be integrated with deep learning for guideline detection and heading angle control. In addition, the robot would be equipped with a shock absorption device and a steering device, so that the robot has more applications.

**Supplementary Materials:** The following supporting information can be downloaded at: https://www.mdpi.com/article/10.3390/agriculture12030428/s1. Video: Autonomous Two Wheel Robot Trailer.

**Author Contributions:** Conceptualization, C.-L.C. and C.-C.Y.; methodology, C.-L.C. and H.-W.C.; software, H.-W.C. and Y.-H.C.; verification, C.-L.C., H.-W.C. and Y.-H.C.; data management, H.-W.C.; writing—manuscript preparation, C.-L.C., H.-W.C. and Y.-H.C.; writing—review and editing, C.-L.C.; visualization, C.-L.C. and Y.-H.C.; supervision, C.-L.C. and C.-C.Y.; project management, C.-L.C.; fund acquisition, C.-L.C. All authors have read and agreed to the published version of the manuscript.

**Funding:** This research was funded by Ministry of Science and Technology (MOST), Taiwan, grant number MOST 109-2321-B-020-004; MOST 110-2221-E-020-019.

**Institutional Review Board Statement:** Not applicable.

**Informed Consent Statement:** Not applicable.

**Data Availability Statement:** The datasets presented in this study are available from the corresponding author on reasonable request.

**Conflicts of Interest:** The authors declare no conflict of interest.

## References

1. McCool, C.; Beattie, J.; Firn, J.; Lehnert, C.; Kulk, J.; Bawden, O.; Russell, R.; Perez, T. Efficacy of mechanical weeding tools: A study into alternative weed management strategies enabled by robotics. *IEEE Robot. Autom. Lett.* **2018**, *3*, 1184–1190. [CrossRef]
2. Hemming, J.; Nieuwenhuizen, A.T.; Struik, L.E. Image analysis system to determine crop row and plant positions for an intra-row weeding machine. In Proceedings of the CIGR International Symposium on Sustainable Bioproduction, Tokyo, Japan, 19–23 September 2011; pp. 1–7.
3. Xiong, Y.; Ge, Y.; Liang, Y.; Blackmore, S. Development of a prototype robot and fast path-planning algorithm for static laser weeding. *Comput. Electron. Agric.* **2017**, *142*, 494–503. [CrossRef]
4. Peruzzi, A.; Martelloni, L.; Frasconi, C.; Fontanelli, M.; Pirchio, M.; Raffaelli, M. Machines for non-chemical intra-row weed control in narrow and wide-row crops: A review. *J. Agric. Eng.* **2017**, *48*, 57–70. [CrossRef]
5. Chang, C.L.; Lin, K.M. Smart agricultural machine with a computer vision-based weeding and variable-rate irrigation scheme. *Robotics* **2018**, *7*, 38. [CrossRef]
6. Chang, C.L.; Xie, B.X.; Chung, S.C. Mechanical control with a deep learning method for precise weeding on a farm. *Agriculture* **2021**, *11*, 1049. [CrossRef]
7. Astrand, B.; Baerveldt, A.J. An agricultural mobile robot with vision-based perception for mechanical weed control. *Auton. Robots* **2002**, *13*, 21–35. [CrossRef]
8. Emmi, L.; Le Flecher, E.; Cadenat, V.; Devy, M. A hybrid representation of the environment to improve autonomous navigation of mobile robots in agriculture. *Precis. Agric.* **2021**, *22*, 524–549. [CrossRef]
9. Opiyo, S.; Okinda, C.; Zhou, J.; Mwangi, E.; Makange, N. Medial axis-based machine-vision system for orchard robot navigation. *Comput. Electron. Agric.* **2021**, *185*, 106153. [CrossRef]
10. Grimstad, L.; From, P.J. The Thorvald II agricultural robotic system. *Robotics* **2017**, *6*, 24. [CrossRef]

11. Reiser, D.; Sehsah, E.-S.; Bumann, O.; Morhard, J.; Griepentrog, H.W. Development of an autonomous electric robot implement for intra-row weeding in vineyards. *Agriculture* **2019**, *9*, 18. [CrossRef]
12. Fue, K.G.; Porter, W.M.; Barnes, E.M.; Rains, G.C. An extensive review of mobile agricultural robotics for field operations: Focus on cotton harvesting. *AgriEngineering* **2020**, *2*, 150–174. [CrossRef]
13. Yin, X.; Du, J.; Geng, D. Development of an automatically guided rice transplanter using RTKGNSS and IMU. *IFAC PapersOnline* **2018**, *51*, 374–378. [CrossRef]
14. Ng, K.M.; Johari, J.; Abdullah SA, C.; Ahmad, A.; Laja, B.N. Performance evaluation of the RTK-GNSS navigating under different landscape. In Proceedings of the 18th International Conference on Control, Automation and Systems (ICCAS), PyeongChang, Korea, 17–18 October 2018; pp. 1424–1428.
15. Mutka, A.; Miklic, D.; Draganjac, I.; Bogdan, S. A low cost vision based localization system using fiducial markers. *IFAC PapersOnline* **2008**, *41*, 9528–9533. [CrossRef]
16. Byun, Y.S.; Kim, Y.C. Localization based on magnetic markers for an all-wheel steering vehicle. *Sensors* **2016**, *16*, 2015. [CrossRef] [PubMed]
17. Chen, Y.-R.; Chao, K.; Kim, M.S. Machine vision technology for agricultural applications. *Comput. Electron. Agric.* **2002**, *36*, 173–191. [CrossRef]
18. Perez, L.; Rodriguez, I.; Rodriguez, N.; Usamentiaga, R.; Garcia, D.F. Robot guidance using machine vision techniques in industrial environments: A Comparative Review. *Sensors* **2016**, *16*, 335. [CrossRef]
19. Kuruvilla, J.; Sukumaran, D.; Sankar, A.; Joy, S.P. A review on image processing and image segmentation. In Proceedings of the 2016 International Conference on Data Mining and Advanced Computing (SAPIENCE), Ernakulam, India, 16–18 March 2016; pp. 198–203. [CrossRef]
20. Malik, M.H.; Zhang, T.; Li, H.; Zhang, M.; Shabbir, S.; Saeed, A. Mature tomato fruit detection algorithm based on improved HSV and watershed algorithm. *IFAC PapersOnline* **2018**, *51*, 431–436. [CrossRef]
21. Wu, X. Review of theory and methods of image segmentation. *Agric. Biotechnol.* **2018**, *7*, 136–141.
22. Preetha, M.M.S.J.; Suresh, L.P.; Bosco, M.J. Image segmentation using seeded region growing. In Proceedings of the 2012 International Conference on Computing, Electronics and Electrical Technologies (ICCEET), Nagercoil, India, 21–22 March 2012; pp. 576–583.
23. Otsu, N. A threshold selection method from gray-level histograms. *IEEE Trans. Syst. Man Cybern.* **1979**, *9*, 62–66. [CrossRef]
24. Haralick, R.M.; Shapiro, L.G. Image segmentation techniques. *Comput. Gr. Image Process.* **1985**, *29*, 100–132. [CrossRef]
25. Marr, D.; Hildreth, E. Theory of edge detection. *Proc. R. Soc. Lond. B Biol. Sci.* **1980**, *207*, 187–217. [CrossRef] [PubMed]
26. Canny, J.A. computational approach to edge detection. *IEEE PAMI* **1986**, *8*, 679–714. [CrossRef]
27. Kaganami, H.G.; Beij, Z. Region based segmentation versus edge detection. In Proceedings of the 2009 Fifth International Conference on Intelligent Information Hiding and Multimedia Signal Processing, Kyoto, Japan, 12–14 September 2009; pp. 1217–1221.
28. Al-Amri, S.S.; Kalyankar, N.; Khamitkar, S. Image segmentation by using edge detection. *Int. J. Comput. Sci. Eng. Technol.* **2010**, *2*, 804–807.
29. Wang, H.; Ying, D. An improved image segmentation algorithm based on OTSU method. *Comput. Simul.* **2011**, *6625*, 262–265.
30. Huang, M.; Yu, W.; Zhu, D. An improved image segmentation algorithm based on the Otsu method. In Proceedings of the 13th ACIS International Conference on Software Engineering, Artificial Intelligence, Networking and Parallel/Distributed Computing, Kyoto, Japan, 8–10 April 2012; pp. 135–139.
31. Huang, C.; Li, X.; Wen, Y. An OTSU image segmentation based on fruitfly optimization algorithm. *Alex. Eng. J.* **2021**, *60*, 183–188. [CrossRef]
32. Wu, Z.; Leahy, R. An optimal graph theoretic approach to data clustering: Theory and its application to image segmentation. *IEEE Trans. Pattern Anal. Mach. Intell.* **1993**, *15*, 1101–1113. [CrossRef]
33. Celebi, M.E.; Kingravi, H.A.; Vela, P.A. A comparative study of efficient initialization methods for the kmeans clustering algorithm. *Expert Syst. Appl.* **2013**, *40*, 200–210. [CrossRef]
34. Dhanachandra, N.; Manglem, K.; Chanu, Y.J. Image segmentation using k-means clustering algorithm and subtractive clustering algorithm. *Procedia Comput. Sci.* **2015**, *54*, 764–771. [CrossRef]
35. Zheng, X.; Lei, Q.; Yao, R.; Gong, Y.; Yin, Q. Image segmentation based on adaptive k-means algorithm. *Eurasip J. Image Video Process.* **2018**, *1*, 68. [CrossRef]
36. Srinivasan, V.; Bhatia, P.; Ong, S.H. Edge detection using a neural network. *Pattern Recognit.* **1994**, *27*, 1653–71662. [CrossRef]
37. Sowmya, B.; Rani, B.S. Colour image segmentation using fuzzy clustering techniques and competitive neural network. *Appl. Soft Comput.* **2011**, *11*, 3170–3178. [CrossRef]
38. Sultana, F.; Sufian, A.; Dutta, P. Evolution of image segmentation using deep convolutional neural network: A survey. *Knowl. Based Syst.* **2020**, *201*, 106062. [CrossRef]
39. Kukolj, D.; Marinovic, I.; Nemet, S. Road edge detection based on combined deep learning and spatial statistics of LiDAR data. *J. Spat. Sci.* **2021**, 1–15. [CrossRef]
40. Xue, J.; Zhang, L.; Grift, T.E. Variable field-of-view machine vision based row guidance of an agricultural robot. *Comput. Electron. Agric.* **2012**, *84*, 85–91. [CrossRef]

41. Li, N.L.; Zhang, C.; Chen, Z.; Ma, Z.; Sun, Z.; Yuan, T.; Li, W.; Zhang, J. Crop positioning for robotic intra-row weeding based on machine vision. *IJABE* **2015**, *8*, 20–29.
42. Pajares, G.; Garcia-Santillan, I.; Campos, Y.; Montalvo, M.; Guerrero, J.M.; Emmi, L.; Romeo, J.; Guijarro, M.; Gonzalez-de-Santos, P. Machine-vision systems selection for agricultural vehicles: A Guide. *J. Imaging* **2016**, *2*, 34. [CrossRef]
43. Shrivakshan, G.T.; Chandrasekar, C. A comparison of various edge detection techniques used in image processing. *Int. J. Comput. Sci. Issues IJCSI* **2012**, *9*, 269.
44. Torii, T.; Kitade, S.; Teshima, T.; Okamoto, T.; Imou, K.; Toda, M. Crop row tracking by an autonomous vehicle using machine vision (part 1): Indoor experiment using a model vehicle. *J. JSAM* **2000**, *62*, 41–48.
45. Bak, T.; Jakobsen, H. Agricultural robotic platform with four wheel steering for weed detection. *Biosyst. Eng.* **2004**, *87*, 125–136. [CrossRef]
46. Leemans, V.; Destain, M.F. Application of the hough transform for seed row localisation using machine vision. *Biosyst. Eng.* **2006**, *94*, 325–336. [CrossRef]
47. Bakker, T.; Wouters, H.; van Asselt, K.; Bontsema, J.; Tang, L.; Muller, J.; van Straten, G. A vision based row detection system for sugar beet. *Comput. Electron. Agric.* **2008**, *60*, 87–95. [CrossRef]
48. Ponnambalam, V.R.; Bakken, M.; Moore, R.J.D.; Gjevestad, J.G.O.; From, P.J. Autonomous crop row guidance using adaptive multi-ROI in strawberry fields. *Sensors* **2020**, *20*, 5249. [CrossRef] [PubMed]
49. Rabab, S.; Badenhorst, P.; Chen, Y.P.P.; Daetwyler, H.D. A template-free machine vision-based crop row detection algorithm. *Precis. Agric.* **2021**, *22*, 124–153. [CrossRef]
50. Ma, Z.; Tao, Z.; Du, X.; Yu, Y.; Wu, C. Automatic detection of crop root rows in paddy fields based on straight-line clustering algorithm and supervised learning method. *Biosyst. Eng.* **2021**, *211*, 63–76. [CrossRef]
51. De Silva, R.; Cielniak, G.; Gao, J. Towards agricultural autonomy: Crop row detection under varying field conditions using deep learning. *arXiv* **2021**, arXiv:2109.08247.
52. Oriolo, G.; De Luca, A.; Vendittelli, M. WMR control via dynamic feedback linearization: Design, implementation, and experimental validation. *IEEE Trans. Control Syst. Technol.* **2002**, *10*, 835–852. [CrossRef]
53. Chwa, D. Robust distance-based tracking control of wheeled mobile robots using vision sensors in the presence of kinematic disturbances. *IEEE Trans. Ind. Electron.* **2016**, *63*, 6172–6183. [CrossRef]
54. Li, J.; Zhang, D.; Ma, Y.; Liu, Q. Lane image detection based on convolution neural network multi-task learning. *Electronics* **2021**, *10*, 2356. [CrossRef]

*agriculture*

MDPI

*Article*

# Design and Experiment of Spiral Discharge Anti-Blocking and Row-Sorting Device of Wheat No-Till Planter

**Yunxiang Li [1,2], Caiyun Lu [1,2,*], Hongwen Li [1,2], Jin He [1,2], Qingjie Wang [1,2], Shenghai Huang [1,2], Zhen Gao [1,2], Panpan Yuan [1,2], Xuyang Wei [1,2] and Huimin Zhan [3]**

1 College of Engineering, China Agricultural University, Beijing 100083, China; liyunxiang@cau.edu.cn (Y.L.); lhwen@cau.edu.cn (H.L.); hejin@cau.edu.cn (J.H.); wangqingjie@cau.edu.cn (Q.W.); huangshenghai@cau.edu.cn (S.H.); b20203070561@cau.edu.cn (Z.G.); yuanpp@cau.edu.cn (P.Y.); weixvyang@cau.edu.cn (X.W.)

2 Key Laboratory of Agricultural Equipment for Conservation Tillage, Ministry of Agricultural and Rural Affairs, Beijing 100083, China

3 Shaanxi Agricultural Machinery Appraisal and Extension Station, Xi'an 710068, China; zhm81871563@gmail.com

* Correspondence: lucaiyun@cau.edu.cn; Tel.: +86-010-6273-7300

**Abstract:** Aiming at the problems of the poor passing capacity of machines and low cleaning rate of seed strip during wheat no-tillage sowing in annual double cropping areas of North China, a spiral discharge anti-blocking and row-sorting device (SDARD) was designed and is reported in this paper. After the straw was cut and chopped by the high-velocity rotating no-till anti-blocking knife group (NAKG), the straw was thrown into the spiral discharging mechanism (SDM) behind the NAKG. The chopped straw was discharged to the non-sowing area to reach the effect of seed strip cleaning through the interaction between the SDM and the row-sorting of straw mechanism (RSM). Based on a theoretical analysis for determining the parameters of crucial components, the quadratic rotation orthogonal combination test method was adopted, and the operating velocity of machines (OVM), the rotary velocity of the spiral shaft (RVSS), and the height of the holding hopper from the ground (HHHG) were selected as the test factors. The straw cleaning rate (SCR) was taken as the test index. The discrete element simulation test was carried out, the regression model of the SCR was established, and parameters optimization and field test were carried out. The results show that the significant order of the three influencing factors on the SCR was HHHG > OVM > RVSS. The optimal combination of operating parameters was that OVM was 5 km/h, RVSS was 80 r/min, and HHHG was 10 mm. Under the optimal parameter combination, the average SCR was 84.49%, which was 15.5% higher than the no-till planter without the device, and the passing capacity of machines was great, which met the agronomic requirements of no-tillage sowing of wheat in annual double cropping areas. This study could provide a reference for the design of no-tillage machines.

**Keywords:** no-tillage sowing; sowing strip cleaning; spiral discharge straw; discrete element simulation; anti-blocking and row-sorting

**Citation:** Li, Y.; Lu, C.; Li, H.; He, J.; Wang, Q.; Huang, S.; Gao, Z.; Yuan, P.; Wei, X.; Zhan, H. Design and Experiment of Spiral Discharge Anti-Blocking and Row-Sorting Device of Wheat No-Till Planter. *Agriculture* **2022**, *12*, 468. https://doi.org/10.3390/agriculture12040468

Academic Editors: Mustafa Ucgul and Chung-Liang Chang

Received: 1 March 2022
Accepted: 22 March 2022
Published: 25 March 2022



## 1. Introduction

Conservation tillage is an advanced agricultural tillage technology that implements no/minimum tillage on farmland covered with crop straw. It can effectively reduce soil wind and water erosion, improve soil fertility level and drought resistance, avoid straw burning, and alleviate ecological pressure. It is an important measure for achieving sustainable agricultural development [1–4]. When winter wheat is no-tillage sown in annual double cropping areas of North China, the straw is chopped and returned to the field, and a large amount of the straw on the soil surface is covered, which causes the no-till planter to be easily blocked during operation. The problems of the poor passing capacity of the machine and poor furrowing effect are severe [5–9], and if the straw or stubble falls on the seeding row during operation, it will affect the contact degree between the seed and the soil and reduce the seed implantation quality [10–15]. Therefore, timely and effective separation of the straw from the planter and its working parts and discharge of the straw residue to the non-sowing area are crucial problems for solving the blockage of the wheat no-till planter and improving the sowing quality [16,17].

Existing research mainly focuses on the anti-blocking performance of no-till planters—that is, the passing capacity of the machines. Wang et al. [18] studied the telescopic lever furrowing and anti-blocking unit, which used the telescopic lever to collect the straw in front of the furrowing opener and release it to the row, which greatly affected anti-blocking ability when furrowing. Lin et al. [19,20] designed and studied an Archimedes spiral type notched disc knife based on the Archimedean spiral, which improved the stubble-breaking performance of the disc knife. Lu et al. [21–24] studied a rotary cutting anti-blocking device with a slide plate pressing straw and the friction characteristics of straw during operation. Compared with a zonal rotary and strip chopping no-till planter, this device had better passing capacity, even emergence, and lower rotary velocity.

Although the currently developed anti-blocking device can meet the requirements of the passing capacity of machines, a part of the straw remains in the sowing strip after the operation, and the cleaning effect of the sowing strip is still difficult to guarantee. Straw falling on the seed row will affect the seed implantation quality, then affect the seed emergence, and even reduce the grain yield. Improving the cleaning effect of seed strips is becoming another research hotspot of no-tillage anti-blocking devices. For example, Yuan et al. [25] studied a straw cleaning device for wide-narrow maize no-tillage sowing strip in the drip irrigation area. They put forward the seed strip cleaning method to meet the agronomic requirements of no-tillage sowing of wide-narrow maize in the drip irrigation area. Yu et al. [26] studied and designed a spiral-split sowing strip cleaning device to solve the problems of straw blockage during no-tillage sowing in Northeast China. These studies mainly focus on the no-tillage sowing of maize in annual once cropping areas in the northeast and northwest, while few studies exist on the seed strip cleaning of wheat no-tillage sowing in annual double cropping areas in the Huang-Huai-Hai. Mainly because the row spacing of wheat no-tillage sowing in annual double cropping areas in the Huang-Huai-Hai is narrow, there is no leisure period between the harvest of the last crop and the sowing of the next crop. The straw has strong deformation resistance during the sowing period, and it is difficult to row-sort the straw and carry out seed strip cleaning.

This paper proposes an SDARD of wheat no-till planter for solving these above problems. It can efficiently discharge the straw to the non-sowing area on both sides of the machine and between the seed strips, while ensuring passing capacity, playing the role of straw cleaning, effectively reducing the furrowing opener's blockage, and concentrating the straw in the non-sowing area. Through theoretical analysis, the parameters of the critical components of the SDARD are determined. The quadratic rotation orthogonal combination test determines the optimal structure and working parameters through a discrete element simulation test. Finally, a field test on the optimized SDARD was conducted to verify the above theoretical analysis and discrete element simulation experiments.

## 2. Materials and Methods

### *2.1. Machine Structure and Working Principle of the SDARD*

#### 2.1.1. Structure of the SDARD

As shown in Figure 1a, the SDARD was installed between the NAKG (1) and the furrowing opener (4) of the wheat no-till planter. Moreover, it was composed of the SDM (2) and the RSM (3). Its overall structure is shown in Figure 1b.

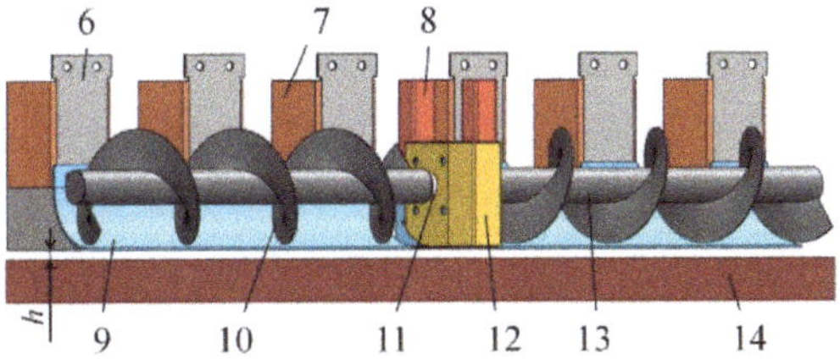


**Figure 1.** SDARD with its specific position. (**a**) Installation position of SDARD of wheat no-till planter; (**b**) the structure of SDARD. Note: 1, NAKG; 2, SDM; 3, RSM; 4, furrowing opener; 5, frame; 6, seed strip baffle; 7, side baffle; 8, support-connecting; 9, holding hopper; 10, spiral blade; 11, rolling bearing; 12, inclined anti-winding plate; 13, spiral shaft; 14, soil. $h$ is the height of the hopper above the ground, mm. $h$ also indicates a gap between the hopper (9) and the soil (14). Moreover, there is no contact. $h > 0$.

The SDM (2) comprised spiral shaft (13), spiral blade (10), holding hopper (9), support-connecting frame (8), rolling bearing (11), and inclined anti-winding plate (12). Wherein the spiral blade (10) was integrally formed with the spiral shaft (13), combined with the helix direction of the spiral blade (10), the two spiral shafts (13) were reversely installed in the rolling bearing (11), fixed on the support-connecting frame (8) through the inclined anti-winding plate (12), and the support-connecting frame (8) was welded on the frame (5). Holding hopper (9) was located directly below the spiral blade (10), and there was a gap between the bottom of the holding hopper (9) and the soil surface (14) (essential position parameter $h$), which was not in contact with the soil.

The RSM (3) comprised seed strip baffle (6) and side baffle (7). Seed strip baffle (6) was installed in front of the furrowing opener (4), and the top was connected with frame (5) through bolts. Both sides of the seed strip baffle (6) were equipped with a side baffle (7) that was integrally formed. Furthermore, the top of the holding hopper (9) was fixed at the bottom of the seed strip baffle (6) by bolts.

2.1.2. Working Principle

The power of the SDARD was provided by the rear output shaft of the tractor. When working, the high-velocity rotating NAKG was achieved by cutting and chopping the soil and straw, reaching the power-driven anti-blocking effect and guaranteeing the passing capacity.

Under the action of high-velocity rotation of the NAKG, most of the straw was thrown directly into the holding hopper, alternatively, to the front side of the seed strip baffle, and fell on the hopper through free-fall motion. The SDM accumulated and transported the straw falling into the hopper to both sides of the planter, playing roles of collection and diversion. When the spiral blade transported the straw in the hopper to the gap between adjacent seed strip baffles, the position corresponded to the no-sowing areas, the straw tended to move backward, and it would fall from the gap to the non-sowing area between the seed strips, achieving the effect of straw-cleaning in the seed strips.

A small part of the straw was directly thrown into the gap between the adjacent seed strip baffles by the NAKG and directly fell, or was rebounded by the side baffle, to the non-sowing area, playing the role of row-sorting the straw.

*2.2. Parameter Design of the SDM*

The relationship between the material bearing limit of the SDM and the straw feeding amounts (SFA) was an important index for evaluating the device's conveying performance stability and straw discharge effect. When designing, in order to ensure that the actual SFA was within the limit of the material bearing capacity of the mechanism, the minimum material bearing capacity of a device should be greater than the maximum SFA to ensure the stable and reliable conveying performance of the device and good straw conveying effect.

Therefore, this research calculated the maximum SFA $Q_s$ according to the actual operating environment of the annual double cropping areas in North China, and then designed and calculated the critical structural dimensions (spiral blade outer diameter $D$, pitch $S$, and HHHG $h$) and operating parameters (RVSS $n$) according to the SFA $Q_s$ (Figures 1 and 2).

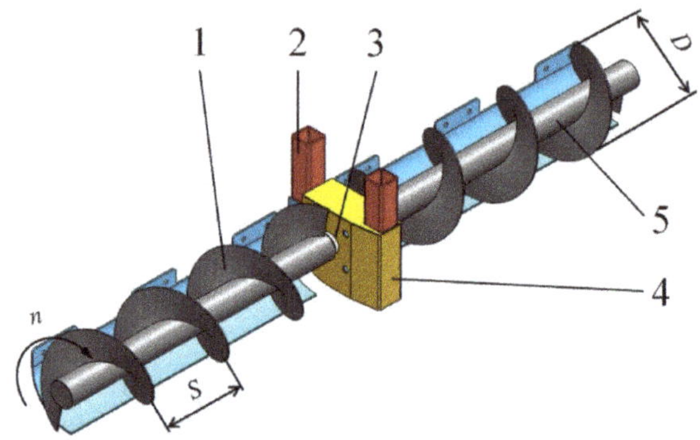


**Figure 2.** Critical parameters of SDM. 1, Spiral blade; 2, support-connecting frame; 3, rolling bearing; 4, inclined anti-winding plate; 6, spiral shaft; 5, holding hopper. $D$ is the outer diameter of the spiral blade, m; $S$ is pitch, m; $n$ is the RVSS, r/min.

2.2.1. The SFA $Q_s$

The SFA $Q_s$ referred to in this paper was the maximum value of the total mass of straw residues thrown into the holding hopper of the SDM by the NAKG in unit time, which was the total mass of disappeared straw in the sowing strip, while these straws were all from stover and stubble of maize on the surface before the operation. Its size was mainly determined by the straw mulching quantity (SMQ), the seed strip cleaning area of no-till planter per-unit time, and the seed strip cleaning rate. The area was determined by the width of machines, operation velocity, and time. Therefore, to obtain quantitative data of SFA, a mathematical model of the SFA and the above-mentioned related factors was constructed.

In order to simplify the model, it was assumed that: (1) the straw was covered evenly on the ground surface before operation; (2) considering the cutting effect of the NAKG on soil and the gap between the bottom of the hopper and soil surface, the soil would be discharged from the gap in preference to the straw under gravity. Thus, the influence of the soil on the SFA was not considered in the modeling process.

As shown in Figure 3, the mathematical model could be described as an ideal process: within a specific time $t$, the total mass $Q_{St}$ of straw thrown into the hopper by the NAKG was equal to the mass difference of straw that disappeared in the sowing strip before and after the operation, that was, $Q_{St} = (Q_1 - Q_2)$. The mass difference per unit time could be determined by SMQ $C_S$, OVM $v$, the number of sowing strips $z$, the total width of sowing strip cleaning $L_1$, and SCR $\lambda$. The conversion from the SMQ $C_S$ to the SFA $Q_S$ was realized.

The mathematical model of the relationship between SFA and SMQ established by the mathematical induction and the quotient-remaining method was as the following equation:

$$Q_s = \frac{Q_{St}}{t} = \frac{Q_1 - Q_2}{t} = \frac{1}{60} C_s v z L_1 \lambda \tag{1}$$

where $Q_s$ is the SFA; $Q_{St}$ is the total feeding mass of straw in the hopper after a specific operation time, kg; $t$ is the operation time of no-till planter, min; $Q_1$ is the mass of straw before sowing strip operation, kg; $Q_2$ is the residual mass of straw after sowing strip operation, kg; $C_S$ is the SMQ, kg/m$^2$; $v$ is the OVM, km/h; $z$ is the number of corresponding sowing strips; $L_1$ is the total cleaning width of the sowing strip, mm; $\lambda$ is the SCR,%.

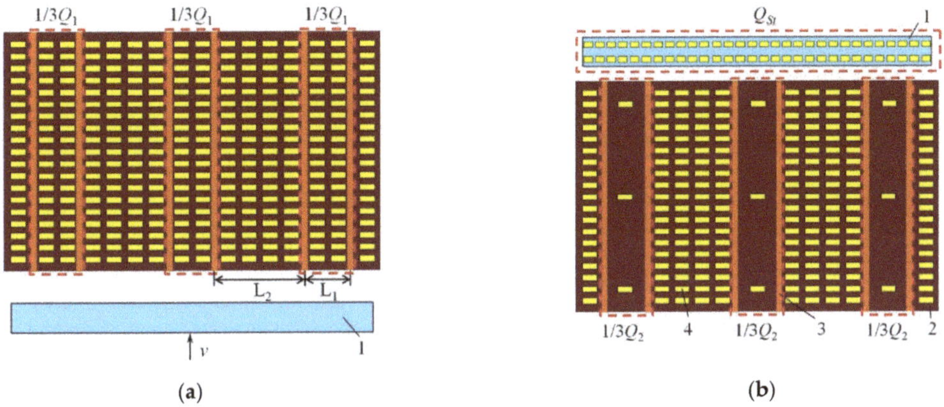


**Figure 3.** Mathematical model of the relationship between the SFA and the SMQ. (**a**) Theoretical model before no-tillage operation; (**b**) theoretical model after a no-tillage operation. Note: 1, the hopper of SDM; 2, no-tillage surface of the field; 3, seed line; 4, straw residue. $Q_1$ is the mass of straw on the sowing strip before an operation, kg; $Q_2$ is the residual mass of straw on the sowing strip after an operation, kg; $Q_{St}$ is the total mass of straw fed into the hopper after a certain period of operation, kg; $v$ is the OVM, km/h; $L_1$ is the total cleaning width of sowing strip, mm; $L_2$ is the total width of non-sowing area, mm.

Through the quadrat method [27], the maximum value of the SMQ $C_S$ on the ground surface of pre-operated maize was 1.72 kg/m$^2$, the typical operating velocity of 2BMQF-6/12A wheat no-till planter was 3~5 km/h, the maximum OVM $v$ was 5 km/h, each side of the SDM corresponded to 3 sowing strips, $z$ was 3, the width $L_1$ of the sowing strip was 150 mm, and the SCR $\lambda$ took the maximum value of 1. According to Equation (1), the SFA $Q_s$ of the SDM was 64.5 kg/min.

2.2.2. Outer Diameter of the Spiral Blade *D*

The outer diameter of the spiral blade was an essential parameter in the design of SDM, which was directly related to the straw discharged capacity and structural size of the mechanism. Referring to the Darnell–Mol solid conveying theory [28] proposed by Darnell and Mol, the calculation equation of the spiral blade outer diameter [29,30] was as follows:

$$D \geq K \cdot \left( \frac{60 Q_s}{\psi \rho C} \right)^{\frac{2}{5}} \tag{2}$$

where $D$ is the outer diameter of the spiral blade, m; $K$ is the material characteristic coefficient; $Q_s$ is the SFA, kg/min; $\psi$ is the filling factor; $\rho$ is the material density, kg/m$^3$; $C$ is the inclination coefficient.

By consulting the literature [29–33], the material characteristic coefficient $K$ was 0.05, filling coefficient $\psi$ took 0.3, straw density $\rho$ took 240 kg/m$^3$, and when the SDM was in the horizontal state during operation, the inclination coefficient $C$ was taken as 1. According to Equation (2), the minimum value of the spiral blade outer diameter $D$ was 0.246 m, and $D$ was 0.250 m after rounding.

2.2.3. Pitch *S*

When the pitch increased, the axial conveying velocity increased, but the conveying stability decreased. The conveying stability was great when the pitch decreased, but the conveying velocity slowed down. The pitch calculation equation was as follows:

$$S = K_1 D \tag{3}$$

where $S$ is the pitch of the spiral blade, m; $K_1$ is the pitch coefficient; $D$ is the outer diameter of the spiral blade, m.

Referring to the transportation machinery design and selection manual [30], when the pitch coefficient $K_1$ was taken as 1, the pitch value $S$ calculated by Equation (3) was 0.250 m.

2.2.4. The HHHG *h*

The hopper should have a suitable height from the ground. On the one hand, it could reduce the disturbance of the hopper to the soil and let part of the soil pass through the gap. The HHHG $h$ should be greater than 0 to avoid contact with the soil surface. On the other hand, combined with the dropping track of straw under the action of NAKG, the value of the HHHG would affect the amount of straw falling into the hopper and also impact the SCR $\lambda$, so the relationship between $h$ and $\lambda$ should be determined according to the simulation test.

2.2.5. The RVSS *n*

When the spiral shaft rotated, the straw was affected by the normal thrust of the spiral blade, and the radial component of the thrust and a frictional force of the spiral blade acting on the straw would carry some straw to rotate around the spiral shaft [31]. When the RVSS was too low, the straw discharge velocity was slow, which was accessible to lead to the blockage of the SDM. When the RVSS increased, the centrifugal force would gradually replace the blade's thrust and play a leading role. When the RVSS exceeded a specific limit value, the straw would be thrown outward due to excessive centrifugal force so that it could not be transported, thus reducing the transportation stability of straw [32,33]. The calculation equation of RVSS was as follows:

$$n \geq \frac{60Q_s}{47D^2 S\psi\rho C} \tag{4}$$

where $n$ is the RVSS, r/min; $Q_s$ is the SFA, kg/min; $D$ is the outer diameter of spiral blade, m; $S$ is the pitch of the spiral blade, m; $\psi$ is the filling factor; $\rho$ is the material density, kg/m$^3$; $C$ is the inclination coefficient.

Substituting the above-mentioned calculated and selected parameters into Equation (4), it was calculated that the minimum RVSS was 73.2 r/min. In order to ensure that its maximum bearing capacity could adapt to the SMQ in annual double cropping areas, the minimum value of $n$ was 80 r/min. Considering that the impact of RVSS $n$ on the SCR was unknown and that the RVSS was the main factor affecting the stability of material transportation [32], follow-up was based on discrete element simulation experiments for analyzing the variation law of $n$ and $\lambda$.

### 2.3. Parameter Design of the RSM

The RSM could discharge the straw residue to the non-sowing area on both sides of the sowing strip and directly prevent the straw from falling into the sowing strip, improving the cleaning effect of the seed strip. The critical parameters of RSM are shown in Figure 4, including width $b_B$, height $h_G$, length $l_G$, and included angle $\theta$.

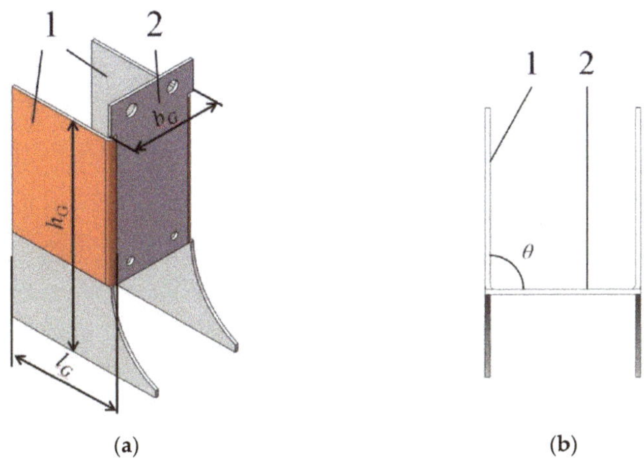


**Figure 4.** Critical parameters of the RSM. (**a**) Oblique view; (**b**) top view. Note: 1, side baffle; 2, seed strip baffle. $b_G$ is the width of the RSM, mm; $h_G$ is the height of the RSM, mm; $l_G$ is the length of the side baffle, mm; $\theta$ is the included angle between the strip baffle and the side baffle.

#### 2.3.1. The Width of the RSM $b_G$

Agronomic conditions had specific requirements for the width of the no-tillage sowing seed strip and the gap between the adjacent sowing strips. When designing the RSM, first of all, it had to be ensured that the straw could not fall into the sowing strip, so the width between the side baffles on both sides of the same RSM had to be greater than the row spacing of narrow rows. Secondly, considering that the row spacing of no-tillage sowing of wheat was narrow and the number of corn stalks in the field at the sowing date was large, there needed to be enough space to place straw, so the spacing between adjacent RSMs had to be maximized.

As shown in Figure 3a, the width and narrow rows of no-tillage wheat in annual double cropping areas in northern China required the row spacing of two adjacent sowing strips to be L = 340 mm, the total cleaning width of sowing strip $L_1 \geq 150$ mm, and the total width of non-sowing area $L_2 \leq L - L_1 = 190$ mm. In order to meet the above requirements, the width of the RSM $b_G = L_1 = 150$ mm was determined.

#### 2.3.2. The Height of the RSM $h_G$

In the RSM, the function of the seed strip baffle and the side baffle was to prevent the straw from falling into the sowing strip from the front and side. If the height of the RSM was short, the straw could fly into the sowing strip across its top or fall into the seed strip along the gap at the bottom, and the cleaning effect of the seed strip would be weakened.

Combined with the installation position of the no-till planter, the optimum range of height threshold of RSM was 350~370 mm. In order to improve the effect of preventing straw from falling into the seed strip, the height of RSM $h_G$ was selected as 370 mm.

2.3.3. The Length of the Side Baffle $l_G$

The side baffle would guide the straw to the non-sowing area. The reasonable length design would avoid the straw blocking the furrowing openers and ensure the seed strip cleaning effect. Considering the above factors, the length of the side baffle $l_G$ was 180 mm.

2.3.4. The Included Angle between the Seed Strip Baffle and the Side Baffle $\theta$

In order to ensure that all straw fell in the non-sowing area and was distributed evenly, the included angle $\theta$ between the seed strip baffle and the side baffle had to be designed. As shown in Figure 5, when $\theta < 90°$, the function area of the side baffle could not completely cover the sowing strip, which increased the probability of straw falling into the seed row. When $\theta > 90°$, most straw was concentrated in the centerline of the non-sowing area, and the uniformity of straw distribution was poor. To sum up, the included angle was taken $\theta = 90°$.

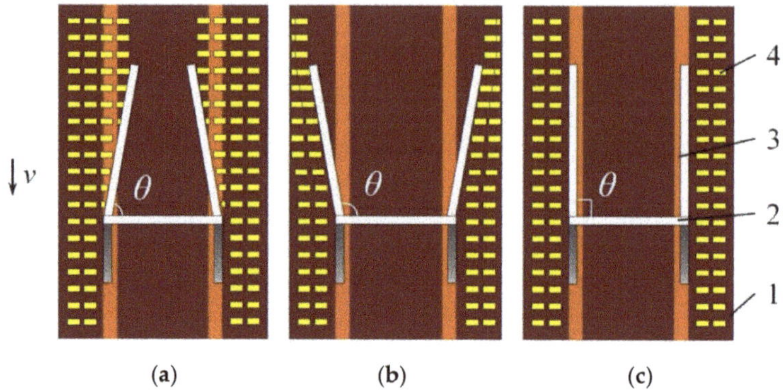


**Figure 5.** Effect of $\theta$ on cleaning effect of seed strip. (**a**) $\theta < 90°$; (**b**) $\theta > 90°$; (**c**) $\theta = 90°$. Note: 1, no-tillage surface of the field; 2, RSM; 3, seed line; 4, straw.

*2.4. Rational and Reasonable Parameters of the SDARD by EDEM Optimization Method*

The discrete element method was often used to study the relationship between working parts and multi-granular media such as soil. In this research, the discrete element simulation software EDEM was used to establish the computer simulation model between the SDARD and its operation object (straw and soil) and to simulate the actual operating conditions of the SDARD. The SCR in the sowing strip was selected as the test index in the simulation process, and OVM, RVSS, and HHHG were taken as the primary test factors. The influence of various factors on the SCR was analyzed through a discrete element simulation test to provide the basis for the optimal structure and operating parameters and to lay the foundation for the subsequent prototype trial production and field test.

2.4.1. Establishment of the Discrete Element Simulation Model

Because the computer's computing power and simulation efficiency were considered, spherical particles with a radius of 7 mm were selected as soil particles. The mechanical relationship model between soil particles was a coupled contact model consisting of the Hertz–Mindlin with bonding model and the Hertz–Mindlin (no slip) model [25,26], and the shear modulus was $1.0 \times 10^6$ Pa, soil density 1850 kg/m$^3$, Poisson's ratio 0.38. They were combined with the actual situation of straw in the field. An extended linear model with a length of 88 mm composed of a diameter of 16 mm and a ball center interval of 4 mm was used as the straw particle model. The Hertz–Mindlin (no slip) model was selected as the mechanical relationship model of the straw model, and the shear modulus was $1.0 \times 10^6$ Pa, straw density 240 kg/m$^3$, Poisson's ratio 0.4, and the basic contact parameters of the discrete element simulation [34,35] are shown in Table 1.

**Table 1.** Parameter settings for EDEM simulation.

| Parameters | Value | Parameters | Value |
|---|---|---|---|
| Coefficient of restitution among soil | 0.20 | Coefficient of rolling friction between soil and steel | 0.04 |
| Coefficient of static friction among soil | 0.40 | Coefficient of restitution among straw | 0.40 |
| Coefficient of rolling friction among soil | 0.31 | Coefficient of static friction among straw | 0.30 |
| Coefficient of restitution between soil and straw | 0.30 | Coefficient of rolling friction among straw | 0.01 |
| Coefficient of static friction between soil and straw | 0.30 | Coefficient of static friction between straw and steel | 0.30 |
| Coefficient of rolling friction between soil and straw | 0.01 | Coefficient of rolling friction between straw and steel | 0.30 |
| Coefficient of restitution between soil and steel | 0.30 | Coefficient of restitution between straw and steel | 0.01 |
| Coefficient of static friction between soil and steel | 0.50 | | |

In EDEM, a virtual soil bin suitable for SDRAD of wheat no-till planter was established. Firstly, the soil particles were generated to make the soil particles settle naturally under gravity, and by appropriately extending the simulation time, it was ensured that stable bonding bonds were formed between soil particles and that they were as static as possible. After the soil was settled, the straw was dropped, and the straw particles were dropped above the soil particles to make the model as close as possible to the actual soil and straw conditions. The actual size of the soil bin (length × wide × height) was set at 4000 mm × 1500 mm × 220 mm, with a 180 mm thick soil layer and 40 mm thick straw layer. Figure 6 shows the built virtual soil bin model.

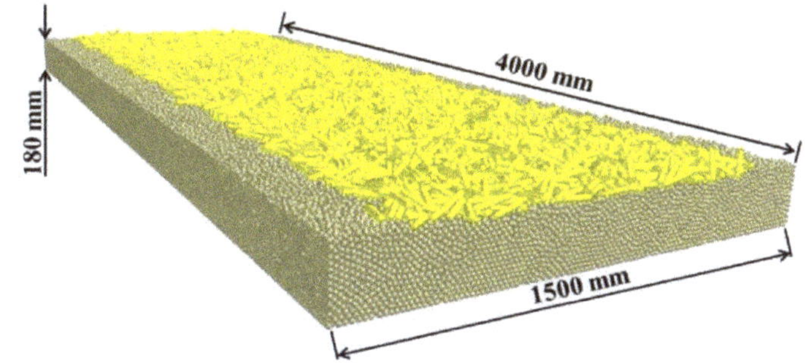


**Figure 6.** Virtual soil bin model.

After the soil bin was established, the SOLIDWORKS 3D model of SDARD with a scale of 1:1 was imported into the EDEM software in a STEP format file. The material property of the SDARD was 45 steel, the density was 7800 kg/m$^3$, the Poisson's ratio was 0.31, and the shear modulus was $7.0 \times 10^{10}$ Pa.

2.4.2. Simulation Process

The SDARD was set at one end of the soil bin and started. In order to ensure the continuity of the operation process of the SDARD and the movement of soil and straw particles, the minimum step size was set as 0.0001 s, the total simulation time was 5.0 s, and the grid size was set as 2.5 times of the minimum soil particle size—i.e., 17.5 mm. The simulation operation process of SDARD is shown in Figure 7.

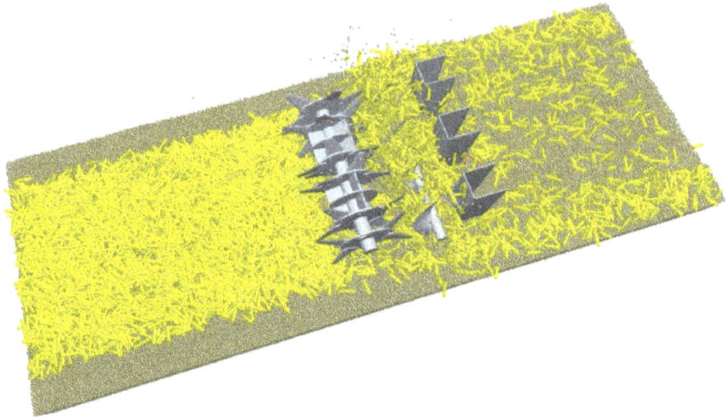

**Figure 7.** Simulation operation process of the SDARD.

### 2.4.3. Simulation Experimental Design

Comprehensively considering the theoretical analysis and results of the design of critical components, taking the SCR as the evaluation index, and selecting OVM, RVSS, and HHHG as the test factors, the quadratic regression combined simulation test was carried out in EDEM. According to the main technical parameters, theoretical analysis, and pre-test of the device, it was determined that the value range of the OVM was 3~5 km/h, the value range of RVSS was 80~160 r/min, and the value range of HHHG was 10~30 mm. The Box–Behnken method in Design Expert 10.0.7 software was used to determine the test factors and codes, as shown in Table 2.

**Table 2.** Test factors and coding.

| Code | Factors | | |
|---|---|---|---|
| | The OVM $X_1$/(km·h$^{-1}$) | The RVSS $X_2$/(r·min$^{-1}$) | The HHHG $X_3$/mm |
| −1 | 3 | 80 | 10 |
| 0 | 4 | 120 | 20 |
| 1 | 5 | 160 | 30 |

## 3. Results and Discussion

The Selection option in the Analyst module of EDEM software was used to establish a Total Mass Sensor grid area (mass sensor) with a length and width of 2000 × 150 mm at the corresponding sowing strip position. The corresponding SCR was obtained by calculating the straw quality change before and after the operation of the SDARD in the area. The local straw cleaning effect of the sowing strip is shown in Figure 8.

The calculation equation of SCR was as follows:

$$\lambda = \frac{m_1 - m_2}{m_1} \times 100\% \tag{5}$$

where $\lambda$ is the SCR, %; $m_1$ is the mass of straw before the operation, kg; $m_2$ is the mass of straw after the operation, kg.

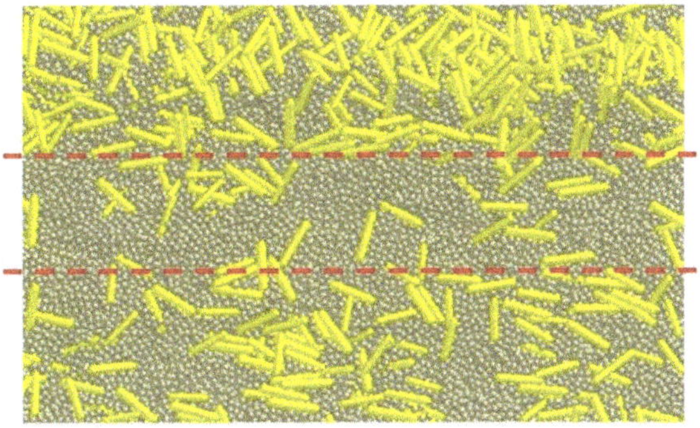

**Figure 8.** Test area of sowing strip straw cleaning effect. Note: The red dotted line area is the sowing strip after the operation of the SDARD.

## 3.1. Regression Model Construction

The test results are shown in Table 3. According to the test data in Table 3, the quadratic regression analysis of the straw cleaning test results of the sowing strip was carried out through the Design Expert software, and a quadratic response surface regression model was established with the test index as the SCR and the test factors as the OVM, the RVSS, and the HHHG; the equation was as follows:

$$y = 88.48 + 2.55X_1 - 1.05X_2 - 3.52X_3 - 0.93X_1X_3 - 0.55X_1^2 - 0.65X_3^2 \quad (6)$$

**Table 3.** Experimental design and results.

| Test No. | Factors | | | | |
|---|---|---|---|---|---|
| | The OVM $X_1$ | The RVSS $X_2$ | The HHHG $X_3$ | SCR $y$/% | Calculated Values $y$ |
| 1 | −1 | −1 | 0 | 83.25 | 86.43 |
| 2 | 1 | −1 | 0 | 87.28 | 91.53 |
| 3 | −1 | 1 | 0 | 80.75 | 84.33 |
| 4 | 1 | 1 | 0 | 85.85 | 89.43 |
| 5 | −1 | 0 | −1 | 83.65 | 87.32 |
| 6 | 1 | 0 | −1 | 91.15 | 94.28 |
| 7 | −1 | 0 | 1 | 78.15 | 82.14 |
| 8 | 1 | 0 | 1 | 81.92 | 85.38 |
| 9 | 0 | −1 | −1 | 88.59 | 92.40 |
| 10 | 0 | 1 | −1 | 86.48 | 90.30 |
| 11 | 0 | −1 | 1 | 82.01 | 85.36 |
| 12 | 0 | 1 | 1 | 79.65 | 83.26 |
| 13 | 0 | 0 | 0 | 84.86 | 88.48 |
| 14 | 0 | 0 | 0 | 84.05 | 88.48 |
| 15 | 0 | 0 | 0 | 86.03 | 88.48 |
| 16 | 0 | 0 | 0 | 84.66 | 88.48 |
| 17 | 0 | 0 | 0 | 84.97 | 88.48 |

The results of the analysis of significant variance (Table 4) show that in the regression model, the $y$ model, which is representative of the SCR, is $p < 0.01$; the regression model is highly significant; and the influence of $X_1$, $X_2$, $X_3$, and $X_{1 \times 3}$ is highly significant ($p < 0.01$). $X_3^2$ has a significant effect ($0.01 < p < 0.05$). The effect of $X_1^2$ is relatively significant ($0.05 < p < 0.1$). The other items are not significant. The significant order of the three influencing factors on the SCR is $X_3 > X_1 > X_2$—that is, the HHHG, the OVM, and the RVSS. The lack of fit test $p > 0.1$, the regression model is highly significant, the lack of fit is insignificant, and the coefficient of determination $R^2$ of the equations was 0.9453, demonstrating that the regression models could explain more than 85% of the response values. Moreover, the regression model is effective.

**Table 4.** Experimental design and results.

| Variation Source | Sum of Squares | Degree Freedom | Mean Square | *F* Values | *p*-Values |
|---|---|---|---|---|---|
| Model | 166.57 | 6 | 27.76 | 86.21 | <0.0001 *** |
| $X_1$ | 52.02 | 1 | 52.02 | 161.54 | <0.0001 *** |
| $X_2$ | 8.82 | 1 | 8.82 | 27.39 | 0.0004 *** |
| $X_3$ | 98.98 | 1 | 98.98 | 307.37 | <0.0001 *** |
| $X_1X_3$ | 3.48 | 1 | 3.48 | 10.80 | 0.0082 *** |
| $X_1{}^2$ | 1.29 | 1 | 1.29 | 4.00 | 0.0733 * |
| $X_3{}^2$ | 1.80 | 1 | 1.80 | 5.58 | 0.0397 ** |
| Residual | 3.22 | 10 | 0.32 | | |
| Lack of fit | 1.16 | 6 | 0.19 | 0.37 | 0.8642 |
| Pure error | 2.06 | 4 | 0.52 | | |
| Total sum | 169.79 | 16 | | | |

Note: *** means highly significant ($p < 0.01$), ** means significant ($0.01 < p < 0.05$), * means relatively significant ($0.05 < p < 0.1$).

### 3.2. Analysis of the Influence of Particular Factor Numerical Change on Indicator

Changes in either factor will have an impact on the results. It can be seen from Equation (6) that when the value range of $X_1$ is 3~5 km/h, $X_1$ and $y$ are positively correlated, so when $X_1$ increases, the value of $y$ will also increase. When the value range of $X_2$ is 80~160 r/min, $X_2$ is negatively correlated with $y$, and when $X_2$ increases, the value of $y$ will decrease slightly. When the value range of $X_3$ is 10~30 mm, $X_3$ and $y$ are also negatively correlated, so when $X_3$ increases, the value of $y$ will decrease sharply. Furthermore, because the coefficient of $X_1$ is 2.55, the coefficient of $X_2$ is $-1.05$, and the coefficient of $X_3$ is $-3.52$. The more significant the absolute value of the coefficient, the more significant the influence on $y$. The significant order is $X_3 > X_1 > X_2$. The conclusion of the above analysis is the same.

### 3.3. Basic Theoretical Analysis of the Influence of Interaction Factors on Indicator

Based on the experimental design and results (Table 4), this study separately analyzed the influences of the interactive factors $X_1$, $X_3$, and $X_1X_3$ on $y$. It can be seen that $X_1$ and $X_3$ have an interactive effect on $y$, and the response surface of the two interactions is shown in Figure 9. $X_2$ is fixed at 120 r/min. When $X_1$ is constant, $y$ decreases with the increase in $X_3$, and the two factors are negatively correlated. When $X_3$ is within 10~30 mm, the smaller the gap between the bottom of the holding hopper and the ground surface and the less straw that can pass directly through the gap when the planter is operating, which can effectively improve the operation effect and $y$ of the planter.

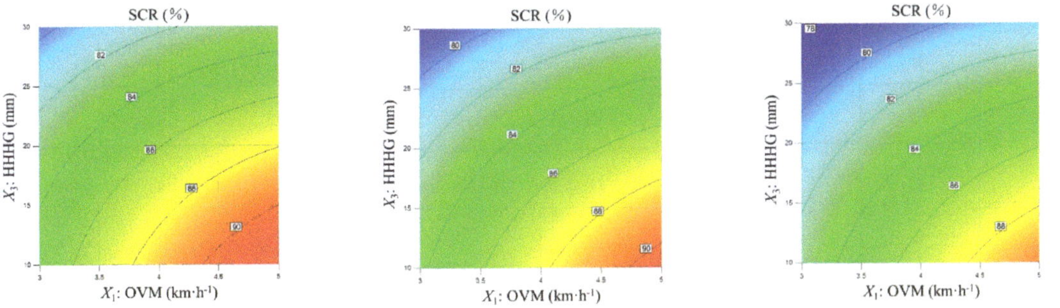


**(a)** **(b)** **(c)**

**Figure 9.** Effects of interaction factors on the SCR. (**a**) $X_2$ = 80 r/min; (**b**) $X_2$ = 120 r/min; (**c**) $X_2$ = 160 r/min.

When $X_3$ is constant, $y$ increases with the increase in $X_1$. Thus, $y$ is positively correlated with $X_1$. Considering a greater $X_1$, the greater the soil cutting pitch during the operation of the NAKG and the fewer times the soil and straw are effectively cut and chopped within the same distance. In addition, the amount of soil beyond a certain depth is more significant than the amount of straw, $X_1$ is increased, the amount of soil thrown into the holding hopper within the same distance is reduced, the decline degree of soil surface level is reduced, the amount of straw that can pass through the gap between the bottom of the holding hopper and the surface is reduced, and $y$ is improved. $X_1$ is low, the amount of soil cut by the NAKG is large, and the decline in the surface soil level will increase the gap, resulting in more straw staying in the sowing strip, which has a specific impact on the straw cleaning effect in the seed strip.

It can be seen from Figure 9 that with the increase in $X_2$ (increasing from 80 to 160 in numerical value), the model shows a gradually falling trend. It is more evident that the vertices in the red area of the model in Figure 9a are above $y$ = 90%, while the red vertices of the model in Figure 9c are below $y$ = 90%. However, with the change in $X_2$, the model's shape is basically unchanged, which shows that the overall trend of the model for the test index $y$ is unchanged, and it also shows that the model is stable and effective. Moreover, as $X_2$ decreases, the area of the red area is getting more minor, and the area of the blue area is getting more evident. Further, the more significant the red area, the higher the value of $y$, the more significant the blue area, the lower the value of $y$.

### 3.4. Discussion

Based on the above analysis, it can be concluded that the relationship between OVM and SCR is such that when OVM gradually increases, SCR will decrease when OVM is at 3~5 km/h—on the premise that OVM interacts with HHHG. It is because the more significant the soil cutting pitch during the operation of the NAKG, the fewer times the soil and straw are effectively cut and chopped within the same distance, the soil level drops less, and only less straw can pass through the gap between the bottom of the holding hopper and the surface. This viewpoint is similar to Yu's [26] discrete element simulation analysis viewpoint.

In terms of RVSS, this research concludes that when RVSS is in the range of 80–160 r/min, RVSS is negatively correlated with SCR, and SCR will decrease with the increase in RVSS. The reason is that when the RVSS increases, the centrifugal force would gradually replace the blade's thrust and play a leading role, the axial conveying effect of the blade is weakened, and the efficiency of conveying straw is reduced, thereby reducing the SCR. This point of view is consistent with Wulantuya's point of view [31–33] in the theoretical analysis.

According to the interaction between HHHG and OVM, when HHHG is at 10~30 mm, there is a negative correlation between SCR and HHHG. The smaller the gap between the bottom of the holding hopper and the ground surface, the less straw can pass directly through the gap when the planter is operating, effectively improving the operation effect and the SCR of the planter.

## 4. Optimization and Verification

### 4.1. Parameter Optimization

In order to obtain the optimal parameter combination of the SDARD, the Optimization module in Design Expert software was used to optimize and solve the above-established regression model. Taking the maximum SCR as the goal, the objective function was established as follow:

$$\begin{cases} \text{Max } y(X_1, X_2, X_3) \\ 3 \leq X_1 \leq 5 \\ 80 \leq X_2 \leq 160 \\ 10 \leq X_3 \leq 30 \end{cases} \tag{7}$$

The optimization results to obtain the parameter values of the three experimental factors were as follows: the OVM was 5 km/h, the RVSS was 80 r/min, the HHHG was 10 mm, and the optimal SCR was 91.72%. The best parameters obtained by optimization were used for simulation experiments in EDEM, and the SCR was 90.87%, which was basically consistent with the optimization results.

### 4.2. Field Test Verification

#### 4.2.1. Field Test Conditions

The test was carried out on 26 October 2021 at the Machines Trial Production Base of the Conservation Tillage Research Center of the Ministry of Agriculture and Rural Affairs of China Agricultural University, Baimasi Town, Luolong District, Luoyang City, Henan Province (112°35′27″ E longitude, 34°43′19″ N latitude, 133 m above sea level, air pressure 1006.4 hPa). The test field was a crop field covered with corn stalks that had been chopped and returned to the field, and the previous crop was autumn maize planted in June of the same year. The soil texture of the test site was brown clay, and the average daily temperature during the test period was 9~21 °C without precipitation. Affected by the continuous rainfall in Zhengzhou, the water content of soil and straw in the field was relatively high. The main parameters of the experimental area are shown in Table 5.

**Table 5.** Main parameters of field tests.

| Items | Parameters | Values |
|---|---|---|
| The straw of the field | Average length/mm | 138 |
| | Length range/mm | 65–220 |
| | Average diameter/mm | 4.32 |
| | Diameter range/mm | 1–12 |
| | Covering thickness/mm | 47 |
| | SMQ/(kg·m$^{-2}$) | 1.63 |
| | Moisture content/% | 68.06 |
| 0–100 mm soil layer | Firmness/kPa | 532 |
| | Moisture content/% | 21.46 |
| | Bulk density/(g·cm$^{-3}$) | 1.49 |
| | Temperature/°C | 16.8 |

#### 4.2.2. Test Method and Result Analysis

The Passing Capacity of No-Till Planter

According to the technical requirements in NY/T 1768–2009 "Technical Specifications for Quality Evaluation of No-till planters", the no-till planter would make a round trip in the test area with a length of 60 m at a conventional operating velocity, and the number of shutdowns caused by blockage during the planter operation was observed and recorded.

No blockage or a slight blockage was considered qualified. The SDARD was installed on a 2BMQF-6/12A no-tillage fertilizer planter. The field test site is shown in Figure 10.

The test results show that the no-till planter with the SDARD operated smoothly without blocking. By observing the actual operation effect, compared with the no-till planter without the SDARD, the passing capacity performance of the planter had been improved.

**Figure 10.** Field verification test.

The SCR of the Sowing Strip

On the field test day, a 0.25 $m^2$ square metal frame and an electronic balance were used to randomly sample and weigh the SMQ in the field by the five-point sampling method [36]. After the operation, the straw in the area of 15 × 15 $cm^2$ of sowing strip was weighed, and the quantity of straw remaining on the strip was calculated. The measurement points were the data of 10 consecutive weighing positions with an interval of 1 m in the stable operation area. The 10 sets of data were then averaged. The SCR in the sowing strip after the operation was calculated by Equation (5).

According to the optimized operation parameters, the OVM was set to 5 km/h, the RVSS was 80 r/min, and the HHHG was 10 mm. The cleaning effect of the seed strip after the operation is shown in Figure 11, and the field test results of the SCR of the sowing strip are shown in Table 6.

(a) (b) (c)

**Figure 11.** Straw cleaning effect in seed strip after the operation. (**a**) The overall effect with SDARD after operation; (**b**) seed strip cleaning effect with SDARD; (**c**) seed strip cleaning effect without SDARD.

**Table 6.** Field test results of straw cleaning effect of sowing strip.

| Parameters | The SCR of Seed Strip Installed with SDARD/% | The SCR of Seed Strip Installed without SDARD/% |
|---|---|---|
| Average | 84.49 | 73.16 |
| Variation | 6.73 | 9.48 |

It can be seen from Table 6 that the average SCR of the planter equipped with the SDARD is 84.49%, which is increased by 15.5% compared with the no-till planter without the SDARD, as shown in Figure 11, meeting the agronomic requirements of no-tillage sowing of wheat in an annual double cropping area in China.

## 5. Conclusions

(1) In this paper, an SDARD was designed. During the operation of the no-till planter, the straw was cut and chopped by the high-velocity rotating NAKG and then thrown to the SDM. Through the interaction between the SDM and the RSM, the chopped straw was transported to the non-sowing area, which could effectively reduce the blockage of the furrowing opener, concentrate the straw in the non-sowing area, and realize the cleaning of the seed strip.

(2) The relationship between the material bearing limit of the SDM and the SFA was an important index for evaluating the conveying performance stability and straw discharge effect of the device. Therefore, the mathematical model of the relationship between SFA and SMQ was established by mathematical induction and quotient-remaining method, which provided a theoretical basis for the parameter design of SDM and RSM. On this basis, the specific numerical values and the ranges of the critical structural dimensions and working parameters of the SDM and the RSM were determined.

(3) Using the quadratic rotation orthogonal combination test method—selecting OVM, RVSS, and HHHG as the test factors and SCR as the test index—the discrete element simulation test was carried out, the regression model of SCR was established, and the parameter optimization and field test were carried out. The results show that the significant factors affecting the SCR in the sowing strip from high to low were HHHG, OVM, and RVSS. The optimal combination of operating parameters was that the OVM was 5 km/h, the RVSS was 80 r/min, and the HHHG was 10 mm. Under the optimal parameter combination, the average SCR of the planter equipped with the SDARD was 84.49%, 15.5% higher than that of the no-till planter without the device. The passing capacity of the machines was great, which meets the agronomic requirements of no-tillage sowing of wheat in the annual double cropping area.

The SDARD designed in this paper might provide a new design idea for the anti-blocking and row-sorting device of no-till planters for wheat in the annual double cropping area in northern China. Further, it had the potential to improve the passing capacity of no-till planters and sowing quality to increase grain production in this area. However, this paper mainly conducted parameters design and discrete element simulation experiments for a new mechanism for determining the rational and reasonable values of critical parameters. However, it is essential to assess how much the indicator of reliability, durability, maintainability has changed compared with the control device. Moreover, it is also imperative and meaningful to evaluate the preservation ability of agricultural machinery for theoretical analysis and further exploration. In addition, we will conduct in-depth research on the above issues to improve further the operational reliability of the anti-blocking and row-sorting device.

**Author Contributions:** Conceptualization, Y.L. and C.L.; methodology, Y.L., C.L., S.H. and P.Y.; software, Y.L., S.H. and Z.G.; writing—original draft preparation, Y.L. and X.W.; writing—review and editing, C.L.; H.L., J.H., Q.W. and H.Z.; funding acquisition, H.L. and C.L. All authors have read and agreed to the published version of the manuscript.

**Funding:** This research was funded by the China Agriculture Research System of MOF and MARA (Grant No.CARS-03), Innovative Research Team in University of China (Grant No.IRT13039), and the 2115 Talent Development Program of China Agricultural University and Chinese Universities Scientific Fund (Grant No.2021TC105).

**Institutional Review Board Statement:** Not applicable.

**Informed Consent Statement:** Not applicable.

**Data Availability Statement:** Not applicable.

**Conflicts of Interest:** The authors declare no conflict of interest.

## Nomenclature

| | |
|---|---|
| SDARD | Spiral discharge anti-blocking and row-sorting device |
| NAKG | No-till anti-blocking knife group |
| SDM | Spiral discharging mechanism |
| RSM | Row-sorting of straw mechanism |
| OVM/$v$ | Operating velocity of machines |
| RVSS/$n$ | Rotary velocity of the spiral shaft |
| HHHG/$h$ | Height of the holding hopper from the ground |
| SCR/$\lambda$ | Straw cleaning rate |
| SFA/$Q_s$ | Straw feeding amounts |
| SMQ/$C_S$ | Straw mulching quantity |
| $Q_1$ | Mass of straw on the sowing strip before an operation |
| $Q_2$ | Residual mass of straw on the sowing strip after an operation |
| $Q_{St}$ | Total mass of straw fed into the hopper after a certain period of operation |
| $L_1$ | Total cleaning width of sowing strip |
| $L_2$ | Total width of non-sowing area |
| $D$ | Outer diameter of the spiral blade |
| $S$ | Pitch |
| $z$ | Number of corresponding sowing strips |
| $K$ | Material characteristic coefficient |
| $\psi$ | Filling factor |
| $\rho$ | Material density |
| $C$ | Inclination coefficient |
| $K_1$ | Pitch coefficient |
| $b_G$ | Width of the RSM |
| $h_G$ | Height of the RSM |
| $l_G$ | Length of the side baffle |
| $\theta$ | Included angle between the strip baffle and the side baffle |
| $m_1$ | Mass of straw before the operation |
| $m_2$ | Mass of straw after the operation |

## References

1. He, J.; Li, H.; Chen, H.; Lu, C.; Wang, Q. Research progress of conservation tillage technology and machine. *Trans. Chin. Soc. Agric. Mach.* **2018**, *49*, 1–19. [CrossRef]
2. Cao, X.; Wang, Q.; Li, H.; He, J.; Lu, C. Design and experiment of active rotating collective straw-cleaner. *Trans. Chin. Soc. Agric. Eng.* **2021**, *37*, 26–34. [CrossRef]
3. Cao, X.; Wang, Q.; Li, H.; He, J.; Lu, C. Combined row cleaners research with side cutter and stubble clean disk of corn no-till seeder. *Trans. Chin. Soc. Agric. Mach.* **2021**, *52*, 9. [CrossRef]
4. Wang, Q.; Cao, X.; Wang, C.; Li, H.; He, J.; Lu, C. Research progress of no/minimum tillage corn seeding technology and machine in northeast black soil regions of China. *Trans. Chin. Soc. Agric. Mach.* **2021**, *52*, 1–15.

5. Jia, H.; Zhao, J.; Jiang, X.; Jiang, T.; Wang, Y.; Guo, H. Design and experiment of anti-blocking mechanism for inter-row no-tillage seeder. *Trans. Chin. Soc. Agric. Eng.* **2013**, *29*, 16–25. [CrossRef]
6. Yao, Z.; Gao, H.; Wang, X.; Li, H.; Li, W. Design and experiment on 2BMX-5 no-till wheat-maize seeder. *Trans. Chin. Soc. Agric. Mach.* **2008**, *39*, 64–68.
7. Fallahi, S.; Raoufat, M.H. Row-crop planter attachments in a conservation tillage system: A comparative study. *Soil Tillage Res.* **2008**, *98*, 27–34. [CrossRef]
8. He, J.; Li, H.W.; Wang, Q.J.; Gao, H.W.; Li, W.Y.; Zhang, X.M.; McGiffen, M. The adoption of conservation tillage in China. *Ann. N. Y. Acad. Sci.* **2010**, *1195*, E96–E106. [CrossRef]
9. He, J.; Zhang, Z.; Li, H.; Wang, Q. Development of small/medium size no-till and minimum-till seeders in Asia: A review. *Int. J. Agric. Biol. Eng.* **2014**, *7*, 1–12. [CrossRef]
10. Siemens, M.C.; Wilkins, D.E.; Correa, R.F. Development and evaluation of a residue management wheel for hoe-type no-till drills. *Trans. ASAE* **2003**, *47*, 397–404. [CrossRef]
11. Tourn, M.; Soza, E.; Botta, G.; Mete, A. Direct corn seedling. Effects of residue clearance on implant efficiency. *Span. J. Agric. Res.* **2003**, *1*, 99–103. [CrossRef]
12. Morris, N.L.; Miller, P.C.H.; Orson, J.H.; Froud-Williams, R.J. The adoption of non-inversion tillage systems in the United Kingdom and the agronomic impact on soil, crops and the environment—A review. *Soil Tillage Res.* **2010**, *108*, 1–15. [CrossRef]
13. Celik, A.; Altikat, S.; Way, T.R. Strip tillage width effects on sunflower seed emergence and yield. *Soil Tillage Res.* **2013**, *131*, 20–27. [CrossRef]
14. Li, H.; Liu, H.; Zhou, J.; Wei, G.; Shi, S.; Zhang, X.; Zhang, R.; Zhu, H.; He, T. Development and first results of a no-till pneumatic seeder for maize precise sowing in Huang-Huai-Hai plain of China. *Agriculture* **2021**, *11*, 1023. [CrossRef]
15. Wang, C.; Li, H.; He, J.; Wang, Q.; Lu, C.; Yang, H. Optimization design of a pneumatic wheat-shooting device based on numerical simulation and field test in rice–wheat rotation areas. *Agriculture* **2022**, *12*, 56. [CrossRef]
16. Zhao, H. *Study on Driven Seedbed-Cleaning and Anti-Blocking Device of Residue Inter-Row Side-Throwing for Minimum till Wheat Seeding*; China Agricultural University: Beijing, China, 2019.
17. Zheng, Z. *Study on Corn Straw Pickup-Chopping and Ditch-Burying Returning Field Machine*; China Agricultural University: Beijing, China, 2017.
18. Wang, Q.; He, J.; Li, H.; Lu, C.; Rasaily, R.G.; Su, Y. Design and experiment on furrowing and anti-blocking unit for no-till planter. *Trans. Chin. Soc. Agric. Eng.* **2012**, *28*, 27–31. [CrossRef]
19. Lin, J.; Li, B.; Li, H. Design and experiment of Archimedes spiral type stubble breaking ditching device and stubble breaking anti blocking device. *Trans. Chin. Soc. Agric. Eng.* **2015**, *31*, 10–19. [CrossRef]
20. Lin, J.; Li, B.; Li, B.; Niu, J.; Qian, W. Parameter optimization and experiment on archimedes spiral type of gap cutting disc. *Trans. Chin. Soc. Agric. Mach.* **2014**, *45*, 118–124. [CrossRef]
21. Lu, C. *Study on Anti-Blocking Technology and Device of Rotary Cutting with Slide Plate Pressing Straw for No-Till Planter*; China Agricultural University: Beijing, China, 2014.
22. Lu, C.; Zhao, C.; Meng, Z.; Wang, X.; Wu, G.; Gao, N. Straw friction characteristic based on rotary cutting anti-blocking device with slide plate pressing straw. *Trans. Chin. Soc. Agric. Eng.* **2016**, *32*, 83–89. [CrossRef]
23. Lu, C.; Li, H.; He, J.; Zhu, H.; Xu, D. Floated support anti-blocking device of wheat no-till seeder. *Trans. Chin. Soc. Agric. Eng.* **2013**, *44*, 52–55, 153. [CrossRef]
24. Lu, C.; He, J.; Li, H.; Wang, Q.; Zhang, X.; Liu, J. Finite element analysis and experiment on anti-blocking device based on support cutting. *Trans. Chin. Soc. Agric. Mach.* **2013**, *44*, 61–66. [CrossRef]
25. Yuan, P.; Li, H.; Jiang, G.; He, J.; Lu, C.; Huang, S. Design and experiment of straw cleaning device for wide narrow maize no-tillage sowing strip in drip irrigation area. *Trans. Chin. Soc. Agric. Mach.* **2021**, *52*, 43–52. [CrossRef]
26. Yu, C.; Wang, Q.; Li, H.; He, J.; Lu, C. Design and experiment of spiral-split sowing strip cleaning device. *Trans. Chin. Soc. Agric. Eng.* **2020**, *51*, 212–219.
27. Zhou, H. *The Application of Quadrat Method in Parameters of Forest Spatial Structure Investigation*; Chinese Academy of Forestry: Beijing, China, 2009. [CrossRef]
28. Darnell, W.H. Solid Conveying in Extruders. *SPE J.* **1956**, *12*, 20–29.
29. Yuan, Q.; Xu, L.; Niu, C.; Ma, S.; Yan, C.; Zhao, S.; Liu, F.; Wang, K. Development of soil-fertilizer mixing layered backfiller for organic fertilizer deep applicator in orchard. *Trans. Chin. Soc. Agric. Eng.* **2021**, *37*, 11–19. [CrossRef]
30. "Transportation Machinery Design and Selection Manual" Committee. *Transport Machinery Design and Selection Manual, the Next Volume*; Chemical Industry Press: Beijing, China, 1999.
31. Wang, C.; Zhao, F.; Wang, X. Experiment and optimization of screw conveyor parameters for rubbing and breaking corn straw. *J. China Agric. Univ.* **2019**, *24*, 115–122. [CrossRef]
32. Wang, C.; Qi, S.; Yan, J.; Wang, J. Test and analysis of performance of screw conveyor for rubbing and breaking corn straw. *Trans. Chin. Soc. Agric. Eng.* **2015**, *31*, 51–59. [CrossRef]
33. *Study on the Mechanism of Screw Conveying Theory for Rubbing and Breaking Corn Straw*; Inner Mongolia Agricultural University: Hohhot, China, 2016.

34. Wang, X.; Hu, H.; Wang, Q.; Li, H.; He, J.; Chen, W. Calibration method of soil contact characteristic parameters based on DEM theory. *Trans. Chin. Soc. Agric. Eng.* **2017**, *48*, 78–85. [CrossRef]
35. Fang, H.; Ji, C.; Ahmed, A.T.; Zhang, Q.; Guo, J. Simulation analysis of straw movement in straw soil rotary blade system. *Trans. Chin. Soc. Agric. Mach.* **2016**, *47*, 60–67. [CrossRef]
36. Ni, H.; Lu, F.; Luo, X.; Tian, H.; Wang, J.; Guan, Y.; Chen, S.; Luo, X.; Zeng, E. Assessment of sampling designs to measure riverine fluxes from the pearl river delta, china to the south china sea. *Environ. Monit. Assess.* **2008**, *143*, 291–301. [CrossRef]

*agriculture*

MDPI

*Article*

# Simulation and Experiment of Spiral Soil Separation Mechanism of Compound Planter Based on Discrete Element Method (DEM)

**Lianjie Han [1], Wei Yuan [1], Jinjin Yu [1], Jiajun Jin [1], Dongshen Xie [1], Xiaobo Xi [1,2,*], Yifu Zhang [1] and Ruihong Zhang [1,2]**

[1] School of Mechanical Engineering, Yangzhou University, Yangzhou 225127, China; mx120210537@stu.yzu.edu.cn (L.H.); mx120210538@yzu.edu.cn (W.Y.); 191203126@stu.yzu.edu.cn (J.Y.); 191205109@stu.yzu.edu.cn (J.J.); 191205125@stu.yzu.edu.cn (D.X.); zyfu@yzu.edu.cn (Y.Z.); 002674@yzu.edu.cn (R.Z.)

[2] Jiangsu Engineering Center for Modern Agricultural Machinery and Agronomy Technology, No. 196, Huayangxilu, Hanjiang District, Yangzhou 225127, China

* Correspondence: xbxi@yzu.edu.cn

**Abstract:** In order to solve the problems of blocking the drainage ditch and reducing the soil flatness caused by soil accumulation when using compound planter with plowshare to ditch, a spiral soil separation mechanism (SSSM) is designed. The SSSM is analyzed. In order to obtain the optimal parameters of the SSSM, based on the discrete element method, the multifactor test is carried out with the embedded depth, pitch, and rotation speed of the spiral blade as the test factors and the soil separation distance and uniformity as the evaluation index. The optimal parameters are the embedded depth 49 mm, pitch 331 mm, and rotation speed of the spiral blade 318 r min$^{-1}$. The field experiment is carried out with these parameters, with soil separation distance 900 mm and standard deviation of soil height 7.8 mm, which is consistent with the simulation results. No blockage of drainage ditch was found, which shows that this device can effectively solve the problem. This study can provide a reference for the design of soil separation equipment using spiral soil separation device.

**Keywords:** compound planter; ditching; soil separation spiral; discrete element method; parameter optimization

**Citation:** Han, L.; Yuan, W.; Yu, J.; Jin, J.; Xie, D.; Xi, X.; Zhang, Y.; Zhang, R. Simulation and Experiment of Spiral Soil Separation Mechanism of Compound Planter Based on Discrete Element Method (DEM). *Agriculture* **2022**, *12*, 511. https://doi.org/10.3390/agriculture12040511

Academic Editors: Mustafa Ucgul and Chung-Liang Chang

Received: 1 March 2022
Accepted: 31 March 2022
Published: 4 April 2022



## 1. Introduction

Wheat is one of the main food crops in the world. Nearly one-third of the world's population takes wheat as the main food [1,2]. China is the largest wheat producer in the world. The sowing of wheat generally goes through the processes of tillage and land preparation, fertilization, sowing, soil covering and pressing, and digging drainage ditches. There are many processes, and the operation of a single machine and tool affects the production efficiency. Therefore, it has become a trend to use the compound planter that can complete multiple processes at one time [3–7]. Soil moisture content has a great impact on the yield of wheat. Too high soil moisture content is not suitable for the growth of wheat. Therefore, it is necessary to dig a drainage ditch in the field to drain the accumulated water in the field in time [8–11]. At present, there are plowshare type, rotary disc type, chain knife type, and other ditching machines. The rotary disc ditching machine can dig the drainage ditch through the milling principle, and can evenly scatter the milled soil to avoid soil accumulation. However, the structure of the machine is complex, which is difficult to apply in the compound planter. The plowshare ditching machine has simple structure and is easy to be integrated with the compound planter. However, there are some problems such as soil accumulation after ditching, resulting in soil falling back, blocking drainage ditches and reducing soil flatness [12–14]. Qin et al. [15] designed double disk opener mechanism for harvesting, ditching, and burying grass, which can complete the ditching operation in the process of rice and wheat harvest. Bao et al. [16] designed a plow-rotary style ditching and ridging device. The device adopts a plowshare ditcher, and uses a bulldozing plate and a soil dividing plate to lift and push the soil in the ditch to the ridge area, but this structure is not suitable for the compound planter of wheat.

Discrete element method is a computer numerical simulation method based on the assumption of discontinuity. The use of discrete element method can improve the design efficiency and reduce the design cost. Therefore, more and more researchers use discrete element method to design agricultural equipment [17–20]. Zhou et al. [21] used EDEM software to study the optimal clearance of optimum gap of the screw extraction device, and the soil covering effect of 3~4 mm clearance is better. Owen et al. [22] used the discrete element method to numerically simulate the periodic boundary conditions of single pitch conveyor, and predicted the performance of screw conveyor according to the changes of particle speed, mass flow, and energy consumption. Pezo et al. [23] studied the effects of screw structure, particle size, and conveyor length on the mixing performance of mixed materials transported by screw conveyor through discrete element simulation. However, the above are the research of closed screw conveyor, while the research on open screw conveyor with soil as the transportation material is rare.

Overall, aiming to solve problems of soil accumulation leading to soil falling down, blocking the drainage ditch and reducing soil flatness in the drainage ditch of compound planter, the spiral soil separation mechanism (SSSM) is designed, the working mechanism is analyzed, and discrete element simulation model is established. Single factor and response surface tests are designed, the optimized parameters of the device are obtained, and field tests are carried out. The simulation results are compared with the field test results to verify the reliability of the model. This study is expected to provide a guide line for the design of soil separation equipment using spiral soil separation device.

## 2. Materials and Methods

### 2.1. General Structure

As shown in Figure 1, the general structure of the compound planter is composed of biaxial rotary tillage mechanism, press grooving roller, seeder, ditcher, spiral soil separation mechanism (SSSM), and other components. The compound planter relates to the tractor through the three-point linkage, which can complete the processes of tillage and land preparation, sowing, soil covering and pressing, and ditching at one time.

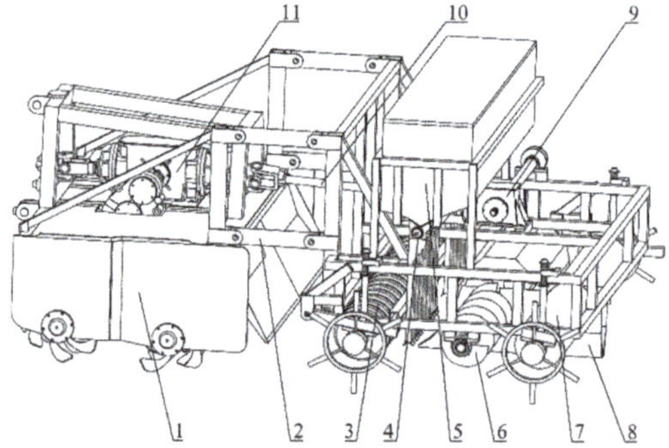


**Figure 1.** General structure of compound planter. 1. Biaxial rotary tillage mechanism. 2. Connecting rod. 3. press grooving roller. 4. Seeder. 5. Seed box. 6. Spiral soil separating mechanism. 7. Repress roller. 8. Ditcher. 9. Steering box. 10. Transmission shaft. 11. Intermediate transmission box.

The SSSM is composed of two spiral shafts symmetrically installed on the bearing seat of the frame, and the spiral blades on the two spiral shafts rotate in the opposite direction.

The SSSM is an active moving part, its power comes from the intermediate transmission box of biaxial rotary tillage mechanism, and the power transmitted through universal joint and commutator. The ditcher is installed in the middle of the spiral shafts and firmly connected to the frame.

The working process of the SSSM is shown in Figure 2. As the ditcher moves forward, the soil in the middle is forced to move to both sides of the ditcher, and banded soil accumulate on both sides of the drainage ditch under shear and lifting action of the ditcher. When the SSSM passes through the banded soil, the soil is pushed to both sides by the conveying action of rotating spiral blades. This function can solve the problem of drainage ditch blockage by preventing the soil from falling back into the drainage ditch, and obtaining higher soil flatness.

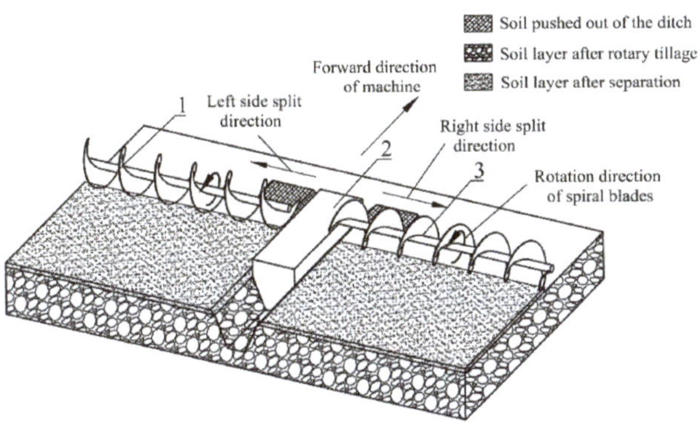


**Figure 2.** Working process of SSSM. 1. Left spiral blades. 2. Ditching. 3. Right spiral blades.

In order to analyze the soil separating mechanization of spiral blade, two soil particles are taken below and in front of the spiral blades respectively (q1 and q2) for mechanical analysis and kinematic analysis. The analysis results are shown in Figure 3.

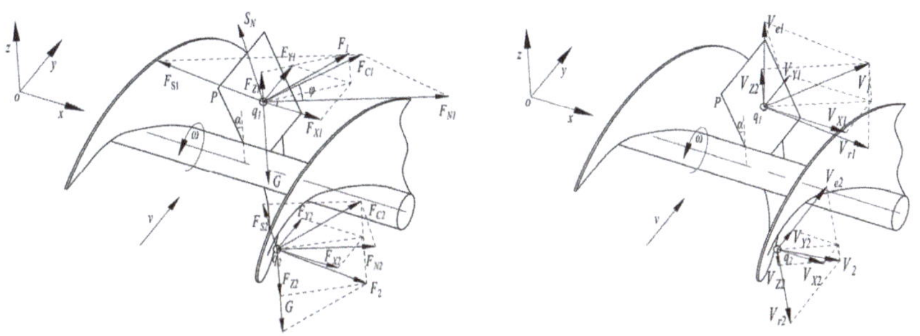

**Figure 3.** Mechanical and kinematic analysis of soil particles. where, $v$ is the forward speed of the compound planter, m s$^{-1}$; $\omega$ is the rotation speed of spiral blade, $r$ min$^{-1}$; $\alpha$ is the spiral angle, rad; $\varphi$ is the friction angle between soil particles and blades, rad; P is the tangent plane of the contact point between the spiral blade and soil particles; $F_S$ is the friction force of soil particles by helical blades, N; $F_N$ is the supporting force of soil particles by spiral blades, N; $F_C$ is the resultant force of leaves on soil particles, N; $F$ is the resultant force on soil particles, N; $F_X$, $F_Y$ and $F_Z$ are the component force of soil particles in the transverse, forward and vertical directions, N; G is the gravity of soil particles, N; $S_N$ is the support of soil particles by the underlying soil, N; *Ve* is the implicated movement of soil particles, m s$^{-1}$; *Vr* is the implicated movement of soil particles, m s$^{-1}$; $V$ is the absolute movement of soil particles, m s$^{-1}$; $V_X$, $V_Y$ and $V_Z$ are the component velocity of soil particles in transverse, forward and vertical directions, m s$^{-1}$.

The soil particle q1 is affected by four forces from different directions, including the support force of the soil particle below ($S_{N1}$), the support force and friction force of the blade ($F_{N1}$, $F_{S1}$) and its gravity ($G$). Under the action of the resultant force of above four forces ($F_1$), q1 moves centrifugally at a speed of $V_1$, and gradually moves away from the central axis of the helical blade.

When q1 is separated from the spiral blade, due to loss of resultant force of the blade to the soil particle ($F_{c1}$) and $S_{N1}$, q1 finally falls to the ground under the action of $G$.

With the rotation and forward movement of the spiral blades, the soil particle q2 is affected by three forces from different directions, including the support force and friction force of the blade ($F_{N2}$, $F_{S2}$), and its gravity ($G$). Under the action of the resultant force of the above three forces ($F_2$), q2 moves centrifugally at a speed of $V_2$. $V_{Z2}$ is the partial velocity of $V_Z$ in the vertical direction, and its velocity direction is downward. When q2 is separated from the spiral blade, it falls to the ground under the action of $G$.

In order to meet the requirements of continuous transportation of soil, the component speed direction of the soil in the $y$ direction and the component force ($F_Y$) direction of the resultant force must be the same as the forward speed direction, as shown in Formula (1). This condition limits the relationship between the soil separation direction, the forward direction of the machine, the rotation direction of the spiral blade, and the helical direction of the spiral blade. As shown in Formula (1).

$$\frac{\vec{v}}{\left|\vec{v}\right|} = \frac{\vec{F_Y}}{\left|\vec{F_Y}\right|} = \frac{\vec{V_Y}}{\left|\vec{V_Y}\right|} \quad (1)$$

The soil separation effect of spiral blade is affected by many factors, including pitch, rotation speed, forward speed, radius, and the embedded depth. While the working speed of the compound planter is relatively constant, which is 3 km $h^{-1}$, and the radius of the spiral blade is not easy to change due to the influence of the structure of the machine. Therefore, this paper only studied the influence of the three factors of the embedded depth, pitch, and rotation speed of the spiral blade on the soil separation effect.

### 2.2. Discrete Element Simulation

In this paper, the operation process of SSSM were simulated by EDEM 2020 (DEM Solutions Ltd., Edinburgh, Scotland, UK) to analyze its soil-separating effect. Due to its symmetrical structure, the discrete element simulation can be carried out only on one side of the SSSM. The simulation model is shown in Figure 4. Similar to many researchers, spherical particles with a radius of 5 mm were used as soil particles in this study [17,19,20]. The soil model was divided into two layers (a, b), where layer a indicated the surface soil with size of 2500 mm × 1600 mm × 40 mm, and layer b indicated the banded soil accumulate on both sides of the drainage ditch with size of 2500 mm × 200 mm × 60 mm. The Hertz–Mindlin with JKR model was selected as the bonding model for the soil particles, and Hertz–Mindlin (no slip) was selected as the contact model between soil and spiral blade. The SSSM model was imported from SolidWorks, the length was 1200 mm, and the diameter of spiral blade was 240 mm. The soil was analyzed in the field test area (33°22′21.5″ N, 118°15′59.6″ E; clay) for the angle of the repose experiment (Figure 5) to determine the relevant parameters of the soil model [24–32], as shown in Table 1.

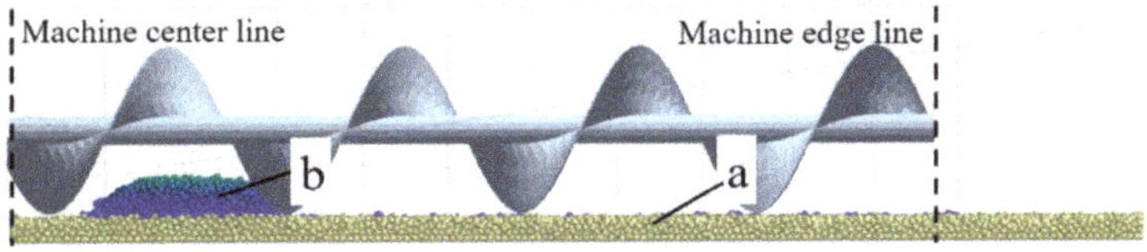


**Figure 4.** Discrete element simulation model diagram.

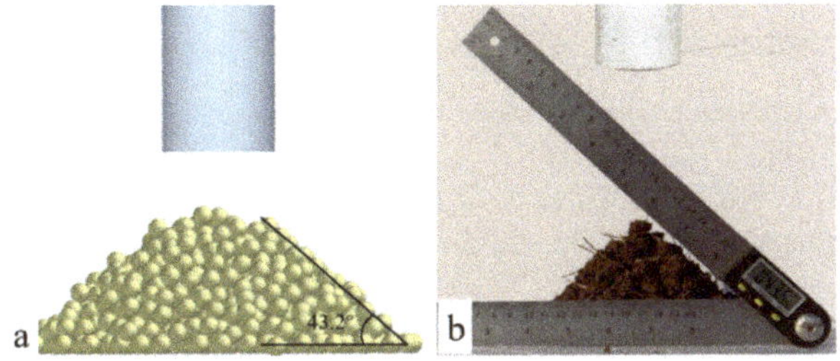


**Figure 5.** (**a**,**b**) Angle of repose experiment of soil.

**Table 1.** Soil parameters for simulation.

| Parameters | Values |
|---|---|
| Soil Poisson's ratio | 0.38 |
| Soil density (kg m$^{-3}$) | 2680 |
| Soil shear modulus | $1.2 \times 10^{6}$ |
| Recovery coefficient between soil particles | 0.31 |
| Static friction factor between soil particles | 0.77 |
| Dynamic friction coefficient between soil particles | 0.08 |
| Steel Poisson's ratio | 0.3 |
| Steel density (kg m$^{-3}$) | 7850 |
| Shear modulus of steel | $7 \times 10^{10}$ |
| Recovery coefficient between soil particles and steel | 0.3 |
| Static friction factor between soil particles and steel | 0.36 |
| Dynamic friction coefficient between soil particles and steel | 0.08 |
| Surface energy (J m$^{-3}$) | 6.1 |

In order to analyze the influence of the embedded depth, pitch, and rotation speed of the spiral blade on the soil separation quality, three groups of single factor tests were designed to analyze the influence of each parameter on the soil separation distance and uniformity. The test parameters are shown in Table 2.

**Table 2.** Single factor experimental design.

| Group | Test Factor | | |
|---|---|---|---|
| | Embedded Depth (mm) | Pitch (mm) | Spiral Rotation Speed (r min$^{-1}$) |
| 1 | 42 | 300 | 130, 200, 270, 340, 410 |
| 2 | 42 | 150, 200, 250, 300, 350 | 270 |
| 3 | 26, 34, 42, 50, 58 | 300 | 270 |

In the single factor test, the influence trend of three parameters on soil separation distance and uniformity can be obtained. In order to obtain the optimal parameters of target soil separation distance and uniformity, a central composite design was carried out for embedded depth, pitch, and rotation speed by using Design-Expert software (version 12). The level of test factors is shown in Table 3.

**Table 3.** Experimental factors codes.

| Test Factor | Symbol | Test Level | | |
|---|---|---|---|---|
| | | −1 | 0 | 1 |
| Embedded depth/(mm) | A | 26 | 42 | 58 |
| Pitch/(mm) | B | 150 | 250 | 350 |
| Spiral rotation speed/(r min$^{-1}$) | C | 130 | 270 | 410 |

## 2.3. *Soil Separation Quality Evaluation Index*

### 2.3.1. Soil Separation Distance

The soil separation distance $D_W$ is used to describe the maximum soil separation distance of the SSSM. As shown in Figure 6, the edge of the strip-shaped soil is the starting point, the outer edge of the soil after the operation of the soil divider is the end point, and the distance between the two edges is $D_W$. Mark the surface soil particles and measure the distance.

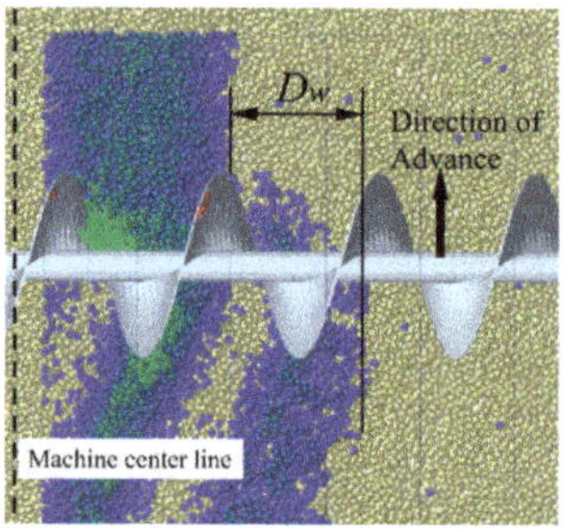


**Figure 6.** Evaluation index of soil separation distance.

### 2.3.2. Soil Uniformity

Soil uniformity $E$ is used to describe the uniformity of soil distribution within the width of machines. The larger the value of $E$, the higher the uniformity of soil distribution. As shown in Figure 7, 10 grids were established in the stable soil division section with size of 1000 mm × 120 mm × 300 mm. The calculation formula of $E$ is as follows:

$$\begin{cases} \overline{n} = \frac{\sum_{i=1}^{10} n_i}{10} \\ U = \sqrt{\frac{\sum_{i=1}^{10} (n_i - \overline{n})^2}{10}} \\ E = 2000 - U \end{cases} \quad (2)$$

where $n_i$ ($I$ = 1,2,···,10) is the number of soil particles in each grid derived from EDEM, $U$ is the overall calculation standard deviation of $n_i$, $\overline{n}$ is the average value of the number of soil particles in ten grids.

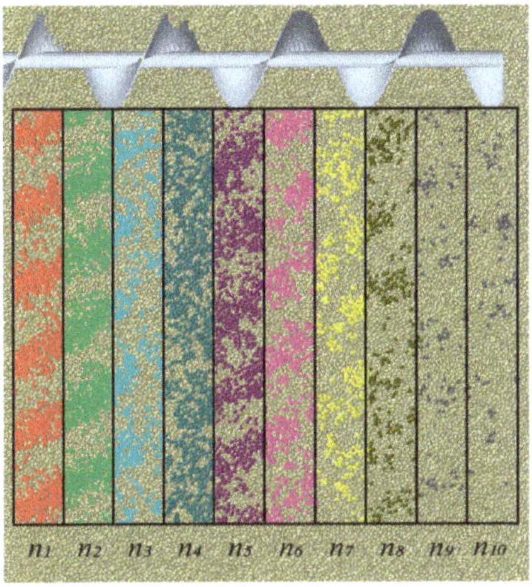


**Figure 7.** Evaluation index of soil uniformity.

## 2.4. Field Experiment

On 19 November 2021, field experiment was carried out in Sihong County, Suqian City, Jiangsu Province (33°22′21.5″ N, 118°15′59.6″ E). The soil type in the experimental area was clay with a moisture content of 21.38%. The test equipment is shown in Figure 8a. The embedded depth, pitch, and rotation speed of spiral blade were 49 mm, 331 mm, and 318 r min$^{-1}$ respectively (the optimal parameters obtained by the regression model).

**Figure 8.** Diagram of test equipment and test site. (**a**) test equipment. (**b**) test site.

In practical operation, the soil uniformity was reflected in the difference of field soil accumulation per unit distance in the direction of operation width. Due to the existence of repress roller, it was difficult to measure the soil accumulation after the operation of compound planter, and there were large errors. Considering that the soil type in the test area was clay and its compressibility was low, this characteristic will cause the area with large soil accumulation to be unable to maintain the same height as other areas after compaction, so the height difference of the soil after compaction was used to reflect the uniformity of soil division. The horizontal line was determined by a laser level (accuracy $\pm$ 1.5°), as shown in Figure 8b. The distance between the horizontal line and the surface soil was measured by a steel ruler, and the result was converted into the soil height value $H$. The starting point of the measurement is the edge of the drainage ditch, the measurement range is 1.5 m, the measurement interval is 10 cm, and the distance between the measurement point and the edge of the drainage ditch is $D_C$. The average value was calculated after ten groups of data were randomly measured in the forward direction of the compound planter.

## 3. Results and Analysis

As shown in Figure 9, the soil particles move outward under the action of spiral blades, and the soil particles at area "a" and "b" were dispersed within the operation width by pushing and scattering of the blades respectively.

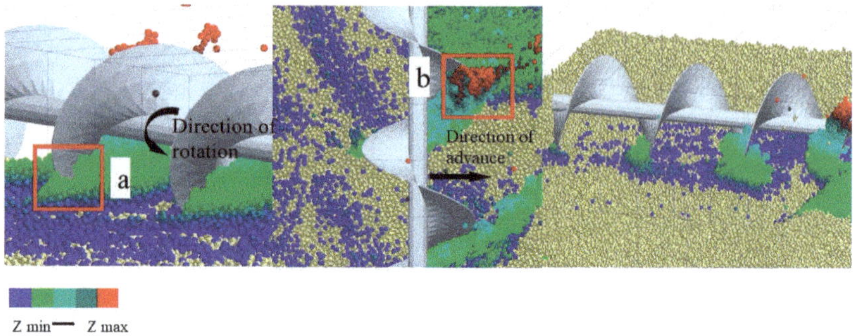


**Figure 9.** Soil separating process. (**a**) soil particles below the spiral blades. (**b**) soil particles in front of the spiral blades.

### 3.1. Single Factor Experiment

(1) Effects of rotation speed on soil separation distance and uniformity

Figure 10 shows the relationship curves between rotation speed and $D_W$, $E$ when the embedded depth was 42 mm and the pitch was 300 mm. When the rotation speed increased from 130 r min$^{-1}$ to 410 r min$^{-1}$, the soil separation distance increased from 254 mm to 956 mm, which increased by 702 mm. When the rotation speed was about 380 r min$^{-1}$, the optimal soil separation distance was 900 mm, and the rotation speed was approximately linear with the separation distance. The soil uniformity increased by 765 from 710 to 1475. When the rotation speed increased from 130 r min$^{-1}$ to 270 r min$^{-1}$, the soil uniformity increased by 590. When the rotation speed increased from 270 r min$^{-1}$ to 410 r min$^{-1}$, the soil uniformity increased by 170, indicating that with the increase of rotation speed, the improvement of rotation speed on soil uniformity gradually weakened.

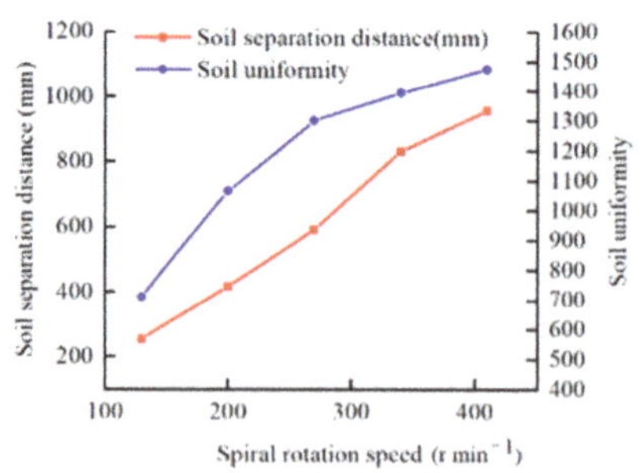


**Figure 10.** Relationship curves between rotation speed and $D_W$, $E$.

(2) Effects of pitch on soil separation distance and uniformity

Figure 11 shows the relationship curves between pitch and $D_W$, $E$ when the soil depth was 42 mm and the rotation speed was 270 r min$^{-1}$. When the pitch increased from 150 mm to 350 mm, the soil separation distance increased from 204 mm to 542 mm, which increased by 338 mm. With the increased of pitch, the effect of pitch on the improvement of soil separation distance gradually weakened. This was because the helix angle of the spiral blade increased with the increase of the pitch, which made $F_{X1}$ and $F_{X2}$ decrease, thus reducing the push ability of the spiral blade to the *x* direction of the soil particles. The uniformity of soil separation increased from 815 to 1323, which increased by 508, similar to the effect of rotation speed on soil uniformity. With the increased of pitch, the improvement of pitch on soil uniformity gradually weakened. The above results showed that increasing the pitch can increase the soil separation distance and soil uniformity, but with the increased of pitch, the increase rate of soil separation distance and soil uniformity became slow.

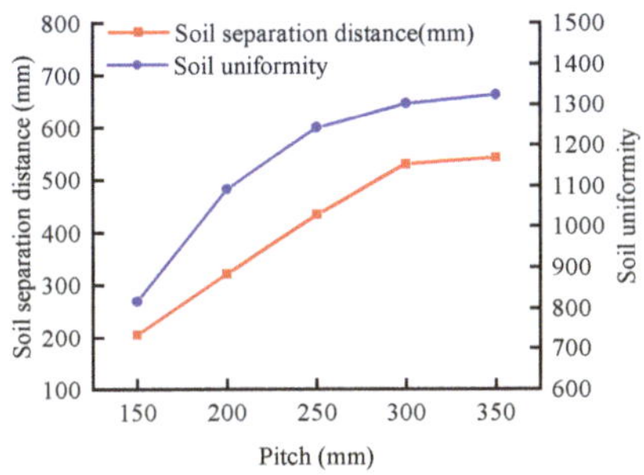


**Figure 11.** Relationship curves between pitch and $D_W$, $E$.

(3) Effects of soil depth on soil separation distance and uniformity

Figure 12 shows the relationship curve of the embedded depth of the blade with the soil separation distance and the soil uniformity when the pitch was 300 mm and the rotation speed was 270 r min$^{-1}$. When the embedded depth increased from 26 mm to 58 mm, the soil separation distance increased from 346 mm to 1350 mm, which increased by 1004 mm. When the embedded depth was about 48 mm, the optimal soil separation distance was 900 mm. With the increased of soil depth, the increase rate of soil separation distance gradually increased. This was because with the increase of the embedded depth, the volume of soil transported by spiral blades increased, and the contact time between soil particles and spiral blades increased, which increased the action time of $F_{X1}$ and $F_{X2}$. More contact time improved the pushing ability of the spiral blade to the soil particles in the x direction, and some soil particles were scattered outside the working width. The soil uniformity increased from 599 to 1915, which increased by 1316, and the embedded depth was approximately linear with the soil uniformity. The above results showed that increasing the embedded depth can effectively improve soil separation distance and uniformity.

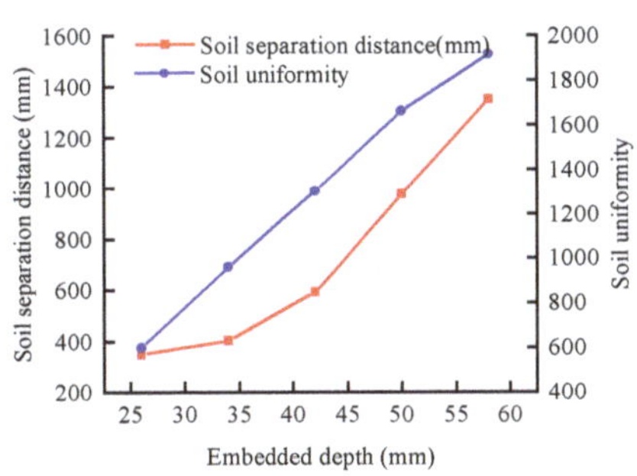


**Figure 12.** Relationship curves between embedded depth and $D_W$, $E$.

### 3.2. Multifactor Experiment

The experimental design and results of response surface analysis method are shown in Table 4. The quadratic polynomial regression model of soil separation distance and soil separation uniformity was obtained by Design-Expert software.

$$D_W = 513.8 + 333A + 165.88B + 229.38C + 83.5AB + 90AC + 21.25BC + 172.72A^2 - 60.03B^2 - 22.02C^2$$
$$E = 1237.4 + 464.5A + 212.63B + 342.13C + 204.25AB + 94.75AC - 20.5BC - 113.7A^2 - 111.45B^2 - 145.45C^2 \quad (3)$$

**Table 4.** Response surface analysis test design and results.

| Test Number | Factor | | | Response Value | |
|---|---|---|---|---|---|
| | Embedded Depth | Pitch | Spiral Rotary Speed | Soil Separation Distance (mm) | Soil Uniformity |
| 1 | −1 | −1 | 0 | 170 | 494 |
| 2 | 0 | −1 | −1 | 132 | 442 |
| 3 | −1 | 1 | 0 | 428 | 597 |
| 4 | 0 | 0 | 0 | 520 | 1243 |
| 5 | 1 | −1 | 0 | 658 | 1019 |
| 6 | 0 | 0 | 0 | 443 | 1324 |
| 7 | 0 | 0 | 0 | 468 | 1152 |
| 8 | 0 | −1 | 1 | 493 | 1180 |
| 9 | 0 | 1 | 1 | 774 | 1478 |
| 10 | −1 | 0 | 1 | 493 | 757 |
| 11 | 0 | 0 | 0 | 562 | 1294 |
| 12 | −1 | 0 | −1 | 159 | 275 |
| 13 | 1 | 0 | −1 | 656 | 1010 |
| 14 | 0 | 0 | 0 | 576 | 1174 |
| 15 | 0 | 1 | −1 | 328 | 822 |
| 16 | 1 | 0 | 1 | 1350 | 1871 |
| 17 | 1 | 1 | 0 | 1250 | 1939 |

The variance analysis of the response values $D_W$ and $E$ is shown in Tables 5 and 6. The $p$-values of the two models are less than 0.01, indicating that the selected factors in the equation are extremely significant. The misfits are greater than 0.05, which are not significant.

**Table 5.** Variance analysis of quadratic term model of soil separation distance.

| Error Source | Sum of Squares ($\times 10^5$) | Freedom | Mean Square ($\times 10^5$) | F Value | $p$ Value |
|---|---|---|---|---|---|
| Model | 17.28 | 9 | 1.92 | 36.24 | <0.0001 *** |
| A | 8.87 | 1 | 8.87 | 167.47 | <0.0001 *** |
| B | 2.20 | 1 | 2.20 | 41.55 | 0.0004 *** |
| C | 4.21 | 1 | 4.21 | 79.46 | <0.0001 *** |
| AB | 0.28 | 1 | 0.28 | 5.27 | 0.0554 |
| C | 0.32 | 1 | 0.32 | 6.12 | 0.0426 * |
| BC | 0.02 | 1 | 0.02 | 0.34 | 0.5776 |
| $A^2$ | 1.26 | 1 | 1.26 | 23.71 | 0.0018 ** |
| $B^2$ | 0.15 | 1 | 0.15 | 2.86 | 0.1344 |
| $C^2$ | 0.02 | 1 | 0.02 | 0.39 | 0.5543 |
| Residual | 0.37 | 7 | 0.05 | | |
| Lack of Fit | 0.24 | 3 | 0.08 | 2.37 | 0.2113 |
| Pure Error | 0.13 | 4 | 0.03 | | |
| Total | 17.65 | 16 | | | |
| $R^2$ = 0.979, Adjusted $R^2$ = 0.9520, Predicted $R^2$ = 0.7730, Adeq Precision = 22.5568 | | | | | |

Note: *, ** and *** indicate significant difference at $p < 0.05$, $p < 0.01$ and $p < 0.001$ levels, respectively.

**Table 6.** Variance analysis of quadratic term model of soil uniformity.

| Error Source | Sum of Squares ($\times 10^5$) | Freedom | Mean Square ($\times 10^5$) | F Value | *p* Value |
|---|---|---|---|---|---|
| Model | 34.47 | 9 | 3.83 | 71.89 | <0.0001 *** |
| A | 17.26 | 1 | 17.26 | 324.01 | <0.0001 *** |
| B | 3.62 | 1 | 3.62 | 67.89 | <0.0001 *** |
| C | 9.36 | 1 | 9.36 | 175.77 | <0.0001 *** |
| AB | 1.67 | 1 | 1.67 | 31.32 | 0.0008 *** |
| C | 0.36 | 1 | 0.36 | 6.74 | 0.0356 |
| BC | 0.02 | 1 | 0.02 | 0.32 | 0.5918 |
| $A^2$ | 0.54 | 1 | 0.54 | 10.22 | 0.0151 * |
| $B^2$ | 0.52 | 1 | 0.52 | 9.82 | 0.0165 * |
| $C^2$ | 0.89 | 1 | 0.89 | 16.72 | 0.0046 ** |
| Residual | 0.37 | 7 | 0.05 | | |
| Lack of Fit | 0.15 | 3 | 0.05 | 0.92 | 0.5069 |
| Pure Error | 0.22 | 4 | 0.06 | | |
| Total | 34.84 | 16 | | | |
| $R^2 = 0.9893$, Adjusted $R^2 = 0.9755$, Predicted $R^2 = 0.9201$, Adeq Precision = 29.0687 | | | | | |

Note: *, ** and *** indicate significant difference at $p < 0.05$, $p < 0.01$ and $p < 0.001$ levels, respectively.

Figure 13 shows the response surface analysis of the relationship between test factors and separation distance. The influence of the embedded depth and rotation speed on soil separation distance was greater than that of pitch, which was consistent with the *p*-values of the embedded depth ($p < 0.0001$), the rotation speed ($p < 0.0001$), and the pitch ($p = 0.0004$) in Table 5. With the increase of the embedded depth, rotation speed, and pitch, the soil separation distance increased. The greater the embedded depth and rotation speed, the greater the soil separation distance.

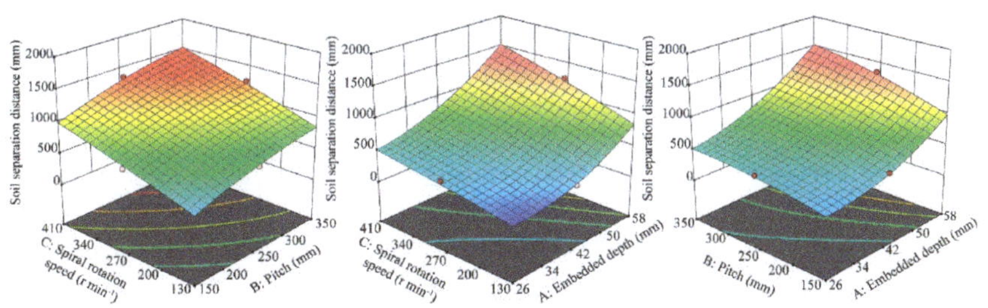


**Figure 13.** Response surface analysis of the relationship between test factors and separation distance.

Figure 14 shows the response surface analysis of the relationship between test factors and soil uniformity. The embedded depth, pitch, and rotation speed had significant indigenous effects on soil uniformity. With the increase of the embedded depth, rotation speed, and pitch, soil uniformity increased significantly. The greater the embedded depth, the greater the increase of soil uniformity.

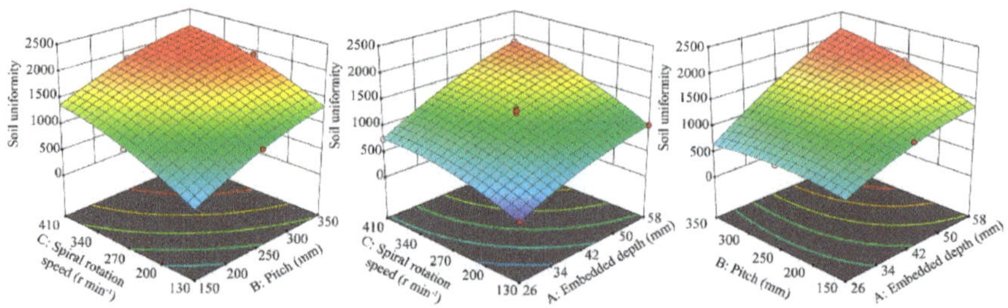


**Figure 14.** Response surface analysis of the relationship between test factors and soil uniformity.

When the soil separation distance of the spiral blade is 900 mm, the soil accumulated on both sides of the drainage ditch covers the operation width of the SSSM, and this distance is the best soil separation distance. The response surface analysis shows that the requirement of soil separation distance could not be the condition of the best soil separation uniformity. With the increase of various test factors, the soil increases, but the excessive soil separation uniformity makes the soil separation distance much larger than the target soil separation distance. Therefore, it was necessary to improve the soil separation uniformity as far as possible under the condition of soil separation distance 900 mm.

The optimization equation was obtained by the Design-Expert software multiobjective optimization method with $D_W$ and $E$ as the optimization objective function 4. The optimal parameters were obtained. Embedded depth was 49 mm, pitch was 331 mm, and rotation speed was 318 r min$^{-1}$. The discrete element simulation model was established with the optimal parameters, the soil separation distance was 917 mm, the soil separation uniformity was 1728. Figure 15 shows the simulation results.

$$\begin{cases} 26 < \mathrm{A} < 58 \\ 150 < \mathrm{B} < 350 \\ 130 < \mathrm{C} < 410 \\ D_{\mathrm{w}} = 900 \\ \max E \end{cases} \quad (4)$$

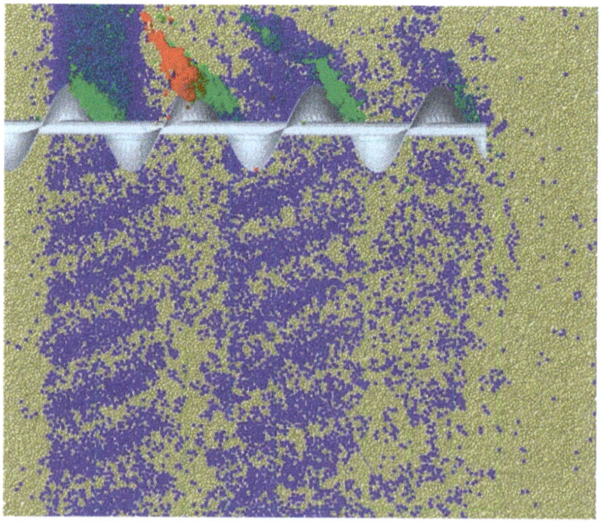

**Figure 15.** Simulation results after optimizing parameters of spiral soil separating device.

### 3.3. Field Experimental Results

The average value of ten groups of data in the field experiment is shown in Figure 16. The position where $D_C$ was equal to 0 mm was the edge of the drainage ditch, and the position where $D_C$ was equal to 200 mm was the edge line of the origin of soil separation distance measurement in the simulation test. Since H was higher in the range of $D_C$ from 0 to 1100 mm than that in the range of $D_C$ from 1200 to 1500 mm, the range of 200 to 1100 mm was the soil separation interval. The soil separation distance in the field test was about 900 mm, which was consistent with 917 mm in the simulation test. In the range of $D_C$ from 0 to 1100 mm, when $D_C = 200$, the maximum $H = 68$ mm, when $D_C = 1100$, the minimum $H = 44$ mm, and the height difference of $H$ was 24 mm. Within this range, the standard deviation of $H$ was 7.8 mm, the level of soil flatness was high, and there was no blockage of drainage ditch, as the Figure 17 shown.

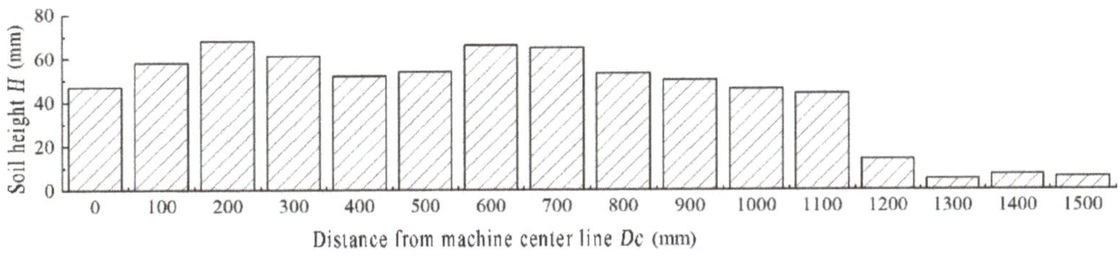


**Figure 16.** Field test results.

**Figure 17.** Drainage ditch.

## 4. Discussion

It is reported that the simplified or compound planting technological process has a positive impact on farming efficiency. Drainage ditches can drain surface waterlogging during wheat growing, especially in areas with excessive rainfall. However, considering the structure and operation characteristics of various ditching machines, the device for digging drainage ditch is not integrated in the general compound planter. Therefore, this study differed from previous ones dramatically. First, SSSM which can be integrated with compound planter was proposed. This device can complete the operation of opening drainage ditch and evenly disperse the accumulated soil. Second, the influence of the structural parameters of the SSSM on optimizing soil particle distribution was studied by discrete element method. The structural parameters have been optimized to achieve a better soil separating effect.

In this paper, Formula (1) was obtained through kinematic and dynamic analysis, which established the relationship between the soil separation direction, the forward direction of the machine, the rotation direction of the spiral blade, and the helical direction of the spiral blade. Yang et al. [33] designed a track filling assembly mounted on wheeled-tractor for paddy fields. The relationship between the soil separation direction, the forward direction of the machine, the rotation direction of the spiral blade, and the helical direction of the spiral blade of the device met Formula (1) was consistent with the research results of this paper. The best spiral blade parameters obtained in this paper were embedded depth = 49 mm, pitch = 331 mm, rotation speed = 318 r $min^{-1}$. This result was not consistent with the study of Yang et al. (depth = 45 mm, pitch = 200 mm, rotation speed = 200 r $min^{-1}$). This was due to the different characteristics of transported materials, transportation volume, forward speed of machines, and the helical direction of the spiral blade, etc. This study can provide a reference for the design of soil separation equipment using spiral soil separation device, and for the design of devices for conveying and leveling materials with spiral blades.

However, there are several limitations in this research. First, it lacks the support of wheat planting experiment. In fact, the effects of ditching and soil separation will affect the growth status and final yield of wheat. Second, there is a lack of research on soil structure parameters after soil separation, such as soil particle distribution, soil porosity, and water stable aggregates, etc. Further study is needed to judge the impact of ditching and soil separation on soil structure. In the future, we will focus on the effects of SSSM on wheat yield and soil structure parameters.

## 5. Conclusions

(1) The design of the SSSM solves the problems encountered in the compound planter ditching process, such as uneven soil, drainage ditch blockage. At the same time, the soil separation mechanism of spiral blades in the soil separation device was analyzed, and the stress analysis and kinematic analysis of the soil separation process were carried out.

(2) The discrete element simulation model of SSSM was established, and the evaluation indexes of soil separation distance and soil uniformity was established to evaluate the operation quality of SSSM, which laid the foundation for the parameter setting of SSSM.

(3) The single factor simulation test showed that there was an approximate linear relationship between the rotation speed and soil separation distance, but the increase of soil separation uniformity decreases with the increase of rotation speed. Increasing pitch can increase soil separation distance and soil separation uniformity. However, with the increase of pitch, the increase of soil separation distance and soil uniformity decreased. Increasing embedded depth can effectively improve soil separation distance and uniformity.

(4) The results of response surface analysis showed that with the increase of embedded depth, rotation speed, and pitch, soil separation distance and soil uniformity also increased. When the embedded depth and rotation speed increase, the soil separation distance also increases. Similarly, the greater the embedded depth, the greater the increase in soil uniformity.

(5) The quadratic regression model of soil separation quality was established, and the following optimization parameters were obtained: the embedded depth was 49 mm, the pitch was 331 mm, and the rotation speed was 318 r $min^{-1}$. The optimized parameters were obtained in the field experiment. The soil separation distance was 900 mm, and the standard deviation of soil flatness was 7.8 mm, which was consistent with the simulation results, and there was no drainage ditch blockage.

**Author Contributions:** Conceptualization, L.H. and R.Z.; methodology, X.X. and J.Y.; software, L.H. and X.X.; validation, W.Y. and J.Y.; formal analysis, L.H.; investigation, W.Y.; resources, J.J. and X.X.; data curation, Y.Z.; writing—original draft preparation, W.Y.; writing—review and editing, J.J. and D.X.; visualization, Y.Z.; supervision, D.X.; funding acquisition, R.Z. All authors have read and agreed to the published version of the manuscript.

**Funding:** This research was funded by Yangzhou University Interdisciplinary Research Foundation for Crop Science Discipline of Targeted Support (yzuxk202007), the Key Research and Development Program of Jiangsu Province (BE2020319), Jiangsu Modern Agricultural Machinery Equipment and

Technology Demonstration and Promotion Project (NJ2021-16), High-end Talent Support Program of Yangzhou University.

**Institutional Review Board Statement:** Not applicable.

**Informed Consent Statement:** Not applicable.

**Data Availability Statement:** Not applicable.

**Conflicts of Interest:** The authors declare no conflict of interest.

## References

1. Liu, H.; Xiong, W.; Mottaleb, K.A.; Krupnik, T.J.; Burgueño, J.; Pequeno, D.N.L.; Wu, W. Contrasting contributions of five factors to wheat yield growth in China by process-based and statistical models. *Eur. J. Agron.* **2021**, *130*, 126370. [CrossRef]
2. Ren, A.-X.; Sun, M.; Wang, P.-R.; Xue, L.-Z.; Lei, M.-M.; Xue, J.-F.; Gao, Z.-Q.; Yang, Z.-P. Optimization of sowing date and seeding rate for high winter wheat yield based on pre-winter plant development and soil water usage in the Loess Plateau, China. *J. Integr. Agric.* **2019**, *18*, 33–42. [CrossRef]
3. Chen, J.; Zheng, M.-J.; Pang, D.-W.; Yin, Y.-P.; Han, M.-M.; Li, Y.-X.; Luo, Y.-L.; Xu, X.; Li, Y.; Wang, Z.-L. Straw return and appropriate tillage method improve grain yield and nitrogen efficiency of winter wheat. *J. Integr. Agric.* **2017**, *16*, 1708–1719. [CrossRef]
4. He, J.; Shi, Y.; Zhao, J.; Yu, Z. Strip rotary tillage with subsoiling increases winter wheat yield by alleviating leaf senescence and increasing grain filling. *Crop J.* **2020**, *8*, 327–340. [CrossRef]
5. Xi, X.; Gao, W.; Gu, C.; Shi, Y.; Han, L.; Zhang, Y.; Zhang, B.; Zhang, R. Optimisation of no-tube seeding and its application in rice planting. *Biosyst. Eng.* **2021**, *210*, 115–128. [CrossRef]
6. Xi, X.; Gu, C.; Shi, Y.; Zhao, Y.; Zhang, Y.; Zhang, Q.; Jin, Y.; Zhang, R. Design and experiment of no-tube seeder for wheat sowing. *Soil Tillage Res.* **2020**, *204*, 104724. [CrossRef]
7. Wu, F.; Zhai, L.-C.; Xu, P.; Zhang, Z.-B.; Baillo, E.H.; Tolosa, L.N.; Kimotho, R.N.; Jia, X.-L.; Guo, H.-Q. Effects of deep vertical rotary tillage on the grain yield and resource use efficiency of winter wheat in the Huang-Huai-Hai Plain of China. *J. Integr. Agric.* **2021**, *20*, 593–605. [CrossRef]
8. Zhang, C.; Liu, J.; Shang, J.; Cai, H. Capability of crop water content for revealing variability of winter wheat grain yield and soil moisture under limited irrigation. *Sci. Total Environ.* **2018**, *631*, 677–687. [CrossRef]
9. Zhang, H.; Li, Y.; Meng, Y.-L.; Cao, N.; Li, D.-S.; Zhou, Z.-G.; Chen, B.-L.; Dou, F.-G. The effects of soil moisture and salinity as functions of groundwater depth on wheat growth and yield in coastal saline soils. *J. Integr. Agric.* **2019**, *18*, 2472–2482. [CrossRef]
10. Wu, S.; Ren, J.; Chen, Z.; Yang, P.; Li, H. Soil moisture estimation based on the microwave scattering mechanism during different crop phenological periods in a winter wheat-producing region. *J. Hydrol.* **2020**, *590*, 125521. [CrossRef]
11. Zhang, Y.; Wang, J.; Gong, S.; Xu, D.; Mo, Y.; Zhang, B. Straw mulching improves soil water content, increases flag leaf photosynthetic parameters and maintaines the yield of winter wheat with different irrigation amounts. *Agric. Water Manag.* **2021**, *249*, 106809. [CrossRef]
12. Aikins, K.A.; Barr, J.B.; Antille, D.L.; Ucgul, M.; Jensen, T.A.; Desbiolles, J.M.A. Analysis of effect of bentleg opener geometry on performance in cohesive soil using the discrete element method. *Biosyst. Eng.* **2021**, *209*, 106–124. [CrossRef]
13. Aikins, K.A.; Antille, D.L.; Ucgul, M.; Barr, J.B.; Jensen, T.A.; Desbiolles, J.M.A. Analysis of effects of operating speed and depth on bentleg opener performance in cohesive soil using the discrete element method. *Comput. Electron. Agric.* **2021**, *187*, 106236. [CrossRef]
14. Barr, J.; Desbiolles, J.; Ucgul, M.; Fielke, J.M. Bentleg furrow opener performance analysis using the discrete element method. *Biosyst. Eng.* **2020**, *189*, 99–115. [CrossRef]
15. Qin, K.; Ding, W.; Fang, Z.; Du, T.; Zhao, S. Design and parameter optimization of double disk opener mechanism for harvest ditch and stalk-disposing machine. *Trans. Chin. Soc. Agric. Eng.* **2017**, *33*, 27–35.
16. Bao, P.; Wu, M.; Guan, C.; Luo, H.; He, Y.; Xiang, W. Design of plow-rotary style ditching and ridging device for rapeseed seeding. *Trans. Chin. Soc. Agric. Eng.* **2017**, *33*, 23–31.
17. Aikins, K.A.; Ucgul, M.; Barr, J.B.; Jensen, T.A.; Antille, D.L.; Desbiolles, J.M.A. Determination of discrete element model parameters for a cohesive soil and validation through narrow point opener performance analysis. *Soil Tillage Res.* **2021**, *213*, 105123. [CrossRef]
18. Barr, J.B.; Desbiolles, J.M.A.; Fielke, J.M.; Ucgul, M. Development and field evaluation of a high-speed no–till seeding system. *Soil Tillage Res.* **2019**, *194*, 104337. [CrossRef]
19. Bahrami, M.; Naderi-Boldaji, M.; Ghanbarian, D.; Ucgul, M.; Keller, T. Simulation of plate sinkage in soil using discrete element modelling: Calibration of model parameters and experimental validation. *Soil Tillage Res.* **2020**, *203*, 104700. [CrossRef]
20. Li, B.; Chen, Y.; Chen, J. Modeling of soil–claw interaction using the discrete element method (DEM). *Soil Tillage Res.* **2016**, *158*, 177–185. [CrossRef]
21. Zhou, Y.; Wang, X.; Yu, Y. Research and Analysis of the Optimum Gap of the Screw Extraction Device Based on EDEM Software. *J. Agric. Mech. Res.* **2017**, *39*, 38–42.

22. Owen, P.J.; Cleary, P.W. Prediction of screw conveyor performance using the Discrete Element Method (DEM). *Powder Technol.* **2009**, *193*, 274–288. [CrossRef]
23. Pezo, M.; Pezo, L.; Jovanović, A.P.; Terzić, A.; Andrić, L.; Lončar, B.; Kojić, P. Discrete element model of particle transport and premixing action in modified screw conveyors. *Powder Technol.* **2018**, *336*, 255–264. [CrossRef]
24. Müller, D.; Fimbinger, E.; Brand, C. Algorithm for the determination of the angle of repose in bulk material analysis. *Powder Technol.* **2021**, *383*, 598–605. [CrossRef]
25. Asaf, Z.; Rubinstein, D.; Shmulevich, I. Determination of discrete element model parameters required for soil tillage. *Soil Tillage Res.* **2007**, *92*, 227–242. [CrossRef]
26. Mak, J.; Chen, Y.; Sadek, M.A. Determining parameters of a discrete element model for soil–tool interaction. *Soil Tillage Res.* **2012**, *118*, 117–122. [CrossRef]
27. Obermayr, M.; Vrettos, C.; Eberhard, P.; Däuwel, T. A discrete element model and its experimental validation for the prediction of draft forces in cohesive soil. *J. Terramechanics* **2014**, *53*, 93–104. [CrossRef]
28. Horabik, J.; Molenda, M. Parameters and contact models for DEM simulations of agricultural granular materials: A review. *Biosyst. Eng.* **2016**, *147*, 206–225. [CrossRef]
29. Obermayr, M.; Dressler, K.; Vrettos, C.; Eberhard, P. Prediction of draft forces in cohesionless soil with the Discrete Element Method. *J. Terramechanics* **2011**, *48*, 347–358. [CrossRef]
30. Ucgul, M.; Fielke, J.M.; Saunders, C. Three-dimensional discrete element modelling of tillage: Determination of a suitable contact model and parameters for a cohesionless soil. *Biosyst. Eng.* **2014**, *121*, 105–117. [CrossRef]
31. Wang, X.; Hu, H.; Wang, Q.; LI, H.; He, J.; Chen, W. Calibration Method of Soil Contact Characteristic Parameters Based on DEM Theory. *Trans. Chin. Soc. Agric. Mach.* **2017**, *48*, 78–85.
32. Chen, C.; Quan, W.; Wu, M.; Zhang, W. Parameter optimization of vertical soil-filling hole-forming parts for rapeseed transplantation based on discrete element method. *J. Hunan Agric. Univ.* **2019**, *45*, 433–439.
33. Yang, W.; Luo, X.; Wang, Z.; Zhang, M.; Zeng, S.; Zang, Y. Design and experiment of track filling assembly mounted on wheeled-tractor for paddy fields. *Trans. Chin. Soc. Agric. Eng.* **2016**, *32*, 26–31.

*agriculture*

MDPI

*Article*

# Regression-Based Correction and I-PSO-Based Optimization of HMCVT's Speed Regulating Characteristics for Agricultural Machinery

**Zhun Cheng [1,*] and Zhixiong Lu [2]**

[1] Department of Vehicle Engineering, Nanjing Forestry University, Nanjing 210037, China
[2] College of Engineering, Nanjing Agricultural University, Nanjing 210031, China; luzx@njau.edu.cn
[*] Correspondence: cz38@njfu.edu.cn

**Abstract:** To improve the speed regulating characteristics of continuously variable transmission for agricultural machinery, in order to meet the engineering and technical requirements of precision agriculture and intelligent agriculture, the paper researches and proposes a method combining the analysis of speed regulating characteristics, regression-based correction, and the improved particle swarm optimization (I-PSO) algorithm. First, the paper analyzes the degree of deviation between the linearization degree and the theoretical value of the speed regulating characteristics of the variable-pump constant-motor system of agricultural machinery according to the measurement results of the bench test. Next, the paper corrects the speed regulating characteristics and compares the regression results based on four models. Finally, the paper proposes a design method for the expected speed regulating characteristics of agricultural machinery and it completes the optimization of speed regulating characteristics and the matching of transmission parameters with the I-PSO algorithm. Results indicate that the speed regulating characteristics of the variable-pump constant-motor system show high linearization (with a coefficient of determination of 0.9775). The theoretical and measured values of the speed regulating characteristics have a certain deviation (with a coefficient of determination of 0.8934). Therefore, correcting the speed regulating characteristics of the variable-pimp constant-motor system is highly necessary. In addition, the second reciprocal function model proposed has the highest correction precision (with a coefficient of determination of 0.9978). The I-PSO algorithm is applicable to the design and application of hydro-mechanical continuously variable transmission (HMCVT) for agricultural machinery. The new method proposed can improve the HMCVT's speed regulating characteristics efficiently and quickly. It also ensures that the speed regulating characteristics are highly consistent with the expected design characteristics (with a mean error of 1.73%). Thus, the research offers a theoretical direction and design basis for the research and development of continuously variable transmission units in agricultural machinery.

**Keywords:** agricultural machinery; HMCVT; correction of characteristics; I-PSO algorithm; parameter match

**Citation:** Cheng, Z.; Lu, Z. Regression-Based Correction and I-PSO-Based Optimization of HMCVT's Speed Regulating Characteristics for Agricultural Machinery. *Agriculture* **2022**, *12*, 580. https://doi.org/10.3390/agriculture12050580

Academic Editors: Mustafa Ucgul and Chung-Liang Chang

Received: 25 March 2022
Accepted: 19 April 2022
Published: 21 April 2022



## 1. Introduction

There is a wide variety of agriculture and forestry machinery. The tractor is one of the most important pieces of agricultural working machinery applied in the fields of agriculture and forestry [1–3]. Moreover, the agriculture working machinery also includes the grain harvester combine, the cotton picker, and so on. The forestry working machinery includes the skidder, the forest fire truck, the harvesting–cultivating combination machine, and so on [4]. The agriculture and forestry machinery generally works in the conditions of severe environments and variable loads [5–8]. Running reliably at the required speed is one of the most basic performance requirements of agriculture and forestry machinery. It requires a wider range of speed changes.

For vehicles (mainly fuel engine vehicles [9,10] and electric vehicles [11–13]), when the engine is determined, the transmission system plays an important role in changing the speed. Agricultural machinery and vehicles have a certain degree of commonality. Tractors are also non-road vehicles. Therefore, the method of analogy can be used to study and analyze the power transmission system. The HMCVT, as a power dividing (composed of the mechanical power and the hydraulic power), continuously variable transmission unit [14], is applied to agricultural and forestry machinery such as tractors [15–17].

The HMCVT receives the power transmitted from the engine and then outputs the power after changing the speed according to the transmission characteristics. Then, the power goes through other transmission systems (with the fixed transmission ratio) and wheels [18–20]. Finally, the running speed of the agriculture and forestry machinery is formed. Therefore, to ensure that the agriculture and forestry machinery runs reliably at the required speed, it is necessary to design the speed regulating characteristics properly. However, designing the speed regulating characteristics of a HMCVT properly requires the correct mastery of the speed regulating characteristics. Meanwhile, the proper design of the speed regulating characteristics is the premise of using the continuously transmission unit.

Recently, a few scholars have conducted studies on the analysis and optimization design of the HMCVT's speed regulating characteristics.

In the studies of the HMCVT's speed regulating characteristic analysis, most scholars use theoretical analysis based on the mechanical transmission principle to derive the relational expression of the HMCVT's speed regulating characteristics (i.e., the relational expression between the displacement ratio of the pump–motor system and the HMCVT's ratio). Xu et al. [21] gave a relational expression between the HMCVT speed ratio and variable-pump constant-motor displacement ratio based on a theoretical analysis. The research pointed out that the relational model could offer a theoretical basis for the transmission scheme determination, parameter matching and performance analysis of the HMCVT. Later, Xu et al. [22] pointed out the requirement of the HMCVT's steady section shift to the working section transmission ratio. Sung et al. [23] researched the speed regulating characteristics of 12 HMCVTs using the network analysis method. Yu et al. [24], after obtaining a theoretical analysis model of speed regulating characteristics, analyzed the linearization degree of the speed regulating characteristics and the change range of the HMCVT's output speed, researched using a prototype test. Li et al. [25] analyzed the relationship between the transmission ratio and displacement ratio using a theoretical model of output speed built in the environment of Matlab. Meanwhile, the research analyzed the hydraulic–mechanical mixed section and the purely hydraulic section separately and pointed out the continuity of the speed regulating characteristics. Xia and Sun [26] derived the change relational expression of speed regulating characteristics based on the working principle of continuously variable transmission. Then, on this basis, they analyzed the working advantages of the HMCVT.

In the research field of HMCVT design, most scholars have used the optimization algorithm for the design and matching adjustment of parameters. Volpe et al. [27] used the difference evolution algorithm, the simulated annealing algorithm, and the simplex method for the optimization design of two types of power-dividing continuously variable transmission. Macor and Rossetti [28,29] set parameters, such as the gear ratio and the planetary gear transmission ratio, as the variables of optimization design, and used the PSO algorithm for the multi-object optimization design. Moreover, the research pointed out that the HMCVT's optimization design is complicated and has strongly nonlinear characteristics. Zhang et al. [30] used the non-dominated sorting genetic algorithm for the optimization of the planetary row and transmission ratio. Some scholars first determined the form of HMCVT transmission scheme (an equal-ratio or equal-difference transmission scheme), and then determined the values of transmission parameters according to the empirical method and formulas. For instance, Ni et al. [31] designed a four-stage HMCVT with the common transmission ratios of sections as 1.88, and verified the continuity of output speed. Zhang et al. [32] and He et al. [33] designed the HMCVT's parameters with the equal-ratio and equal-difference principles, respectively. Moreover, Liu et al. [34] used the theoretical analysis formulas of kinematics and mechanics combined with the Newton–Raphson method for parameter matching. Cheng et al. [35] performed a mechanism analysis of the HMCVT's speed regulating characteristics, composed of the multi-planetary-row compound transmission, and designed a non-equal-ratio-transmission HMCVT using the improved genetic algorithm. The research matched the required range of tractor working speed with the change range of each HMCVT section transmission ratio. This helped to improve the flexibility of the HMCVT's speed regulating characteristics design.

To sum up, using the theoretical relational expression of speed regulating characteristics to design the HMCVT is common. However, the method depends on the precision of the model of the speed regulating characteristics. The HMCVT is composed of the mechanical system (generally composed of the fixed-shaft gear pair and the planetary gear train) and the hydraulic system (generally composed of the pump and the motor). In the process of transmission, it can be considered that the actual transmission characteristics are consistent with the theoretical transmission characteristics of the mechanical system. However, the hydraulic system is greatly affected by the environment, thus causing a certain deviation between the actual and theoretical transmission characteristics. Therefore, only when the transmission characteristics of the mechanical and hydraulic systems are both correct, the HMCVT's speed regulating characteristics have certain precision. In particular, it is difficult to improve the precision of the HMCVT's speed regulating characteristics model in the research, development, and design stage. Generally, the comparative verification of the speed regulating characteristics can only be done after the HMCVT is made, thus resulting in increasing research and development costs and duration. In addition, current studies have a serious deficiency in this respect.

To solve the problem described above, the paper proposes a regression-based correction method and an I-PSO-based optimization method for the HMCVT's speed regulating characteristics. The paper mainly researches the following three aspects. Firstly, the paper describes a test of the speed regulating characteristics of the hydraulic system using the test bench of a variable-pump constant-motor system. Then, based on the measurement results of the test, the paper analyzes the linearization degree of the measured output speed and the deviation degree between the measured and theoretical values. The method of least squares is used for the regression analysis of the speed regulating characteristics of the variable-pump constant-motor system. The regression model is considered as the correction model of the HMCVT hydraulic system's speed regulating characteristics. Secondly, according to agricultural machinery's working speed requirement, engine characteristics, and the corrected model of the HMCVT's speed regulating characteristics, the paper uses the I-PSO algorithm for the optimization design of the speed regulating characteristics and matching of transmission parameters. Thirdly, using the coefficient of determination $R^2$ and the mean absolute percentage error *MAPE* as evaluation indexes, the paper compares before and after optimization and before and after correction of the speed regulating characteristics. Comparison results show that the corrected model of the speed regulating characteristics of the hydraulic system has high precision; the speed regulating characteristics based on the I-PSO algorithm are highly consistent with the expected characteristics and the matching of the transmission design parameters has a good result. According to the method proposed, it is only necessary to test the output speed of the pump–motor system chosen to complete the optimization design of the speed regulating characteristics and the matching of the transmission parameters of the HMCVT effectively in the research, development, and design stage.

## 2. Materials and Methods

### 2.1. Working Principle of Tractor HMCVT Researched

Figure 1 shows the transmission scheme of the tractor HMCVT researched in the paper. The continuously variable transmission unit has eight gear pairs (suppose that their transmission ratios are $i_1$, $i_2$, $i_3$, $i_4$, $i_5$, $i_6$, $i_7$, and $i_8$, respectively), two planetary rows $P_1$ and

$P_2$ (suppose that the parameters of planetary rows are $k_1$ and $k_2$, respectively), three wet clutches $C_0$, $C_1$, and $C_2$, and one variable-pump constant-motor system.

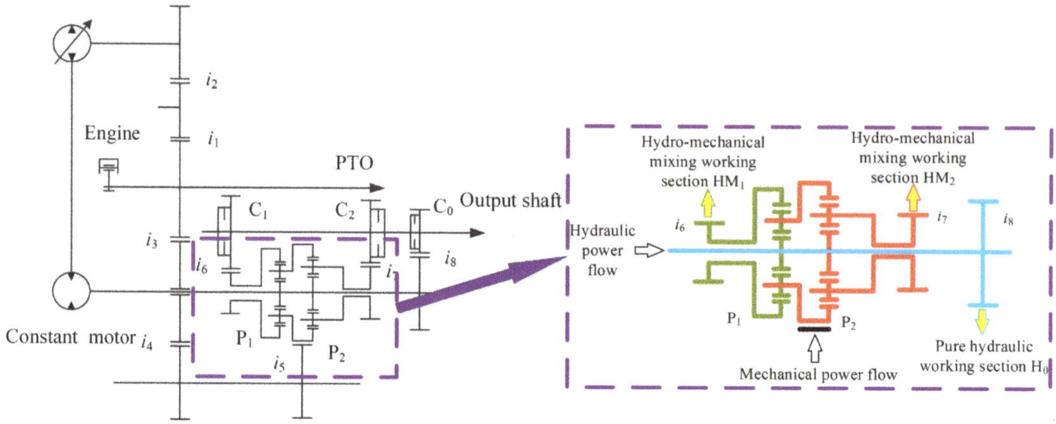


**Figure 1.** Transmission scheme of HMCVT for tractor researched in the paper.

When a tractor runs normally and works (corresponding to the HMCVT hydraulic–mechanical-power mixed working mode), the overall power of the transmission system is output from the engine and then input into the variable-pump constant-motor system (the hydraulic power flow) and the planetary gear train (the mechanical power flow). The divided power converges through the planetary gear train and then is output through $P_1$'s gear ring or $P_2$'s planetary carrier. When a tractor starts up (corresponding to the purely hydraulic working mode of HMCVT), the overall power of the transmission system is output from the engine and then input into the variable-pump constant-motor system directly, and drives gear pair $i_8$ directly to offer the power to the output end. The core part of the continuously variable unit (i.e., the power confluence mechanism) of the tractor is the Simpson planetary gear transmission mechanism and has three working sections (including the purely hydraulic working section $H_0$ and the hydraulic–mechanical mixed working sections $HM_1$ and $HM_2$). Figure 1 also shows the power output routes of working sections, in which sections $H_0$, $HM_1$, and $HM_2$ are shown in blue, green, and red, respectively.

The calculation formulas for the speeds of the confluence mechanism planetary carrier, gear ring, and sun gear are as follows [36,37]:

$$k_1 n_r = (1 + k_1) n_c - n_s \tag{1}$$

in which $n_r$ is the output speed of the gear ring; $n_c$ is the input speed of the planetary carrier, and $n_s$ is the input speed of the sun gear.

Theoretically, the transmission relationship between the variable pump and constant motor is as follows [38]:

$$n_m = n_p \varepsilon \tag{2}$$

in which $n_m$ is the speed of the constant motor; $n_p$ is the working speed of the variable pump, and $\varepsilon$ is the displacement ratio of the variable-pump constant-motor system.

Combining Equation (1) with Equation (2), we then obtain the relational expression of the 3-section HMCVT's speed regulating characteristics (i.e., the relational expression between the displacement ratio of the pump–motor system and the transmission ratio of HMCVT) as follows [39]:

$$i_{H_0} = -\frac{i_1 i_2 i_8}{\varepsilon} \tag{3}$$

$$i_{HM_1} = \frac{i_1 i_2 i_3 i_4 i_5 i_6 k_1}{\varepsilon i_3 i_4 i_5 + i_1 i_2 (1 + k_1)} \quad (4)$$

$$i_{HM_2} = \frac{i_1 i_2 i_3 i_4 i_5 i_7 (1 + k_2)}{i_1 i_2 k_2 - \varepsilon i_3 i_4 i_5} \quad (5)$$

### *2.2. Build the 'Variable-Pump and Constant-Motor' Test Bench of HMCVT for Tractor*

Figure 2 shows the variable-pump constant-motor test bench built for the HMCVT for a tractor.

**Figure 2.** Test bench of 'variable-pump and constant-motor' system. Note: ① Engine (DEUTZ TCD2013L062V); ② Gear box combined by the 'variable-pump and constant-motor' system; ③ Speed torque sensor of ZJ-5000A model; ④ Auxiliary gearbox; ⑤ Electrical eddy current dynamometer of DW250 model; ⑥ Hydraulic system (realizing lubrication, cooling, and other functions).

The test bench uses the speed torque sensors of the ZJ-2000A model and ZJ-5000A model of Lanling Jiangsu, China. Table 1 gives the ranges of speed and torque.

**Table 1.** Related parameters of motor.

| Model | Measurement Range |
|---|---|
| ZJ-2000A | Rated Torque: 2000 N · m<br>Working Speed: 0–3000 r/min |
| ZJ-5000A | Rated Torque: 4000 N · m<br>Working Speed: 0–5000 r/min |

The test bench uses the variable pump of the Linde HPV-02 model with 55 $cm^3$/rev displacement and 75 kW continuous working power, and the constant motor of the Linde HMF-02 model with 55 $cm^3$/rev displacement and 93 kW continuous working power.

### *2.3. Test of Speed Regulating Characteristics for 'Variable-Pump and Constant-Motor' System*

The test aims to measure the speed regulating characteristics of the variable-pump constant-motor system in the real environment, and offers measured data to the studies on the linearization degree of the motor output speed, the deviation degree between the measured value and theoretical value of output speed, and the correction of the speed regulating characteristics based on regression.

The test requires us to fix the variable pump's input speed and record the constant motor's output speed by changing displacement ratio $\varepsilon$.

According to Equation (2), the change range of the absolute value of the displacement ratio is 0~1. Therefore, to cover the change range of displacement ratio $\varepsilon$ completely, the test chooses displacement ratios of 0.2, 0.25, 0.3, 0.375, 0.5, 0.625, 0.75, and 1 (in total, 8 groups of tests), and considers the average of stable constant motor output speeds in each group of tests as the output speed in the current working condition. In addition, the research records the input speed data of the variable pump to analyze and determine whether the volatility of the input speed is reasonable. The analysis aims to alleviate the influence of the input speed volatility on the system's output speed.

### *2.4. The Analysis Method for Speed Regulating Characteristics of 'Variable-Pump and Constant-Motor' System*

The paper calculates and obtains $\overline{n}_p$ (the average input speed of variable pump) and $\sigma_p$ (the standard deviation) to determine whether the volatility of the input speed is reasonable. The calculation formulas of average input speed $\overline{n}_p$ and standard deviation $\sigma_p$ are as follows:

$$\overline{n}_p = \frac{1}{N_p}\sum_{i=1}^{N_p} n_{pi} \tag{6}$$

$$\sigma_p = \sqrt{\frac{1}{N_p}\sum_{i=1}^{N_p} (n_{pi} - \overline{n}_p)^2} \tag{7}$$

in which $N_p$ is the total number of measured data and $n_{pi}$ is the value of the $i$th measured datum.

The higher the linearization degree of the HMCVT's speed regulating characteristics, the better the adjustability of the tractor's driving speed [24]. According to the expression of the speed regulating characteristics of the HMCVT used, i.e., Equations (3)–(5), the HMCVT's output speed and displacement ratio $\varepsilon$ have a linear variation relationship theoretically. The gear system transmission is reliable, and its theoretical and actual transmission characteristics are basically consistent. Therefore, the linearization degree of the variable-pump constant-motor system's speed regulating characteristics plays a decisive role in the linearization degree of the HMCVT's speed regulating characteristics.

The paper proposes to perform linear fitting to the motor output speed based on the least squares method and using $R^2$, the coefficient of determination of linear fitting, as the determination basis of the linearization degree. The following is the calculation formula of $R^2$:

$$R^2 = 1 - \frac{\sum_{i=1}^{N} [y_{measured} - y_{ideal}]^2}{\sum_{i=1}^{N} [y_{ideal} - (\sum_{i=1}^{N} y_{ideal})/N]^2} \tag{8}$$

in which $y_{ideal}$ is the theoretical value, $y_{measured}$ is the measured value of the bench test, and $N$ is the total number of data to be fitted.

The paper uses the coefficient of determination $R^2$ to evaluate the degree of consistency between the measured and theoretical output speeds of the pump–motor system.

### *2.5. The Regression-Based Correction Method of HMCVT's Speed Regulating Characteristics for Tractor*

According to Equation (2), the relational expression of $i_{pm}$, the transmission ratio of the variable-pump constant-motor system (i.e., $n_p/n_m$, the ratio of the input speed of pump and the output speed of motor), and the displacement ratio is as follows:

$$i_{pm} = 1/\varepsilon \tag{9}$$

According to the variation trend of measured data, the research uses the linear and nonlinear least squares methods for the regression analysis of the speed regulating characteristics of the variable-pimp constant-motor system. The polynomial regression [40,41] has good generalization ability and strong applicability in the field of engineering technology. In addition, according to Equations (2) and (9), the research adds items to the numerator and the denominator based on the original theoretical model to correct the model. To sum up, the paper uses four models in the regression analysis, including the 2-order polynomial model, the 3-order polynomial model, and two ε reciprocal function models. The four models' forms are as follows.

The 2-order polynomial model of speed regulating characteristics:

$$i_{pm} = a_1\varepsilon^2 + a_2\varepsilon + a_3 \tag{10}$$

The 3-order polynomial model of speed regulating characteristics:

$$i_{pm} = a_1\varepsilon^3 + a_2\varepsilon^2 + a_3\varepsilon + a_4 \tag{11}$$

The first reciprocal function model of speed regulating characteristics:

$$i_{pm} = a_1/(\varepsilon + a_2) \tag{12}$$

The second reciprocal function model of speed regulating characteristics:

$$i_{pm} = (a_1\varepsilon + a_2)/(\varepsilon + a_3) \tag{13}$$

The paper uses the coefficient of determination $R^2$ to evaluate and determine the final variation expression of the speed regulating characteristics of the variable-pump constant-motor system, and substitutes the final form of $i_{pm}$ into Equations (3)–(5) to obtain the corrected change law of the speed regulating characteristics of the tractor HMCVT. Figure 3 shows the technical route of the correction method proposed.

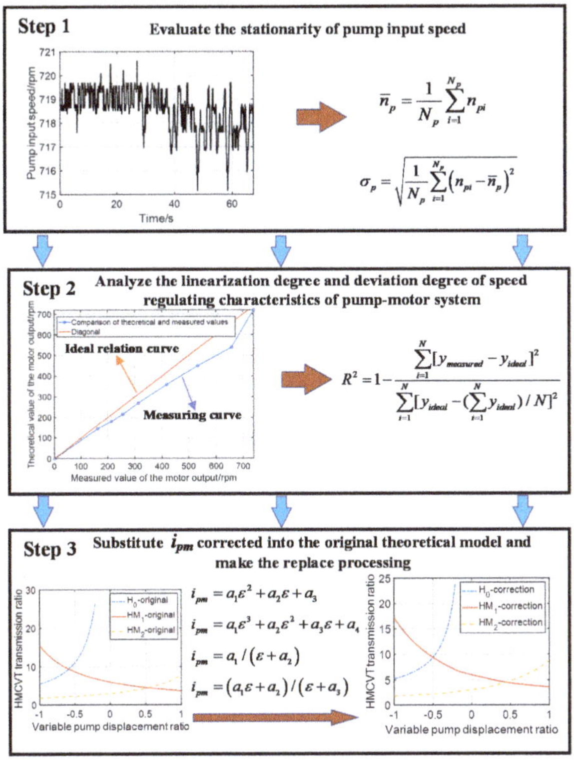


**Figure 3.** The technical route of correction method proposed for speed regulating characteristics.

*2.6. The I-PSO-Based Optimization Design Method of HMCVT's Speed Regulating Characteristics for Tractor*

Figure 4 shows the original speed regulating characteristics of the HMCVT researched. The HMCVT of this model has three sections ($H_0$, $HM_1$, and $HM_2$), of which the speed regulating characteristics intersect mutually. Moreover, there are two synchronous section-shift transmission ratios. The design aims to ensure the continuity of the transmission ratio of the HMCVT.

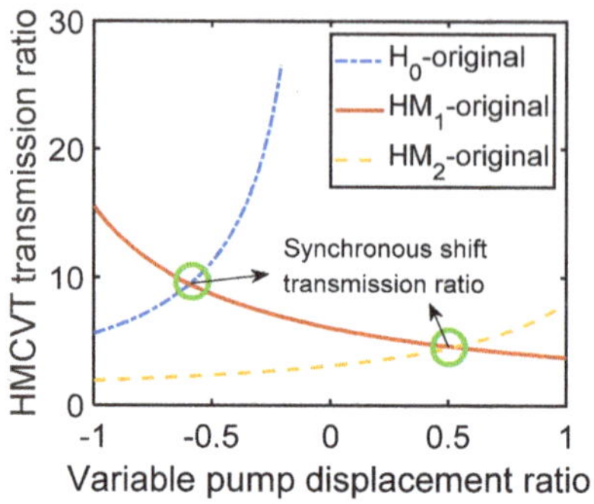


**Figure 4.** The original transmission characteristics of HMCVT researched.

To make the HMCVT system easier to control, the two synchronous section-shift transmission ratios should be distributed in the marginal area with the displacement ratio ε of −1~1 as much as possible because, in this case, the displacement ratio ε has a wider range of change. If the actually required displacement ratio ε has a small change range, the requirement to control precision is high. For instance, suppose that the designed HMCVT transmission ratio $i_{cvt}$ has a change range of 1~11; then, if displacement ratio ε changes in the range of −1~1, ε changing by 0.1 corresponds to $i_{cvt}$ changing by 0.5, on average; if displacement ratio ε changes in the range of −0.5~0.5, ε changing by 0.05 corresponds to $i_{cvt}$ changing by 0.5 on average. This indicates that the wider displacement ratio ε's available range is, the more applicable it is to practical engineering.

The paper sets the tire radius of the tractor as 0.9 m and the other transmission system's transmission ratio as 9. Figure 5 shows the external characteristics of the diesel engine used.

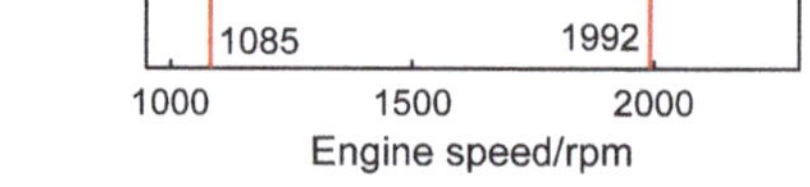


**Figure 5.** External characteristic curve of diesel engine used.

In Figure 5, the values of torque in the positions marked with the left and right red lines (i.e., the speeds of the diesel engine are 1085 and 1992 r/min, respectively) correspond to 90% of the maximum torque of the diesel engine. The design in the paper uses Wang's analysis of the research results of Resch and Renius for reference [42]. In the whole life cycle of the tractor, the proportion of time for which the tractor works in the speed section of 4~20 km/h is approximately 76~93% (in which the proportion of time in 4~12 km/h is 61~68% and the proportion of time in 12~20 km/h is approximately 15~25%).

Therefore, the paper designs section $HM_1$ as the farmland working section of the tractor and section $HM_2$ as the driving and transportation section of the tractor. The design can also avoid the frequent switch problem of the wet clutch. To sum up, the transmission ratios of the sections of the tractor HMCVT are shown in Table 2.

**Table 2.** Transmission ratios of HMCVT's sections for tractor.

| Section | Vehicle Speed (km/h) | Transmission Ratio |
|---|---|---|
| $HM_1$ | 4~20 | 4.34~14.14 |
| $HM_2$ | 20~40 and above | 2.17 and below~4.34 |

Section $H_0$ is the starting section of the tractor, so, in the design, it should have a synchronous section-shift transmission ratio with section $HM_1$, and the minimum vehicle speed corresponding to the synchronous section-shift transmission ratio should be equal to or greater than 4 km/h.

Moreover, when the displacement ratio $\varepsilon = 0$, in section $HM_1$, only the gear system is transmitting the power, so the power transmission efficiency is the maximum in this case. To ensure that section $HM_1$ fully plays its role in the case where the displacement ratio $\varepsilon$ is 0, the paper considers the speed of the diesel engine corresponding to 90% of maximum torque as the critical speed for calculation, and calculates the range of vehicle speed corresponding to the critical speed. Meanwhile, we further calculate the proportion of the vehicle speed range in 4~12 km/h, and take the HMCVT transmission ratio corresponding to 80% of the maximum proportion as the designed value of the expected transmission ratio of section $HM_1$ in the case where displacement ratio $\varepsilon$ is 0. The corresponding calculation formula (engine output speed through the HMCVT and other transmission systems to cause the tire to produce translation speed) of the tractor's speed and the HMCVT transmission ratio is as follows (0.377 is the coefficient used for unit conversion):

$$u_a = 0.377\frac{r_d n_e}{i_0 i_{cvt}} \tag{14}$$

in which $u_a$ is the running speed of the tractor, $r_d$ is the radius of the wheel, $n_e$ is the working speed of the engine, and $i_0$ is the overall transmission ratio of the other transmission system.

According to the calculation result, when displacement ratio $\varepsilon = 0$, the value range of $i_{cvt}$, the HMCVT transmission ratio corresponding to section $HM_1$, is 5.37~7.82.

According to Equations (3)–(5), the HMCVT's speed regulating characteristics have a nonlinearity characteristic and many characteristic parameters (i.e., many transmission parameters to be designed). We classify the transmission parameters of the HMCVT researched and obtain the following seven transmission parameters to be designed: $i_1$ and $i_2$; $i_3$, $i_4$, and $i_5$; $i_6$; $i_7$; $i_8$; $k_1$; and $k_2$. If we use the enumeration method to match the transmission parameters and optimize the speed regulating characteristics (suppose that the precision of the transmission parameter of the mechanical part is 0.01), it is necessary to carry out $350^5 + 150^2 = 5.25 \times 10^{12}$ times of calculation and matching processes (the range of gear transmission ratio is 0.5~4 and the value range of planetary row's characteristic parameter is 2.5~4). In addition, there are constraint conditions for the transmission parameters in the matching process. Meanwhile, a great amount of work is required if replanning the HMCVT's speed regulating characteristics to match the working requirements of other agricultural machinery. The reasons above cause the difficulty in optimizing the speed regulating characteristics using the enumeration method. The heuristic intelligent optimization algorithm has obvious advantages in performance optimization, parameter matching, and identification, and has been used to accurately solve a series of complex engineering problems [43–46].

Therefore, the paper proposes using the I-PSO algorithm for the optimization of the speed regulating characteristics of the tractor HMCVT and the matching of the transmission parameters. As for the I-PSO algorithm used, the paper uses the I-PSO algorithm process [47,48], proposed in previous research and verified for engineering applications, for reference.

The optimization of the speed regulating characteristics of the HMCVT is a multi-object optimization problem. The objective function *fitness* of the I-PSO algorithm proposed and used in the research is as follows:

$$fitness = \left|i_{\mathrm{HM}_1}(\varepsilon=-1) - 14.14\right| + \left|i_{\mathrm{HM}_1}(\varepsilon=1) - 4.34\right| + \left|i_{\mathrm{HM}_2}(\varepsilon=1) - 4.34\right| \quad (15)$$

in which $i_{\mathrm{HM}_1}(\varepsilon=-1)$ and $i_{\mathrm{HM}_1}(\varepsilon=1)$ are the transmission ratios of the HMCVT working in section $\mathrm{HM}_1$ with the displacement ratios of $-1$ and 1, respectively, and $i_{\mathrm{HM}_2}(\varepsilon=1)$ is the transmission ratio of the HMCVT working in section $\mathrm{HM}_2$ with the displacement ratio of 1.

According to Equation (15), the paper considers the sum of three sub-objective functions with equal weight as the overall objective function. Each term of a sub-objective function considers the absolute value of error between the design value and expected value as the calculation formula. Meanwhile, the constraint conditions of parameter matching are as follows:

$$i_{\mathrm{HM}_2}(\varepsilon=-1) \le 2.17 \quad (16)$$

$$5.37 \le i_{\mathrm{HM}_1}(\varepsilon=0) \le 7.82 \quad (17)$$

$$i_{\mathrm{H}_0}(\varepsilon=c) = i_{\mathrm{HM}_1}(\varepsilon=c) \ge 8.48 \quad (18)$$

in which $i_{\mathrm{H}_0}(\varepsilon=c)$ is the transmission ratio of the HMCVT working in stage $\mathrm{H}_0$ when the displacement ratio is equal to $c$.

To sum up, Figure 6 shows the optimization process of the speed regulating characteristics based on the I-PSO algorithm proposed.

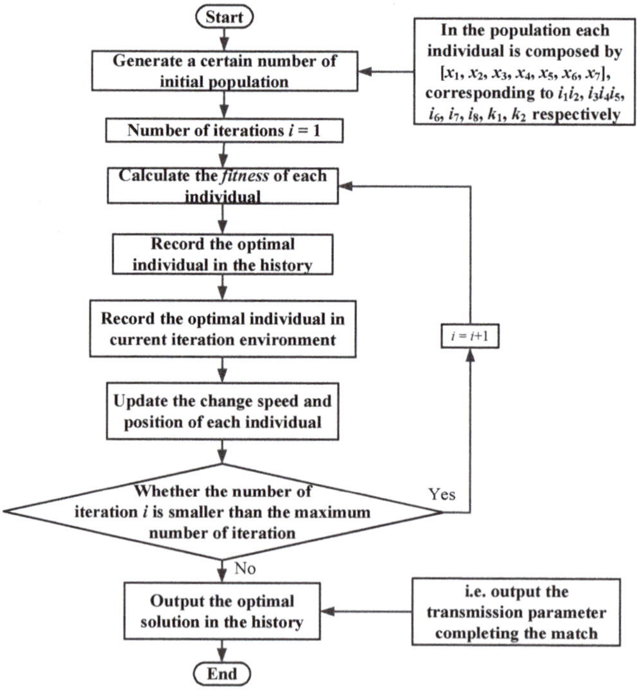


**Figure 6.** Flow diagram of optimization for speed regulating characteristics based on the I-PSO algorithm.

### 2.7. The Comparative Analysis Method of Speed Regulating Characteristics

To compare the optimization results of the speed regulating characteristics better, the paper uses three comparative analysis methods.

First, we calculate the errors of $i_{\mathrm{HM}_1}(\varepsilon=-1)$, $i_{\mathrm{HM}_1}(\varepsilon=1)$, and $i_{\mathrm{HM}_2}(\varepsilon=1)$ of the speed regulation of the tractor HMCVT after optimization with their expected values to verify the optimization result of the speed regulating characteristics. Meanwhile, we check whether the transmission ratios of $i_{\mathrm{HM}_2}(\varepsilon=-1)$, $i_{\mathrm{HM}_1}(\varepsilon=0)$ and the intersection position of section $\mathrm{H}_0$ and section $\mathrm{HM}_1$ are in the ranges specified by Equations (16)–(18).

Next, we compare and analyze the difference in speed regulating characteristics before and after optimization by calculating and generating figures.

Finally, based on the optimization results, we compare and analyze the difference in the speed regulating characteristics before and after correction by calculating and generating figures.

The research uses the coefficient of determination $R^2$ and the mean absolute percentage error *MAPE* for the analysis and evaluation of different types of speed regulating characteristics. See Equation (8) for the calculation formula of the coefficient of determination $R^2$.

The following is the calculation formula of the mean absolute percentage error *MAPE*.

$$MAPE=\frac{1}{n}\sum_{i=1}^{n}\left|\frac{y_{design}-y_{ideal}}{y_{ideal}}\right|\times 100\% \tag{19}$$

in which $y_{ideal}$ is the ideal expected value of the speed regulating characteristics of the HMCVT of the tractor, $y_{design}$ is the designed result of the speed regulating characteristics of the HMCVT of the tractor, and $n$ is the total number of data of the speed regulating characteristics.

## 3. Results and Discussion

### 3.1. The Results and Analysis of Speed Regulating Characteristics for 'Variable-Pump and Constant-Motor' System

Figures 7 and 8 show the test results of the speed regulating characteristics of the variable-pump constant-motor system. In the test, $\overline{n}_p$ (the mean working speed of variable pump) is 718.57 r/min and $\sigma_p$ (standard deviation) is 0.88 r/min. The results indicate that in the whole test process of the speed regulating characteristics, the variable pump's working speed remains essentially stable, with low volatility. The variable pump's working speed ultimately will not affect the constant motor's output speed (i.e., it will not affect the speed regulating characteristics of the pump–motor system).

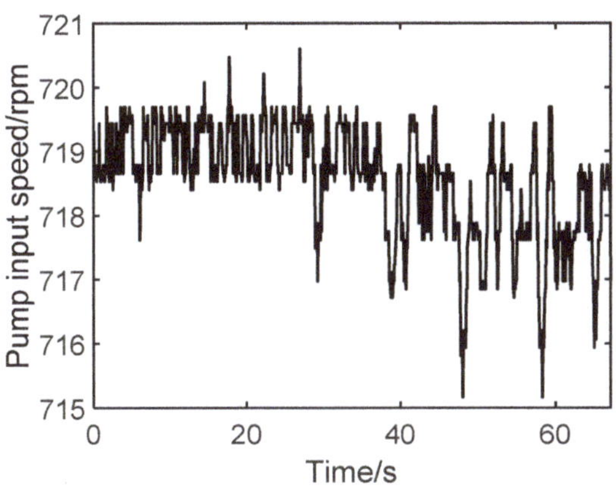


**Figure 7.** Measured input speed of variable pump.

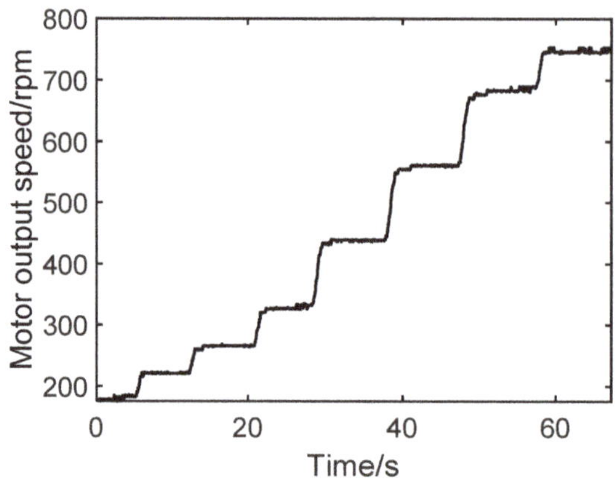


**Figure 8.** Measured output speed of constant motor.

In Figure 8, the output speed of the constant motor (showing the law of step rising) corresponds to the cases where the displacement ratio ε is 0.2, 0.25, 0.3, 0.375, 0.5, 0.625, 0.75, and 1, respectively, from left to right. We use the least squares method for the linear fitting of the variation characteristics of displacement ratio ε, and obtain the following result:

$$n_m = 760.90\varepsilon + 30.79 \tag{20}$$

$R^2$, the coefficient of determination of linear fitting, is 0.9775. The result shows that the variable-pump constant-motor system researched has a high degree of linearization, consistent with the original transmission law of the variable-pump constant-motor system. Meanwhile, it indicates that the variation characteristics between the tractor's speed and displacement ratio ε also show a linear relationship in the adjustment of the system.

Figure 9 shows the comparison between the measured and theoretical values of the output speed of the variable-pump constant-motor system.

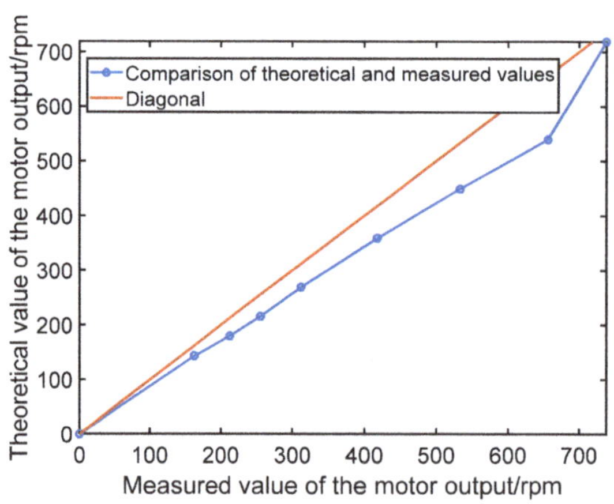


**Figure 9.** The comparison of theoretical and measured values for output speed of pump–motor system.

From Figure 9, we can see that the coefficient of determination $R^2$ between the measured and theoretical values is 0.8934. The result shows that the variation characteristics of the measured and theoretical values are basically consistent, but there is still a certain deviation. Therefore, correcting the model of the speed regulating characteristics of the variable-pump constant-motor system is particularly important.

### *3.2. The Results and Analysis of Regression-Based Correction of HMCVT's Speed Regulating Characteristics*

We use the least squares method for the regression analysis of Equations (10)–(13). Figure 10 shows the results.

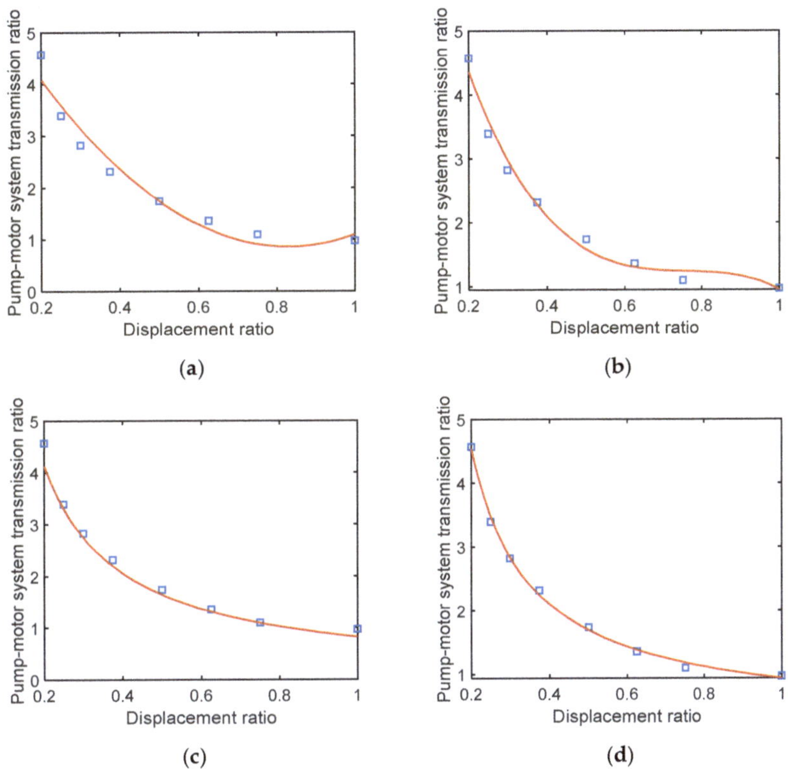


**Figure 10.** Correction results of speed regulating characteristics of variable-pump constant-motor system: (**a**) 2-order polynomial correction model; (**b**) 3-order polynomial correction model; (**c**) the first type of reciprocal correction model; (**d**) the second type of reciprocal correction model.

Table 3 shows the values and coefficients of determination $R^2$ of the parameters of the four correction models.

**Table 3.** The comparison of parameter values of coefficients of determination $R^2$ for four correction models.

| Type of Model | $a_1$ | $a_2$ | $a_3$ | $a_4$ | $R^2$ |
|---|---|---|---|---|---|
| 2-order Polynomial Correction Model | 8.1710 | −13.5400 | 6.4600 | - | 0.9544 |
| 3-order Polynomial Correction Model | −17.5100 | 39.7700 | −30.2700 | 8.9590 | 0.9848 |
| First Type of Reciprocal Correction Model | 0.8182 | −0.01756 | - | - | 0.9953 |
| Second Type of Reciprocal Correction Model | 0.2611 | 0.6309 | −0.04965 | - | 0.9978 |

According to Figure 10 and Table 3, the second reciprocal correction model has the proper number of parameters and the highest coefficient of determination. Therefore, we choose the second reciprocal correction model as the model of the speed regulating characteristics of the variable-pump constant-motor system. The research also shows that the correction based on the original law model produces a good result.

Therefore, the modified models of the speed regulating characteristics of the HMCVT of the tractor are shown as follows and the variation law of characteristics is shown in Figure 11.

$$i_{H_0} = -\frac{i_1 i_2 i_8 (0.2611\varepsilon + 0.6309)}{(\varepsilon - 0.04965)} \tag{21}$$

$$i_{HM_1} = \frac{i_1 i_2 i_3 i_4 i_5 i_6 k_1}{i_3 i_4 i_5 (\varepsilon - 0.04965)/(0.2611\varepsilon + 0.6309) + i_1 i_2 (1 + k_1)} \tag{22}$$

$$i_{HM_2} = \frac{i_1 i_2 i_3 i_4 i_5 i_7 (1 + k_2)}{i_1 i_2 k_2 - i_3 i_4 i_5 (\varepsilon - 0.04965)/(0.2611\varepsilon + 0.6309)} \tag{23}$$

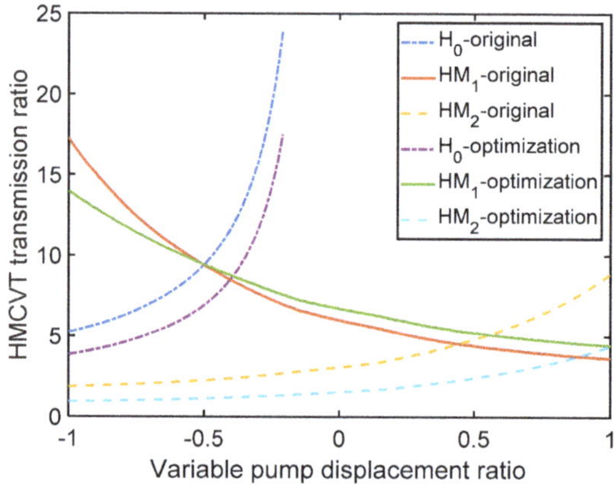


**Figure 11.** The comparison of speed regulating characteristics of transmission parameters before and after optimization.

### 3.3. The Results and Analysis of I-PSO-Based Optimization for Speed Regulating Characteristics

Figure 12 shows the iteration evolution curve of the optimization of the speed regulating characteristics based on the I-PSO algorithm.

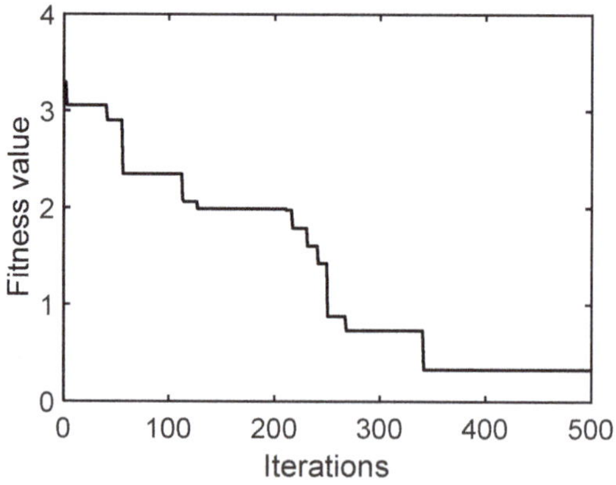


**Figure 12.** I-PSO-based iteration evolution curve.

According to Figure 12, for the strongly nonlinear problem, the I-PSO algorithm produces a good optimization result. From the perspective of the overall iteration process, the I-PSO algorithm's optimization result can evolve continuously. It indicates that the I-PSO algorithm can avoid the problem of prematurity well.

After the optimization, $i_{\mathrm{HM}_1}(\varepsilon = -1)$ has an error of 1.08% compared with the expected value 14.14; $i_{\mathrm{HM}_1}(\varepsilon = 1)$ has an error of 2.32% compared with the expected value 4.3; $i_{\mathrm{HM}_2}(\varepsilon = 1)$ has an error of 1.79% compared with the expected value 4.34. This indicates that the optimization results of the speed regulating characteristics are consistent with the expected results.

After the optimization, $i_{\mathrm{HM}_1}(\varepsilon = 1)$ and $i_{\mathrm{HM}_2}(\varepsilon = 1)$ have an error of 0.51%; the transmission ratio characteristics of section $\mathrm{H}_0$ and section HM1 have an intersecting point when displacement ratio $\varepsilon$ is $-0.39$. This shows that after the optimization design, the tractor HMCVT's speed regulating characteristics are continuous, without any point of discontinuity.

After the optimization, $i_{\mathrm{HM}_2}(\varepsilon = -1) = 0.95 < 2.17$. The results show that the highest driving speed of the tractor after optimization meets the requirement of >40 km/h. Moreover, after the optimization design, $i_{\mathrm{HM}_1}(\varepsilon = 0) = 6.74$, and the transmission ratio value in the intersection position of section $\mathrm{H}_0$ and section $\mathrm{HM}_1$ is 8.71. The optimization result completely meets the design requirement.

Figure 11 shows the comparison results of the speed regulating characteristics of the HMCVT of the tractor before and after optimization.

According to Figure 11, after the optimization, the position of the intersection point of section $\mathrm{HM}_1$ and section $\mathrm{HM}_2$ moves to the position with a larger displacement ratio and finally stops in the position with the displacement ratio of 1. This improves the available range of the displacement ratio, completely consistent with the design expectation.

Based on the optimization result, Figure 13 shows the comparison results of the speed regulating characteristics before and after correction.

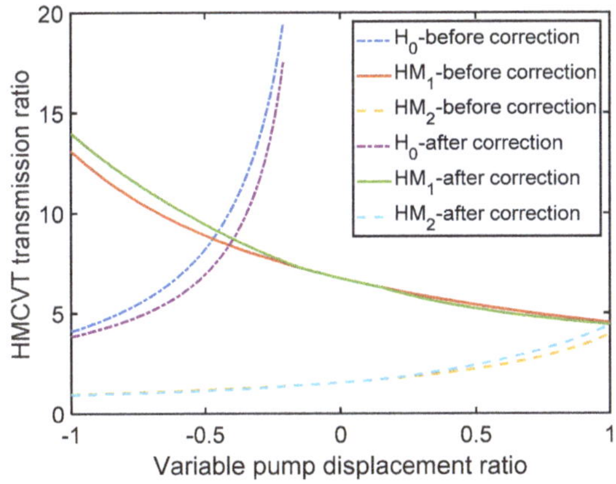


**Figure 13.** The comparison of speed regulating characteristics before and after correction.

According to Figure 13, $R^2$, the coefficient of determination, of the speed regulating characteristics of section $\mathrm{H}_0$ (choosing the transmission data with the displacement ratio of $-1 \leq \varepsilon \leq -0.4$), section $\mathrm{HM}_1$ (choosing the transmission data with the displacement ratio of $-1 \leq \varepsilon \leq -0.5$), and section $\mathrm{HM}_2$ (choosing the transmission data with the displacement ratio of $0.5 \leq \varepsilon \leq 1$) is 0.5139, 0.6257, and 0.6351, respectively, before and after correction; *MAPE* is 13.69%, 6.95%, and 10.24%, respectively. Meanwhile, section $\mathrm{HM}_1$ and section $\mathrm{HM}_2$ have no intersection point before correction. This shows that the speed regulating characteristics of the HMCVT are not even continuous. To sum up, if we use the speed regulating characteristics in ideal conditions for optimization design, the design results have certain errors compared with the actual transmission characteristics.

## 4. Conclusions

The paper proposes an analysis method for the linearization degree and deviation degree of the speed regulating characteristics of a variable-pump constant-motor system. The linear variation relationship between the actual output speed and displacement ratio of the tractor with the HMCVT is verified using the test bench results. Moreover, the research points out that the correction based on measured results is necessary for the speed regulating characteristics of the variable-pump constant-motor system.

The second reciprocal correction model proposed can be built easily (with only three parameters to be determined) and is highly consistent with the measured value of the speed regulating characteristics (the coefficient of determination is close to 1), so it should replace the original model of the speed regulating characteristics of the variable-pump constant-motor system. Moreover, the research shows that the correction of the original law model produces a good improvement result.

The method proposed for the optimization of the speed regulating characteristics and transmission parameter matching of agricultural machinery HMCVT is highly effective. Using the method combined with the I-PSO algorithm, the optimization design of the speed regulating characteristics and transmission parameter matching for the HMCVT can be completed effectively, merely by testing the output speed of the pump–motor system of the chosen model in the research, development, and design stage.

**Author Contributions:** Methodology, Z.C.; software, Z.C.; validation, Z.C.; investigation, Z.C.; resources, Z.L.; writing—original draft preparation, Z.C.; writing—review and editing, Z.C. and Z.L.; supervision, Z.L.; project administration, Z.C. and Z.L. All authors have read and agreed to the published version of the manuscript.

**Funding:** This research was funded by the National Natural Science Foundation of China (grant number: 52105063), National Key Research and Development Plan (2016YFD0701103), and Metasequoia Teacher Research Start-Up Fund of Nanjing Forestry University (163106061).

**Institutional Review Board Statement:** Not applicable.

**Informed Consent Statement:** Not applicable.

**Data Availability Statement:** The data presented in this study are available on demand from the corresponding author or first author at (luzx@njau.edu.cn or chengzhun38@163.com).

**Acknowledgments:** The authors thank the National Natural Science Foundation of China (grant number: 52105063), National Key Research and Development Plan (2016YFD0701103), and Metasequoia Teacher Research Start-Up Fund of Nanjing Forestry University (163106061) for funding. We also thank the anonymous reviewers for providing critical comments and suggestions that improved the manuscript.

**Conflicts of Interest:** The authors declare no conflict of interest.

## References

1. Liu, Z.; Zhang, G.; Chu, G.; Niu, H.; Zhang, Y.; Yang, F. Design matching and dynamic performance test for an HST-based drive system of a hillside crawler tractor. *Agriculture* **2021**, *11*, 466. [CrossRef]
2. Bulgakov, V.; Aboltins, A.; Ivanovs, S.; Holovach, I.; Nadykto, V.; Beloev, H. A mathematical model of plane-parallel movement of the tractor aggregate modular type. *Agriculture* **2020**, *10*, 454. [CrossRef]
3. Kalinichenko, A.; Havrysh, V.; Hruban, V. Heat recovery systems for agricultural vehicles: Utilization ways and their efficiency. *Agriculture* **2018**, *8*, 199. [CrossRef]
4. Sun, S.; Zhang, S.; Li, Y.; Wu, J.; Chu, J. Studies of several large-scale forestry operating vehicles at home and abroad and prospect of vehicle type design. *J. Beijing For. Univ.* **2019**, *41*, 154–166.
5. Lu, L.; Zhou, Y.; Li, H.; Wang, Y.; Yin, Y.; Zhao, J. Electro-hydraulic Shift Quality of Power Shift Transmission of Heavy Duty Tractor. *Trans. Chin. Soc. Agric. Mach.* **2020**, *51*, 550–556, 602.

6. Xi, Z.; Zhou, Z.; Zhang, M.; Cao, Q. Shift Characteristics and Control Strategy of Powershift Transmission on Tractor. *Trans. Chin. Soc. Agric. Mach.* **2016**, *47*, 350–357.
7. Wang, J.Y.; Xia, C.G.; Fan, X.; Cai, J.Y. Research on transmission characteristics of hydromechanical continuously variable transmission of tractor. *Math. Probl. Eng.* **2020**, *2020*, 6978329. [CrossRef]
8. Jenane, C.; Bashford, L.L. Tractive performance of a mechanical front-wheel assist tractor as related to forward speeds. *J. Agric. Eng. Res.* **2000**, *77*, 221–226. [CrossRef]
9. Tian, J.; Zeng, Q.K.; Wang, P.; Wang, X.Q. Active steering control based on preview theory for articulated heavy vehicles. *PLoS ONE* **2021**, *16*, e0252098. [CrossRef]
10. Xu, X.M.; Zhang, L.; Jiang, Y.P.; Chen, N. Active Control on Path Following and Lateral Stability for Truck-Trailer Combinations. *Arab. J. Sci. Eng.* **2019**, *44*, 1365–1377. [CrossRef]
11. Tian, J.; Tong, J.; Luo, S. Differential steering control of four-wheel independent-drive electric vehicles. *Energies* **2018**, *11*, 2892. [CrossRef]
12. Tian, J.; Wang, Q.; Ding, J.; Wang, Y.Q.; Ma, Z.S. Integrated control with DYC and DSS for 4WID electric vehicles. *IEEE Access* **2019**, *7*, 124077–124086. [CrossRef]
13. Zhou, W.L.; Zheng, Y.P.; Pan, Z.J.; Lu, Q. Review on the Battery Model and SOC Estimation Method. *Processes* **2021**, *9*, 1685. [CrossRef]
14. Qian, Y.; Cheng, Z.; Lu, Z.X. Bench testing and modeling analysis of optimum shifting point of HMCVT. *Complexity* **2021**, *2021*, 6629561. [CrossRef]
15. Bao, M.X.; Ni, X.D.; Zhao, X.; Li, S. Research on the HMCVT gear shifting smoothness of the four-speed self-propelled cotton picker. *Mech. Sci.* **2020**, *11*, 267–283. [CrossRef]
16. Zhang, M.; Wang, J.; Wang, J.; Guo, Z.; Guo, F.; Xi, Z.; Xu, J. Speed changing control strategy for improving tractor fuel economy. *Trans. Chin. Soc. Agric. Eng.* **2020**, *36*, 82–89.
17. Zhang, G.Q.; Zhang, H.T.; Ge, Y.Y.; Qiu, W.; Xiao, M.H.; Xu, X.M.; Zhou, M.H. Mechanical efficiency of HMCVT under steady-state conditions. *Shock Vib.* **2021**, *2021*, 4275922. [CrossRef]
18. Wan, L.R.; Dai, H.Z.; Zeng, Q.L.; Sun, Z.Y.; Tian, M.Q. Characteristic analysis and co-validation of hydro-mechanical continuously variable transmission based on the wheel loader. *Appl. Sci.* **2020**, *10*, 5900. [CrossRef]
19. Zhu, Z.; Gao, X.; Cao, L.L.; Cai, Y.M.; Pan, D.Y. Research on the shift strategy of HMCVT based on the physical parameters and shift time. *Appl. Math. Model.* **2016**, *40*, 6889–6907. [CrossRef]
20. Baek, S.M.; Kim, W.S.; Kim, Y.S.; Baek, S.Y.; Kim, Y.J. Development of a simulation model for HMT of a 50 kw class agricultural tractor. *Appl. Sci.* **2020**, *10*, 4064. [CrossRef]
21. Xu, L.; Zhou, Z.; Zhang, M.; Li, Y. Characteristics analysis of hydro-mechanical continuously variable transmission of tractor. *J. China Agric. Univ.* **2006**, *11*, 70–74.
22. Xu, L.; Zhou, Z.; Peng, Q.; Wang, B. Drive Scheme Design and Characteristic Analysis of Multi-range Hydro-mechanical CVT. *China Mech. Eng.* **2012**, *23*, 2641–2645.
23. Sung, D.; Hwang, S.; Kim, H. Design of hydromechanical transmission using network analysis. *Proc. Inst. Mech. Eng. Part D J. Automob. Eng.* **2005**, *219*, 53–63. [CrossRef]
24. Yu, J.; Wu, C.; Hu, Y.; Mou, J. Characteristic analysis of a new compound HMCVT. *J. Jiangsu Univ. Nat. Sci. Ed.* **2016**, *37*, 507–511.
25. Li, J.; Liu, L.; Xiao, M.; Wang, T.; Wang, X.; Zhang, H. Research on dynamic characteristics of hydro-mechanical continuously variable transmission. *J. Mech. Strength* **2017**, *39*, 14–19.
26. Xia, Y.; Sun, D.Y. Characteristic analysis on a new hydro-mechanical continuously variable transmission system. *Mech. Mach. Theory* **2018**, *126*, 457–467. [CrossRef]
27. Volpe, S.S.; Carbone, G.; Napolitano, M.; Sedoni, E. Design optimization of input and output coupled power split infinitely variable transmissions. *J. Mech. Des.* **2009**, *131*, 111002. [CrossRef]
28. Macor, A.; Rossetti, A. Optimization of hydro-mechanical power split transmissions. *Mech. Mach. Theory* **2011**, *46*, 1901–1919. [CrossRef]
29. Rossetti, A.; Macor, A. Multi-objective optimization of hydro-mechanical power split transmissions. *Mech. Mach. Theory* **2013**, *62*, 112–128. [CrossRef]
30. Zhang, Q.; Sun, D.Y.; Qin, D.T. Optimal parameters design method for power reflux hydro-mechanical transmission system. *Proc. Inst. Mech. Eng. Part D J. Automob. Eng.* **2019**, *233*, 585–594. [CrossRef]
31. Ni, X.; Zhu, S.; Ouyang, D.; Chang, Y.; Wang, G.; Nguyen, W.T. Design and experiment of hydro-mechanical CVT speed ratio for tractor. *Trans. Chin. Soc. Agric. Mach.* **2013**, *44*, 15–20.
32. Zhang, P.; Ni, X.; Mei, W.; Peng, X. Design and characteristic analysis of hydro-mechanical continuous variable transmission of cotton picker. *Mach. Des. Manuf.* **2017**, *10*, 64–66, 70.
33. He, C.; Lang, P.; Kang, M.; Zhang, H. Transmission design and force analysis of HMCVT for high power tractor. *J. Mech. Transm.* **2018**, *42*, 54–59.
34. Liu, F.X.; Wu, W.; Hu, J.B.; Yuan, S.H. Design of multi-range hydro-mechanical transmission using modular method. *Mech. Syst. Signal Process.* **2019**, *126*, 1–20. [CrossRef]

35. Cheng, Z.; Lu, Z.; Qian, J. A new non-geometric transmission parameter optimization design method for HMCVT based on improved GA and maximum transmission efficiency. *Comput. Electron. Agric.* **2019**, *167*, 105034. [CrossRef]
36. Cheng, Z. I-SA algorithm based optimization design and mode-switching strategy for a novel 3-axis-simpson dual-motor coupling drive system of PEV. *World Electr. Veh. J.* **2021**, *12*, 221. [CrossRef]
37. Yu, H.S.; Zhang, T.; Ma, Z.T.; Wang, R.P. Torsional vibration analysis of planetary hybrid electric vehicle driveline. *Trans. Chin. Soc. Agric. Eng.* **2013**, *29*, 57–64.
38. Hu, B.; Xu, P.; Gao, X.; Wang, Z. Matching calculation of hydrostatic transmission system of small agricultural loader. *Tract. Farm Transp.* **2014**, *41*, 22–24.
39. Cheng, Z.; Zheng, S.; Qian, Y.; Lu, Z.; Zhang, H. Based on improved SA and GA a new method for optimizing transmission parameters of automotive HMCVT. *J. Mech. Strength* **2020**, *42*, 61–66.
40. Tao, H.; Cao, W. Principle and application of polynomial regression and response surface analysis. *Stat. Decis.* **2020**, *36*, 36–40.
41. Zhang, D.; Zhang, X.; Yang, X.; Hou, X. Parameter analysis of vehicle-pedestrian accidents in untypical contact state based on orthogonal tests and polynomial regression analysis. *J. Shanghai Jiaotong Univ.* **2019**, *53*, 55–61.
42. Wang, G. Study on Characteristics, Control and Fault Diagnosis of Tractor Hydro-Mechanical CVT. Ph.D. Thesis, Nanjing Agricultural University, Nanjing, China, 2014.
43. Xu, X.M.; Lin, P. Parameter identification of sound absorption model of porous materials based on modified particle swarm optimization algorithm. *PLoS ONE* **2021**, *16*, e0250950. [CrossRef] [PubMed]
44. Chang, C.C.; Zheng, Y.P.; Yu, Y. Estimation for battery state of charge based on temperature effect and fractional extended kalman filter. *Energies* **2020**, *13*, 5947. [CrossRef]
45. Wang, H.; Zheng, Y.P.; Yu, Y. Joint estimation of soc of lithium battery based on dual kalman filter. *Processes* **2021**, *9*, 1412. [CrossRef]
46. Li, Y.J.; Ma, Z.S.; Zheng, M.; Li, D.X.; Lu, Z.H.; Xu, B. Performance analysis and optimization of a high-temperature PEMFC vehicle based on particle swarm optimization algorithm. *Membranes* **2021**, *11*, 691. [CrossRef] [PubMed]
47. Cheng, Z.; Lu, Z. Semi-empirical model for elastic tyre trafficability and methods for the rapid determination of its related parameters. *Biosyst. Eng.* **2018**, *174*, 204–218. [CrossRef]
48. Cheng, Z.; Lu, Z. A novel efficient feature dimensionality reduction method and its application in engineering. *Complexity* **2018**, *2018*, 2879640. [CrossRef]

*agriculture*

*Article*

# Parameters Optimization and Test of an Arc-Shaped Nail-Tooth Roller-Type Recovery Machine for Sowing Layer Residual Film

**Zhiyuan Zhang [1,2,3], Jingbin Li [1,2,3,*], Xianfei Wang [1,2,3], Yongman Zhao [1,2,3], Shuaikang Xue [1,2,3] and Zipeng Su [1,2,3]**

1 College of Mechanical and Electrical Engineering, Shihezi University, Shihezi 832003, China; zhangzy@stu.shzu.edu.cn (Z.Z.); wangxf@shzu.edu.cn (X.W.); zhrym@shzu.edu.cn (Y.Z.); 20192009026@stu.shzu.edu.cn (S.X.); szp@stu.shzu.edu.cn (Z.S.)
2 Xinjiang Production and Construction Corps Key Laboratory of Modern Agricultural Machinery, Shihezi 832003, China
3 Industrial Technology Research Institute of Xinjiang Production and Construction Corps, Shihezi 832003, China
* Correspondence: lijingbin@shzu.edu.cn

**Abstract:** The aim of this paper is to optimize the working parameters of the arc-shaped nail-tooth roller-type recovery machine for sowing layer residual film. Firstly, the tooth roller device of the residual film recovery machine is designed, and the main working parameters affecting the operation of the machine and the value range of each parameter are determined through the analysis of the operation process. Secondly, virtual simulation technology is used to establish a virtual simulation model of the interaction process between the tooth roller device and soil. At the same time, taking the soil-hilling quantity as the index, we build a quadratic regression mathematical model with three factors—the forward speed, rotation speed, and working depth—using the Box–Behnken method. Consequently, the analysis of the simulation results show that the order of the most significant factors is working depth, rotation speed, and forward speed. The optimal combination of working parameters are as follows: a forward speed of 4.5 km/h, a rotation speed of 43.2 r/min, and a working depth of 100.0 mm. Meanwhile, the predicted value of the soil-hilling quantity is 23.1 kg. Finally, we carried out field tests using the optimal combination parameters; the results show that the normal residual film collection rate is 66.8%, the soil-hilling quantity is 24.2 kg, and the relative error between the test value and the predicted value is 4.8%. This indicates that the devised DEM simulation model can be used to predict the operational performance of the tooth roller device in the working process. This study provides a reference that can be used in the planning and boundary enhancement of agricultural machinery and equipment.

**Keywords:** residual film recovery machine; DEM; virtual simulation; parameter optimization; agricultural machinery

**Citation:** Zhang, Z.; Li, J.; Wang, X.; Zhao, Y.; Xue, S.; Su, Z. Parameters Optimization and Test of an Arc-Shaped Nail-Tooth Roller-Type Recovery Machine for Sowing Layer Residual Film. *Agriculture* **2022**, *12*, 660. https://doi.org/10.3390/agriculture12050660

Academic Editors: Mustafa Ucgul and Chung-Liang Chang

Received: 5 April 2022
Accepted: 30 April 2022
Published: 3 May 2022



## 1. Introduction

Cotton planting has embraced the established model for film laying and tube laying in Xinjiang, which has reached 100%. In the meantime, the recovery rate of farmland residual film is just 60%, causing a significant amount of residual film to remain in the field, causing genuine white contamination of farmland; this has turned into an unmistakable issue that limits the green advancement of agriculture [1–3]. Mechanized residual film recuperation has developed into an unavoidable pattern in the treatment of residual film contamination [4,5]. In light of the radial plate arc-shaped nail-tooth roller structure, an arc-shaped formed nail-tooth roller-type planting layer residual film recovery machine (hereafter alluded to as the residual film recovery machine) was designed by our group. Field tests have shown that the average recovery rate of the residual film recovery machine is 67%, which can effectively recover residual film from the seeding layer. Nonetheless, the unmatched and unreasonable operation parameters of the residual film recovery machine have led to the phenomenon of hilling, subsequently resulting in the poor operational effect of the machine [6]. Therefore, the identification of sensible working parameters is important to improve the working efficiency of the residual film recovery machine.

With the widespread application of computer technology in agricultural production, the discrete element method (DEM) is also widely used in the simulation of the working parts of agricultural machinery, such as the process simulation of seeders and subsoilers, and the results can provide a reference for field tests [7]. Zhang [8] and Jiang et al. [9] conducted a simulation test study on the operating performance of the rotating drum device and spiral impurity cleaning device of the residual film recovery machine using DEM, and verified the reliability of the DEM approach combined with field tests. Wang [10], Tong [11], and Hang et al. [12] used DEM to study the influence of a subsoiler on soil tillage force and disturbance characteristics during operation. Dai et al. [13] concentrated on the working presentation of a crosswise belt-type whole plastic film-ridging residual corn seeder on double ridges by combining DEM simulation technology with field experiments, which further verified the superiority of the DEM. Hence, the combination of simulation technology and DEM has commonly been applied in the exploration of the cooperation law between the soil and soil contact of agricultural machinery, and has accomplished great outcomes. Compared with the traditional method of depending on an indoor soil bin test bench or genuine field test to optimize the operating parameters, the virtual simulation test using the DEM can not only shorten the test cycle and decrease the research and development work costs, but can also naturally and precisely control and measure the physical quantities of the machine, such as operation speed, operation depth, load, and soil resistance. Hence, the DEM is frequently utilized in the design, performance testing, and product optimization of agricultural machinery and equipment.

In view of the past research, this paper principally concentrates on the influence of various working parameters on the quality of soil hilling when the residual film recovery machine works using theoretical analysis, DEM simulation, and field testing. We obtain the scope of the key variables that influence the efficiency of the tooth roller device's activity by theoretical analysis. Moreover, we use the DEM to set up the simulation model between the soil and the tooth roller device. In addition, we find the optimum working parameters of the tooth roller device through response surface methodology, and validate the accuracy of the simulation results using the indoor soil trough test. This review can give significant hypothetical premise and specialized help for the design and parameter optimization of the key components of agricultural machinery equipment such as the residual film recovery machine.

## 2. Materials and Methods

### 2.1. *Structure of Residual Film Recovery Machine*

The residual film recovery machine fundamentally comprises the traction unit, the rack, the drive system, the tooth roller device, the film stripping unit, the depth-limiting wheel, and the film collection unit, as displayed in Figure 1. The proposed residual film recovery machine was trailed by a working tractor. The power take-off (PTO) shaft of the work tractor gives power to the drive system, and at that point, the drive framework distributes the power to the tooth roller device and the film-stripping unit to pick and strip residual films effectively.

### 2.2. *Design and Analysis of Tooth Roller Device*

#### 2.2.1. Design of Tooth Roller Device

The tooth roller device is the critical part of the residual film recovery machine, which mostly comprises the arc-shaped nail-tooth, base, flange, shaft head, central tube, and radial plate, as shown in Figure 2. The compelling working width is 1600 mm, the most extreme rotating diameter is 950 mm, and the distance between the two spiral plates is 236 mm. The bases are uniformly distributed in the peripheral direction of the tooth roller device. There are 12 or 13 arc-shaped nail-teeth on each row of holders. The distance between two adjacent teeth is 100 mm, and the circumferential arc-shaped nail-teeth are arranged in staggered mode.

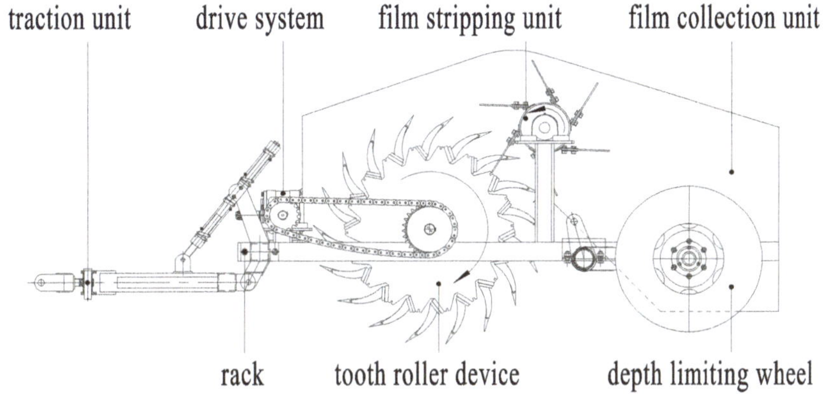


**Figure 1.** Structure of the residual film recovery machine.

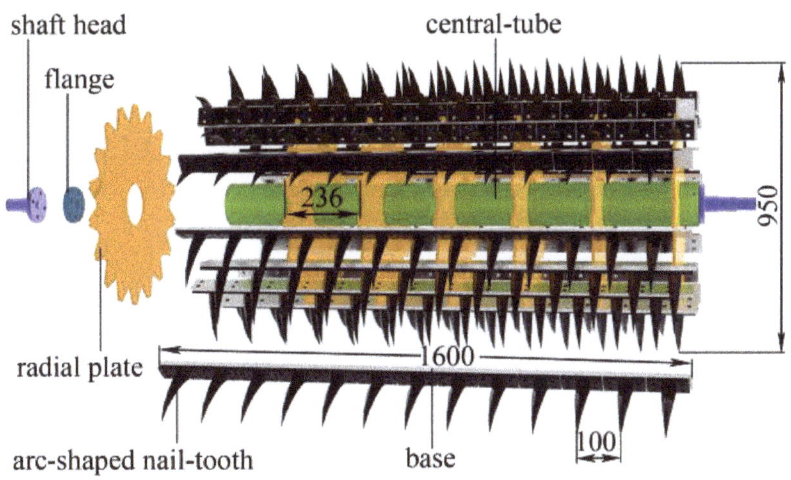


**Figure 2.** Structure diagram of the tooth roller device.

2.2.2. Analysis of Film-Picking Operation of Arc-Shaped Nail-Teeth

To finish the picking activity, the nail-teeth pierce through the residual film with the tooth tip when the machine is working. Meanwhile, the residual film is moved along with the arc-shaped nail-teeth to isolate it from the soil. Therefore, the residual film in the soil was taken as the research object for force analysis, and the rectangular coordinate system *o-xy* established with the *o* point, as shown in Figure 3.

The forces can be decomposed along the $x$-axis and $y$-axis, as follows:

$$\begin{cases} f + F_N = F_Z + F_L \cdot \cos\theta \\ F_L \cdot \sin\theta = mg \cdot \sin\delta + f_L \\ F_N = Mg \cdot \cos\delta \\ F_L = (M+m) \cdot v^2/R \\ v = 2\pi Rn \end{cases} \tag{1}$$

where $mg$ is simply the gravity of the residual film, N; $Mg$ is the gravity of soil on the film, N; $F_Z$ is the force on the residual film when the arc-shaped nail-tooth is tied, N; $F_N$ is the pressure of the soil on the residual film while it is attached with the curve-formed nail-tooth, N; $F_L$ is the force on the residual film when the tooth roller device rotates, N; $f$ is the frictional resistance of the residual film to the arc-shaped nail-tooth during operation, N; $f_L$ is the frictional resistance of soil to residual film during operation, N; $\theta$ and $\delta$ are the angles, rad; and $n$ is the rotation speed, r/min.

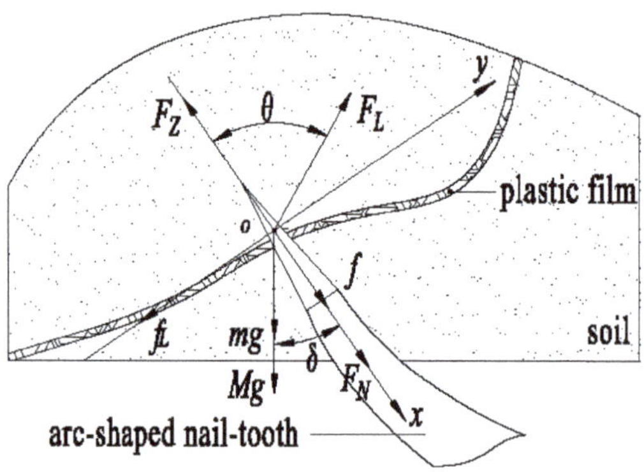


**Figure 3.** Schematic diagram of residual film force analysis during film-picking operation.

According to the empirical formula [14], the work used by the tooth roller device to rotate once is:

$$\begin{cases} Q = \frac{2\pi M_0}{zShb} + \frac{0.1F}{hb} \\ M_0 = F_Z \cdot R \end{cases} \tag{2}$$

where $Q$ is the work, J; $z$ is the number of arc-shaped nail-teeth, pcs; $S$ is the penetration pitch between adjacent arc-shaped nail-teeth, mm; $h$ is the working depth, mm; $b$ is the width of the arc-shaped nail-tooth, mm; $F$ is the traction force of the machine, N; $M_0$ is the torque, N m; and $R$ is the radius of the tooth roller device, mm.

Since the mulching film gravity $mg$ is very tiny, it can be ignored. In combination with (1) and (2):

$$Q = \frac{2\pi\left[4\pi^2R^2n^2M \cdot (\sin\theta + \cos\theta) - (Mg \cdot \cos\delta - f - f_L) \cdot R\right] + 0.1zSF}{zShb} \tag{3}$$

From Equation (3), we can see that when the structural parameters of the tooth roller device are fixed, the work used in the operation is related to the working parameters of the tooth roller speed, the quality of the soil on the film, and the working depth.

### 2.2.3. Determination of Tooth Roller Device Operation Parameters

If the rotation speed goes much faster when the tooth roller device is working, the disturbed soil is thrown forward, and the residual film on the tooth roller device moves with the soil. This is not helpful for the film-picking activity of the device. In contrast, if the rotation speed goes much slower, the soil generated by the disturbance will cause soil hilling in front of the residual film recovery machine, thus reducing the operating efficiency of the device and increasing energy consumption during the machine's activity. However, in our field test of the arc-shaped nail-tooth roller-type sowing layer residual film recovery machine, we found that an appropriate amount of soil hilling was conducive to the film-picking activity of the model. Henceforth, it was of extraordinary importance to look for sensible operation parameters to further develop the efficiency of the tooth roller device's operation.

The motion of the tooth roller device during its operation is compound motion, wherein the rotational motion around the center axis of the flanged drum is relative, and the motion along with the tractor is implicated motion. Thus, the motion track of the arc gear binding vertex *D* during the rotation of the tooth roller device is the trochoid, as shown in Figure 4.

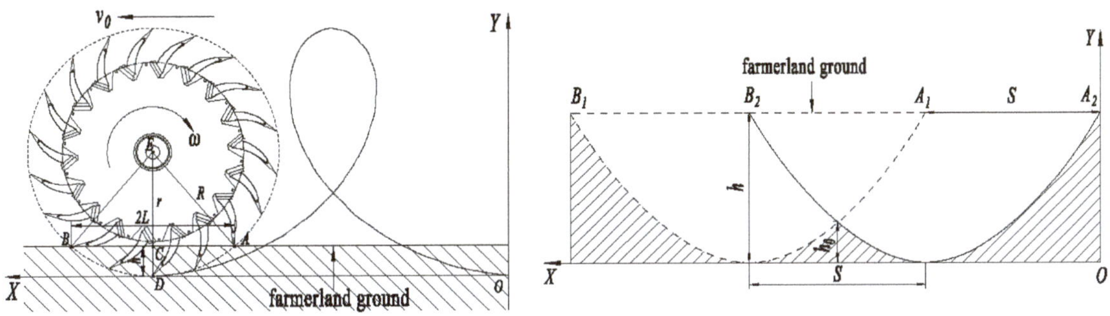


(a) (b)

**Figure 4.** Motion analysis of the tooth roller device. (**a**) Movement trail of the arc-shaped nail-tooth; (**b**) analysis of continuous work process.

In Figure 4a, the trochoid is the motion trajectory *OD* of the tooth tip *D* on the tooth roller device. The rectangular coordinate system *O-XY* is established with the point *O* when tooth tip *D* enters the deepest soil as the origin, the straight line *OD* as the *X*-axis, and the straight line opposite to *OD* going through *O* as the *Y*-axis. Then, at that point, the trajectory equation of the arc-shaped nail-tooth tip motion is:

$$\begin{cases} x = v_0 \cdot t + R \cdot \sin(\omega t) \\ y = R - R \cdot \cos(\omega t) \\ \lambda = v/v_0 > 1 \end{cases} \quad (4)$$

where $R$ is the maximum rotation radius of the tooth roller device, mm; $\omega$ is the angular velocity of rotation, rad/s; $v$ is the moving speed of the tooth tip D, m/s; $v_0$ is the forward speed of the tractor, m/s; $t$ is the movement time, s; and $\lambda$ is the speed ratio.

The field test results of the residual film recovery machine in the early stage showed that when the working velocity was $v_0$ = 4–6 km/h, the residual film recovery machine performed well and could complete the recovery of the residual film. Therefore, the forward speed of the machine is settled as $v_0$ = 4–6 km/h, $\lambda$ = 3–10, and we could obtain the rotational angular velocity of the gear roller device by taking them into Equation (4): $\omega$ = 2.14–14.43 rad/s. According to the relationship between the angular velocity of the circular motion and the velocity when the rigid body rotates, the rotation speed of the tooth roller device can be obtained as $n$ = 20.43–137.80 r/min.

The tooth roller device ought to guarantee that the residual film inside the extent of the sowing layer has been obtained, and guarantee that the residual film in the sowing layer was not missed during operation. Thus, the number of teeth embedded into the soil was one of the key factors that influenced operational efficiency. In addition, related studies show that in order to guarantee that the residual film in the soil does not leak when picked up, at least one arc-shaped nail-tooth remained in the soil when the first tooth was unearthed [1].

In Figure 4a, $AE$ is the maximum rotation radius of the nail-tooth roller on the tooth tip $D$, $AE = R$, mm; $EC$ is the radius of the tooth roller device, $EC = r$, mm; $AB$ is the distance between the relative track of the tooth tip $D$ relative to the rotation center $E$ of the tooth roller device and the ground, and $AC = BC = L$, mm; and $CD$ is the operating depth, $CD = h$, mm. Then, in $Rt\Delta ACE$, there is:

$$\begin{cases} R^2 = L^2 + r^2 \\ r = R - h \end{cases} \tag{5}$$

In Figure 4b, when the circumferential adjacent arc-shaped nail-teeth pierce through the films, there will be a protruding part in the soil. The protruding height $h_0$ is the distance from the joint of the trajectories of the arc-shaped nail-teeth to the maximum penetration depth of the nail-teeth. The cotton-sowing depth in Xinjiang is 25–35 mm, so the effective working depth of an arc-shaped nail-tooth is 140 mm. Then:

$$35 \text{ mm} \leq h - h_0 \leq 140 \text{ mm} \tag{6}$$

The area of residual plastic film in the Xinjiang cotton field is mostly 400–2500 mm$^2$ [15,16]. To simplify the calculation model, it can be considered that the shape of the residual film is square, and its side length is $l$ = 20–50 mm. Then, the numbers of teeth $N$ in the circumferential soil acquired by associating Equations (5) and (6) is:

$$N_{\text{min}} = L_{\text{min}} / l_{\text{max}} = 3.57 \tag{7}$$

That is, the number of teeth in the circumferential soil $N_{\text{min}} = 3$. When the circumferential adjacent nail-teeth successively penetrate the soil at time $t$, the distance of the device in the forward direction is the penetrating pitch $S$:

$$S = v_0 \cdot t = L/N \tag{8}$$

According to the structure of the tooth roller device, the interval time of the successively penetrated circumferential adjacent arc-shaped nail teeth is:

$$t = \frac{2\pi}{z\omega} \tag{9}$$

where $z$ is the number of rows of installed circumferential arc-shaped nail-teeth, row.

As indicated by Equations (4)–(9) above, it tends to be acquired that $z$ = 5.9–19.8, working depth $h$ = 43.8–140.0 mm. In this paper, according to the geometric characteristics of residual film in farmland, $z = 18$ was chosen to improve the picking efficiency of the residual film recovery machine and increase the number of wrapping films with arc-shaped nail-teeth.

### 2.3. DEM Simulation Model Establishment

#### 2.3.1. Simulation Modeling

The particle contact model is a fundamental basis for the accuracy of the calculation results of the simulation analysis model. To accurately simulate the interaction behavior between the soil and tools in the actual production process, the soil particle radius was set to 2.5 mm and the soil particle contact model selected was the Hertz–Mindlin with the JKR model by referring to references [8,9,17,18]. To speed up the simulation calculation, one-third of the total length of the tooth roller device was selected as the geometry simulation model; the 3D modeling was carried out with SolidWorks 2016 software (Dassault Systèmes S.E., Massachusetts, Concord, MA, USA) in the proportion 1:1, and was saved in the format of "-.step" and imported into the EDEM 2020 software

(DEM Solutions Ltd., Edinburgh, Scotland, UK). As indicated by the underlying size of the geometry simulation model, the size of the particle factory was established as $L \times W \times H$ = (4000 × 700 × 250) mm. The simulation model of the soil particles and geometry are displayed in Figure 5.

#### 2.3.2. Simulation Parameter Settings

The simulation fixed time step was set to $6.12 \times 10^{-5}$ s, the simulation time was set to 14.00 s, the simulation data storage interval was set to 0.05 s, and the mesh cell size of the geometry simulation model was set to 3.00 mm. According to the requirements of the simulation parameters in the virtual simulation test as well as the literature [8,9,19–22], the intrinsic parameters (density, Poisson's ratio, and shear modulus) and contact parameters (recovery coefficient, static friction coefficient, and dynamic friction coefficient) of the soil and geometry in this study were determined, as shown in Table 1.

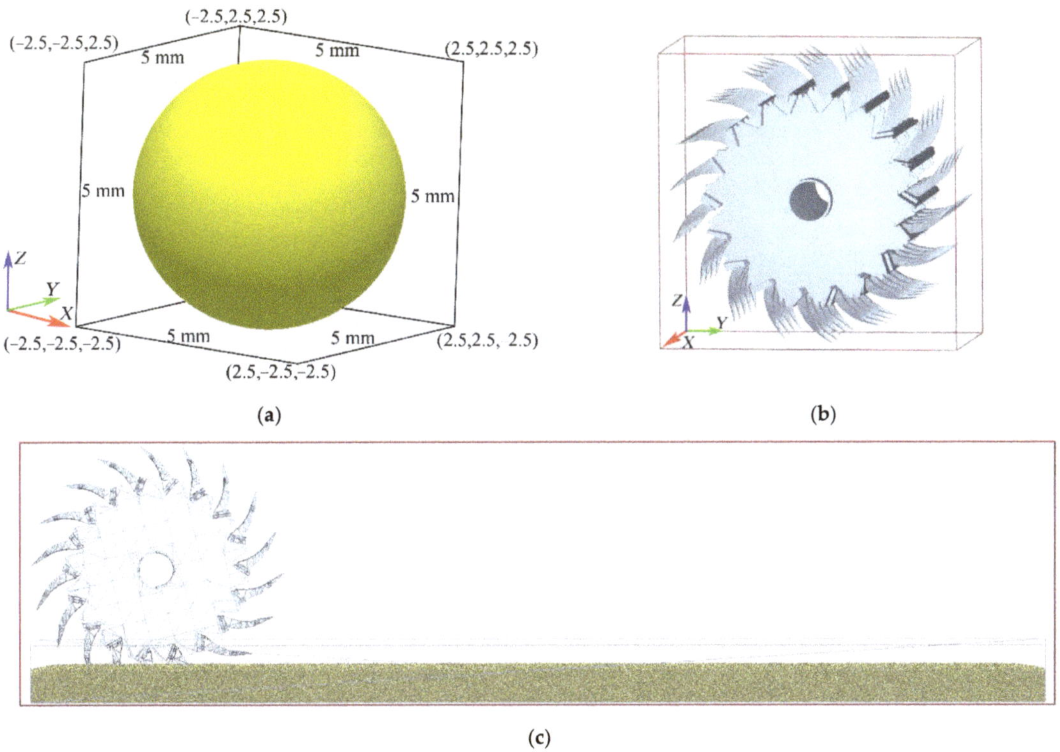


**Figure 5.** Particle and geometric simulation model. (**a**) Particle model of the soil; (**b**) geometric model of the tooth roller device; (**c**) simulation model of the EDEM.

**Table 1.** Simulation parameter settings of soil particles and geometry.

| Item | Parameter | Value |
|---|---|---|
| Soil particles | Poisson's ratio | 0.40 |
| | Shear modulus/MPa | $1.09 \times 10^6$ |
| | Density/(kg·m$^{-3}$) | 1400.00 |
| Arc-shaped nail-tooth | Poisson's ratio | 0.30 |
| | Shear modulus/MPa | $7.90 \times 10^{10}$ |
| | Density/(kg·m$^{-3}$) | 7850.00 |
| Particle—Particle | Recovery coefficient | 0.20 |
| | Static friction coefficient | 0.40 |
| | Dynamic friction coefficient | 0.30 |
| Particle—Arc-shaped nail-tooth | Recovery coefficient | 0.30 |
| | Static friction coefficient | 0.40 |
| | Dynamic friction coefficient | 0.10 |

#### 2.3.3. Design of Simulation Test

In the field experiments, the research group found that an appropriate soil-hilling phenomenon was conducive to the film-picking activity of the drum device. Therefore, the quality of soil accumulated on the ground (hereafter referred to as the soil-hilling quantity) was determined as the test response index. In addition, according to the analysis results in 2.2.2 and 2.2.3, the main operating parameters that affect the operation of the device were selected as the forward speed of the machine, the rotation speed of the nail-tooth roller, and the working depth, and the value range of each parameter was 4–6 km/h, 20.43–137.80 r/min, and 43.8–140.0 mm, respectively. Therefore, we used the forward speed of the machine $v_0$, the rotation speed of the nail-tooth roller $n$, and the working depth $h$ as the simulation test factors, and the quality of the soil hilling $Y$ as the testing indicators. The Box–Behnken test module method of Design-Expert V8.0.6.1 software (Stat-Ease Inc., Minneapolis, MN, USA) was used to design a central combined test with three factors and three levels, as shown in Table 2.

**Table 2.** Horizontal coding table of experimental factors.

| Levels | Factors | | |
|---|---|---|---|
| | $v_0$/km·h$^{-1}$ | $n$/r·min$^{-1}$ | $h$/mm |
| Upper level (1) | 6.00 | 140.00 (137.80) | 140.00 |
| Zero level (0) | 5.00 | 80.00 | 90.00 |
| Lower level (−1) | 4.00 | 20.00 (20.43) | 40.00 (43.75) |

## 3. Results and Discussion

### 3.1. Regression Analysis of Test Results

The Box–Behnken central combination tests comprised 17 groups, including 12 groups of analysis factors and 5 groups of zero-point estimation error tests. The analysis factor tests were repeated three times in each group, and the results were calculated as an arithmetic average. The experimental design and results are shown in Table 3.

**Table 3.** Experimental design and results.

| No. | Factors and Levels | | | Response Index |
|---|---|---|---|---|
| | $A$/km·h$^{-1}$ | $B$/r·min$^{-1}$ | $C$/mm | $Y$/kg |
| 1 | 4.00 | 20.00 | 90.00 | 8.14 |
| 2 | 6.00 | 20.00 | 90.00 | 22.89 |
| 3 | 4.00 | 140.00 | 90.00 | 16.21 |
| 4 | 6.00 | 140.00 | 90.00 | 27.11 |
| 5 | 4.00 | 80.00 | 40.00 | 6.84 |
| 6 | 6.00 | 80.00 | 40.00 | 16.46 |
| 7 | 4.00 | 80.00 | 140.00 | 34.37 |
| 8 | 6.00 | 80.00 | 140.00 | 49.02 |
| 9 | 5.00 | 20.00 | 40.00 | 3.26 |
| 10 | 5.00 | 140.00 | 40.00 | 10.51 |
| 11 | 5.00 | 20.00 | 140.00 | 36.48 |
| 12 | 5.00 | 140.00 | 140.00 | 39.37 |
| 13 | 5.00 | 80.00 | 90.00 | 25.99 |
| 14 | 5.00 | 80.00 | 90.00 | 25.87 |
| 15 | 5.00 | 80.00 | 90.00 | 26.64 |
| 16 | 5.00 | 80.00 | 90.00 | 25.63 |
| 17 | 5.00 | 80.00 | 90.00 | 26.87 |

Note: $A$ is the forward speed of machinery, km/h; $B$ is the rotating speed of nail-tooth roller, r/min; C is working depth, N; $Y$ is the soil-hilling quantity, kg.

We conducted multiple linear quadratic regression analyses on the simulation test results in Table 3 using the Design-Expert V8.0.6.1 software. The results are shown in Table 4.

The value of the model for the soil-hilling quality $Y$ was $p < 0.0001$, the value of the lack-of-fit term $p > 0.1$, the second-order response model determination coefficient $R^2 = 99.90\%$, the correction determination coefficient $R^2{}_{adj} = 99.77\%$, and the prediction determination coefficient $R^2{}_{pred} = 99.07\%$. This showed that the regression model is extremely significant, and could predict and optimize the soil-hilling quality $Y$ of the tooth roller device under different working parameters. According to the regression analysis of the variance of the results, multiple regression fitting analysis was conducted on the test results. Equation (10) is the mathematical regression model of the true value of each influencing factor on the response value.

**Table 4.** Variance analysis of regression models.

| Index | Difference Source | Sum of Squares | Degree of Freedom | Mean Square | $F$-Value | $p$-Value |
|---|---|---|---|---|---|---|
| Quantity of soil hilling/kg | Model | 2432.62 | 9 | 270.29 | 778.21 | <0.0001 |
| | $A$ | 311.50 | 1 | 311.50 | 896.86 | <0.0001 |
| | $B$ | 62.89 | 1 | 62.89 | 181.06 | <0.0001 |
| | $C$ | 1865.69 | 1 | 1865.69 | 5371.59 | <0.0001 |
| | $AB$ | 3.71 | 1 | 3.71 | 10.67 | 0.0137 |
| | $AC$ | 6.33 | 1 | 6.33 | 18.21 | 0.0037 |
| | $BC$ | 4.75 | 1 | 4.75 | 13.68 | 0.0077 |
| | $A^2$ | 11.78 | 1 | 11.78 | 33.91 | 0.0006 |
| | $B^2$ | 148.56 | 1 | 148.56 | 427.73 | <0.0001 |
| | $C^2$ | 19.37 | 1 | 19.37 | 55.78 | 0.0001 |
| | Residual | 2.43 | 7 | 0.35 | | |
| | Lake of Fit | 1.31 | 3 | 0.44 | 1.56 | 0.3304 |
| | Pure Error | 1.12 | 4 | 0.28 | | |
| $R^2 = 99.90\%$; $R^2{}_{\text{adj}} = 99.77\%$; $C.V. = 0.41\%$; $R^2{}_{\text{pred}} = 99.07\%$ | | | | | | |

Note: $p < 0.05$ (highly significant); $0.05 < p < 0.1$ (significant); $p > 0.1$ (not significant).

$$
\begin{aligned}
Y = {} & -79.36 + 21.98 \cdot A + 0.42 \cdot B + 0.05 \cdot C \\
& -0.02 \cdot AB + 0.03 \cdot AC - 3.63 \times 10^{-4} \cdot BC \\
& -1.67 \cdot A^2 - 1.65 \times 10^{-3} \cdot B^2 + 8.58 \times 10^{-4} \cdot C^2
\end{aligned} \quad (10)
$$

*3.2. Analysis of Interaction between Experimental Factors and Indicators*

3.2.1. Influence of Forward Speed and Rotation Speed on Soil-Hilling Quantity

As shown in Figure 6, when the working depth $h$ is at the level of 0 ($h$ = 90.0 mm), the interaction between the machine forward speed $v_0$ and the rotation speed $n$ can have an important impact on the soil-hilling quantity $Y$. As the forward speed $v_0$ of the machine tool increases, the soil-hilling quantity $Y$ increases and its change becomes slower. Meanwhile, the rotation speed $n$ increases, the soil-hilling quantity $Y$ increases at first, and then decreases, and its change tends to be faster. This indicates that the influence of rotation speed $n$ on the soil-hilling quantity $Y$ is more significant than the influence of the forward speed of the machine $v_0$.

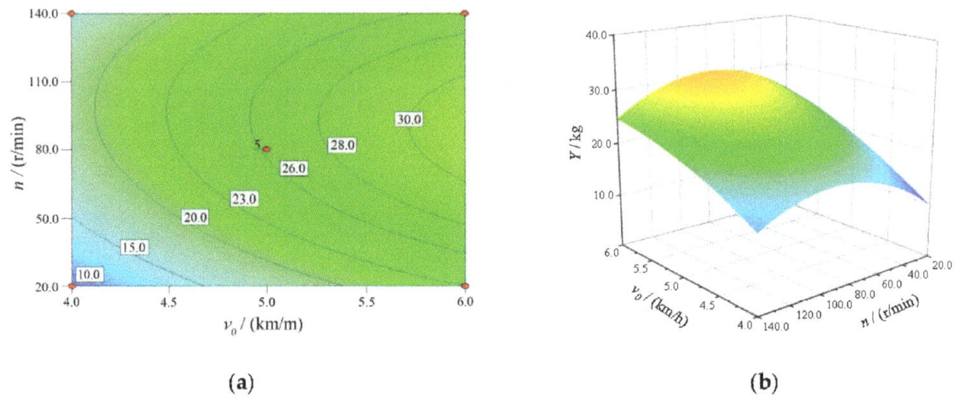


**Figure 6.** Influence of the forward speed and rotation speed on the soil-hilling quantity. (**a**) Contour map; (**b**) response surface diagram.

3.2.2. Influence of Forward Speed and Working Depth on Soil-Hilling Quantity

As shown in Figure 7, when the rotation speed $n$ is at the level of 0 ($n$ = 80.0 r/min), the interaction between the operating speed $v_0$ and the working depth $h$ affects the soil-hilling quantity $Y$. According to the figure, the soil-hilling quantity $Y$ increases and its variation range slows when the forward speed $v_0$ increases. As the working depth $h$ increases, the soil-hilling quantity $Y$ increases correspondingly, and its variation range becomes faster. This shows that the effect of working depth $h$ on the soil-hilling quantity $Y$ is more significant than that of the working speed $v_0$.

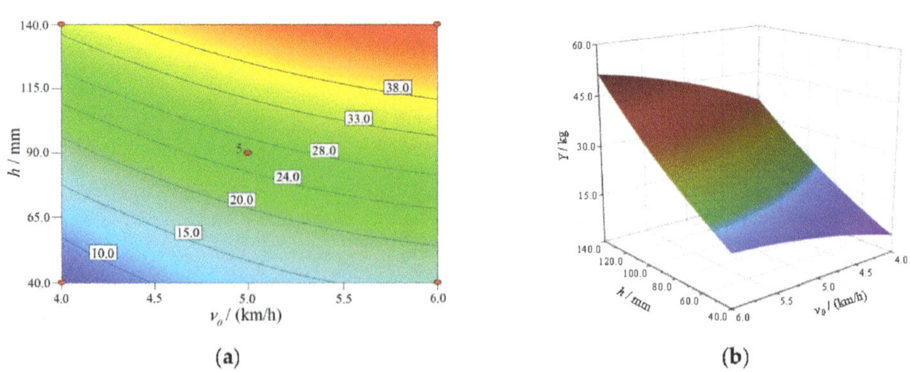


**Figure 7.** Influence of the forward speed and working depth on the soil-hilling quantity. (**a**) Contour map; (**b**) response surface diagram.

3.2.3. Influence of Rotation Speed and Working Depth on Soil-Hilling Quantity

As shown in Figure 8, the forward speed of the machine $v_0$ is at the level of 0 ($v_0$ = 5.0 km/h), the interaction between rotation speed n and working depth h also affects the soil-hilling quantity $Y$. It can be seen that when the working depth $h$ increases, the soil-hilling quantity $Y$ increases rapidly and its change range becomes faster; when the rotation speed $n$ increases, the soil-hilling quantity $Y$ begins increasing and then decreases with a slower change range. This shows that the impact of the working depth $h$ on the soil-hilling quantity $Y$ is more significant than that of the rotation speed $n$.

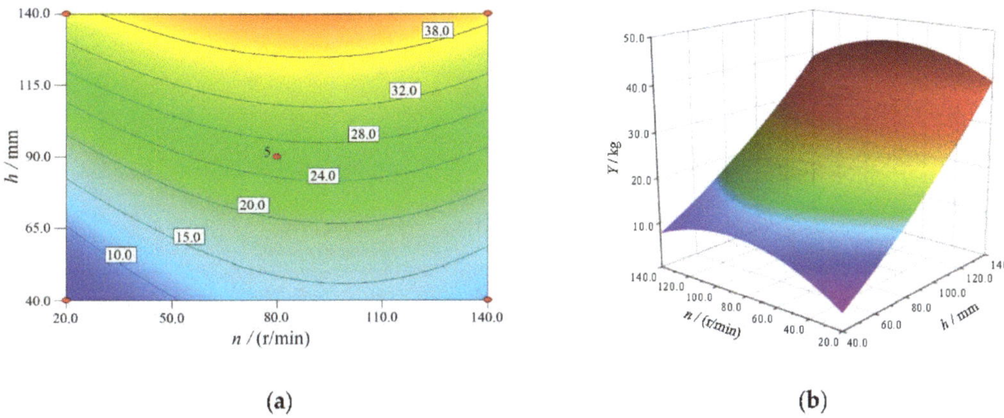


**Figure 8.** Influence of the forward speed and working depth on the soil-hilling quantity. (**a**) Contour map; (**b**) response surface diagram.

In conclusion, the order of significance in terms of the impact on the soil-hilling quantity $Y$ is as follows: working depth $h$, rotation speed $n$, forward speed $v_0$.

## 3.3. *Parameter Optimization and Verification*

### 3.3.1. Parameter Optimization

To obtain the combination of optimal parameters of the tooth roller device, we used the optimization module of Design-Expert V8.0.6.1 software [23,24]. The constraint conditions of the test factors and test indexes were:

$$\begin{cases} Y_{\min} = F(A,B,C) \\ s.t \begin{cases} A \in [4.0, 5.0] \\ B \in [40.0, 60.0] \\ C \in [80.0, 120.0] \end{cases} \end{cases} \tag{11}$$

The advancement results showed that the combination with the highest satisfaction degree was selected as the optimal combination of parameters: the forward speed $v_0$ is 4.5 km/h, the rotation speed $n$ is 43.2 r/min, the working depth $h$ is 100.0 mm, and the predicted value of the soil-hilling quantity $Y$ = 23.1 kg.

### 3.3.2. Field Verification Test

Figure 9 was the field test process. A field performance test of the prototype was carried out referring to the methods specified in the Agricultural Machinery Promotion and Appraisal Outline of the Ministry of Agriculture and Rural Affairs of the People's Republic of China, DG/T 149-2021 "film collector". The parameter combination of the forward speed of the machine, the rotation speed of the nail-tooth roller, and the working depth were 4.5 km/h, 43.2 r/min, and 100.0 mm, respectively. The working width of the machine was 1600 mm. All of the above tests were repeated five times, and the results were calculated as an arithmetic average, as displayed in Table 5.

**(a)** **(b)**

**Figure 9.** Residual film recovery machine test. (**a**) The test process; (**b**) the test results.

**Table 5.** Measurement results of the field test.

| **Items** | **Soil-Hilling Quantity** | | | | | | **Residual Films Collection Rate** |
|---|---|---|---|---|---|---|---|
| | **MAX/kg** | **MIN/kg** | **AVG/kg** | **C.V/%** | **PPV/kg** | **RE/%** | **AVG/%** |
| Value | 24.8 | 23.3 | 24.2 | 2.3 | 23.1 | 4.8 | 66.8 |

It can be seen from Table 5 that the soil-hilling quantity of the field test is 24.2 kg, the relative error is 4.8% compared with the predicted value, and the coefficient of variation of the five experimental results is 2.3% < 15%. The average residual films collection rate is 66.8%. The results show that the field test results are dependable, and the established EDEM simulation model could be used to predict the operation of the tooth roller device in the working process.

### *3.4. Discussion*

The results show that the EDEM simulation model established in this paper can be used to predict the operation performance of the tooth roller device in the working process, and the research can provide important theoretical and methodological support for the design and parameter optimization of agricultural machinery equipment, which is consistent with the research results of Liu [25], Pan [26], Jia [27], and Hu et al. [28]. In addition, Luo et al. [29] designed a chaining screen tillage residual film recovery machine, and the test results showed that the residual film recovery rate was 85.07%. Our research group [30] designed a nail-toothed roller residual film recovery device in the early stage, and the residual film recovery rate was 70.56%, higher than the residual film recovery rate in this study. Guo et al. [31] designed a residual film recovery machine for the plough layer, and the field test results showed that the residual film recovery rate was 55.04%, which was lower than that in this study. A comprehensive analysis of the reason is that all types of residual film recovery machine are in the stage of research and development, and the structure of the key operating components is the main reason for the different residual film recovery rates. At the same time, in the field tests of the prototype, the research group [32] found that the area of residual film debris, soil type, soil moisture content, and region could all affect the operational effect of the residual film recovery machine.

## 4. Conclusions

In this paper, given the working qualities of the residual film recovery machine, the theoretical analysis method was used to determine the range of key operating parameters of the tooth roller device. Then, EDEM 2020 software was used to set up the physical model of a virtual simulation experiment. The Box–Behnken test technique of the Design-Expert V8.0.6.1 software was used to establish the quadratic regression model with the soil-hilling quantity Y as the test index, taking the forward speed $v_0$, the rotation speed $n$, and the working depth $h$ as the test factors. Furthermore, the order of significant factors that influence the test indexes was determined, and the optimal working parameters of the device were obtained as follows: the forward speed $v_0$ = 4.5 km/h, the rotation speed $n$ = 43.2 r/min, the working depth $h$ = 100.0 mm, and the soil-hilling quantity $Y$ = 23.1 kg. Finally, field tests of the prototype were carried out using the best parameter combination. The results showed that the average collection rate of residual films is 66.8%, the soil-hilling quantity $Y$ is 24.2 kg, and the relative error is 4.8%. This indicates that the EDEM simulation model we established can be used to predict the operational performance of the tooth roller device in the working process, and the optimal combination of working parameters can be used as the basis for the actual operation of the residual film recovery machine.

**Author Contributions:** Conceptualization, methodology, data curation, formal analysis, writing—original draft, writing—review and editing, Z.Z. and J.L.; investigation, X.W.; data curation, Y.Z.; funding acquisition, J.L. and X.W.; validation, S.X.; supervision, Z.S. All authors have read and agreed to the published version of the manuscript.

**Funding:** This research was funded by the National Natural Science Foundation of China (52175240; 51765057) and the Autonomous Region Postgraduate Research Innovation Project of Shihezi University in 2022 (XJ2022G011).

**Institutional Review Board Statement:** Not applicable.

**Informed Consent Statement:** Not applicable.

**Data Availability Statement:** All data are presented in this article in the form of figures and tables.

**Conflicts of Interest:** The authors declare that they have no conflict of interest or financial conflicts to disclose. This article does not contain any studies with human or animal subjects performed by any of the authors.

## References

1. Chen, X.; Zhang, J.; Li, J.; Li, C.; Yang, Y. Design and Finite Element Analysis of Film Pick-up Components of Nail Tooth Drum Residual Film Recovery Device. *J. Agric. Mech. Res.* **2021**, *43*, 26–32.
2. Zhang, H.; Chen, X.; Yan, L.; Yang, S. Design and test of master-slave straw returning and residual film recycling combine machine. *Trans. Chin. Soc. Agric. Eng.* **2019**, *35*, 11–19.
3. Zhao, Y.; Chen, X.; Wen, H.; Zheng, X.; Niu, Q.; Kang, J. Research status and prospect of control technology for residual plastic film pollution in farmland. *Trans. Chin. Soc. Agric. Mach.* **2017**, *48*, 1–14.
4. Xue, S.; Chen, X.; Li, J.; Wang, X.; Zhang, Z. Design of and Experiment on a Film Removal Device of an Arc-Toothed Residual Film Recovery Machine before Sowing. *Appl. Sci.* **2021**, *11*, 8551. [CrossRef]
5. Liang, R.; Chen, X.; Zhang, B.; Meng, H.; Jiang, P.; Peng, X.; Kan, Z.; Li, W. Problems and countermeasures of recycling methods and resource reuse of residual film in cotton fields of Xinjiang. *Trans. Chin. Soc. Agric. Eng.* **2019**, *35*, 1–13.
6. You, J.; Zhang, B.; Wen, H.; Kang, J.; Song, Y.; Chen, X. Design and test optimization on spade and tine combined residual plastic film device. *Trans. Chin. Soc. Agric. Mach.* **2017**, *48*, 1–9.
7. He, Y.; Wu, M.; Xiang, W.; Yan, B.; Wang, J.; Bao, P. Application Progress of Discrete Element Method in Agricultural Engineering. *Chin. Agric. Sci. Bull.* **2017**, *33*, 133–137.
8. Zhang, H.; Yan, L.; Chen, X.; Jiang, D.; Yang, S. Simulation and test of film surface cleaning roller of residual film collector. *Int. Agric. Eng. J.* **2019**, *28*, 257–267.
9. Jiang, D.; Chen, X.; Yan, L.; Mo, Y.; Yang, S. Design and Experiment on Spiral Impurity Cleaning Device for Profile Modeling Residual Plastic Film Collector. *Trans. Chin. Soc. Agric. Mach.* **2019**, *50*, 137–145.

10. Wang, X.; Li, P.; He, J.; Wei, W.; Huang, Y. Discrete element simulations and experiments of soil-winged subsoiler interaction. *Int. J. Agric. Biol. Eng.* **2021**, *14*, 50–62. [CrossRef]
11. Tong, J.; Jiang, X.; Wang, Y.; Ma, Y.; Li, J.; Sun, J. Tillage force and disturbance characteristics of different geometric-shaped subsoilers via DEM. *Adv. Manuf.* **2020**, *8*, 392–404. [CrossRef]
12. Hang, C.; Huang, Y.; Zhu, R. Analysis of the movement behavior of soil between subsoilers based on the discrete element method. *J. Terramech.* **2017**, *74*, 35–43. [CrossRef]
13. Dai, F.; Guo, W.; Song, X.; Shi, R.; Zhao, W.; Zhang, F. Design and test of crosswise belt type whole plastic-film ridging-residual corn seeder on double ridges. *Int. J. Agric. Biol. Eng.* **2019**, *12*, 88–96.
14. Bernacki, H.; Haman, J.; Kanafojski, C. *Agriculture Machines, Theory and Construction*; China Machine Press: Beijing, China, 1985; Volume 1, pp. 171–195.
15. Hu, C.; Wang, X.; Chen, X.; Tang, X.; Zhao, Y.; Yan, C. Current situation and control strategies of residual film pollution in Xinjiang. *Trans. Chin. Soc. Agric. Eng.* **2019**, *35*, 223–234.
16. He, H.; Wang, Z.; Guo, L.; Zheng, X.; Zhang, J.; Li, W.; Fan, B. Distribution characteristics of residual film over a cotton field under long-term film residual and drip irrigation in an oasis agroecosystem. *Soil Tillage Res.* **2018**, *180*, 194–203. [CrossRef]
17. Liang, R.; Chen, X.; Jiang, P.; Zhang, B.; Meng, H.; Peng, X.; Kan, Z. Calibration of the simulation parameters of the particulate materials in film mixed materials. *Int. J. Agric. Biol. Eng.* **2020**, *13*, 29–36. [CrossRef]
18. Wang, X.; Hu, H.; Wang, Q.; Li, H.; He, J.; Chen, W. Calibration Method of Soil Contact Characteristic Parameters Based on DEM Theory. *Trans. Chin. Soc. Agric. Mach.* **2017**, *48*, 78–85.
19. Dai, F.; Song, X.; Zhao, W.; Zhang, F.; Ma, H.; Ma, M. Simulative Calibration on Contact Parameters of Discrete Elements for Covering Soil on Whole Plastic Film residual on Double Ridge. *Trans. Chin. Soc. Agric. Mach.* **2019**, *50*, 49–56, 77.
20. Ucgul, M.; Sannders, C.; Fielke, J. Discrete element modelling of top soil burial using a full scale mouldboard plough under field conditions. *Biosyst. Eng.* **2017**, *160*, 140–153. [CrossRef]
21. Zhao, S.; Wang, J.; Chen, J.; Yang, Y.; Tan, H. Design and Experiment of Fitting Curve Subsoiler of Conservation Tillage. *Trans. Chin. Soc. Agric. Mach.* **2018**, *49*, 82–92.
22. Yang, Y.; Wen, B.; Ding, L.; Li, L.; Chen, X.; Li, J. Soil particle modeling and parameter calibration for use with discrete element method. *Trans. ASABE* **2021**, *64*, 2011–2023. [CrossRef]
23. Ge, Y.; Zhang, L.; Gu, J.; Fu, W.; Zhu, R.; Zhang, H. Parameter optimization and experiment of dual roller harvesting device for safflower. *Trans. Chin. Soc. Agric. Eng.* **2015**, *31*, 35–42.
24. Shi, G.; Li, J.; Kan, Z.; Ding, L.; Ding, H.; Zhou, L.; Wang, L. Design and Parameters Optimization of a Provoke-Suction Type Harvester for Ground Jujube Fruit. *Agriculture* **2022**, *12*, 409. [CrossRef]
25. Liu, L.; Ma, C.; Liu, Z. EDEM-based Parameter Optimization and Experiment of Full-layer Fertilization Shovel for Strip Subsoiling. *Trans. Chin. Soc. Agric. Mach.* **2021**, *52*, 74–83.
26. Pan, J.; Chen, F.; Hu, J.; Yue, R.; Yao, M.; Li, J. Simulation and Optimization Design of Finger-clamping Seedling Picking Claw Based on EDEM-RecurDyn. *Trans. Chin. Soc. Agric. Mach.* **2022**. Available online: https://kns.cnki.net/kcms/detail/11.1964.s.20220316.2211.008.html (accessed on 26 March 2022).
27. Jia, L.; Jiang, Y.; Sun, J.; Fu, T.; Jiang, L.; Wang, J. EDEM-based simulation and test of disc blade rollers of disc spring tooth combination type paddy grader. *Acta Agric. Univ. Jiangxiensis* **2021**, *43*, 1406–1414.
28. Hu, T.; Hu, J.; Bi, Y.; Yang, M.; Li, M.; Yang, J. Design of Sea-buckthorn roller screen based on EDEM numerical. *J. Chin. Agric. Mech.* **2021**, *42*, 122–129.
29. Luo, K.; Yuan, P.; Jin, W.; Yan, J.; Bai, S.; Zhang, C.; Zhang, X. Design of chain-sieve type residual film recovery machine in plough layer and optimization of its working parameters. *Trans. Chin. Soc. Agric. Eng.* **2018**, *34*, 19–27.
30. Chen, X.; Chen, X.; Li, J.; Li, C.; Yang, Y. Design and test of nail-teeth roller-type residual film recovery device before sowing. *Trans. Chin. Soc. Agric. Eng.* **2020**, *36*, 30–39.
31. Guo, W.; He, X.; Wang, L.; Zhao, P.; Hu, C.; Hou, S.; Wang, X. Development of a comb tooth loosening and pneumatic stripping plough layer residual film recovery machine. *Trans. Chin. Soc. Agric. Eng.* **2020**, *36*, 1–10.
32. Wang, X.; Zhang, Z.; Zhao, Y.; Xue, S.; Li, J. Design and test of arc-shaped film binding nail teeth. *J. Chin. Agric. Mech.* **2021**, *42*, 24–29.

agriculture

MDPI

*Article*

# Optimization Design Based on I-GA and Simulation Test Verification of 5-Stage Hydraulic Mechanical Continuously Variable Transmission Used for Tractor

**Zhun Cheng [1,*], Yuting Chen [1], Wenjie Li [1], Pengfei Zhou [1], Junhao Liu [1], Li Li [1], Wenjuan Chang [1] and Yu Qian [2]**

[1] Department of Vehicle Engineering, Nanjing Forestry University, Nanjing 210037, China; edward_2022@126.com (Y.C.); wy13813938850@163.com (W.L.); zhoupengfei0328@163.com (P.Z.); junhao12378988@163.com (J.L.); maxleegk@163.com (L.L.); c15251893977@163.com (W.C.)

[2] College of Engineering, Nanjing Agricultural University, Nanjing 210031, China; m18763892368@163.com

* Correspondence: cz38@njfu.edu.cn

**Abstract:** To improve the characteristics of a tractor power transmission system, make the transmission system not only meet the geometric principle but also meet the working speed requirements of the tractor, realize the continuous adjustment of the speed ratio, and solve the power interruption caused by the change of speed ratio. This paper not only proposes a new type of five-stage hydraulic mechanical continuously variable transmission (HMCVT), but also proposes a geometric design method based on I-GA. In this paper, the intersection position of each working section is introduced in the optimization design process, and six wet clutches, one brake and two planetary rows are used to achieve a flexible geometric design. The results show that the HMCVT proposed in this paper can effectively match the required speed of the tractor. After optimization, the transmission ratio characteristics are consistent with the target characteristics (the average error is about 3.27% and the common ratio is about 1.81); the simulation results based on Simulation X are highly consistent with the theoretical design values (the MAPE of 36 simulation experiments is about 0.72%); and the maximum speed of the tractor is 41.62 km/h. The proposed HMCVT design scheme and optimization design method provide a direct basis for the research and development of the composite power transmission system of tractors or other vehicles and agricultural machinery.

**Keywords:** hydro-mechanical continuously variable transmission; tractor; optimization design; simulation experiments; I-GA; geometric principle

**Citation:** Cheng, Z.; Chen, Y.; Li, W.; Zhou, P.; Liu, J.; Li, L.; Chang, W.; Qian, Y. Optimization Design Based on I-GA and Simulation Test Verification of 5-Stage Hydraulic Mechanical Continuously Variable Transmission Used for Tractor. *Agriculture* **2022**, *12*, 807. https://doi.org/10.3390/agriculture12060807

Academic Editors: Mustafa Ucgul and Chung-Liang Chang

Received: 26 April 2022
Accepted: 31 May 2022
Published: 2 June 2022



## 1. Introduction

As an off-road vehicle, a tractor plays an important role in agricultural machinery [1,2]. The working conditions of tractors are complex, mainly for rotary tillage, ditching, bulldozing, ploughing, harrowing, sowing, cultivation, harvesting, road transportation and other work [3]. The complex operating conditions have a clear requirement for the power source and transmission system of the tractors, that is, that the vehicle has a wide speed range. Whether electrical motor (for example, battery electric vehicles mainly use power batteries [4,5] and power systems [6]) or internal combustion engine, centralized drive systems are commonly used in tractors. For example, Li et al. [7] proposed a dual-input coupling powertrain system (DICPS), which provides centralized power through two motors, and then transmits power to the driving wheel by the cooperation of two brakes and gearboxes; Chen et al. [8] studied the parameter-matching and optimization design of a dual-motor centralized drive system; Li et al. [9] and Xia and Sun [10] studied the characteristics of the variable-speed transmission system driven by a single engine. The transmission system had a planetary gear mechanism and the power was converged and then output. The speed ratio variation characteristics of a variable speed system is one of the decisive factors for the reasonable adjustment of tractor speed. In addition, when the tractor speed ratio changes, it is required that the power does not break as much as possible and the whole process be smooth [11]. The energy consumption is also related to the variation range of speed ratio for the power transmission system and the number of speed ratios that can be changed. To sum up, it is of great significance to study the design of a tractor transmission system.

As a kind of continuously variable transmission (CVT) [12,13], HMCVT can achieve many beneficial effects under the condition of reasonable design. Transmission ratio characteristics can be designed to vary continuously [14]. Power does not interrupt when the transmission ratio changes. The power transmission is smooth and stable during the change of transmission ratio. This can be well-matched with the speed requirements of tractor operations. At present, the characteristics analysis and control of the power-split CVT system have been studied to some extent. For example, Ince and Guler [15] analyzed the power flow and efficiency characteristics of the power-split CVT. Based on bench test and stepwise regression analysis, Qian et al. [16] studied the method of reducing shift impact on HMCVT. Xiao et al. [17] analyzed the characteristics of a new type of tractor HMCVT, and conducted the speed ratio matching research based on tractor fuel economy combined with simulations and tests.

At present, for the design of HMCVT, many studies first discuss the transmission scheme, and then design the parameters of each component in the transmission scheme based on geometric principle. In addition, the tractor's transmission system is often designed with a geometric transmission scheme. Ju et al. [18] pointed out that the transmission ratio of a tractor stepped-transmission system should be designed according to the geometric principle. The study calculated the transmission system of various types of tractors, and concluded that these tractors basically conform to geometric principle. According to geometric principle, Ni et al. [19] designed the forward working section of a tractor HMCVT as 4 sections, and the transmission common ratio of each working section is 1.88. He et al. [20] designed the tractor HMCVT with the variation range of the transmission ratio from 0.533 to 6.667, and calculated and matched each transmission parameter based on geometric principle in design. However, the design flexibility of the traditional geometric principle is insufficient, and the calculation of matching transmission parameters by solving equations is cumbersome. The solving process often depends on the initial assumptions, and limits the matching degree with the required speed of tractor operation.

In order to solve the above problems, expand the variation range of tractor's speed ratio and improve the flexibility of transmission parameter optimization design, this paper proposes a new transmission scheme of HMCVT for tractors and an optimization design method combining the intersection position of each working section and I-GA. According to the optimization design method proposed in this paper, the optimal matching of transmission parameters can be effectively realized by introducing the intersection position of each working section. It plays a positive role in the decision-making stage of transmission parameters based on geometric principle. In addition, this paper builds a tractor HMCVT simulation model based on Simulation X software and conducts corresponding verification. This paper provides considerable reference for the research and development of tractor (or other agricultural machinery, engineering vehicles, and so on) power transmission systems.

## 2. Materials and Methods

### 2.1. A Five-Stage HMCVT Designed in This Paper

The five-stage HMCVT is mainly composed of a variable displacement pump, fixed displacement motor, fixed axis gears (transmission ratio of gear pair is $i_1$, $i_R$, $i_3$, $i_4$, $i_5$, $i_6$, $i_p$ and $i_m$, respectively), wet clutches ($C_V$, $C_R$, $C_1$, $C_2$, $C_3$ and $C_4$, respectively), planetary gear sets ($P_1$ and $P_2$, transmission characteristic parameters are $k_1$ and $k_2$), and brake $B_1$, as shown in Figure 1.

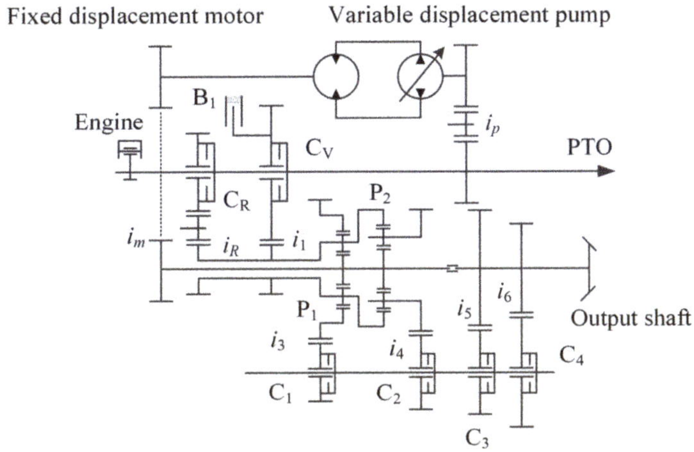


**Figure 1.** Schematics of five-stage HMCVT proposed in this paper.

### 2.2. Working Principle of HMCVT

The HMCVT designed in this paper has five forward sections, one pure hydraulic working section ($H_0$ section) and four hydraulic power flow and mechanical power flow coupling working section ($HM_1$~$HM_4$ section). The schematic diagram of power flow in different stages is shown in Figure 2. The red line in Figure 2 represents the transmission route of the power flow in the current working mode.

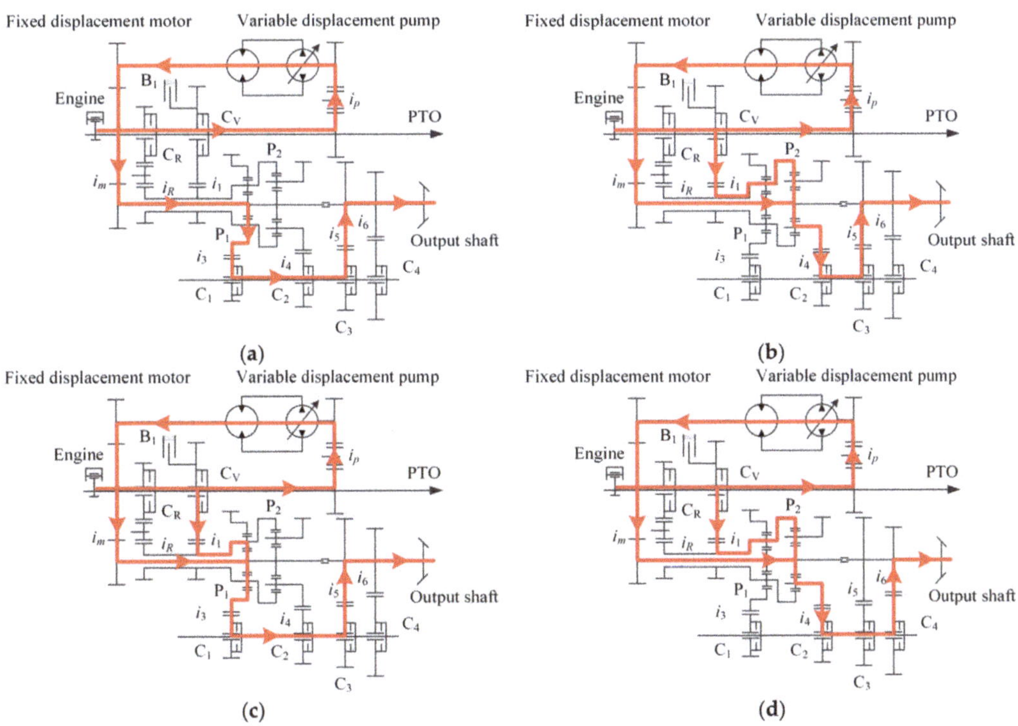


**Figure 2.** *Cont.*

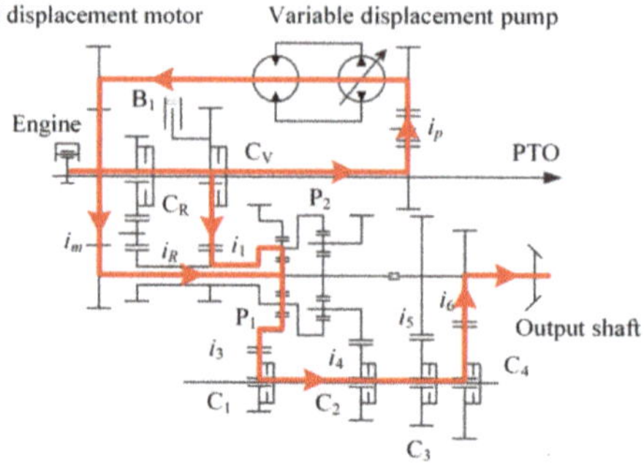


(e)

**Figure 2.** The schematic diagram of power flow in different stages. (**a**) $H_0$ section; (**b**) $HM_1$ section; (**c**) $HM_2$ section; (**d**) $HM_3$ section; and (**e**) $HM_4$ section.

In the $H_0$ section mode, the carrier of planetary gear set $P_1$ is locked because of the brake $B_1$ working. The engine output power is input to the sun gear of the planetary gear set $P_1$ only through a variable displacement pump, fixed displacement motor, front gear pair of pumps and rear gear pair of motors. The power is then input to the output shaft by the ring gear of planetary gear set $P_1$, gear pair $i_3$ (wet clutch $C_1$ engaged) and $i_5$ (wet clutch $C_3$ engaged).

In $HM_1$~$HM_4$ section mode, the engagement of wet clutch $C_V$ leads to the engine output power input to the carrier of planetary gear set $P_1$ and the ring gear of $P_2$ through gear pair $i_1$. The other part of the engine output power is input to the variable displacement pump through the gear pair $i_p$, and then input to the gear sun of planetary gear set $P_1$ and $P_2$ through the fixed displacement motor and gear pair $i_m$ (the gear sun of planetary gear set $P_1$ and $P_2$ is designed to be fixed on the same drive shaft). Finally, the power is input to the output shaft through the gear pair $i_3$ (or $i_4$) and gear pair $i_5$ (or $i_6$) by the engagement and separation of wet clutch $C_1$ (or $C_2$) and $C_3$ (or $C_4$).

In addition, with the same working principle of the forward sections, the HMCVT designed in this paper has 5 backward sections (wet clutch $C_R$ is engaged in this state). The specific working modes logic of HMCVT proposed in this paper is shown in Table 1 (taking forward sections as an example, if it is the backward stages, only $C_V$ is changed to $C_R$).

**Table 1.** Working modes logic.

| Working Mode | List of Brake and Wet Clutches | | | | | |
|---|---|---|---|---|---|---|
| | $C_V$ | $B_1$ | $C_1$ | $C_2$ | $C_3$ | $C_4$ |
| $H_0$ section | × | • | • | × | • | × |
| $HM_1$ section | • | × | × | • | • | × |
| $HM_2$ section | • | × | • | × | • | × |
| $HM_3$ section | • | × | × | • | × | • |
| $HM_4$ section | • | × | • | × | × | • |

Note: • represents that the wet clutch (or brake) is working, × represents that the wet clutch (or brake) is out of working.

## 2.3. Variation Characteristics of HMCVT Transmission Ratio

According to the speed calculation formula of planetary carrier, ring gear and sun gear of the planetary gear mechanism [21], the variation characteristics of the HMCVT transmission ratio at each working section are derived as follows:

$$i_{H_0} = \frac{i_p i_m k_1 i_3 i_5}{\varepsilon} \tag{1}$$

$$i_{\text{HM}_1} = -\frac{(1+k_2)i_p i_m i_1 i_4 i_5}{k_2 i_p i_m + \varepsilon i_1} \tag{2}$$

$$i_{\text{HM}_2} = \frac{k_1 i_p i_m i_1 i_3 i_5}{-(1+k_1)i_p i_m + \varepsilon i_1} \tag{3}$$

$$i_{\text{HM}_3} = -\frac{(1+k_2)i_p i_m i_1 i_4 i_6}{k_2 i_p i_m + \varepsilon i_1} \tag{4}$$

$$i_{\text{HM}_4} = \frac{k_1 i_p i_m i_1 i_3 i_6}{-(1+k_1)i_p i_m + \varepsilon i_1} \tag{5}$$

In the formula, $i_{\text{H}_0}$, $i_{\text{HM}_1}$, $i_{\text{HM}_2}$, $i_{\text{HM}_3}$ and $i_{\text{HM}_4}$ are the transmission ratios of $\text{H}_0$, $\text{HM}_1$, $\text{HM}_2$, $\text{HM}_3$ and $\text{HM}_4$, respectively.$\varepsilon$ is the displacement ratio.

### *2.4. Matching of Tractor Speed and HMCVT Transmission Ratio*

Considering the variety of tractor working types (low speed working conditions mainly include rotary tillage, ditching, bulldozing and so on; medium–high speed working conditions mainly include plowing, harrowing, sowing, cultivation, harvesting, road transportation and so on), the tractor working types are often divided by driving speed. According to the introduction of this paper, the design of a tractor transmission system generally adopts the geometric principle. Based on the research of Chen [22], Ni et al. [19] and Sun [23], this paper sets the speed of a tractor at low-speed operating conditions (corresponding to $\text{H}_0$ section) at 0–4 km/h, at medium–low-speed operating conditions (corresponding to $\text{HM}_1$ section) at 4–7.2 km/h (the ratio is 1.80), at medium-speed operating conditions (corresponding to $\text{HM}_2$ section) at 7.2–12 km/h (the ratio is 1.67), at medium–high-speed operating conditions (corresponding to $\text{HM}_3$ section) at 12–20 km/h (the ratio is 1.67) and at high-speed operating conditions (corresponding to the $\text{HM}_4$ section) at 20–35 km/h (the ratio is 1.75). The ratio of speed variation range of each section above is approximate (the ratio is about 1.72).

The working speed range of the tractor engine is set to 750–2200 rpm, the rolling radius of the driving wheel is 0.976 m, and the total transmission ratio of other transmission mechanisms is 26.866. The calculation formula of tractor speed is as follows. Speed calculation formula reference [7,24,25]. Because the research is oriented to the conceptual design stage of the novel HMCVT proposed in this paper, the research ignores the wheel slip.

$$u_a = 0.377\frac{r_d n_e}{i_0 i_{cvt}} \tag{6}$$

where $u_a$ is the driving speed of tractor;$r_d$ is the radius of wheels; $n_e$ is the working speed of engine; $i_0$ is the total transmission ratio of other transmission mechanism; and $i_{cvt}$ is the transmission ratio of HMCVT.

Based on the speed design of geometric principle and combined with Equation (6), the target transmission ratio range of each working section of HMCVT is matched.

### *2.5. Optimal Design Method Based on Geometric Principle and Considering the Intersection Position of Each Working Section*

The standard design method based on geometric principle makes the ratio of the maximum transmission ratio and the minimum transmission ratio of each HMCVT working section approximately equal, so as to match the transmission design parameters (the transmission ratio of each gear pair and the characteristic parameters of planetary gear mechanism) by solving the equation. The equation to be solved is as follows:

$$\frac{\max(i_{\text{HM}_1})}{\min(i_{\text{HM}_1})} \approx \frac{\max(i_{\text{HM}_2})}{\min(i_{\text{HM}_2})} \approx \frac{\max(i_{\text{HM}_3})}{\min(i_{\text{HM}_3})} \approx \frac{\max(i_{\text{HM}_4})}{\min(i_{\text{HM}_4})} \approx q \tag{7}$$

where $\max(i_{HM})$ is the maximum transmission ratio of HM section; $\min(i_{HM})$ is the minimum transmission ratio of HM section; and $q$ is the common ratio.

Apparently, the value of $\max(i_{HM})$ and $\min(i_{HM})$ is critical. According to Equations (1)–(5), the transmission ratio of each HMCVT working section is a monotonic function of displacement ratio $\varepsilon$. For HMCVT, the variation characteristics of the transmission ratio at each working section intersect each other (Figure 3). This makes the tractor transmission ratio change continuously and reduces the vibration and shock during shift. Therefore, the effective section of transmission ratio for each HMCVT section is the area in the intersection of left and right ends (for example, in Figure 3, the effective section for the $HM_2$ section is marked). In summary, the transmission ratio of each HMCVT working section is not necessarily the maximum or minimum when the displacement ratio $\varepsilon = \pm 1$. Uncertainty of the intersection position of the HMCVT sections leads to difficulty in solving Equation (7) (the traditional geometric design method often assumes that when the displacement ratio $\varepsilon = \pm 1$, the transmission ratio of each HMCVT working section has the maximum or minimum value).

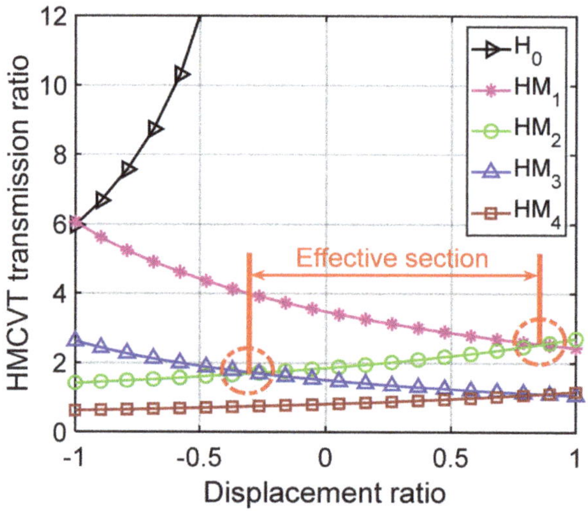


**Figure 3.** The transmission ratio characteristic and effective section of each working mode.

In this paper, a new optimization design method based on geometric principle is proposed, which considers the intersection position of each working section. Based on the original design individuals to be optimized (the transmission ratio of each gear pair and the characteristic parameters of a planetary gear mechanism), this method introduces four intersection position individuals (the five-stage HMCVT proposed in this paper has four intersections). The intersection position individual is the value of the displacement ratio $\varepsilon$ at the intersection of each HMCVT working stage. The individuals to be optimized are:

$$X = \left[ i_p i_m \ i_1 \ i_3 \ i_4 \ i_5 \ i_6 \ k_1 \ k_2 \ j_1 \ j_2 \ j_3 \ j_4 \right] \tag{8}$$

where $j_{1\sim4}$ is the x-coordinate of the intersection of $H_0$ and $HM_1$, $HM_1$ and $HM_2$, $HM_2$ and $HM_3$, and $HM_3$ and $HM_4$ (the value of the displacement ratio $\varepsilon$).

The optimization design objective function $fitness$ of the proposed method in this paper is:

$$fitness(X) = \sum_{m=1}^{4} \left( \left| i_{j_m} - z_m \right| + \left| i'_{j_m} - z_m \right| \right) \tag{9}$$

where $j_m$ is the displacement ratio at $m$ intersection; $z_m$ is the design target value at $m$ intersection; $i_{j_m}$ is the transmission ratio at $j_m$ for the HMCVT low-speed section; and $i'_{j_m}$ is the transmission ratio at $j_m$ for the HMCVT high-speed section.

In the optimization process, the HMCVT's minimum transmission ratio corresponding to the maximum speed of the tractor as the constraint condition, the flow chart of the proposed method in this paper is shown in Figure 4.

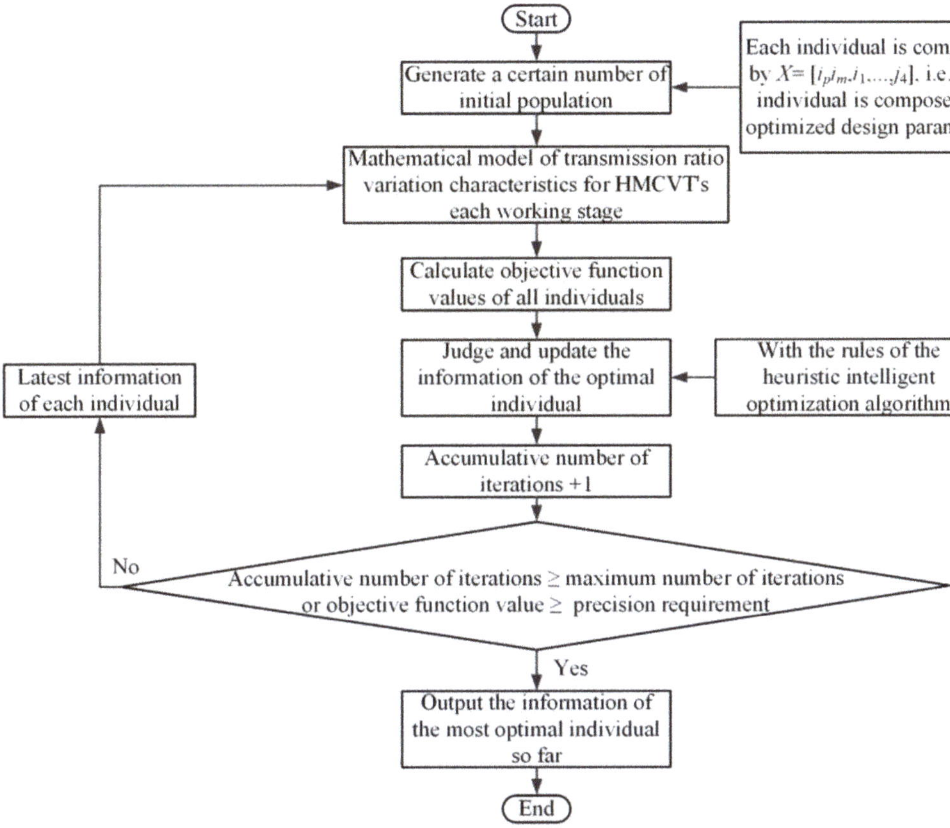


**Figure 4.** The flow chart of the method proposed in this paper.

### 2.6. Algorithm for Optimization Design

The proposed HMCVT in this paper has a complex structure and many parameters to be optimized. The objective function of this optimization problem is nonlinear. The heuristic optimization algorithm [25–28] is suitable for solving a series of complex engineering problems. The genetic algorithm (GA) is combined with the optimization design method proposed in this paper to complete the matching of transmission ratio parameters. GA is an evolutionary algorithm with the survival of the fittest as the basic rule of iterative computation and evolution. It is suitable for multi-parameter and multi-objective engineering optimization problems. To further improve the effect of parameter optimization design, this paper uses I-GA, which has been proposed and verified in previous studies [29]. The I-GA improves the convergence speed and accuracy of the algorithm mainly from several aspects. The I-GA screen and delete the super individuals in the population. Adaptive changes of the population size, crossover and mutation probabilities are introduced into the I-GA.

The range of population size is set as 1000–2000. The rest of the algorithm parameters are consistent with the reference [29]. The values of $z_m$ can be seen in Section 3.1 in the paper, which are 7.53, 4.18, 2.51 and 1.51, respectively.

*2.7. Simulation Test Platform Based on Simulation X Software*

Simulation X is a multidisciplinary and widely used simulation software. Its modular modeling method is suitable for the research object of this paper. The simulation model of a five-stage HMCVT is built by the built-in planetary gear model, friction load model, fixed axis gear model, clutch model and engine model (Figure 5).

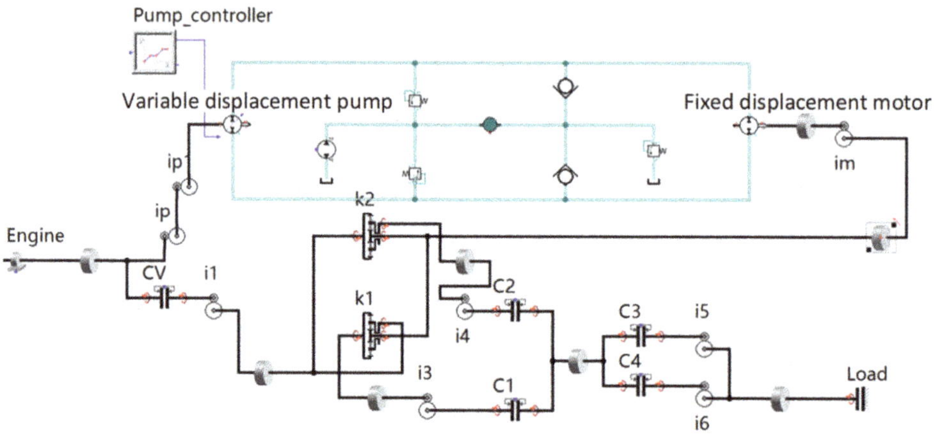


**Figure 5.** The simulation model of an HMCVT based on Simulation X.

## 3. Results and Discussion

*3.1. Matching Results of Tractor Speed and HMCVT Transmission Ratio*

The variation range of the target transmission ratio at each HMCVT working section and its corresponding relationship with tractor speed are shown in Figure 6.

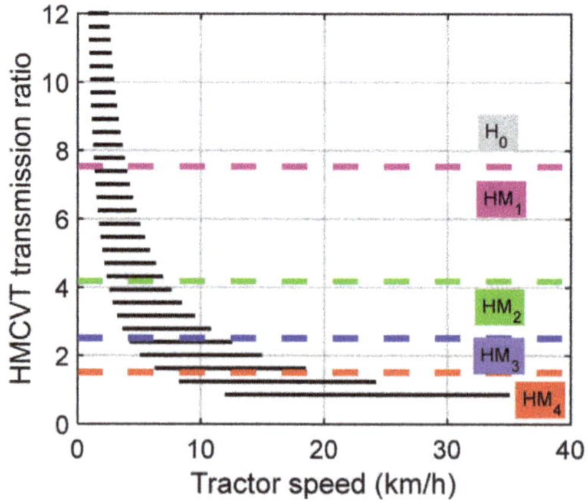


**Figure 6.** The variation range of target transmission ratio and its corresponding relationship with tractor speed.

According to the calculation results of geometric principle, the target transmission ratios of each HMCVT working section are: 7.53~+∞ ($H_0$ section), 4.18~7.53 ($HM_1$ section), 2.51~4.18 ($HM_2$ section), 1.51~2.51 ($HM_3$ section) and 0.86~1.51 ($HM_4$ section), respectively.

### *3.2. Results of Transmission Ratio Optimization Design*

The transmission parameters of the five-stage HMCVT are optimized by the proposed method. Since the number of decision variables is 12, the design freedom of the expanded objective function is 8. The number of decision variables is greater than the design degrees of freedom of the objective function. Therefore, multiple feasible solutions will be generated after optimization iteration. This paper draws on the design ideas of Kisssoft. There are many feasible solutions in mechanical system design. In this paper, the optimization design method is independently executed three times to obtain three feasible solutions. The results of the three feasible solutions are compared and one of them was selected as the final design scheme of this paper. The results are shown in Figure 7 and Table 2.

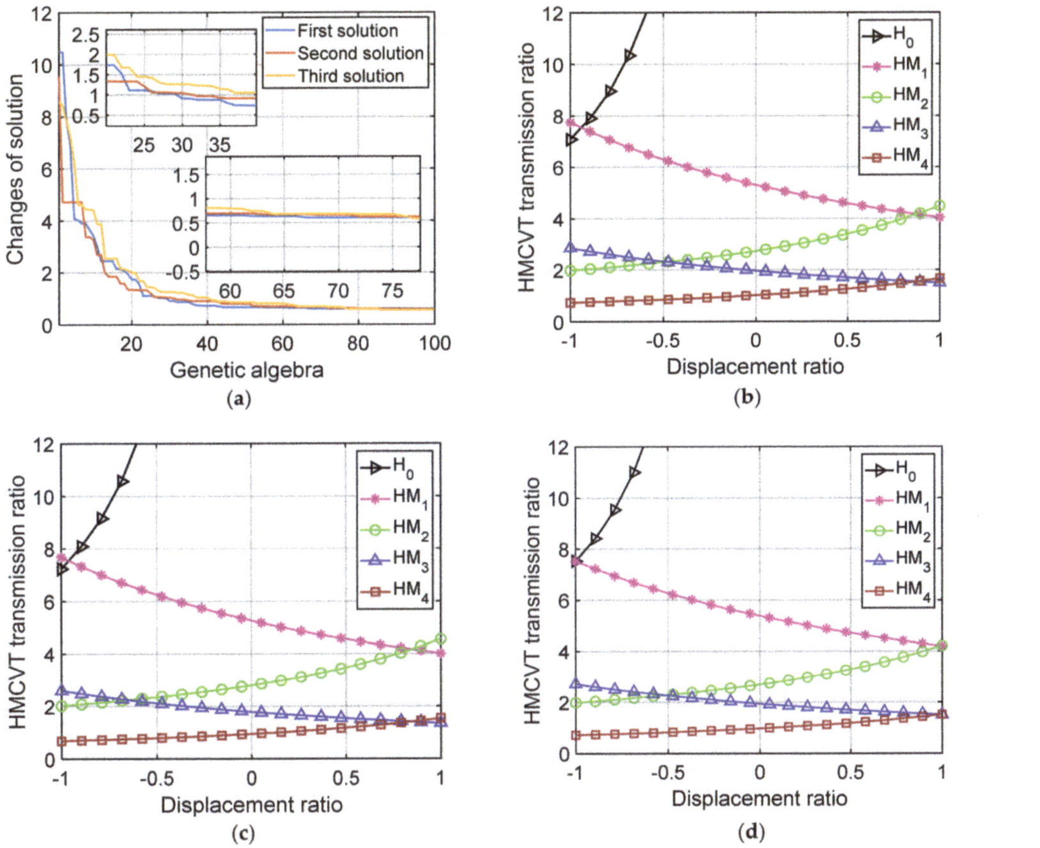


**Figure 7.** Optimization results of the proposed method. (**a**) Iterative evolution curve; (**b**) characteristics obtained by the first solution; (**c**) characteristics obtained by the second solution; (**d**) characteristics obtained by the third solution.

**Table 2.** Comparison of multiple optimization results.

| **Serial Number** | $i_{pim}$ | $i_1$ | $i_R$ | $i_3$ | $i_4$ | $i_5$ | $i_6$ |
|---|---|---|---|---|---|---|---|
| I | 2.40 | 2.50 | 2.50 | 0.60 | 0.56 | 2.92 | 1.08 |
| II | 0.80 | 0.88 | 0.88 | 2.77 | 2.64 | 1.76 | 0.59 |
| III | 1.18 | 1.27 | 1.27 | 1.16 | 1.22 | 2.76 | 1.00 |
| Serial Number | $k_1$ | $k_2$ | $j_1$ | $j_2$ | $j_3$ | $j_4$ | Mean error |
| I | 1.68 | 3.31 | −0.94 | 0.87 | −0.48 | 0.88 | 2.62% |
| II | 1.85 | 3.50 | −0.96 | 0.85 | −0.65 | 0.85 | 4.69% |
| III | 2.00 | 3.79 | −1.00 | 1.00 | −0.52 | 1.00 | 2.49% |

From Figure 7 and Table 2, the three optimization results converge to the same minimum. The application of I-GA in this project has good effect. The algorithm can continuously decline during the iterative process (the initial optimization speed of the algorithm is fast and the optimization speed tends to be gentle in the middle and late stages of the algorithm, but it still has the ability of continuous decline). This effectively avoids the premature phenomenon of the algorithm. After the optimization of three transmission parameters, the HMCVT's transmission characteristics are consistent with the target characteristics (the mean error is 2.62%, 4.69% and 2.49%, respectively). And the minimum HMCVT transmission ratio is less than 0.86. The maximum speed of the tractor meets the design requirements.

In the first optimization, the ratios of the transmission ratio's effective section of each HMCVT working section are 1.81, 1.80, 1.50 and 2.10, respectively. In the second optimization, the ratios of the transmission ratio's effective section of each HMCVT working section are 1.76, 1.86, 1.59 and 2.04, respectively. In the third optimization, the ratios of the transmission ratio's effective section of each HMCVT working section are 1.80, 1.82, 1.52 and 2.11, respectively. The three optimization results are 1.80, 1.81 and 1.81, respectively; the variances are 0.06, 0.04 and 0.06, respectively (the data fluctuation is small). Therefore, the designed HMCVT basically meets the geometric design requirements.

According to the three optimization design results, the gear pair transmission parameters $i_{pim}$, $i_1$, $i_3$, $i_4$ and $i_5$ can be designed in a wide range of changes. The $i_6$ mainly works at mid–high-speed and high-speed state of the tractor, and its design range is small and the value is small. The characteristic parameters of the planetary gear that output by the ring gear are smaller ($k_1 \leq 2$), while the characteristic parameters of the planetary gear that output by the planetary carrier are larger ($k_2 \geq 3$). Compared with the results of the first and second optimization, the results of the third optimization have a wider range of the displacement ratio's utilization intervals (the utilization ratio of the displacement ratio's interval of $HM_2$ and $HM_3$ is 76%, while the interval utilization ratio of the other three working sections is 100%, and the average utilization ratio of the whole working section is 90.4%). Meanwhile, the results of the first optimization and the second optimization are both partial utilization of the displacement ratio's intervals (the average utilization rates of the whole working section are 82.8% and 85.8%, respectively). In addition, the third optimization result has the smallest error. In summary, the third optimization result is selected as the final design result.

### *3.3. Simulation Test Verification Based on Simulation X*

The $H_0$ section is a pure hydraulic working section (the hydraulic power flow transmitted only by variable displacement pump and fixed displacement motor in the tractor transmission system), and basically only acts when the tractor starts. The backward-working principle of the HMCVT is consistent with the forward-working principle, and only one idler wheel is added for the transmission. Therefore, the simulation experiment in this paper verifies the transmission ratio characteristics of the hydraulic power flow and mechanical power flow coupling working sections ($HM_1$, $HM_2$, $HM_3$ and $HM_4$ sections) when the tractor moves forward. Each working section divides the displacement ratio into nine levels (the displacement ratios are −1, −0.75, −0.5, −0.25, 0, 0.25, 0.5, 0.75 and 1, respectively) and carries on each level simulation test one by one, with a total of 36 groups of simulation tests. The verification results are shown in Figure 8.

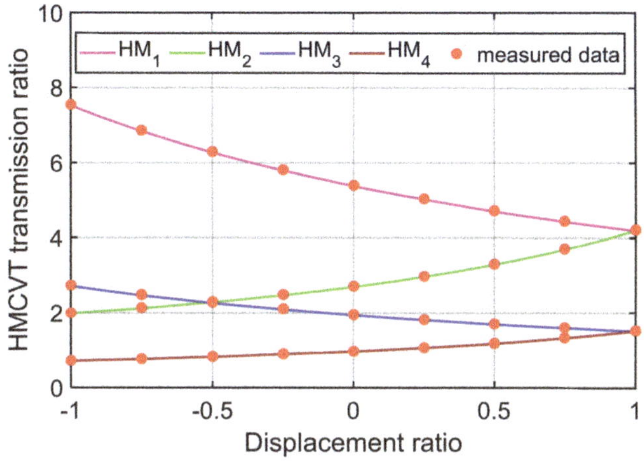


**Figure 8.** Verification results of transmission ratio characteristics.

According to Figure 8, the *MAPE* (mean absolute percentage error) of transmission ratio characteristics at the $HM_1$, $HM_2$, $HM_3$ and $HM_4$ sections is 0.41%, 0.57%, 0.85% and 1.06%, respectively. The *MAPE* of all transmission ratio data (a total of 36 simulation tests) is 0.72%. Then set the displacement ratio of the variable displacement pump to −1; clutch $C_V$, $C_1$ and $C_4$ engaged (HMCVT is in the state of minimum transmission ratio); the engine speed is equally divided into six parts within the working speed range (750, 1040, 1330, 1620, 1910 and 2200, respectively). The simulation results based on Simulation X are shown in Figure 9.

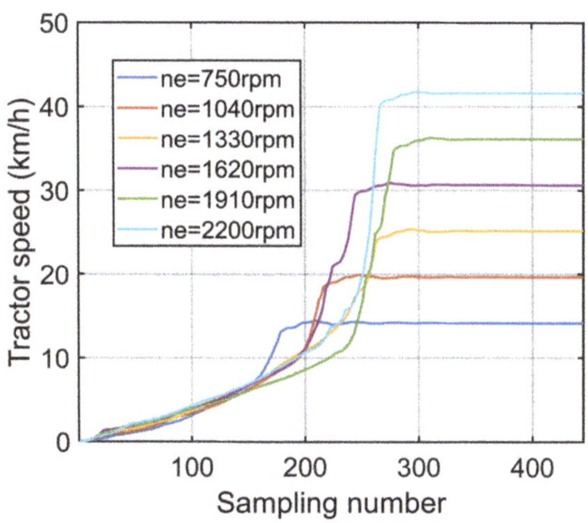


**Figure 9.** Simulation results of tractor speed.

According to Figure 9, when the engine is at the maximum speed of 2200 rpm, the maximum speed of the tractor reaches 41.62 km/h, which meets the design requirements of the maximum speed of 35 km/h.

## 4. Conclusions

In this paper, a five-stage HMCVT is proposed for tractors, and a geometric design method for transmission ratio characteristics is proposed based on the intersection position of each working section and I-GA. After the optimization design of transmission parameters, the transmission ratio characteristics of the five working sections change continuously, and there is no discontinuity of transmission ratio characteristics.

The utilization rate of the displacement ratio's change interval is high and the average utilization rate of the whole working section is 90.4%. The five-stage HMCVT proposed in this paper meets the demand for tractor speed. The optimization design method proposed in this paper is based on I-GA that can flexibly complete the matching of transmission parameters according to geometric principle. The intersection position information of each working section is introduced in the optimization process of transmission parameters, and the problem of the standard method's solving difficulty caused by the uncertainty of intersection position is solved effectively. The HMCVT transmission characteristics are highly consistent with the target characteristics, the average error of the three solutions is about 3.27%, and the common ratio is about 1.81.

The simulation experiment based on Simulation X verifies the correctness of the design in this paper. The average error of 36 simulation tests is about 0.72%. The maximum driving speed of the tractor is greater than 35 km/h, which meets the design requirements.

**Author Contributions:** Methodology, Z.C.; software, Z.C., J.L., Y.C. and W.L.; validation, Z.C., Y.C.; investigation, Z.C., Y.C., L.L., P.Z. and W.C.; resources, Z.C. and W.L.; writing—original draft preparation, Z.C.; writing—review and editing, Z.C., Y.C. and Y.Q.; supervision, Z.C.; and project administration, Z.C. All authors have read and agreed to the published version of the manuscript.

**Funding:** This research was funded by the National Natural Science Foundation of China (grant number: 52105063).

**Institutional Review Board Statement:** Not applicable.

**Informed Consent Statement:** Not applicable.

**Data Availability Statement:** The data presented in this study are available on demand from the corresponding author or first author at (cz38@njfu.edu.cn).

**Acknowledgments:** The authors thank the National Natural Science Foundation of China (grant number: 52105063) for funding. We also thank the anonymous reviewers for providing critical comments and suggestions that improved the manuscript.

**Conflicts of Interest:** The authors declare no conflict of interest.

## References

1. Liu, Z.; Zhang, G.; Chu, G.; Niu, H.; Zhang, Y.; Yang, F. Design Matching and Dynamic Performance Test for an HST-Based Drive System of a Hillside Crawler Tractor. *Agriculture* **2021**, *11*, 466. [CrossRef]
2. Park, Y.-J.; Kim, S.-C.; Kim, J.-G. Analysis and verification of power transmission characteristics of the hydromechanical transmission for agricultural tractors. *J. Mech. Sci. Technol.* **2016**, *30*, 5063–5072. [CrossRef]
3. Sun, J.B.; Chu, G.P.; Pan, G.T.; Meng, C.; Liu, Z.J.; Yang, F.Z. Design and performance test of remote control omnidirectional leveling hillside crawler tractor. *Trans. Chin. Soc. Agric. Mach.* **2021**, *52*, 358–369.
4. Zhou, W.; Zheng, Y.; Pan, Z.; Lu, Q. Review on the Battery Model and SOC Estimation Method. *Processes* **2021**, *9*, 1685. [CrossRef]
5. Wang, H.; Zheng, Y.; Yu, Y. Lithium-Ion Battery SOC Estimation Based on Adaptive Forgetting Factor Least Squares Online Identification and Unscented Kalman Filter. *Mathematics* **2021**, *9*, 1733. [CrossRef]
6. Tian, J.; Wang, Q.; Ding, J.; Wang, Y.; Ma, Z. Integrated Control with DYC and DSS for 4WID Electric Vehicles. *IEEE Access* **2019**, *7*, 124077–124086. [CrossRef]
7. Li, T.; Xie, B.; Li, Z.; Li, J. Design and Optimization of a Dual-Input Coupling Powertrain System: A Case Study for Electric Tractors. *Appl. Sci.* **2020**, *10*, 1608. [CrossRef]

8. Chen, Y.; Xie, B.; Du, Y.; Mao, E. Powertrain parameter matching and optimal design of dual-motor driven electric tractor. *Int. J. Agric. Biol. Eng.* **2019**, *12*, 33–41. [CrossRef]
9. Li, J.; Liu, L.; Xiao, M.; Wang, T.; Wang, X.; Zhang, H. Research on dynamic characteristics of hydro-mechanical con-tinuously variable transmission. *J. Mech. Strength* **2017**, *39*, 14–19.
10. Xia, Y.; Sun, D. Characteristic analysis on a new hydro-mechanical continuously variable transmission system. *Mech. Mach. Theory* **2018**, *126*, 457–467. [CrossRef]
11. Lu, L.; Zhou, Y.; Li, H.; Wang, Y.; Yin, Y.; Zhao, J. Electro-hydraulic Shift Quality of Power Shift Transmission of Heavy Duty Tractor. *Trans. Chin. Soc. Agric. Mach.* **2020**, *51*, 550–556, 602.
12. Macor, A.; Rossetti, A. Optimization of hydro-mechanical power split transmissions. *Mech. Mach. Theory* **2011**, *46*, 1901–1919. [CrossRef]
13. Rossetti, A.; Macor, A. Multi-objective optimization of hydro-mechanical power split transmissions. *Mech. Mach. Theory* **2013**, *62*, 112–128. [CrossRef]
14. Cheng, Z.; Zheng, S.; Qian, Y.; Lu, Z.; Zhang, H. Based on improved SA and GA a new method for optimizing transmission parameters of automotive HMCVT. *J. Mech. Strength* **2020**, *42*, 61–66.
15. Ince, E.; Guler, M.A. Design and analysis of a novel power-split infinitely variable power transmission system. *J. Mech. Des.* **2019**, *141*, 054501. [CrossRef]
16. Qian, Y.; Cheng, Z.; Lu, Z. Bench Testing and Modeling Analysis of Optimum Shifting Point of HMCVT. *Complexity* **2021**, *2021*, 6629561. [CrossRef]
17. Xiao, M.; Zhao, J.; Wang, Y.; Zhang, H.; Lu, Z.; Wei, W. Fuel economy of multiple conditions self-adaptive tractors with hydromechanical CVT. *Int. J. Agric. Biol. Eng.* **2018**, *11*, 102–109. [CrossRef]
18. Ju, C.; Wang, Y.Q.; Dong, W.H.; Yang, Y.P.; Zhai, Y. Research on adjacent transmission ratio distribution of tractor range transmission. *Tract. Farm Transp.* **2016**, *43*, 28–33.
19. Ni, X.D.; Zhu, S.H.; Ouyang, D.Y.; Chang, Y.L.; Wang, G.M.; Nguyen, W.T. Design and experiment of hydro-mechanical CVT speed ratio for tractor. *Trans. Chin. Soc. Agric. Mach.* **2013**, *44*, 15–20.
20. He, C.; Lang, P.; Kang, M.; Zhang, H. Transmission design and force analysis of HMCVT for high power tractor. *J. Mech. Transm.* **2018**, *42*, 54–59.
21. Yu, H.; Zhang, T.; Ma, Z.; Wang, R. Torsional vibration analysis of planetary hybrid electric vehicle driveline. *Trans. Chin. Soc. Agric. Eng.* **2013**, *29*, 57–64.
22. Chen, Y.N. *Study on Design and Drive Control Methods of Powertrain for Electric Tractor*; China Agricultural University: Beijing, China, 2018.
23. Sun, X.W. *Design and Characteristic Research of Transmission System of Tractor Full Power Shift Transmission*; Jilin University: Changchun, China, 2019.
24. Li, D.; Xu, B.; Tian, J.; Ma, Z. Energy Management Strategy for Fuel Cell and Battery Hybrid Vehicle Based on Fuzzy Logic. *Processes* **2020**, *8*, 882. [CrossRef]
25. Li, Y.; Ma, Z.; Zheng, M.; Li, D.; Lu, Z.; Xu, B. Performance Analysis and Optimization of a High-Temperature PEMFC Vehicle Based on Particle Swarm Optimization Algorithm. *Membranes* **2021**, *11*, 691. [CrossRef] [PubMed]
26. Xu, X.; Lin, P. Parameter identification of sound absorption model of porous materials based on modified particle swarm optimization algorithm. *PLoS ONE* **2021**, *16*, e0250950. [CrossRef]
27. Chang, C.; Zheng, Y.; Yu, Y. Estimation for Battery State of Charge Based on Temperature Effect and Fractional Extended Kalman Filter. *Energies* **2020**, *13*, 5947. [CrossRef]
28. Wang, H.; Zheng, Y.; Yu, Y. Joint Estimation of SOC of Lithium Battery Based on Dual Kalman Filter. *Processes* **2021**, *9*, 1412. [CrossRef]
29. Cheng, Z.; Lu, Z.; Qian, J. A new non-geometric transmission parameter optimization design method for HMCVT based on improved GA and maximum transmission efficiency. *Comput. Electron. Agric.* **2019**, *167*, 105034. [CrossRef]

*agriculture*

MDPI

*Article*

# The Method of Calculating Ploughshares Durability in Agricultural Machines Verified on Plasma-Hardened Parts

**Alexandr Gulyarenko [1] and Michał Bembenek [2,*]**

[1] Faculty of Technology, S. Seifullin Kazakh Agro Technical University, A. Moldagulova Street, 29a-302, Nur-Sultan 010000, Kazakhstan; gulyarenko@mail.ru

[2] Faculty of Mechanical Engineering and Robotics, AGH University of Science and Technology, A. Mickiewicza 30, 30-059 Kraków, Poland

* Correspondence: bembenek@agh.edu.pl

**Abstract:** Reliability consists of four components: failure-less operation, maintainability, durability, and preservation ability. For different machines and different conditions of operation, different combinations of these properties, and differences in how they are balanced and proportioned are essential. For tractors, the most important aspect of reliability is maintainability, while for agricultural machines, durability is most important. Using the example of a ploughshare, the issue of increasing the durability has been studied; a method for calculating the durability of a ploughshare for various types of soils has been described. The use of plasma hardening of the surface of a 65G-steel ploughshare has been proposed; the effectiveness of plasma hardening of soil-cutting parts and its economic feasibility have been proved. Due to hardening to a depth of 1–1.8 mm, the service life of parts increases by 2–3 times; moreover the downtime of expensive machine-tractor units for replacing worn-out parts is reduced.

**Keywords:** plough; ploughshares; durability calculation method; agricultural machine; wear; plasma-hardening surface

**Citation:** Gulyarenko, A.; Bembenek, M. The Method of Calculating Ploughshares Durability in Agricultural Machines Verified on Plasma-Hardened Parts. *Agriculture* **2022**, *12*, 841. https://doi.org/10.3390/agriculture12060841

Academic Editor: Mustafa Ucgul

Received: 22 May 2022
Accepted: 9 June 2022
Published: 10 June 2022



## 1. Introduction

The applied and economic significance of the development of agriculture is obvious. The food security of the country depends on it. Nowadays, the leading positions in the development of agriculture are held by China, India, and the United States [1]. At the same time, the basis of agriculture is crop production. In turn, the profitability of crop production largely depends on the efficiency of using machine-tractor units (MTU). Modern high-performance and high-tech MTUs are the primary tool for crop production. The effectiveness of using MTUs depends on a large number of factors. However, no matter how powerful and technological they are, the effectiveness of their use is primarily determined by their reliability. At the same time, the MTU consists of two main parts, the tractor and the process machine; therefore, it is evident that the reliability of the MTU comprises two components, i.e., the reliability of the tractor and the reliability of the process machine. So, in this case, the reliability components will not have the same effect on the tractor and the process machine. For a tractor, the principal factors influencing the efficiency of use are the reliability components, i.e., reliability and maintainability. Their influence on the efficiency of using MTUs and ways of managing the reliability indicators of tractors have already been investigated, and recommendations have already been given for equipping MTUs with tractors of different levels of reliability [2–8].

To solve the issue of increasing the efficiency of using MTUs in a comprehensive way, it is necessary to investigate the issue of increasing the reliability of the process machine. Soil-cutting machines will be considered process machines in this study, since they perform the leading and most energy-consuming operations in crop production [9–11].

The most important aspect of improving the technical level of soil-cutting machines is considered to be increasing the service life of their tools [9–16]. In this case, it is the indicators of durability that will be of paramount importance [17–21]. Since it is the durability of the tools of the process machine that will affect the machine-tractor unit as a whole, the very durability of the soil-cutters will have a significant impact on the increase in energy costs (fuel consumption), the observance of agrotechnical requirements (yield), and even the reliability of the tractor itself [22,23]. Therefore, the qualitative increase in these indicators can be achieved only by identifying the main reason for their decrease. In this case, both a tractor and an agricultural machine are needed to perform a process operation, and if this operation is cutting the soil with a tool, then, accordingly, the most obvious way to qualitatively improve the indicators is to study the cutting process and its optimisation [24–26]. As a result of intense abrasive wear, the geometry of the cutting part and the overall dimensions of the tools change [27–29], and therefore it is necessary to increase the hardness of the working bodies using various methods of hardening [30–37], and, at the same time, to develop reliable methods for calculating durability [38,39].

Thus, our study combines three key aspects:

1. The relevance of work for agriculture, in particular, for crop production, since the forced frequent replacement of parts of the working bodies leads to a decrease in labour productivity and an increase in processing costs [40–42]. For example, as calculations show, based on the existing resources and the prices of parts of the plough tools, every 100 hectares of ploughing required monetary costs of at least USD 70 only for their replacement and at least four person-hours of labour costs. These figures reach about USD 85 million in Kazakhstan and an additional need for about three thousand machine operators. Therefore, the relationship between durability and maintainability is also obvious, i.e., the less durable the machine, the higher its maintainability should be since frequent replacements of tools require labour and time, which again leads to costs and non-compliance with agrotechnical requirements.

2. The proposed method for calculating the durability of the ploughshares will allow the durability of the share for different types of soil and different hardness of the plow surface to be calculated. In the existing methods, the nominal parameters of the ploughshares are used in the calculations and only soil indicators vary [19,43–46].

In addition, studies into the hardness of the ploughshare surface (operational control by ultrasonic method, depth measurement and structure analysis), as well as field comparative tests of hardened and nominal ploughs in identical conditions (installed on the same unit) will confirm the correctness of the durability calculation method and predict the resource of the plough share in surface hardness and soil type.

3. The proposed method of manual plasma hardening has a number of advantages in comparison with existing hardening methods. One example of this is the method used in the USA for argon-arc surfacing of petrochemical fittings with hard-alloy stellite [47]. Due to its brittleness, this cobalt alloy cannot be drawn into a wire, so continuous feeding into the arc is carried out only by blowing it in the form of a powder. However, the powder, when injected, scatters, deposited on the tip of the tungsten electrode, and quickly disables it. These problems are being solved, and stellites and methods of their application are still being improved, but in our opinion, any methods of spraying or surfacing cannot be used in this case, since an increase in thickness inevitably leads to an increase in the resistance of the soil-cutting organ, and this is, again, a violation of agricultural requirements, increased load, breakdowns, excessive fuel consumption, etc. At the same time, there are a large number of hardening methods precisely due to spraying and surfacing [20,48–50]. This direction is still relevant and has been developing since the first half of the 20th century; however, the main disadvantage of these methods has been and will be the consumption of the sprayed or deposited substance. The high cost of these hardening methods and the increase in the thickness of soil-cutting methods make them unacceptable for our study. At the same time, there are methods for hardening parts, but they are also not acceptable, since when a fully hardened part becomes hard, at the same time, it becomes too brittle [51]. Due

to the heterogeneity of the soil, chips appear, while excessively plastic parts undergo plastic deformation, while wear also increases. Thus, to solve our problem, a hardening method is required that allows the hardness of the surface layer of the metal to be increased, and at the same time, allows the elasticity and plasticity of the soil-cutting part to be maintained, and all without increasing the thickness.

Consequently, reliability is paramount for a tractor, and durability is paramount for a process machine, and only after that, comes maintainability in the case of breakdown or wear. In this regard, the issue of the development and production of high-quality and long-life soil-cutters, ensuring compliance with agrotechnical requirements during treatment, which are long-living and competitive in terms of their cost, is quite acute. The method of calculating the durability of the plough blade presented in the article will make it possible to make comparative calculations of durability for different types of soil [14,18,19,52–57]. Moreover, the method of plasma hardening [58–66] of soil-cutting working bodies used by us will increase the durability of the plough compared to serial samples and experimentally confirm the correctness of the calculations.

## 2. Material and Methods

### *2.1. Calculation Methodology Model*

In the general case, the service life of the tools can be represented as a function of the following main varying parameters:

$$T = f\,(I, m, p, v, \eta_1, \eta_2 \ldots \eta_{\mathrm{n}}) \tag{1}$$

where $T$ is the service life, h (ha); $I$ is the wear resistance of the tool material, h/g (h/mm); $m$ is the wear capacity of the soil, g/h (mm/h); $p$ is the soil pressure on the working surface of the tool, MPa; $v$ is the speed of movement of the tool relative to the soil, km/h; and $\eta_1$, $\eta_2 \ldots \eta_n$ are the factors characterising the change in the main parameters depending on the condition of the soil, the composition of the material of the tools, and the modes of their heat treatment, the design parameters of the tools, etc.

It is possible to manage the service life of the tools if the general pattern of ensuring their performance and the nature of wear in the soil are known. Many papers are devoted to the establishment of such patterns and the development of recommendations for determining the intensity of wear and predicting the service life of tools. However, their practical application is constrained because they do not fully take into account those complex dependencies that exist in the process of abrasive wear. Notably, it was revealed that the relative wear resistance of materials and the wear capacity of the abrasive medium (soil) are not constant values. They vary depending on the pressure of the abrasive medium on the tool. The lack of a reasonably simple methodology for determining the wear rate and the service life of tools hinders the development and justification of using new materials and technologies when hardening tools to increase their service life. These circumstances have led to the fact that modern ploughs today use ploughshares, the design parameters and materials developed more than 40 years ago. However, their operation modes have changed significantly, i.e., the processing speed, the weight of the machines, and, consequently, and the compaction of soils within the processing period, especially when harvesting, have all increased. All this leads to an increase in the load on the tools and, accordingly, their wear rate.

A characteristic feature of the soil-cutting tools is a relatively large area of contact with the cultivated soil. In this case, the loads on individual sections of the working surface differ significantly. For example, the ploughshare has the most significant pressure on the tip and significantly less pressure on the blade. In this regard, the wear rate of different sections is not the same. Consequently, the tools are rejected due to the wear rate on one relatively small section, while the rest of the sections have a significant residual life.

Let us consider the methodology for predicting the service life on the example of a ploughshare, depending on the types of soils, materials of which it is made and which are used to harden it, and changes in some design parameters.

The intensity of wear of tools has been studied in the field, and research materials of other authors have been used [12–15,19]. Consequently, a mathematical expression of the abrasive wear rate of the tools has been developed, depending on several parameters. In the general case, wear rate, cm, of the most wear-prone section is calculated as follows:

$$W = k_{ref}\,\frac{m\;\eta_1\;p\;v_p\;t}{\varepsilon_{ref}\;\eta_2\;\chi} \tag{2}$$

where $k_{ref}$ is the factor of proportionality of the wear of the reference sample under reference conditions: $k_{ref}$ = 0.016 cm/(MPa·km); m is the relative wear capacity of the soil (in terms of particle size distribution) at the reference pressure of the abrasive medium (quartz); $\eta_1$ is the factor that takes into account the change in the relative wear capacity of the soil depending on pressure; $p$ is the pressure of the soil (abrasive medium) on the most wear-prone section of the tool, MPa; $v_p$ is the forward speed of the tool, km/h; $t$ is the operating time of the tool, h; $\varepsilon_{ref}$ is the relative wear resistance of the material under reference test conditions; $\eta_2$ is the factor that takes into account the change in the relative wear resistance of the material depending on the pressure; and $\chi$ is the ratio of the speed of movement of the soil layer on the surface of the tool to the speed of the tool.

The following formula can determine the durability, h, of the tool:

$$T = \frac{W_{crit}\;\varepsilon_{ref}\;\eta_2\;\chi}{k_{ref}\;m\;\eta_1\;p\;v_p} \tag{3}$$

where $W_{crit}$ is the limiting wear rate of the most wear-prone section of the tool, cm.

Steel 45 with a hardness of 90 HRB (or 180 HB) is taken as reference material. The following are taken as reference wear conditions: pressure $p_{ref}$ = 0.1 MPa; abrasive medium is quartz particles with a size of 0.16÷0.32 microns; relative wear capacity of the abrasive medium m = 1; $v_p$ = 1 km/h.

The values of the relative wear capacity of soils are given in Table 1.

**Table 1.** Relative wear capacity of soils [19].

| Soil Type | Relative Wear Capacity of the Soil, *m* |
|---|---|
| Sandy soil | 0.87 |
| Sandy-loam soil | 0.62 |
| Loamy soil: | |
| light | 0.42 |
| medium | 0.32 |
| heavy | 0.22 |
| Clayey soil: | |
| light | 0.15 |
| medium | 0.10 |
| heavy | 0.06 |
| Quartz particles with a size of 0.16 ÷ 0.32 μm | 1.0 |

The dependence of the relative wear resistance of steels, of which the tools of soil-cutting machines are made, on their chemical composition and hardness is presented in the form of an empirical equation [16]:

$$\varepsilon = 0.24X_1 + 0.07X_2 + 0.11X_3 - 3.54, \tag{4}$$

where ε is the relative wear resistance of steel (the standard is steel 45 with a hardness of 90 HRB, the abrasive medium is quartz with particles of 0.16 ÷ 0.32 microns in size, the pressure of the abrasive medium is $p$ = 0.33 MPa); $X_1$ is carbon content, %; $X_2$ is chromium content, %; and $X_3$ is hardness, HRC.

Permanent alloying elements in raw, untreated steels, manganese and silicon, have a positive effect on some characteristics of steels but not on their wear resistance [67–69]. The content of elements such as tungsten, molybdenum, and vanadium in steels increases wear resistance above 60 HRC. At lower hardness, their effect on wear resistance is minor [9,30–32].

Therefore, these elements are not included in the equation. The following empirical formulas determine the values of the correction factors $\eta_1$ and $\eta_2$:

$$\eta_1 = 9.6p - 0.04 \tag{5}$$

$$\eta_2 = 1.75p + 0.825 \tag{6}$$

If the value of the relative wear resistance of steel at the reference pressure is unknown, then it is determined by the following formula:

$$\varepsilon_{ref} = \frac{\varepsilon}{\eta_2} \tag{7}$$

where ε is the relative wear resistance of steel at a pressure of $p$ = 0.33 MPa (see Equation (4)).

Considering that the load on the ploughshare tip and the intensity of its wear vary significantly from the exact parameters of the blade, the durability of the ploughshare is calculated according to two criteria, i.e., wear rate of the tip and wear rate of the blade.

The permissible wear rate of the tip is determined by the difference between the original $H$ (Figure 1) and the limiting, $H_{crit}$, tip height. The permissible wear rate of the blade is also determined by the difference between the initial, $h$, and the permissible, $h_{crit}$, blade width or the proper blade thickness, $a$.

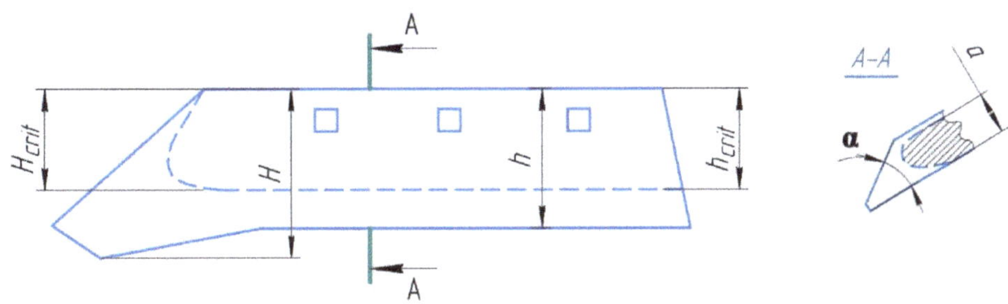


**Figure 1.** The rejecting parameters of the ploughshare.

The durability of the share according to the cultivated area in hectares (ha), according to the wear of the tip:

$$T_{tip} = \frac{\varepsilon_{ref}\ \eta_2\ \chi\ A\ (H{-}H_{crit})}{k_{ref}\ \ m\ \ \eta_1\ \ p\ \ v_p} \tag{8}$$

where $A$ is the performance of the plough body, ha/h; $H$-$H_{crit}$ is the limiting wear rate of the tip in height, cm.

The durability of the ploughshare, ha, according to the wear rate of the blade:

$$T_{bla} = \frac{\varepsilon_{ref}\ \eta_2\ \chi\ A\ (h{-}h_{crit})}{k_{ref}\ \ m\ \ \eta_1\ \ p\ \ v_p} \tag{9}$$

where $h - h_{crit}$ is the limiting wear rate of the blade in width, cm.

In most cases, the ploughshares are rejected not as per the wear rate of the blade in width but as per the limiting thickness of the blade.

The durability of the ploughshare blade as per the limiting thickness:

$$T_{bla} = \frac{(a-b)\varepsilon_{ref}\ \eta_2\ \chi\ A}{k_{ref}\ \ m\ \ \eta_1\ \ p\ \ v_p\ \ \tan\alpha} \tag{10}$$

where $a$ is the limiting thickness of the ploughshare blade for specific ploughing conditions, cm; $b$ is the initial thickness of the new ploughshare blade, cm; $\alpha$ is the angle of sharpening the ploughshare.

As can be seen from Equations (8)–(10), the durability of the ploughshare is directly proportional to the relative wear resistance of the material. It is inversely proportional to the wear capacity of the soil, the pressure of the abrasive medium, the speed of the plough, and the angle of sharpening the blade. The larger the sharpening angle, the faster the blade will reach its limiting thickness and will be rejected due to its poor penetration.

The maximum total pressures acting on the tip and the blade of the ploughshare can be determined by the following empirical relationships [19,70]:

$$p_{bla} = (0.012 \div 0.016)(1 + 0.028v_p)(1 + 0.01\beta)\left(1.45 + H + 0.5H^{1.5}\right) \tag{11}$$

$$p_{tip} = (0.06 \div 0.065)(1 + 0.028v_p)(1 + 0.01\beta)\left(1.45 + H + 0.5\mathrm{H}^{1.5}\right) \tag{12}$$

where $p_{bla}$ is the pressure on the ploughshare blade, MPa; $p_{tip}$ is the pressure on the ploughshare tip, MPa; $v_p$ is the speed of movement of the tool, km/h; $\beta$ is the angle of inclination of the ploughshare to the bottom of the furrow, °; and $H$ is the soil hardness, MPa.

The proper thickness (mm) of the ploughshare blade, at which a constant ploughing depth is provided, can be determined by the following empirical equation:

$$a = 8 - H \tag{13}$$

For example, the durability of a serial 65G steel ploughshare without additional hardening will we calculated.

The calculation will be carried out using the following ploughing conditions:

- Types of soils: sandy, light-loamy, and light-clayey;
- Soil hardness: $H = 1\text{MPa}$, $H = 3\text{MPa}$, $H = 5\text{MPa}$;
- Ploughing speed: $v_p$ = 10 km/h;
- Performance of the plough body: $A$ = 0.35 ha/h;
- The angle of inclination of the ploughshare to the bottom of the furrow, $\beta$ = 30°.

Serial ploughshare parameters:

- Relative wear resistance, ε = 1.28 for 65G steel and at the reference pressure of the abrasive medium;
- Original tip blade thickness, $b$ = 2 mm;

1. Limiting wear rate of the tip in height, $H_{crit} = 6.8\text{cm}$;

- Limiting blade thickness at $H = 5\text{MPa}$, $a$ = 3 mm; at $H$ = 3 MPa, $a$ = 5 mm; at $H$ = 1 MPa, $a$ = 7 mm.

The 7 mm limitation of the blade thickness is due to the limitation of the ploughshare wear rate in width; and blade sharpening angle, α = 8°.

The calculation results are shown in Table 2.

According to the data in Table 2, the service life of serial ploughshares varies from 2.06 to 9.83 ha on sandy soils (depending on their hardness). With a hardness of 5 MPa, the ploughshares will be rejected according to the maximum thickness of the blade. The service life of the tip is greater than that of the blade. To increase the service life of the blade, for example, to 2.95 ha, its sharpening angle should be reduced. Consequently, the potential for blade wear rate is increased without compromising ploughshare performance.

**Table 2.** The calculation results.

| Parameter | Parameter Values on Soil | | | | | | | | |
|---|---|---|---|---|---|---|---|---|---|
| | Sandy | | | Loamy (Light) | | | Clayey (Light) | | |
| | Hardness, MPa | | | | | | | | |
| | 1 | 3 | 5 | 1 | 3 | 5 | 1 | 3 | 5 |
| Soil pressure on the tip, $p_{tip}$, MPa | 0.48 | 0.82 | 1.24 | 0.48 | 0.82 | 1.24 | 0.48 | 0.82 | 1.24 |
| Soil pressure on the blade, $p_{bla}$, MPa | 0.12 | 0.27 | 0.31 | 0.12 | 0.27 | 0.31 | 0.12 | 0.27 | 0.31 |
| Serial ploughshare service life, ha: | | | | | | | | | |
| tip | 9.8 | 4.85 | 2.95 | 20.7 | 10.0 | 6.03 | 58.8 | 28.3 | 16.7 |
| blade | 28.5 | 4.6 | 2.06 | 60.4 | 10.1 | 4.42 | 166.2 | 28.4 | 12.5 |
| The ratio of the service lives of the blade and the tip | 2.9 | 0.95 | 0.69 | 2.9 | 1.01 | 0.73 | 2.82 | 1.01 | 0.74 |
| The service life of the test ploughshare of 65G steel, ha: | | | | | | | | | |
| tip | 14.02 | 6.94 | 4.15 | 29.7 | 14.6 | 8.66 | 81.4 | 40.4 | 23.9 |
| blade | 79.3 | 12.35 | 3.31 | 160.0 | 26.2 | 6.9 | 450.0 | 72.1 | 19.3 |
| The ratio of the service lives of the blade and the tip | 5.65 | 1.78 | 0.79 | 5.38 | 1.79 | 0.8 | 5.52 | 1.78 | 0.8 |

With a sandy soil hardness of 3 MPa, the service lives of the tip and the blades are 4.85 and 4.6 ha, respectively. That is, the ploughshare is subject to wear almost evenly.

On loamy soils with a hardness of 5 MPa, the service life of the tip exceeds the service life of the blade. The ploughshare will be rejected because it is out of plough. With a soil hardness of 3 MPa, the service lives of the tip and the blade are equal to 10.1 ha. That is, the ploughshare is subject to wear evenly. With a hardness of 1 MPa on loamy soils, the service life of a serial ploughshare is 20.7 ha. In this case, the tip is primarily exposed to wear. The residual life of the blade when rejecting the ploughshare will be about 40 ha.

On clay soils, the service life of the serial ploughshare, depending on the hardness of the soil, will vary from 12.5 ha with a hardness of 5 MPa to 58.8 ha with a hardness of 1 MPa. In the latter case, when the ploughshare is rejected, its blade will be underutilised by about 100 ha of ploughing, i.e., by hardening the tip part, the ploughshare service life of about 160 ha can be achieved.

As practice shows in most cases, the hardness of sandy and light loamy soils at a depth of 20 ÷ 30 cm is 2.2 ÷ 2.8 MPa. This means that the service life of ploughshares made of 65G steel without hardening for such soils will be 7 ÷ 14 ha. By hardening the tip of these ploughshares only, it is possible to bring their service life up to 26 ÷ 36 ha, respectively. Therefore, by hardening the tip of a 65G steel ploughshare using plasma hardening, it is possible to achieve at least a 2.6-fold increase in its service life compared to a non-hardened serial ploughshare. When ploughing medium and heavy loamy soils, the difference in the service lives of serial and test ploughshares will be much more significant.

Let us consider the possibilities of increasing the durability of the ploughshare by hardening the tip, the blade, or both, proceeding from ensuring their equal wear resistance. In the general case, to ensure equal wear resistance of the blade and the tip of the ploughshare, the required relative wear resistance can be determined based on the equality of the durability:

$$\frac{(H - H_{crit})\varepsilon_{ref}^{tip}\,\eta_2^{tip}}{\eta_1^{tip}\,p_{tip}} = \frac{(a - b)\varepsilon_{ref}^{bla}\,\eta_2^{bla}}{\eta_1^{bla}\,p_{bla}\tan\alpha} \quad (14)$$

where from:

$$\varepsilon_{ref}^{tip} = \frac{(a-b)\varepsilon_{ref}^{bla}\,\eta_2^{bla}\eta_1^{tip}p_{tip}}{(H-H_{crit})\eta_2^{tip}\eta_1^{bla}p_{bla}\tan\alpha} \quad (15)$$

where $\varepsilon_{ref}^{tip}$ and $\varepsilon_{ref}^{bla}$ are the relative wear resistance of the tip and the blade, respectively; $\eta_1^{tip}$ and $\eta_1^{bla}$ are correction factors that take into account the change in the wear capacity of soils, respectively, on the tip and the blade; $\eta_2^{tip}$ and $\eta_2^{bla}$ are correction factors that take into account the change in the relative wear resistance of materials, respectively, of the tip and the blade; and $p_{tip}$ and $p_{bla}$ are soil pressure on the tip and the blade, respectively.

### 2.2. Methods of Confirming the Results of Calculations by Experiment

Experience shows that the hardening of structural steels to such a depth is achievable using surface plasma treatment (hardening) technology. Let us also note that to ensure tribotechnical properties (increase in wear resistance and decrease in the friction coefficient), which provide the required durability of the parts in the friction units, the thickness of the hardened layer of over 1.0–1.8 mm is not required. Since only the friction surface is subject to wear, in this case, as mentioned above, hardening of the entire part will increase the brittleness of the part. In addition, it is not economically feasible, and would involve an unreasonable increase in the thickness of the hardened layer. The depth of the hardened layer turned out to be sufficient; this was confirmed by the first results of field tests. To obtain test samples of hardened parts, a UDGZ-200 (Russtankom, Ekaterinburg, Russian Federation) plasma-hardening unit, which allowed a hardened layer depth of 0.5 to 2.0 mm and a width of 7–15 mm to be obtained, was used. Plasma hardening was performed with the following parameters: nozzle diameter 11 mm, argon flow 15 L/min and arc length 15–20 mm at a current of 150 A. Before the hardening, the sample had to be properly prepared: recommended roughness $R_z$ < 16 μm, cleaning from soil, grease, paint and rust is necessary. To remove the paint, the grinding disc NC-22,23-G40-D125 and an angle grinder P.I.T 61808 PRO (speed 9600 rpm) were used; the same equipment were used to remove rust.

Low-alloyed structural 65G steel was used for ploughshare prototypes test with the following chemical composition (GOST 14959–2016): 0.62–0.70% C; 0.90–1.20% Mn; 0.17–0.37% Si; < 0.035% P; <0.035% S; <0.25% Cr; <0.20% Ni; and <0.25% Cu.

Macroscopic cross-section image was made with a Keyence VHX-7000 microscope (Keyence, Osaka, Japan). Metal microstructure studies were carried out on an optical research microscope Axio Observer D1m Carl Zeiss (Carl Zeiss AG, Obrekochen, Germany) designed to study the phase composition and structural features of the treated steel at a magnification from ×100 to ×1000.

To assess the most important indicators of reliability and durability over a long period of operation, various methods of non-destructive testing are usually used [35], including the ultrasonic method. Operational control of the hardness of the hardened surface was carried out using the UZIT-3 device (Introtest, Ekaterinburg, Russian Federation). The device does not require special setting or training and allows for on-line control of metal hardness by ultrasonic method without damaging the surface within HRC from 20 to 70, HB from 80 to 450.

The field tests were conducted with 2 mounted plows PLN-8-35 (Almaz, Barnaul, Russian) used together with 2 tractors Kirovec K744 p2 (Peterburgskiy Traktornyy Zavod, Petersburg, Russian). The operating speed of the machine-tractor unit was 10 km/h. The test was conducted on light loamy soil.

## 3. Results

Obviously, the greater the total thickness of the tip (chisel), the worse the penetration ability. The use of plasma hardening of soil-cutting tools will increase the wear resistance, which is especially important. There will be no need to increase the thickness. Moreover, it is crucial to determine the ideal thickness of the hardened layer. Experience shows that the efficiency of the plough body of a three-four-body non-reversible plough with a specific weight within the range of 110 ÷ 150 kg per body is relatively high if the thickness of the tip (chisel) is under 14 mm. A sufficient level of operability of eight-nine-body non-reversible ploughs and almost all reversible ploughs with a specific weight of 220 ÷ 480 kg per body is ensured with a tip (chisel) thickness of 16 ÷ 20 mm.

Let us calculate the rational thickness of the hardened layer, $h_{plaz}$, depending on the relative wear resistance:

$$h_{plaz} = \frac{c\varepsilon_{ref}^{tip}}{\varepsilon_{ref.plaz}^{tip} + \varepsilon_{ref}^{tip}} \tag{16}$$

where $\varepsilon_{ref}^{tip}$ and $\varepsilon_{ref.plaz}^{tip}$ are the relative wear resistance of the primary and hardened layers, respectively; and $c$ is the thickness of the tip.

Based on the total thickness of the tip, $c$ = 12 mm, let us perform the calculations for ploughshares made of 65G steel to work on light loamy soils with a hardness of 3 MPa. The values of the relative wear resistance of the steel and the hardened layer are reduced to the wear conditions at a pressure of $p_{ref}$ = 0.1 MPa. The calculation results are given in Table 3.

**Table 3.** The rational thickness of the hardened layer and the potential service life of the ploughshare tip on the example of 65G steel (ε = 1.28).

| Hardened Material, Its Relative Wear Resistance | The Thickness of the Hardened Layer, mm | The Relative Wear Resistance of the Tip $\varepsilon_{ref}^{tip}$ | The Potential Life of the Tip, Ha |
|---|---|---|---|
| 65G steel, $\varepsilon_{ref.plaz}^{tip} = 8$ | 1.6 | 1.28 | 31.6 |

After that, the influence of the quality of surface preparation for hardening was experimentally validated. The difference between these surfaces is shown in Figure 2.

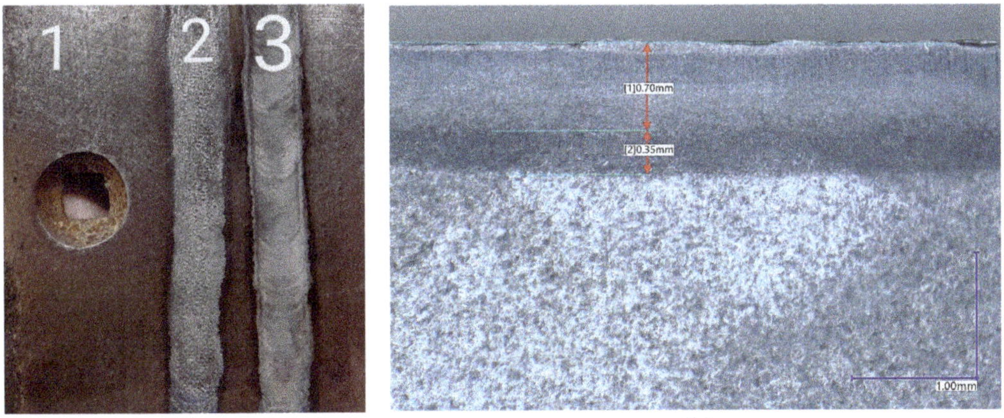


(**a**) (**b**)

**Figure 2.** The effect of surface pretreatment: (**a**) the view, (**b**) the cross-section of hardening line 2; 1—non-hardened surface; 2—surface hardened without additional treatments; 3—surface preliminarily cleaned from surface marks (rust, scratches, etc.) using a grinding machine before hardening.

At the same time, further repeated testing of the surface hardness with the UZIT-3 ultrasonic hardness tester (Figure 3) showed that there was no significant difference in the hardness of samples 2 and 3. Surface pre-treatment after hardening before ultrasonic hardness testing was not carried out, since there is no surface damage if the hardening mode is chosen correctly. At the same time, the thickness of the plow share in the section was 8–12 mm.

(**a**) (**b**) (**c**) (**d**)

**Figure 3.** Operational hardness test: (**a**,**b**) surface hardened without additional treatments; (**c**,**d**) surface preliminarily cleaned from surface marks (rust, scratches, etc.) using a grinding machine before hardening, 1—non-hardened surface; 2—surface hardened without additional treatments; 3—surface preliminarily cleaned from surface marks (rust, scratches, etc.) using a grinding machine before hardening.

Consequently, hardening can be carried out without pretreatment; the only difference is the surface roughness; it changed in the second sample since it had not been pretreated. The difference is not significant, since, during the operation of the unit, the roughness of surfaces 2 and 3 are levelled within the first minutes of operation. Moreover, within the framework of the study, the study of changes in the chemical composition of steel after hardening and metallographic studies of changes in the metal structure was carried out [60–64,71].

The steel structure is shown in Figure 4. As shown in previous studies [60] on the surface, there is a microfusion zone, whose chemical composition corresponds to the steel composition with a carbon content of 0.65%. Upon rapid cooling, it transforms into acicular martensite with a fineness of needles amounting to 5–15 μm. This zone is followed by a zone of austenite transformed into martensite. In the microstructure of these layers, there is a small amount of retained austenite close to 20%, which depends on the depth of the hardened layer. This zone is followed by a layer of troostite, where the microhardness decreases and depends on the volumetric content of the occurring phases, and then sorbitol appears in the structure. The zone of sorbite location is determined by the central zones of the former austenite grains. It is characterised by a lower dispersion level of ferrite and cementite components therein, compared to the dispersion level of troostite, and exhibits a lower microhardness level. The microhardness in this zone also depends on the volumetric amount of the occurring phases. Furthermore, as one goes deeper into the sample, ferrite appears at the junction between the boundaries of former austenite grains, and the amount thereof exhibits a gradual increase. The structure remains ferrite-sorbitic and then smoothly transforms into a ferrite-pearlitic structure. The total microhardness decreases to the initial level. The initial structure represents a mixture of ferrite and pearlite grains with a volume fraction of each amounting to 50% [60].

The results of the study confirmed the operational measurements; the depth of the hardened layer was 1–1.8 mm, at which point the indicators obtained by the UZIT-3 ultrasonic hardness tester were confirmed, i.e., the hardness increased 3-fold. Table 4 shows the hardness measurement results and the calculated averages of the HB indicators.

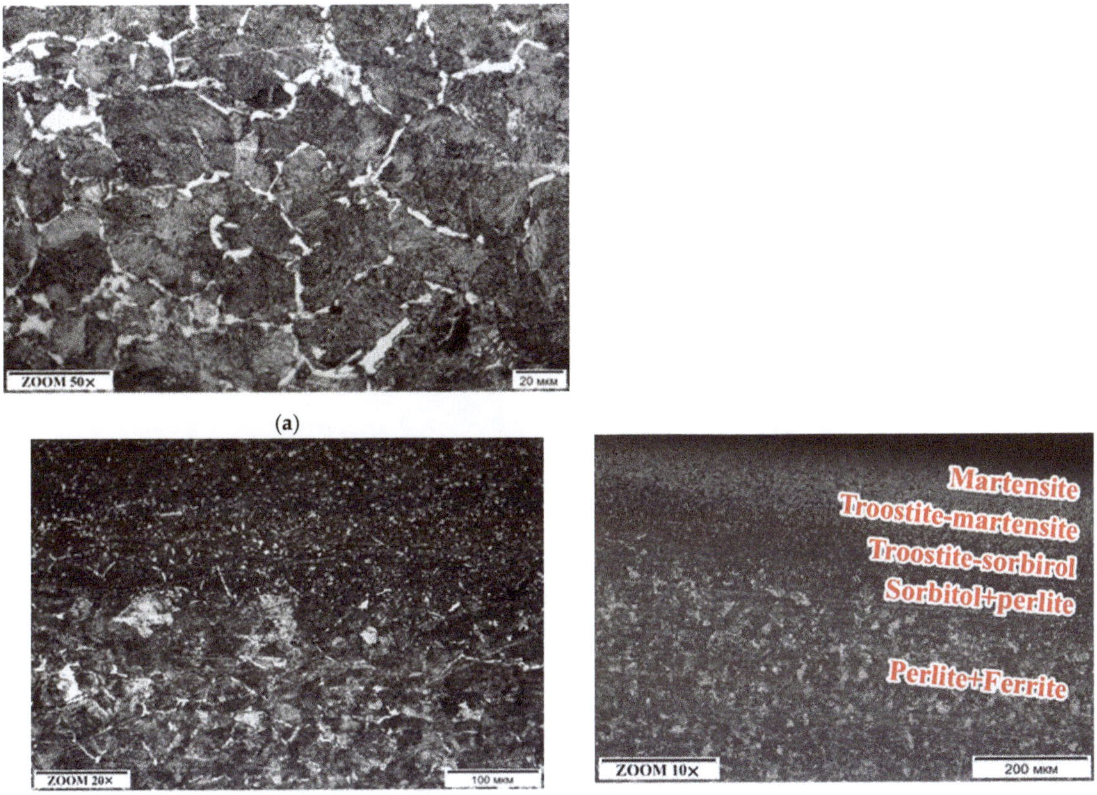


(a)

(b)

(c)

**Figure 4.** The steel structure: (**a**) non-hardened surface; (**b**) surface hardened without additional treatments; (**c**) surface preliminarily cleaned from surface marks (rust, scratches, etc.) using a grinding machine before hardening.

**Table 4.** Hardness measurement results and calculated average values of indicators on the HRC scale.

| | Hardness on HRC Scale before Hardening, Steel 65G (Serial Sample) | Hardness on HRC Scale after Hardening, Steel 65G |
|---|---|---|
| The average of middle 10 measurnet (without 5 max. and 5 min.) | 18.2 | 53.2 |
| Standard deviation | 0.70 | 2.64 |
| Variance | 0.49 | 6.97 |

As can be seen from Tables 2 and 3, when ploughing light loamy soils, the service life of the tip of the serial ploughshare after plasma hardening increases in proportion to the increase in hardness 3-fold, from 10.0 ha to 31.6 ha. The preliminary results of field tests have shown that wear occurs evenly, that is, the blade and tip wear evenly.

Moreover, field tests have confirmed an increase in the service life of the ploughshares. Plasma hardening was applied to the test ploughshares made of 65G steel, which were installed together with the serial ones. After plowing 20.5 ha, serial shares fell into disrepair (resource was exhausted), the residual resource of hardened shares was about 20 hectares. The appearance of the plasma-taped ploughshare is shown in Figure 5.

**Figure 5.** The appearance of the plasma-hardened ploughshare.

In addition, during field tests, it was noticed that sharper plasma-hardened ploughshares have a positive effect on the unit's performance, and this is noticeable not only in the relatively rapidly worn-out serial samples, but also in the ploughshares modified with various claddings, since any cladding leads to thickening. At the same time, as shown in Figure 5, plasma hardening does not change the geometry of the plough at all and, at the same time, it makes it 2–3 times more durable depending on soil conditions.

## 4. Discussion

The test results confirmed the validity of the proposed methodology for predicting the durability of the tools of soil-cutting agricultural machines, as well as the efficiency of using the UDGZ-200 plasma-hardening unit for their treatment. At the same time, currently, surfacing of various hardening materials (T-590 electrode, OSI-6 electrode, FBH-6–2 plasma surfacing, X12 steel, IH300H9F6 wear-resistant cast iron, TK-G corundum ceramics, WK-20 hard alloy) is widespread instead of plasma hardening and even a transition from a homogeneous stabilised blade to a two-layer equally resistant blade is being observed [30–34,36,37]. In our opinion, this significantly reduces the penetration ability and leads to a violation of agro-requirements when performing process operations and, consequently, to a loss of crop yields. Moreover, these alloys are next-order higher in cost than plasma hardening. At the same time, plasma hardening is not a difficult operation in terms of technology and can be carried out directly in an agricultural enterprise.

Therefore, the hardening of heavy-duty soil-cutting parts may significantly increase the profitability of agricultural production. Due to hardening to a depth of 1–1.8 mm, the service life of parts increases 2–3-fold depending on soil conditions; the downtime of expensive machine-tractor units for replacing worn-out parts is reduced. The sharpness of the blades of soil-cutting machines determines not only the fuel consumption rate and the reliability of tractors but also the performance of the MTUs as a whole, and, therefore, the observance of agrotechnical terms, requirements, and yield, and consequently, the profit of the agricultural enterprise and the profitability of crop production.

**Author Contributions:** Conceptualisation, M.B. and A.G.; formal analysis, M.B. and A.G.; methodology, M.B. and A.G.; investigation, M.B. and A.G.; writing—original draft A.G.; writing—review and editing, M.B. All authors have read and agreed to the published version of the manuscript.

**Funding:** This research was funded by the Science Committee of the Ministry of Education and Science of the Republic of Kazakhstan. The paper has been completed within the framework of project No. AP08052699 "The Development and Creation of an Experimental Site for Hardening Heavy-Duty Parts of Soil-Cutting Machines Using Innovative Plasma Technology".

**Institutional Review Board Statement:** Not applicable.

**Informed Consent Statement:** Not applicable.

**Data Availability Statement:** The data presented in this study are available upon request from the corresponding author.

**Acknowledgments:** We would like to thank Jan Pawlik for providing cross-section image.

**Conflicts of Interest:** The authors declare no conflict of interest.

## References

1. Seok, J.H.; Moon, H. Agricultural Exports and Agricultural Economic Growth in Developed Countries: Evidence from OECD Countries. *J. Int. Trade Econ. Dev.* **2021**, *30*, 1004–1019. [CrossRef]
2. Kipriyanov, F.A.; Medvedeva, N.A.; Medvedeva, S.V. Ensuring the Operational Reliability of the Tractor Fleet. *IOP Conf. Ser. Earth Environ. Sci.* **2019**, *315*, 062015. [CrossRef]
3. Miodragović, R.; Tanasijević, M.; Mileusnić, Z.; Jovančić, P. Effectiveness Assessment of Agricultural Machinery Based on Fuzzy Sets Theory. *Expert Syst. Appl.* **2012**, *39*, 8940–8946. [CrossRef]
4. Shilovsky, V.N.; Pitukhini, E.A.; Skobtsov, I.G. Technique for Improving the Organization of Maintenance of Transport and Technological Vehicles. *IOP Conf. Ser. Earth Environ. Sci.* **2021**, *666*, 062089. [CrossRef]
5. Beaton, A.J.; Dhuyvetter, K.C.; Kastens, T.L.; Williams, J.R. Per unit costs to own and operate farm machinery. *J. Agric. Appl. Econ.* **2005**, *37*, 131–144. [CrossRef]
6. Galiev, I.; Khafizov, C.; Adigamov, N.; Khusainov, R. Increase of Efficiency of Tractors Use in Agricultural Production. *Eng. Rural Dev.* **2018**, *17*, 23–25. [CrossRef]
7. Gulyarenko, A.A. Calculation Method of the Reasonable Reliability Level Based on the Cost Criteria. *J. Mach. Manuf. Reliab.* **2018**, *47*, 96–103. [CrossRef]
8. Babchenko, L.A.; Gulyarenko, A.A. Data Control for Reliability of Agricultural Tractors. *J. Mach. Manuf. Reliab.* **2020**, *49*, 900–906. [CrossRef]
9. Starovoytov, S.I.; Akhalaya, B.K.; Sidorov, S.; Mironova, A. The trend of tillage equipment development. *AMA Agric. Mech. Asia Afr. Lat. Am.* **2020**, *51*, 77–81. Available online: https://www.agriculturalmechanization.com/app/download/14479928935/VOL.51_NO.3_SUMMER_2020.pdf?t=1611645201 (accessed on 10 May 2022).
10. Ranjbarian, S.; Askari, M.; Jannatkhah, J. Performance of Tractor and Tillage Implements in Clay Soil. *J. Saudi Soc. Agric. Sci.* **2017**, *16*, 154–162. [CrossRef]
11. Mohamed, K.G.; Rezakalla, A.I. Reliability-based design optimization strategy for soil tillage equipment considering soil parameter uncertainty. *Adv. Eng. Res.* **2016**, *16*, 136–147. [CrossRef]
12. Vladutoiu, L.; Vladut, V.; Voiculescu, I.; Matache, M.; Radu, O.; Biris, S.; Voicea, I.; Paraschiv, G.; Atanasov, A.; Usenko, M. Increasing Agricultural Machinery Active Parts Durability by Hardening. *Aktualni Zadaci Meh. Poljopr. Actual Tasks Agric. Eng.* **2015**, *43*, 153–164. Available online: https://www.researchgate.net/profile/Valentin-Vladut/publication/274286917_INCREASING_AGRICULTURAL_MACHINERY_ACTIVE_PARTS_DURABILITY_BY_HARDENING/links/57dfaaa808ae72d72eac3f3d/INCREASING-AGRICULTURAL-MACHINERY-ACTIVE-PARTS-DURABILITY-BY-HARDENING.pdf (accessed on 10 May 2022).

13. Owsiak, Z. Wear of Spring Tine Cultivator Points in Sandy Loam and Light Clay Soils in Southern Poland. *Soil Tillage Res.* **1999**, *50*, 333–340. [CrossRef]
14. Motorin, V.A.; Gapich, D.S.; Borisenko, I.B.; Kurbanov, D.B. Simulation of the Wear of the Working Bodies of Chisel Plows. *J. Frict. Wear* **2020**, *41*, 71–77. [CrossRef]
15. Kostencki, P.; Stawicki, T.; Białobrzeska, B. Durability and Wear Geometry of Subsoiler Shanks Provided with Sintered Carbide Plates. *Tribol. Int.* **2016**, *104*, 19–35. [CrossRef]
16. Dickey, E.C.; Shelton, D.P.; Meyer, G.; Fairbanks, K. Determining crop residue cover with electronic image analysis. In Proceedings of the Eleventh International Congress on Agricultural Engineering, Dublin, Ireland, 4–8 September 1989; pp. 3056–3064.
17. Bialobrzeska, B.; Kostencki, P. Abrasive Wear Characteristics of Selected Low-Alloy Boron Steels as Measured in Both Field Experiments and Laboratory Tests. *Wear* **2015**, *328–329*, 149–159. [CrossRef]
18. Mironov, D.A.; Sidorov, S.A.; Liskin, I.V. Strength and Durability Characteristics of Soil-Cutting Working Tools. *Agric. Mach. Technol.* **2019**, *13*, 39–43. [CrossRef]
19. Erokhin, M.N.; Novikov, V.S. Forecasting Durability of Working Elements Of Designed Soil-Cultivating Machines. *Vestn. Fed. State Educ. Inst. High. Prof. Educ. Mosc. State Agroengineering Univ. Named After V.P. Goryachkin* **2017**, *6*, 56–62. [CrossRef]
20. Jankauskas, V.; Katinas, E.; Pusvaškis, M.; Leišys, R. A Study of the Durability of Hardened Plough Point. *J. Frict. Wear* **2020**, *41*, 78–84. [CrossRef]
21. Katsitadze, J.; Kapanadze, I.; Kutelia, G.; Bidzinashvili, I. Research on Reliability of Plows Operating in Mountainous Conditions. *Mach. Technol. Mater.* **2016**, *10*, 7–9. Available online: https://stumejournals.com/journals/mtm/2016/3/7.full.pdf (accessed on 10 May 2022).
22. Tajbakhsh, O.; Karparvarfard, S.H. Estimation of the fuel consumption rate of tractor due to different aspect ratios of chisel plow blade. *Iran. J. Biosyst. Eng.* **2015**, *46*, 201–208. [CrossRef]
23. ALkhafaji, A.J.; Almosawi, A.A.; Alqazzaz, K.M. Performance of Combined Tillage Equipment and It's Effect on Soil Properties. *Int. J. Environ. Agric. Biotechnol.* **2018**, *3*, 799–805. [CrossRef]
24. Aulin, V.; Warouma, A.; Lysenko, S.; Kuzyk, A. Improving of the Wear Resistance of Working Parts Agricultural Machinery by the Implementation of the Effect of Self-Sharpening. *Int. J. Eng. Technol.* **2016**, *5*, 126. [CrossRef]
25. Selig, E.T.; Nelson, R.D. Observations of Soil Cutting with Blades. *J. Terramechanics* **1964**, *1*, 32–53. [CrossRef]
26. Siemens, J.; Weber, J.; Thornburn, T. Mechanics of soil as influenced by model tillage tools. *Trans. ASAE* **1965**, *8*, 1–0007. [CrossRef]
27. Natsis, A.; Petropoulos, G.; Pandazaras, C. Influence of Local Soil Conditions on Mouldboard Ploughshare Abrasive Wear. *Tribol. Int.* **2008**, *41*, 151–157. [CrossRef]
28. Nalbant, M.; Palali, A.T. Effects of different material coatings on the wearing of plowshares in soil tillage. *Turk. J. Agric. For.* **2011**, *35*, 215–223. [CrossRef]
29. Er, U.; Par, B. Wear of Plowshare Components in SAE 950C Steel Surface Hardened by Powder Boriding. *Wear* **2006**, *261*, 251–255. [CrossRef]
30. Rempp, A.; Widmann, M.; Killinger, A.; Gadow, R. Advanced Ceramic/Metal Polymer Multilayered Coatings for Industrial Applications. *Key Eng. Mater.* **2012**, *533*, 133–144. [CrossRef]
31. Oliveira, T.G.d.; Costa, A.R.d. Influence of microstructure on micro-abrasive wear resistance of alloys Fe-Cr-C and Fe-Cr-C-Nb. *Matéria* **2019**, *24*, 1. [CrossRef]
32. Wang, J.; Lu, J.; Xing, X.; Zhou, Y.; Liu, S.; Qi, X.; Yang, Q. Effects of B Contents on the Microstructure and Wear Resistance of Hypereutectic Fe-Cr-C Hardfacing Alloy Coating. *Mater. Res. Express* **2019**, *6*, 1065h2. [CrossRef]
33. Latypov, R.; Serov, A.; Serov, N.; Chekha, O. Technology of Hardening Plowshares by Electrocontact Welding Using Waste from Tool Production. In *Robotics, Machinery and Engineering Technology for Precision Agriculture*; Springer: Singapore, 2021; pp. 197–203. [CrossRef]
34. Titov, N.V.; Kolomeichenko, A.V.; Logachev, V.N.; Kravchenko, I.N.; Litovchenko, N.N. Investigation of the Hardness and Wear Resistance of Working Sections of Machines Hardened by Vibroarc Surfacing Using Cermet Materials. *Weld. Int.* **2015**, *29*, 737–739. [CrossRef]
35. Remshev, E.Y.; Danilin, G.A.; Vorob'eva, G.A.; Silaev, M.Y. Ensuring the Operational Reliability of Elastic Elements by Acoustic Methods. *Metallurgist* **2015**, *59*, 225–228. [CrossRef]
36. Kornilovich, S.A.; Redreev, G.V.; Vorobyov, D.A.; Zakharov, S.V. Improvement of Adhesion Strength of the Sprayed Material. *IOP Conf. Ser. Mater. Sci. Eng.* **2019**, *582*, 012042. [CrossRef]
37. Królicka, A.; Szczepański, Ł.; Konat, Ł.; Stawicki, T.; Kostencki, P. The Influence of Microstructure on Abrasive Wear Micro-Mechanisms of the Claddings Produced by Welding Used in Agricultural Soil. *Materials* **2020**, *13*, 1920. [CrossRef]
38. Borawski, A.; Szpica, D.; Mieczkowski, G.; Awad, M.M.; Shalaby, R.M.; Sallah, M. Proposition of Abrasion Resistance Determining Method of Materials Used for Production of Plowshares. In Proceedings of the International Scientific Conference Engineering for Rural Development, Jelgava, Latvia, 26–28 May 2021. [CrossRef]
39. Swanson, P.; Vetter, A. The measurement of abrasive particle shape and its effect on wear. *ASLE Trans.* **1985**, *28*, 225–230. [CrossRef]
40. Taha, S.; Taha, F.; Nekolayevech, Z. Estimating economical costs for operating agricultural machines and equipment during plowing operations. *Iraqi J. Agric. Sci.* **2016**, *47*, 385–390. [CrossRef]

41. Kanaev, A.; Nukeshev, S.; Zhusin, B.; Milokumov, V. The Experience and the Problems in Improvement of Technical Level and Economic Effectiveness of Agricultural Machinery. *Sci. Rev.* **2012**, *2*, 21–27. Available online: https://kazatu.edu.kz/assets/i/science/sr2012_I10_05.pdf (accessed on 10 May 2022).
42. Molenhuis, J. *Budgeting Farm Machinery Costs*; Ontario Ministry of Agriculture, Food and Rural Affairs: Guelph, ON, Canada, 2001. Available online: https://files.ontario.ca/omafra-budgeting-farm-machinery-costs-20-007-en-2020-05-12.pdf (accessed on 10 May 2022).
43. Krivtsov, A. Forecasting the nature of the formation of two-layer soil-cutting profiles in the process of wear. *Agro Panorama* **2005**, *2*, 24–28. Available online: https://rep.bsatu.by/bitstream/doc/11383/1/krivcov-a-v-prognozirovanie-haraktera-formoobrazovaniya-dvuhslojnyh-pochvorezhushchih-profilej-v-processe-iznashivaniya.pdf (accessed on 10 May 2022).
44. Sidorov, S.A.; Mironov, D.A.; Tsench, Y.S.; Mironova, A.V. Evaluation of wear resistance and resource of two-layer hardened soil-cutting working bodies in various soil conditions. *Eng. Technol. Syst.* **2020**, *30*, 699–710. [CrossRef]
45. Orekhova, G. Forecasting the durability of plow shares in the treatment of soils of different degrees of moisture. *Bull. Bryansk State Agric. Acad.* **2020**, *2*, 28–32. Available online: https://cyberleninka.ru/article/n/prognozirovanie-dolgovechnosti-pluzhnyh-lemehov-pri-obrabotke-pochv-raznoy-stepeni-vlazhnosti (accessed on 10 May 2022).
46. Korshunov, V.Y. Calculation of the wear rate of deposited materials during the restoration of tillage implements. *Rural Mach. Oper.* **2020**, *1*, 32–33. Available online: https://libryansk.ru/files/projectimage/agro/text/korshunov_sm_2020_01.pdf (accessed on 10 May 2022).
47. Gray, R.D. *Alloys and AutomobilesThe Life of Elwood Haynes*; Indiana Historical Society: Indianapolis, IN, USA, 1979.
48. Sidorov, S.; Mironov, D.; Khoroshenkov, V.; Khlusova, E. Surfacing methods for increasing the service life of rapidly wearing working tools of agricultural machines. *Weld. Int.* **2016**, *30*, 808–812. [CrossRef]
49. Kostencki, P.; Stawicki, T.; Królicka, A.; Sędłak, P. Wear of cultivator coulters reinforced with cemented-carbide plates and hardfacing. *Wear* **2019**, *438*, 203063. [CrossRef]
50. Lvov, P. A Method of Surfacing Wear-Resistant Alloys. 1951. Available online: https://rusneb.ru/catalog/000224_000128_0000092002_19510101_A1_SU/ (accessed on 10 May 2022).
51. Erokhin, M.; Novikov, V. Increasing the strength and wear resistance of the plowshare. *Agroengineering* **2008**, *3*, 100–107. Available online: https://cyberleninka.ru/article/n/povyshenie-prochnosti-i-iznosostoykosti-lemeha-pluga (accessed on 10 May 2022).
52. Katsitadze, J.; Pawlowski, T.; Phutkaradze, Z. Plowshares Wear Investigation of the Plows Working in Mountainous Conditions Using Statistical Probabilistic Modeling. *J. Agric. Sci. Technol. A* **2018**, *8*, 253–257. [CrossRef]
53. Dzhabborov, N.I.; Dobrinov, A.V.; Jabborov, P.N. Research and Modeling of the Wear Process of Parts of the Soil Tillage Working Implements. *IOP Conf. Ser. Earth Environ. Sci.* **2021**, *699*, 012038. [CrossRef]
54. Ozhegov, N.; Dzhabborov, N.; Dobrinov, A.; Fedkin, D. Estimation of wear rate of hard faced ploughshares. In Proceedings of the Environmentally Friendly Agriculture and Forestry for Future Generations, Saint Petersburg, Russia, 26–28 February 2015; pp. 191–193. Available online: https://elibrary.ru/item.asp?id=24861069 (accessed on 10 May 2022).
55. Buylov, V.N.; Lyulyakov, I.V.; Makarov, S.A.; Danilin, A.V.; Chekmarev, V.V. Forecasting of the Soil Processing Units Working Bodies Resource. *IOP Conf. Ser. Earth Environ. Sci.* **2020**, *422*, 012115. [CrossRef]
56. Bartenev, I.; Pozdnyakov, E. Wearing ability of soils and its effect on the durability of the working bodies of tillage machines. *J. For. Eng.* **2013**, *3*, 114–123. [CrossRef]
57. Myalenko, V.I. Modeling of Reliability When Designing Soil Cutting Parts of Agricultural Machines. *Tract. Agric. Mach.* **2015**, *82*, 44–47. [CrossRef]
58. Korotkov, V.A. Strengthening of Steel by Plasma Quenching and Carbonitriding. *Russ. Eng. Res.* **2019**, *39*, 234–236. [CrossRef]
59. Kanaev, A.; Gulyarenko, A.; Bogomolov, A.; Sarsembaeva, T. Analysis of mechanisms for hardening constructional steels by structure parameters. *Metalurgija* **2022**, *61*, 241–243. Available online: https://hrcak.srce.hr/clanak/381655 (accessed on 10 May 2022).
60. Kanaev, A.T.; Gulyarenko, A.A.; Sarsembaeva, T.E.; Ayazbaeva, A.B. Structure Formation under Plasma-Assisted Hardening of Thin-Walled Low-Weight Parts. *Steel Transl.* **2021**, *51*, 582–586. [CrossRef]
61. Yang, L.J. Plasma Surface Hardening of ASSAB 760 Steel Specimens with Taguchi Optimisation of the Processing Parameters. *J. Mater. Processing Technol.* **2001**, *113*, 521–526. [CrossRef]
62. Korotkov, V.A. Impact of Plasma Hardening on the Wear Resistance of 38XC Steel. *J. Frict. Wear* **2017**, *38*, 302–304. [CrossRef]
63. Korotkov, V.A. Properties and Commercial Application of Manual Plasma Hardening. *Met. Sci. Heat Treat.* **2016**, *58*, 449–454. [CrossRef]
64. Konstantinova, M.; Balanovskiy, A.; Gozbenko, V.; Kargapoltsev, S.; Karlina, A.; Shtayger, M.; Guseva, E.; Kuznetsov, B. Application of plasma surface quenching to reduce rail side wear. In *IOP Conference Series: Materials Science and Engineering*; IOP Publishing: Bristol, UK, 2019; p. 012146. [CrossRef]
65. Yang, Y.; Yan, M.; Zhang, S.; Guo, J.; Jiang, S.; Li, D. Diffusion behavior of carbon and its hardening effect on plasma carburized M50NiL steel: Influences of treatment temperature and duration. *Surf. Coat. Technol.* **2018**, *333*, 96–103. [CrossRef]
66. Wu, B.; Li, J.; Ye, J.; Tan, J.; Liu, L.; Song, J.; Chen, X.; Pan, F. Work hardening behavior of Ti particle reinforced AZ91 composite prepared by spark plasma sintering. *Vacuum* **2021**, *183*, 109833. [CrossRef]

67. Krawczyk, J.; Bembenek, M.; Pawlik, J. The Role of Chemical Composition of High-Manganese Cast Steels on Wear of Excavating Chain in Railway Shoulder Bed Ballast Cleaning Machine. *Materials* **2021**, *14*, 7794. [CrossRef]
68. Bembenek, M.; Krawczyk, J.; Pańcikiewicz, K. The Wear on Roller Press Rollers Made of 20Cr4/1.7027 Steel under Conditions of Copper Concentrate Briquetting. *Materials* **2020**, *13*, 5782. [CrossRef]
69. Krawczyk, J.; Bembenek, M.; Frocisz, Ł.; Śleboda, T.; Paćko, M. The Effect of Sandblasting on Properties and Structures of the DC03/1.0347, DC04/1.0338, DC05/1.0312, and DD14/1.0389 Steels for Deep Drawing. *Materials* **2021**, *14*, 3540. [CrossRef]
70. Orynbekov, D.; Sarsembayeva, T.; Kanaev, A.; Gulyarenko, A. Nanoparticle-strengthened-martensitic Surface Layered Constructed Steel by Plasma Hardening Rout. *J. Nanostruct.* **2021**, *11*, 814–824. [CrossRef]
71. Saburkin, D.A. *Increasing the Durability of the Plowshare by Improving Its Design and Technological Parameters*; Moscow State Agroengineering University: Moscow, Russia, 2008. Available online: https://static.freereferats.ru/_avtoreferats/01003440483.pdf (accessed on 10 May 2022).

*agriculture*

MDPI

*Article*

# Research on Wet Clutch Switching Quality in the Shifting Stage of an Agricultural Tractor Transmission System

**Yuting Chen [1], Zhun Cheng [1,*] and Yu Qian [2]**

[1] Department of Vehicle Engineering, Nanjing Forestry University, Nanjing 210037, China
[2] College of Engineering, Nanjing Agricultural University, Nanjing 210031, China
* Correspondence: cz38@njfu.edu.cn

**Abstract:** In order to improve the working quality of wet clutch switching in an agricultural tractor, in this paper, we took a power shift system composed of multiple wet clutches as the research object for full-factorial performance measurement, multi-factor analysis of the degree of influence, establishment of a single evaluation index model, formation of a comprehensive evaluation index, and formulation of adjustable factor control strategies. We studied the simulation test platform of an agricultural tractor power transmission system based on the SimulationX software and obtained 225 sets of sample data under a full-use condition. Partial least squares and range analysis were applied to comprehensively analyze the influence of multiple factors on the working quality of wet clutches. In this paper, we proposed a modeling method for a single evaluation index of the wet clutch (combined with polynomial regression and tentative method, the goal is determined in the form of a model with the maximum coefficient of determination) and two control strategy optimization methods for the wet clutch adjustable factors, i.e., Method 1 (integrated optimization) and Method 2 (step-by-step optimization), both methods were based on an improved genetic algorithm. The results showed that oil pressure, flow rate, and load had significant effects on the dynamic load characteristics (the degrees were 0.38, −0.44, and −0.63, respectively, with a negative sign representing an inverse correlation); rate of flow and load had significant effects on speed drop characteristics (the degrees were −0.56 and 0.73, respectively). A multivariate first-order linear model accurately described the dynamic load characteristics ($R^2$ = 0.9371). The accuracy of the dynamic load characteristic model was improved by 5.5037% after adding the second-order term and interaction term of oil pressure. The polynomial model containing the first-order oil pressure, first-order flow rate, second-order flow rate, and interaction terms could explain the speed drop characteristics, with an $R^2$ of 0.9927. If agricultural tractors operate under medium and large loads, the oil pressure and flow rate in their definitional domains should be small and large values, respectively; if operating under small loads, both oil pressure and flow rate should be high. When the wet clutch dynamic load and speed drop characteristics were improved, the sliding friction energy loss also decreased synchronously (the reduction could reach 70.19%).

**Keywords:** simulation; quality improvement; improved genetic algorithm; full-factorial test; single evaluation index modeling method; control strategy

**Citation:** Chen, Y.; Cheng, Z.; Qian, Y. Research on Wet Clutch Switching Quality in the Shifting Stage of an Agricultural Tractor Transmission System. *Agriculture* **2022**, *12*, 1174. https://doi.org/10.3390/agriculture12081174

Academic Editors: Mustafa Ucgul and Chung-Liang Chang

Received: 18 July 2022
Accepted: 5 August 2022
Published: 7 August 2022



## 1. Introduction

A tractor is a widely used vehicle in agricultural operations [1–3], which has some differences in driving speed and load when performing ploughing, rotary tillage, or transportation tasks [4]. Therefore, there are clear requirements for the coordination of the power source and transmission system when operating agricultural tractors. Regarding power sources, an agricultural tractor is similar to road driving vehicles, mainly using a "power battery [5,6] motor [7–9] system" or an internal combustion engine system [10]. The combined use of a variable speed transmission system can further improve the working performance of an agricultural tractor (mainly the power performance and economic performance).

A transmission gearbox is the core device of a vehicle transmission system that changes the transmission ratio to achieve acceleration and torque reduction or deceleration and increased torque. The advanced variable speed transmission systems in agricultural tractors (these new systems are formed based on the development of computer technology, electronic control technology, and hydraulic technology [11]) are mainly composed of power shift transmissions [12,13] and power-split continuously variable transmissions (CVT) [14,15]. These advanced transmission systems can change their operating mode when changing transmission ratios. During the process of changing the operating mode, there is a need for a reliable mechanical system with a smooth switching process, and the power cannot be interrupted during the switching process. To meet the above requirements, wet clutches are widely used in the advanced transmission systems of agricultural tractors to achieve transmission gearbox switching [16].

At present, there is increased interest in research on agricultural tractor wet clutches. Cheng et al. [17,18], Sun et al. [19], and Li et al. [20] have conducted corresponding design, parameter matching, and performance analyses of a tractor with hydraulic mechanical continuously variable transmission (HMCVT) and wet clutches. Current research on wet clutches has focused on improving clutch performance, specifically, mainly including a wet clutch working quality analysis and control strategy formulation. For example, Qian et al. [21] conducted an orthogonal test with five factors and four levels, combined with the SimulationX software and stepwise regression to study the shift quality of wet clutches of heavy tractors. They developed a mathematical model of wet clutch performance based on an oil pressure domain of 2~6 MPa, a flow domain of 3~6 L/min, and a load torque domain 200~641 Nm. Ni et al. [22] also used an orthogonal test with five factors and four levels (but the factors and levels were different from that of the Qian et al. [21] study) combined with a bench test. They conducted 16 trials and used range analysis to obtain the optimal oil pressure and flow during clutch operation. Wang et al. [23] conducted an orthogonal test with four factors and three levels and a single-factor test and mainly relied on the analysis of variance and a range analysis. Stockinger et al. [24] analyzed the friction performance of a multiple sheet wet clutch. The research of Raikwar et al. [25] mainly used the MATLAB software to build a tractor vehicle simulation model, including a wet clutch model and analyzed smooth clutch engagement, reduced transmission shock, and operator discomfort.

Overall, currently, the number of studies on wet clutches in agricultural tractors is relatively small. The existing studies have mainly conducted simulation tests or actual tests. However, most of the studies have had a relatively small number of trial samples. Most studies have explored the scheme of improving the working quality of a wet clutch by means of an orthogonal test combined with the analysis of variance or a range analysis. However, it is difficult to apply these methods for in-depth analyses. In addition, most studies have tended to use the revolution or load torque working conditions of an agricultural tractor as a factor in orthogonal tests. However, in fact, agricultural tractors can or need to work under varied working conditions. Therefore, it is much more important to study the switching quality of agricultural tractors under different working conditions. There is a lack of research on evaluation index modeling related to the switching quality of a wet clutch under the full-use conditions of an agricultural tractor.

To solve the abovementioned problems, in this paper, we have focused on five parts (full factorial simulation tests, analysis of the degree of influence of factors, model establishment of a single evaluation index, model establishment of a comprehensive evaluation index, and control strategy optimization and formulation). The value and innovation of this research include: Relatively large sample data are used to investigate the working quality of an agricultural tractor wet clutch; partial least squares, a range analysis, polynomial regression, analytic hierarchy process, and the heuristic intelligent optimization algorithm are comprehensively used for systematic research and analysis; two optimization and formulation methods for the wet clutch control strategies for an agricultural tractor under full-use conditions are proposed; the modeling method of a work quality evaluation index model is proposed. In this paper, we provide a basis and valuable reference for the design, performance evaluation, performance estimation, control, and performance improvement of a wet clutch for agricultural machinery.

## 2. Materials and Methods

### *2.1. Agricultural Tractor Transmission System with Wet Clutch*

In this paper, we studied an advanced transmission system for an agricultural tractor. See Figure 1 for the transmission scheme. The transmission system is a hydraulic mechanical CVT (HMCVT). It has five working segments, and each working segment can realize the continuous change of its own transmission ratio within a certain range.

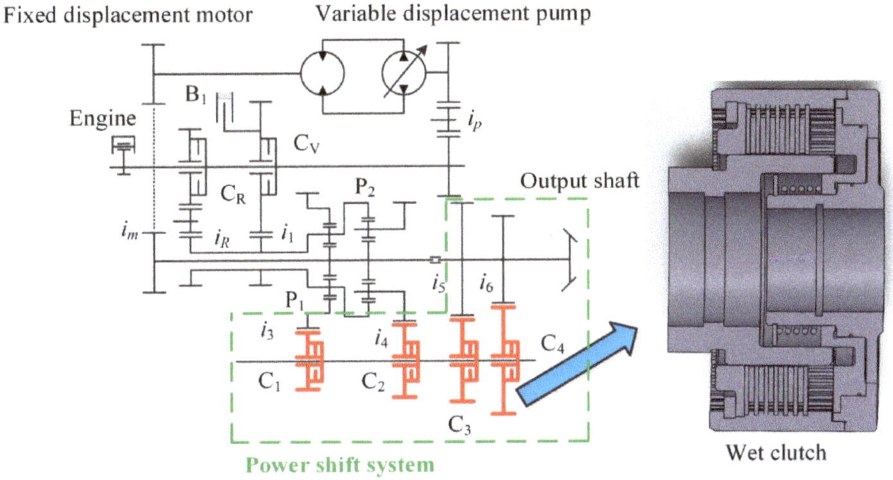


**Figure 1.** HMCVT for agricultural tractors, as studied in the paper.

The "power shift system" shown in Figure 1, can also be a core device for other types of power shift transmission for agricultural tractors. Therefore, in this paper, we focused on a hydraulic and mechanical power shunt CVT transmission, and also on a power shift transmission.

The HMCVT mainly consisted of two planetary gear devices ($P_1$ and $P_2$), six wet clutches ($C_V$, $C_R$, $C_1$, $C_2$, $C_3$, and $C_4$), one brake ($B_1$), one pump-motor system (the working mechanism involved a variable pump that controlled the quantitative motor to change the revolution of the motor output shaft), and multiple fixed gears (the gear secondary transmission ratio was set to $i_1$, $i_R$, $i_3$, $i_4$, $i_5$, $i_6$, $i_p$, and $i_m$ respectively). The working parameters and working principles of the system were referred to in a previous study [18].

### *2.2. Wet Clutch Switch Simulation Test Platform Based on the SimulationX Software*

Simulation is a core technology that is widely used in engineering research. Simulation results in a variety of engineering fields have been verified for their accuracy and effectiveness (for example, the research works of Talati et al. [26], Torshizian et al. [27], Aliakbari et al. [28], etc.). The SimulationX software has been verified by several previous studies in clutch performance simulation tests (for example, Lu et al. [29] showed that the maximum error of the simulation test results of the wet clutch and the actual test results did not exceed 6%, and the average error did not exceed 5%; Wang et al. [30] showed that the relative error of the simulation test results of the wet clutch and the actual test results was less than 10%.).Therefore, in this paper, we studied various models built using the SimulationX software to build the HMCVT gear switching simulation test platform and used the abovementioned studies in the literature as references (see Figure 2 for the schematic diagram of the simulation test platform). The simulation test platform mainly included a pump-motor system model (the pump was a variable pump, and its displacement ratio was adjustable), a fixed shaft gear model, a motor model, a load simulation model, a wet clutch model, a clutch oil filling pressure simulation model, and a planetary gear model.

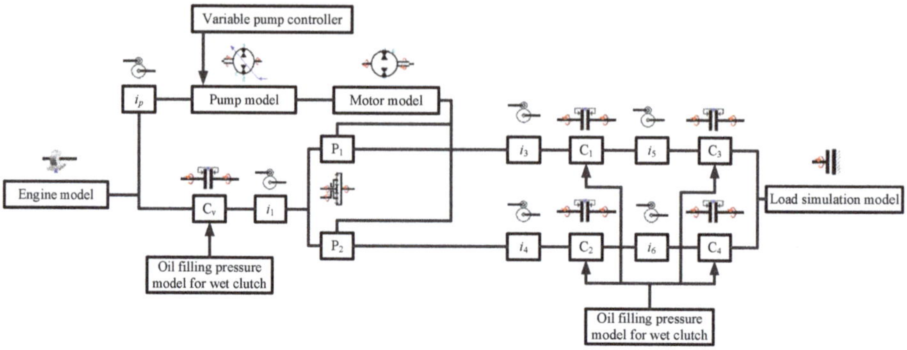


**Figure 2.** Schematic diagram of the simulation test platform used in this paper.

### 2.3. Full-Factorial Test Design under Full-Use Conditions of Agricultural Tractors

The goal of this research was to explore the performance change characteristics of a wet clutch when the power shift transmission of an agricultural tractor was shifted under various operating conditions. The optimal control strategies of oil pressure and flow rate were formulated for optimal working quality of the clutch. Therefore, in this paper, we took the switch of an HM1 segment bit to HM2 segment bit in a five-segment HMCVT as an example. During the gear switch, the $C_2$ wet clutch was disconnected and the $C_1$ wet clutch combined.

In this paper, the total mass of the agricultural tractor was set to be 3000 kg; the working revolution range of the diesel engine was 800–2200 rpm; the radius of the driving wheel was 0.976 m, and the rolling resistance coefficient was assumed to be 0.2. The maximum load condition of the agricultural tractor was the plough industry condition. The plough resistance of an agricultural tractor is usually estimated using the following formula [31]:

$$F_p = r_1 zbhk \tag{1}$$

where $r_1$ is the instability coefficient, which is used to characterize fluctuations in the plowing resistance when an agricultural tractor is ploughing in a field; $z$ is the number of ploughshares, and the value in this study is 5; $b$ is the tillage width of a single plough body, which is 40 cm in this study; $h$ is the tillage depth, which is 18 cm in this study; $k$ is the soil specific resistance, which is 60 kPa in this study.

Since agricultural tractors travel slowly when plowing, in this study, the air resistance of the agricultural tractor was ignored. The load resistance torque at the output end of the agricultural tractor transmission is (refer to vehicle driving resistance equation [32]):

$$T_{out} = (F_p + F_f)r_d/i_0 = (F_p + mgf)r_d/i_0 \tag{2}$$

where $T_{out}$ is the load resistance torque at the output end of the transmission; $F_f$ is the rolling resistance of the agricultural tractor; $m$ is the total mass of the agricultural tractor; $g$ is the acceleration of gravity; $f$ is the rolling resistance coefficient; $r_d$ is the driving force radius; $i_0$ is the total gear ratio of the drive train, excluding the transmission.

Combining Equations (1) and (2), it could be calculated that the maximum load resistance torque at the output end of the agricultural tractor transmission was about 1000 Nm. Therefore, the full-use conditions of the agricultural tractor studied in this paper were: the working revolution of the engine varied from 800 to 2200 rpm (that is, the working revolution of the input end of the transmission varied from 800 to 2200 rpm); the load torque variation range of the transmission output end was 0~1000 Nm.

In this paper, the working revolution of the input end and the output end load torque of the HMCVT were divided into three levels in their respective definition domains. Atotal of nine use conditions for an agricultural tractor were studied in this paper (see Table 1).

**Table 1.** Factor level combination of the agricultural tractor full-use conditions.

| HMCVT Output Load Torque Nm | HMCVT Input Working Revolution rpm | | |
|---|---|---|---|
| | 800 | 1500 | 2200 |
| 200 | Low speed small load | Medium speed small load | High speed small load |
| 600 | Low speed medium load | Medium speed medium load | High speed medium load |
| 1000 | Low speed large load | Medium speed large load | High speed large load |

Under the working conditions, the working parameters (oil filling pressure and filling flow) of the wet clutches were divided into five levels within their respective domain range. Then, the horizontal combination of oil pressure was 2, 3, 4, 5, and 6 MPa; the horizontal combination of flow was 2, 3, 4, 5, and 6 L/min. Therefore, 25 sets of simulation tests were required for each working condition of the agricultural tractor. In summary, the number of full-factorial test groups used in this study was 225.

### 2.4. Analysis Method for the Degree of Influence of Working Condition Factors and Adjustable Factors

In this paper, the working revolution of the HMCVT input end and the output end load torque were the working condition factors studied, that is, the agricultural tractor always worked at a certain revolution and torque working conditions.

The oil filling pressure and flow rate of a wet clutch were the adjustable factors studied, that is, the operation of the agricultural tractor could be adjusted and controlled under any working condition.

We used partial least squares (PLS) [33,34] and a range analysis (RA) [35] to analyze the degree of influence of factors (including working conditions and adjustable factors) for the results of the full-factorial tests.

The calculation formula for the range differential analysis is as follows [35]:

$$R_i = \max(\overline{X}_i) - \min(\overline{X}_i) \tag{3}$$

where $R_i$ is the range of the $i_{th}$ factor, $\overline{X}_i$ is the set of sample data means of all levels of the $i_{th}$ factor.

The analysis process for the degree of influence of factors used in this study is as follows:

**Step 1.** Calculate and analyze the degree of influence for 225 sets of test results on working condition factors (HMCVT input working speed and output load torque) by using PLS and RA, respectively.
**Step 2.** Calculate and analyze the degree of influence for 225 sets of test results on working condition factors (oil filling pressure for wet clutch) by using PLS and RA, respectively.
**Step 3.** Compare the analysis results of the PLS and RA, and draw common conclusions.

### 2.5. Selection of a Single Evaluation Index and the Model-Building Method

There were a number of indicators in the performance evaluation of the wet clutches, among which the most important were the evaluation indicators related to the output end speed and torque. The wear dissipation energy (i.e., sliding friction work) was another important physical quantity. The calculation formula of sliding friction work is as follows [30]:

$$W_c = \int_{t_1}^{t_2} T_c(t)|\Delta\omega(t)|dt \tag{4}$$

where $W_c$ is the energy loss of sliding friction, $t_1$ is the start time of clutch engagement, $t_2$ is the end time of clutch engagement, $T_c(t)$ is the torque transmitted by the clutch, $\Delta\omega(t)$ is the revolution difference between the master and driven ends of the clutch.

From Equation (4), the evaluation indicators related to the revolutions and torque also had the ability to reflect the sliding friction size during the wet clutch bonding process. Therefore, in this paper, the physical quantity related to the output end revolutions and the torque was selected as the evaluation index, including the speed drop and the dynamic load, respectively.

The speed drop formula is calculated as follows [30]:

$$J = |\omega_\infty - \omega_{\min}| \tag{5}$$

where $J$ is the speed drop, which is dimensionless; $\omega_\infty$ is the output revolution of the transmission in steady state after shifting; $\omega_{\min}$ is the minimum output revolution of the transmission during the clutch switching process.

The formula of the dynamic load is as follows [30]:

$$K = T_{c\_\max}/T_\infty \tag{6}$$

where $K$ is the dynamic load, dimensionless; $T_{c\_\max}$ is the maximum torque at the output end of the transmission during the clutch switching process; $T_\infty$ is the output torque in the steady state after the transmission is shifted, and this physical quantity is basically determined by the output end load.

Polynomial regression models are widely used in the engineering field. In this paper, we proposed a polynomial regression-based modeling method for wet clutch performance evaluation indicators (speed drop and dynamic load). This method combined the observation of 225 sets of full-factorial simulation test data and used the tentative method of a heuristic intelligent optimization algorithm (such as the artificial fish school algorithm). For the full working conditions of the simulation test, the estimation model of the evaluation index (speed drop and dynamic load) was determined by testing several types of polynomial models and taking the maximum target of the coefficient. The test process took the multivariate first-order linear regression model as the first test model, and then added the second-order terms and interaction terms of each independent variable one by one to form a new test model. The final model form was determined by comparing the dependent coefficients of the previous and the subsequent tentative model. A flow chart of this modeling method is presented in Figure 3.

### 2.6. Method for Establishing a Comprehensive Evaluation Index of Wet Clutch Working Quality

In this paper, the weighting coefficient method was used to establish a comprehensive evaluation index by combining two single evaluation indexes (speed drop and dynamic load). According to the literature [36], the analytic hierarchy process (AHP) was used to determine the weighting coefficient of two single evaluation indexes (i.e., speed drop and dynamic load). The mathematical expression of the comprehensive evaluation index is as follows:

$$CEI = w_1 K/K_{\max} + w_2 J/J_{\max} \tag{7}$$

where $CEI$ is the comprehensive evaluation index of the working quality of the wet clutch; $w_1$ and $w_2$ are the weighting coefficients of dynamic load and speed drop, respectively, while $w_1 + w_2 = 1$; $K_{max}$ and $J_{max}$ are the maximum values of dynamic load and speed drop in the test sample data, respectively. In this way, the original index was converted into a dimensionless index to assimilate the order of magnitude of the two single evaluation indexes.

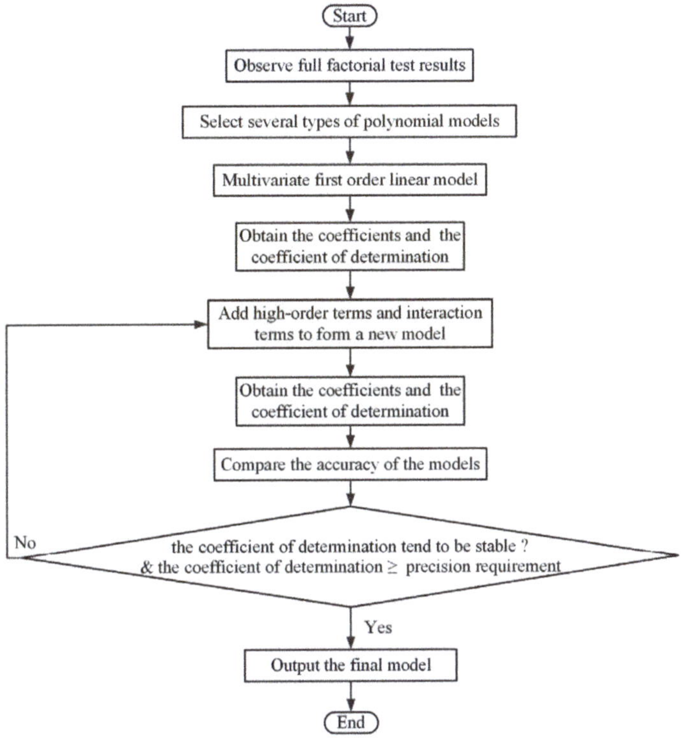


**Figure 3.** Flow chart of the performance evaluation index modeling method proposed in this paper.

The layers of the wet clutch switching quality are shown in Figure 4.

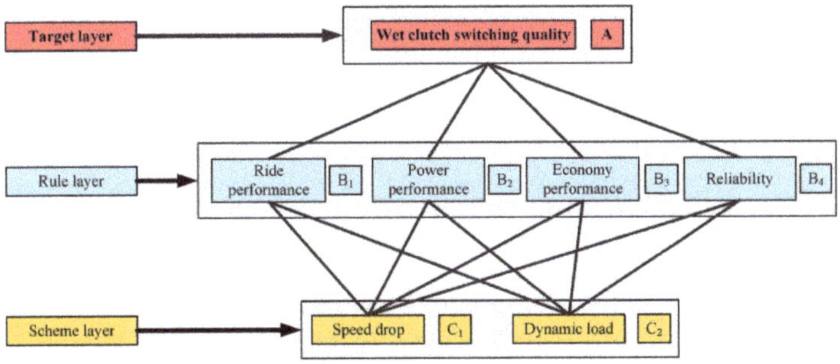


**Figure 4.** Layers of the wet clutch switching quality.

*2.7. Acquisition Method for the Optimal Control Strategy*

In this paper, we proposed two methods to obtain the optimal control strategy of the wet clutch adjustable factors (i.e., oil filling pressure and flow rate) with the optimization objective of the minimum comprehensive evaluation index. Consistent results were obtained between the two optimal control strategies.

**Method 1:** Integrated optimization based on an improved genetic algorithm (I-GA)

Method 1 used an I-GA to optimize the comprehensive evaluation index considering oil pressure and flow as a whole. Heuristic intelligent optimization algorithms are widely used for optimization in the engineering field. A number of studies [37–40] have shown that they can be used to effectively solve engineering problems. The I-GA reference used in this study had been previously studied [41], and its effect in engineering applications had been verified. The flow of this method is shown in Figure 5.

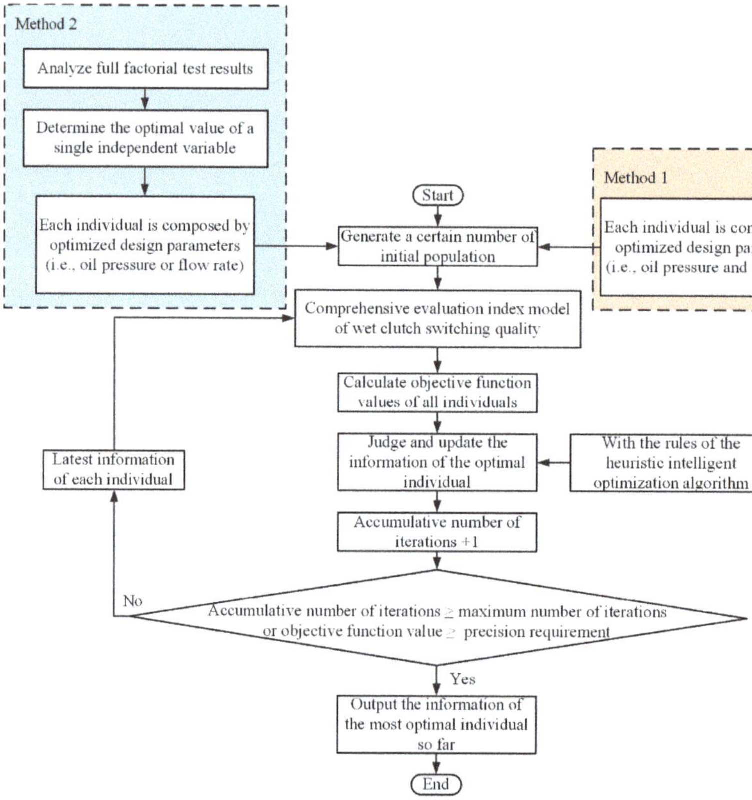


**Figure 5.** Flow chart of Methods 1 and 2.

**Method 2:** I-GA-based step-by-step optimization

In each working condition of the agricultural tractor, Method 2 first analyzed the full-factorial test data to obtain the optimal adjustable factor test level combination of dynamic load and speed drop, respectively. The optimal tuning interval for a single independent variable was also determined to hopefully reduce the dimension of the decision variables during the optimization process. Secondly, when the decision variable dimension was reduced, the comprehensive evaluation index model was combined to form a new optimization objective function. Finally, the remaining decision variables were optimized by using the same I-GA from Method 1. The flow of Method 2 is shown in Figure 5.

## 3. Results and Discussion

### *3.1. Full-Factorial Simulation Test Results*

The full-factorial simulation test results of the working quality of wet clutch switching in an agricultural tractor variable transmission system are shown in Figure 6 for 225 sets of test sample data. The meanings of sample data numbers in Figure 6 are shown in Table 2.

**Table 2.** The meanings of sample data numbers in Figure 6.

| Sample Number | Oil Pressure (MPa) | Flow Rate (L/min) |
|---|---|---|
| 1~5 | 2 | 2~6 |
| 6~10 | 3 | 2~6 |
| 11~15 | 4 | 2~6 |
| 16~20 | 5 | 2~6 |
| 21~25 | 6 | 2~6 |

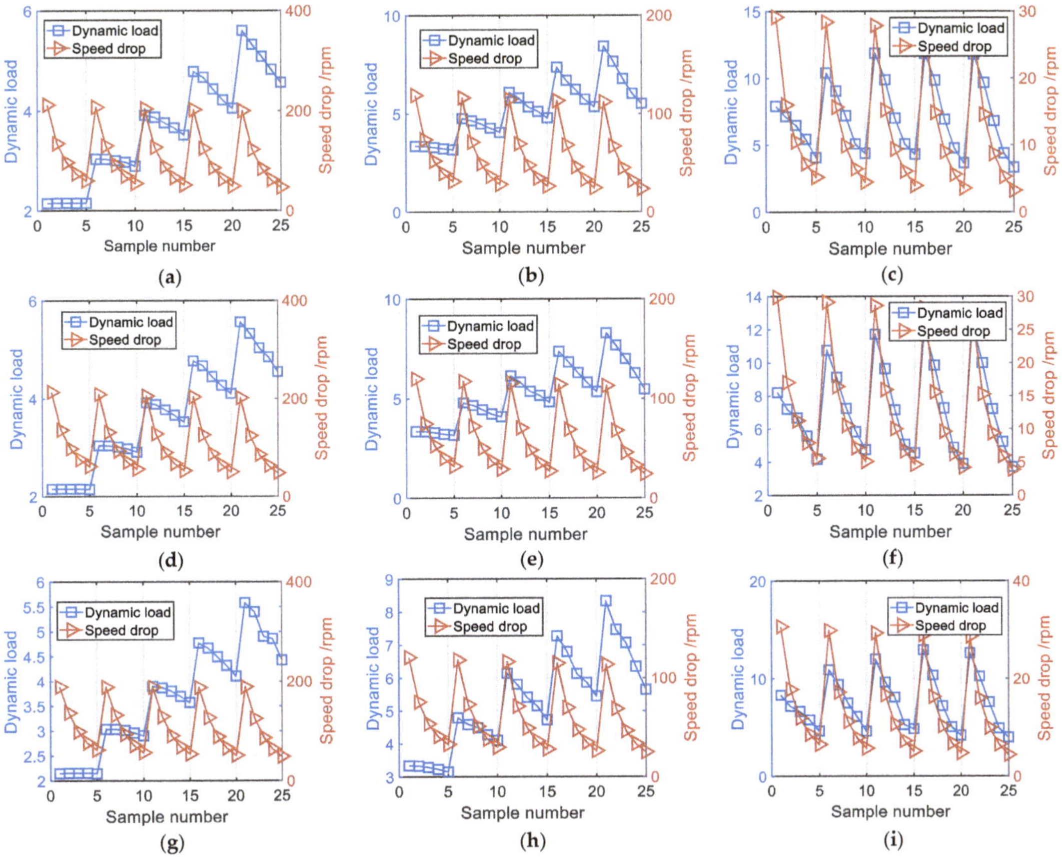


**Figure 6.** Full-factorial simulation test results: (**a**) High-speed and large-load working condition; (**b**) high-speed and medium-load working condition; (**c**) high-speed and small-load working condition; (**d**) medium-speed and large-load working condition; (**e**) medium-speed and medium-load working condition; (**f**) medium-speed and small-load working condition; (**g**) low-speed and large-load working condition; (**h**) low-speed and medium-load working condition; (**i**) low-speed and low-load working condition.

From Figure 6, there were obvious differences between the dynamic load characteristics and the speed drop characteristics of the agricultural tractor transmission system when the gears were switched. However, the difference between the two tended to decrease with load. In particular, the dynamic load characteristics and the speed drop characteristics were similar under small load conditions. Dynamic load was obviously affected by oil pressure, the flow rate, and the load, and the dynamic load characteristics under the different oil pressures, flow rates, an loads varied significantly. The speed drop was obviously affected by the flow and load, that is, the speed drop characteristics under different flow and load were significantly different.

### *3.2. Analysis of the Degree of Influence of Working Condition Factors and Adjustable Factors*

The results of the influence of working conditions and adjustable factors using PLS are shown in Figure 7.

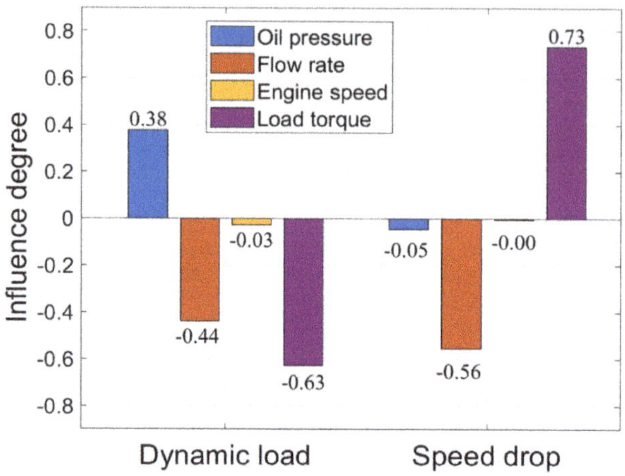


**Figure 7.** Calculation results based on PLS.

The calculation results based on PLS showed that: (1) The factors with significant influence on dynamic load characteristics were oil pressure, flow rate, and load (from small to large). Among them, oil pressure was positively correlated with dynamic load, and flow rate and load were inversely correlated with dynamic load. (2) The factors that had a significant influence on the speed drop characteristics were flow and load (from small to large). Among them, flow rate was inversely correlated with speed drop, and load was positively correlated with speed drop.

The results of the influence of the RA on working conditions and adjustable factors are shown in Table 3.

**Table 3.** Calculation results based on the range analysis.

| Factor | Oil Pressure | Flow Rate | Engine Speed | Load Torque |
|---|---|---|---|---|
| Dynamic load range | 2.60 | 2.90 | 0.17 | 3.69 |
| Speed drop range (rpm) | 6.97 | 86.44 | 0.96 | 94.86 |

The calculated results of the degree of influence based on the RA were highly consistent with the results of the PLS analysis.

### 3.3. Establishment and Analysis of a Single Evaluation Index Model

According to the performance evaluation index modeling method proposed in Section 2.5 of this paper, there were three feasible forms to study the available dynamic load characteristic models and speed drop characteristic models. The polynomial models of the three feasible forms are shown below.

Model 1:

$$Q = a_0 + a_1 P + a_2 F \quad (8)$$

where $Q$ is a single evaluation index, namely dynamic load $K$ or speed drop $J$; $a_0$~$a_2$ are the coefficients in Model 1; $P$ is the clutch oil filling pressure; $F$ is the clutch oil filling flow.

Model 2

$$Q = b_0 + b_1 P + b_2 F + b_3 P^2 + b_4 PF \quad (9)$$

where $b_0$~$b_4$ are the coefficients of each item in Model 2.

Model 3

$$Q = c_0 + c_1 P + c_2 F + c_3 PF + c_4 F^2 \quad (10)$$

where $c_0$~$c_4$ are the coefficients of each item in Model 3.

The dynamic load characteristic models and speed drop characteristic models of the nine tractor working conditions are shown in Figure 8 (the test group numbers from 1 to 9 are: high-speed and large-load working condition, high-speed and medium-load working condition, high-speed and small-load working condition, medium-speed and large-load working condition, medium-speed and medium-load working condition, medium-speed and small-load working condition, low-speed and large-load working condition, low-speed and medium-load working condition, low-speed and small-load working condition).

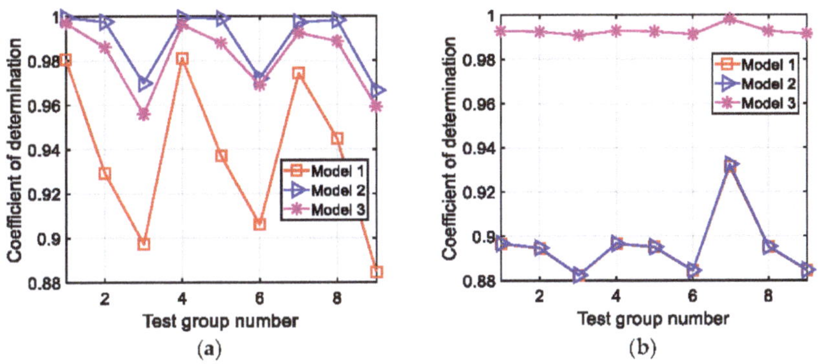


**Figure 8.** Accuracy comparison of three feasible form models: (**a**) Dynamic load; (**b**) speed drop.

Combined with the forms of the three models and Figure 8, the new forms formed by adding second-order terms and interaction terms (i.e., Model 2 and Model 3) continued to improve the accuracy on the basis of Model 1(specifically, Model 2 improved by 5.5037% and Model 3 improved by 4.7105%). Model 2 had the highest accuracy (the mean coefficient of determination $R^2$ was 0.9887) and the smallest variance (the variance of nine tractor operating conditions was 0.0002). Therefore, Model 2 had the highest match degree with the dynamic load characteristics.

For speed drop characteristics, Model 1 and Model 2 had similar accuracy (where the mean accuracy of Model 1 was 0.8955 and Model 2 was 0.8958). This suggested that the multivariate first-order linear model had a limited matching degree with the speed drop properties. Moreover, the second-order term of the oil filling pressure had little effect on improving the model accuracy. Model 3 had the highest match with speed drop characteristics (mean accuracy is 0.9927). The first-order term of oil pressure, the first-order term of flow, the second-order term of flow, and the interaction term had the ability to accurately explain the speed drop characteristics. As compared with Model 1 and Model 2, the accuracy of Model 3 was improved by 10.86% and 10.82%, respectively.

Taking the agricultural tractor in high-speed and large-load conditions as an example, the dynamic load characteristic model and speed drop characteristic model are shown in Figure 9.

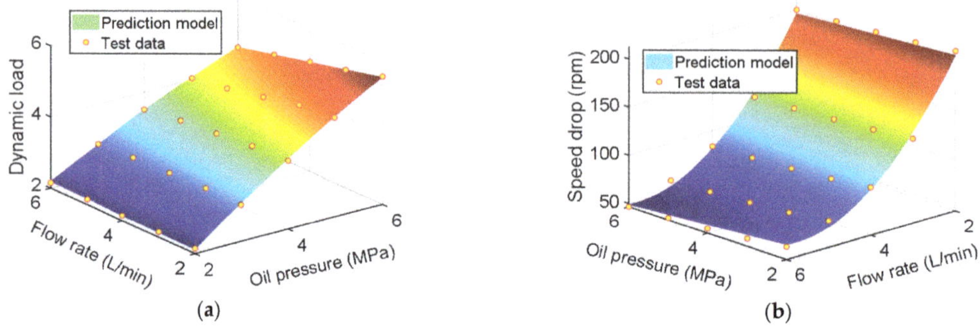


**Figure 9.** Example of the establishment result of single evaluation index model: (**a**) Dynamic load; (**b**) speed drop.

### 3.4. Establishment of Comprehensive Evaluation Indicators

The mutual factor weight matrix for each level of wet clutch performance was derived from expert opinion [36]. The mutual factor weights of rule layer B are shown in Table 4, and the mutual factor weights of the scheme layer to the rule layer are shown in Table 5.

**Table 4.** Mutual factor weights of the rule layer B.

| A | $B_1$ | $B_2$ | $B_3$ | $B_4$ |
|---|---|---|---|---|
| $B_1$ | 1 | 1/2 | 5 | 4 |
| $B_2$ | 2 | 1 | 5 | 3 |
| $B_3$ | 1/5 | 1/5 | 1 | 1/5 |
| $B_4$ | 1/4 | 1/3 | 5 | 1 |

**Table 5.** Mutual factor weights of scheme layer c for rule layer B.

| B | $B_1$ | | $B_2$ | | $B_3$ | | $B_4$ | |
|---|---|---|---|---|---|---|---|---|
| | $C_1$ | $C_2$ | $C_1$ | $C_2$ | $C_1$ | $C_2$ | $C_1$ | $C_2$ |
| $C_1$ | 1 | 1/2 | 1 | 2 | 1 | 1/3 | 1 | 1/2 |
| $C_2$ | 2 | 1 | 1/2 | 1 | 3 | 1 | 2 | 1 |

In summary, the $w_1$ and $w_2$ of the comprehensive evaluation index obtained by the AHP method were 0.4752 and 0.5248, respectively.

### 3.5. Control Strategy Formulation and Comparison of Adjustable Factors

According to the two control strategy formulation methods proposed in this paper (see Section 2.7), an I-GA was used to optimize the nine working conditions of an agricultural tractor to minimize the comprehensive evaluation index. The iterative evolution curves for Method 1 are shown in Figure 10.

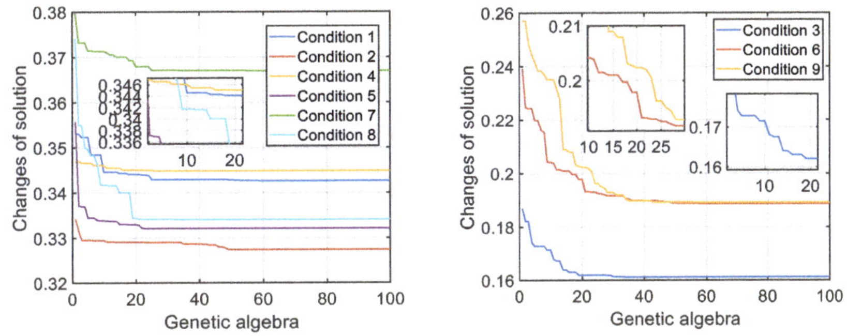


(**a**) (**b**)

**Figure 10.** Iterative evolution curves of the I-GA (Method 1): (**a**) Medium and high-load working condition; (**b**) small-load working condition.

The control strategies of adjustable factors during wet clutch switching of optimized agricultural tractors are shown in Table 6.

**Table 6.** Adjustable factor control strategy for wet clutch switching.

| Operating Conditions of Agricultural Tractors | Method 1 | | Method 2 | |
|---|---|---|---|---|
| | Oil Pressure | Flow Rate | Oil Pressure | Flow Rate |
| High-speed and large-load | 2 | 5.75 | 2 | 6 |
| High-speed and medium-load | 2 | 5.80 | 2 | 6 |
| High-speed and small-load | 6 | 6 | 6 | 6 |
| Medium-speed and large-load | 2 | 5.76 | 2 | 6 |
| Medium-speed and medium-load | 2 | 5.82 | 2 | 6 |
| Medium-speed and small-load | 6 | 6 | 6 | 6 |
| Low-speed and large-load | 2 | 6 | 2 | 6 |
| Low-speed and medium-load | 2 | 5.82 | 2 | 6 |
| Low-speed and small-load | 6 | 6 | 6 | 6 |

The full-factorial test data were first analyzed according to Method 2. As compared with the results of 225 sets of test data using the enumeration method, it could be found that the dynamic load and speed drop had relative minimum values when the flow rate was 6 L/min. Therefore, it was determined that the oil filling flow of the wet clutch should be controlled and adjusted to 6 L/min under the full-use condition of am agricultural tractor.

The research combines the whole sample data at the flow level of 6 L/min to obtain the new model after dimension reduction. The new model after dimensionality reduction is shown in Figure 11.

The control strategy was optimized based on the new model after I-GA and dimensionality reduction, and the results are shown in Table 6.

According to Table 6 and the observations of the full-factorial test data, a larger flow rate was beneficial to the speed drop and dynamic load characteristics. The oil pressure regulation strategy was affected by the operating conditions of the agricultural tractor. When the load on the agricultural tractor was small, small oil pressure helped to improve the working quality of the wet clutch. When the load of the agricultural tractor was large, large oil pressure helped to improve the working quality of the wet clutch. The results of the control strategies of Method 1 and Method 2 were highly consistent.

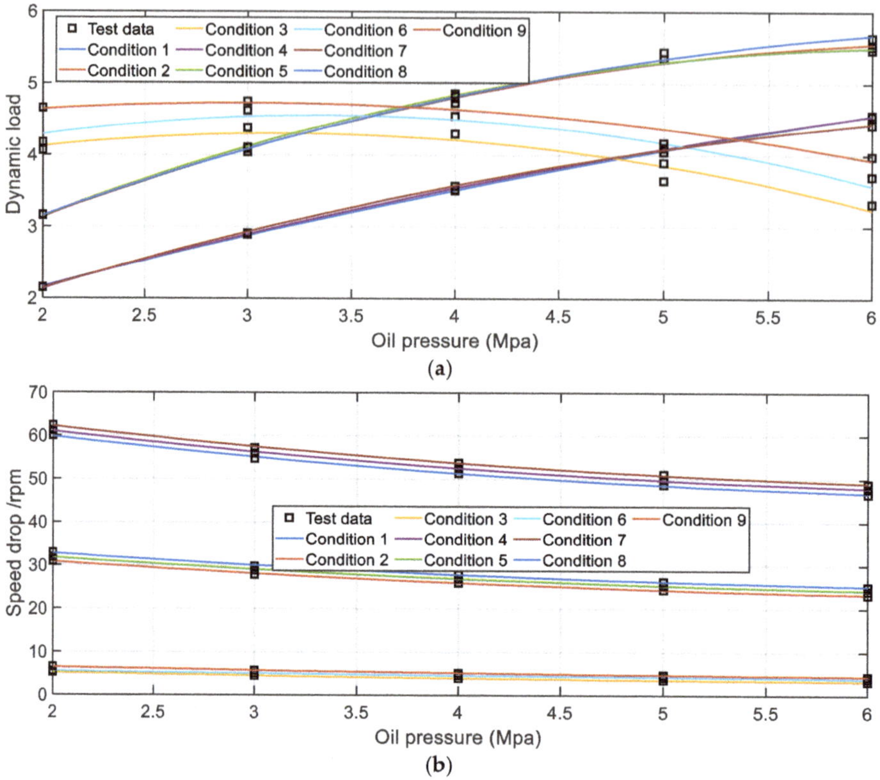


**Figure 11.** The new model of evaluation index after dimensionality reduction: (**a**) Dynamic load; (**b**) speed drop.

When the dynamic load and speed drop characteristics were improved, the sliding change was further analyzed. Taking the agricultural tractor operation at high-speed and large-load conditions as an example, the maximum sliding value (oil pressure and flow within the respective domain) was 2.49 kJ, and the average value of all test combinations of adjustable factors was 0.63 kJ. Applying the control strategies, the sliding work decreased by 70.19%.

## 4. Conclusions

Wet clutches are often used in advanced transmission systems for agricultural tractors. Under the action of different factors (mainly including speed, torque, oil pressure, flow, etc.) there are significant differences in wet clutch switching quality. In order to improve the operating characteristics of an agricultural tractor wet clutch, in this paper, we studied the working quality of wet clutch switching under the full-use condition of an agricultural tractor. The effects of four factors (oil pressure, flow rate, engine speed, and load torque) on dynamic load and speed drop are 0.38, −0.44, −0.03, −0.63 and −0.05, −0.56, −0.00, 0.73, respectively. Model 2 should be used for dynamic load characteristics (the mean of $R^2$ is 0.9887). Model 3 should be used for the speed drop characteristics (the mean of $R^2$ is 0.9927).

Combined with the two wet clutch adjustable factor control strategies proposed here (based on an I-GA), the agricultural tractor's adjustable factors (oil pressure and flow rate) need not change at high speed, medium speed, or low speed. If an agricultural tractor operates under medium and large loads, oil pressure should be at a smaller value, while flow rate should be at a larger value in their definitional domains, respectively. When an agricultural tractor is under a small-load condition, oil pressure and flow rate should take larger values in each definitional domain. In addition, the research results in this paper show that improving the dynamic load and speed drop can also effectively reduce the sliding friction of a wet clutch.

This study (mainly including the influence of various factors on quality, the establishment of an evaluation index model, and the formulation of control strategies) provides a basis and valuable reference for the design, performance evaluation, performance estimation, as well as control and performance improvement of wet clutches in agricultural machinery.

**Author Contributions:** Methodology, Y.C. and Z.C.; software, Z.C. and Y.C.; validation, Z.C. and Y.C.; investigation, Y.C., Y.Q. and Z.C.; resources, Z.C.; writing—original draft preparation, Z.C. and Y.C.; writing—review and editing, Z.C., Y.Q. and Y.C.; supervision, Z.C.; and project administration, Z.C. All authors have read and agreed to the published version of the manuscript.

**Funding:** This research was funded by the National Natural Science Foundation of China (grant number: 52105063).

**Institutional Review Board Statement:** Not applicable.

**Informed Consent Statement:** Not applicable.

**Data Availability Statement:** The data presented in this study are available on demand from the corresponding author or first author at (edward_2022@126.com or cz38@njfu.edu.cn).

**Acknowledgments:** The authors thank the National Natural Science Foundation of China (grant number: 52105063) for funding. We also thank the anonymous reviewers for providing critical comments and suggestions that improved the manuscript.

**Conflicts of Interest:** The authors declare no conflict of interest.

## References

1. Kalinichenko, A.; Havrysh, V.; Hruban, V. Heat recovery systems for agricultural vehicles: Utilization ways and their efficiency. *Agriculture* **2018**, *8*, 199. [CrossRef]
2. Bulgakov, V.; Aboltins, A.; Ivanovs, S.; Holovach, I.; Nadykto, V.; Beloev, H. A mathematical model of plane-parallel movement of the tractor aggregate modular type. *Agriculture* **2020**, *10*, 454. [CrossRef]
3. Liu, Z.; Zhang, G.; Chu, G.; Niu, H.; Zhang, Y.; Yang, F. Design Matching and Dynamic Performance Test for an HST-Based Drive System of a Hillside Crawler Tractor. *Agriculture* **2021**, *11*, 466. [CrossRef]
4. Sun, J.B.; Chu, G.P.; Pan, G.T.; Meng, C.; Liu, Z.J.; Yang, F.Z. Design and performance test of remote control omnidirectional leveling hillside crawler tractor. *Trans. Chin. Soc. Agric. Mach.* **2021**, *52*, 358–369.
5. Zhou, W.L.; Zheng, Y.P.; Pan, Z.J.; Lu, Q. Review on the Battery Model and SOC Estimation Method. *Processes* **2021**, *9*, 1685. [CrossRef]
6. Wang, H.; Zheng, Y.P.; Yu, Y. Lithium-Ion Battery SOC Estimation Based on Adaptive Forgetting Factor Least Squares Online Identification and Unscented Kalman Filter. *Mathematics* **2021**, *9*, 1733. [CrossRef]
7. Li, T.H.; Xie, B.; Li, Z.; Li, J.K. Design and optimization of a dual-input coupling powertrain system: A case study for electric tractors. *Appl. Sci.* **2020**, *10*, 1608. [CrossRef]
8. Tian, J.; Wang, Q.; Ding, J.; Wang, Y.Q.; Ma, Z.S. Integrated control with DYC and DSS for 4WID electric vehicles. *IEEE Access* **2019**, *7*, 124077–124086. [CrossRef]
9. Chen, Y.N.; Xie, B.; Du, Y.F.; Mao, E.R. Powertrain parameter matching and optimal design of dual-motor driven electric tractor. *Int. J. Agric. Biol. Eng.* **2019**, *12*, 33–41. [CrossRef]
10. Gao, H.S.; Xue, J.L. Modeling and economic assessment of electric transformation of agricultural tractors fueled with diesel. *Sustain. Energy Technol. Assess.* **2020**, *39*, 100697. [CrossRef]
11. Yin, Y.F.; Lu, L.Q.; Zhao, J.; Gao, J.H.; Li, D.F. Application status and trend of tractor full-power shift transmission technology. *Tract. Farm. Transp.* **2019**, *46*, 1–5.
12. Xia, G.; Chen, J.S.; Tang, X.W.; Zhao, L.F.; Sun, B.Q. Shift quality optimization control of power shift transmission based on particle swarm optimization-genetic algorithm. *Proc. Inst. Mech. Eng. Part. D-J. Automob. Eng.* **2022**, *236*, 09544070211031132. [CrossRef]
13. Fu, S.H.; Gu, J.H.; Li, Z.; Mao, E.R.; Du, Y.F.; Zhu, Z.X. Pressure Control Method of Wet Clutch for PST of High-power Tractor Based on MFAPC Algorithm. *Trans. Chin. Soc. Agric. Mach.* **2020**, *51*, 367–376.

14. Ince, E.; Guler, M.A. On the advantages of the new power-split infinitely variable transmission over conventional mechanical transmissions based on fuel consumption analysis. *J. Clean. Prod.* **2020**, *244*, 118795. [CrossRef]
15. Ince, E.; Guler, M.A. Design and analysis of a novel power-split infinitely variable power transmission system. *J. Mech. Des.* **2019**, *141*, 054501. [CrossRef]
16. Cheng, Z.; Lu, Z.X. Research on Dynamic Load Characteristics of Advanced Variable Speed Drive System for Agricultural Machinery during Engagement. *Agriculture* **2022**, *12*, 161. [CrossRef]
17. Cheng, Z.; Lu, Z.X. Regression-Based Correction and I-PSO-Based Optimization of HMCVT's Speed Regulating Characteristics for Agricultural Machinery. *Agriculture* **2022**, *12*, 580. [CrossRef]
18. Cheng, Z.; Chen, Y.T.; Li, W.J.; Zhou, P.F.; Liu, J.H.; Li, L.; Chang, W.J.; Qian, Y. Optimization Design Based on I-GA and Simulation Test Verification of 5-Stage Hydraulic Mechanical Continuously Variable Transmission Used for Tractor. *Agriculture* **2022**, *12*, 807. [CrossRef]
19. Sun, X.X.; Lu, Z.X.; Chen, Y. Lightweight design of hydro-mechanical continuously variable transmission box based on weight optimization. *J. Hunan Agric. Univ. (Nat. Sci.)* **2022**, *48*, 363–369.
20. Li, J.; Zhai, Z.Q.; Song, Z.S.; Fu, S.H.; Zhu, Z.X.; Mao, E.R. Optimization of the transmission characteristics of an HMCVT for a high-powered tractor based on an improved NSGA-II algorithm. *Proc. Inst. Mech. Eng. Part. D-J. Automob. Eng.* **2022**, 09544070211067961. [CrossRef]
21. Qian, Y.; Cheng, Z.; Lu, Z.X. Study on stepwise regression optimization of shift quality of heavy-duty tractor HMCVT based on five factors. *J. Nanjing Agric. Univ.* **2020**, *43*, 564–573.
22. Ni, X.D.; Zhu, S.H.; Zhang, H.J.; Chang, Y.L.; Ouyang, D.Y.; Wang, G.M. Experiment of shift quality factors for hydro-mechanical CVT. *Trans. Chin. Soc. Agric. Mach.* **2013**, *44*, 29–34.
23. Wang, G.M. Study on Characteristics, Control and Fault Diagnosis of Tractor Hydro-Mechanical CVT. Ph.D. Thesis, Nanjing Agricultural University, Nanjing, China, 2014.
24. Stockinger, U.; Groetsch, D.; Reiner, F.; Voelkel, K.; Pflaum, H.; Stahl, K. Friction behavior of innovative carbon friction linings for wet multi-plate clutches. *Forsch. Im Ing.-Eng. Res.* **2021**, *85*, 115–127. [CrossRef]
25. Raikwar, S.; Tewari, V.K.; Mukhopadhyay, S.; Verma, C.R.B.; Rao, M.S. Simulation of components of a power shuttle transmission system for an agricultural tractor. *Comput. Electron. Agric.* **2015**, *114*, 114–124. [CrossRef]
26. Talati, H.; Aliakbari, K.; Ebrahimi-Moghadam, A.; Farokhad, H.K.; Nasrabad, A.E. Optimal design and analysis of a novel variable-length intake manifold on a four-cylinder gasoline engine. *Appl. Therm. Eng.* **2022**, *200*, 117631. [CrossRef]
27. Torshizian, M.R.; Aliakbari, K.; Ghonchegi, M. Failure Analysis of Ductile Iron Differential Housing Spline in 4WD Passenger Car. *Int. J. Met.* **2021**, *15*, 587–601. [CrossRef]
28. Aliakbari, K.; Nejad, R.M.; Mamaghani, T.A.; Pouryamout, P.; Asiabaraki, H.R. Failure analysis of ductile iron crankshaft in compact pickup truck diesel engine. *Structures* **2022**, *36*, 482–492. [CrossRef]
29. Lu, K.; Lu, Z.X.; Cheng, Z.; Zheng, S.Q. Study on influence rules of clutch parameters on HMCVT shift performance. *Mech. Sci. Technol. Aerosp. Eng.* **2019**, *38*, 1695–1701.
30. Wang, G.M.; Zhang, X.H.; Zhu, S.H.; Zhang, H.J.; Tai, J.J.; Nguyen, V. Shift performance of tractor hydraulic power-split continuously variable transmission. *Trans. Chin. Soc. Agric. Mach.* **2015**, *46*, 7–15.
31. Cheng, Z.; Zhou, H.D.; Lu, Z.X. A Novel 10-Parameter Motor Efficiency Model Based on I-SA and Its Comparative Application of Energy Utilization Efficiency in Different Driving Modes for Electric Tractor. *Agriculture* **2022**, *12*, 362. [CrossRef]
32. Li, D.X.; Xu, B.; Tian, J.; Ma, Z.S. Energy Management Strategy for Fuel Cell and Battery Hybrid Vehicle Based on Fuzzy Logic. *Processes* **2020**, *8*, 882. [CrossRef]
33. Xia, L. Analysis of partial least squares modeling and multi-collinearity ability. *Agro Food Ind. Hi-Tech* **2017**, *28*, 885–889.
34. Xu, Q.S.; Liang, Y.Z.; Shen, H.L. Generalized PLS regression. *J. Chemom.* **2001**, *15*, 135–148. [CrossRef]
35. Cheng, Z.; Chen, Y.T.; Li, W.J.; Liu, J.H.; Li, L.; Zhou, P.F.; Chang, W.J.; Lu, Z.X. Full Factorial Simulation Test Analysis and I-GA Based Piecewise Model Comparison for Efficiency Characteristics of Hydro Mechanical CVT. *Machines* **2022**, *10*, 358. [CrossRef]
36. Lu, K. Hydraulic Mechanical Continuously Variable Transmission Shift Clutch Design and Research on Quality of Shifting Process. Master's Thesis, Nanjing Agricultural University, Nanjing, China, 2019.
37. Xu, X.M.; Lin, P. Parameter identification of sound absorption model of porous materials based on modified particle swarm optimization algorithm. *PLoS ONE* **2021**, *16*, e0250950.
38. Chang, C.C.; Zheng, Y.P.; Yu, Y. Estimation for battery state of charge based on temperature effect and fractional extended kalman filter. *Energies* **2020**, *13*, 5947. [CrossRef]
39. Wang, H.; Zheng, Y.P.; Yu, Y. Joint estimation of soc of lithium battery based on dual kalman filter. *Processes* **2021**, *9*, 1412. [CrossRef]
40. Li, Y.J.; Ma, Z.S.; Zheng, M.; Li, D.X.; Lu, Z.H.; Xu, B. Performance analysis and optimization of a high-temperature PEMFC vehicle based on particle swarm optimization algorithm. *Membranes* **2021**, *11*, 691. [CrossRef]
41. Cheng, Z.; Lu, Z.; Qian, J. A new non-geometric transmission parameter optimization design method for HMCVT based on improved GA and maximum transmission efficiency. *Comput. Electron. Agric.* **2019**, *167*, 105034. [CrossRef]

*agriculture*

MDPI

*Article*

# Design and Optimization of a Soil-Covering Device for a Corn No-Till Planter

**Yuanle Geng [1], Xianliang Wang [1,*], Xiaokang Zhong [1], Xiangcai Zhang [1], Kun Chen [1], Zhongcai Wei [1], Qingxin Lu [1], Xiupei Cheng [1] and Mingtao Wei [2]**

[1] School of Agricultural Engineering and Food Science, Shandong University of Technology, Zibo 255000, China
[2] Weifang Letian Agricultural Science & Technology Co., Ltd., Weifang 261102, China
* Correspondence: wxl1990@sdut.edu.cn

**Abstract:** Aiming to solve the problems of easy generation of clods, poor soil fluidity and poor soil-covering effect in the no-tillage seeding operation mode, this study has designed a soil-covering device with a soil-closing function for the chisel-type furrow opener of a corn no-tillage planter. Theoretical analysis and design of the soil dividing plate and the soil-covering disc were carried out, and the key parameters were determined. Then, the soil-covering device with soil-closing function and the soil movement model was constructed in Altair EDEM, and the spatial soil particle mobility was analyzed. The numerical simulation method was combined with the center plane composite response test scheme to carry out the simulation test, and the optimal working parameter combination was obtained. The operating speed was 6.35 km·h$^{-1}$, the inclination angle of the soil-covering disc was 60°, the opening angle of the soil-covering disc was 70°, and the inclination angle of the soil dividing plate was 40°. A comparative test of the soil-covering effect and performance was carried out, and the results showed that the designed device was better than the traditional soil-covering device in terms of the consistency of soil-covering thickness and the stability of seed position deviation. The relative errors of the soil bin experiment results and the simulation optimization results for the cover soil thickness and seed position deviation were 5.7% and 11.1%, respectively. The operation effect of the designed soil-covering device with soil-closing function meets the requirements of soil covering under no-tillage conditions and provides a basis for the research and development of soil-covering suppression technology and devices under no-tillage seeding mode.

**Keywords:** no-tillage; soil cover; discrete element; soil-covering thickness; seed offset

**Citation:** Geng, Y.; Wang, X.; Zhong, X.; Zhang, X.; Chen, K.; Wei, Z.; Lu, Q.; Cheng, X.; Wei, M. Design and Optimization of a Soil-Covering Device for a Corn No-Till Planter. *Agriculture* **2022**, *12*, 1218. https://doi.org/10.3390/agriculture12081218

Academic Editors: Mustafa Ucgul and Chung-Liang Chang

Received: 27 June 2022
Accepted: 11 August 2022
Published: 13 August 2022



## 1. Introduction

No-till seeding technology is one of the core technologies of conservation tillage, which can not only reduce wind erosion and water erosion, and improve soil fertility and drought resistance, but also have the advantages of water conservation and moisture conservation, cost-saving and efficiency improvement, and soil fertility [1,2]. However, under the continuous no-tillage mode, the surface soil is hard, and the surface is uneven [3]. During the operation of the seeder, the chisel-type furrow opener has problems such as easy generation of clods, poor soil fluidity and a poor soil-covering effect [4], which seriously affect the sowing quality. Therefore, it is of great significance to analyze the flow characteristics of no-till soil during the operation of the chisel-type furrow opener and optimize the soil-covering device of the no-till planter to improve the quality of no-till sowing and promote conservation tillage.

The law interaction between soil and tillage components is the basis for designing and selecting soil tillage components. The discrete element method (DEM) can be used to study the flow of particulate matter and the behavior of soil fragmentation and is widely considered in the research of soil-contacting components of agricultural machinery [5,6].

Scholars have conducted extensive research on the interaction process between soil and tillage equipment based on DEM [7–11], and these studies have confirmed the superiority of discrete element simulation of the tillage process. Compared with the traditional theoretical derivation and experimental analysis, the DEM has the advantages of saving time and labor, low cost, and visualization of results; compared with the numerical simulation method (such as the finite element method), the DEM has the model fidelity high and high accuracy of simulation results. Moreover, the movement trajectory of the grains can be tracked in the discrete element to analyze the soil particles' force and movement during the tillage process.

Presently, scholars' research on soil-covering devices mainly focuses on the shape and structure design and precise soil-covering control of soil-covering devices. In terms of the shape and structure design of the soil-covering device, Zhang et al. [12] designed a dryland auger-type precise soil-covering device, which improved the operation accuracy and soil-covering effect. Zhang et al. [13] designed a crushing soil-covering device based on bionics, which improved the crushing performance. In order to solve the problems of excessive resistance and poor soil-covering effect during the straw returning process, Lin et al. [14] designed a spiral soil-covering device. Liu et al. [15] designed a new double-layer disc-type soil-covering device, which remarkably affects soil covering, improves the sowing accuracy and reduces the displacement of seeds in the process of covering soil. At the same time, some scholars optimize the parameters of the soil-covering device through simulation analysis to improve the soil-covering effect [16–18].

Regarding precise soil cover control, that is, to achieve accurate control and real-time adjustment of the amount and thickness of soil cover. Li et al. [19] designed a cover soil-suppression linkage monitoring device to adjust the amount of cover soil in real-time. Chao et al. [20] designed an automatic adjustment device for the amount of soil covered by a transplanter to realize the automatic adjustment and control of the amount of soil covered on the membrane of the transplanter. At the same time, some scholars mainly conducted research on the control of sowing depth of precision seeding technology [21–24]. In general, the existing research on the soil-covering device mainly focuses on the depth of soil-covering and active control of the sowing depth, and there are few studies on the effect of the soil-covering device and the soil flow movement state of the soil-covering process and the displacement of the seeds after touching the soil under the no-tillage seeding mode.

Therefore, to improve the soil-covering effect of the opener of the seeder when working on the surface of no-tillage sowing, a soil-covering device with a soil-closing function was designed in this paper, and the discrete element method was used to carry out a four-factor and three-level regression combination test. The simulation model determines the optimal parameter combination of the soil-closing device and uses the soil bin experiment results to verify it to provide a theoretical basis and technical support for the precision no-tillage seeder to achieve the consistency of seeding depth and the development of the soil-covering device.

## 2. Materials and Methods

### *2.1. Design of Soil-Covering Device with Soil-Closing Function*

#### 2.1.1. Overall Design

The soil-covering device with a soil-closing function is composed mainly of a trench opener, a share shaft, a seed metering device, a soil dividing plate, and soil-covering discs. The overall structure is shown in Figure 1. The soil dividing plate is mounted on both sides of the retaining plate through the fixing holes, and the connecting frame is installed on the soil retaining plate through the chute. The two earth-covering discs are fixedly installed on the connecting frame, and the discs' inclination angle and opening angle can be changed by adjusting the fixed position of the bolts in the positioning holes. The biggest difference between the device we designed and optimized and the traditional soil-covering device is the presence or absence of the soil-dividing plate.

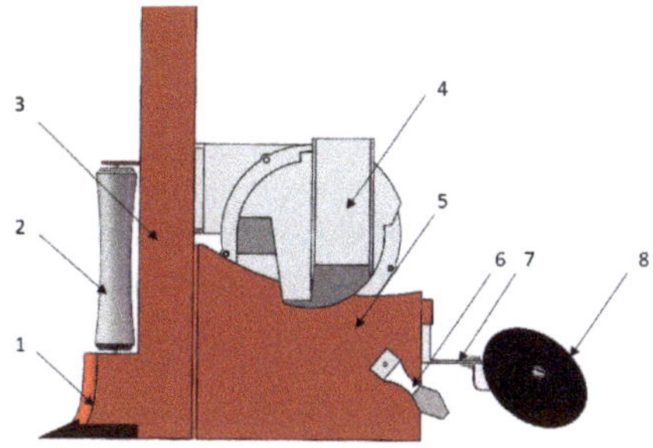


**Figure 1.** Schematic diagram of the overall structure of the soil-covering device with soil-closing function: 1—opener; 2—anti-blocking roller; 3—share shaft; 4—seed metering device; 5—retaining plate; 6—dividing plate; 7—connection frame; 8—covering disc.

2.1.2. Working Process

During the operation, the trench is first opened by the opener, and the front-end soil dividing plate first covers a small amount of finely divided soil on the surface of the seeds to realize the contact between the seeds and the loose and finely divided soil and complete the first covering of soil. Then, the loose soil formed by the opener after the operation is actively covered back into the seed furrow formed after ditching along the inner side of the covering disc by the soil-covering discs at the rear end to complete the second covering of the seeds. The soil covering of the device is completed in two parts. The first step is to cover a small part of the soil through the soil dividing plate to reduce the soil disturbance. The second step is to cover the soil through the discs to meet the soil-covering requirements. By adjusting the design parameters such as the opening angle, the round inclination angle and the inclination angle of the soil dividing plate, the amount of soil-covering can be adjusted, thereby changing the thickness of the soil covering of the seeds.

2.1.3. Design of Key Components

According to the soil-covering device's structure and soil dynamics model with soil-closing function, the analysis part of the soil-covering device is mainly divided into two parts: the soil dividing plate and the soil-covering discs.

When the opener is working, the soil on both sides will flow into the seed ditch by itself with the advance of the opener, which realizes the function of self-covering to a certain extent. The thickness of its self-covering soil is related to parameters such as the natural angle of repose φ of the soil particles and the distance x between the two discs. The relationship between them is shown in Figure 2. From reference [25], it is concluded that the expression of the opener after self-covering is:

$$H = H_1 - 7.245x_1^{0.367} \quad (1)$$

where $H$—self-covering soil thickness of the opener (mm); $H_1$—ditching depth (mm); $x_1$—ditching width (mm).

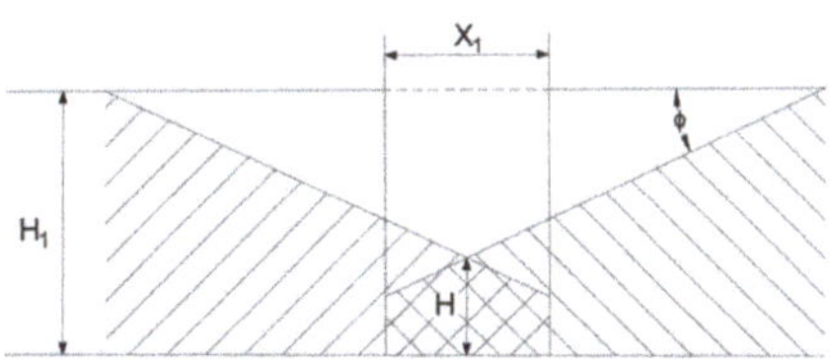


**Figure 2.** Self-covering depth of chisel-type furrow opener.

The width of the opener used in this study is 50 mm, and the penetration depth of the opener is 50 mm. By substituting Equation (1), the thickness of the self-covering soil of the opener is about 20 mm. It can be concluded that the self-covering effect of the opener is not significant enough to meet the requirements of seed thickness, so it is necessary to design the soil-covering device behind the opener.

The soil-dividing plate is installed on both sides of the rear end of the opener, and its basic function is to cover a small amount of small soil particles on the surface of the seeds to complete the first soil cover [26]. In order to enhance the diversion effect of the soil, the soil dividing plate is bent inward at an angle, and when the opener moves forward, the inner side pushes the finely divided soil on both sides of the seed ditch so that the soil slips relative to the soil dividing plate and falls into the ditch. Under the action of soil separation on the outer side of the soil separation plate, the soil block falls back to the seed furrow with a delay so that the finely divided soil is preferentially contacted with the seeds, and then the larger soil particles and soil blocks are covered on the top surface. In order to explore the structural design of the soil dividing plate, the force analysis of the soil dividing plate is carried out, refer to Figure 3.

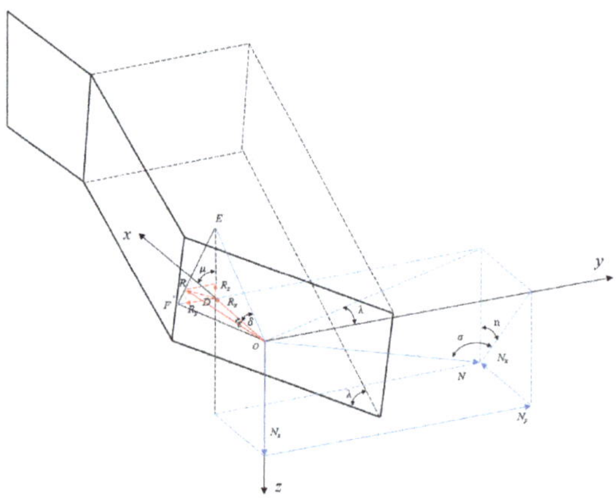


**Figure 3.** Force analysis diagram of dividing plate.

Select any point o of the soil-contacting part on the soil dividing plate as the research object, and establish a space rectangular coordinate system, as shown in Figure 3. The positive direction of the *x*-axis is the forward direction of the soil dividing plate, the normal stress of soil particles to the soil-dividing plate is σ, and the friction force is *R*. The triangle OE′F′ is the micro-element plane of the point O on the plane, and the friction force of the soil-dividing plate is analyzed based on the three-sided wedge OD′E′F′ [27]. On the triangle OE′F′, the friction force *R* is perpendicular to the straight line E′F′, and the angle ε between the friction force R and the *x*-direction is the minimum value. According to the resultant vector projection theorem, the relationship between the normal stress *N* and the friction force *R* projected in the x-direction is:

$$\begin{cases} \sigma_x = N\cos \text{n} \sin\delta \\ R_x = R\cos\varepsilon \end{cases} \quad (2)$$

In the formula, $\sigma_x$ is the component force of N in the x-direction, and the unit is *N*. $R_x$ is the component force of the friction force *R* in the x-direction, and the unit is *N*. n is the angle between the normal stress N and the XOZ plane, the unit is (°). $\delta$ is the inclination angle of the soil dividing plate; the unit is (°).

From the geometric relationship, we can obtain:

$$\begin{cases} \tan\sigma = \frac{\sin\delta}{\tan\lambda} \\ \tan\varepsilon = \tan\delta\cos\mu \\ \tan\mu = \frac{\tan\delta}{\tan\lambda} \end{cases} \tag{3}$$

$$R_{fx} = \frac{N\tan\lambda\sin\delta}{\sqrt{\tan^2\lambda + \sin^2\delta}} + N\tan\varphi\sqrt{\frac{\tan^2\lambda\cos^2\delta + \sin^2\delta}{\tan^2\lambda + \sin^2\delta}} \tag{4}$$

It can be seen from Equation (4) that the normal stress $N$, the friction angle $\varphi$, the element line angle $\lambda$ and the inclination angle $\delta$ jointly determine the size of the working resistance $R_{fx}$ of the soil dividing plate.

The structural design of the soil dividing plate is shown in Figure 4. In the figure, $\angle\lambda$ is the inclination angle of the soil dividing plate, and $\angle\eta$ is the inner angle of the soil dividing plate. In order to ensure that the seeds fall in the seed furrow, the distance between the width of the opener and the inflection point of the soil-dividing plate must meet [26]:

$$60 \leq H_1 + 2l_2 \leq 80 \tag{5}$$

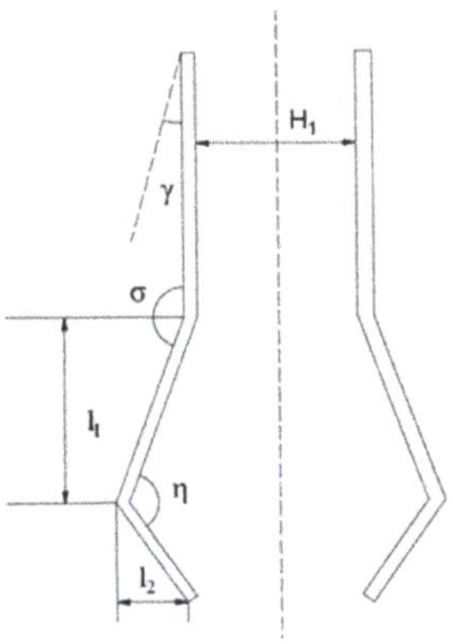


**Figure 4.** Structural design drawing of dividing plate.

The width $H_1$ of the opener designed in this paper is 40 mm. According to Equation (5), the value range of $l_2$ is 10–20 mm. In this paper, $l_2$ = 15 mm. According to the geometric relationship of the structure, the distance $l_1$ from the soil dividing plate to the inflection point is:

$$l_1 = \frac{l_2}{\tan(180^\circ - \sigma)} \tag{6}$$

According to Equation (6), $l_1$ is 44 mm. In order to ensure the smooth backfilling of the soil, the inclination angle λ of the soil dividing plate should be much larger than the friction angle between the soil and the shovel body.

The basic function of the soil-cover disc is to cover the soil on both sides of the opener to the top of the seeds and achieve a certain thickness of soil cover. In order to increase the amount of soil cover, two soil-cover discs were installed on the soil-cover frame to cover the seeds. The soil-covering discs also belong to the disc harrow category, and the relevant design parameters are calculated according to the empirical formula of concave disc design. The formula for calculating the diameter of the covering disc is as follows:

$$D = KB \tag{7}$$

In the formula: $D$ is the diameter of the soil-covering disc, the unit is mm; $K$ is the diameter-depth ratio coefficient; $B$ is the designed soil depth, and the unit is mm. The designed soil-covering device is mainly aimed at the no-tillage sowing of corn. The sowing depth of corn is generally 30–50 mm, and the $K$ value of the disc is generally 4–6 [24], so the diameter of the disc is designed to be 125 mm.

When the covering disk is cut into the soil, the force analysis is shown in Figure 5. During the soil-covering process, the soil-covering device moves forward with the speed $V$, $P$ is the power of the tractor, and the pulling force is $F_1$. $\angle\alpha$ is the inclination angle of the soil-covering device, $\angle\beta$ is the opening angle of the soil-covering device, and the friction force $F_f$ (excluding the force between the soils). The force $F_A$ is:

$$F_A = F_N = \frac{P}{V} \tag{8}$$

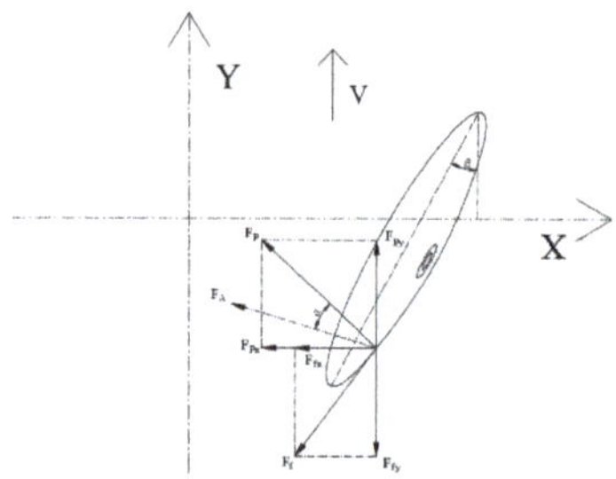


**Figure 5.** Force Analysis of Disc (the top view).

$F_P$ is the component force of $F_A$ on the plane XOY surface:

$$F_P = F_A \cos a \tag{9}$$

The friction force is:

$$F_f = \mu F_N \tag{10}$$

The force analysis of the covering disk along the $y$-axis direction is as follows:

$$F = F_1 + F_{py} - F_{fy} \tag{11}$$

In the formula: $F$ is the resultant force on the covering disk in the y-axis direction; $F_{py}$ is the y-axis component of the resultant pressure on the soil on the disk; $F_{fy}$ is the y-axis component of the friction force between the soil and the disk.

Figure 6 is a schematic diagram of the grounding situation of the earth-covered disc, and the calculation formula of its grounding area is:

$$S = \frac{S_{ABC}}{\cos\beta} \tag{12}$$

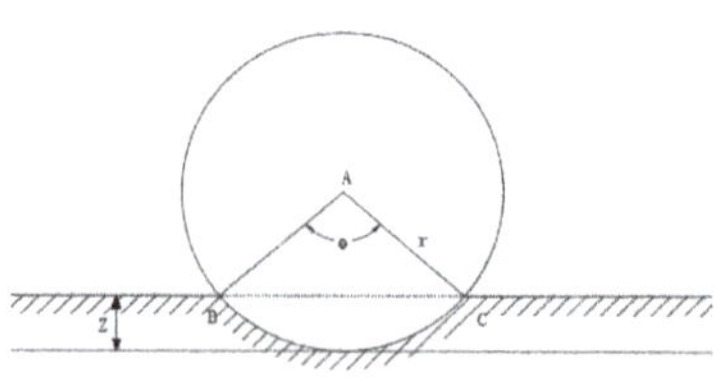

**Figure 6.** Schematic diagram of the effective area of the disc grounding.

The area of the arc in the formula is:

$$S_{ABC} = r^2 \cdot \left(\frac{\pi\theta}{360}\right) - \frac{1}{2} l_{AB}(r - z_0) \quad (13)$$

where:

$$l_{AB} = 2r\sin\left(\frac{\theta}{2}\right) \quad (14)$$

$$\theta = 2\arccos\left(\frac{r - Z_0}{r}\right) \quad (15)$$

It can be seen from Equations (8) and (10) that the effect of soil covering is related to the structural parameters of the soil-covering device, soil conditions, and forward speed, and the specific influence law needs to be further studied through experiments.

### *2.2. Altair EDEM Simulation Analysis*

#### 2.2.1. Soil Particle Model

Relevant research shows that the no-till soil structure is mostly cemented into granules and small clumps; the soil is biased towards viscosity [28]. Combined computational efficiency and simulation effects. In this study, the soil particle model was set up as spherical particles with a particle size of 5 mm [29]. Particle-to-particle contact is selected to Hertz-Mindin with bonding contact model simulation [30,31]. All particles can be bonded together by setting up a bonding contact in the particle contact model to generate particle aggregates of other shapes.

#### 2.2.2. Corn Grain Model

The brand of corn seeds used in this study was "Jade 188"; we randomly selected 100 corn seed pellets and measured their three-axis dimensions with a vernier caliper. The average of the measured corn seed height, width, and thickness was selected as the size of the corn pellet model. The length of the corn seeds was13 mm, the width was 9.5 mm, and the thickness was 4.5 mm.

A maize seed simulation model was established using the spherical particle polymer method [32–35]. First, we created a 3D model of corn seeds in Solid works. Then, we imported the corn seed 3D model into Altair EDEM, filled it with a large number of spherical particles and obtained a corn seed model. The corn grain model is shown in Figure 7.

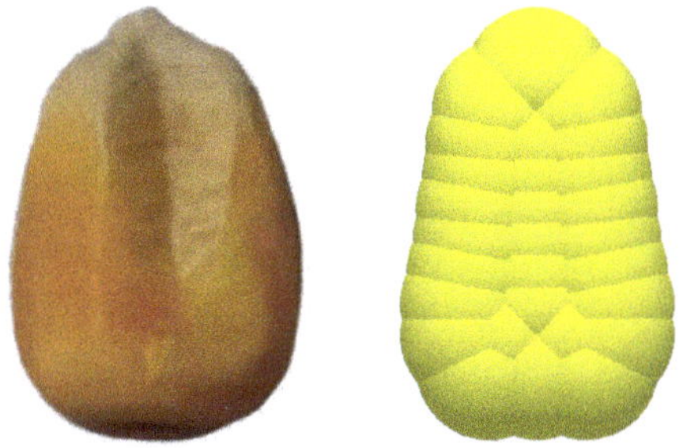

(a) (b)

**Figure 7.** Corn grain model: (**a**) Actual corn seed; (**b**) Corn Seed Simulation Model.

#### 2.2.3. Altair EDEM Simulation Parameters

Discrete element simulation parameters mainly include material parameters and contact mechanics parameters. Material parameters mainly include soil-covering device, soil, the density of corn seeds, Poisson's ratio, and shear modulus [36–38]. The contact mechanics parameters mainly include soil-to-soil, the recovery coefficient between the covering device and the soil, the static friction coefficient, the rolling friction coefficient, etc. [39,40]. The remaining parameters are measured by rest angle, the inclined plate test and other physical and mechanical tests. The discrete element simulation time step was taken as $1 \times 10^{-5}$ s, and the acceleration of gravity was 9.8 $m{\cdot}s^{-2}$. The remaining main parameters are shown in Table 1.

**Table 1.** Basic Parameters of Discrete Element Models for Granular Materials and Covering Devices.

| **Parameter** | **Granular Material** | | | | **Covering Device (Steel)** |
|---|---|---|---|---|---|
| | **Soil** | | **Corn** | | |
| density/($kg{\cdot}m^{-3}$) | 2600 | | 1197 | | 7800 |
| Poisson's ratio | 0.5 | | 0.4 | | 0.3 |
| Modulus of shear/Pa | $2 \times 10^7$ | | $1.37 \times 10^8$ | | $7.9 \times 10^{10}$ |
| | soil-soil | soil-corn | soil-steel | corn-corn | corn-steel |
| Recovery factor | 0.2 | 0.068 | 0.1 | 0.182 | 0.456 |
| coefficient of friction | 0.25 | 0.698 | 0.2 | 0.2 | 0.436 |
| Rolling coefficient of friction | 0.04 | 0.155 | 0.02 | 0.0782 | 0.059 |

### 2.2.4. Altair EDEM Simulation

In order to meet the working conditions of simulation and reduce the experiment of simulation, the trough size (length × width × height) set in this study is 3000 mm × 1000 mm × 220 mm. Set the depth of the opener to 50mm. The total simulation time is 10 s, of which 0~5 s is the formation time of the bottom soil particles. During this period, 360,000 soil particles that meet the sowing conditions are generated, and the soil particles settle naturally for 1 s. The formation time of topsoil particles is 6~8 s; set the particle generation mode to random generation, and particle sizes are randomly generated in the range of 0.5 to 1.5 times the original particle radius. In order to make the small size particles fill the crevices of the large size soil particles as soon as possible and avoid forming too loose soil, the number of topsoil particles is set to 200,000. Soil particles settle naturally for 1 s. Apply Solid works 2021 software (Dassault Systemes, Vélizy-Villacoublay, France) to create geometric simulation models of coverers with different parameters. To facilitate the observation of the effect of the job and speed up the simulation time, compress the connecting parts of the cover when the simulation model is established, save the geometric simulation model in IGS format and import it into Altair EDEM.

A pellet generating module for corn seeds is set at the lower part of the seed meter, sets the total amount of grains generated for seeds to 150, and sets the pellet generation module forward speed of the corn seed to be the same as the forward speed of the cover device, maximum simulation of sowing and soil-covering process. At 9 s, the soil-covering device began to move, and the simulation ended at 12s. The simulated geometry of the trough and cover device is shown in Figure 8.

### 2.2.5. Particle Motion Mechanism Analysis

Soil particles located at the front of the opener during the soil-covering device with soil-closing function simulation are picked up by the tip of the shovel, and the soil particles picked up pass through the tip of the shovel. It moves along the retainer under the action of its weight, and surrounding soil, the diversion effect of the soil dividing plate is filled into the trench, after that, the soil is covered by the soil-covering discs, and the entire trenching and covering movement is completed. Figure 9a,c shows the vector distribution of soil particle velocity. Figure 9b,d shows the conventional double-disc covering device's soil particle velocity vector distribution. As seen from the comparison chart, soil-covering devices with soil-closing functions can enhance soil fluidity during no-till operations and cover the seed surface with fine soil fragments.

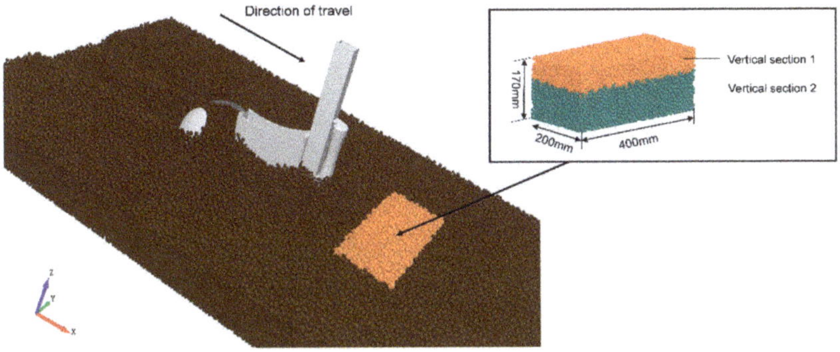


**Figure 8.** Simulation model of cover device and trough.

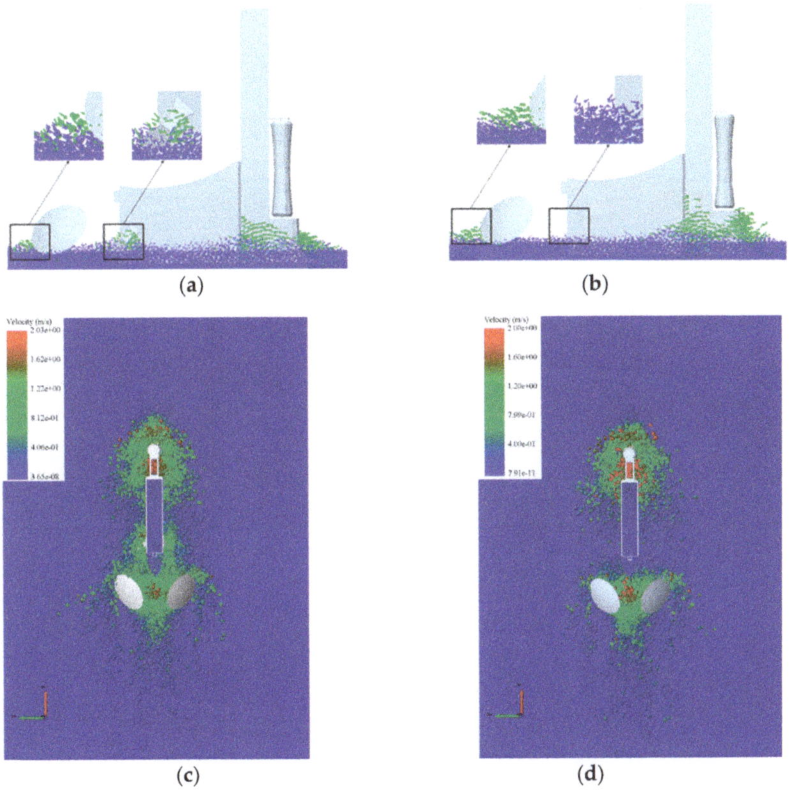

**Figure 9.** Comparison map of vector distribution of soil particle velocity: (**a**) Side view of the soil-covering device with soil-closing function from; (**b**) Side view of the Traditional double-disc soil-covering device; (**c**) Top view of the soil-covering device with soil-closing function from quarter view; (**d**) Top view of the traditional double disc soil-covering device.

As can be seen from the cover renderings of Figure 10, most of the smaller soil particles are located at the seed location when the soil is covered after the soil-covering process, while the larger soil particles cover the surface of the smaller soil particles. Therefore, it is shown that the soil-covering device with soil-closing function first uses the diversion effect of the soil-dividing plate to cover the soil once during the soil-covering process and then uses the surface soil to complete the secondary cover is basically in line with the original design intention.

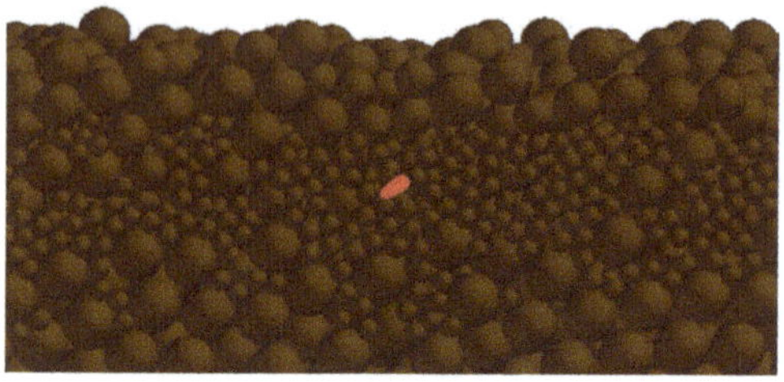

**Figure 10.** Effect drawing of covering soil. Red grains represent corn seed.

In order to explore the movement law of soil under the soil-covering device with a soil-closing function, it is necessary to ensure that the selected soil particles are approximately equal in the direction of the opener and the spacing of each soil particle. Five soil particles within the tillage depth range that meet the conditions were selected, and the selected soil particle distribution is shown in Figure 11. A soil particle is selected at each of the five locations for tracking its three-dimensional movement.

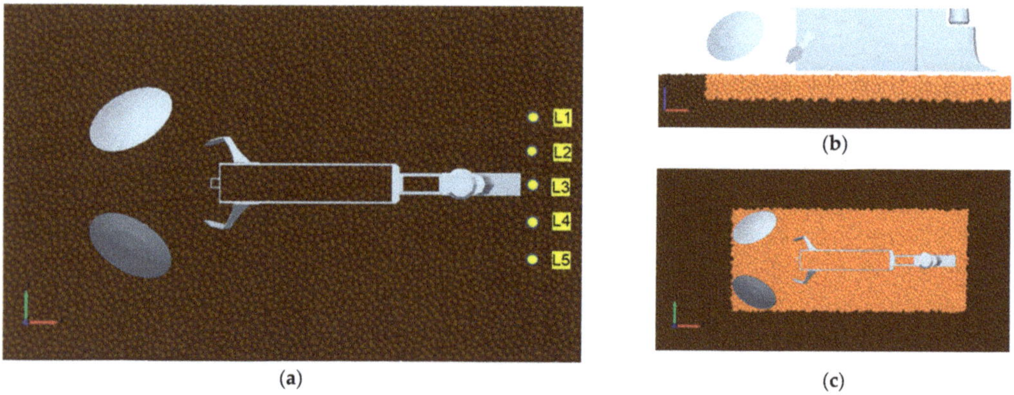


**Figure 11.** Particle distribution of soil: (**a**) Schematic diagram of soil particle distribution, L1–L5 were soil particles selected in sequence.; (**b**) Cross-sectional view of the soil covering a range of the soil-covering device; (c) Top view of the soil covering a range of the soil-covering device.

Solving the trench covering process, soil particles' horizontal, lateral and vertical motion displacement curves are obtained, as shown in Figure 12. By comparing the displacement curves of soil particles at the corresponding position in Figure 12, it is found that the soil particles move backward towards the bottom of the trench as the opener breaks the soil under the dynamic diversion of the soil coverer. As the tip of the shovel gradually approaches the selected soil particles, the soil particles collide under the action of the tip, and the soil particles begin to produce a motion trajectory in an obliquely upward direction. When the tip of the shovel passes through the soil particles, the particles move along the soil dividing plate, creating a diagonal downward movement into the seed ditch. After the soil particles enter the seed ditch, they are gently covered and stop moving under the action of the soil-covering disk, completing the entire soil-covering process.

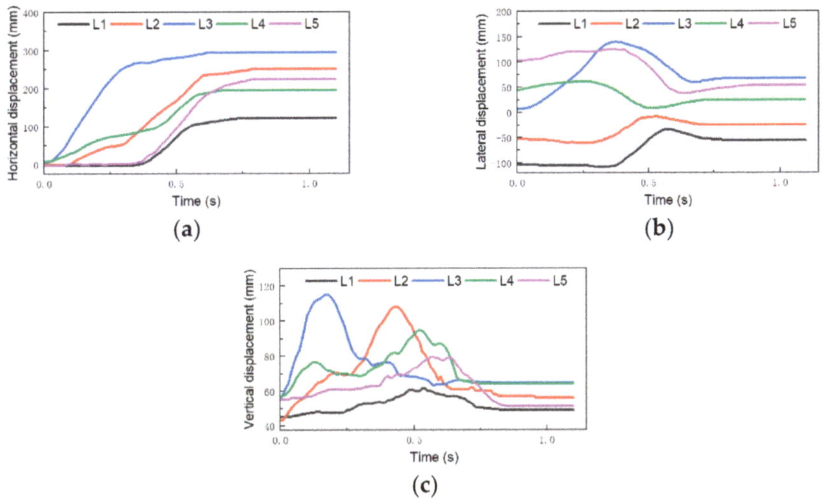


**Figure 12.** The motion curves of soil particles at different locations during tillage: (**a**) movement curve of soil particles along the x-direction; (**b**) movement curve of soil particles along the y-direction; (**c**) Movement curve of soil particles along the x-direction.

## 3. Results

### 3.1. Simulation Test of Soil-Covering Device

The soil-covering device is one of the important working parts of the precision seeding machine. It is used to cover the seeds in the seed trench and achieve the purpose of suitable thickness and uniform soil cover. The performance of the soil-covering device mainly depends on its structural parameters and motion parameters, reducing the positional deviation of the seeds moving with the soil flow, and improving the precision seeding theory. Therefore, in this study, the thickness of the seed covering soil H and the seed offset X were selected as the evaluation indicators of the performance of the covering soil device.

Seed cover thickness refers to the thickness of the soil cover above the seeds. After ditching, select the center of the seed as the measurement reference point, measure the thickness H of the soil covered above the seed, measure five times, and take the average of the results. The soil section after the simulation operation is analyzed through the clipping module in the Altair EDEM post-processing, and the thickness H of the seed covering soil is measured with the scale that comes with the software.

Seed offset refers to the positional deviation of seeds following soil flow after they touch the soil. The $x$-axis and $y$-axis coordinates of the corn seeds are exported in the Altair EDEM post-processing interface, and the seed offset x is calculated according to the $y$-axis coordinates of the corn seeds.

### 3.2. Quadratic Orthogonal Rotational Regression Combined Experimental Design

In order to further study the influence of the soil shunt cover's structural parameters on the seed cover's thickness and the seed's displacement, the quadratic orthogonal rotation regression combination test was carried out using the Altair EDEM 2020 simulation software (DEM Solutions, Edinbrugh, UK). Four key parameters (X1, X2, X3, X4) that affect the soil-covering effect are selected as factors. The thickness of seed covering soil and seed displacement were selected as the test indicators. The quadratic orthogonal rotation regression combination test with four factors and three levels was designed by Design-expert 10 software (Stat-Ease; Minneapolis, MN, USA) to study the relationship between the key factors and the test indicators. Set the depth into the soil to 50mm, select the diameter D of the disc to be 125mm, and the opener width to 50mm. The orthogonal test factors and levels are shown in Table 2 [41,42].

**Table 2.** Factors and Levels of Quadratic Orthogonal Rotational Regression Combination Test of Operation Effect of Soil-Covering Device.

| Factor Level | Working Speed ($km \cdot h^{-1}$) | The Opening Angle of the Disc (°) | The Inclination Angle of the Disc (°) | The Inclination Angle of the Soil Dividing Plate (°) |
|---|---|---|---|---|
| −1 | 3 | 60 | 65 | 40 |
| 0 | 5 | 65 | 70 | 45 |
| 1 | 7 | 70 | 75 | 50 |

### 3.3. Simulation Test Results and Analysis

The test results are shown in Table 3. Among them, X1, X2, X3, and X4 are the coding values of operating speed, disc opening angle, disc inclination angle, and soil-dividing plate inclination angle, respectively.

**Table 3.** Scheme and Results of Quadratic Orthogonal Rotation Regression Combination Test of Soil-covering Device Operation Effect.

| Test Serial Number | Factor | | | | Soil-Covering Thickness /$Y_1$(mm) | Seed Offset /$Y_2$(mm) |
|---|---|---|---|---|---|---|
| | Working Speed /$x_1$($km \cdot h^{-1}$) | The Opening Angle of the Disc /$x_2$(°) | The Inclination Angle of the Disc /$x_3$(°) | The Inclination Angle of the Dividing Plate /$x_4$(°) | | |
| 1 | 7 | 70 | 70 | 45 | 47.5 | 31.2 |
| 2 | 3 | 65 | 70 | 40 | 41.2 | 24.3 |
| 3 | 5 | 65 | 70 | 45 | 52.4 | 25.4 |
| 4 | 7 | 60 | 70 | 45 | 52.1 | 21.6 |
| 5 | 5 | 65 | 65 | 40 | 51.5 | 18.5 |
| 6 | 7 | 65 | 70 | 50 | 44.3 | 39.5 |
| 7 | 3 | 60 | 70 | 45 | 36.8 | 13.8 |
| 8 | 3 | 65 | 65 | 45 | 45.2 | 14.2 |
| 9 | 7 | 65 | 70 | 40 | 45.3 | 14.7 |
| 10 | 5 | 65 | 70 | 45 | 52.2 | 25.0 |
| 11 | 5 | 65 | 75 | 50 | 40.8 | 32.0 |
| 12 | 3 | 65 | 75 | 45 | 31.4 | 13.3 |
| 13 | 7 | 65 | 75 | 45 | 42.8 | 35.9 |
| 14 | 7 | 65 | 65 | 45 | 48.2 | 37.5 |
| 15 | 5 | 65 | 65 | 50 | 52.6 | 35.0 |
| 16 | 5 | 60 | 70 | 40 | 46.8 | 18.2 |
| 17 | 3 | 65 | 70 | 50 | 36.5 | 10.2 |
| 18 | 5 | 70 | 70 | 50 | 51.0 | 21.4 |
| 19 | 5 | 65 | 70 | 45 | 48.2 | 18.1 |
| 20 | 5 | 70 | 70 | 40 | 52.3 | 25.6 |
| 21 | 5 | 70 | 75 | 45 | 44.3 | 28.6 |
| 22 | 5 | 60 | 65 | 45 | 41.9 | 21.9 |
| 23 | 5 | 65 | 70 | 45 | 51.2 | 16.6 |
| 24 | 5 | 60 | 70 | 50 | 46.6 | 20.2 |
| 25 | 5 | 70 | 65 | 45 | 49.5 | 31.4 |
| 26 | 5 | 60 | 75 | 45 | 40.8 | 19.7 |
| 27 | 5 | 65 | 70 | 45 | 50.5 | 18.3 |
| 28 | 5 | 65 | 75 | 40 | 42.8 | 15.6 |
| 29 | 3 | 70 | 70 | 45 | 45.0 | 17.7 |

### 3.3.1. Regression Equation

Design-Expert10 software was used to analyze the experimental data, and the F test was performed under the confidence level $\alpha = 0.05$. The partial sum of squares and degrees of freedom of insignificant factors and interactions were incorporated into the remaining sum of squares, and the analysis was performed again [42]. Ensure that all factors reach a significant or extremely significant level and obtain the optimized regression equation:

$$\begin{aligned}Y_1 = 50.90 + 3.66x_1 + 2.05x_2 - 3.83x_3 - 0.68x_4 - 3.20x_1x_2 + 2.10X_1X_3 + 0.93x_1x_4 \\ -1.03x_2x_3 - 0.28x_2x_4 - 0.78x_3x_4 - 5.80x_1^2 - 1.01x_2^2 - 3.86x_3^2 - 1.38x_4^2\end{aligned} \tag{16}$$

$$\begin{aligned}Y_2 = 2.07 + 0.72x_1 + 0.34x_2 - 0.11x_3 + 0.35x_4 + 0.14x_1x_2 - 0.018x_1x_3 + 0.97x_1x_4 \\ -0.015x_2x_3 - 0.15x_2x_4 - 0.0025x_3x_4 + 0.048{x_1}^2 + 0.016{x_2}^2 + 0.42{x_3}^2 + 0.064{x_4}^2\end{aligned} \tag{17}$$

### 3.3.2. Analysis of Variance of the Regression Equation

It can be seen from Tables 4 and 5 that the regression models of the two indicators are extremely significant ($p < 0.01$), indicating that the selected model is appropriate, and there is a model-determined relationship between the test indicators and the test factors. At the same time, the lack of fit terms was not significant ($p > 0.1$), indicating that the model was suitable, no uncontrolled factors influenced the indicators, and the fitting effect of the model was good. From the regression model in Table 4, it can be seen that the proportion of the influence of the thickness of the covering soil is the inclination angle of the disc, the working speed, the opening angle of the disc, and the inclination angle of the soil dividing plate. The regression model in Table 5 shows that the proportion of the influence of seed offset is the operating speed, the inclination angle of the soil dividing plate, the disc opening angle and the disc inclination angle.

**Table 4.** Analysis of variance for regression model of covering soil thickness.

| Project | Squares | df | Square | F | P |
|---|---|---|---|---|---|
| Model | 734.42 | 14 | 52.46 | 7.11 | 0.0004 |
| $x_1$ | 162.07 | 1 | 162.07 | 21.95 | 0.0004 |
| $x_2$ | 50.46 | 1 | 50.43 | 6.83 | 0.0204 |
| $x_3$ | 176.33 | 1 | 176.33 | 23.89 | 0.0002 |
| $x_4$ | 5.47 | 1 | 5.47 | 0.74 | 0.4040 |
| $x_1x_2$ | 40.96 | 1 | 40.96 | 5.55 | 0.0336 |
| $x_1x_3$ | 17.64 | 1 | 17.64 | 2.39 | 0.1445 |
| $x_1x_4$ | 3.42 | 1 | 3.42 | 0.46 | 0.5070 |
| $x_2x_3$ | 4.20 | 1 | 4.20 | 0.57 | 0.4631 |
| $x_2x_4$ | 0.30 | 1 | 0.30 | 0.041 | 0.8425 |
| $x_3x_4$ | 2.40 | 1 | 2.40 | 0.33 | 0.5774 |
| ${x_1}^2$ | 217.89 | 1 | 217.89 | 29.52 | <0.0001 |
| ${x_2}^2$ | 6.60 | 1 | 6.60 | 0.89 | 0.3606 |
| ${x_3}^2$ | 96.56 | 1 | 96.56 | 13.08 | 0.0028 |
| ${x_4}^2$ | 12.19 | 1 | 12.19 | 1.65 | 002197 |
| Residual | 103.35 | 14 | 7.38 | | |
| Lack of Fit | 91.87 | 10 | 9.19 | 3.20 | 0.1367 |
| Pure Error | 11.48 | 4 | 2.87 | | |
| Cor Total | 837.77 | 28 | | | |

**Table 5.** Regression model analysis of Variance for Seed Offset.

| Project | Squares | df | Square | F | P |
|---|---|---|---|---|---|
| Model | 14.37 | 14 | 1.03 | 4.10 | 0.0063 |
| $x_1$ | 6.29 | 1 | 6.29 | 25.13 | 0.0002 |
| $x_2$ | 1.37 | 1 | 1.37 | 5.46 | 0.0349 |
| $x_3$ | 0.15 | 1 | 0.15 | 0.60 | 0.4524 |
| $x_4$ | 1.43 | 1 | 1.43 | 5.70 | 0.0316 |
| $x_1x_2$ | 0.081 | 1 | 0.08 | 0.324 | 0.5780 |
| $x_1x_3$ | 0.0012 | 1 | 0.0012 | 0.0049 | 0.9452 |
| $x_1x_4$ | 3.78 | 1 | 3.78 | 15.10 | 0.0016 |
| $x_2x_3$ | 0.001 | 1 | 0.0009 | 0.0035 | 0.9530 |
| $x_2x_4$ | 0.0961 | 1 | 0.0961 | 0.384 | 0.5456 |
| $x_3x_4$ | $2.5 \times 10^{-5}$ | 1 | $2.5 \times 10^{-5}$ | $9.98 \times 10^{-5}$ | 0.9922 |
| $x_1^2$ | 0.015 | 1 | 0.015 | 0.060 | 0.8102 |
| $x_2^2$ | 0.0016 | 1 | 0.0016 | 0.0063 | 0.9379 |
| $x_3^2$ | 1.14 | 1 | 1.14 | 4.55 | 0.0510 |
| $x_4^2$ | 0.027 | 1 | 0.0268 | 0.107 | 0.7482 |
| Residual | 3.51 | 14 | 0.25 | | |
| Lack of Fit | 2.81 | 10 | 0.28 | 1.61 | 0.3431 |
| Pure Error | 0.70 | 4 | 0.17 | | |
| Cor Total | 17.88 | 28 | | | |

#### 3.3.3. Response Surface Analysis of Each Test Factor to Test Index

The influence rule of factor interaction and test index is shown in Figure 13. It can be seen from Figure 13a that when the opening angle of the disc is constant, the thickness of the covering soil increases sharply at first and then tends to be gentle; when the working speed is constant, the thickness of the covering soil increases with the increase in the opening angle of the disc. It can be seen from Figure 13b that when the working speed is constant, the thickness of the covering soil changes little with the increase of the inclination angle of the soil dividing plate; when the inclination angle of the soil dividing plate is constant, the thickness of the covering soil increases with the increase of the working speed. It can be seen from Figure 13c that when the inclination angle of the disc is constant, the relationship between the seed offset and the working speed tends to be linear and, at the same time, increases with the increase in the inclination angle of the disc; when the working speed is constant, the seed offset first decreases with the inclination angle of the disc increase after. It can be seen from Figure 13d that when the inclination angle of the disc is constant, the relationship between the seed displacement and the inclination angle of the soil dividing plate tends to be linear and increases with the increase in the inclination angle of the soil dividing plate; when the inclination angle of the soil dividing plate is constant, the seed displacement increases with the changing trend of the inclination angle of the disk is small, and it first decreases and then increases with the increase in the inclination angle of the disk.

#### 3.3.4. Optimization of Structural Parameters

Through the optimization module in Design-expert, the optimal design is carried out with the covering soil thickness of 50 mm and the seed offset as small as possible. The optimal parameter combination is as follows; the operating speed was 6.35 km·h. The opening angle was 60°, the inclination angle of the disc was 70°, and the inclination angle of the soil plate is 40°; at this time, the thickness of the covering soil was 48.24 mm, and the seed offset was 10.67 mm, which meets the agronomic requirements of precision seeding operations [43].

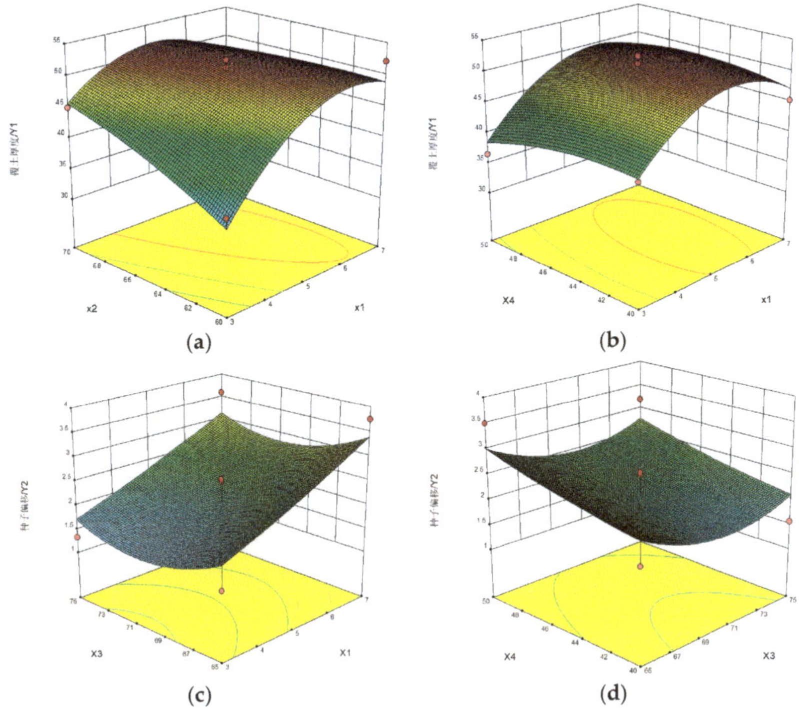


**Figure 13.** The interaction between the factors affects the response surface of the test index: (**a**) x3 and x4 are 70° and 45°, respectively, the relationship between the thickness of the covering soil and X1 X2; (**b**) X2 and X3 are 65° and 70°, respectively, the relationship between the thickness of the covering soil and X1 X4; (**c**) 2 and X4 are 65° and 45°, respectively, the relationship between the seed offset and X1 X3; (**d**) X1 and X2 are 3 km·h$^{-2}$ and 65°, respectively, the relationship between the seed offset and X3 X4.

### 3.4. *Field Test of the Soil-Covering Device*

In order to test the working performance of the soil-covering device, the working performance test was carried out in the soil tank test base of the School of Agricultural Engineering and Food Science, Shandong University of Technology (Zibo, China). In order to test the performance of the soil shunt cover device compared with the traditional soil cover device, a comparative test was carried out. The biggest difference between the device we designed and optimized and the traditional soil-covering device is the presence or absence of the soil-dividing plate. Therefore, we conducted experiments by adding and removing the soil dividing plate.

#### 3.4.1. Comparative Test

The performance test was carried out in the Agricultural Machinery Equipment Laboratory of Shandong University of Technology, as shown in Figure 14. Test conditions: the length × width of the soil tank was 8 × 1.5 m, the soil bulk density of 0–10 cm was 1522 Kg·m$^{-3}$, and the average moisture content was 15%. Test equipment included a soil tank test bench, corn no-tillage seeding monomer, TZS-IIWX soil moisture and temperature measuring instrument, TYD-2 soil compaction instrument, and ruler.

**Figure 14.** Equipment for soil tank test: 1—motor; 2—opener; 3—seed metering device; 4—dividing plate; 5—covering disc.

After installing the designed soil shunt type soil cover and the traditional disc type soil cover device, respectively, on the corn no-tillage seeding unit to carry out the field test operation, the section of the soil groove was manually intercepted. On the section made, we used a brush to gently find the seeds on the section and used a ruler to measure the distance H from the seeds to the soil surface, where H was the soil cover thickness. At the same time, we marked the landing position of the seeds during seeding. After the soil covering was completed, we used a ruler to measure the distance between the marked point and the actual position of the seeds to obtain the offset of the seeds and measured five times at different position.

It can be seen from Tables 6 and 7 that the difference in the soil-covering thickness of the soil shunt type covering device was 48.3–52.4 mm, that of the traditional covering device was 38.6–56.4 mm, and that of the combined soil-covering device was 12~20 mm, and the difference of seed offset of traditional soil-covering device was 25~53 mm. The test results show that the soil-covering device with a soil-closing function is better than the traditional soil-cover device. At the same time, the soil-covering device with a soil-closing function has the function of returning finely crushed soil, and the rationality of the design is also verified. The test results provide a guarantee for future tests.

**Table 6.** The measurement results of the difference in the thickness of the covering soil at the sampling point.

| Types of Soil-Covering Device | Unit | Test Serial Number | Sampling Point | | | | |
|---|---|---|---|---|---|---|---|
| | | | 1 | 2 | 3 | 4 | 5 |
| soil-covering device with soil-closing function | mm | 1 | 53.1 | 52.6 | 50.6 | 62.2 | 55.2 |
| | | 2 | 48.3 | 39.3 | 42.5 | 46.7 | 52.4 |
| | | 3 | 52.2 | 53.0 | 49.7 | 48.3 | 38.2 |
| | | average value | 51.2 | 48.3 | 47.6 | 52.4 | 48.6 |
| Traditional double-disc soil-covering device | mm | 1 | 43.6 | 50.4 | 59.3 | 65.8 | 41.2 |
| | | 2 | 40.5 | 38.6 | 58.5 | 62.8 | 51.2 |
| | | 3 | 31.7 | 37.9 | 48.7 | 40.6 | 45.0 |
| | | average value | 38.6 | 42.3 | 55.5 | 56.4 | 45.8 |

**Table 7.** Measurement results of seed offset difference at the sampling point.

| Types of Soil-Covering Device | Unit | Test Serial Number | Sampling Point | | | | |
|---|---|---|---|---|---|---|---|
| | | | 1 | 2 | 3 | 4 | 5 |
| soil-covering device with soil-closing function | mm | 1 | 22 | 24 | 24 | 15 | 12 |
| | | 2 | 14 | 18 | 16 | 7 | 12 |
| | | 3 | 12 | 18 | 14 | 14 | 21 |
| | | average value | 16 | 20 | 18 | 12 | 15 |
| Traditional double-disc soil-covering device | mm | 1 | 30 | 63 | 31 | 54 | 48 |
| | | 2 | 32 | 15 | 54 | 39 | 57 |
| | | 3 | 13 | 36 | 23 | 66 | 33 |
| | | average value | 25 | 38 | 36 | 53 | 46 |

3.4.2. Performance Test

The field experiment was carried out according to the optimal parameter combination obtained by the simulation experiment, the verification experiment was repeated three times, and five test points were taken each time. The experimental verification results and simulation optimization results are shown in Figure 15.

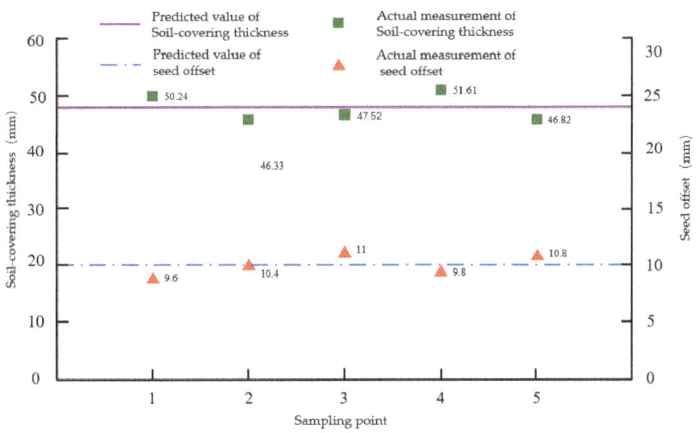


**Figure 15.** Test verification diagram.

It can be seen from Figure 15 that the difference between the simulation results and the results measured in the field test is small, the maximum relative error of the covering soil thickness is 5.7%, and the maximum error of the seed offset is 11.1%, which shows the reliability of the experimental optimization results, and also verifies the feasibility of the simulation.

## 4. Discussion

The changing trend of the response surface analysis shown in Figure 11 indicates that the higher the working speed, the better the soil-breaking performance of the opener, and the greater the soil disturbance. However, an excessively high working speed will bring about excessive power consumption; thus, it is generally advisable to use a working speed that can ensure the covering soil thickness Y1 = 50 mm, 10% floating up and down. Moreover, the working speed is too high, the soil cannot be completely covered in a short time, resulting in poor soil-covering effect, thus reducing the covering soil thickness Y1, so the working speed of the planter should not be too high.

Considering the changes in the trend of the evaluation index Y2 in Figure 11, we concluded that with the increase in working speed, the fluidity of soil increases, which leads to the increase in seed offset Y2, which causes the uniformity of seed arrangement to decrease. In addition, the smaller the inclination angle x4 of the soil dividing plate is, the more stable the soil fluidity of the seed furrow will be, resulting in the smaller seed offset Y2; but if the inclination angle x4 of the soil dividing plate is too small, the amount of flowing soil at the soil dividing plate will increase, and the soil dividing plate cannot withstand a large amount of soil in a short time. This will easily cause the soil to block at the soil dividing board, resulting in a decrease in the thickness of the covering soil Y1, and the thickness of the covering soil cannot be guaranteed.

Further in-depth analysis showed that the change rate of the response surface of the evaluation indicator Y2 in the direction of the working speed and the inclination angle of the soil-dividing plate is better than the change rate of the response surface in the direction of the opening angle of the disc and the inclination angle of the disc. It shows that in this interaction, the influence of the working speed x1 and the inclination angle of the soil-dividing plate x4 on the seed offset Y2 is more significant. x1 and x4 were the dominant factors affecting the inter-row straw mulching Y2.

## 5. Conclusions

(1) A soil-covering device with a soil-closing function that can realize the separation of clod and fine soil and the angle of soil covering can be adjusted is designed, the key parameters of the device were designed, and the interaction model between the soil-covering device and the soil was established.

(2) The quadratic rotation regression combination orthogonal was used to conduct the simulation test. The primary and secondary order of the factors affecting the performance of the soil-covering device was obtained through variance analysis and response surface analysis: speed, disc opening angle, soil dividing plate inclination angle, and disc inclination angle. The optimal combination was as follows: the operating speed was 6.35 $km \cdot h^{-1}$, the disc opening angle was 60°, the disc inclination angle was 70°, and the soil plate inclination angle was 40°. Currently, the covering soil thickness was 48.24 mm, and the seed offset is10.667 mm.

(3) Through comparative experiments, it was determined that the soil-covering device with a soil-closing function could separate soil blocks and enhance the performance of soil fluidity. The results show that the thickness of soil cover was 48.3–52.4 mm, and the seed offset was 12–20 mm. The seed offset was small, and the soil covering was uniform, which meets the agronomic requirements of corn planting.

**Author Contributions:** Conceptualization, Y.G. and X.W.; methodology, X.W.; software, Y.G. and X.Z. (Xiaokang Zhong); experiment, K.C. and Y.G.; data curation, Y.G. and X.Z. (Xiaokang Zhong); writing—original draft preparation, Y.G. and X.Z. (Xiaokang Zhong); writing—review and editing, K.C. and Q.L.; visualization, X.Z. (Xiangcai Zhang); supervision, Z.W.; project administration, X.W.; funding acquisition, M.W. and X.C. All authors have read and agreed to the published version of the manuscript.

**Funding:** This work was supported financially by the National Natural Science Foundation of China (Grant No. 32101631).

**Institutional Review Board Statement:** Not applicable.

**Informed Consent Statement:** Not applicable.

**Data Availability Statement:** Not applicable.

**Acknowledgments:** The authors would like to thank the teachers and supervisors for their technical support. We would also like to acknowledge the assistance provided by brothers and sisters during the tests. Finally, we are grateful to the editor and anonymous reviewers for providing helpful suggestions to improve the quality of this paper.

**Conflicts of Interest:** The authors declare no conflict of interest.

## References


1. Wang, Q.; Cao, X.; Wang, C.; Li, H.; He, J.; Lu, C. Research Progress of No/Minimum Tillage Corn Seeding Technology and Machine in Northeast Black Land of China. *Trans. Chin. Soc. Agric. Mach.* **2021**, *52*, 1–15.
2. Wang, Q.; Xu, Q.; Lu, C.; Li, H.; He, J.; Wang, Q. Research status and development of key technologies for no-tillage seeding intellectualization. *J. South China Agric. Univ.* **2021**, *42*, 27–35.
3. Somerville, P.D.; May, P.B.; Livesley, S.J. Effects of deep tillage and municipal green waste compost amendments on soil properties and tree growth in compacted urban soils. *J. Environ. Manag.* **2018**, *227*, 365–374. [CrossRef] [PubMed]
4. Hargreaves, P.R.; Baker, K.L.; Graceson, A.; Bonnett, S.; Ball, B.C.; Cloy, J.M. Soil compaction effects on grassland silage yields and soil structure under different levels of compaction over three years. *Eur. J. Agron.* **2019**, *109*, 125916. [CrossRef]
5. Liu, L.; Ma, C.; Liu, Z. EDEM-based Parameter Optimization and Experiment of Full-layer Fertilization Shovel for Strip Subsoiling. *Trans. Chin. Soc. Agric. Mach.* **2021**, *10*, 74–83.
6. Yan, B.; Wu, G.; Fu, W.; Gao, N.; Meng, Z.; Zhu, P. Influencing Factors of Corn Implantation Distribution for High-Planting Based on EDEM. *Trans. Chin. Soc. Agric. Mach.* **2020**, *51*, 47–54.
7. Chen, Y.; Munkholm, L.J.; Nyord, T. A discrete element model for soil–sweep interaction in three different soils. *Soil Tillage Res.* **2013**, *126*, 34–41. [CrossRef]
8. Ucgul, M.; Saunders, C. Simulation of tillage forces and furrow profile during soil-mouldboard plough interaction using discrete element modelling. *Biosyst. Eng.* **2020**, *190*, 58–70. [CrossRef]
9. Tekeste, M.Z.; Balvanz, L.R.; Hatfield, J.L.; Ghorbani, S. Discrete element modeling of cultivator sweep-to-soil interaction: Worn and hardened edges effects on soil-tool forces and soil flow. *J. Terramechanics* **2019**, *82*, 1–11. [CrossRef]
10. Hang, C.; Gao, X.; Yuan, M.; Huang, Y.; Zhu, R. Discrete element simulations and experiments of soil disturbance as affected by the tine spacing of subsoiler. *Biosyst. Eng.* **2018**, *168*, 73–82. [CrossRef]
11. Barr, J.; Desbiolles, J.; Ucgul, M.; Fielke, J.M. Bentleg furrow opener performance analysis using the discrete element method. *Biosyst. Eng.* **2020**, *189*, 99–115. [CrossRef]
12. Zhang, H.; Chen, J.; Song, Y.; Han, D.; Ji, C. Design of Auger Screw Precise Soil Covering Device for Dry Land. *Xinjiang Agric. Mech.* **2021**, *5*, 21–22.
13. Zhang, Z.; Sun, X.; Jin, Z.; Bing, Z.; Sun, J.; Tong, J. Design and Test of Crushing Bionic Soil Covering Device of Soybean Seeder. *Trans. Chin. Soc. Agric. Mach.* **2018**, *49*, 34–40, 73.
14. Lin, J.; Meng, F.; Gao, W. Optimum Design and Experimental Study of Soil Covering Device of Straw Deep-buried Returning Machine. *J. Shenyang Agric. Univ.* **2019**, *50*, 694–702.
15. Liu, X.; Jin, L.; Zhao, Y.; Guo, H.; Wang, J. Experiment research of double-deck disk coverer based on uniform design. *J. Chin. Agric. Mech.* **2016**, *37*, 26–28.
16. Song, L.; Yang, X.; Li, Q. Simulation study on precision seeding disc covering device based on SPH's mesh less method. *J. Chin. Agric. Mech.* **2016**, *37*, 29–32, 42.
17. Gou, W.; Ma, R.; Fan, G.; Yang, W.; Liu, L.; Lei, X. Parameter optimization of interlink covering device in relay-cropping system. *Trans. CSAE* **2011**, *27*, 33–37.
18. Wang, J. *The Research of Position Control after Seed Contacting Soil in the Process of Soil Covering and Rolling with Precision Seeder*; Jilin University: Jilin, China, 2012.
19. Li, Y.; Meng, P.; Geng, D.; He, K.; Meng, F.; Jiang, M. Intelligent System for Adjusting and Controlling Corn Seeding Depth. *Trans. Chin. Soc. Agric. Mach.* **2016**, *47*, 62–68.
20. Chao, B.; Yuang, P.; Han, C.; You, J.; Gao, J. Design of Automatic Regulating Device for Soil Cover of Transplanter. *J. Agric. Mech. Res.* **2022**, *44*, 54–58.
21. Nielsen, S.K.; Nørremark, M.; Green, O. Sensor and control for consistent seed drill coulter depth. *Comput. Electron. Agric.* **2016**, *127*, 690–698. [CrossRef]
22. Suomi, P.; Oksanen, T. Automatic working depth control for seed drill using ISO 11783 remote control messages. *Comput. Electron. Agric.* **2015**, *116*, 30–35. [CrossRef]
23. Kiani, S.; Kamgar, S.; Raoufat, M. Automatic on-line depth control of seeding unit using a non-contacting ultrasonic sensor. *Can. Soc. Bioeng.* **2010**, *6*, 13–17.
24. Kirkegaard Nielsen, S.; Munkholm, L.J.; Lamandé, M.; Nørremark, M.; Skou-Nielsen, N.; Edwards, G.T.C.; Green, O. Seed drill instrumentation for spatial coulter depth measurements. *Comput. Electron. Agric.* **2017**, *141*, 207–214. [CrossRef]
25. Chinese Academy of Agricultural Mechanization Sciences. *Agricultural Machinery Design Manual*; Chinese Agricultural Sientech Press: Beijing, China, 2007; pp. 343–389.
26. Liu, X.; Wang, X.; Chen, L.; Zhang, C.; Liu, W.; Ding, Y. Design and experiments of layered and quantitative fertilization device for rapeseed seeder. *Trans. CSAE* **2021**, *37*, 1–10.
27. Zhao, S.; Liu, H.; Yang, C.; Yang, L.; Gao, L.; Yang, Y. Design and Discrete Element Simulation of Interactive Layered Subsoiler with Maize Straw Returned to Filed. *Trans. Chin. Soc. Agric. Mach.* **2021**, *52*, 75–87.
28. Wang, Y. *Simulation Analysis of Structure and Effect of the Subsoiler Based on DEM*; Jilin Agricultural University: Jilin, China, 2014.
29. Ma, Y.; Wang, A.; Zhao, J.; Hao, J.; Li, J.; Ma, L.; Zhao, W.; Wu, Y. Simulation analysis and experiment of drag reduction effect of convex blade subsoiler based on discrete element method. *Trans. CSAE* **2019**, *35*, 16–23.

30. Potyondy, D.O.; Cundall, P.A. A bonded-particle model for rock. *Int. J. Rock Mech. Min. Sci.* **2004**, *41*, 1329–1364. [CrossRef]
31. Zhao, S.; Liu, H.; Tan, H.; Cao, X.; Zhang, X.; Yang, Y. Design and performance experiment of opener based on bionic sailfish head curve. *Trans. CSAE* **2017**, *33*, 32–39.
32. Shi, L.; Wu, J.; Sun, W.; Zhang, F.; Sun, B.; Liu, Q.; Zhao, W. Simulation test for metering process of horizontal disc precision metering device based on discrete element method. *Trans. CSAE* **2014**, *30*, 40–48.
33. Markauskas, D.; Ramírez-Gómez, Á.; Kačianauskas, R.; Zdancevičius, E. Maize grain shape approaches for DEM modelling. *Comput. Electron. Agric.* **2015**, *118*, 247–258. [CrossRef]
34. Barrios, G.K.P.; de Carvalho, R.M.; Kwade, A.; Tavares, L.M. Contact parameter estimation for DEM simulation of iron ore pellet handling. *Powder Technol.* **2013**, *248*, 84–93. [CrossRef]
35. Masson, S.; Martinez, J. Effect of particle mechanical properties on silo flow and stresses from distinct element simulations. *Powder Technol.* **2000**, *109*, 164–178. [CrossRef]
36. Wang, Y.; Liang, Z.; Zhang, D.; Cui, T.; Shi, S.; Li, K.; Yang, L. Calibration method of contact characteristic parameters for corn seeds based on EDEM. *Trans. CSAE* **2016**, *32*, 36–42.
37. Cui, T.; Liu, J.; Yang, L.; Zhang, D.; Zhang, R.; Lan, W. Experiment and simulation of rolling friction characteristic of corn seed based on high-speed photography. *Trans. CSAE* **2013**, *29*, 34–41.
38. Wang, J.; Tang, H.; Wang, Q.; Zhou, W.; Yang, W.; Shen, H. Numerical simulation and experiment on seeding performance of pickup finger precision seed-metering device based on EDEM. *Trans. CSAE* **2015**, *31*, 43–50.
39. Wang, X.; Zhong, X.; Geng, Y.; Wei, Z.; Hu, H.; Geng, D.; Zhang, X. Construction and parameter calibration of the nonlinear elastoplastic discrete element model for no-tillage soil compaction. *Trans. CSAE* **2021**, *37*, 100–107.
40. Huang, Y.; Hang, C.; Yuan, M.; Wang, B.; Zhu, R. Discrete Element Simulation and Experiment on Disturbance Behavior of Subsoiling. *Trans. Chin. Soc. Agric. Mach.* **2016**, *47*, 80–88.
41. Shi, Y.; Yu, H.; Jiang, Y.; Wang, X.; Chen, H.; Liu, H. Optimization of Strip Fertilization Planter for Straw Throwing and Paving. *Agriculture* **2022**, *12*, 613. [CrossRef]
42. Liu, H.; Hang, J.; Chen, J.; Lu, J.; Zhao, S. Performance Simulation and Experiment on Rigid Press Wheel for Hilly Area. *Trans. Chin. Soc. Agric. Mach.* **2018**, *49*, 114–122.
43. Zhou, L. *DEM-Based Modelling of Maize Seeds and the Simulation Analysis and Experimental Study of the Seed-Sowing*; Jilin Agricultural University: Jilin, China, 2021.

*agriculture*

*Article*

# Improvement of Straw Throwing Performance of Harvester Based on Matching Header Width

**Jiang Wang [1,2], Xiaoyan Wang [1,2,*], Hongwen Li [1,2], Caiyun Lu [1,2], Jin He [1,2], Qingjie Wang [1,2], Di Liu [1,2], Bo Deng [1,2] and Meiyu Zhang [1,2]**

[1] College of Engineering, China Agricultural University, Beijing 100083, China
[2] Key Laboratory of Agricultural Equipment for Conservation Tillage, Ministry of Agricultural and Rural Affairs, Beijing 100083, China
* Correspondence: 00525@cau.edu.cn; Tel.: +86-010-6273-7300

**Abstract:** Aiming at the problem wherein a straw crushing and throwing device (SCTD) installed in a rice combine harvester (RCH) has a small throwing width and does not match the harvesting width, this paper proposes an improved plan for installing wind blades (WB) and optimizing the parameters of the deflector. The structural parameters of the WB were determined, and static analysis was carried out. The influence of the number of WB on the airflow field of the crushing chamber (CC) was studied by CFD simulation. The movement of the straw after entering the throwing device (TD) was analyzed. It was determined that the factors affecting the throwing width under the condition of a certain straw speed were the installation angle of the deflector (IAD), and the arc length of the deflector (ALD) through the bench test. The optimal combination of deflectors parameters was determined to match the width of the harvester header. When the straw feeding speed was 4 kg/s, and the straw moisture content was 33.80%, the optimal parameters are that the ALD was 400 mm and the IAD was 9°. The matching degree with the header width (4.50 m) is 98.44%. This study can effectively increase the straw throwing width and create conditions for the smooth implementation of straw returning to the field.

**Keywords:** rice combine harvester; throwing device; wind blades; fluid analysis; deflector optimization; throwing width

**Citation:** Wang, J.; Wang, X.; Li, H.; Lu, C.; He, J.; Wang, Q.; Liu, D.; Deng, B.; Zhang, M. Improvement of Straw Throwing Performance of Harvester Based on Matching Header Width. *Agriculture* **2022**, *12*, 1291. https://doi.org/10.3390/agriculture12091291

Academic Editors: Mustafa Ucgul and Chung-Liang Chang

Received: 24 July 2022
Accepted: 20 August 2022
Published: 23 August 2022



## 1. Introduction

Straw throwing is an essential part of straw returning [1,2], and its operation quality affects the effect of straw returning and whether subsequent operations can be successfully implemented [3–6]. Compared with the straw pulverizer matching the tractor [7–11], the SCTD installed in the RCH has the characteristics of high operation efficiency and low operation cost [12,13], both of which have broad prospects for various applications. Due to the large amount of straw in the operation of rice straw throwing, a problem often emerges wherein the width of the straw throwing is small and the width of the header does not match it size, which causes excessively uneven throwing. During the subsequent burial and return to the field, the straw is more likely to block the working parts [14], and the decomposition rate of the straw will also decrease [15].

Scholars at home and abroad have researched the problem of throwing width. Schillinger [16] installed a fan on the original harvester, and this fan was designed with a two-way hose. The hoses were arranged at the straw discharge and grain bran outlets. At the same time, the acceleration of the straw was completed by discharging strong airflow from the air outlet of the hose, thereby increasing the throwing width of the straw. However, this improvement plan dramatically changed the structure of the harvester, and the fan also increased the power consumption of the harvester. Christian [17] developed a double-shaft vertical pulverizer for increasing the width of the harvester header, which can realize the integration of straw and powder throwing (that is, the straw was also thrown at the same time). The width can reach 12 m, but the operation stability was poor, the power consumption was high, and it was still in its initial test stage. Lu Jian [18] designed a double helix auger throwing mechanism suitable for the half-feed harvester. During the mechanism's operation, the straw can be evenly spread in the working area of the harvester, but the work efficiency was low and the throwing width was small. This was not suitable for axial harvesters. Wang Xin [19] analyzed the influence of the IAD on the throwing width through field experiments but did not systematically study the interaction between the deflector itself and the installation parameters. Most of the existing research on the throwing width is to add a throwing device, which significantly changes the structure of the harvester and has high power consumption, which does not apply to the improvement of the throwing performance of the existing rice combine harvester.

In this paper, the harvester's dynamic model of straw throwing is constructed, and the factors affecting the throwing width are analyzed. The improvement plan for installing WB and optimizing the deflector is proposed, and the relevant parameters are determined. The influence of the WB installation parameters on the airflow in the crushing chamber is studied, and the related parameters of the deflector are determined through optimization tests.

## 2. Materials and Methods

### *2.1. Machine Structure and Working Principle of the SCTD*

#### 2.1.1. Structure of the SCTD

The structure and operation effects of the whole machine are shown in Figure 1a. The structure of the SCTD comprised grass receiving plate (2), the crushing device (4), the TD (5), and pulley (3). The crushing device included a crushing chamber (8), crushing knife shaft (9), first-level fixed knife (11), Second-level fixed knife (10) and the moving knife combination (12); the throwing device included the throwing shell (7) and the deflector (6); the movable knife assembly mainly includes the movable knife blade (13) and the wind blade (14), and the structure is shown in Figure 1b.

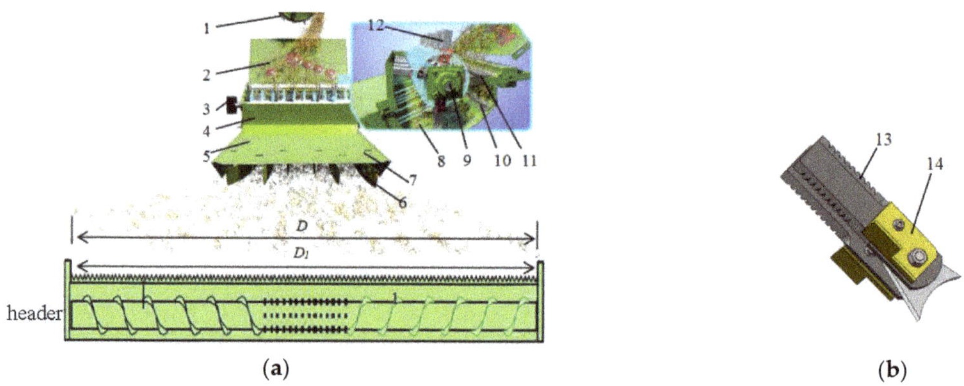


**Figure 1.** Structure diagram of the SCTD. (**a**) Overall structure and renderings; (**b**) structure diagram of moving knife combination. Note: (1) threshing drum; (2) grass receiving plate; (3) pulley; (4) crushing device; (5) TD; (6) baffle; (7) throwing shell; (8) CC; (9) crushing knife shaft; (10) second-level fixed knife; (11) first-level fixed knife; (12) moving knife combination; (13) moving knife; (14) WB; $D$ is the throwing width, m; $D_1$ is the header width, m.

#### 2.1.2. Working Principle

During operation, the power of the SCTD was transmitted from the RCH to the crushing knife shaft through the belt. The operation process can be divided into three stages. (1) Straw feeding stage: the straw come out of the straw discharge port of the threshing device and slid down along the straw collecting plate to the inlet of the CC. (2) Straw crushing stage: when the straw was fed into the CC of the crushing device, the cutting of the straw was completed with the cooperation of the crushing movable and fixed knives. (3) Straw throwing stage: the crushed straw was thrown to the ground under the action of the TD and its gravity.

### 2.2. Design and Installation of Wind Blades

The induced draft fan had been widely used in factories and mining areas as ventilation and dust removal equipment. Its main working principle was to use the fan blade-shaped working parts to rotate at high speed to accelerate the airflow and generate negative pressure to exhaust air [20]. The blade can be installed on the moving knife by drawing on the working principle of the induced draft fan so that when the moving knife rotated at high speed, a higher wind speed was generated so that the straw could be discharged from the crushing device at a higher speed.

#### 2.2.1. Determination of Structure Size

The structure of the WB shown in Figure 2 includes a windward side and a mounting side. The windward side was perpendicular to the blade face of the moving knife. As shown in Figure 2a, both the wind blade mounting surface and the moving blade surface are provided with connecting holes. The main working face generates high-speed airflow. Since the moving blade was made of 65 Mn steel, the welding and tensile properties were poor, and the WB should not be welded or formed at one time. Therefore, a mounting side was designed to install the windward side of the WB on the blade face of the moving knife, and the specific connection method was the bolt locking connection. As shown in Figure 2a, the WB mounting side and the moving knife surface provide connecting holes.

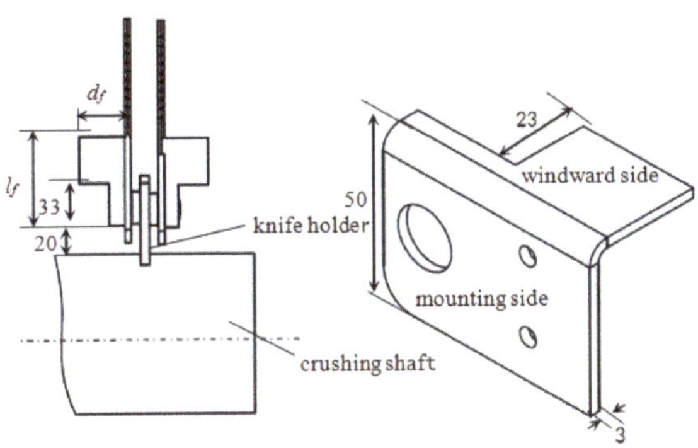

(a)

(b)

**Figure 2.** WB structure and physical pictures. (**a**) WB structure; (**b**) WB installation. $l_f$ is the length of the windward side, mm; $d_f$ is the width of the windward side, mm.

In order to avoid affecting the crushing effect, the length of the windward side ($l_f$) should not overlap with the effective length of the blade of the moving knife. According to the actual test, the $l_f$ was designed to be 70 mm. The width of the windward face ($d_f$) was designed to consider the swing of the moving knife, and the size was 33 mm. At the same time, in order to avoid the phenomenon of knife jamming when the moving knife combination starts and stops, through the indoor installation test, the structure of the windward side was designed, as shown in Figure 2a, using a 70 mm × 35 mm rectangle to cut a 33 mm × 23 mm rectangle. The thickness was designed to be 3 mm. The length of the mounting side should be consistent with the length of the windward side. In order to better connect with the moving knife, the width of the datum plane should be consistent with the width of the moving knife, which was 50 mm. The installation diagram of the wind blade test is shown in Figure 2b.

#### 2.2.2. Establishment of Finite Element Model of WB

The existence of WB can generate airflow under high-speed rotation, but at the same time, the airflow and straw would also impact the WB. In order to avoid fatigue and stress damage during the operation of the designed WB, it is necessary to obtain the windward stress of the WB during the crushing operation. Therefore, carrying out a static finite element analysis of WB was necessary.

In this paper, Solidworks was used to build a three-dimensional model of a WB, and the model was saved in step format and then opened in ANSYS Workbench. The material of the WB was structural steel Q235. After consulting relevant literature [21], the material performance parameters are shown in Table 1. According to the table content, the model's material properties were defined. The model mesh was divided by tetrahedral mesh [22]. Figure 3 shows the model effect after division. A total of 19,522 nodes were divided, and 35,383 nodes were obtained. The number of meshes meets the analysis requirements.

**Table 1.** Q235 material properties.

| Parameters | Value |
|---|---|
| Mass density | $7.85 \times 10^3$ kg/m$^3$ |
| Young's modulus | 210 GPa |
| Poisson's ratio | 0.3 |
| Tensile strength | 450 MP |
| Yield strength | 235 MP |

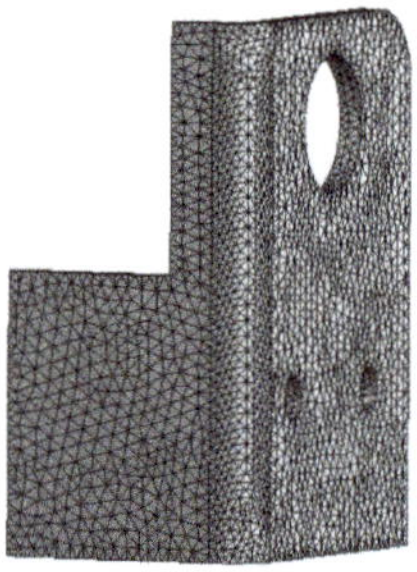

**Figure 3.** The effect of model meshing.

Since the mounting side of the WB was fixedly connected with the moving knife by bolts, a fixed constraint was imposed on the mounting side to analyze the ultimate stress of the windward side of the WB. Before calculating the load pressure, it was assumed that no vortex occurred during the crushing operation and that the angle of the windward side of the straw collision was the same. When the WB rotated at high speed, it would be subjected to air resistance and straw impact force. Reference [23] showed that the air resistance $F_k$ of an object was related to the area of the object, the speed of airflow, and the air resistance coefficient at the airflow speed. Its expression is as follows:

$$F_k = C_d S_1 v_d^2 \quad (1)$$

where $C_d$ is the air resistance coefficient; $S_1$ is the windward area of the object, m$^2$; $v_d$ is the relative velocity of the airflow and the windward surface, m/s. According to a literature review, it can be seen that $C_d$ tends to decrease slowly with the increase in speed. According to this research, $C_d$ is selected as 0.80 in this paper.

When calculating the average action force Fc of straw on wind blades, it can be understood that the product of action force and action time equals the change of straw momentum. The formula is:

$$F_c \Delta t = \Delta m v_j \quad (2)$$

where $\Delta t$ is the collision time, s; $\Delta m$ is the mass of broken straw, kg; $v_j$ is the velocity of the straw after the collision, m/s. Refer to the relevant literature [24] for further information. $\Delta m$ value is $3.12 \times 10^{-4}$ kg; $\Delta t$ value is $1.75 \times 10^{-3}$ s.

The size of the windward side area $S_1$ of the WB is $2.31 \times 10^{-3}$ m$^2$, and the positional velocity $v_f$ of the center point of the windward side can be expressed as:

$$v_f = \frac{2\pi R_f}{60} \times n \quad (3)$$

where $R_f$ is the turning radius of the center point of the windward side, m, and the value is 0.125; $n$ is the rotational speed of the cutter shaft, r/min, which was taken as 2600 r/min in this paper. Bringing in the above formula can obtain $v_f$. The speed of the straw after the collision can be considered to be the same speed as the WB; the WB and the airflow are in relative motion; the WB speed can replace the airflow speed, that is, $v_f = v_j = v_d$. Put the obtained $v_f$ values into Equations (1) and (2), respectively, to obtain the air resistance $F_k$ as 892.81 N, and the average force $F_c$ of the straw on the WB is 391.87 N. As shown in Figure 4, these two forces are added on the WB's windward side.

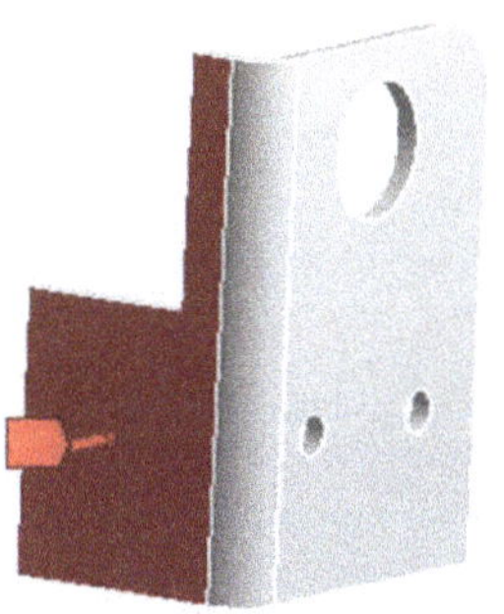

**Figure 4.** Add load.

2.2.3. Pre-Processing of CFD Simulation

Driven by the crushing knife shaft, the WB would generate a complex airflow field and pressure field in the CC, and the characteristics of the flow field would affect the airflow in the CC and thus the movement of the straw. In this chapter, the CFD analysis method was used to study the influence of different installation parameters of WB on the flow field characteristics of the CC to reveal the influence of different installation parameters of WB on the flowability and throwing characteristics of straw.

In order to improve the simulation efficiency, the 3D model established by Solidworks needed to be simplified, and the structures or details that had little effect on the simulation accuracy were omitted, such as bearings, bolts, tool seats, and openings. At the same time, a motion split surface was established to facilitate the extraction and segmentation of subsequent fluid domains. The simplified CC structure is shown in Figure 5. We saved the 3D model in Parasolid format and imported it into Design Modeler to extract the CC fluid domain. The fluid domain was meshed in Meshing, and the boundaries were named. In order to improve the mesh quality and reduce the simulation time, after the mesh model was imported into Fluent, and the mesh was optimized through the mesh polyhedron conversion function that comes with the software. The optimized polyhedron mesh is shown in Figure 6.

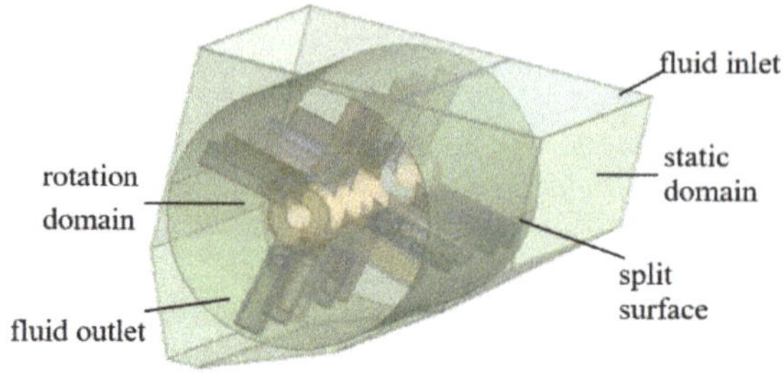


**Figure 5.** Simplified CC structure.

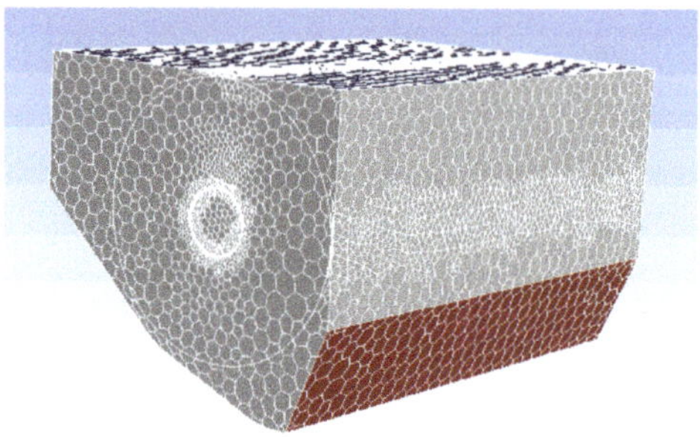

**Figure 6.** Polyhedral mesh after optimization.

The fluid domain between the pulverizer shell and the split surface was set as the static domain, the motion domain of the pulverizing knife shaft was set as the rotation domain, and the rotational speed of the rotation domain was set as 2600 r/min. During the simulation, the rotation domain would drive the static domain, and Changes in airflow were generated. The inlet and outlet boundary conditions of the CC were set to pressure type, and the pressure was standard atmospheric pressure. Due to the phenomenon of airflow fluctuation and eddy currents in the process of rotating motion, the realizable k-ε model with high reliability was used to simulate the turbulent flow in the CC [25]; the calculation method adopted was 'Coupled'. After the simulation calculation was completed, the results were processed in CFD-Post, and three sections, P1, P2, and P3, were selected and established (as shown in Figure 7); the fluid properties of each section were analyzed. Section P1 was located on the left side of the tool axis and passed through the end face of the tool axis; section P2 was located at the center of the tool axis and parallel to the P1 plane; section P3 was located on the right side of the tool axis and passed through the end face of the tool axis.

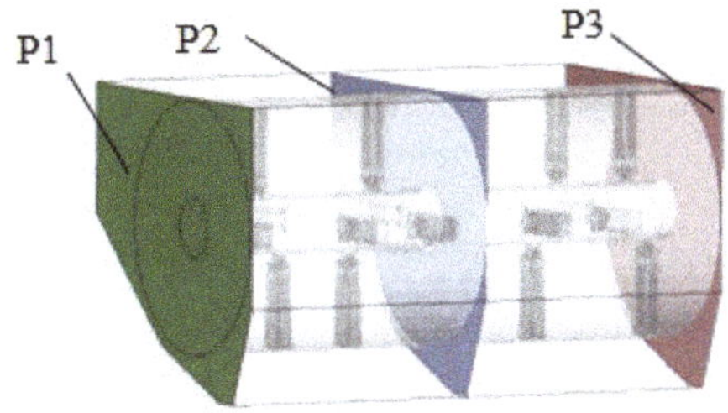


**Figure 7.** Selection of CC section. Section P1 was located on the left side of the tool axis and passed through the end face of the tool axis; section P2 was located at the center of the tool axis and parallel to the P1 plane; section P3 was located on the right side of the tool axis and passed through the end face of the tool axis.

*2.3. Motion Analysis of Straw Entering the TD*

Since the crushing device and the TD were closely connected, it can be considered that the straw entered the TD immediately after being discharged from the crushing device. After the straw entered the throwing device, its motion can be divided into an AB stage (non-guided stage) and Bb stage (guided throwing stage), as shown in Figure 8. Motion analysis of these two processes is performed below.

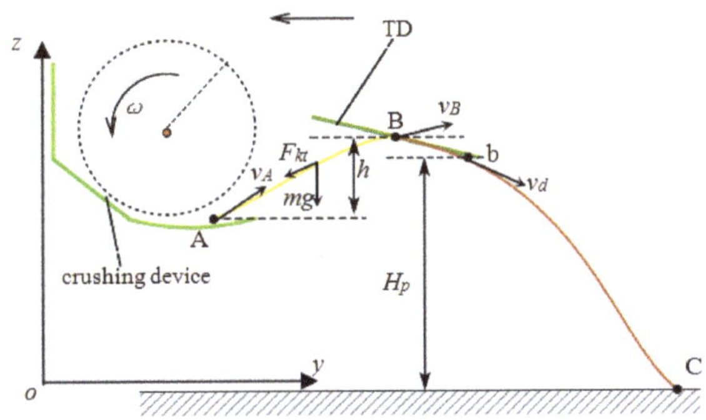


**Figure 8.** Schematic diagram of straw entering into the throwing device.

2.3.1. Straw Non-Guided Stage

The crushed straw was discharged from point A of the outlet of the crushing device under the action of centrifugal force and airflow, entered the TD, and then reached point B of the TD under the action of its impulse airflow and gravity. This process was the non-guided stage of straw. At this stage, the airflow velocity gradient was slight, and it can be considered that the straw does not rotate, so the influence of its rotation on its force was ignored [26]. Since the volume of straw was small, the flow velocity of the air flowing around it can be considered almost unchanged, and the fluid pressure difference on both sides of the straw was slight, so the displacement of the straw caused by the pressure difference can be ignored [27].

The designed WB can accelerate the airflow through high-speed rotation. It can be considered that the airflow does positive work on the movement of the straw. After the crushing device worked stably, the straw would reach the maximum value in a short period and then maintain a certain speed to reach point A. In attenuation, the airflow would do negative work on the straw in the AB stage, so it was considered that the airflow velocity

$v_{d1}$ at point A was less than or equal to the straw velocity $v_A$ at this point. Then, the kinetic equation of straw in the AB stage is

$$-(\Delta mg + F_{kt}) = m\frac{d^2h}{dt^2} \tag{4}$$

where $t$ is the time when the straw leaves point A, s; $\Delta m$ is the straw weight, g; $F_{kt}$ is the air resistance when the straw leaves point A at time $t$, N; $h$ is the height difference between point A and point B, mm.

When the straw moved in the AB stage, the air resistance $F_{kt}$ received can be expressed as:

$$\begin{cases} F_{kt} = C_d S_j v_{dt}^2 \\ S_j = \frac{1}{4}\pi d_j^2 \\ d_j = \left(\frac{6V_j}{\pi}\right)^{\frac{1}{3}} \\ V_j = \frac{1}{4}\pi \overline{d_j}^2 l_j \end{cases} \tag{5}$$

where $S_j$ is the windward area of the straw, $m^2$; $d_j$ is the equivalent volume diameter of the straw, m; $V_j$ is the equivalent straw volume, $m^3$; $\overline{d_j}$ is the average diameter of the cross-section of the straw, m; $l_j$ is the average length of the straw after crushing, m; $v_{dt}$ is the relative velocity of airflow and straw, m/s.

The initial boundary conditions of the AB segment of the straw movement process are $\frac{dh}{dt} = v_A - v_{d1}, h = 0$. Then the height difference of the straw moving in the AB segment can be expressed as:

$$h = -\frac{\Delta mg + F_{kt}}{2\Delta m}t^2 + (v_A - v_{d1})t \tag{6}$$

When the WB was added, the throwing speed of the straw at point A would increase, and the air resistance after being discharged from the crushing device would also increase. It can be seen that h will be smaller according to Formula (5); that is, the collision point B of the straw would move backwards since the guide of the TD was the process of throwing down, which would increase the collision angle of the straw at point B. The increase in the collision angle would reduce the loss of kinetic energy [28].

### 2.3.2. Straw Guided Throwing Stage

The straw contacted and collided with the TD at point B. Since the flow velocity of the straw was fast, the collision occurred instantaneously, so the beating and displacement generated in the process can be ignored. After passing point B, the straw entered the deflector area and was thrown under the guidance of the deflector. As shown in Figure 9, the force on the straw in this process had its gravity $\Delta mg$, the centrifugal force $F_l$ guided by the deflector, the friction force $F_f$ with the deflector, and the air resistance $F_k'$. It can be seen that the straw was subject to more resistance in this process, and there was a specific energy loss. Assuming that the kinetic energy of the straw thrown at point b is $E_b$, it can be expressed as:

$$E_b = E_0 - E_p - E_z \tag{7}$$

where $E_0$ is the kinetic energy when the straw enters the deflector, J; $E_p$ is the collision kinetic energy loss when the straw is in contact with the deflector, J; $E_z$ is the kinetic energy loss of the straw when it drains and slips, J.

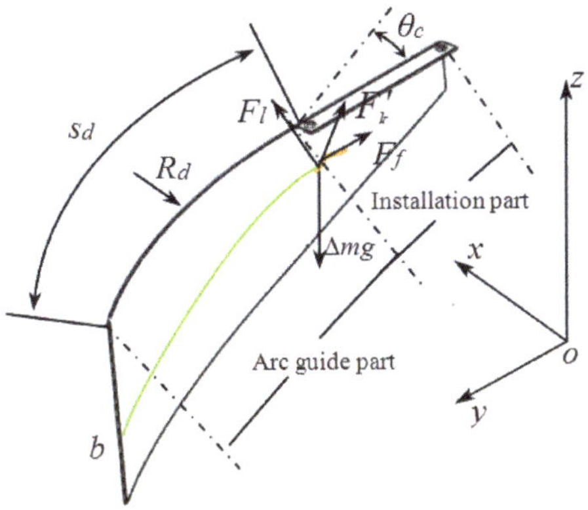


**Figure 9.** Analysis of the structure of the deflector and the movement of straw.

Assuming that the speed of the straw entering the deflector was $v_c$, the kinetic energy of the straw at this time can be expressed as:

$$E_0 = \frac{1}{2}\Delta m v_c^2 \quad (8)$$

When it was considered that the collision between the straw and the deflector was utterly inelastic, the energy loss during the collision can be expressed as:

$$E_b = \frac{1}{2}\Delta m v_c^2 \sin\varepsilon \quad (9)$$

where is the collision angle between the straw and the deflector, °. The energy loss of straw when the deflector slips can be expressed as:

$$E_c = \int_0^{s_d} F_d ds = \int_0^{s_d} (F_f + F_k')ds \quad (10)$$

Substituting Equations (8)–(10) into Equation (7), we can get:

$$E_b = \frac{1}{2}\Delta m v_c^2 - \frac{1}{2}\Delta m v_c^2 \sin\varepsilon - \int_0^{s_d} (F_f + F_k')ds \quad (11)$$

The kinetic energy of straw throwing would affect the throwing width. We can discuss the kinetic energy loss in the Bb stage from the above formula. It can be seen that the kinetic energy loss in this process was related to the collision angle $\varepsilon$ and the ALD $s_d$. When the collision angle $\varepsilon$ increased, the kinetic energy loss increased, and the collision angle $\varepsilon$ was related to the IAD (the angle between the installation section of the deflector and the $y$-axis) $\theta_c$. When the arc length $s_d$ of the deflector was longer, the friction of the straw at this stage and the negative work carried out by the airflow were greater. According to the above analysis, the kinetic energy loss of straw in the TD was represented by two collisions and the loss of resistance. In order to achieve a better throwing width under the condition that the installation angle of the TD remained unchanged, two factors, the IAD $\theta_c$ of the deflector and the ALD $s_d$, needed to be optimized.

### 2.4. Determination of the Structural Parameters of the Deflector

As shown in Figure 9, the structure of the deflector can be divided into an installation part and an arc guide part. The installation part mainly connects the deflector with the throwing shell of the throwing device; it has two mounting holes, which can be adjusted within 0°~15° of the IAD by cooperating with the strip holes on the shell plate. The primary dimension parameters of the arc guide part were the arc radius and arc length. The TD's basic structure and the deflectors' arrangement are shown in Figure 10. The deflectors were divided into three types and a total of six according to different sizes. Since the outermost deflector was directly related to the throwing width of the TD, this paper only optimized the structure of the outermost deflectors 1 and 6. Considering the overall size of the TD, the radius $R_d$ of the arc guide part of the deflector should not be too small, which would cause the straw to be thrown out not smoothly. If it were too large, the velocity component in the $y$-axis direction would be small when the straw was thrown, which was not conducive to increasing the throwing width. According to the actual test and the mapping of the original deflector, the radius Rd of the arc guide part of the improved deflector was determined to be 650 mm. According to the analysis in Section 2.3, it can be seen that if the ALD were too long, it would cause more energy loss of straw. However, the guide effect was not apparent if it was too short. According to deflector installation and diversion requirements, the improved deflector arc length sd was designed to be 360 mm~420 mm. The installation size and width of the improved deflector are the same as the original deflector.

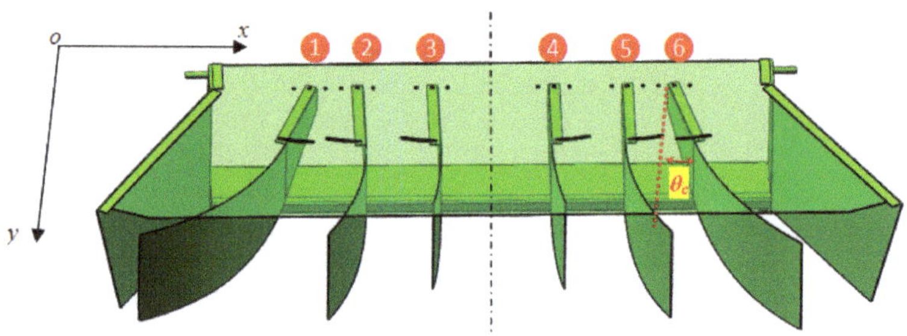


**Figure 10.** The structure of the throwing device and the installation diagram of the deflector. 1–6 represent the deflectors at different installation positions, 1, 6 are the outermost deflectors, and 2–4 are the middle position deflectors.

### 2.5. Parameter Optimization of the Deflector on the Test Bench

In order to further determine the appropriate combination of ALD and IAD when the TD threw rice straw, an optimization test was carried out on the SCTD. In order to avoid the influence of uncertain factors and a complex measurement environment on the measurement results in the field test, the optimization test was carried out on a test bench.

The optimization test of the test bench was carried out in October 2020 at the Harbin Songbei Machinery Factory, Harbin, China. The structural design of the test bench is shown in Figure 11, which mainly includes a straw conveyor belt, a safety box, a motor, a support bench, and SCTD. Through continuous motion, the straw drive belt was used as the straw feeding structure to feed the straw into the crushing device. According to the feed rate of the RCH header and the ratio of grass to grain, it can be determined that the feed rate of straw is 4 kg/s. The existence of the safety box is intended to avoid the unsafe factor when the SCTD operates at high speed. The motor serves as the power source of the crushing device, and the support bench provides a stable support environment for the test bench. The test straw was the unpulverized straw in the nearby farmland with a moisture content of 33.80%, which was collected and used later. The speed of the pulverizer was designed to be 2600 r/min, and the WB was all installed. After being thrown, the straw can be received by the pre-laid roll cloth, and the roll cloth can move backwards at a certain speed under the drive of the reel. This process can simulate the forward speed of the harvester during the field test to avoid repeated spreading of the straw after throwing. The significance of the two factors was analyzed by test. The parameter combination optimization was carried out with the width of the harvester header as the optimization target [29,30]. The test factors and levels are shown in Table 2. After the test, the throwing width ($D$) was measured and evaluated.

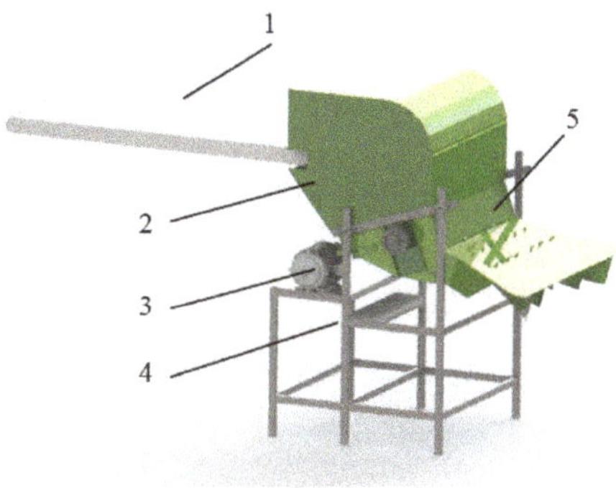

**Figure 11.** Structure diagram of the test bench. Note: (1) straw conveyor belt; (2) safety box; (3) motor; (4) support stand; (5) SCTD.

**Table 2.** Test factors and levels.

| Level | $s_d$/mm | $\theta c$/° |
|---|---|---|
| $+\gamma$ | 360 | 0 |
| 1 | 370 | 2 |
| 0 | 390 | 7.5 |
| $-1$ | 410 | 13 |
| $-\gamma$ | 420 | 15 |
| $\Delta i$ | 20 | 5.5 |

## 3. Results and Analysis

### *3.1. WB Statics Results and Analysis*

The WB model was solved, and the stress, strain, and deformation cloud diagrams are shown in Figure 12. Different colors in the figure represent different effects after being subjected to force. Blue indicated a smaller value, and red indicated a more significant value. From Figure 12a,b, it can be seen that the maximum stress and strain on the windward side of the WB are 43.73 MPa and 0.023, respectively, which were less than the yield strength of the material. At the same time, it was found that the bending angle of the windward side has a stress concentration phenomenon under the load; a triangular transition can be appropriately made at this position when machining parts to avoid damage in the actual work. Figure 12c showed that the maximum deformation of the windward side under load occurs at one end away from the mounting side. The maximum deformation is 0.0044 mm, which can meet the actual strength requirements.

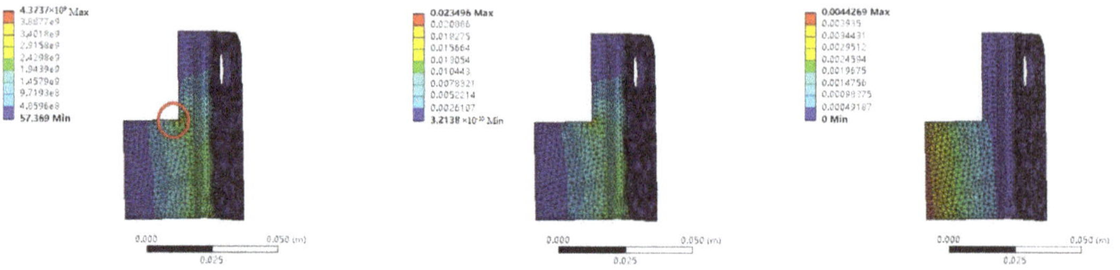

(a) (b) (c)

**Figure 12.** Stress, strain, deformation cloud map. (**a**) Stress analysis cloud diagram; (**b**) strain analysis cloud diagram; (**c**) total deformation analysis cloud diagram.

### 3.2. *Analysis of Installation Parameters of WB Based on CFD*

#### 3.2.1. The Influence of the Installation Direction of the WB on the Surrounding Airflow

The WB had two installation forms when it cooperated with the moving knife, forward installation and reverse installation. The installation structure is shown in Figure 13. When the forward installation was adopted, the windward side of the WB was consistent with the rotating front end surface of the moving knife. When the reverse installation was used, the windward side of the WB was consistent with the rotating rear end surface of the moving knife.

In order to more directly and accurately study the airflow conditions around the WB in different installation directions, the rotation model with the WB installed was further simplified, as shown in Figure 14. Forward and reverse installation can be achieved by changing the rotation of the crushing knife shaft in the model. Figure 15 showed the cloud diagram of the surrounding airflow when the WB were installed forward and reverse. The maximum flow velocity of the fluid was 64.35 m/s when the forward installation was adopted and 62.78 m/s when the reverse installation was adopted, and the difference in flow velocity was insignificant. It can be seen from the figure that the streamline in the rotation range of the WB was relatively uniform when the forward installation was adopted. During the reverse installation, there was a significant difference in the number of streamlines before and after the WB, which may cause irregular movement of the straw during the crushing process. The partial enlarged image showed multiple vortex areas formed between the windward side and the moving knife when the reverse installation was adopted. During the actual crushing operation, the local vortex area would cause the straw to flow unsmoothly so that the straw can hang around the WB. Compared to reverse installation, when the WB was installed in the forward direction, the airflow movement near the windward side was relatively smooth since the windward side does not form a fluid motion space with the moving blade. Therefore, according to the above analysis, the forward installation was adopted when the WB and the moving knife were installed to achieve a better wind effect.

**Figure 13.** Installation form of wind blades. (**a**) Forward installation; (**b**) reverse installation.

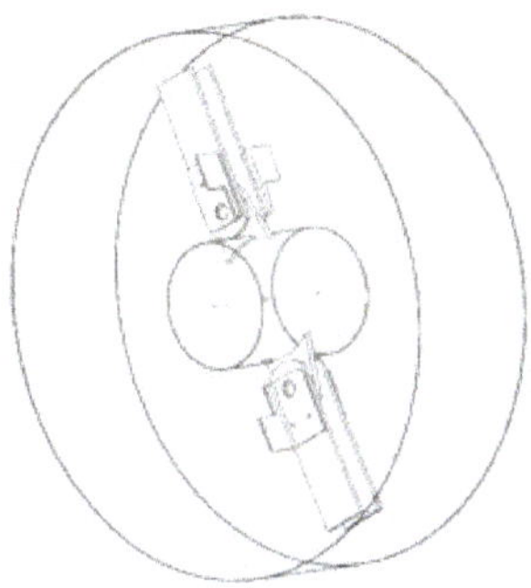

**Figure 14.** Rotation model of wind blade.

**Figure 15.** The surrounding airflow cloud diagram when the WB was installed in the forward and reverse directions. (**a**) Forward installation; (**b**) reverse installation.

3.2.2. The Influence of the Number of WB on the Flow Field

The number of WB installed affects the velocity and pressure of the flow field in the CC, thus affecting the performance of the SCTD [27]. In this paper, referring to the number of moving knives, the installation quantities ($E$) of WB were set to 0, 20 (half the number of installations), and 40 (all installations), respectively. The installation of WB and moving knives adopted forward installation. When $E$ was 20, the uniformity of airflow distribution was considered, and the spaced distribution of WB was adopted. The structure diagram is shown in Figure 16.

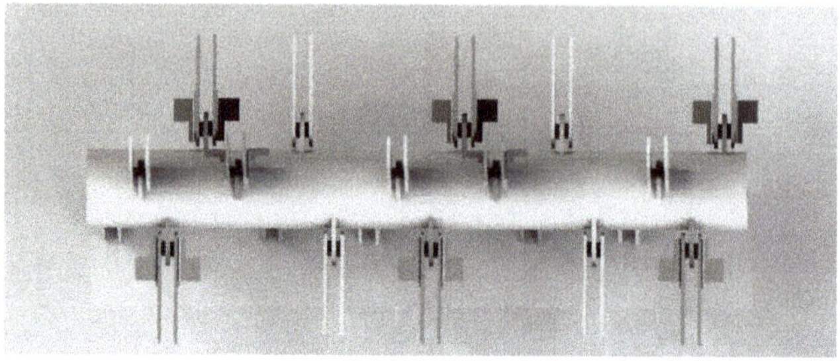

**Figure 16.** Distribution of wind blades when $E = 20$.

Pressure Change at the Inlet

When the crushing knife shaft rotated at high speed, a negative pressure zone would be formed at the inlet of the CC [25]. The larger the value of the negative pressure zone, the more conducive to feeding straws, which can effectively avoid the problem of straw jam and poor feeding. Figure 17 shows the pressure cloud diagram at the inlet under the installed number of different WB. It is known that whether the WB was installed or not would form a negative pressure area at the inlet of the CC. However, when the number of installed WB increases, the high negative pressure area of the inlet begins to increase. When $E = 0$, the maximum negative pressure at the chamber inlet was 14.59 pa, when $E = 20$, it was 17.94 pa, and when $E = 40$, it was 21.86 pa. Therefore, it can be seen that with the increased number of installed WB, the absolute value of the negative pressure at the feeding inlet of the CC also gradually increased. It can be seen from the above analysis that the WB can improve the negative pressure effect at the inlet of the CC, and the negative pressure effect at the inlet was more significant as the number of installed WB increased.

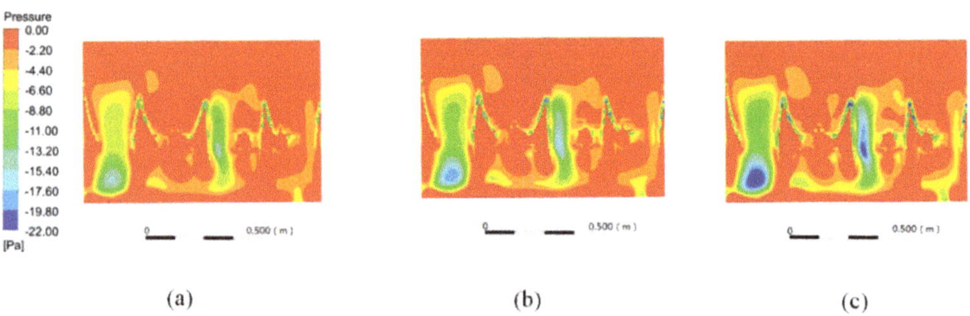


**Figure 17.** Pressure cloud diagram at the inlet under different WB installation numbers. (**a**) $E = 0$; (**b**) $E = 20$; (**c**) $E = 40$.

Analysis of Airflow Fluidity in CC

Figure 18 was the velocity streamline diagram of P1, P2 and P3 sections under different WB installation numbers. It can be seen from the figure that under the same installation parameters, the streamline diagrams between different sections were also different. The airflow streamlines of the P1 and P3 sections on both sides were relatively smooth and stable, and the airflow streamlines of the middle P2 section were relatively fluctuating. The main reason was that the airflow in the CC had a specific axial movement phenomenon., which made the airflow field between the shafts more complicated than the airflow field near the side plate so that the airflow would fluctuate to a certain extent.

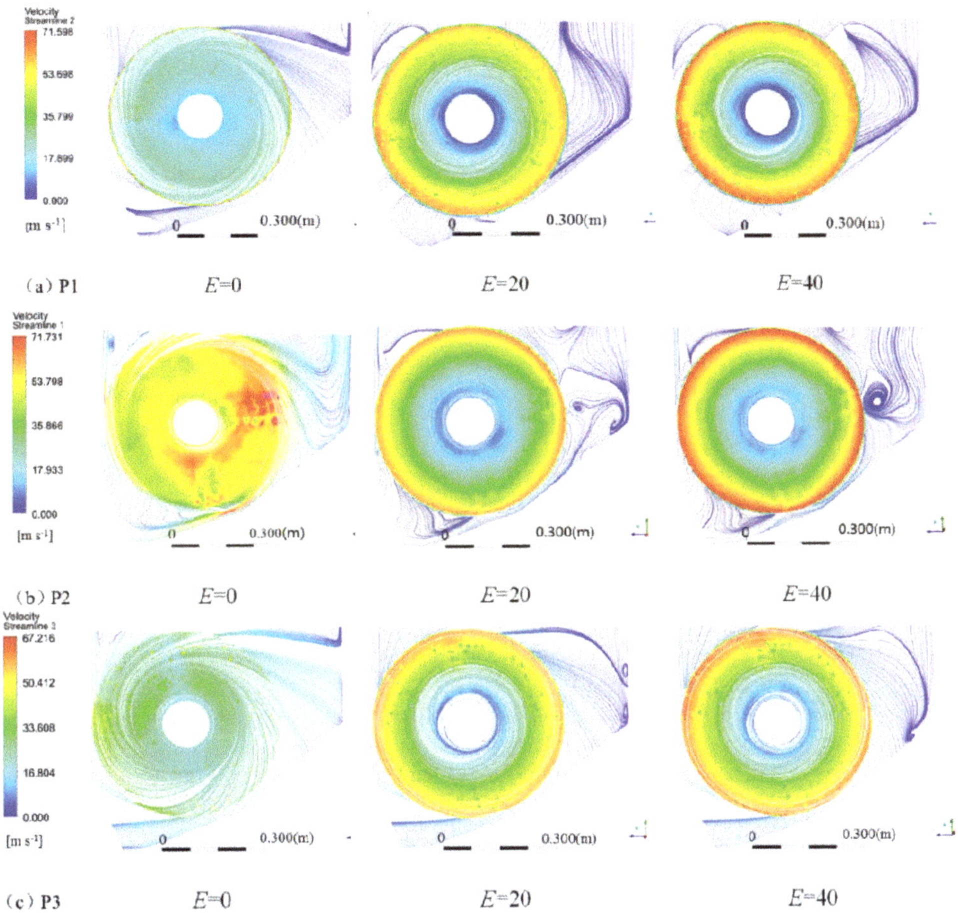


**Figure 18.** Streamline diagram of each section under different number of wind blades installed. (**a**) P1; (**b**) P2; (**c**) P3.

It can be seen from Figure 18a that when the number of installed WB increased, the fluid velocity in the rotating domain on the P1 section increased gradually. When $E$ = 0, the maximum flow velocity of the fluid in the P1 section was 55.18 m/s, $E$ = 20 was 62.81 m/s, and $E$ = 40 was 71.60 m/s. When $E$ = 0, fluid chasm and eddy currents appeared at the inlet of the CC. When the installed number of WB increased from 0 to 20, the velocity streamlines chasm at the inlet of the CC, and the eddy current in the upper right corner disappeared. However, the fluid fluctuation in the CC increased, the streamline density at the outlet decreased, and the streamline area increased. When the installed number of WB increased from 20 to 40, the fluid streamline fluctuation in the CC increased. The streamlined area and density at the outlet did not change significantly. The installation of WB increased the fluid velocity in the left area of the CC but also caused fluid instability, which means that when the WB was not installed, the straw was more concentrated and thrown away from the CC along the bottom plate. The straw was thrown from the entire outlet after installation. Therefore, the number of WB significantly influenced the fluid flow in the left area of the CC.

It can be seen from Figure 18b that when the number of installed WB increased, the fluid velocity in the rotating domain on the P2 section increased gradually. When $E = 0$, the maximum flow velocity of the fluid in the P2 section was 64.82 m/s, $E = 20$ was 65.77 m/s, and $E = 40$ was 71.73 m/s. When $E = 0$, a vortex phenomenon appeared in the fluid at the inlet of the CC. When the number of installed WB increased from 0 to 20, the vortex on the left side of the CC inlet disappeared, and the vortex on the right side still existed. A new vortex appeared on the right side of the rotating domain, and the fluid dynamics at the outlet of the CC remained unchanged. When the installed number of WB increased from 20 to 40, the vortex at the inlet of the CC still existed, the vortex on the right side of the rotating domain was enhanced, and the fluid dynamics at the outlet remained unchanged. The installation of WB increased the fluid velocity in the middle area of the CC but also increased the fluid instability, which had little effect on the outlet fluid flow. Therefore, the number of WB significantly influenced the fluid flow in the middle area of the CC.

It can be seen from Figure 18c that when the number of installed WB increased, the fluid velocity in the rotating domain on the P3 section increased gradually. When $E = 0$, the maximum flow velocity of the fluid in the P3 section was 49.97 m/s, $E = 20$ was 64.59 m/s, and $E = 40$ was 67.22 m/s. When $E = 0$, fluid chasm and eddy currents appeared at the inlet of the CC. When the number of WB installed was increased from 0 to 20, the vortex of the inlet disappeared. The vortex on the right side still existed and moved down to the right side of the rotating domain, and the fluid dynamics at the outlet of the CC remained unchanged. When the installed number of WB increased from 20 to 40, the vortex in the CC continued to move down, but the range decreased, and the fluid dynamics at the outlet remained unchanged. The installation of WB increased the fluid velocity in the right area of the CC but also increased the fluid instability to a certain extent and had little effect on the outlet fluid flow. Therefore, the number of WB significantly influenced the fluid flow in the right area of the CC.

When a fluid chasm occurred at the inlet of the CC, the fluid would concentrate on the left side of the inlet, which was not conducive to the direct feeding of straw into the CC. The above analysis found that installing WB can improve the fluid chasm phenomenon; the more installations, the better the effect. At the same time, increasing the number of WB may reduce the stability of the chamber's fluid flow and generate local eddy currents, which would have a particular impact on the flow in the CC. Whether the effect affects the operation performance of the device still needs to be verified by tests.

Variation of Exit Velocity

The fluid velocity at the outlet of the CC directly determined the kinetic energy when the straw was thrown, thus affecting the throwing characteristics of the device. Figure 19 shows the velocity cloud diagram at the outlet under different WB installation numbers. It can be seen from the figure that with the increase in the number of WB, the high-velocity area at the outlet gradually increased. When $E = 0$, the maximum fluid velocity at the outlet was 9.91 m/s; when $E = 20$, the maximum fluid velocity at the outlet was 10.59 m/s, and when $E = 40$, the maximum fluid velocity at the outlet was 11.46 m/s. During simulation, set ten monitoring points at equal intervals on the lower edge of the exit. The average flow velocity at the outlet of different WB numbers was 5.08 m/s, 8.57 m/s and 10.19 m/s, respectively. When $E = 20$ and $E = 40$, compared with $E = 0$, the maximum export speed increased by 6.86% and 15.74%, and the average speed increased by 68.70% and 100.59%. Therefore, installing WB significantly improved the fluid velocity at the outlet, which increased gradually with the number of installations.

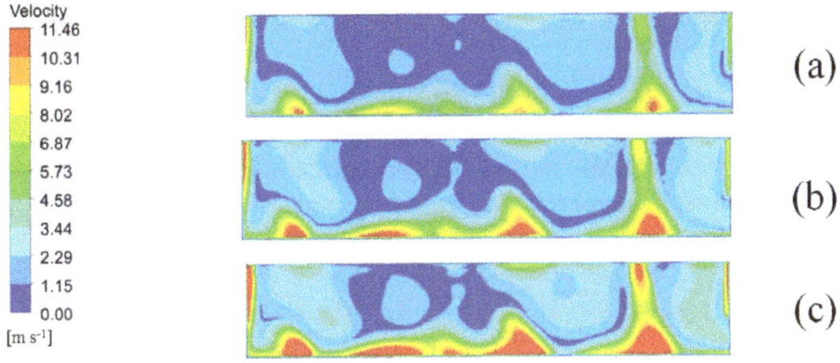


**Figure 19.** Velocity cloud diagram at the outlet with different WB installed. (**a**) $E = 0$; (**b**) $E = 20$; (**c**) $E = 40$.

Through the flow field analysis of the inlet, outlet, and chamber of the CC, it can be known that installing WB benefits the feeding and throwing of straw; at the same time, it will give the straw a higher discharge speed. Considering a large amount of rice straw and the significant difference between the throwing width of the combined harvester's original SCTD and the header width, a better straw feeding effect and higher throwing speed are required. Therefore, the number of WB should be fully installed, $E = 40$.

### *3.3. Analysis and Optimization of Test Results*

The test effect diagram is shown in Figure 20, and the test results are shown in Table 3. The test data were processed by the software Design-expert 13.0 [31,32].

**Figure 20.** Test effect diagram.

**Table 3.** Test results.

| Test No. | $s_d$/mm | $\theta_c$/° | $D$/m |
|---|---|---|---|
| 1 | −1 | −1 | 4.05 |
| 2 | 1 | −1 | 4.58 |
| 3 | −1 | 1 | 4.49 |
| 4 | 1 | 1 | 4.39 |
| 5 | $-\gamma$ | 0 | 4.11 |
| 6 | $+\gamma$ | 0 | 4.71 |
| 7 | 0 | $-\gamma$ | 3.88 |
| 8 | 0 | $+\gamma$ | 4.38 |
| 9 | 0 | 0 | 4.41 |
| 10 | 0 | 0 | 4.43 |
| 11 | 0 | 0 | 4.34 |
| 12 | 0 | 0 | 4.29 |
| 13 | 0 | 0 | 4.45 |

Through the analysis of test data, the variance table of straw throwing width ($D$) is shown in Table 4. According to the content in the table, the experimental model was significant ($p < 0.01$). Among the main factors in the test, the ALD ($s_d$) had a more significant impact on the measurement index. In the interaction term, the ALD ($s_d$) and the IAD ($\theta_c$) significantly impacted the test index. The two factors had primary and secondary effects on the throwing width. The order was $x_1 > x_2$. After excluding the insignificant items in the variance analysis, the analysis was carried out again. The results are shown in 4. The multiple regression fitting of the test results was carried out, and the quadratic regression equation of each factor and the test index was obtained as follows:

$$Y = 4.41 + 0.1598x_1 + 0.1196x_2 - 0.1575x_1x_2 - 0.1053x_2^2 \tag{12}$$

where $x_1$ is the ALD, mm; $x_2$ is the IAD, °.

**Table 4.** Variance of throwing width.

| Variation Source | | Sum of Squares | Degree Freedom | Mean Square | *F* Values | *p* Values |
|---|---|---|---|---|---|---|
| Model | | 0.51/0.50 | 5/4 | 0.10/0.12 | 7.98/9.92 | 0.0083 ***/0.0034 *** |
| $x_1$ | | 0.20/0.20 | 1/1 | 0.20/0.20 | 16.06/16.33 | 0.0051 ***/0.0037 *** |
| $x_2$ | | 0.11/0.11 | 1/1 | 0.11/0.11 | 9.00/9.15 | 0.0199 **/0.0164 ** |
| $x_1$ $x_2$ | | 0.099/0.010 | 1/1 | 0.099/0.0010 | 7.80/7.93 | 0.0268 **/0.0226 ** |
| $x_1^2$ | | 0.011 | 1 | 0.011 | 0.87 | 0.3821 |
| $x_2^2$ | | 0.070/0.079 | 1/ | 0.070/0.079 | 5.48/6.27 | 0.0517 */0.0367 ** |
| Residual | Lack of fit | 0.071/0.082 | 3/4 | 0.024/0.021 | 5.29/4.59 | 0.0706/0.0847 |
| | Pure error | 0.018/0.018 | 4/4 | 0.0045/0.0045 | | |
| Total sum | | 0.60/0.60 | 12/12 | | | |

Note: The numbers after "/" represent the test analysis results after excluding insignificant items; "***" means extremely significant ($p < 0.01$); "**" means significant ($0.01 < p < 0.05$); "*" means significant ($0.05 < p < 0.1$).

According to the regression analysis results, we used Design-expert13.0 software to draw the response surface of the influence of factor interaction on the test index, as shown in Figure 21. It can be seen from the figure that when the ALD and IAD increased, the throwing width first increased and then decreased.

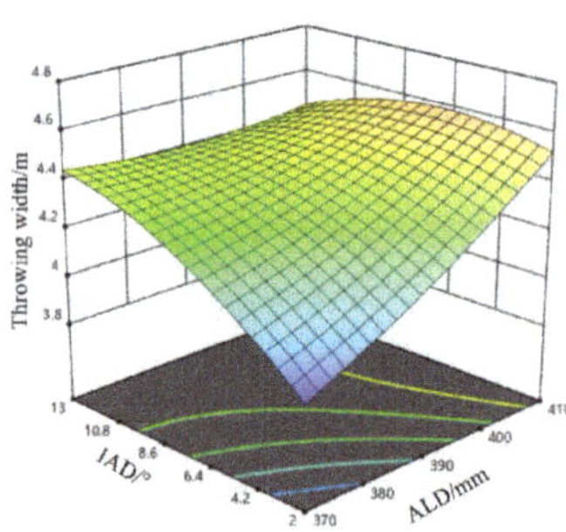


**Figure 21.** Two-factor response surface plot.

In order to obtain the optimal combination of the two parameters, the optimization module in the Design-expert13.0 software was used to solve the regression equation of the throwing width, and the optimization constraints and objectives were selected according to the constraints of factors and actual operation needs [33]. The constraint function is as follows.

$$\begin{cases} Y = 4.5 \\ 360 \leq x_1 \leq 420 \\ 0 \leq x_2 \leq 15 \end{cases} \tag{13}$$

After solving, the result was that the ALD was 401.05 mm, and the IAD was 9.21°. For the convenience of testing and adjustment, the factors were rounded: the ALD was 400 mm and the IAD was 9° after rounding.

### 3.4. Test Verification

In order to verify the reliability of the optimization, a test bench verification test was carried out on the optimized parameter combination in November 2020, and the operating condition test method was consistent with the previous design. The ALD is designed to be 400 mm, and the IAD was 9°. The test was carried out three times, the results are shown in Table 5. T The average value of the test results was divided by the harvester's header width (4.5 m) to obtain the matching degree.

**Table 5.** The effect of throwing width after SCTD optimization.

| Items | Test Results | Average Value | Matching Degree/% |
|---|---|---|---|
| Throwing width D/m | 4.43<br>4.39<br>4.45 | 4.42 | 98.44% |

With the optimized SCTD, the throwing width had significantly been improved to 4.42 m, and the matching degree with the John Deere c100 combine harvester's header width can reach 98.44%, with a high matching degree. Improvement was due to the fact that the WB designed in this chapter increased the speed of the straw when it left the crushing device. On this basis, the related parameters of the deflector of the TD were matched and optimized, thus ensuring the matching degree of straw throwing. During the whole experiment process, the improved device ran smoothly, and there was no straw clogging problem, which showed that increasing the eddy current generated by the WB in the CC had little effect on the smooth operation of the device.

## 4. Conclusions

In order to improve the matching degree with the header width of the combine harvester, this paper innovatively designed the WB, optimized the relevant parameters of the deflector, and improved the throwing performance of the SCTD.

(1) The structural design of the WB is divided into a windward side and a mounting side. The finite element carried out the static analysis of the WB. It was determined that the maximum stress and strain and the maximum deformation on the windward side were 43.73 MPa and 0.023 and 0.0044 mm, respectively, which could meet the material work requirements. Based on CFD, the influence of WB installation parameters on the CC fluid was analyzed. The study found that with the number of WB installed, the negative pressure effect at the CC's inlet increased, and the velocity at the outlet increased. However, the fluid stability in the CC tended to deteriorate. There would be eddy currents around the windward side during reverse installation. Finally, the analysis determined that the number of WB to be installed was 40, and the forward installation was adopted.

(2) The non-guided stage and the guided throwing stage of straw entering the TD were analyzed. The analysis showed that the increased straw speed in the non-guided stage would make the collision point of the straw and the TD move backwards, increasing the collision angle and reducing the collision kinetic energy loss. In the guided throwing stage, the IAD ($\theta_c$) and the ALD ($s_d$) affected the speed direction and kinetic energy loss when the straw was thrown, thereby affecting the throwing width of the straw.

(3) To determine the appropriate combination of IAD ($\theta_c$) and ALD ($s_d$) when throwing rice straw, an optimization test of the SCTD was carried out. The results showed that when the deflector's IAD and ALD increased, the throwing width first increased and then decreased. The optimization took the width of the harvester header as the goal to solve the simulation equation. It obtained the optimal solution of the parameters: an ALD of

400 mm and an IAD of 9°. It was verified through tests that the throwing width under the optimal parameter combination was 4.42 m, and the matching degree with the width of the harvester header was 98.44%. The throwing effect was good and met the actual needs. At the same time, it was found in the test that the adjustment of the deflector of the TD was complicated, and the throwing width was easily affected by environmental changes (wind direction changes, terrain changes). The following research should further improve these problems.

**Author Contributions:** Conceptualization, J.W. and X.W.; methodology, J.W.; software, J.W.; validation, J.W., D.L. and M.Z.; formal analysis, B.D.; investigation, J.W.; resources, X.W.; data curation, J.W.; writing—original draft preparation J.W.; writing—review and editing, H.L. and Q.W.; visualization, J.W.; supervision, J.H. and C.L.; project administration, X.W.; funding acquisition, X.W. All authors have read and agreed to the published version of the manuscript.

**Funding:** This research was funded by Chinese Academy of Agricultural Sciences: 2016YFD030009-039; And the APC was funded by Chinese Universities Scientific Fund: 2021TC011.

**Institutional Review Board Statement:** Not applicable.

**Informed Consent Statement:** Not applicable.

**Data Availability Statement:** Not applicable.

**Conflicts of Interest:** The authors declare no conflict of interest.

## References

1. He, J.; Li, H.; Chen, H.; Lu, C.; Wang, Q. Research progress of conservation tillage technology and machine. *Trans. Chin. Soc. Agric. Mach.* **2018**, *49*, 1–19. [CrossRef]
2. Zhang, Z.; McHugh, A.D.; Li, H.; Ma, S.; Wang, Q.; He, J.; Zhen, K. Global overview of research and development of crop residue management machinery. *Appl. Eng. Agric.* **2017**, *33*, 329–344. [CrossRef]
3. Sun, N.; Dong, W.; Wang, X.; Li, H.; Meng, Y.; Wei, Z. Comprehensive effect of rice harvesting straw treatment methods in northeast rice region. *Trans. Chin. Soc. Agric. Mach.* **2020**, *51*, 69–77. [CrossRef]
4. Botta, G.F.; Tolón-Becerra, A.; Lastra-Bravo, X.; Hidalgo, R.; Rivero, D.; Agnes, D. Alternatives for handling rice (*Oryza sativa* L.) straw to favor its decomposition in direct sowing systems and their incidence on soil compaction. *Geoderma* **2015**, *239–240*, 213–222. [CrossRef]
5. Allmaras, R.R.; Douglas, C.L.; Rasmussen, P.E.; Baarstad, L.L. Distribution of small grain residue produced by combines. *Agron. J.* **1985**, *77*, 730–734. [CrossRef]
6. Siemens, M.C.; Wilkins, D.E. Effect of residue management methods on no-till drill performance. *Appl. Eng. Agric.* **2006**, *22*, 51–62. [CrossRef]
7. Jia, H.; Jiang, X.; Guo, M.; Liu, X.; Wang, L. Design and experiment of V-L shaped smashed straw blade. *Trans. Chin. Soc. Agric. Eng.* **2015**, *31*, 28–33. [CrossRef]
8. Zhen, Z.; He, J.; Li, H.; Diao, P.; Wang, Q.; Zhang, X. Design and experiment of straw-chopping device with chopping and fixed knife supported slide cutting. *Trans. Chin. Soc. Agric. Mach.* **2016**, *47*, 108–116. [CrossRef]
9. Liu, P.; He, J.; Lou, S.; Wang, Y.; Zhang, Z.; Lin, H. Design and experiment for dynamic supporting type maize straw chopping retention device with different rotation speeds of disc blade. *Trans. Chin. Soc. Agric. Mach.* **2021**, *52*, 41–50. [CrossRef]
10. Han, X.; Zhang, J.; Gao, Y.; Zhang, J.; Yu, B.; Geng, A. Design and test of saw disc type straw crushing and returning machine. *J. Chin. Agric. Mech.* **2017**, *38*, 7–11. [CrossRef]
11. Sun, N.; Wang, X.; Li, H.; He, J.; Wang, Q.; Wang, J.; Liu, Z.; Wang, Y. Design and experiment of differential sawing rice straw chopper for turning to field. *Trans. Chin. Soc. Agric. Eng.* **2019**, *35*, 267–276. [CrossRef]
12. Wang, J.; Wang, X.; LI, H.; He, J.; Lu, C.; Liu, D. Design and experiment of rice straw chopping device for agitation sliding cutting and tearing. *Trans. Chin. Soc. Agric. Mach.* **2021**, *52*, 28–40. [CrossRef]
13. Chen, L.; Wang, L.; Zhang, J.; Zhang, J.; He, C. Design of 1JHSX-34 straw crusher for whole-feeding combine harvesters. *Trans. Chin. Soc. Agric. Eng.* **2011**, *27*, 28–32. [CrossRef]
14. Zhang, M. *Research on Anti-Blocking Device of Straw Returning Plow*; China Agricultural University: Beijing, China, 2020.
15. Sun, N.; Dong, W.; Wang, X.; Li, H.; Peng, X.; Wang, J. Performance of straw returning equipment under different mechanized straw returning pattern in northeast rice area. *Trans. Chin. Soc. Agric. Mach.* **2018**, *49*, 68–75. [CrossRef]
16. Schillinger, W.F.; Smith, T.A.; Schafer, H.L. Chaff and straw spreader for a plot combine. *Agron. J.* **2008**, *100*, 398–399. [CrossRef]
17. Korn, C.; Fehrmann, J.; Herlitzius, T.; Flanhardt, M.; Acimas, A. Development of a straw chopper for combines for increased working width. *Landtechnik* **2012**, *67*, 11–16.

18. Lu, J.; Liao, M.; Zhao, M.; Cao, L.; Wang, Y. Research on rice and wheat straw throwing and returning machine for half-feed combine harvester. *Agriculture* **2014**, *40*, 13–16. [CrossRef]
19. Wang, X.; Jin, C.; Xu, J.; Ni, Y. Study and design optimization on the angle adjustments of the straw spreader panel in a combined wheat harvester. *J. Agric. Mach. Res.* **2019**, *8*, 127–135. [CrossRef]
20. Li, Y.; Li, C.; Zhang, J. Analysis of the influence of different adjustment methods on the economy of induced draft fan. *J. Eng. Therm. Energy Power.* **2021**, *36*, 39–43. [CrossRef]
21. Chen, D. *Mechanical Design Handbook*, 6th ed.; Chemical Industry Press: Beijing, China, 2017.
22. Zhu, H. *ANSYS 15.0 Geometric Modeling and Meshing Practical Guide*; People Post Press: Beijing, China, 2014.
23. Wu, X.; Li, H.; Shi, Z. Simulation of air drag coefficient of pig model based on OpenFOAM. *J. China Agric. Univ.* **2021**, *26*, 197–205. [CrossRef]
24. Tutar, M.; Oguz, G. Large eddy simulation of wind flow around parallel buildings with varying configurations. *Fluid Dyn. Res.* **2002**, *31*, 289–315. [CrossRef]
25. Zhang, Z. *Research on Corn Straw Chopping and Spreading Machine Design and Dynamic Characteristic of Straw*; China Agricultural University: Beijing, China, 2018.
26. Ren, B.; Zhong, W.; Jin, B. Study on the Drag of a Cylinder-Shaped Particle in Steady Upward Gas Flow. *Ind. Eng. Chem. Res.* **2011**, *50*, 7593–7600. [CrossRef]
27. Fang, M.; Yu, Z.; Zhang, W.; Liu, W.; Bie, Z.; Song, J. Analysis and experiments of the movement process for the shredded material of disc knife chaff cutter. *Trans. Chin. Soc. Agric. Eng.* **2021**, *37*, 76–84. [CrossRef]
28. Wu, F.; Xu, H.; Gu, W.; Chen, Y.; Shi, L.; Hu, Z. Improvement of straw transport device for straw-smashing back-throwing type multi-function no-tillage planter. *Trans. Chin. Soc. Agric. Eng.* **2017**, *33*, 18–28. [CrossRef]
29. Ge, Y. *Experiment Design Method and Application of Design-Expert Software*; Harbin Institute of Technology Press: Harbin, China, 2014.
30. Wang, Y.; Rong, G.; Li, H.; Wang, Q.; He, J.; Lu, C. Design and parameter optimization of vertical driving-type surface rotary tillage machine. *Trans. Chin. Soc. Agric. Eng.* **2019**, *35*, 38–47. [CrossRef]
31. Zhang, H.; Chen, X.; Yan, L.; Yang, S. Design and test of master-slave straw returning and residual film recycling combine machine. *Trans. Chin. Soc. Agric. Eng.* **2019**, *35*, 11–19. [CrossRef]
32. Shi, N.; Chen, H.; Wei, Z.; Hou, S.; Zhou, Z.; Wang, X. Design and experiment of stalk returning proportion adjusting device for corn original stubble. *Trans. Chin. Soc. Agric. Eng.* **2020**, *36*, 11–22. [CrossRef]
33. Luo, W.; Hu, Z.; Wu, F.; Gu, F.; Xu, H.; Chen, Y. Design and optimization for smashed straw guide device of wheat clean area planter under full straw field. *Trans. Chin. Soc. Agric. Eng.* **2019**, *35*, 1–10. [CrossRef]

agriculture

MDPI

Article

# Design of Cotton Recovery Device and Operation Parameters Optimization

Hezheng Wang [1], Silin Cao [2,*], Yongrui Liu [1], Yuxin Yang [3], Xiangyu Meng [1] and Peng Ji [1]

1 College of Mechanical and Electrical Engineering, Shihezi University, Shihezi 832000, China
2 Mechanical Equipment Research Institute, Xinjiang Academy of Land Reclamation Sciences, Shihezi 832000, China
3 Institute of Agricultural Mechanization, Xinjiang Academy of Agricultural Sciences, Urumqi 830091, China
* Correspondence: caosilin@shzu.edu.cn

**Abstract:** This research aims to optimize the working parameters of the sawtooth-type recovery device for cotton fallen on the ground to enhance cotton's recovery effect. Firstly, the cotton-picking mechanism and cotton unloading mechanism of the cotton recovery device were designed. The movement trajectory of the serrated tooth end of the designed device, the cotton non-missing picking condition, and the cotton unloading condition were noted. Secondly, virtual simulation technology developed a model of the interaction process between the picking equipment and the soil. To determine the optimal combination of operating parameters for the recovery device, a three-factor, three-level response surface optimization test was conducted using Box–Behnken's central combination method with operating machine speed, spacing between serrated discs, and serrated disc speed as the test factors, and the picking and impurity rate as the test indexes. In addition, a response surface regression model was developed to analyze the effects of the selected factors on the recovery unit, and each factor was optimized. When the picking and impurity rates were 79.09% and 35.12%, respectively, the optimal operating speed of the machine was 0.96 m/s, the spacing of the serrated discs was 40 mm, and the speed of the serrated discs was 68 rpm. The relative error between the experimental findings and the theoretical optimized values was less than 5%, and the optimized working parameters were reliable. This study can provide a reference for the device used to recover cotton that has fallen to the ground.

**Keywords:** cotton recovery device; EDEM; virtual simulation; parameters optimization; agricultural machinery

**Citation:** Wang, H.; Cao, S.; Liu, Y.; Yang, Y.; Meng, X.; Ji, P. Design of Cotton Recovery Device and Operation Parameters Optimization. *Agriculture* **2022**, *12*, 1296. https://doi.org/10.3390/agriculture12091296

Academic Editors: Mustafa Ucgul and Chung-Liang Chang

Received: 31 July 2022
Accepted: 20 August 2022
Published: 24 August 2022



## 1. Introduction

Cotton is a globally significant crop that can be used as the textile industry's raw material. In developing countries, cotton production is essential for economic growth and improving living conditions [1–3]. There are cotton-growing regions throughout Asia, Africa, the United States, Europe, and Oceania, but they are located primarily in Asia and the Americas [4]. China is the world's largest consumer and the second-largest cotton producer, and cotton production is vital to its national economy [5–7]. China's leading cotton-producing region is in Xinjiang. In recent years, the rate of cotton harvesting by machine has steadily increased, and mechanical cotton harvesting has grown popular [8–11]. The quality guideline for cotton picker operations mandates a loss rate of less than 5 percent for mechanical harvesting [12]. In practice, however, cotton falls to the ground due to the effects of defoliation and ripening, cotton-picking machine head contact, weather, and numerous other causes. According to early studies, approximately 3 percent of cotton falls unharvested yearly. According to statistics [13], in 2021, the cultivated cotton area in Xinjiang was around $2.51 \times 10^6$ hm$^2$, and the yield reached $5 \times 10^6$ tons. Approximately $1.5 \times 10^5$ tons of cotton are left in the cotton field after a 3% loss, and the amount of cotton that has fallen is ample. Consequently, one of the challenges of cotton resource exploitation is the effective recovery of cotton distributed over the ground.

Currently, the existing ground recovery technologies for collecting cotton fallen on the ground are primarily pneumatic, mechanical, and mechanical combinations in the research and development stage [14]. Lawrence A. Lehman et al. [15] invented a gripper-type fallen cotton pickup device equipped with a flexible belt with transverse grooves that open and close when the belt runs through the circular surface of the front roller, thereby grasping the fallen cotton in the furrow while wedging in a significant amount of soil and broken branches. The CXC-1.2-type horizontal spindle picker manufactured in the former Soviet Union was fitted with an airflow picker at the rear of the machine. This airflow picker collected cotton that touched the ground through the suction pipe installed on the fan. However, cotton leaves are easier to inhale, increasing the number of impurities in cotton. Jianlong et al. [16,17] invented an air-suction-type picking machine for cotton fallen on the ground based on the principle of negative pressure suction picking; here, the front-end comb teeth scrape the cotton off the stalks as the machine advances, while the blowing ports on both sides of the cotton collection platform blow the cotton fallen on the ground to the center of the platform and then suction the cotton into the collection box using the suction tube. The machine's structure is simple, but the impurities, such as cotton sticks and leaves, are scraped to the ground's surface, making subsequent cleaning more difficult. Li-Guang et al. [18] invented a pickup roller with retractable pickup teeth for cotton that has fallen to the ground. When the pickup teeth move to the surface, they extend and hook cotton, and when they hook and cling onto the back of the pickup roller, they retract, releasing the cotton. Its simplistic structure will pick up mulch and drip irrigation tape and wrap it around the drum, causing machine failure.

Regarding existing technologies, a pneumatic cotton recovery method inhales a significant amount of crop waste, cotton stalks, soil, and other impurities during operation. As a result, cotton includes a high rate of impurities and thus reduces the unit's capacity to run continuously. In addition, the uneven surface of the field makes it impossible to recover the majority of cotton that has fallen to the ground using a mechanical approach.

Aiming at the characteristics of the cotton machine harvesting in Xinjiang (e.g., the enormous volume of cotton fallen on the ground and lack of appropriate recovery equipment) and based on the structure and characteristics of an existing cotton recovery device, a sawtooth-type recovery device was designed. Through theoretical analysis and field testing to determine the machine's reasonable structure and operating parameters, the influence of different factors on the effect of cotton recovery was explored, and a better combination of parameters was obtained. This study provides a theoretical and technological reference for developing cotton recovery equipment.

## 2. Materials and Methods

### 2.1. Structure and Operating Principle of the Cotton Recovery Device

Figure 1 shows the structure of the sawtooth-type recovery device for fallen cotton, which consists mainly of the frame, straw guard, power transmission system, sawtooth roll-tie cotton-picking mechanism, cotton unloading mechanism, cotton collection box, and other components. The tractor places the device on the ground via the suspension, and the ground wheel for each group of cotton-picking parts advances under traction after touching the ground. The forward process achieves topographical adaptation according to the difference in the ground surface undulation. The transmission system includes a hydraulic motor, sprocket, and other components. The device is powered by the tractor's rear output shaft, and the hydraulic motor drives the sawtooth roll-tie-type cotton-picking mechanism and the unloading mechanism. The main technical parameters of the device are shown in Table 1.

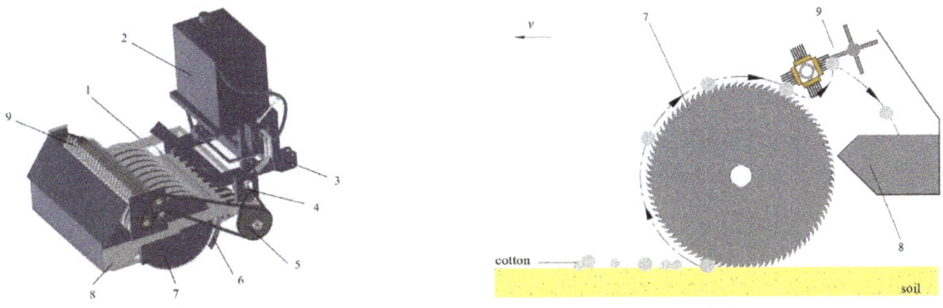


**Figure 1.** Cotton recovery device: 1. frame; 2. hydraulic device; 3. hangers; 4. hydraulic motor; 5. sprockets; 6. straw shield; 7. sawtooth roll-tie cotton-picking mechanism; 8. cotton collection box; and 9. cotton unloading mechanism. (**a**) Structure diagram of the cotton recovery device. (**b**) Schematic diagram of the principle of the cotton recovery device.

**Table 1.** Main technical parameters of the cotton recovery device.

| Parameter | Value |
|---|---|
| Structure form | Traction type |
| Overall dimension (length × width × height)/mm | 1670 × 1240 × 1550 |
| Engine rated power/kW | 51.5 |
| Effective working width/mm | 1000 |
| Picking rate/% | ≥75 |
| Impurity rate/% | 40≤ |

When the device is in operation, the serrated disk rapidly rotates clockwise; the end of the teeth of the serrated disk first contacts the cotton fallen on the ground and gradually deepens as the machine advances; the serrated teeth completely hook the cotton fiber and turn backward; and the cotton boll shell, cotton peach, and large cotton rod are thrown out under the action of the centrifugal force. Since the friction coefficient of the cotton differs between the brush roller and the serrated disc, the cotton on the serrated disc can be brushed off when the hooked cotton continues to rotate backward and comes into contact with the brush roller rotating at reverse high speed [19,20]. In addition, the wind force generated by the brush roller at high-speed rotation blocks part of the impurities from entering the collection box. Subsequently, the stripping roller rolls the cotton wound on the brush roller to the collection box at the frame's rear to complete the picking up of cotton fallen on the ground.

### 2.2. *Design and Analysis of the Major Components*

#### 2.2.1. Design of the Sawtooth Roll-Tie-Type Cotton-Picking Mechanism

The principal mechanism of the device is a sawtooth roll-tie-style cotton-picking mechanism. The cotton-picking mechanism comprises ten groups of picking components, including a guard ring, ground wheel, serrated disc, and straw shield (Figure 2). Its primary purpose is picking up, hooking, and transporting dropped cotton to the unloading location. The operational width of the cotton-picking section is 1000 mm, the diameter of the ground wheel is 610 mm, and the sawtooth disc circumference has 90 saw teeth evenly spaced. In addition, the serrated disc's top circle and tooth root circle diameters are 600 and 560 mm, respectively.

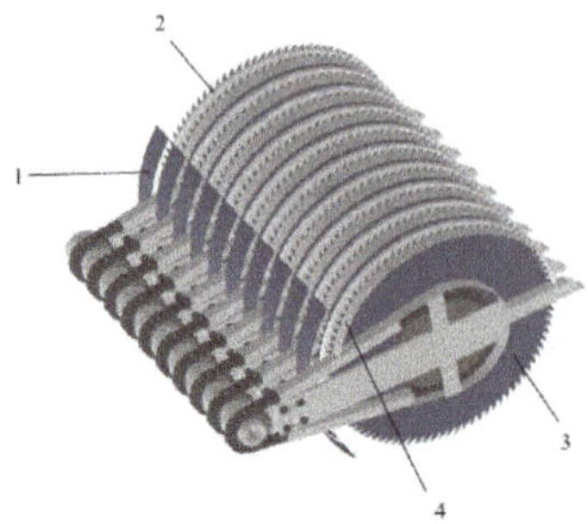


**Figure 2.** Sawtooth roll-tie-type cotton-picking mechanism: 1. straw shield; 2. ground wheel; 3. serrated disc; 4. guard ring.

2.2.2. Analysis of the Motion Characteristics of the Serrated Disc

When the device is operating, the circumferential line speed below the serrated disc is the same as the forward direction. The trajectory of the saw tooth end $E$ point is depicted in Figure 3, with the machine moving forward at a constant velocity $v$, the forward direction for the $x$-axis being positive and the $y$-axis being vertically upward, and the saw tooth disk rotating at a constant angular velocity $\omega$.

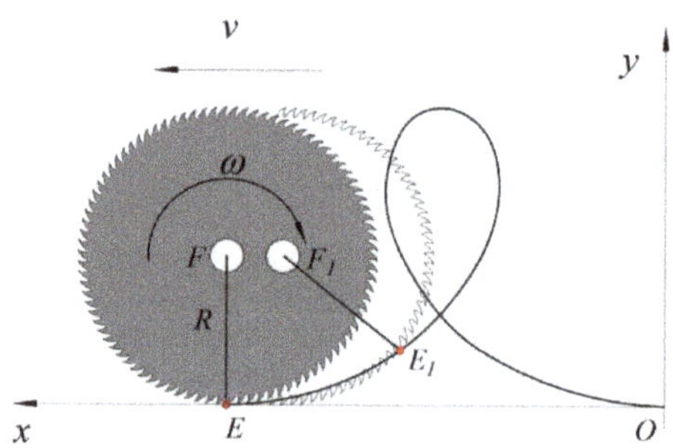


**Figure 3.** Motion trajectory of the serrated tooth end.

The serrated disc moves from point $F_1$ to point $F$ after time $t$, while the sawtooth tooth endpoint $E_1$ rotates to point $E$. If the angle through which the sawtooth tooth end rotates $\alpha(\alpha = \omega t)$, then the sawtooth tooth end $E$ point's equation of motion is:

$$\begin{cases} x = R\sin\omega t + v_0 t \\ y = R - R\cos\omega t \end{cases} \quad (1)$$

The first-order derivative of the above system of equations with respect to time $t$ yields the sawtooth end velocity.

$$\begin{cases} v_x = \frac{dx}{dt} = \omega R\cos(\omega t) + v \\ v_y = \frac{dy}{dt} = \omega R\sin(\omega t) \end{cases} \quad (2)$$

where $R$ is the radius of rotation of the serrated disc, mm; $F$ and $F_1$ are the center of rotation of the serrated disk at different moments.

According to Equation (2), it can be shown that the backward horizontal component speed of the machine is determined by the magnitude of $v_x$, and the backward horizontal component speed ensures the effect of the saw teeth hooking the cotton. When $\omega Rcos(\omega t) + v = 0$, we can obtain:

$$\frac{\omega R}{v} = -\frac{1}{\sin(\omega t)} \tag{3}$$

$$v_1 = \omega R \tag{4}$$

Let the ratio of the serrated disc's circumferential linear velocity $v_1$ to the device's forward velocity $v$ be:

$$\lambda = \frac{v_1}{v} \tag{5}$$

According to Equation (2), when $\lambda < 1$, the serrated tooth end backward velocity is 0, which cannot complete the picking operation; when $\lambda > 1$, the serrated disk rotation at the same time relative to the ground in the picking area has backward velocity; in this movement, the landed cotton is hooked up with the serrated upward movement. Therefore, the value of $\lambda$ can pick up the landing cotton, and thus has an important impact; if the value is too small, it is easy to miss the phenomenon; if the value is too large, the impact of the serrated teeth on the landing cotton increases, resulting in the fracture of cotton fibers, which only hook up a small part, plus the difficulty of picking up. According to the results of the preliminary field test, when the machine operating speed is set at $v$ = 0.6~1.2 m/s, $\lambda$ = 2.1~2.8, the device operating performance is good and can complete the cotton-picking operation. Bringing v and λ into Equation (3), the sawtooth disc speed can be initially determined as $n$ = 40.11~106.95 rpm.

2.2.3. Analysis of the Conditions of Flooring Cotton without Missing Pickup

To achieve continuity when picking cotton, there should be no gap between the motion trajectories formed by the tooth ends of adjacent serrations; that is, adjacent serrations are required to just tie into the initial point $E$ when the tooth end of the previous serration has not left the end point of picking cotton $E_2$. The serrations arrive at the initial point $E$ of tying into the cotton; at this time, the initial phase angle between $L_{OE}$ and $Y$-axis is $\theta$. The time elapsed from the movement of the serrated disk axis $O$ to $O_1$ is $t_1$, and the time elapsed from the movement to $O_2$ is $t_2$. According to the analysis presented in Figure 4,

$$h \leq R(1 - \cos \alpha) \tag{6}$$

where $h$ is the distance between the intersection of two adjacent serrations in the trajectory of the landing cotton from the deepest point of the serration into the cotton (mm).

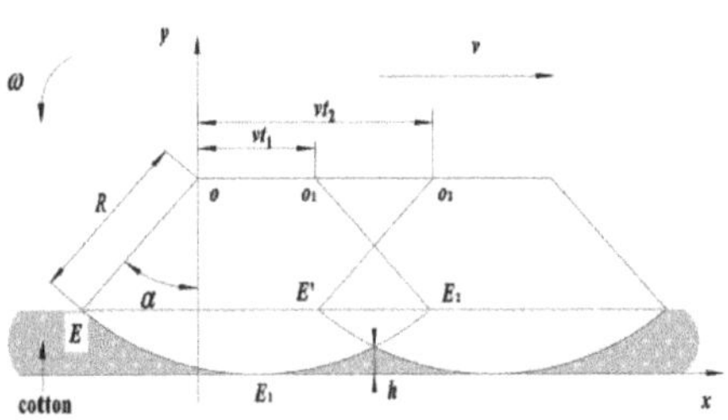


**Figure 4.** Adjacent serrated tooth end motion trajectory. $v$ is the operating speed of the machine (m/s); $\omega$ is the angular speed of the serrated disc, (rad·s$^{-1}$); $O$, $O_1$, and $O_2$ are the centers of rotation of the serrated disc at different moments; $E$ is the initial point of the serrated tooth end into the cotton; $E_1$ is the deepest point of the serrated tooth end into the cotton; $E_2$ is the point of the serrated tooth end out of the cotton; $E'$ is the point of the adjacent serrated tooth end into the cotton; $vt_1$ is the displacement of the rotary center of the previous serrated tooth in time $t_1$ (mm); $vt_2$ is the displacement of the rotary center of the latter saw tooth in time $t_2$ (mm).

When $E_2$ and $E_1$ coincide, the criterion for continual collection of cotton fallen to the ground is satisfied, and the critical condition can be attained as:

$$L_{o_1o_2} = 2R\sin\alpha \tag{7}$$

where $L_{O_1O_2}$ is the previous saw teeth leaving the end of the pick cotton point immediately after the next saw teeth just into the initial point of tying cotton serrated disc axis horizontal movement distance (mm).

Figure 4 illustrates the parameters that must be met to achieve continuity while picking up cotton:

$$2R\sin\alpha \geq v(t_2 - t_1) \tag{8}$$

One must enter the cotton from the tip of the serration to pick up the cotton; the angle of the sawtooth end is rotated for $2\alpha$, and the time required is $t_1 = 2\alpha/\omega$. When the number of saw teeth on the serrated disc is set to $z$, the adjacent serrated angle can be obtained as $\theta = 2\pi/z$. Subsequently, the next saw teeth turn the angle after the serrated tooth end is located at point $E$; the required time is $t_3 = 2\pi/z\omega$, where $t_3 = t_2 - t_1$. Thus, $t_1$, $t_2$ appear in Equation (8), and finishing can be obtained as:

$$\omega \geq \frac{\pi v}{zR\sin\alpha} \tag{9}$$

In Equations (6)–(9), $t_1$ is the time required for the serrated tooth end to move from $E$ to $E_2$ (s); $t_2$ is the required time for adjacent serrated tooth ends to hook the cotton individually (s); $t_3$ is the time required for the latter serrated tooth end to move to point $E$ until the previous serrated tooth leaves point $E_2$ (s); and $\alpha$ is the initial phase angle of the serrated tooth end from immediately hooking the cotton (°).

Using the sine theorem, we can write:

$$\frac{R}{\sin 90^\circ} = \frac{R-H}{\cos\alpha} \tag{10}$$

According to the structure and dimensional data of the sawtooth-type picking device, $R$ = 300 mm was selected as the radius of the rotation of the serrated disk. The maximum depth of the serrated teeth $H$ was set to 20 mm, and the initial phase angle $\alpha$ = 21.04° was calculated using Equation (10). Under the condition of continuously picking up cotton, the number of saw teeth on the same serrated disk was $z$ = 90; machine forward speed $v$ was set in the range of 2–5 km/h (0.56–1.39 m/s) according to the field situation; $\omega$ was calculated as 0.198 rad/s. In other words, the serrated disc speed of 1.89 rpm is sufficient to meet the demands of continuous cotton plucking.

### 2.3. Design of the Cotton Unloading Mechanism

To take off cotton that is picked up by the serrated disc and send it to the collection box, a cotton unloading mechanism was designed for the structural characteristics of the sawtooth roll-tie cotton-picking mechanism; its structural schematic is depicted in Figure 5. The unloading mechanism consists of brush and stripping rollers. The brush roller comprises a drive shaft and a brush strip, with the brush strip having dimensions of 930 mm in length and 60 mm in width. At 40 mm in length, the bristle is composed of a nylon substance with good wear resistance and tensile strength. Four brush strips are installed at equal intervals around the perimeter of the drive shaft, and each brush strip is bolted to the drive shaft. A cotton removal shaft and spikes are fitted on the cotton removal shaft of the cotton removal roller. Here, the cotton removal shaft has a 20 mm diameter. In addition, the spikes are circumferentially placed on the drive shaft, and there are four rows with 22 or 23 spikes in each row, with a 30 mm distance between spikes.

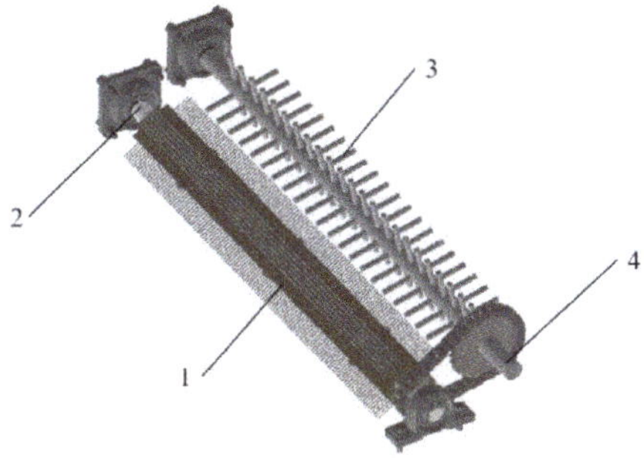


**Figure 5.** 1. Brush strip; 2. drive shaft; 3. spikes; and 4. unloading the cotton shaft.

A hydraulic motor drives the cotton unloading and cotton-picking mechanisms to rotate in opposing directions via chain drive and reversing gear during operation. When the tooth end of the serrated tooth hooks the cotton to the position of contact with the brush roller, the force of the brush roller is sufficient to overcome the resistance of the serrated tooth and brush the cotton off the serrated tooth; this is because the friction coefficient between the brush and the cotton fiber is more significant than that between the cotton fiber and the serrated tooth [21].

Figure 6 depicts the force sketch of cotton while unloading cotton. The equilibrium force condition of cotton is as follows:

$$\begin{cases} F_f + G\cos\theta = F_{\omega_1} \\ F + G\sin\theta = F_N \\ G = mg \\ F_{\omega_1} = 4\pi^2 n_1^2 mR \end{cases} \quad (11)$$

where $m$ is the mass of cotton fallen on the ground on a single serrated type (g); $g$ is the acceleration of gravity (9.8 $\mathrm{m/s^2}$); $n_1$ is the rotational speed of the serrated disk (rpm); $R$ is the instantaneous rotation radius of the serrated disk (mm).

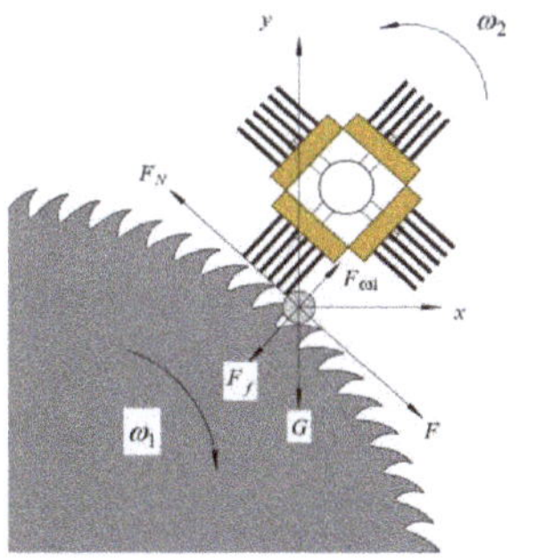


**Figure 6.** Force analysis of cotton. $\omega_1$ and $\omega_2$ are the rotational angular speed of the serrated disk and brush roller, respectively (rad/s); $F_f$ is the force of the serrated teeth on the cotton (N); $F$ is the force of the brush roller on the cotton (N); $\theta$ is the angle between the force of the brush roller on the cotton and its own gravity (°); $F_1$ is the centrifugal force on the cotton (N); $F_N$ is the support force of the serrated teeth on the cotton (N); and $G$ is the gravity of the cotton (N).

The force of the brush roller on the cotton and the combined centrifugal force on the cotton on the serrated disk must be larger than the force of the serrated teeth on the cotton for the brush rollers to brush down the cotton from the cotton-picking mechanism smoothly. According to Equation (11), the condition necessary for the cotton to detach from the serration is:

$$n_1 = \frac{1}{2\pi}\sqrt{\frac{F_f + mg\cos\theta}{mR}} \quad (12)$$

Using the universal testing equipment, the force $F_f$ of the saw teeth on the cotton was measured to be approximately 2.2 times the cotton's specific gravity, or $F_f$ = 2.2 $mg$. In addition, $R$ = 300 mm was chosen as the rotational radius of the serrated disk. When $cos\theta = 0$, the minimum rotation speed of the cotton out of the serrated disc was $n_1$ = 80.96 rpm. If the speed is poor, the picking efficiency will be low, and the operating requirements will not be reached. Under centrifugal force, cotton is easily dislodged from the sawtooth disk if the speed is extremely high. Combining theoretical analysis with the actual operation, the initial serrated disk speed was set as 60 rpm. According to [22], the brush roll surface's linear speed is typically 1.5–2 times that of the serrated tooth end. The spinning radius of the brush roller is 85 mm, and its computed speed range is 317.64~423.53 rpm. In conjunction with the preliminary test, the final brush roller speed determination is 400 rpm. Since the speed of the serrated disc cannot exceed 80.96 rpm, the speed ratio of the serrated disc to the brush roll exceeds 1:5, and the brush roll will brush over the serrated teeth many times without missing the cotton.

### *2.4. Discrete Element Modeling of the Motion Process of the Cotton-Picking Mechanism*

#### 2.4.1. Modeling and Parameter Setting of the Simulation Model

Due to the loose nature of the soil, the cotton-picking mechanism will unavoidably come into contact with the ground. An EDEM simulation test was performed to further analyze the force on the device's serrated tooth end during operation. To precisely simulate the interaction between the picking device and the soil throughout the operation, the soil particle radius was set to 3 mm. The Hertz–Mindlin with bonding model was used to build the contact model of the soil particles in EDEM 2018 (DEM Solutions Ltd. Edinburgh, Scotland, UK). In order to accelerate the simulation calculation, we simplified the cotton-picking mechanism while retaining the serrated disc and ground wheel. We then used SolidWorks 2018 software (Dassault Systèmes S.E., Massachusetts, Concord, MA, USA) to build the simplified 3D model structure of the cotton-picking mechanism according to the 1:1 ratio, saved it as an ".x t" format file, and imported it into EDEM 2018 software.

The model of the soil trough was created in EDEM software with the dimension $L \times W \times H$ = (2500 × 1000 × 250) mm. Figure 7 depicts the beginning state of the EDEM simulation model. The cotton-picking mechanism is placed at the soil trough's upper right end. In order to imitate the actual motion law of the cotton-picking mechanism, the serrated disk is configured to rotate clockwise at 60 rpm with a constant speed to the left of 0.9 m/s and a sinking amount of 20 mm.

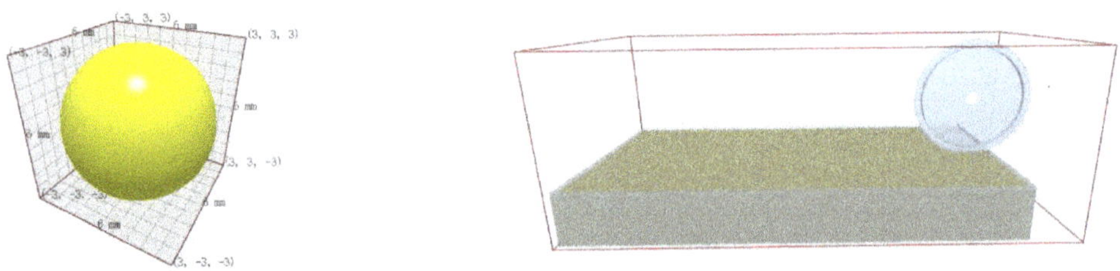


(**a**) (**b**)

**Figure 7.** Particle and geometric simulation model. (**a**) Particle model of the soil; (**b**) simulation model of the EDEM.

The simulation time step is set to $1.5 \times 10^{-6}$, the simulation time is set to 2 s, and the grid cell size is three times the minimum soil particle size. The contact models were selected from soil–soil and soil–cotton-picking mechanisms. The main parameters included contact parameters (soil recovery coefficient, static friction coefficient, and dynamic friction coefficient) and intrinsic parameters (density, Poisson's ratio, and shear modulus). The data of the main parameters of the discrete element method test model were obtained by the method of calibration and optimization of the discrete element parameters of clay loam soil from the stacking test [23–25], and the relevant parameters are shown in Table 2.

**Table 2.** Simulation parameter settings of soil particles and geometry.

| Item | Parameter | Value |
|---|---|---|
| Soil particles | Poisson's ratio | 0.30 |
| | Shear modulus/Pa | 5 × 107 |
| | Density/(kg·m$^{-3}$) | 2600.00 |
| Cotton-picking mechanism | Poisson's ratio | 0.30 |
| | Shear modulus/Pa | 7.90 × 1010 |
| | Density/(kg·m$^{-3}$) | 7850.00 |
| Particle–Particles | Recovery coefficient | 0.21 |
| | Static friction coefficient | 0.68 |
| | Dynamic friction coefficient | 0.27 |
| Particle–Cotton-picking mechanism | Recovery coefficient | 0.54 |
| | Static friction coefficient | 0.53 |
| | Dynamic friction coefficient | 0.13 |

2.4.2. Analysis of Simulation Results

Figure 8 depicts the simulation result of the force on the serrated disk in the EDEM software once the simulation is complete. The result indicates that the maximum force exerted on the serrated disk upon entry into the soil is 201.88 N.

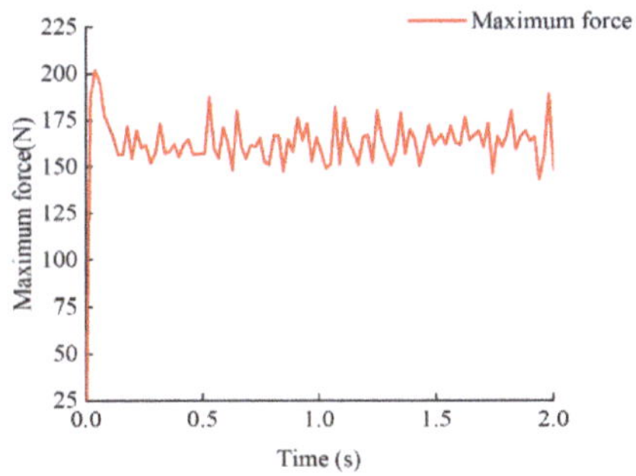


**Figure 8.** Variation of force on serrated disc with time.

The strain and von Mises stress of the serrated tooth end under maximum force are analyzed using ANSYS software (ANSYS, Inc., Canonsburg, PA, USA). The load on the serrated tooth end is set to 201.88 N under the boundary condition of complete constraint at the center of the serrated disk, and the force is set opposite to the advancing direction of the serrated disk to bend it, and the force area is the working area of the serrated tooth end. In order to simplify the computation, the serrated tooth tip is assumed to experience a uniform load. Figure 9 illustrates the results. At the end of the saw tooth, there is a maximum strain of 0.023063 mm. The most significant equivalent force arises at the root of the serrated tooth, and the equivalent force is 64.677 MPa, which is much less than the yield strength of the serrated disk, 345 MPa. Therefore, the serrated disk's structural strength meets the design's requirements.

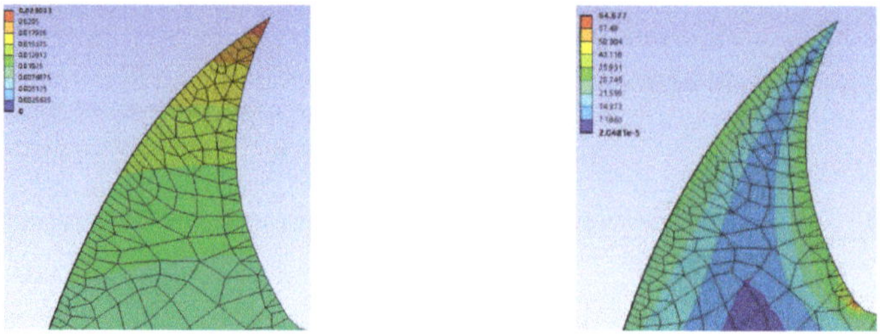

(a) (b)

**Figure 9.** Analysis results. (**a**) Serrated disc tooth strain model; (**b**) Serrated disc tooth stress model.

## 2.5. Test Materials

To test the reliability and operational efficiency of the cotton recovery device and to determine the optimal operating parameters, a field test was conducted at the Shihezi University test site in Shihezi, Xinjiang, on 3 October 2021.

The test was conducted in a cotton field after the cotton picker operation; the cotton variety was Xin Luzao 68. Cotton was planted in wide and narrow rows using a dense planting pattern (660 mm + 100 mm). The test plot was flat and ditchless, and the drip irrigation belt was recovered. The rated power of the employed tractor (KEER KE704B) was 51.5 kW. Test equipment included an electronic balance (HZ-C5033, the maximum weighing = 500 g, the minimum sense = 0.001 g), LJD15 meter ruler (range = 100 m), photoelectric tachometer (AR926, measurement range = 2.5–99,999 rpm), stopwatch, label paper, and encapsulated experimental bag.

## 2.6. Test Methods

Before machine operation, a test area of 50 m in length and 1.0 m in width was chosen in the test cotton field. Five points were chosen at random as testing points in the test area, with each testing point measuring 5 m in length. As the test result for this test area, the average value of the picking rate of the five testing points was used [26]. Before the test, five cotton field reference points with the same area as the test points were randomly selected, and all the fallen cotton in the reference points was manually recovered, and the average value of their masses was recorded as $M_1$. After the machine operation, the mass of the missed cotton at the test points was manually recovered and cleaned up as $M_2$, and the picking rate was calculated using Equation (13).

At the end of every test, five cotton samples weighing no less than 2000 g were picked randomly from various positions in the cotton collection box, concentrated, and thoroughly mixed. Then, five samples weighing 1000 g each were extracted. Manual removal of stalks, broken leaves, boll shells, and other impurities from the samples was conducted. The lint cotton was separated using the test gin while coarse impurities were collected. Lint cotton was then separated from cotton and impurities using a cotton impurity separator; the impurity rate of cotton was calculated using Equation (14). Finally, the impurity rate of each test was determined by averaging the results of five samples [27].

$$\eta_1 = \frac{M_1 - M_2}{M_1} \tag{13}$$

$$\eta_2 = \frac{M_d + M_c + M_x}{M_y} \tag{14}$$

where $\eta_1$ is the picking rate of cotton (%); $\eta_2$ is the impurities rate of cotton (%); $M_1$ is the total mass of cotton in the testing points (g); $M_2$ is the mass of cotton left in each testing point (g); $M_d$ is the mass of impurities, such as stalks, broken leaves, and boll shells picked out manually (g); $M_c$ is the mass of impurities separated from the sample using the test gin (g); $M_x$ is the mass of impurities separated from the lint in the sample using the cotton impurity separator (g); $M_y$ is the mass of the sample (g).

Based on the structure parameters and operating characteristics of the device, the three critical parameters of machine operating speed, sawtooth disc speed, and sawtooth disc spacing were identified as the primary influencing factors of this test. A single-factor test was conducted to determine the effect of different levels of the same factor on the picking rate $\eta_1$ and the impurity rate $\eta_2$. Figure 10 illustrates the single-factor test's results. When the operating speed of the machine increased, the picking rate showed a trend of first increasing and then decreasing, and the overall impurity rate showed an increasing trend; when the spacing of the serrated disc increased, the picking rate showed a continuous decreasing trend, and the impurity rate first decreased and then increased; when the speed of the serrated disc increased, the picking rate first increased and then decreased, and the overall impurity rate showed an increasing trend.

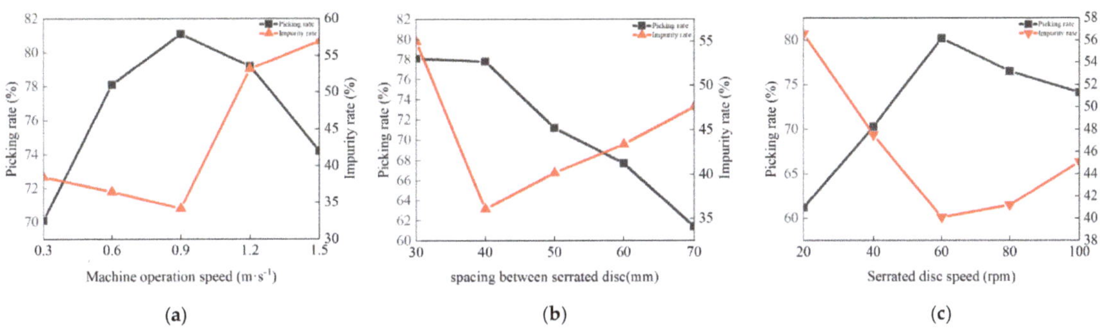


**Figure 10.** Single factor test. (**a**) The effect of machine operation speed on the operating effect. (**b**) The effect of spacing between serrated discs on the operating effect. (**c**) The effect of the serrated disc speed on the operating effect.

Therefore, 0.6, 0.9, and 1.2 m/s were selected as the factor levels of the operating machine speed, 40, 50, and 60 mm were selected as the factor levels of spacing between the serrated discs, and 40, 60, and 80 rpm were selected as the factor levels of the serrated disc speed, based on the premise of ensuring the picking rate and taking into account the impurity rate. Table 3 details the experiment with considered factors and level coding.

**Table 3.** Test factors and levels.

| Coded Value | Machine Operation Speed $X_1$ (m·s$^{-1}$) | Spacing between Serrated Discs $X_2$ (mm) | Serrated Disc Speed $X_3$ (rpm) |
|---|---|---|---|
| −1 | 0.6 | 40 | 40 |
| 0 | 0.9 | 50 | 60 |
| 1 | 1.2 | 60 | 80 |

## 2.7. Test Results

Seventeen groups were administered a three-factor, three-level test based on the Box–Behnken test principle [28]. Table 4 shows the experimental protocol and results.

**Table 4.** Test plans and results.

| Test | $X_1$ | $X_2$ | $X_3$ | $\eta_1$ | $\eta_2$ |
|---|---|---|---|---|---|
| 1 | 0 | 1 | 1 | 78.1 | 40.1 |
| 2 | 0 | 0 | 0 | 81.4 | 36.5 |
| 3 | −1 | 1 | 0 | 78.1 | 60.3 |
| 4 | −1 | 0 | −1 | 71.1 | 48.3 |
| 5 | 1 | 0 | −1 | 74.7 | 57.4 |
| 6 | 0 | 0 | 0 | 80.6 | 37.6 |
| 7 | 0 | 0 | 0 | 79.9 | 34.7 |
| 8 | 1 | −1 | 0 | 79.3 | 53.6 |
| 9 | 0 | 0 | 0 | 81.3 | 33.8 |
| 10 | 1 | 1 | 0 | 81.2 | 54.3 |
| 11 | −1 | −1 | 0 | 76.3 | 39.7 |
| 12 | 0 | 1 | −1 | 78.3 | 39.4 |
| 13 | 1 | 0 | 1 | 77.5 | 59.6 |
| 14 | 0 | 0 | 0 | 81.2 | 35.7 |
| 15 | 0 | −1 | −1 | 74.4 | 36.4 |
| 16 | 0 | −1 | 1 | 79.3 | 33.8 |
| 17 | −1 | 0 | 1 | 76.3 | 53.1 |

## 3. Results and Discussion

As shown in Table 5, we used Design Expert 13.0 software (Stat-Ease Inc., Minneapolis, MN, USA) to analyze the test results and the multiple regression fit. Table 5 shows the results of the picking rate and impurity rate variance analyses. The significance of the regression equations of $\eta_1$ and $\eta_2$ on $X_1$, $X_2$, and $X_3$ was tested.

**Table 5.** Analysis of variance of regression equation.

| Source of Variation | *DOF* | Picking Rate $\eta_1$/% | | | Impurity Rate $\eta_2$/% | | |
|---|---|---|---|---|---|---|---|
| | | **Sum of Squares** | ***F*** | **Significant Level *p*** | **Sum of Squares** | ***F*** | **Significant Level *p*** |
| Models | 9 | 130.47 | 30.89 | <0.0001 ** | 1514.95 | 26.68 | 0.0001 ** |
| $X_1$ | 1 | 14.85 | 31.64 | 0.0008 ** | 69.03 | 10.94 | 0.0130 * |
| $X_2$ | 1 | 5.12 | 10.91 | 0.0131 * | 117.05 | 18.55 | 0.0035 ** |
| $X_3$ | 1 | 20.16 | 42.96 | 0.0003 ** | 3.25 | 0.5153 | 0.4961 |
| $X_{1\times2}$ | 1 | 0.0025 | 0.0053 | 0.9439 | 99.00 | 15.69 | 0.0055 ** |
| $X_{1\times3}$ | 1 | 1.44 | 3.07 | 0.1233 | 1.69 | 0.2679 | 0.6207 |
| $X_{2\times3}$ | 1 | 6.50 | 13.85 | 0.0074 * | 2.72 | 0.4315 | 0.5322 |
| $X_1^2$ | 1 | 24.05 | 51.24 | 0.0002 ** | 1180.61 | 187.12 | <0.0001 ** |
| $X_2^2$ | 1 | 0.2325 | 0.4954 | 0.5043 | 0.7785 | 0.1234 | 0.7357 |
| $X_3^2$ | 1 | 54.27 | 115.62 | <0.0001 ** | 20.29 | 3.22 | 0.1160 |
| Residual | 7 | 3.29 | | | 44.16 | | |
| Lack of fit | 3 | 1.70 | 1.43 | 0.3954 | 35.31 | 5.32 | 0.0701 |
| Pure error | 4 | 1.59 | | | 8.85 | | |
| Total | 16 | 133.75 | | | 1559.12 | | |

Note: ** means highly significant ($p < 0.01$), and * means significant ($0.01 \leq p < 0.05$).

(1) Establishment of the regression equation and significance analysis of the picking rate

The variance analysis for the picking rate indicated that in this regression model, $X_1$, $X_3$, $X_1^2$, and $X_3^2$ had an extremely significant impact on the picking rate model. $X_2$ and $X_{2\times3}$ had a more significant impact on the picking rate model. The significance of the influence of each variable on the pickup rate was in the following order, from more to less significant: the serrated disc speed, the machine operation speed, and the spacing between serrated discs. After eliminating the insignificant factors, the quadratic regression equation of each variable on the picking rate was obtained [29], as shown in Equation (15):

$$\eta_1 = 80.88 + 1.36X_1 + 0.8000X_2 + 1.59X_3 - 1.27X_2X_3 - 2.39X_1^2 - 3.59X_3^2 \quad (15)$$

(2) Establishment of the regression equation and significance analysis of the impurity rate

The variance analysis for the impurity rate indicated that the $X_2$, $X_{1\times 2}$, and $X_1{}^2$ had an extremely significant impact on the impurity rate model. $X_1$ had a significant impact on the impurity rate model. The significance of the influence of each variable on the impurity rate was in the following order, from more to less significant: the spacing between serrated discs, the machine operation speed, and the serrated disc speed. After eliminating the insignificant factors, the quadratic regression equation of each variable on the impurity rate was obtained as shown in Equation (16):

$$\eta_2 = 35.66 + 2.94X_1 + 3.83X_2 - 4.98X_1X_2 + 16.74X_1^2 \quad (16)$$

### 3.1. Response Surface Analysis

Using the 3D-surface response surface plots created by Design Expert 13.0.1.0, we analyzed the interactions between the parameters that affect the pickup and impurity rates of the cotton fallen on the ground (Figure 11).

(a) (b)

**Figure 11.** Field test of the experimental device. (**a**) Field test. (**b**) Device operation effect.

(1) Analysis of the influence of the picking rate

Figure 12a demonstrates that when $X_3$ is fixed at 60 rpm and $X_1$ is increased, the picking rate increases, then decreases, and the rate of decline becomes more gradual. When $X_2$ increases, the picking rate increases gradually, with a moderate degree of change. Figure 12b shows that when $X_2$ is fixed at 50 mm, $X_1$ and $X_3$ are increased, and the picking rate of cotton fallen on the ground increases initially and then decreases. Figure 12c indicates that when $X_1$ controls 0.9 m/s, the pickup rate gradually grows as $X_2$ increases, with a relatively flat amplitude of change; when $X_3$ increases, the pickup rate gradually climbs and drops slowly.

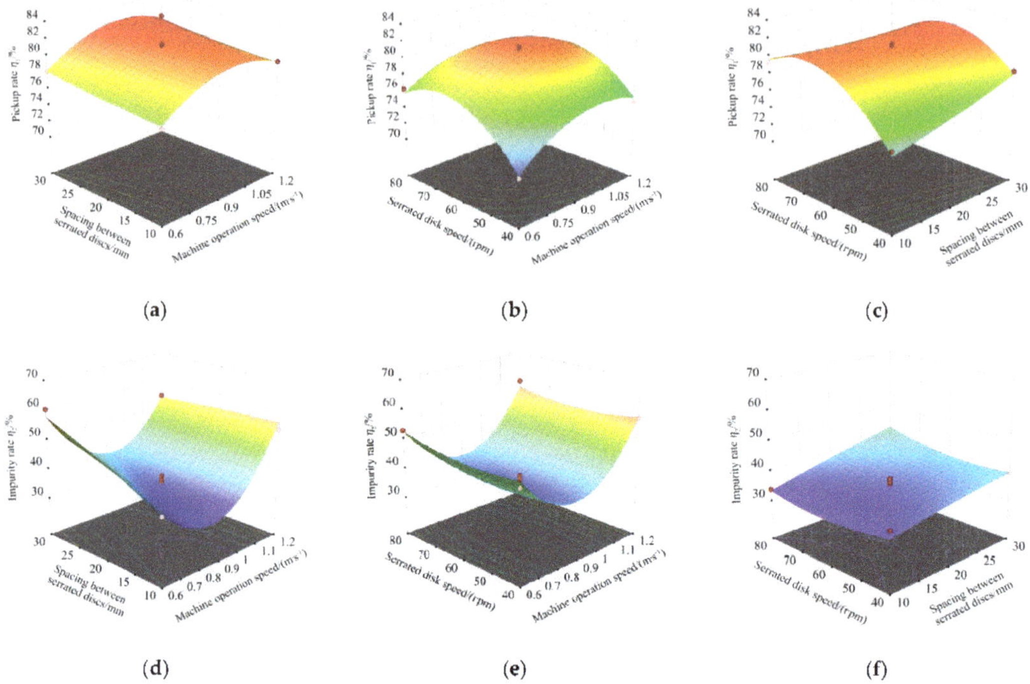


**Figure 12.** Effects of the interaction of various factors on the picking rate and trash content of cotton fallen on the ground. (**a**). $\eta_1$ = ($X_1$, $X_2$, 60); (**b**). $\eta_1$ = ($X_1$, 50, $X_3$); (**c**). $\eta_1$ = (0.9, $X_2$, $X_3$); (**d**). $\eta_2$ = ($X_1$, $X_2$, 60); (**e**). $\eta_2$ = ($X_1$, 50, $X_3$); (**f**). $\eta_2$ = (0.9, $X_2$, $X_3$).

Indeed, the faster the machine speed, the greater the weight of cotton hooked by the serrated teeth per unit of time, and the easier it is to fall off from the teeth, resulting in a lower pickup rate. When the distance between serrated discs is too tiny, adjacent saw discs quickly pick up the same piece of cotton, reducing the pickup rate. When the speed of the serrated disk increases, it is easy to tear the cotton when it touches the cotton so that only a tiny part of the serrated end is hooked, increasing the difficulty of picking [30]. Simultaneously, the centrifugal force on the landed cotton hooked by the serrated teeth increases accordingly, and the cotton fallen on the ground that is not hooked tightly is easily thrown away from the serrated disc.

(2) Analysis of the influence of the impurity rate

Figure 12d indicates that when $X_3$ is held constant at 60 rpm and $X_1$ increases, the impurity rate drops and then increases with a significant trend. In addition, the rate of impurity gradually increases as $X_2$ rises. Figure 12e demonstrates that when $X_2$ is fixed at 50 mm, and $X_1$ and $X_3$ increase, the impurity rate increases and subsequently declines. Figure 12f demonstrates that when $X_1$ is regulated to 0.9 m/s, the impurity rate grows gradually, with only minor variations. In addition, as $X_3$ grows, the impurity rate reduces gradually and then rises gradually, with changes that are likewise relatively slow.

The reason for this is that the operating speed of the machine and the cotton falling to the ground are gradually matched to the optimal levels. In fact, as the speed increases, the amount of cotton picked by the device per unit of time increases, resulting in a high rate of impurities. When the spacing between serrated discs increases, the picking efficiency rises and more cotton enters the unloading mechanism, resulting in a relative increase in the impurity rate. Moreover, the faster the speed of the serrated disc, the greater the impurities on the hooked cotton due to the centrifugal force. Thus, impurities that adhere to the cotton can be easily thrown away. As the rotational speed increases, the conveying capacity of the serrated disc increases, preventing the majority of impurities from being thrown out and gathered directly, resulting in a high rate of impurities.

### 3.2. Parameter Optimization and Test Validation

In order to maximize the performance of the whole machine, the prototype's influencing factors were optimized. Using Design Expert 13.0 software, the model was optimized and analyzed based on the operating conditions, performance requirements, and above-mentioned analysis results [31,32]. The software's optimization conditions are set to dual-objective equal-weight optimization, and the constraints are as follows:

$$\begin{cases} \eta_{1max} = F(X_1, X_2, X_3) \\ \eta_{2min} = F(X_1, X_2, X_3) \\ s.t \begin{cases} X_1 \in [0.6, 1.2] \\ X_2 \in [40.0, 60.0] \\ X_3 \in [40.0, 80.0] \end{cases} \end{cases} \quad (17)$$

The objective function was optimized and solved. The optimal parameter combination was as follows: the optimized machine operating speed was 0.958 m/s, the optimized spacing between serrated discs was 40 mm, and the optimized serrated disc speed was 67.603 rpm. The predicted values of the cotton pickup and impurity rates were 81.01 and 33.79%, respectively. To verify the accuracy of the prediction model, a verification test was conducted, as shown in Figure 12. We took the operating machine speed of 0.96 m/s, the spacing between serrated discs of 40 mm, and the speed of the serrated disk at 68 rpm. We conducted five verification tests and took the average value. The results are shown in Table 6.

**Table 6.** Comparison between the optimum theoretical and test results.

| Parameter | Picking Rate $\eta_1$/% | Impurity Rate $\eta_2$/% |
|---|---|---|
| Theoretical optimization value | 81.01 | 33.79 |
| Test average | 79.09 | 35.12 |
| Relative error | 2.37 | 3.79 |

The difference between the experimental value and the theoretical optimization value of the model was less than 5% after verification. Therefore, the optimization model is reasonable and is capable of meeting operation requirements.

## 4. Conclusions

1. For cotton machine harvesting in Xinjiang, there was no suitable mechanism to recover cotton fallen on the ground. In this study, a sawtooth-type recovery device was designed to recover cotton fallen on the ground and efficiently unload it. The device consists of a sawtooth roll-tie-type cotton-picking mechanism, cotton unloading mechanism, cotton collection box, and other parts. The primary design parameters were determined using the analysis of the motion of the serrated discs, the cotton non-missing picking condition, and the cotton unloading condition.
2. EDEM simulated the process of the cotton-picking mechanism movement. The maximum force on the tooth end of the serrated teeth was obtained during the working process. Then, ANSYS analysis of strain and stress on the tooth end of the serrated teeth was carried out to verify that the structural strength of the serrated disc meets the design requirements.
3. Considering the machine operating speed, spacing between serrated discs, and serrated disc speed as the experimental factors, the picking and impurity rates of the

cotton fallen on the ground were used as the test indicators. Additionally, the response surface data were analyzed using Design Expert software, and multiple fittings obtained the regression equation of the picking and impurity rates. The influence of the interaction of various factors on the picking and impurity rates was determined.

4. Experimental tests on the device proved that when the optimized machine operating speed was 0.96 m/s, the spacing between serrated discs was 40 mm, and the speed of the serrated disc was 68 rpm. In addition, the picking and impurity rates of the cotton fallen on the ground were 79.09 and 35.12%, respectively. The optimized operating parameters were verified experimentally. Relative errors between the experimental results and optimized theoretical values of the picking and impurity rates were 2.37 and 3.79%, respectively, relatively small. Thus, the model was highly reliable.

**Author Contributions:** Conceptualization, methodology, writing—original draft, H.W. and S.C.; writing—review and editing, S.C. and Y.Y.; software, H.W. and Y.L.; investigation, H.W. and Y.L.; data curation, X.M. and P.J.; funding acquisition, S.C.; validation, Y.Y. All authors have read and agreed to the published version of the manuscript.

**Funding:** This research was funded by Corps Young and Middle-aged Leading Talents Program Project (2018CB011).

**Institutional Review Board Statement:** Not applicable.

**Informed Consent Statement:** Not applicable.

**Data Availability Statement:** All relevant data presented in the article are stored according to institutional requirements and, as such, are not available online. However, all data used in this manuscript can be made available upon request to the authors.

**Conflicts of Interest:** The authors declare no conflict of interest.

## References

1. Jie, Z. The Research On Environment Cost in Cotton Production in Xinjiang. Master's Thesis, Tarim University, Tarim, China, 2014.
2. Abdelraheem, A.; Esmaeili, N.; O'Connell, M.; Zhang, J. Progress and perspective on drought and salt stress tolerance in cotton. *Ind. Crops Prod.* **2019**, *130*, 118–129. [CrossRef]
3. ShuXun, Y.; Shuli, F.; Hantao, W.; Hengling, W.; Chaoyou, P. Progresses in Research on Cotton High Yield Breeding in China. *Sci. Agric. Sin.* **2016**, *49*, 3465–3476. [CrossRef]
4. Wang, Y.C.; Peng, S.B.; Huang, J.L.; Zhang, Y.L.; Feng, L.; Zhao, W.Q.; Qi, H.K.; Zhou, G.S.; Deng, N.Y. Prospects for cotton self-sufficiency in China by closing yield gaps. *Eur. J. Agron.* **2022**, *133*, 126437. [CrossRef]
5. Khan, N.; Han, Y.; Xing, F.; Feng, L.; Wang, Z.; Wang, G.; Yang, B.; Fan, Z.; Lei, Y.; Xiong, S. Plant Density Influences Reproductive Growth, Lint Yield and Boll Spatial Distribution of Cotton. *Agronomy* **2020**, *10*, 14. [CrossRef]
6. Aslam, S.; Khan, S.H.; Ahmed, A.; Dandekar, A.M. The tale of cotton plant: From wild type to domestication, leading to its improvement by genetic transformation. *Am. J. Mol. Biol.* **2020**, *10*, 91–127. [CrossRef]
7. Beibei, Z.; Wei, G.; Jianyu, C.; Kangguo, M.; Ying, M.; Lin, H. Assessment of the Biomass Energy Use Potential of Cotton Byproducts in China. *Cotton Sci.* **2016**, *28*, 384–391. [CrossRef]
8. Tingguan, C.; Hongwen, Z.; Lei, W.; Longchang, Z.; Jun, W.; Jianxin, L.; Yanqing, G. Optimization and experiments of picking head transmission system of horizontal spindle type cotton picker. *Trans. Chin. Soc. Agric. Eng.* **2020**, *36*, 18–26. [CrossRef]
9. Cheng, W.M.; Wang, L. How to reduce the cost of cotton cultivation? A Case Study of Xinjiang, China. *Custos E Agronegocio Line* **2019**, *15*, 458–470.
10. Xiuru, L.; Xiaoyue, J.; Jiahui, N. The Present Situation and Prospects of Cotton Industry Development in China. The Present Situation and Prospects of Cotton Industry. *Sci. Agric. Sin.* **2018**, *51*, 26–36. [CrossRef]
11. Chengling, Z.; Tianhui, L.; Jianyu, L. Detection of Impurity Rate of Machine-Picked Cotton Based on Improved Canny Operator. *Electronics* **2022**, *11*, 974. [CrossRef]
12. *GB/T 21397-2008*; Cotton Harvester. Agriculture Machinery of Standardization Administration of China. China Standard Press: Beijing, China, 2018.
13. Statistical Bureau of the People's Republic of China. *China Statistical Yearbook*; China Statistics Press: Beijing, China, 2021.
14. Lindsey, M.; Charles, S.C.; Joe, E. Mechanical gleaning and ginning of ground-loss cotton in the Yazoo-Mississippi Delta. 1969. Available online: https://scholarsjunction.msstate.edu/cgi/viewcontent.cgi?article=1074&context=mafes-bulletins (accessed on 20 March 2022).

15. Lehman, L.A. Ground Cotton Retriever with Dual Cleaning Means. 1985. Available online: https://patents.google.com/patent/US4497088 (accessed on 5 March 2022).
16. Jianlong, H.; Xiaolong, H. Air Suction Type Picking Machine for Cotton Falling on Ground. 2022. Available online: https://patents.google.com/patent/CN215380014U (accessed on 8 March 2022).
17. Shi, G.; Li, J.; Kan, Z.; Ding, L.; Ding, H.; Zhou, L.; Wang, L. Design and Parameters Optimization of a Provoke-Suction Type Harvester for Ground Jujube Fruit. *Agriculture* **2022**, *12*, 409. [CrossRef]
18. Guangli, S.; Liyong, S.; Kun, L. Machine for Picking Cotton Falling to Ground. 2021. Available online: https://patents.google.com/patent/CN214102380U (accessed on 8 March 2022).
19. Changlin, C.; Lei, S.; Yutong, Z.; Senlin, M.; Yongfei, S.; Mingsen, H. Design and experiment of the MQZ-4A type site cotton pre-processor. *J. Chin. Agric. Mech.* **2015**, *36*, 17–19. [CrossRef]
20. Huang, M.; Shi, L.; Zhang, Y.; Chen, C.; Sun, Y.; Xie, Q.; Kong, F. Optimization experiment of machine-mounted seed cotton pre-treatment apparatus for cotton stripper harvester. *Trans. Chin. Soc. Agric. Eng.* **2016**, *32*, 21–29. [CrossRef]
21. Hu, W.; Chen, X.; Wang, J.; Li, Y. Construction and finite element simulation of the saw type gin stand using fractal theory. *Text. Res. J.* **2020**, *90*, 2755–2768. [CrossRef]
22. Le, S. Cleaning performance of modified cylinder cleaners. *J. Cotton Sci.* **2006**, *10*, 273–283.
23. Zhong, W.Q.; Yu, A.B.; Liu, X.J.; Tong, Z.B.; Zhang, H. DEM/CFD-DEM Modelling of Non-spherical Particulate Systems: Theoretical Developments and Applications. *Powder Technol.* **2016**, *302*, 108–152. [CrossRef]
24. Wu, T.; Huang, W.F.; Chen, X.S.; Ma, X.; Han, Z.Q.; Pan, T. Calibration of discrete element model parameters for cohesive soil considering the cohesion between particles. *J. South China Agric. Univ.* **2017**, *38*, 93–98. [CrossRef]
25. Xianlinang, W.; Hong, H.; Qingjie, W.; Hongwen, L.; Jin, H.; Wanzhi, C. Calibration method of soil contact characteristic parameters based on DEM theory. *Trans. Chin. Soc. Mach.* **2017**, *48*, 78–85.
26. Jianhua, X.; Yuxin, Y.; Silin, C.; Yi, Z.; Yabin, Z.; Weibin, M. Design and experiments of rake type surface residual film recycling machine with guide chain. *Trans. Chin. Soc. Agric. Eng.* **2020**, *36*, 76–86. [CrossRef]
27. Mingsen, H.; Lei, S.; Yutong, Z.; Changlin, C.; Yongfei, S.; Qing, X.; Fanting, K.; Chongyou, W. Revised design and experiments on brush-rolling cotton harvester. *Trans. Chin. Soc. Agric. Eng.* **2017**, *33*, 41–47. [CrossRef]
28. Sohail, Y.; Parag, B.; Nemeshwaree, B.; Giorgio, R. Optimizing Organophosphorus Fire Resistant Finish for Cotton Fabric Using Box-Behnken Design. *Int. J. Environ. Res.* **2016**, *10*, 313–320. [CrossRef]
29. Xuejun, Z.; Jiaqiang, L.; Zenglu, S.; Wei, J.; Jinshan, Y.; Mengjie, Y. Design and parameter optimization of reverse membrane and soil separation device for residual film recovery machine. *Trans. Chin. Soc. Agric. Eng.* **2019**, *35*, 46–55. [CrossRef]
30. Wen, H.; Di, W.; Xiaochuan, C.; Jun, W.; Yong, L. Finite element simulation of cotton serrated ginning state based on cottonseed modeling. *J. Text. Res.* **2020**, *41*, 27–32. [CrossRef]
31. Wei, Y.; Zhichao, H.; Nu, W.; Hongbo, X.; Zhaoyan, Y.; Xinxing, Z. Parameter optimization and experiment for plastic film transport mechanism of shovel screen type plastic film residue collector. *Trans. Chin. Soc. Agric. Eng.* **2017**, *33*, 17–24. [CrossRef]
32. Lili, S.; Zhichao, H.; Fengwei, G.; Feng, W.; Penglai, W. Design and parameter optimization on teeth residue plastic film collector of ridged peanut. *Trans. Chin. Soc. Agric. Eng.* **2017**, *33*, 8–15. [CrossRef]

*agriculture*

*Article*

# Test and Simulation Analysis of the Working Process of Soybean Seeding Monomer

**Dongxu Yan [1], Tianyue Xu [2], Jianqun Yu [3], Yang Wang [3], Wei Guan [4], Ye Tian [5] and Na Zhang [3,*]**

1 Hua Lookeng Honors College, Changzhou University, Changzhou 213164, China
2 College of Engineering and Technology, Jilin Agriculture University, Changchun 130022, China
3 School of Biological and Agricultural Engineering, Jilin University, Changchun 130022, China
4 School of Mechanical and Aerospace Engineering, Jilin University, Changchun 130025, China
5 Center of Industry and Technology, Hebei University of Technology Petroleum, Chengde 067000, China
* Correspondence: zna18@mails.jlu.edu.cn; Tel.: +86-13844941571

**Abstract:** Soybean seeding monomers can realize the process of opening, seed throwing, covering, and compacting when they work. Due to the complexity of their working process, the relevant process cannot be analyzed by the discrete element method (DEM) alone. The DEM coupled with the multi-rigid body dynamics method (MBD) can solve the above problem, and the simulation analysis of the above process is realized by coupling the EDEM software with RecurDyn software. The changes in the position of soybean seed particles before and after covering and compacting are analyzed. The results show that when the working speed of the seeding monomer increases, the distance along the vertical direction of the soybean seed particles after covering gradually increases, and the distance along the horizontal direction gradually decreases. The effect of different working speeds of seeding monomer on the opening situation and the variation in seed particle positions is studied. The results show that the ditch angle gradually decreases as the working speed of the seeding monomer increases. The distribution of seed particle spacing is also analyzed. The above tests are simulated, and the results show a high agreement between the simulation and test results, proving the accuracy of the coupling method. This paper applies the coupling method for the first time to the simulation of the seeding monomer. This method can be applied not only to the analysis of the sowing process of soybean seeding monomers, but also be applied to the analysis of other machinery working processes, such as the tillage process, the sieving process, the planting and harvesting processes of crops, etc. It also deepens the application of the discrete element method in the field of agriculture.

**Keywords:** DEM; MBD; coupled simulation; seeding; soybean; seed–soil

**Citation:** Yan, D.; Xu, T.; Yu, J.; Wang, Y.; Guan, W.; Tian, Y.; Zhang, N. Test and Simulation Analysis of the Working Process of Soybean Seeding Monomer. *Agriculture* **2022**, *12*, 1464. https://doi.org/10.3390/agriculture12091464

Academic Editors: Mustafa Ucgul and Chung-Liang Chang

Received: 23 August 2022
Accepted: 12 September 2022
Published: 14 September 2022



## 1. Introduction

Currently, the discrete element method (DEM) has become a common method for analyzing the contact interaction between particles and between particles and mechanical components. It has been widely used in fields such as agricultural engineering [1–9].

Wang [10] analyzed the simulation and test results of the sieving process of a pendulum screen based on the coupled DEM and multi-body dynamics (MBD) algorithms, using AgriDEM software developed independently by the Digital Design Laboratory of the College of Biological and Agricultural Engineering, Jilin University.

Yuan et al. [11] established a self-excited vibration deep loosening machine–soil system model based on the coupling algorithm of DEM and MBD. AgriDEM software was used to simulate and analyze the interaction between mechanical components and soil particles, and the accuracy of the model was verified.

Xu [12] simulated the working process of the coverer and roller based on the coupling algorithm of DEM and MBD, using EDEM software coupled with ADAMS software.

The feasibility and applicability of the coupling method were verified by comparing the displacement of seeds during simulation and testing.

The above analysis shows that the use of different software coupling methods to analyze the field of agricultural machinery has been initially applied. However, for the sowing monomer, when it works under the action of the traction machine, the processes of opening the furrow, throwing the seed, covering, and compacting can be realized. The whole process is a complex mechanical movement. It has not yet been analyzed and studied in depth.

In this paper, test analyses of the covering, compacting, and ditching of seeding monomers are performed. On this basis, an in-depth study of these problems is carried out by means of simulation analysis. The simulation results are compared with the test results to prove the feasibility of the coupling method and the validity of the simulation parameters. It provides a reference for analyzing and optimizing agricultural machinery components such as the opener, cover and roller. The method also offers the possibility for the simulation and analysis of other agricultural machines.

## 2. Composition and Structure of the Seeding Monomer

This paper takes the King Helen soybean seeding monomer as the test object, whose 2D perspective view is shown in Figure 1. The seeding monomer mainly consists of the opener, seed box, seedmeter, coverer, and roller through a combination of connecting devices. The main working parts of the sowing unit are described below.

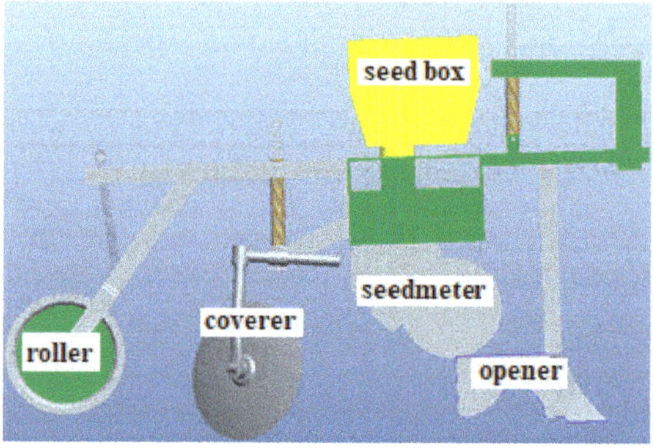


**Figure 1.** Two-dimensional perspective view of the seeding monomer.

### 2.1. Opener

The function of the opener is mainly to open the seed furrow, guide the seeds into the seed furrow and make the wet soil cover the seeds when the seeding monomer is working. Its main structure types are hoe shovel type, wide wing shovel type, arrow shovel type, core share type, slide knife type, double disc type, single disc type, etc. The opener on the seeding monomer of this paper is the core-share type opener, whose structure is shown in Figure 2.

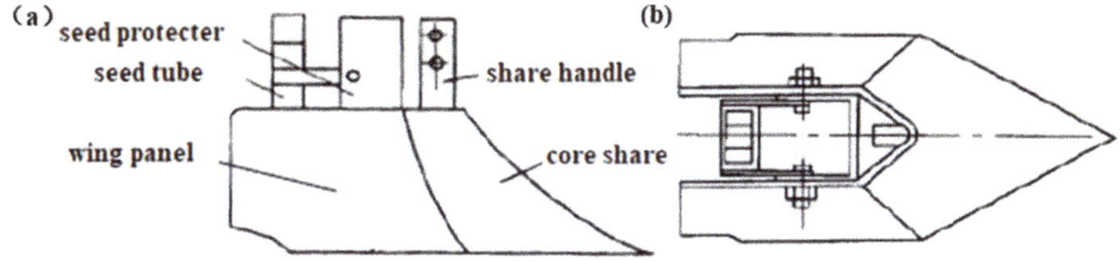


**Figure 2.** (**a**)The front view and (**b**) top view of core-share opener [13].

## 2.2. Vertical Type-Hole Wheel Seedmeter

The vertical type-hole wheel seedmeter mainly consists of seed cylinder, seed rower body, type-hole wheel, seed scraper and seed guard, etc.; see Figure 3.

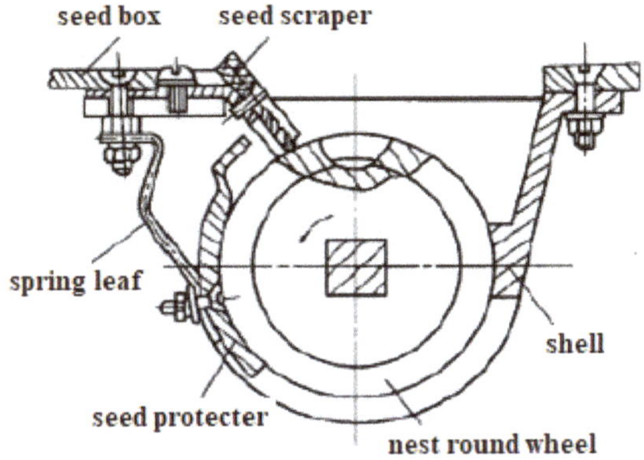


**Figure 3.** Structure diagram of the type-hole wheel seedmeter [13].

### 2.2.1. Type-Hole Wheel

The shape of the profile hole can be cylindrical, conical, and hemispherical. In order to reduce damage to the seed when filling and scraping, generally, the type-hole is paired with a front groove, a tail groove, or a chamfer. Its diameter and depth with the size of the soybean seed are sized to adapt.

In this paper, the seedmeter is a double-row type-hole wheel with a diameter of 200 mm, a hemispherical hole shape, a hole diameter of 8.5 mm, and 100 holes. When the type-hole wheel rotates, the seeds are distributed inside each hole, and the seeds are broadcast with the rotation of the wheel. See Figure 4.

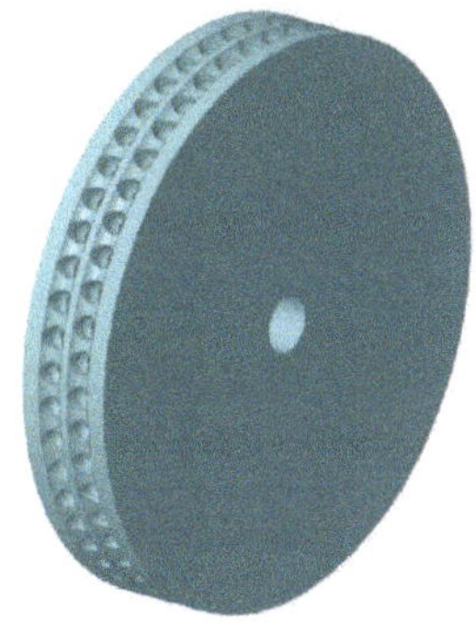

**Figure 4.** The double-row type-hole wheel.

2.2.2. Seed Scraper

The seed scraper used on the type-hole wheel seed meter has various forms, and the common ones include brush, rubber seed scraping tongue, brush seed scraping wheel, rubber brush seed wheel, steel knurled seed scraping wheel and so on. The installation position of the seed scraper is indicated by the β angle, generally 22~45°. The seed scraper in this paper is a rubber seed scraper whose installation angle β is 30°. In order to avoid damaging the seeds, the seed scraper cannot be installed vertically above the type-hole wheel, so it should be installed with a certain safety angle α. In this paper, angle α is 10°, see Figure 5.

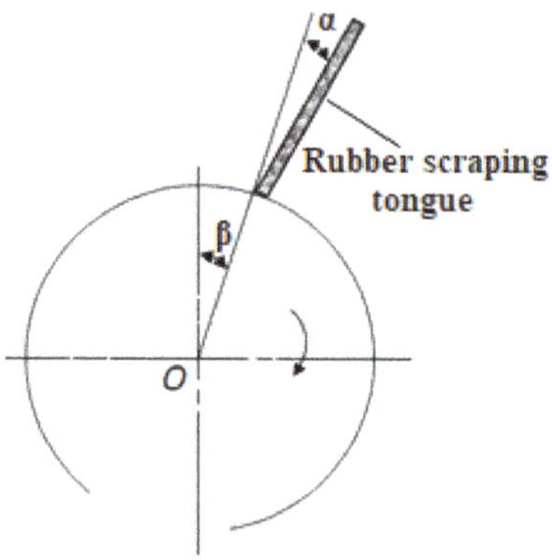


**Figure 5.** The installation angle of rubber scraping tongue [13].

2.2.3. Seed Guard Plate

The seed guard is used to ensure that the seed in the type-hole does not fall out from the type-hole during movement so that it reaches the seed-throwing place and is accurately put into the seed furrow. The material of the seed guard is made of iron, plexiglass or foam, and some seed guards are part of the seedmeter shell. In this paper, the seed guard is made of metal tin, which is connected with the seedmeter shell.

*2.3. Coverer*

After the seeds fall into the bottom of the furrow, the furrow opener will cover the seeds with a shallow layer of soil, and then it needs to be covered to make it reach a certain depth of coverage. The type of cover in this paper is a double-disc-type eight-character cover, as shown in Figure 6. The angle of tension of the double-disc cover can be adjusted. When the cover tension angle was varied from 50° to 70°, the congestion of the disc was relieved as the disc tension angle increased, indicating that the larger the tension angle of the cover disc was, the less the soil was disturbed [14]. Based on the above study, the tension angle of the overburden disc is 60° during the simulation.

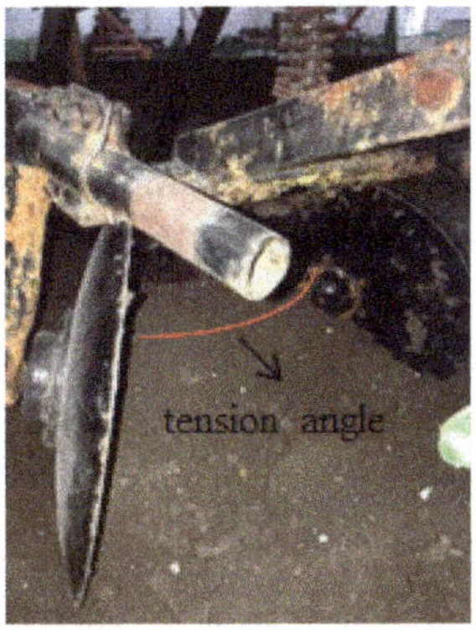


**Figure 6.** Double-disc-type eight-character cover.

### 2.4. Compacting Roller

Compaction is very necessary after covering soil. The main structure types of compacting roller are: a cylindrical compacting roller, concave and convex compacting rollers, a conical compound compacting roller, rubber ring compacting roller, wide compacting roller, narrow rubber compacting roller, etc. The roller of the seeding monomer in this paper is a cylindrical compacting roller with a diameter of 260 mm and a working width of 1000 mm, as shown in Figure 7.

**Figure 7.** Cylindrical compacting roller.

## 3. Covering and Compacting Test

The tests are conducted in the soil tank test bed of the agricultural laboratory of Jilin University. The soil type is a sandy loam with a moisture content of 18 $\pm$ 2%. The soil tank test vehicle is connected to the seeding monomer; see Figure 8a. To construct the seed furrow artificially, two sets of seed furrows with a height of 60 mm, a width of 150 mm and a length of 4000 mm are constructed on the surface of the seed bed, and the seed trench appears in the middle, as shown in Figure 8b. The seeds are sown by hand with a distance of 300 mm between every two seeds. The place of the seeds is marked on a string on the outside of the seed trench; see Figure 8c,d. The vertical distance of the seeds is also measured, as shown in Figure 8e.

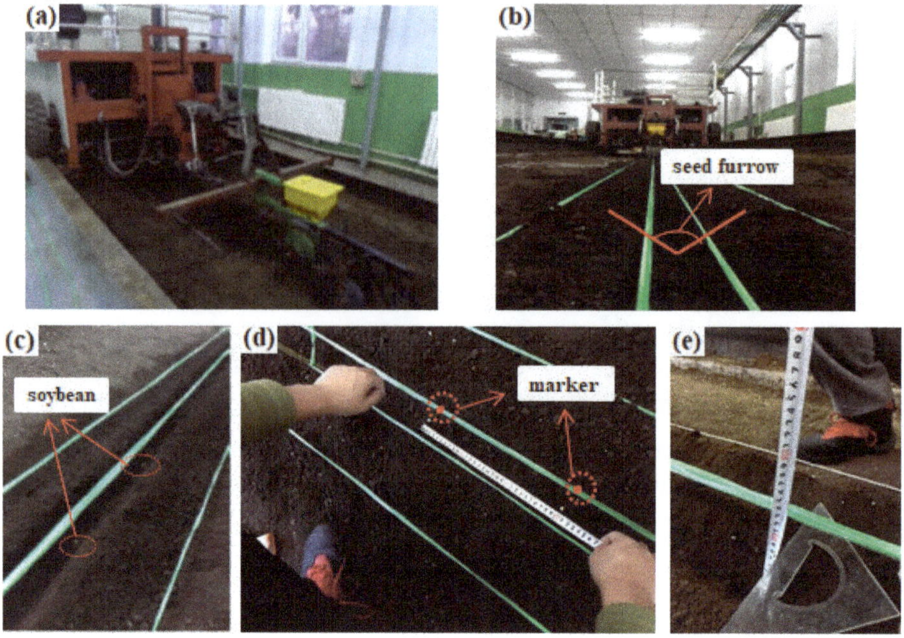


**Figure 8.** The pictures of the covering and compacting test process, (**a**) the seeding monomer, (**b**) the seed furrow, (**c**) soybean seed particles seeded in the seed furrow, (**d**) the measurement of seed particle spacing and (**e**) the measurement of vertical distance of soybean seed particles.

The covering test is carried out as follows: firstly, only the cover of the sowing unit is operated, the covering disc is tensioned at an angle of 60°, and the test vehicle is driven at a speed of 0.75 m/s, 1.11 m/s and 1.47 m/s. After covering, the soil covering the seeds is removed, and the horizontal and vertical distances of the seeds are measured. The change in the position of the soybean seed particles after mulching is analyzed, and the test data are recorded.

The compacting test was carried out as follows, with the cover and compactor of the sowing monomer working simultaneously and the cover disc tensioned at an angle of 60°. After covering and compacting the seeds, the soil was removed from the top of the seeds, and the horizontal and vertical distances of the seeds were measured. The change to the position of the soybean seed particles after compacting was analyzed, and the test data were recorded.

## 4. Covering and Compacting Simulation

To simulate the working process of the seeding monomer during the covering and compacting test, the coupling between EDEM and RecurDyn needs to be used. The simulation setup steps are as follows: import the STL file of the 3D diagram into RecurDyn software; set the material property of the part to "Steel"; and then set the connection method between the rigid bodies involved in the seeding monomer (see Table 1).

**Table 1.** Connection between rigid bodies in the seeding monomer.

| Constraint Number | Constraint Object | Constraint Type |
|---|---|---|
| 1 | opener–ground | sliding pair |
| 2 | seeding wheel–shaft | revolute pair |
| 3 | coverer–shaft | revolute pair |
| 4 | cross beam–back beam | revolute pair |
| 5 | back beam–roller | revolute pair |

In order to realize the coupling of the two software types, it is necessary to set up the corresponding settings in EDEM (Version 2018). After opening the EDEM software, it is necessary to select RecurDyn Coupling in the coupling option of EDEM. Next, the wall files are imported into EDEM. The height of the soil tank is set according to the height of the seeding monomer. After analysis, the dimensions of the soil tank are determined as 4820 mm$\times$ 820 mm $\times$ 420 mm. The soybean seed variety in this paper is SN42 and has an ellipsoidal shape, the DEM model of soybean seed particles was established using the 13-sphere model, and a population of soybean seed particles was generated in the simulation according to a normal distribution [15,16]. The soil particle shapes are sphere-like and triangle-like. In order to save simulation time, the soil particle model was simulated with a particle size of 10 mm and a population of soil particles was generated according to a uniform distribution. The physical and mechanical parameters of soybean seed particles, soil particles and the material of the seeding monomer (galvanized steel) are determined by the author's period research, as shown in Table 2 [15,16].

**Table 2.** Physical and mechanical parameters of the simulated materials [15,16].

| Material | Density, kg/m$^3$ | Poisson's Ratio | Elasticity Modulus, Pa |
|---|---|---|---|
| Soybean | 1257 | 0.4 | $7.60 \times 10^8$ |
| Galvanized steel | 7850 | 0.3 | $7.90 \times 10^{11}$ |
| Soil | 1844 | 0.25 | $1.00 \times 10^6$ |

In order to save calculation time, the radius of soil particles is taken as 10 mm for the simulation [17], and about 200,000 soil particles are generated according to the uniform size. The contact model is the Edinburgh model [18–21].

At the time of simulation, the soybean variety is SN42. The parameters of the material interactions during the simulation are measured. In this paper, the particle size becomes larger during simulation, and the surface energy between particles is adjusted to 4 by simulation analysis. The specific parameters are shown in Table 3 [15,16].

**Table 3.** Parameters of material interactions [15,16].

| Parameter | SN42–SN42 | SN42–Soil | Soil–Soil |
|---|---|---|---|
| Coefficient of restitution | 0.627 | 0.75 | 0.6 |
| Coefficient of static friction | 0.2 | 0.254 | 0.9 |
| Coefficient of rolling friction | 0.02 | 0.011 | 0.7 |
| Constant pull-off force, N | 0 | 0 | 0 |
| Surface energy, J/m$^2$ | 0 | 0.5 | 4 |
| Contact Plasticity Ratio | 0.35 | 0.35 | 0.35 |
| Slope Exp | 1 | 1.5 | 1.5 |
| Tensile Exp | 1 | 1 | 1 |
| Tangential stiff Multiplier | 0.67 | 0.67 | 0.67 |

The coefficient of restitution, the coefficient of static friction, and the coefficient of rolling friction between the soil particles and the soil tank and the soybean seed particles and the soil tank have little effect on the simulation results in this paper. Therefore, the values are taken by the system as default.

Figure 9 shows a screenshot of the simulation interface of the two software types at different simulation moments for a seeding monomer working speed of 1.47 m/s.

(a)

(b)

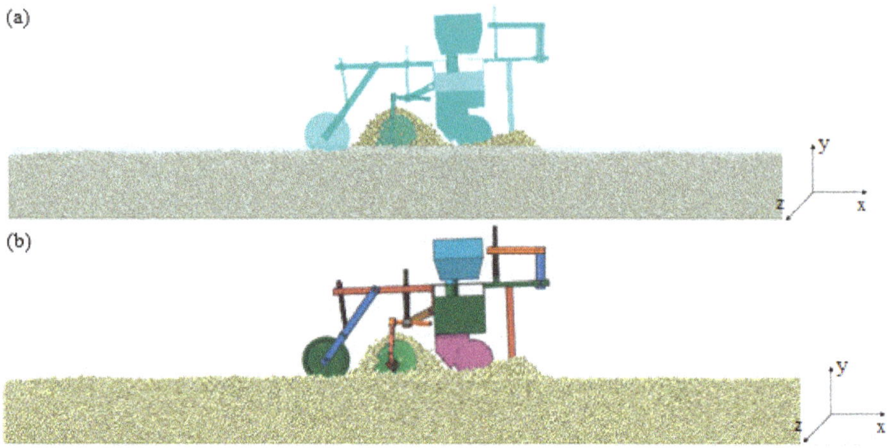


**Figure 9.** Screenshot of (**a**) EDEM simulation and (**b**) RecurDyn simulation at t = 2 s.

## 5. Analysis of the Results

Analysis and statistics of the changes in the position of soybean seed particles after covering and compacting, the opening of the furrow opener and the distribution of grain spacing at different working speeds were carried out.

### *5.1. Analysis of Covering and Compacting Results*

The change in the position of soybean seed particles after covering is compared with the test results. When conducting the simulation, the +X direction was set as the horizontal driving direction of the seeding monomer, the *Z*-axis direction corresponded to the lateral displacement of the test, and the *Y*-axis direction corresponded to the vertical direction.

The analysis process for the variation in horizontal and vertical displacements of a seed particle is taken as an example. The position of the soil particle without covering is shown in Figure 10a. Figure 10b shows the position of the soybean seed particle when it has been covered with soil. Figure 10c shows the position of the soybean seed particles in the soil after compacting. Zooming in on the images, it can be clearly observed that the soybean seed particles are significantly displaced in the *X*-axis direction, as shown in Figure 10d. Five soybean seed particles are selected to obtain information on the position of soybean seed particles in the X, Y and Z directions, and the test data are analyzed.

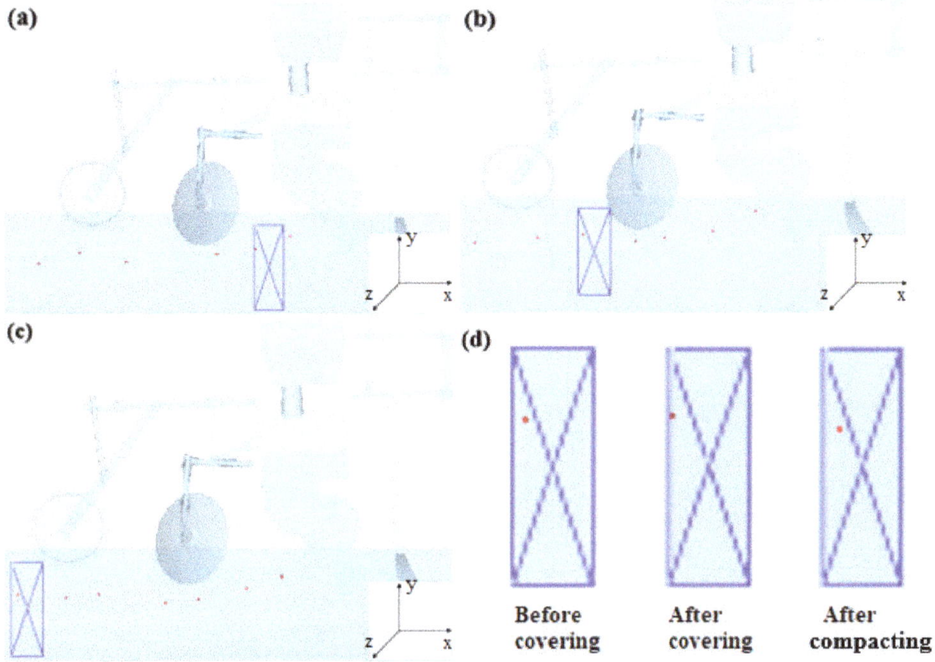


**Figure 10.** Position of soybean seed particles in the soil (**a**) before covering, (**b**) after covering, (**c**) after compacting and (**d**) partially enlarged view.

#### 5.1.1. Analysis of Covering Test Results

The simulation and test results [14] of the change in the vertical and horizontal position of soybean seed particles before and after covering at different working speeds of the seeding monomer are shown in Figure 11. The bars in the figure are the error band resulting from the processing of the replicate test and simulation results and are expressed as the standard deviation. The analysis shows that when the working speed of the seeding monomer increases, the distance along the vertical direction of the soybean seed particles after covering gradually increases, and the distance along the horizontal direction gradually decreases. The simulation results have the same trend as the test results as a whole.

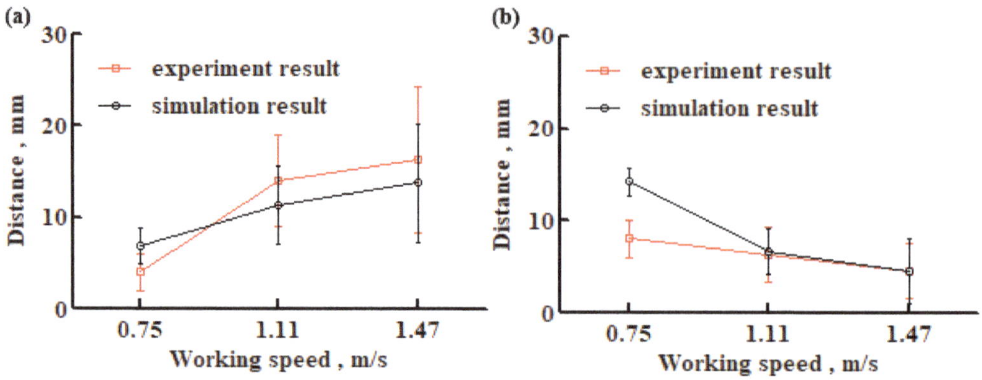


**Figure 11.** The relationship between the change in the position of soybean seed particles after covering and the working speed of the seeding monomer of (**a**) vertical and (**b**) horizontal.

5.1.2. Analysis of the Results of the Compacting Test

The simulation and test results [14] of the change in the vertical and horizontal position of soybean seed particles before and after compacting at different working speeds of the seeding monomer are shown in Figure 12. Analysis shows that with the increase in the working speed of the seeding monomer, the change in the position of soybean seed particles in both horizontal and vertical directions after compacting shows a trend of gradually becoming larger. When the working speed of the seeding monomer is 1.11 m/s, the horizontal and vertical position changes in soybean seed particles differ the most from the test results, and the difference between them and the test is 1.69 mm and 1.33 mm, respectively, which is a relatively small value. Therefore, it can be considered that the overall simulation results and the test results are relatively close to each other.

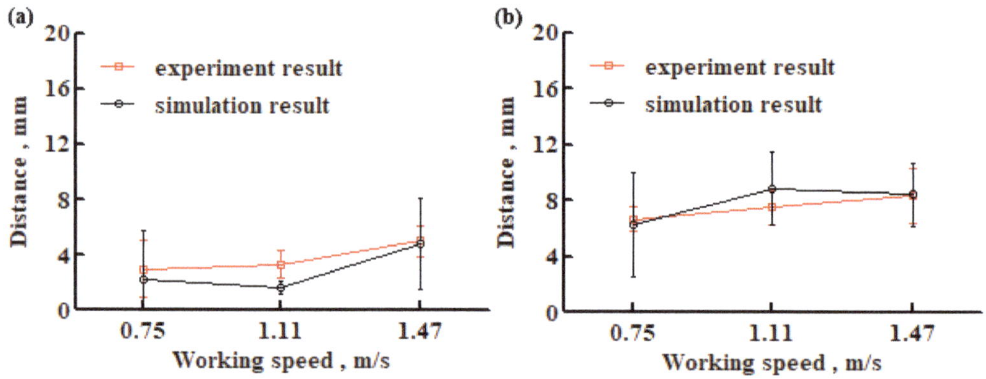


**Figure 12.** The relationship between the change in the position of soybean seed particles after compacting and the working speed of the seeding monomer of (**a**) vertical and (**b**) horizontal.

The change in the position of soybean seed particles in the vertical direction after covering and after compacting at different working speeds is shown in Figure 13. It can be seen from the analysis that the change in the vertical direction of soybean seed particles in the soil after covering and after compacting is not significant. The maximum difference between the two is 3.7 mm when the working speed of the seeding monomer is 0.75 m/s. The minimum difference between the two is 1.77 mm when the working speed of the seeding monomer is 1.11 m/s.

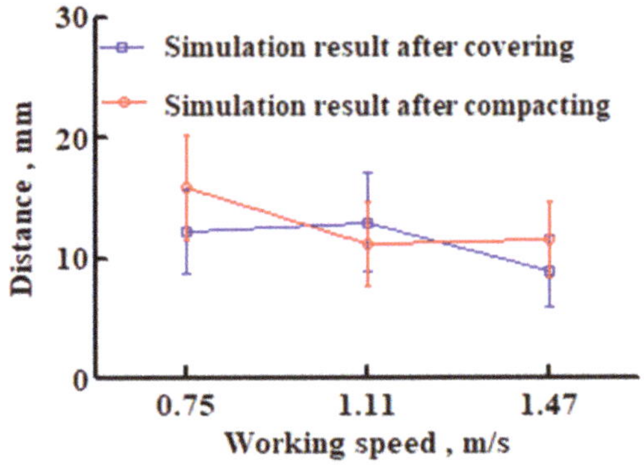


**Figure 13.** The change in the position of soybean seed particles in the vertical direction after covering and after compacting at different working speeds.

The comprehensive analysis shows that the faster the working speed of the seeding monomer, the greater the change in the position of the soybean seed particles that occurs. The reason for this is that the faster the seeding monomer works in contact with the soil particles, the more kinetic energy the soil particles will have, which means that the soil particles will be displaced more when they come to rest. The seeding monomer coupling simulation results have the same trend as the test results, and the parameters chosen for the coupling simulation are reasonable.

*5.2. Analysis of the Open Furrow Results*

The ditch angle θ of the opener when the working speed of the seeding monomer is 1.47 m/s, as shown in Figure 14. The red particle in the figure is the SN42 soybean seed particle that is falling down.

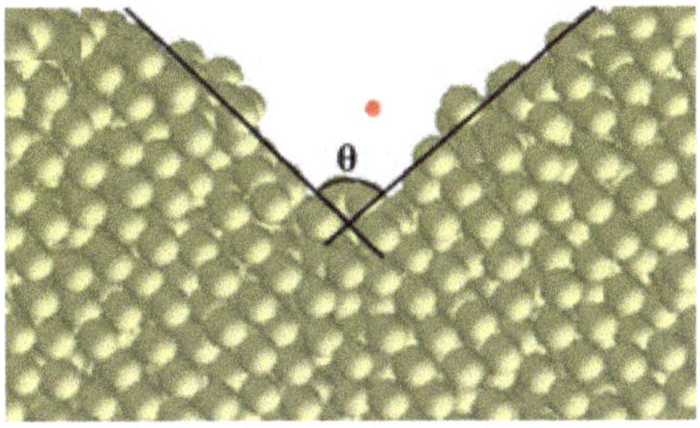


**Figure 14.** The ditch shape diagram at a seeding monomer working speed of 1.47 m/s.

The ditch angle is analyzed at the simulation moment of 1 s, 2 s and 3 s when the working speed of the seeding monomer is 0.75 m/s, and the average value is obtained. The ditch angle of the soil at working speeds of 0.75 m/s, 1.11 m/s and 1.47 m/s are analyzed separately, and the results are shown in Figure 15. The analysis shows that the ditch angle tends to decrease gradually as the working speed of the seeding monomer increases within the range studied in this paper. When the working speed is 0.75 m/s, the maximum ditch angle is 94.5°. When the working speed of the seeding monomer is 1.47 m/s, the ditch angle is the smallest, and the value is 90.31°.

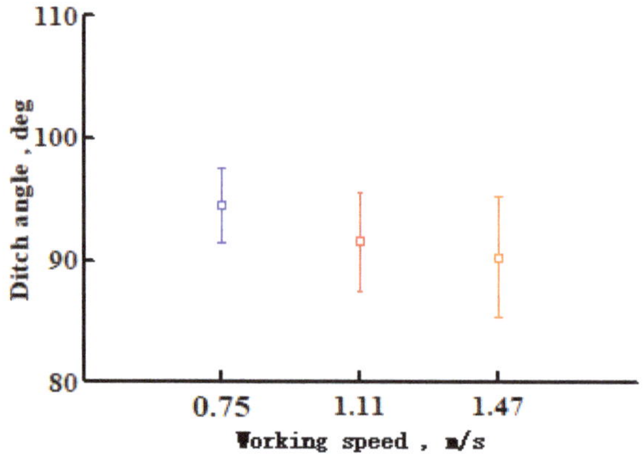


**Figure 15.** The ditch angle of the soil at different working speeds.

From the above analysis, it can be seen that the ditch angle gradually decreases as the working speed of the seeding monomer increases. The likely reason for this is that the faster the seeding unit works in contact with the soil particles, the greater the kinetic energy of the soil particles, which results in more soil particles being distributed on the outside of the seed furrow.

### *5.3. Seed Spacing Analysis*

In the simulation, a 10 mm × 10 mm × 10 mm particle factory is set up at the seed discharge port of the seedmeter, and the soybean seed particles are generated in the particle factory at a rate of 10 per s. The speed of the particle factory is the same as that of the seeding monomer. Therefore, the theoretical seed spacing for soybean seeding, corresponding to the forward speed of the seeding monomer, was 75 mm, 110 mm, and 147 mm for 0.75 m/s, 1.11 m/s, and 1.47 m/s, respectively.

The distribution of soybean seed particles in the soil after covering and compacting the seeds at different working speeds is shown in Figures 16–18.

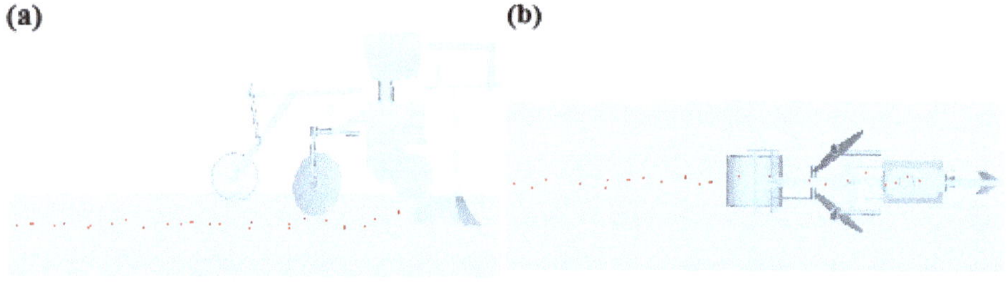


**Figure 16.** Distribution of soybean seed particles in the soil in (**a**) XOY plane and (**b**) XOZ plane at a working speed of 0.75 m/s.

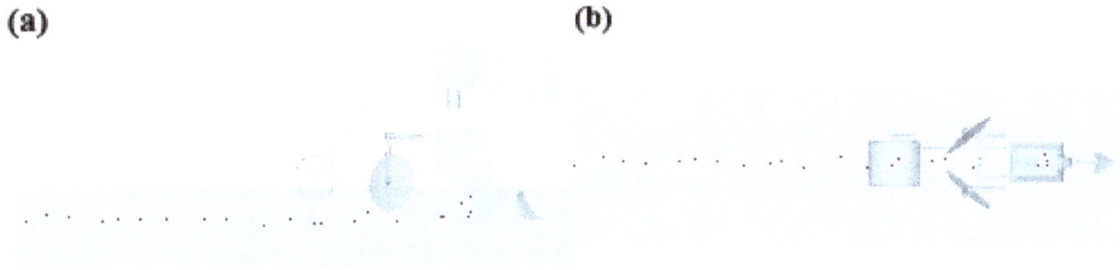


**Figure 17.** Distribution of soybean seed particles in the soil in (**a**) XOY plane and (**b**) XOZ plane at a working speed of 1.11 m/s.

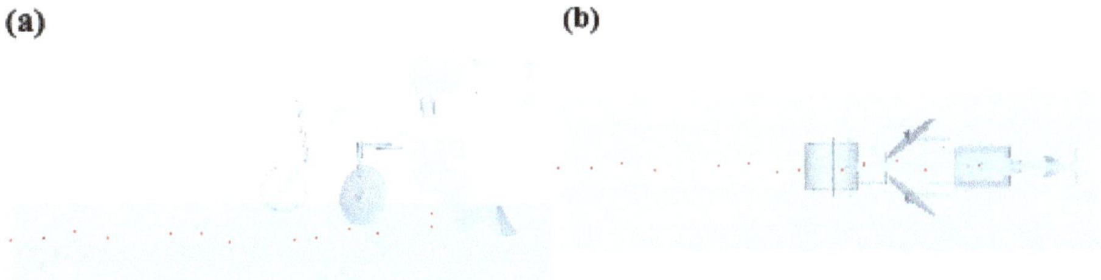


**Figure 18.** Distribution of soybean seed particles in the soil in (**a**) XOY plane and (**b**) XOZ plane at a working speed of 1.47 m/s.

Figure 19 shows the difference between the theoretical seed spacing and the simulation results at different working speeds. For the sake of comparison, the concept of relative error is defined. The relative error is the ratio of the difference between the mean value of the simulation results and the theoretical value. When the working speed is 0.75 m/s, the difference between the simulation result and the theoretical seed spacing is only 0.48 mm, with a relative error of 0.6%. When the working speed is 1.11 m/s, the difference between the simulation result and the theoretical seed spacing is the largest, and the difference is 6.44 mm, with a relative error of 5.8%. When the working speed is 1.47 m/s, the difference between the simulation result and the theoretical seed spacing is 3.42 mm, and the relative error is 2.3%. The comprehensive analysis shows that the simulation results are basically consistent with the theoretical grain distance at different working speeds.

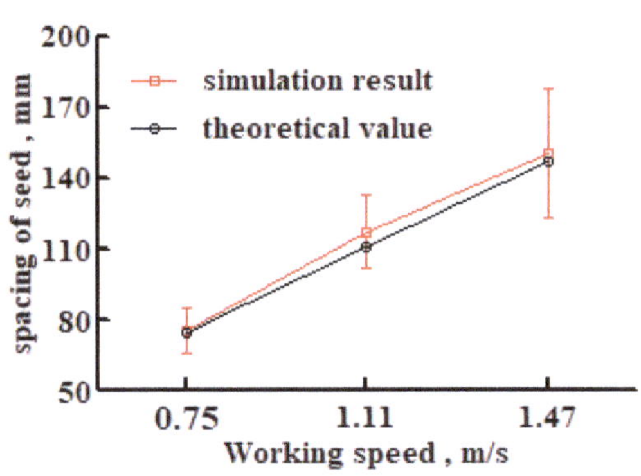


**Figure 19.** The relationship between seed spacing and working speed.

## 6. Conclusions

In this paper, tests on the covering and compacting processes of seeding monomers are carried out, and the above processes are simulated and analyzed by using the DEM coupled with MBD. Simulation analysis is also carried out for the open furrow situation and the uniformity of seed spacing. In the future, DEM and MBD coupling will be widely used for the optimal design of earth-touching machine components. The specific conclusions are as follows:

(1) The EDEM and RecurDyn software are coupled to simulate the process of opening, seed throwing, covering, and compacting of the seeding monomer. The comparison of simulation and test results shows that with the increase in working speed, after covering, the position change in soybean seed particles in the vertical direction gradually increases and in the horizontal direction gradually decreases.

(2) After compacting, the position change of soybean seed particles in the vertical and horizontal directions gradually increases. The simulation is basically consistent with the trend of the test results. Meanwhile, after the experience of covering the soil and then compacting it, the change in the vertical direction of soybean seed particles is not significant.

(3) The ditching angle of the soil gradually decreases when the working speed of the seeding monomer increases. The average seed spacing of the simulation is basically the same as the theoretical spacing at different working speeds of the seeding monomer.

(4) In the simulations of this paper, the boundary conditions of the mechanical components have been ignored, which may produce certain errors and, therefore, will need to be taken into account in future research work. The influence of different soil particle models on the simulation results also needs further study.

**Author Contributions:** Conceptualization, D.Y.; methodology, D.Y.; validation, D.Y. and W.G.; investigation, W.G. and resources, J.Y. and W.G.; writing—original draft preparation D.Y. and N.Z.; writing—review and editing, Y.W. and N.Z.; supervision, Y.T. and T.X.; project administration, J.Y.; funding acquisition, J.Y. All authors have read and agreed to the published version of the manuscript.

**Funding:** The authors are grateful to the National Natural Science Foundation of China (No. 52130001) for the financial support of this work.

**Data Availability Statement:** Not applicable.

**Conflicts of Interest:** The authors declare no conflict of interest.

## References

1. Van Zeebroeck, M.; Tijskens, E.; Dintwa, E.; Kafashan, J.; Loodts, J.; De Baerdemaeker, J.; Ramon, H. The discrete element method (DEM) to simulate fruit impact damage during transport and handling: Model building and validation of DEM to predict bruise damage of apples. *Postharvest Biol. Technol.* **2006**, *41*, 85–91. [CrossRef]
2. Kafashan, J.; Van Zeebroeck, M.; Ramon, H.; Tijskens, B. A novel approach to a realistic discrete element modelling (DEM) in 3D. *Commun. Agric. Appl. Biol. Sci.* **2007**, *72*, 205–208. [PubMed]
3. Kafash, J.; Van Liedekerke, P.; Ramon, H.; Tijskens, B. An Approach to represent realistic particles of bulk assembly in three-dimensional-DEM simulations and applications. *Commun. Agric. Appl. Biol. Sci.* **2011**, *76*, 33–36.
4. Kafashan, J.; Tijskens, B.; Ramon, H. Shape modelling of fruit by image processing. *Commun. Agric. Appl. Biol. Sci.* **2005**, *70*, 161–164. [PubMed]
5. Obermayr, M.; Vrettos, C.; Eberhard, P.; Däuwel, T. A discrete element model and its experimental validation for the prediction of draft forces in cohesive soil. *J. Terramech.* **2014**, *53*, 93–104. [CrossRef]
6. Yu, Y.J.; Fu, H.; Yu, J.Q. DEM-based simulation of the corn threshing process. *Adv. Powder Technol.* **2015**, *26*, 1400–1409. [CrossRef]
7. Kafashan, J.; Wiącek, J.; Noorhazlinda Abd, R.; Gan, J. Two-dimensional particle shapes modelling for DEM simulations in engineering: A review. *Granul. Matter* **2019**, *21*, 80. [CrossRef]
8. Kafashan, J.; Wiacek, J.; Ramon, H.; Mouazen, A.M. Modelling and simulation of fruit drop tests by discrete element method. *Biosyst. Eng.* **2021**, *212*, 228–240. [CrossRef]
9. Yan, D.X.; Yu, J.Q.; Wang, Y.; Zhou, L.; Sun, K.; Tian, Y. A Review of the Application of Discrete Element Method in Agricultural Engineering: A Case Study of Soybean. *Processes* **2022**, *10*, 1305. [CrossRef]
10. Wang, Y.; Yu, J.Q.; Yu, Y.J. Validation of a coupled model of discrete element method with multibody kinematics to simulate the screening process of a swing-bar sieve. *Powder Technol.* **2019**, *346*, 193–202. [CrossRef]
11. Yuan, J.; Yu, J.Q. Analysis on Operational Process of Self-excited Vibrating Subsoiler Based on DEM-MBD Coupling Algorithm. *Trans. Chin. Soc. Agric. Mach.* **2020**, *51*, 17–24.

12. Xu, T.Y.; Yu, J.Q.; Yu, Y.J.; Wang, Y. A modelling and verification approach for soybean seed particles using the discrete element method. *Adv. Powder Technol.* **2018**, *29*, 3274–3290. [CrossRef]
13. Chinese Academy of Agricultural Mechanization Sciences. *Agricultural Machinery Design Manual*; China Agricultural Science and Technology Press: Beijing, China, 2007.
14. Xu, T.Y.; Zhang, R.X.; Wang, Y.; Jiang, X.M.; Feng, W.Z.; Wang, J.L. Simulation and Analysis of the Working Process of Soil Covering and Compacting of Precision Seeding Units Based on the Coupling Model of DEM with MBD. *Processes* **2022**, *10*, 1103. [CrossRef]
15. Yan, D.X.; Yu, J.Q.; Zhang, N.; Tian, Y.; Wang, L. Test and Simulation Analysis of Soybean Seed Throwing Process. *Processes* **2022**, *10*, 1731. [CrossRef]
16. Yan, D.X.; Yu, J.Q.; Wang, Y.; Zhou, L.; Tian, Y.; Zhang, N. Soil Particle Modeling and Parameter Calibration Based on Discrete Element Method. *Agriculture* **2022**, *12*, 1421. [CrossRef]
17. Gürsoy, S.; Chen, Y.; Li, B. Measurement and modelling of soil displacement from sweeps with different cutting widths. *Biosyst. Eng.* **2017**, *161*, 1–13. [CrossRef]
18. Janda, A.; Ooi, J.Y. DEM modeling of cone penetration and unconfined compression in cohesive solids. *Powder Technol.* **2016**, *293*, 60–68. [CrossRef]
19. Thakur, S.C.; Morrissey, J.P.; Sun, J.; Chen, J.F.; Ooi, J.Y. Micromechanical analysis of cohesive granular materials using the discrete element method with an adhesive elasto-plastic contact model. *Granul. Matter* **2014**, *16*, 383–400. [CrossRef]
20. Ma, Y.J.; Wang, A.; Zhao, J.G.; Hao, J.; Li, J.; Ma, L.; Zhao, W. Simulation analysis and experiment of drag reduction effect of convex blade subsoiler based on discrete element method. *Trans. Chin. Soc. Agric. Eng.* **2019**, *35*, 16–23.
21. Wang, X.L.; Hu, H.; Wang, Q.G.; Li, H.; He, J.; Chen, W. Calibration Method of Soil Contact Characteristic Parameters Based on DEM Theory. *Trans. Chin. Soc. Agric. Mach.* **2017**, *48*, 78–85.

*agriculture*

MDPI

*Article*

# Measurement Method of Collision Restitution Coefficient between Corn Seed and Soil Based on the Collision Dynamics Theory of Mass Point and Fixed Surface

**Zhen Gao [1,2], Caiyun Lu [1,2,*], Hongwen Li [1,2], Jin He [1,2], Qingjie Wang [1,2], Shenghai Huang [1,2], Yunxiang Li [1,2] and Huimin Zhan [3]**

1 College of Engineering, China Agricultural University, Beijing 100083, China
2 Scientific Observing and Experiment Station of Arable Land Conservation (North Hebei), Ministry of Agricultural and Rural Affairs, Beijing 100083, China
3 Shaanxi Agricultural Machinery Appraisal and Extension Station, Xi'an 710068, China
* Correspondence: lucaiyun@cau.edu.cn; Tel.: +86-010-6273-7300

**Citation:** Gao, Z.; Lu, C.; Li, H.; He, J.; Wang, Q.; Huang, S.; Li, Y.; Zhan, H. Measurement Method of Collision Restitution Coefficient between Corn Seed and Soil Based on the Collision Dynamics Theory of Mass Point and Fixed Surface. *Agriculture* **2022**, *12*, 1611. https://doi.org/10.3390/agriculture12101611

Academic Editors: Mustafa Ucgul and Chung-Liang Chang

Received: 23 August 2022
Accepted: 26 September 2022
Published: 4 October 2022



**Abstract:** The collision restitution coefficient (CRC) is the essential parameter of the discrete element method (DEM) to study the interaction mechanism between corn seed and soil. The accuracy of its measurement results is the criticalness to ensuring simulation accuracy. In the current study, the CRC between corn seed and soil is mainly measured by the method of corn seed colliding with soil. However, since the soil is a granular body, the essence of the collision of corn seed with soil is the collision between one corn seed and a quantity of soil particles, resulting in inaccurate measurement results soil particles. Therefore, based on the collision dynamics theory of mass point and fixed surface, this paper proposed a measurement method of CRC to study the interaction mechanism between corn seed and soil. This paper analyzed the influence of mass point selection on the measurement results of CRC in the collision process. The measurement method was determined by taking soil particle as mass point and corn seed as fixed surface. To verify the feasibility of this method, a CRC measurement system was established based on high-speed camera technology. The soil particle was dropped at the heights of 10, 15, 20, and 25 cm to collide with the corn seed. The separation velocity and approaching velocity of the soil particle was measured to obtain the CRC between the corn seed and soil. Taking the CRC as the essential parameter of discrete element simulation, the discrete element simulation experiment (DESE) and the soil bin experiment (SBE)of corn seed free fall impact on soil were carried out, and the experiment results were compared and analyzed. In the experiment, the falling height (40, 50, 60, 70 cm) of corn seed and the touching position of the corn seed (positions a, b, c, d) were taken as the influencing factors. The movement state of corn seed and the separation approaching velocity ratio (SAVR) $\xi$ during the collision were taken as the indexes. The results showed that when the positions of corn seed touched the soil was the same, with the increase of the falling height, the SAVR $\xi$ obtained from the DESE and the SBE showed a downward trend. When the falling height was the same, and the corn seed touched the soil at positions b, c, and d, the motion state in the DESE was basically the same as that in the SBE. The relative error of the SAVR between the DESE and the SBE was less than 10%, which proves the feasibility of the method proposed in this paper. The measurement method of the CRC between corn seed and soil proposed in this study provided a theoretical research basis for clarifying the interaction mechanism between corn seed and soil and solving the seed bouncing in high-speed precision sowing.

**Keywords:** corn seed; soil; collision restitution coefficient; EDEM

## 1. Introduction

The mechanization level of corn sowing has reached 90%. During the sowing operation, corn seeds are discharged from the seed metering device and fall into the seedbed through the seed guide pipe. During the implantation process, the corn seeds collide with the soil and bounce on the seedbed, resulting in a high variation coefficient of sowing depth and a obviously difference in seed implantation position, which seriously affects the operation quality of the corn planter [1]. Studying the movement state of seed after collided with soil and clarifying the mechanism of seed-soil interaction is exceedingly essential to solve the seed bouncing on the seedbed and improve the quality of precision planting.

Currently, the primary methods to study the mechanism of corn seed and soil interaction are field experiments and SBE. Limited by the observation technology, operating environment, and the instantaneous completion of seed colliding with soil during seed dropping, it is tough to obtain the movement change process caused by seed colliding with soil during field experiments [2]. The SBE can't effectively obtain the force change and energy change when the seed collide with the soil, and it's complicated to analyze the seed soil interaction mechanism [3]. The DEM is widely used in particle motion analysis. It can obtain the movement velocity, force, energy loss, and other information of the working object at each movement time. It is an effective means to study the mechanism of seed-soil interaction. When using the DEM, the appropriate parameter setting is the basis for studying the interaction between materials, materials, and mechanical devices [4], which is of great significance for accurately simulating the actual physical behavior of macro granular materials [5]. Among them, the CRC, as the essential parameter for studying the problem of seed impacting soil, is of great significance for analyzing the mechanism of seed-soil interaction [6,7].

The CRC is a parameter reflecting the deformation recovery ability of objects during a collision. At present, scholars have carried out a lot of research on the CRC of agricultural materials. The existing research usually uses the impact experiment to measure CRC [8]. Feng et al. [9] measured and analyzed the CRC of potato when it collided with 65Mn steel, rubber, soil block, and other materials for potato tubers during harvest. Wang et al. [10] measured the elastic properties of processed tomato seeds when they collided with rubber, plexiglass, Q235, and other materials. Lu et al. [11] measured and analyzed the CRC of peanut pods when they collided with Q235, aluminum alloy, plexiglass, rubber, white sand (moisture content is 17.5%). Huang et al. [12] measured and analyzed the CRC of rape seeds when they collided with Q235, aluminum alloy, plexiglass, rubber, and other materials. Liu et al. [13] measured and analyzed the CRC between micro potato and steel plate. Zhou et al. [14] measured the CRC between three different shapes of corn seed and the galvanized wall, and analyzed the influence of corn seed's varying collision positions, collision velocity, and collision angles on the CRC. Sui et al. [15] analyzed the collision process between soybean and soil through high-speed camera technology, corrected the CRC between soybean seed and soil by comparing the experiment and discrete element simulation results, and calibrated the contact parameters between soybean and soil.

The above analysis shows that most of the research on the CRC of agricultural materials focuses on the measurement of the CRC between crop seeds with a regular shape (processing tomato seeds, peanut pods, micro potatoes, soybeans, etc.) and agricultural machinery equipment materials (Q235, aluminum alloy, plexiglass, rubber, etc.) with large volume and relatively flat surface. There are few studies on collision between soil and other particulate matter. In the study of the CRC between corn seed and soil, it is mainly measured by the method of corn seed colliding with soil. However, since the soil is a granular body, the essence of the collision of corn seed with soil is the collision between one corn seed and a quantity of soil particles, resulting in inaccurate measurement results soil particles. To this end, this paper proposes a method for measuring the CRC between corn seed and soil during collision based on the mass point fixed surface collision kinetics theory. Through the analysis of the collision process, the mass point and the fixed surface in the collision process of corn seed and soil were determined, and the measurement method of the CRC was clarified. In the cause of providing a theoretical research basis for defining the interaction mechanism between corn seed and soil and solving the problem of seed bouncing, we compare the results of DESE and SBE to analyze its feasibility.

## 2. Theoretical Analysis of CRC

According to the collision kinetics theory of mass point and fixed surface, the two interaction objects that collide with each other can be regarded as mass point and fixed surface, respectively. The mass point collides with the fixed surface at a certain velocity. According to the relative position relationship between the mass point and the fixed surface, the collision process can be divided into the approach stage and the separation stage, as shown in Figure 1.

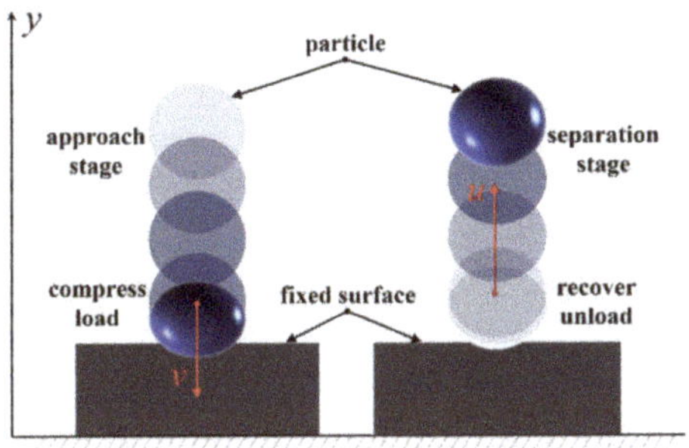


**Figure 1.** Analysis of collision process.

In the approach stage, the distance between the mass point and the fixed surface is continuously reduced, and the mass point contacts the fixed surface with the velocity $v$, which compresses and loads the fixed surface. The mass point is subjected to the reverse force of the fixed surface, and the velocity rapidly decreases to zero. At this stage, the kinetic energy of the mass point is loaded into the mass point fixed surface system, which is absorbed and converted into elastic strain energy and plastic strain energy.

In the separation stage, the distance between the mass point and the fixed surface gradually increases, and the mass point and the fixed surface begin to recover elastic deformation. The elastic strain energy is unloaded and converted into the kinetic energy of the mass point. The mass point velocity increases rapidly from zero and reaches the maximum value $u$ when it comes out of contact with the fixed surface. The plastic strain energy generated by the kinetic energy loading of the mass point in the approach stage is absorbed and converted into internal energy by the mass point and the fixed surface in the separation stage, resulting in plastic deformation [16–18].

The CRC is the primary parameter representing the energy change when the mass point collides with the fixed surface. It is only related to the material characteristics of the collision object. It reflects the ability of the object to recover the deformation during the collision. The more significant the CRC, the lower the energy loss, and the easier it is to recover to the initial state. At present, the CRC is mainly obtained by measuring the ratio of velocity (Newton measurement method), impulse (Poisson measurement method), and energy (strong measurement method) at the time of separation and approach [19]. Since it is difficult to directly obtain the change of impulse and energy in the actual measurement experiment, the Newton measurement method is adopted in this paper. In other words, the CRC is obtained by measuring the separation velocity and approach velocity of the mass point and calculating the SAVR.

## 3. Measurement Method of CRC between Corn Seed and Soil

Based on the kinetics theory of mass point fixed surface collision, the Newton method was used to measure the CRC between corn seed and soil. The criticalness lay in the selection of mass points and fixed surfaces. Therefore, the collision processes of corn seed and soil particles were analyzed, respectively, to obtain the optimal measurement method of the CRC of corn seed and soil.

### *3.1. Analysis of Collision Process when Taking Corn Seed as Mass Point*

According to the above theoretical analysis of the collision process, the collision model between corn seed and soil was established using corn seed as the mass point and the soil as the fixed surface. The soil was a granular system composed of *n* soil particles. There was a small cohesion between the particles, and the particles would be separated under the action of external forces. The macroscopic performance was the plastic deformation of the soil [20–22]. Considering that the shape of corn seed was complex, and irregular compared with that of wheat, soybean, peanut, and other crops, there were also obviously differences in the shape of corn seed of different varieties. The maximum average sphericity of Yuyu 30 was 76.01%, while the maximum average sphericity of Yudan 9953 was only 65.73% [23–25]. In this paper, only the bottom surface (opposite to the basal fruit stalk) and the side surface of the corn seed touching the soil were taken as examples for analysis, as shown in Figures 2 and 3.

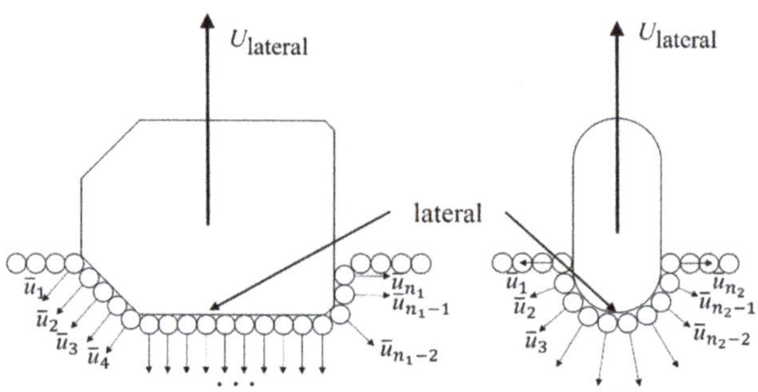


**Figure 2.** Instantaneous impulse Analysis when the side of corn seed touches soil.

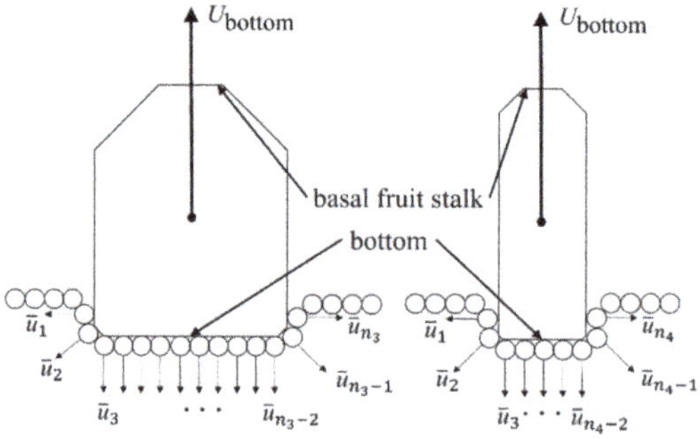


**Figure 3.** Instantaneous impulse Analysis when the bottom of corn seed touches soil.

According to the impulse theorem, the impact impulse acting on the *i*th soil particle is divided into external impact impulse $\overline{S}_i{}^{(e)}$ and internal impact impulse $\overline{S}_i{}^{(i)}$, then:

$$\overline{S}_i{}^{(e)} + \overline{S}_i{}^{(e)} = m_i\overline{u}_i - m_i\overline{v}_i (i = 1, 2, 3, \cdots, n) \quad (1)$$

For the particle system composes of $n$ soil particles, there were:

$$\sum_{i=1}^{n} \overline{S}_i^{(e)} + \sum_{i=1}^{n} \overline{S}_i^{(i)} = \sum_{i=1}^{n} m_i \overline{u}_i - \sum_{i=1}^{n} m_i \overline{v}_i \tag{2}$$

where $m_i$ is the mass of the $i$th soil particle; $\overline{u}_i$ is the velocity of the center of mass at the end of collision of the $i$th soil particle; $\overline{v}_i$ is the velocity of the center of mass of the $i$th soil particle at the beginning of collision, and the soil is stationary at the beginning of the collision, $\overline{v}_i = 0$; $\overline{S}_i^{(e)}$ is the external impact impulse on the $i$th soil particle; $\overline{S}_i^{(i)}$ is the internal impact impulse on the $i$th soil particle.

Since the internal impact impulse appeared as a pair, $\sum_{i=1}^{n} \overline{S}_i^{(i)} = 0$. Let the total mass of soil particles moving after collision be $M$, $\overline{u}_s$ and $\overline{v}_s$ is the average velocity of soil particle centroid at the end of collision and at the beginning of collision, where $\overline{v}_s = 0$, obtained from the theorem of mass center motion and impulse theorem:

$$\begin{cases} S = S_1 + S_2 \\ S = NV \\ S_1 = \sum_{i=1}^{n} \overline{S}_i^{(e)} \\ \sum_{i=1}^{n} \overline{S}_i^{(e)} = M\overline{u}_s - M\overline{v}_s \\ S_2 = NU \end{cases} \tag{3}$$

The simultaneous solution shows that:

$$U = V - \frac{M}{N}\overline{u}_s \tag{4}$$

where $S$ is the total impulse of the system; $S_1$ is the impulse absorbed by soil particles; $S_2$ is the impulse of corn seed collision and separation stage; $N$ is the quality of corn seed; $V$ is the maximum velocity before the corn seed contacts the soil; $U$ is the maximum velocity of corn seed separated from the soil.

When measuring the CRC, it was found that if the corn seed was taken as the mass point, the corn seed collides with the soil, which was essentially the contact collision between the corn seed and multiple soil particles. In the actual experiment, it was complicated to analyze the interaction between a single soil particle and corn seed during the collision process, and the obtained corn seed approach velocity $v$ and separation velocity $u$ could not be applied to the measurement of CRC [26].

Moreover, the movement state of corn seed in the separation stage is completed by the joint action of multiple soil particles. As shown in Figures 2 and 3, the number and velocity of soil particles are different at the side and bottom when they collide with the soil. The impulse of soil particles' absorption and transformation is different, and the movement state of corn seed in the separation stage is different, resulting in the difference in the measured SAVR. In addition, the CRC obtained according to the Newton measurement method is different.

### 3.2. *Analysis of Collision Process when Taking Soil Particles as Mass Point*

As shown in Figure 4, the collision model takes soil particles as mass points and corn seed as fixed surfaces. According to the conservation of momentum and the conservation of mechanical energy in the collision process system, the kinetic energy changes in the two stages are showed in Equations (5) and (6).

$$\Delta E_1 = \frac{1}{2}m_1(v_{10}^2 - v_{11}^2) + \frac{1}{2}m_2(v_{20}^2 - v_{22}^2) \tag{5}$$

$$\Delta E_2 = \frac{1}{2}m_1(v_{1t}^2 - v_{11}^2) + \frac{1}{2}m_2(v_{2t}^2 - v_{22}^2) \tag{6}$$

Equation (7) is obtained according to the momentum conservation of the collision system.

$$m_1v_{10} + m_2v_{20} = m_1v_{1t} + m_2v_{2t} \tag{7}$$

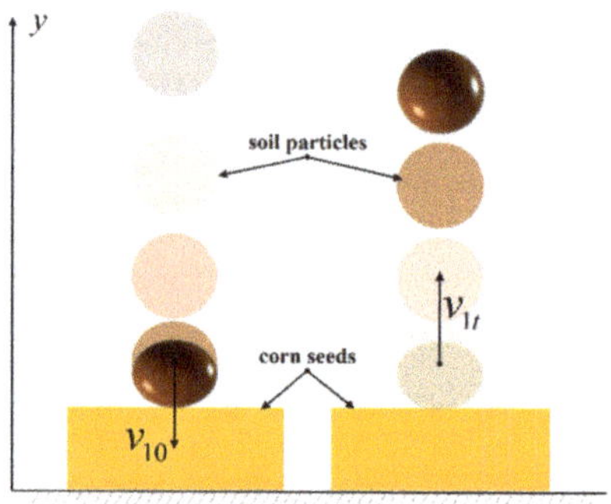


**Figure 4.** Process analysis of soil particle colliding with corn.

Equation (8) is obtained.

$$\begin{cases} \Delta E_1 = \frac{\frac{1}{2}m_1m_2(v_{10}-v_{20})^2}{m_1+m_2} \\ \Delta E_2 = \frac{\frac{1}{2}m_1m_2(v_{2t}-v_{1t})^2}{m_1+m_2} \end{cases} \tag{8}$$

where $\Delta E_1$ is kinetic energy change value in compression stage; $\Delta E_2$ is kinetic energy change value in the recovery stage, J; $m_1$ is the mass of the mass point, g; $m_2$ is the mass of the fixed surface, g; $v_{10}$ is the velocity before particle collision, m/s; $v_{20}$ is the velocity of the fixed surface before the collision, m/s; $v_{1t}$ is the velocity of the particle after the collision, m/s; $v_{2t}$ is the velocity of fixed surface after the collision, m/s; $v_{11}$ is the velocity of mass point when separation, m/s; $v_{22}$ is the velocity of fixed surface when separation, m/s.

where $v_{20}$, $v_{2t}$, $v_{22}$ both are 0 and $m_2$ is much larger than $m_1$. Accordingly:

$$\begin{cases} \Delta E_1 = \frac{1}{2}m_1{v_{10}}^2 \\ \Delta E_2 = \frac{1}{2}m_1{v_{1t}}^2 \end{cases} \tag{9}$$

According to the definition of CRC, then:

$$CRC = \sqrt{\frac{\Delta E_2}{\Delta E_1}} = \frac{v_{1t}}{v_{10}} \tag{10}$$

According to the above analysis, taking soil particles as mass point could effectively avoid the simultaneous contact between corn seed and multiple soil particles and realized the contact between single collision objects, which meets the mass point fixed surface collision kinetics theory requirements. According to the above analysis, the CRC could be obtained by measuring the separation velocity $v_{1t}$ and the approaching velocity $v_{10}$ of the mass point.

### 3.3. Measurement Method

According to the above analysis, the paper adopted the method of measuring the CRC with soil particle as mass point and corn seed as fixed surface. In the cause of obtaining the CRC between corn seed and soil, a CRC measurement system was built based on high-speed camera technology to measure the separation velocity and approach velocity of soil particles. To ensure the accuracy of the experiment system, the following requirements shall be met:

(1) The experiment was conducted in a relatively closed environment to avoid the influence of the airflow field in the experiment space on the measurement results.
(2) To obtain the CRC between corn seed and soil, improve the contrast and ensure the shooting effect, a simple CRC experiment bench was built with a white background plate.
(3) The white background plate was vertically placed on the desktop to obtain the actual distance of soil particle movement. The vertical distance between the high-speed camera and the soil particle movement plane in all experiments should be kept consistent.
(4) Drawn a straight line parallel to the desktop on the white background plate to facilitate the experiment and subsequent high-speed camera data analysis.The main experiment equipment was the revealer 5f01 high-speed camera produced by Hefei Fuhuang Junda high tech Information Technology Co., Ltd. (Hefei, China), with a shooting frame rate of 1000 fps and an exposure time of 998 μs, using high-speed video target tracking measurement software to process image data.

The specific operation steps during the experiment were:

(1) Fix the corn seed on the tabletop of the experiment bench, and the side is close to the background plate so that the surface is relatively smooth and flat, and the large plane with uniform distribution of horny endosperm is upward as the fixing surface.
(2) Open the high-speed camera control software to make the high-speed camera ready for shooting. Adjust the focus, exposure rate, white balance and other parameters of the high-speed camera according to the images collected by the control software. Make the white background plate and corn seed in the center of the video acquisition area, and ensure the best definition of the picture. Adjust the position of the lighting device to ensure that shadows do not cover the movement range of soil particles.
(3) To avoid the influence of sweat and grease secreted by fingers on the properties of soil particles, during the experiment, the soil particles were clamped with tweezers, and the tweezers were loosened at the straight positions of different heights to make them fall naturally and collide with corn seed. In the cause of capturing the complete falling process, trigger the high-speed camera shooting command before releasing the tweezers.

The collision separation and approach velocity of soil particles were analyzed and obtained by the high-speed video target tracking measurement software. After the shooting, the specific methods for processing the obtained video files were as follows:

(1) Coordinate system is set at the starting point of soil particle falling;
(2) Calibrate the size in the drawing, measure the scale in the drawing, and calculate the corresponding ratio between the actual size and the picture size: $L_{\text{Actual}}$:$L_{\text{Figure}} = k$:1 is used for conversion of experiment data. If the camera position changes during the experiment, it needs to be recalibrated;
(3) After the calibration is completed, the soil particle is set as the tracking target, and the target tracking module is used to obtain the movement trajectory of the soil particles, the separation velocity $v_{1t}$ of the soil particles after the collision, and the approaching velocity $v_{10}$ of the soil particles before the collision. The CRC is obtained from Equation (10).

## 4. Experiment Verification

To verify the feasibility of the method proposed in this paper, taking soil particle as mass point to collide with corn seed, measured the CRC between soil particle and corn seed. The DESE and the SBE were carried out with the corn seed as the mass point and the soil bin as the fixed surface, compared the movement state and SAVR of the corn seed.

*4.1. Measurement of CRC*

The experiment was carried out to measure the CRC between corn seed and soil. The moisture content of corn seed was 12.8%, the weight of a thousand seeds was 351 g, the average density was 1.197 kg/m$^3$, the shape was complete, and the surface was flat. To reduce the influence of corn seed' uneven surface on CRC measurement results, the large surface with a uniform and flat distribution of horny endosperm was used as the fixed surface. During the experiment, the soil was first treated with a sieve. After natural air drying, the soil particles were taken as the mass points. The soil bulk density was 1346 kg/m$^{3,}$ and the particle size range was 2–3 mm. It should be noted that according to the mass point fixed surface collision kinetics theory, the CRC is independent of the weight of the mass point. However, in the actual measurement, multiple particles were bonded together due to the little cohesion between soil particles in the natural state. Soil particles with large size were formed by the bonding of several small particles with little cohesion between them. In actual measurements, dispersion occurred after collision, which was inconsistent with the particle-fixed surface collision dynamics theory. Soil particles with small size, whose motion process was affected by air resistance close to its own gravity, greatly influence the motion state, and could not be used to measure the CRC. Therefore, the diameter of soil particles should not be too large or too small.

Using the measurement method of the CRC proposed above, the soil particles are clamped to fall freely at the heights of 10, 15, 20, and 25 cm to collide with corn particles, as shown in Figure 5. The experiment was repeated several times. The high-speed camera recorded the collision process, and the approaching velocity $v_{10}$ and separation velocity $v_{1t}$ were analyzed. Since the surface of corn seed is not an ideal plane, the direction of soil particles and corn will have a divergent shift after the collision, resulting in the measured separation velocity being smaller than the ideal state. In this paper, the maximum SAVR in the normal direction of the contact was taken as the CRC. Sorted out the experiment data, measured the collision approaching velocity and separation velocity of soil particles, calculated the SAVR, and took the maximum value as the corn seed soil CRC = 0.607.

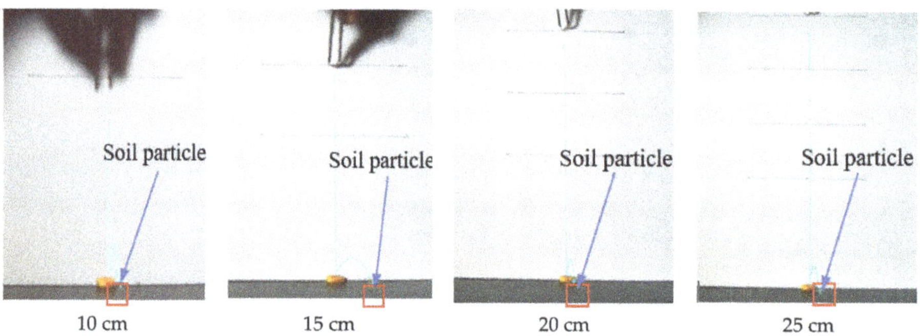


**Figure 5.** Image analysis of high-speed photography.

*4.2. Verification of Measurement Results*

Based on the CRC measured in the previous paper, we carried out the DESE and SBE of the seed soil interaction mechanism. We compared and analyzed the movement state of corn seed falling at different heights and touching the soil at different positions, as well as the SAVR $\xi$ of corn seed colliding the soil bin, to verify the accuracy of the CRC measured by the measurement method proposed in the article.

4.2.1. Seed Impact Soil DESE

Measure the size of corn seed used in the experiment, established a three-dimensional model of corn seed, as shown in Figure 6, and save as Step format was imported into

EDEM2018 software as a template. Considering the influence of calculation speed and accuracy, we establish a multi-ball model of corn seed with 11 particles, as shown in Figure 7.

A cylindrical soil bin with a diameter of 60 mm and a height of 10 mm was established to generate soil particles in the soil bin. The soil particle diameter was set to 1 mm to improve the accuracy and ensure the calculation speed. A polygon plate with a diameter of 60 mm was charged at the top of the soil bin, and linear motion is added. The plate motion compresses the soil bin downward by 2 mm, and the compression rate of the discrete element soil model is 20%. The contact model of soil particles was the Hertz-Mindlin non-sliding contact model. The contact model between corn seed and soil was linear spring model [27–29]. The values of characteristic velocity were showed in Table 1. The discrete element simulation model parameters of corn seed and soil were shown in Table 2, and the discrete element simulation contact parameters of corn seed and soil were shown in Table 3.

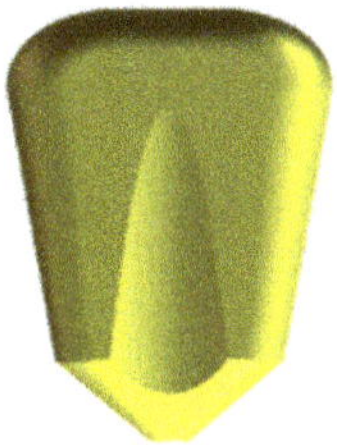

**Figure 6.** Three-dimensional model of corn seed.

**Figure 7.** Discrete element simulation model.

**Table 1.** Values of Characteristic Velocity.

| Number | Falling Height (mm) | Characteristic Velocity (mm/s) |
|---|---|---|
| 1 | 400 | 2857.1 |
| 2 | 500 | 3130.5 |
| 3 | 600 | 3429.3 |
| 4 | 700 | 3704.1 |

**Table 2.** Discrete element simulation model parameters of corn seed and soil.

| Parameter | Corn | | | | Soil | | | |
|---|---|---|---|---|---|---|---|---|
| | Particle Density $\rho$/(kg/m$^3$) | Particle Radius $r$/mm | Poisson's Ratio $v$ | Shear Modulus $G$/(MPa) | Particle Density $\rho$/(kg/m$^3$) | Particle Radius $r$/mm | Poisson's Ratio $v$ | Shear Modulus $G$/(MPa) |
| value | 1197 | 2.4~2.5 | 0.40 | $1.37 \times 10^6$ | 1346 | 0.5 | 0.40 | $1 \times 10^6$ |

Two factor and four level experiments were carried out. The factor levels were shown in Table 4. The positions of corn seed touched the soil are divided into four positions according to the shape characteristics of corn seed, that is, basal fruit stalk a, lateral surface b, large plane c, and bottom surface d, as shown in Figure 8.

**Table 3.** Discrete element simulation contact parameters of corn seed and soil.

| Parameter | Corn-Corn | | | Soil-Soil | | | Soil-Corn | | |
|---|---|---|---|---|---|---|---|---|---|
| | Coefficient of Static Friction $\mu s_1$ | Coefficient of Rolling Friction $\mu r_1$ | CRC $e_1$ | Coefficient of Static Friction $\mu s_2$ | Coefficient of Rolling Friction $\mu r_2$ | CRC $e_2$ | Coefficient of Static Friction $\mu s_3$ | Coefficient of Rolling Friction $\mu r_3$ | CRC $e_3$ |
| value | 0.431 | 0.0782 | 0.182 | 0.2 | 0.4 | 0.3 | 0.22 | 0.727 | 0.607 |

**Table 4.** Experiment factor level.

| Level | Touch Soil Position | Falling Height/(mm) |
|---|---|---|
| 1 | a | 400 |
| 2 | b | 500 |
| 3 | c | 600 |
| 4 | d | 700 |

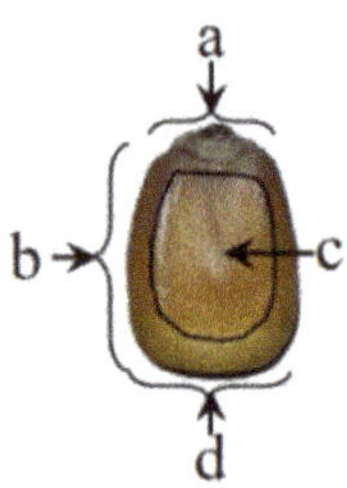


**Figure 8.** The position of seed touching the soil.

We set a particle factory at the height of 400, 500, 600, and 700 mm from the soil model to make the seed fall in free fall motion, and saved the movement state data of corn seed. The corn seed approaching velocity $V_0$ and separation velocity $V_t$ were saved and the SAVR $\xi_1$ was calculated automatically, the calculation equation is shown in Equation (11). Each group of experiments was repeated five times to obtain the mean value.

$$\xi_1 = \frac{V_t}{V_0} \tag{11}$$

4.2.2. Soil Bin Experiment

We built a collision experiment platform to make corn seed fall into the soil. The corn seed used had the same variety and moisture content as the corn seed used in measurement the CRC. The soil used was taken from the same plot as the soil particles used to measure the CRC.

Over the aluminum soil bin, the soil was screened with a 2 mm aperture sieve so that the soil particles fell vertically into the aluminum soil bin. The soil particles above the top edge of the soil bin were scraped so that the soil surface was flush with the bin edge. The universal material machine compresses the soil in the soil bin, and the compression height is 20% of the height of the soil bin so that the compression ratio of the actual soil bin was consistent with that of the discrete element model. Placed the prepared soil bin on the support platform to make the corn seed fall freely at the positions of 400, 500, 600, and 700 mm from the soil bin's soil surface, and conducted the experiment that corn seed collide soil. We used the high-speed camera of revealer 5f01 to capture the moments when the corn seed collide, approach, touch the soil and collide and separate. We analyzed the movement process of the corn seed through the high-speed video target tracking measurement software, obtained the approach velocity $v_0$ and separation velocity $v_t$ of the corn seed when the contact positions are a, b, c, d, and calculated the SAVR $\xi_2$, the calculation equation is shown in Equation (12).

$$\xi_2 = \frac{v_t}{v_0} \quad (12)$$

4.2.3. Comparison and Analysis of Experiment Results

(1) Comparative Analysis of the Movement State of Corn Seed

Aiming at the process of collision, approaching, and separation between corn seed and soil bin, the SBE and DESE under the same experiment conditions were selected to analyze the movement state of corn seed. The experiment conditions were 400 mm height drop and soil contact at position c. We compared and analyzed the movement state of corn seed and drawn the curve of seed velocity and height with time, as shown in Figure 9. The positive value of velocity represents the direction of seed falling close to the soil. The negative value of velocity represents the direction of seed rebound and soil separation. The corn seed starts to fall at 0 s and reach the maximum velocity of falling at about 0.283 s. The collision approaches velocity $v_0$, and collides with the soil. The movement state changes, causing a sudden change in velocity. After the corn seed moves to the lowest point, it starts to rebound, the elastic deformation between particles recovers, and the elastic strain energy is converted into the potential energy of particles. The velocity of corn seed begins to increase and reaches the maximum value $v_t$ when it separated from soil particles, that is, the collision separation velocity is $v_t$.

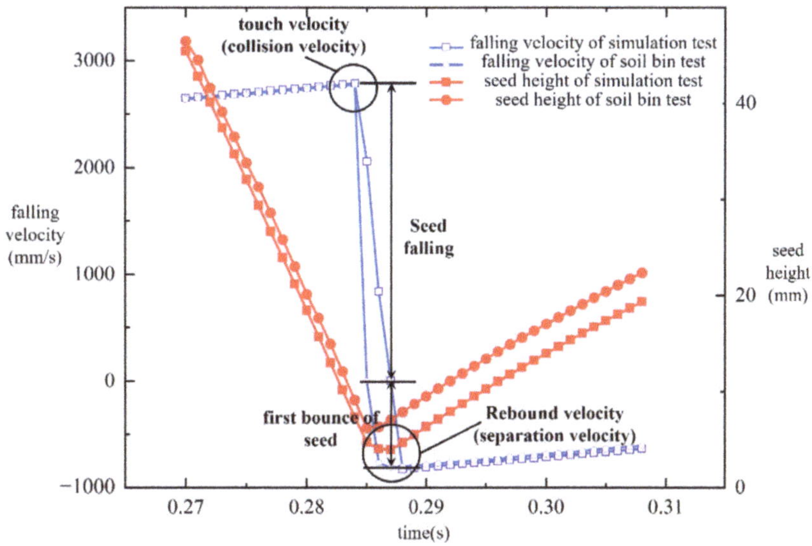


**Figure 9.** Velocity and height of single seed in the DESE and the SBE.

When the falling height and contact position of corn seeds in the DESE and the SBE are the same, the movement state of seed is compared. The movement state of seeds in the free fall stage is consistent, and the movement state of seeds after collision and separation is slightly different, but the difference is seldom. The separated corn seed showed a compound motion state of rotating motion with the contact side as the axis and parabola motion with the contact point as the starting point. The motion state of three groups (a, b, c) of simulation and real soil bin comparison experiments arbitrarily selected are shown in Figure 10.

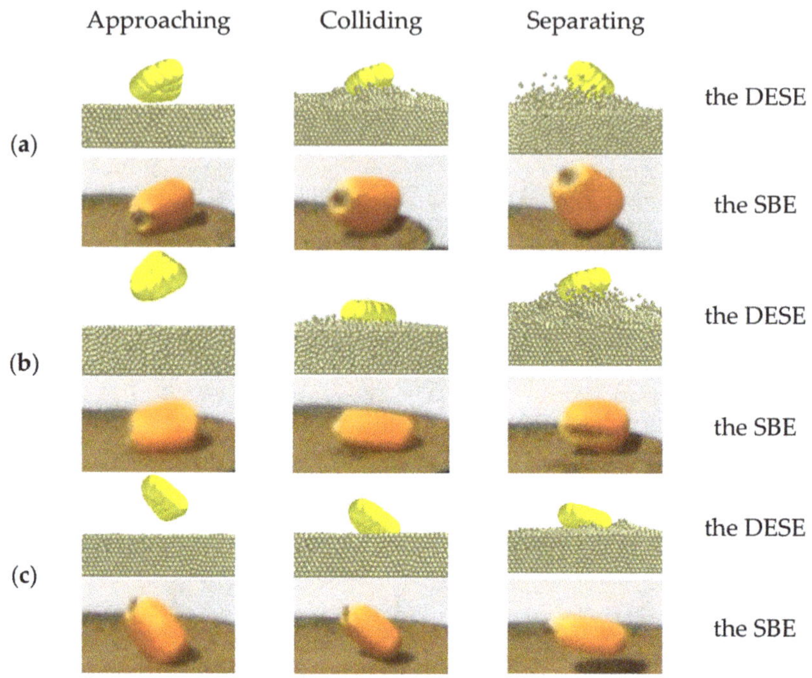


**Figure 10.** Comparison of movement status. (**a–c**) represent three groups of comparative tests, in each group of tests the falling height and contact position of corn seeds in the DESE and the SBE are the same, to compare the movement state of seeds in the DESE and the SBE.

(2) Comparative Analysis of SAVR $\xi$

The SAVR $\xi$ of the DESE and the SBE are shown in Table 5. The comparative analysis shows that the SAVR $\xi$ decreases with the increase of the falling height. The reason was that, according to the impulse theorem, with the drop height increase, the corn seeds' movement velocity before the collision contact increased. The increase of collision velocity led to the increase of collision impulse and collision force, and the deformation of contact position. The increase of deformation at the contact position leads to the increase of the contact area between corn seed and soil and energy loss due to contact friction. The energy loss led to the relative increase of the energy lost by the system, the reduction of the strain energy released in the unloading stage, the change of the corn seed movement state, and the relative reduction of the collision separation velocity. Therefore, the SAVR decreased with the increase of height, the SAVR $\xi$ decreased with the increase of height. Except for position a, the relative error between the SBE and the DESE was within 10%, which confirms the reliability of the proposed method. As for the reason of the obviously error in position a, it was that the material properties of corn seed in the simulation experiment are all horny endosperms. In contrast, the characteristics of the basal fruit stalk materials of actual corn seed are different from those of horny endosperm materials. Therefore, there is an obviously difference in the SAVR between the SBE and the DESE in the simulation and the actual experiment. In the future research, the CRC between position A, and soil should be accurately measured, and different CRC should be used according to different materials in the EDEM to improve the accuracy of the simulation.

**Table 5.** Comparison of results of the SBE and the DESE.

| Falling Height/cm | Touch Soil Position | The SBE $\xi_2$ | The DESE $\xi_1$ | Relative Error/% |
|---|---|---|---|---|
| 40 | a | 0.169 ± 0.00676 | 0.07 ± 0.00245 | 58.6 |
| | b | 0.172 ± 0.00516 | 0.168 ± 0.009408 | 2.3 |
| | c | 0.27 ± 0.0189 | 0.294 ± 0.014994 | 8.9 |
| | d | 0.15 ± 0.009 | 0.137 ± 0.007124 | 8.7 |
| 50 | a | 0.226 ± 0.0113 | 0.049 ± 0.002744 | 78.3 |
| | b | 0.153 ± 0.006579 | 0.16 ± 0.004 | 4.6 |
| | c | 0.253 ± 0.013156 | 0.261 ± 0.003915 | 3.2 |
| | d | 0.137 ± 0.007535 | 0.123 ± 0.004305 | 6.1 |
| 60 | a | 0.146 ± 0.00949 | 0.054 ± 0.001836 | 63 |
| | b | 0.142 ± 0.0071 | 0.145 ± 0.004495 | 2.1 |
| | c | 0.208 ± 0.00728 | 0.206 ± 0.00206 | 2.9 |
| | d | 0.116 ± 0.0029 | 0.121 ± 0.00605 | 4.3 |
| 70 | a | 0.183 ± 0.008235 | 0.06 ± 0.0021 | 67.2 |
| | b | 0.135 ± 0.007425 | 0.145 ± 0.003625 | 7.4 |
| | c | 0.175 ± 0.0042 | 0.182 ± 0.005642 | 4 |
| | d | 0.134 ± 0.005762 | 0.125 ± 0.0065 | 6.7 |

## 5. Conclusions

(1) Through the analysis of the collision process of corn seed colliding with soil, it was found that when using corn seed colliding with soil to measure the CRC, it was essentially a collision of one surface and multiple points. A large number of soil particles determined the movement state of corn seed after collision and separation. It was complicated to obtain the interaction relationship between a single soil particle and corn seed, and it could not be applied to measurement the CRC.

(2) A method for measuring the CRC by taking soil particles as the mass point and taking corn seed as the fixed surface was proposed. The analysis of the impact process shown that the contact between corn seed and multiple soil particles could be avoided, and the CRC of corn seed and soil could be obtained intuitively. Using soil particles to fall and collide with corn seed, the CRC between corn and soil was 0.607.

(3) The results of the verification experiment shown that the movement state of corn seed in the DESE and the SBE were basically the same when they fall at the same height and touch the soil at the same position; In the DESE and the SBE, the error of the SAVR was within 10% except for the position a, which verifies the accuracy of the proposed method.

**Author Contributions:** Conceptualization, Z.G. and C.L.; methodology, Z.G. and S.H.; software, Z.G., S.H. and Y.L.; writing—original draft preparation, Z.G. and C.L.; writing—review and editing, Z.G., C.L., H.L., J.H., Q.W. and H.Z.; funding acquisition, C.L. and H.L. All authors have read and agreed to the published version of the manuscript.

**Funding:** This research was funded by the China Agriculture Research System of MOF and MARA: CARS-03; the 2115 Talent Development Program of China Agricultural University and Chinese Universities Scientific Fund: 2021TC105.

**Institutional Review Board Statement:** Not applicable.

**Informed Consent Statement:** Not applicable.

**Data Availability Statement:** Not applicable.

**Conflicts of Interest:** The authors declare no conflict of interest.

## References

1. Horabik, J.; Beczek, M.; Mazur, R.; Parafiniuk, P.; Ryżak, M.; Molenda, M. Determination of the restitution coefficient of seeds and coefficients of visco-elastic Hertz contact models for DEM simulations. *Biosyst. Eng.* **2017**, *161*, 106–119. [CrossRef]

2. Badua, S.A. Control System Response for Seed Placement Accuracy on Row Crop Planters. Ph.D. Thesis, Kansas State University, Manhattan, KS, USA, 2020.
3. Yan, B.; Wu, G.; Fu, W.; Gao, N.; Meng, Z.; Zhu, P. Influencing factors of corn implantation distribution for high-height planting based on EDEM. *Trans. Chin. Soc. Agric. Mach.* **2020**, *51*, 47–54. [CrossRef]
4. Han, S.; Qi, J.; Kan, Z.; Li, Y.; Meng, H. Parameters calibration of discrete element for deep application of bulk manure in Xinjiang orchard. *Trans. Chin. Soc. Agric. Mach.* **2021**, *52*, 101–108. [CrossRef]
5. Zeng, Z.; Ma, X.; Cao, X.; Li, Z.; Wang, X. Critical review of applications of discrete element method in agricultural engineering. *Trans. Chin. Soc. Agric. Mach.* **2021**, *52*, 1–20. [CrossRef]
6. Wang, C.; Liu, Q.; Ma, L.; Li, L. Cottonseed particle motion law in 3-DOF hybrid vibration screen surface. *Trans. Chin. Soc. Agric. Eng.* **2015**, *31*, 49–56. [CrossRef]
7. Müller, G.T.; Huang, K. Influence of the liquid film thickness on the coefficient of restitution for wet particles. *Phys. Rev. E* **2016**, *93*, 429–437. [CrossRef] [PubMed]
8. Hastie, D.B. Experimental measurement of the coefficient of restitution of irregular shaped particles impacting on horizontal surfaces. *Chem. Eng. Sci.* **2013**, *101*, 828–836. [CrossRef]
9. Feng, B.; Sun, W.; Shi, L.; Sun, B.; Zhang, T.; Wu, J. Determination of restitution coefficient of potato tubers collision in harvest and analysis of its influence factors. *Trans. Chin. Soc. Agric. Eng.* **2017**, *33*, 50–57. [CrossRef]
10. Wang, L.; Ran, B.; Kan, Z.; Kan, Z. Experiment of restitution coefficient of processing tomato seeds in collision models. *J. Jiangsu Univ.* **2015**, *36*, 48–52. [CrossRef]
11. Lu, Y.; Wu, N.; Wang, B.; Yu, Z.; Lin, D.; Hu, Z. Measurement and analysis of peanuts' restitution coefficient in point-to-plate collision mode. *J. China Agric. Univ.* **2016**, *21*, 111–118. [CrossRef]
12. Huang, X.; Zha, X.; Pan, H.; Zong, W.; Chen, H. Measurement and analysis of rapeseeds' restitution coefficient in point-to-plate collision model. *Trans. Chin. Soc. Agric. Eng.* **2014**, *30*, 22–29. [CrossRef]
13. Liu, W.; He, J.; Li, H.; Li, X.; Zheng, K.; Wei, Z. Calibration of simulation parameters for potato minituber based on EDEM. *Trans. Chin. Soc. Agric. Mach.* **2018**, *49*, 125–135, 142 . [CrossRef]
14. Zhou, W. The Physical and Mechanical Properties Research of Corn Grain. Ph.D. Thesis, Northeast Agricultural University, Harbin, China, 2015.
15. Sui, J. Research on Experiment and Simulation Analysis of Soybean Collision with Soil. Ph.D. Thesis, Jilin University, Changchun, China, 2016.
16. Zhang, G.; Xiang, X.; Tang, H. Field test and numerical calculation of restitution coefficient of rockfall collision. *Chin. J. Rock Mech. Eng.* **2011**, *30*, 1266–1273.
17. Yu, X.; Cui, Y.; Chen, F.; Hu, G. Measurement of coefficient of restitution and gravitational acceleration by using bouncing ball. *Coll. Phys.* **2010**, *29*, 35–36. [CrossRef]
18. Qin, Z.; Lu, Q. Analysis of impact process model based on restitution coefficient. *J. Dyn. Control* **2006**, *4*, 294–298.
19. Liu, Y.; Zong, W.; Ma, L.; Huang, X.; Li, M.; Tang, C. Determination of three-dimensional collision restitution coefficient of oil sunflower grain by high-speed photography. *Trans. Chin. Soc. Agric. Eng.* **2020**, *36*, 44–53. [CrossRef]
20. Li, G. *Advanced Soil Mechanics*; Tsinghua University Press: Beijing, China, 2013.
21. Ren, L. *Soil Adhesion Mechanics*; China Machine Press: Beijing, China, 2011.
22. Zeng, D. *Mechanical Soil Dynamics*; Beijing Science and Technology Press: Beijing, China, 1990.
23. Zhang, F.; Zhao, W.; Han, Z.; Liu, J.; Yang, X.; Dai, F. Experimental researches on mechanical properties of corn kernels. *Chin. Agric. Mech.* **2010**, *3*, 75–78. [CrossRef]
24. Shi, Z.; Cheng, H.; Li, J.; Feng, J. Characteristic parameters to identify varieties of corn seeds by image processing. *Trans. Chin. Soc. Agric. Eng.* **2008**, *24*, 193–195.
25. Hao, J.; Yang, J.; Du, T. A study on basic morphologic information and classification of maize cultivars based on seed image process. *Sci. Agric. Sin.* **2008**, *41*, 994–1002.
26. Peter, E.; Hu, B. *Eberhard Modern Contact Dynamics*; Southeast University Press: Shanghai, China, 2003.
27. Xie, F.; Wu, Z.; Wang, X.; Liu, D.; Wu, B.; Zhang, Z. Calibration of discrete element parameters of soils based on unconfined compressive strength test. *Trans. Chin. Soc. Agric. Eng.* **2020**, *36*, 39–47. [CrossRef]
28. Wang, X. Study on the Evaluation of Soil Compaction by Agricultural Tire and Combination Shovel Loose Soil Technology. Ph.D. Thesis, China Agricultural University, Beijing, China, 2018.
29. Wang, Y.; Liang, Z.; Zhang, D.; Cui, D.; Shi, S.; Li, K.; Yang, L. Calibration method of contact characteristic parameters for corn seeds based on EDEM. *Trans. Chin. Soc. Agric. Eng.* **2016**, *32*, 36–42. [CrossRef]

agriculture

MDPI

Article

# Design and Operating Parameters Optimization of the Hook-and-Tooth Chain Rail Type Residual Film Picking Device

Silin Cao [1,2], Jianhua Xie [1,3,*], Hezheng Wang [4], Yuxin Yang [5], Yanhong Zhang [1], Jinbao Zhou [1] and Shihua Wu [1]

1 College of Mechanical and Electrical Engineering, Xinjiang Agricultural University, Urumqi 830052, China
2 Mechanical Equipment Research Institute, Xinjiang Academy of Land Reclamation Sciences, Shihezi 832000, China
3 Xinjiang Key Laboratory of Intelligent Agricultural Equipment, Urumqi 830052, China
4 College of Mechanical and Electrical Engineering, Shihezi University, Shihezi 832000, China
5 Institute of Agricultural Mechanization, Xinjiang Academy of Agricultural Sciences, Urumqi 830091, China
* Correspondence: 20212109027@stu.shzu.edu.cn

**Abstract:** We attempt to solve the current problems of high impurity content and the poor reliability of chain harrow type residual film recovery machines when picking up residual film. This study makes a device for picking up residual film with a hook-and-tooth chain rail. First, we conducted an analysis of the trajectory of the tip movement of the hooked teeth in the designed picking device, with the condition that the residual mulch does not miss the picking, and the force conditions of the residual mulch in the conveying process and the collection process are presented. Secondly, to determine the optimal working parameters of the picking device, a three-factor, three-level response surface optimization test was conducted with the machine forward speed, hook tooth entry depth, and chain harrow input speed as the test factors and the residual film picking rate and the impurity rate of residual film as the test indexes. In addition, a response surface regression model was developed to analyze the effects of the selected factors on the picking device. When the forward speed of the machine was 1.62 m/s, the hook tooth entry depth was 38.51 mm, and the input speed of the chain harrow was 241.42 rpm, the pickup rate and the impurity rate of the residual film were 88.27% and 9.96%, respectively. Finally, the simulation test was carried out under optimal working conditions, with the maximum force of the hook teeth being 60.7 N, the maximum deformation being 31.42 mm, and the maximum stress being 215.33 Mpa. This study can be used as a guide to further improve the design of the residual film recovery machine.

**Keywords:** residual film recovery device; EDEM; virtual simulation; response surface regression model; parameters optimization

**Citation:** Cao, S.; Xie, J.; Wang, H.; Yang, Y.; Zhang, Y.; Zhou, J.; Wu, S. Design and Operating Parameters Optimization of the Hook-and-Tooth Chain Rail Type Residual Film Picking Device. *Agriculture* **2022**, *12*, 1717. https://doi.org/10.3390/agriculture12101717

Academic Editors: Mustafa Ucgul and Chung-Liang Chang

Received: 26 September 2022
Accepted: 15 October 2022
Published: 18 October 2022



## 1. Introduction

Mulching technology has been widely used in the production of a variety of crops since its introduction to China in the 1970s due to its ability to increase temperature and conserve moisture, save water and drought, and improve quality and yield [1,2]. For more than 40 years, Xinjiang has used mulch cultivation technology in cotton fields as an important commercial cotton production base in China, which has greatly contributed to the development of the local agricultural economy [3]. In addition, as the amount of film applied and years of use increase, the "white pollution" caused by large amounts of residual film not only seriously affects agricultural production, but also poses a great threat to the safety and health of the agricultural environment. Residual film on farmland has become one of the most important elements that affects the agricultural environment, destroys soil structure, reduces the quality of cultivated land, affects the growth of crops, and causes crop yield reduction, which poses a huge threat for the sustainable development of agriculture [4].

Developed countries use film with a thickness of 0.02 mm or greater, covering a shorter period of time while maintaining greater strength and not easily breaking when recovered, which is generally recovered in the form of rolls, with the main working parts being the starting shovel and rolls. Lavor invented a film recycling machine that removed soil from the film with a brush before rolling it up for collection [5]. R.L. Parish created a device that controls the flow of the hydraulic motor in order to regulate its speed and match the speed of the mulcher roll line to the speed of the recycler [6]. The film recycler was designed by ROCCA to transport the film to the film rollers at the back of the machine using a chain conveyor that lifts the film and separates it from the soil on its surface [7].

To address the problem of film pollution, related research institutions in China developed over a hundred different types of film recovery machines, which can be classified as telescopic roller type, shovel type, wheel-tooth type, chain rake type, and other major film picking devices based on different film collection principles [8–10]. Among these are the chain rake collection principles, which include high-speed film transfer to the collection device via the bullet teeth, chain rake damage that is easy to replace, a strong ability to remove soil and row miscellaneous, and a higher residual film recovery rate [11]. Guo Wensong created the 4CML-1000 chain harrow type residual film recovery machine, which can perform film collection, transfer, and unloading operations in real time. The residual film recovery rate is higher, but the forward speed of the machine is only 2.542 km/h, resulting in low operational efficiency. [12]. Xie Jianhua designed a surface residual film recovery machine with guided chain rake tines to solve the problem of a chain tine type film picking device for film removal that requires bending of the conveyor chain and uses a rotating film removal device with guided rake tines to enhance the film removal effect [13]. Moreover, the machine is only designed for 2.05 m cotton fields, and differences in planting conditions and farming operations affect edge film compaction during harvesting, making the effect of edge film lifting uncertain.

Based on the dwarf dense planting pattern of Xinjiang machine harvested cotton and the characteristics of an existing chain harrow type residual film recovery machine, a hook-and-tooth chain rail type residual film picking device was designed in this paper based on the structure and characteristics of existing recovery devices. The reasonable structure and operation parameters of the machine were determined through theoretical analysis and field tests, and the influence of different factors on the effect of residual film recovery was explored. As a result, a better combination of parameters was obtained to provide reference for further optimization of the design of the residual film recovery machine.

## 2. Materials and Methods

### *2.1. Structure and Working Principle of Residual Film Recycling Machine*

The residual film recycling machine is mainly composed of a hook-up device, a power transmission system, a straw crushing device, a hook-and-tooth chain rail type residual film picking device, and a film storage and launching device, as shown in Figure 1.

When the machine is activated, the tractor's rear output shaft is connected to the machine drive system and power is transmitted to the straw shredding device. The shredding knife rotates at high speed to shred the straw, and the shredded straw is thrown to the winch in a centrifugal action. The rotating winch throws the straw to the ground. In the film picking device, the hook and tine assembly's curved side conveyor chain drives the passive shaft under the chain rake frame for repeated movement. Picked residual film is conveyed upward to the de-filming shaft, which scrapes it into the film collection box. When unloading film, a hydraulic control system opens the side door and drives the fork arm to compact and push out the film.

#### 2.1.1. The Main Components and Working Principle of The Picking Device

The executive component of the film pick-up operation is the hook and tooth chain rail type residual film picking device, which is made up of a chain rake frame, an articulated short shaft, a flap for discharging debris and preventing mulch from sucking back parts, a de-filming shaft, a curved edge conveyor chain, a guide beam, a hook and tooth assembly, a chain rake upper shaft assembly, and a lower shaft assembly. The guide beam is attached to both sides of the chain rake frame's inner edge, the curved side conveyor chain is attached to the upper part of the guide beam, and the articulated short shaft is symmetrically arranged on both sides of the curved edge conveyor chain and fixed on the curved edge conveyor chain bend plate using bolt sets. The articulated short shaft is inserted into the hook tooth mounting shaft from both ends, and a hook tooth assembly is arranged for every 6 links. The film removal axis is installed on the outer side of the chain rake upper shaft assembly, and the mounting plate is welded to the chain rake frame, located below the film removal axis to connect the picking device to the entire machine. Figure 2 shows the installation position of each component of the picking device. Table 1 shows the mechanism's main operating parameters.

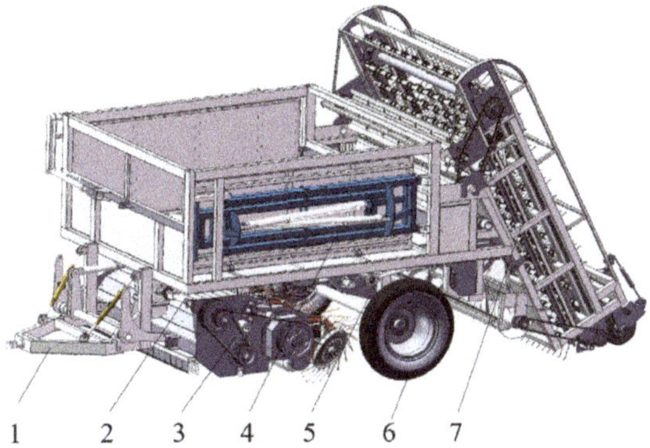


**Figure 1.** The residual film recycling machine: 1. Traction frame; 2. Power transmission system; 3. Straw crushing device; 4. Mulch storage and pushing out device; 5. Machine frame; 6. Wheels; 7. Hook-and-tooth chain rail type residual film picking device.

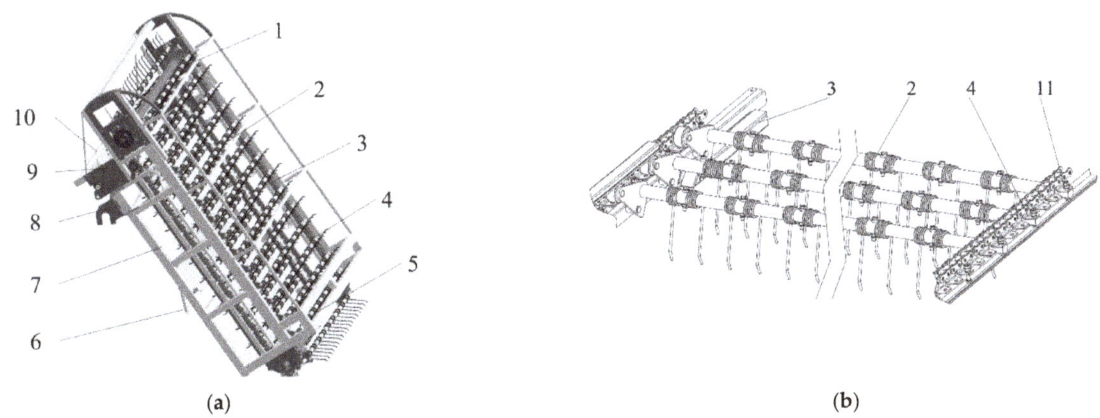


**Figure 2.** Hook-and-tooth chain rail type film picking device: 1. Chained rake frame upper shaft assembly; 2. Hook tooth assembly; 3. Guide beam; 4. Curved edge conveyor chain; 5. Chain rake frame lower shaft assembly; 6. Flap for discharging debris; 7. Chain rake frame; 8. Mounting plate; 9. De-filming shaft; 10. Prevention of mulch sucking back parts; 11. Articulated short shaft. (**a**) the main components of the picking device, (**b**) Pick up film chain rake structure schematic diagram.

**Table 1.** The main working parameters of a picking device.

| Main Parameters of the Picking Device | Value |
|---|---|
| Structure form | Traction type |
| Dimension (length × width × height)/mm | 2240 × 880 × 3170 |
| Matched power/kW | 88.20 |
| Operating speed/(km·h$^{-1}$) | 4–8 |
| Effective operating width/mm | 1875 |
| Residual film pick-up rate/% | ≥85 |
| Residual film impurity rate/% | ≤12 |

When the picking device operates, power is input by the machine's rear drive shaft, and the power input sprocket rotates the upper shaft of the chain rake, while the curved side conveyor chain synchronously drives the hook tooth installation shaft and the lower shaft of the chain rake installed on it. Hook teeth in the machine travel forward into the soil to pick up film, with the installation axis running from bottom to top to transport the residual film. The impurities in the residual film are picked up by shaking off the chain drive and fall down to the two-way spiral discharge winch at the discharge flap, where the size of the opening of the discharge flap can be controlled by the adjusting screw. In the upward conveying process, the articulated short shaft supports the hook tooth assembly, the limiting bracket prevents axial flinging in the film conveying, and the guideway beams on both sides ensure the motion track. When the hook teeth transfer the residual film to the top of the film removal area, the hook teeth assembly in the chain rake frame shift slide and limit bracket together, completing the hook teeth installation axis shift flip, and changing the tooth tip direction. The film scraper on the film release shaft now rotates in the opposite direction to scrape off the residual film on the hook teeth. A serrated baffle is welded to the front end of the film shaft to prevent the scraped out residual film from being hung up again by the hook teeth, ensuring that the residual film can fall smoothly to the film storage device after scraping out.

### 2.1.2. Hook Tooth Assembly

The hook tooth assembly is made up of a roller, a guiding side plate, a limiting short shaft, a hook tooth mounting shaft, a hook tooth, a threaded fastener, and other components. Figure 3 shows the structure.

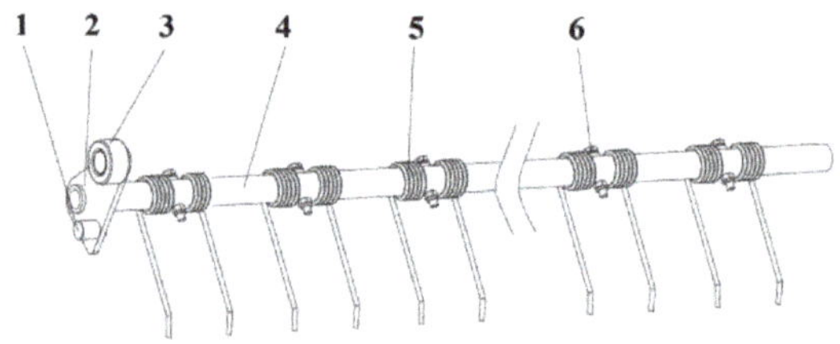


**Figure 3.** A schematic diagram of the hook and tooth assembly structure: 1. Limit short shaft; 2. Guide side plate; 3. Roller; 4. Hook tooth mounting shaft; 5. Hook tooth; 6. Threaded fastener.

In Figure 2, the chain rake frame structure characteristics can be seen. In picking up film teeth movement to the nearest position, direction changing chute will top up the guide side plate connected to the short axis so that the hook tooth assembly around the two ends of the hook tooth articulated short axis produce deflection, and the hook tooth assembly installed hook tooth rod and vertical direction angle becomes smaller, as in the hook tooth assembly off the film area rotation, as shown in Figure 4.

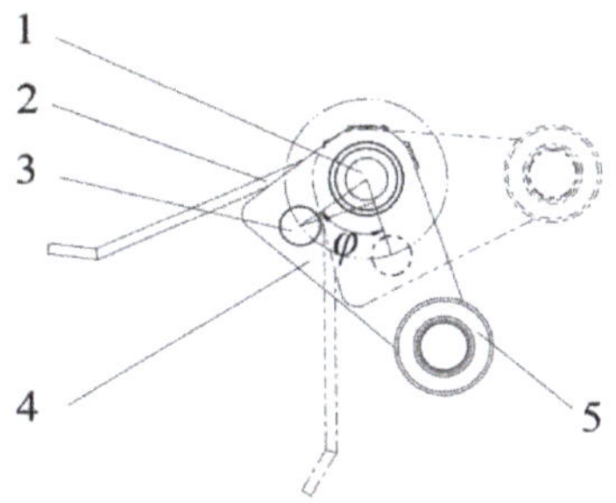


**Figure 4.** A schematic diagram of the rotation of the decapping area of the hook tooth assembly: 1. Hook tooth mounting shaft; 2. Hook tooth; 3. Short shaft; 4. Side guide plate; 5. Roller.

Then the effective lenght of the change of direction slide is

$$L_a = R_a cos\varphi \tag{1}$$

where $L_a$ is the effective length of the deflection chute, (mm); $R_a$ is the distance from the center of the hook tooth mounting shaft to the center of the short shaft, (mm); $\varphi$ is the rotation angle of the hook tooth assembly, (°).

The design hook tooth installation axis is 58 mm away from the center of the short axis, the rotation angle of the hook tooth assembly in the de-filming area is 60°, and $L_a$ = 29 mm can be obtained from an equation based on the characteristics of the guiding hook tooth assembly (1). If the total number of guiding hook teeth designed to rotate in the de-filming area is greater than two, the total number of guiding hook teeth for variable rotation in the de-filming area is 2.5 groups, and the distance between two adjacent rake teeth installation axis of rotation is 190 mm, then the effective length of the directional slide is 475 mm.

2.1.3. Pick-Up Film Hook Teeth

The hook tooth is the most critical operational component, and its performance has a direct effect on picking quality [14]. The ability of the hook tooth to support and transport residual film during operation is critical to the hook tooth's reliability. The hook tooth's working phase is divided into above-ground and below-ground phases, with the below-ground phase being the most complicated, as it will collide violently with the soil and produce violent vibrations when it enters the soil. To improve the self-adaptive ability of picking parts, the hook teeth are made of 65 Mn spring steel, which has good flexibility and impact resistance. By taking the entire calculation, the hook tooth diameter is 8 mm, the distance between the hook tooth teeth is 100 mm, the distance between the hook tooth axis and the hook tooth bend is 170 mm, the distance between the hook tooth bend and the tooth tip is 30 mm, and the distance between the hook tooth axis and the tooth tip is 200 mm. Furthermore, the hook tooth tip is bent in the direction of rotation to reduce residual film slippage. By conducting field pre-tests, mainly considering the deformation and damage of the hook teeth into the soil after being impacted by the load and the residual film picking effect, it was found that the best picking effect could be obtained when the angle of the hook teeth tip was 130°~160°, and there was no deformation and damage of the hook teeth, which was taken as 150° in this design, as shown in Figure 5.

According to the hook tooth force as shown in Figure 5a; with the hook tooth into the soil by the soil resistance $F_f$ after a certain twist, the angle of the twist is $\varphi_1$, and in order to ensure the efficiency of the residual film pick-up, at this time the hook tooth operation should meet the conditions of

$$k\frac{Ed^4\varphi_1}{3667\text{Dn}} \geq PF_fL \tag{2}$$

where $k$ is the number of hook teeth needed to pick up the residual film per unit area; $E$ is the modulus of elasticity (Pa); $d$ is the diameter of the hook teeth (mm); $\varphi_1$ is the angle of twist of the hook teeth into the soil (°); $L$ is the distance from the center of rotation of the hook teeth to the tip of the teeth (mm); $D$ is the diameter of the torsion spring (mm); $n$ is the effective number of turns of the torsion spring; $F_f$ is the resistance of the hook teeth into the soil (N); $P$ is the number of hook teeth per row.

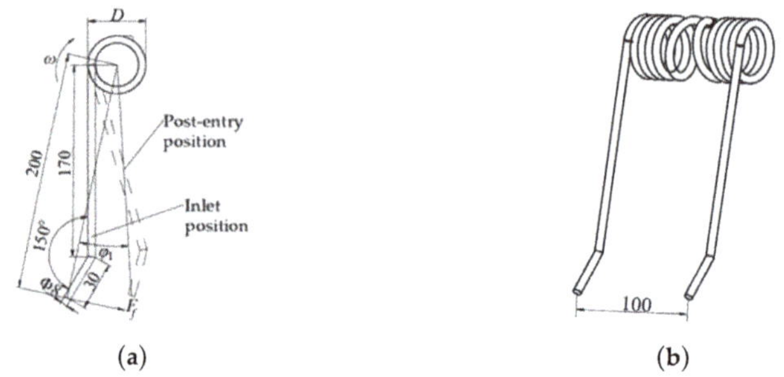


**Figure 5.** A schematic diagram of the hook tooth structure: (**a**) Diagram of the influence of resistance of hook teeth into the soil; (**b**) Axonometric drawing.

The picking device is designed to pick up the width of a film of six rows, so the condition of the number of teeth per row is

$$P = \frac{2S + 3Z}{D_1} + 1 \tag{3}$$

where $D_1$ is the distance between hook teeth in the same row (mm); $S$ is the distance between wide rows of cotton stalks (mm); $Z$ is the distance between narrow rows of cotton stalks (mm).

According to Equations (2) and (3), if the material characteristics of the hook teeth and the middle diameter of the spring are known, whether the residual film can be transported in the clockwise direction according to the operation requirements is related to the effective number of turns of the spring and the spacing of the hook teeth. According to the previous test, the effective number of spring turns is generally 4–6 turns, and this design requires 6; according to the planting pattern of one film six rows (660 + 100 mm) in Xinjiang, the spacing of cotton stalks in wide rows $S$ is 660 mm, the spacing of cotton stalks in narrow rows $Z$ is 100 mm, the spacing of hook teeth in the same row $D_1$ is 200 mm, and the number of hook teeth in each row $P$ is 9.1 by substituting the above formula; then, 9 hook teeth or 9 half hook teeth are arranged on each of the hook tooth mounting shafts.

### 2.1.4. Hook Tooth Arrangement

According to the width of cotton planting in Xinjiang, the measured width after sowing is 1850~1900 mm, and the length of the hook tooth installation axis is designed to be 1875 mm. The film picking device adopts uniform film picking, and in order to avoid stress concentration in the process of film picking, the overall arrangement of hook teeth is staggered [15–17]. The same hook tooth installation axis is arranged with 9 hook teeth or 9 half hook teeth, and the lateral spacing of hook teeth is 200 mm; to ensure the strength of the pick-up film chain rake structure using the type 20A curved edge conveyor chain, the chain rake upper and lower axis assembly sprocket utilizing a 20A 16 tooth sprocket is employed. According to the side guide plate structure design, every 6 links required a set of guide hook teeth assemblies, and the two phase hook teeth installation axis distance was 190 mm. According to the height of the transport design, the pick-up film device to install the total number of guide rake tooth assemblies was 33 groups. The hook teeth spread arrangement is as shown in Figure 6.

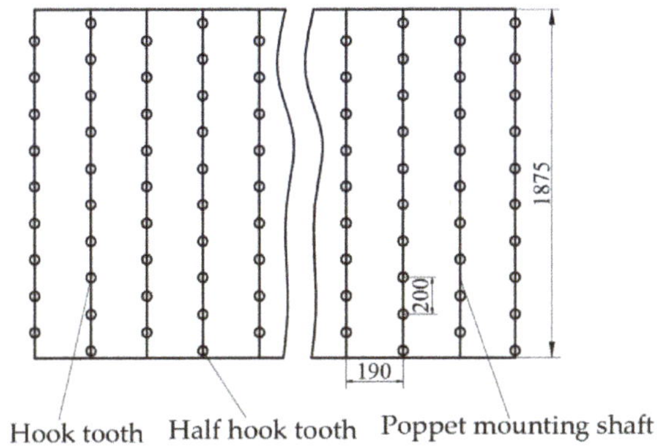


**Figure 6.** The spread layout of the film picking teeth.

## *2.2. Analysis of Picking Process of Hook-Tooth Chain Rail Film Picking Device*

### 2.2.1. Determination of Inclination Angle of Hook Teeth

The process of picking up the film is shown in Figure 7. The chain rake conveyor chain drives the hook assembly to complete the film picking, upward transportation, and reversing rotation. A guide beam is set at the chain rake frame on the right side of the tool direction, and the fixed guide side plate, roller, and limit short shaft on the right side of the hook gear assembly are matched to meet the position and attitude control of the pick-up teeth, so as to realize the limit operation of the hook gear assembly. When picking up the film, the roller is limited by the driven shaft of the chain rake to ensure the angle of the hook tooth in the soil. When conveying, the roller is constrained in the chain rake chain bending plate and the changing slideway, and the hook teeth move upward while hanging the residual mulch film without falling off. When the hook-tooth assembly moves to the peeling area, the hook-tooth assembly rotates, the straight rod of the hook-tooth changes into a vertical direction, and the tooth tip is situated downward. The lifting effect of the hook-tooth group on the residual film is weakened, and the residual film transported by the hook-tooth is unloaded under the action of the scraper, falling off smoothly to the film collecting box.

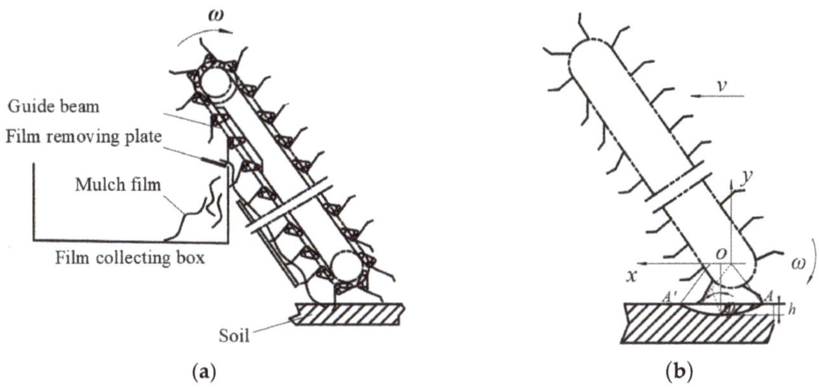


**Figure 7.** Working process of the film pickup device. (**a**) Picking film process diagram, (**b**) Motion diagram of the pick-up hook tooth.

Based on the geometric relationship of the parts, using the cosine theorem, when the depth of the pickup hook reaches the maximum h, the angle $\varphi$ between the $L_{OA}$ and the vertical direction is:

$$\varphi = \arccos\left\{\frac{L_s^2 + L_{OA}^2 - \left(\frac{D}{2} + E\right)}{2L_s^2 L_{OA}}\right\} \tag{4}$$

where $\varphi$ is the angle between the straight line of the center of the hook assembly and the tooth tip and the vertical direction (°); $L_s$ is the distance between rake axis and tooth tip, (mm); $D$ is the chain rake driven shaft sprocket indexing circle diameter, (mm); $E$ is the distance from the center of the crochet assembly to the center of the chain rake chain roller (mm); $L_{OA}$ is the distance from the center of chain rake driven shaft to the tip of hook tooth (mm).

The dip angle $\alpha$ of the hook tooth into soil obtained by the sine theorem is:

$$\alpha = \arcsin\left(\frac{L_{OA} - h}{L_{OA}}\right) \tag{5}$$

According to the previous test combined with the existing data, it can be seen that the hook tooth needs to meet $\varphi < 10°$ when it reaches the maximum depth into the soil. Ignoring the influence of the deformation of the hook tooth, the diameter of the sprocket dividing circle is designed to be $D$ = 243.25 mm, and the distance between the center of the hook tooth assembly and the center of the chain rake chain roller is $E$ = 40.8 mm. The angle $\varphi$ between the straight line and the vertical direction of the center of the hook tooth assembly and the tooth tip is 5°, which can be substituted into the formula to obtain $L_{OA}$ = 359 mm. The maximum embedded depth $h$ of the hook tooth is 60 mm, and the embedded angle $\alpha$ of the hook tooth is 56.4°.

#### 2.2.2. Motion Analysis of Film-Picking Hook Teeth

As the main residual film picking part, the film picking hook tooth has an important influence on the effect of film picking operation. Therefore, the motion process of the film picking hook tooth is analyzed, and the value range of the chain rake shaft speed is determined. The curved edge conveyor chain drives the pick-up hook gear assembly to pick up the film, and the absolute motion of the pick-up hook gear is the synthesis of the forward motion of the machine and the motion of the pick-up hook gear itself. The motion trajectory of the pick-up hook gear is shown in Figure 8. Taking the rotation center O of the lower shaft of the conveyor chain rake as the coordinate origin, the forward direction of the machine is the $x$-axis, and the vertical direction is the $y$-axis to establish the rectangular coordinate system. The machine runs at a constant speed with the forward speed $v$, and the lower shaft of the chain rake rotates with the angular velocity. The trajectory equation of the hook tooth tip is

$$\begin{cases} x = vt + L_{OA}\cos\omega t \\ y = L_{OA}\sin\omega t \end{cases} \tag{6}$$

where $L_{OA}$ is the rotation radius of the hook (mm).

The velocity equation of the hook tooth tip can be obtained by taking the derivative of Formula (1) with respect to time

$$\begin{cases} v_x = v - L_{OA}\omega\sin\omega t \\ v_y = L_{OA}\omega\cos\omega t \end{cases} \tag{7}$$

where $v_x$ and $v_y$ are the speed of the hook tip in the $x$ and $y$ directions (m/s).

As can be seen from Figure 8, the absolute motion direction of the picking film hook tooth is the tangential direction of the trajectory line. Only when $v_x > 0$ does the picking film hook tooth have a horizontal forward speed to complete the picking film operation. At this time, the motion trajectory of the hook tooth is a trochoidal line. When $v_x = v - L_{OA}w\sin wt = 0$:

$$\frac{\omega L_{OA}}{v} = \frac{1}{sin\omega t} \tag{8}$$

On account of:

$$v_1 = \omega L_{OA} \tag{9}$$

Substituting Formula (9) into Formula (8) has

$$\frac{v_1}{v} = \frac{1}{sin\omega t} \tag{10}$$

where $v_1$ is the linear velocity of the hook tip (m/s).

Let the ratio of the linear speed $v_1$ of the pick-up hook tooth to the forward speed $v$ of the machine be $\lambda$. According to the Formulae (7) and (10), it can be seen that when $\lambda > 1$, the pick-up hook tooth has a horizontal forward speed, and the pick-up chain rake rotation will pick up the residual film upward transport. If $\lambda$ is too small, it is easy to make the residual film leak too much and reduce the work efficiency. If $\lambda$ is too large, it is easy for the phenomenon of residual film return to appear. According to the preliminary experiment, when $\lambda$ is 2.8~3.4, the test effect is better. At this time, the chain rake shaft speed can be preliminarily determined to be 210~258 r/min.

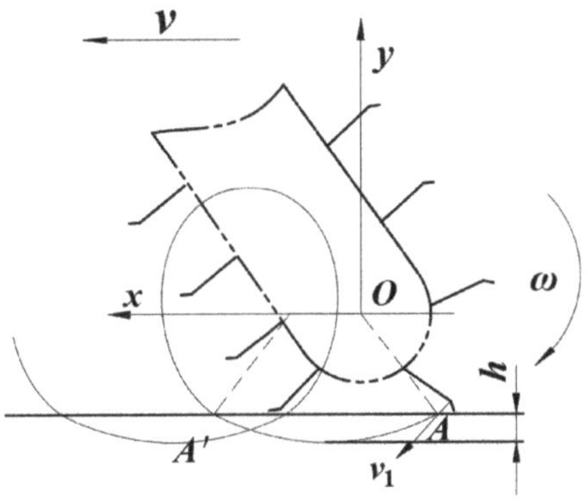


**Figure 8.** The motion trajectory of the pickup hook tooth.

2.2.3. Reliable Conditions for Residual Film Pickup

In order to realize the continuous picking up of residual film, it is necessary to analyze the conditions of the non-leakage picking up of residual film. When the machine is working, the picking film hook teeth arranged on the chain rake stand pick up the residual film in turn. The motion trajectory of the two adjacent rows of picking film hook teeth on the chain rake stand is shown in Figure 9. As the machine moves forward, the former picking film hook tooth is buried at point $A$ at time $t_0$, and the $A'$ point is unearthed at time $t_1$. The adjacent latter picking film hook tooth is buried at point $B$ at time $t_2$, and the $B'$ point is unearthed at time $t_3$. In this process, the distance of the lower shaft axis of the chain rake is $S_0$. In order to realize that the residual film does not leak during the picking process, the following conditions need to be met:

$$S_0 \leq S_a = S_b \tag{11}$$

where $S_a$ is the distance from the previous pickup hook tooth to the lower shaft of the unearthed chain rake (mm); $S_b$ is the distance between the adjacent rear pick-up film hook teeth and the lower shaft of the unearthed chain rake (mm).

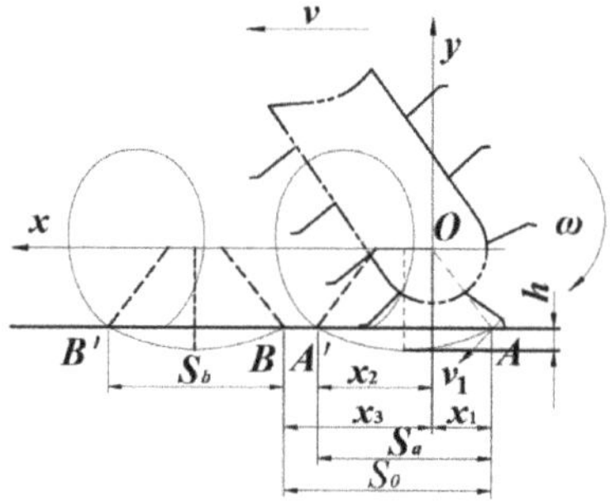


**Figure 9.** The motion track of two adjacent film-picking hook teeth on the conveying chain rake.

Substituting $y = -L_{OA} + h$ into Equation (6), we can get:

$$\begin{aligned} sin\omega t &= \tfrac{L_{OA}-h}{L_{OA}} \\ \cos\omega t &= \pm\tfrac{(2L_{OA}-h^2)^{\frac{1}{2}}}{L_{OA}} \end{aligned} \quad (12)$$

Substituting Formula (12) into Formula (6), we can get:

$$x_1 = vt_0 - (2L_{OA} - h^2)^{\frac{1}{2}} \quad (13)$$

$$x_2 = vt_1 + (2L_{OA} - h^2)^{\frac{1}{2}} \quad (14)$$

$$x_3 = vt_2 + (2L_{OA} - h^2)^{\frac{1}{2}} \quad (15)$$

It can be known from Figure 9:

$$S_a = S_b = x_1 + x_2 \quad (16)$$

$$S_0 = x_1 + x_3 \quad (17)$$

Substituting Formulas (13) and (14) into Formulas (16) and (13), and Formula (15) into Formula (17):

$$S_a = v(t_0 + t_1) = \frac{v(\pi - 2\alpha)}{\omega} \quad (18)$$

$$S_0 = v(t_0 + t_2) = \frac{v\pi}{2\omega} \quad (19)$$

Substituting Formulas (18) and (19) into Formula (11):

$$\frac{v\pi}{2\omega} \leq \frac{v(\pi - 2\alpha)}{\omega} \quad (20)$$

Simplify Formula (20)

$$v \geq \frac{4\alpha}{\pi} \quad (21)$$

According to our analysis:

$$v = \frac{v_1}{\lambda} = \frac{\pi n L_{OA}}{30\lambda} \quad (22)$$

Substituting Formula (22) into Formula (21):

$$n \geq \frac{120\alpha\lambda}{\pi^2 L_{OA}} \quad (23)$$

The maximum value of $\lambda$ is 3.4, and the values of $\lambda$, $\alpha$ and $L_{OA}$ are substituted into Formula (23). The condition where the residual film is not missed in the picking process is $n \geq 118.7$ r/min.

#### 2.2.4. Force Analysis of Residual Film in Conveying Process

The residual film is provoked by the picking film hook tooth, and the stress condition when it is transported upward is shown in Figure 10.

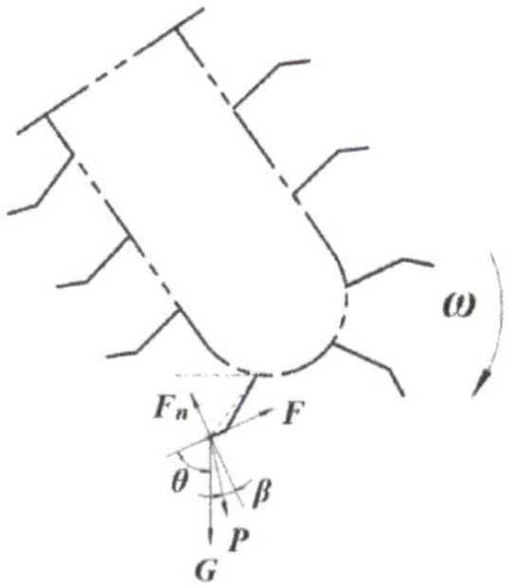


**Figure 10.** A force analysis diagram of residual film in the conveying process.

The actual force of the picking film hook tooth is affected by the soil, stubble on the residual film viscous pressure, and wind resistance, in addition to the gravity of the residual film, the friction of the hook tooth to the residual film, and the centrifugal inertia force during the movement. Because soil and stubble viscous pressures in response to residual film and wind resistance are random forces, their size and direction cannot be determined. As a result, the viscous pressure of soil and stubble on the residual film, as well as wind resistance, are not considered in the dynamic analysis of pickup hook teeth.

When the residual film is in equilibrium on the hook teeth, the equilibrium equation is:

$$\begin{cases} mg\cos\theta + P\sin\beta = f \cdot F_n \\ mg\sin\theta + P\cos\beta = F_n \\ P = m\omega^2 R \end{cases} \tag{24}$$

When the friction force of the residual film is greater than the centrifugal force of the residual film, the residual film does not fall during the lifting process:

$$f \cdot mg\sin\theta + f \cdot m\omega^2 R\cos\beta \geq mg\cos\theta + m\omega^2 R\sin\beta \tag{25}$$

where $F_n$ is the support force of residual film (N); $m$ is the mass of residual film (kg); $R$ is the instantaneous radius of the center of gravity of the residual film (mm); $G$ is the acceleration of gravity (m/s$^2$); $f$ is the friction coefficient of the pickup hook tooth and residual film; $P$ is the centrifugal force (N); $\omega$ is the chain rake shaft rotation angle (rad/s); $\theta$ is the residual film in the direction of the gravity angle (°); $\beta$ is the angle between the support force and centrifugal force (°).

The film collecting effect of the film collecting device is related to the movement speed of the chain rake, as shown by Formula (25). When the chain rake speed is low, the residual film is subjected to little force and is easily slid down under the action of continuous film kneading and pulling, resulting in film leakage, which is not conducive to residual film collection and transportation. When the chain rake speed is high, the picking up film hook tooth force can instantaneously film up, but there is a risk of film being torn. When the speed increases to a certain value, there is a risk of film being broken, but the subsequent picking up film hook teeth, having staggered into the soil area overlap, is conducive to picking up film operations.

### 2.3. Analysis of the De-Filming Process

The condition of residual film separation occurs when the residual film moves to the de-filming area under the action of centrifugal force to overcome the chain tooth rake adsorption winding role and separation. Because of the inertia force of the film movement and the softness and easy adsorption of the film, the machine adopts the rotating scraper to remove the film. The scraper works close to the middle or end of the hook teeth, and there is elastic and plastic deformation between the hook teeth and the scraper in the contact process [18].

As shown in Figure 11, in the process of de-filming, when the residual film held by the hook moves to the de-filming area, the motion of the hook tooth assembly is deflected, the hook tooth tends to be vertical, and the film hanging on the hook tooth is relaxed and slips off under the action of gravity. At this time, the scraper produces a reverse scraping or slapping effect on the hook tooth assembly, and the film is scraped off the hook tooth and falls into the film box for collection after the de-filming operation.

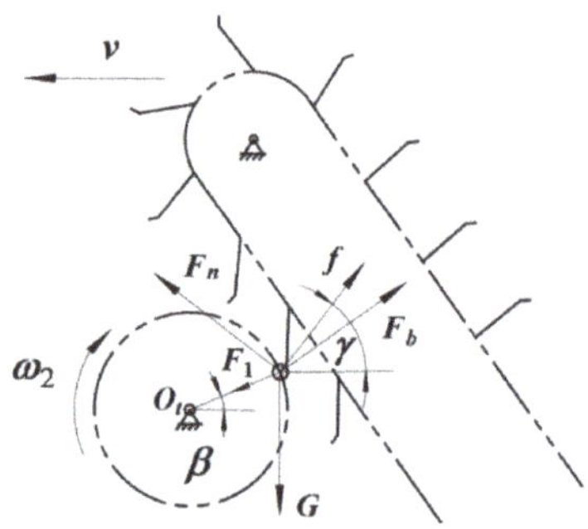


**Figure 11.** A schematic diagram of the forces on the residual film during the de-filming process.

The force analysis of the residual film separation moment is shown in Figure 11. The electrostatic adsorption force between the film and the scraper is very small, so its influence is ignored. When the hook tooth assembly is tilted to convey the residual film, its rotation radius tends to infinity, so $F_b$ tends to 0. In the process of film removal, when the film removal device and the hook tooth assembly move relative to each other, the condition that the residual film is separated from the hook tooth at this time is

$$\begin{cases} F_1\cos\beta + F_n\sin\gamma > f\cos\gamma \\ F_n\cos\gamma + f\sin\gamma < F_1\sin\beta + G \end{cases} \tag{26}$$

where $F_1 = m\omega_2{}^2 r_t$ and $G = mg$.

Collation is obtained.

$$\begin{cases} m\,\omega_{22} r_t\cos\beta + m\,g\cos\gamma\sin\gamma > \mu m\,g\cos^2\gamma \\ m\,g\cos^2\gamma + \mu m\,g\cos\gamma\sin\gamma < m\,\omega_{22} r_t\sin\beta + m\,g \end{cases} \tag{27}$$

where $F_1$ is the force of the stripping device on the hook tooth (N); $F_n$ is the support force of the hook tooth on the residual film (N); $m$ is the mass of the residual film carried by the hook tooth ($g$); $G$ is the gravity of the residual film (N); $r_t$ is the rotation radius of the stripping device m. $g$ is the acceleration of gravity, m/s$^2$; $\omega_2$ is the angular velocity of rotation of the de-filming axis rad/s; $f$ is the friction force between the residual film and the hook tooth (N); $F_b$ is the centripetal force generated by the motion of the residual film (N); $\beta$ is the angle between the center of the de-filming axis and the line connecting the contact point and the horizontal direction (°); $\gamma$ is the angle between the hook tooth rod and the vertical direction during the de-filming operation (°).

### 2.4. *Field Tests*

#### 2.4.1. Test Materials

The picking device is driven by a John Deere tractor (tractor model 6A-1024, rated power 88.2 kW). The test instruments consist of an electronic balance (model: YP2002 electronic balance, range: 0–500 g, accuracy: 0.001 g), a measuring tape (range: 50 m), a Topun soil moisture tester (model: TZS-1K-G), a Topun soil compactness tester (model: TJSD-750-2), a cling bag, a stopwatch, label paper, and a marker pen.

The film width of the test cotton field was 2050 mm, the film thickness was 0.01 mm, and the longitudinal film length was approximately 340 m. The trial field's film surface was flat, the drip irrigation tape had been recovered prior to the trial, and there were a few broken remnants of film, a few boll husks, leaves, and soil on the film surface. The grey desert soil moisture content was 19.29%, and the grey desert soil firmness was 5182.32 kPa.

#### 2.4.2. Test Methods

Field tests were carried out on a picking device with reference to the national standard GB/T 25412-2010 Residual Film Recycler. There were 17 test areas, and the length of each was 100 m, while the width was one film width (2.05 m). 3 points were randomly selected from each test area as test points, each test point was 10 m in length, and the average value of the 3 test points was taken as the test result for that stroke. The residual film pick-up rate $\mu_1$ and the residual film impurity rate $\mu_2$ were determined as test indicators in conjunction with the actual operation. Use Formula (28) to calculate the residual film pick-up rate and Formula (29) to calculate the residual film impurity rate. Before the test, the total mass $G_1$, of the film laid at the test site was calculated by measuring the film mass of one square meter by electronic balance. After the test, the residual film left at each detection point was collected, and the total mass of the residual film $G_2$, was measured by electronic balance. The total mass of residual film impurities $G_3$, was measured by electronic scale. The impurities in the residual film of the cleaning machine were manually picked up, and the mass $G_4$, of the impurities was measured by an electronic balance.

$$\mu_1 = \frac{G_1 - G_2}{G_1} \times 100\% \tag{28}$$

$$\mu_2 = \frac{G_4}{G_3} \times 100\% \tag{29}$$

where $\mu_1$ is the residual film pick-up rate %; $\mu_2$ is the residual film impurity rate %; $G_1$ is the total mass of film laid at the test points in the test plot g; $G_2$ is the mass of residual film left at each test point g; $G_3$ is the total mass of film impurities g; $G_4$ is the mass of impurities in the film impurities g.

The machine advancing velocity, the depth of hook tooth, and the chain rake input speed are the three main factors influencing the effect of residual film picking and residual film separation, according to the structure and working principle of the picking device. The machine advancing velocity of the implement is determined by the tractor's travel speed, and the best velocity of the implement is 1.5 m/s, according to the subject group's research. Given that the actual situation in the field differs from that on the experimental bench, the range is set at 0.5 m/s. Hook teeth into the soil depth will affect the machine's operational effect to some extent; according to the structural design of the machine tail depth limit wheel, the hook teeth minimum depth should be 20 mm, the maximum depth should be 62 mm, and the rake teeth should be set into the soil depth level at 20, 40, and 60 mm. The previous theoretical analysis determined that the chain rake input speed is 210, 234, and 258 rpm.

In this paper, a three-factor, three-level response surface approach was used to design the experiment for the operation of the picking device in the field, and the experimental factors were coded as shown in Table 2.

**Table 2.** The test factors and levels.

| Coded Value | Machine Advancing Velocity $X_1$ ($m{\cdot}s^{-1}$) | Depth of Hook Tooth $X_2$ (mm) | Chain Rake Input Speed $X_3$ (rpm) |
|---|---|---|---|
| −1 | 1.0 | 20 | 210 |
| 0 | 1.5 | 40 | 234 |
| 1 | 2.0 | 60 | 258 |

### 2.4.3. Test Results

The experiment was designed using Design Expert 12.0 software (Stat-Ease Inc., Minneapolis, MN, USA), and 17 trial areas were arranged, with 1 trial in each, for a total of 17 sets of trials. Calculate the residua film pick-up rate using Formula (28), the residual film impurity rate using Formula (29), and the specific test scheme and results are shown in Table 3.

**Table 3.** The test plans and results.

| Test | $X_1$ | $X_2$ | $X_3$ | $\mu_1$ | $\mu_2$ |
|---|---|---|---|---|---|
| 1 | 0 | −1 | −1 | 84.1 | 13.4 |
| 2 | 1 | 1 | 0 | 83.8 | 13.2 |
| 3 | 1 | 0 | 1 | 85.2 | 11.6 |
| 4 | 0 | 0 | 0 | 87.6 | 10.5 |
| 5 | 0 | 0 | 0 | 88.5 | 9.8 |
| 6 | 1 | −1 | 0 | 82.8 | 13.6 |
| 7 | 0 | 0 | 0 | 88.1 | 10.2 |
| 8 | 1 | 0 | −1 | 80.7 | 15.3 |
| 9 | 0 | 1 | −1 | 82.3 | 14.1 |
| 10 | −1 | 0 | −1 | 82.4 | 14.2 |
| 11 | 0 | 1 | 1 | 84.8 | 12.6 |
| 12 | 0 | 0 | 0 | 88.2 | 10.3 |
| 13 | 0 | 0 | 0 | 87.9 | 10.3 |
| 14 | 0 | −1 | 1 | 85.3 | 11.4 |
| 15 | −1 | 0 | 1 | 78.2 | 16.4 |
| 16 | −1 | −1 | 0 | 81.3 | 14.7 |
| 17 | −1 | 1 | 0 | 79.2 | 16.1 |

## 3. Results and Discussion

As shown in Table 4, we used Design Expert software to analyze the test results and the multiple regression fit. Table 4 shows the results of the residua film pick-up rate and residua film impurity rate variance analyses. The significance of the regression equations of $\mu_1$ and $\mu_2$ on $X_1$, $X_2$, and $X_3$ was tested.

(1) Establishment of the regression equation and significance analysis of the residual film pick-up rate

According to the model ANOVA, the order of significance of the test factors affecting the residual film pick-up rate is $X_1 > X_3 > X_2$. The *p*-value of 0.0763 for the lack of fit of the residue picking rate was not significant, indicating that the regression model was valid and accurate, and that it could be used to analyze and predict the residue picking rate [19]. After removing the non-significant term, the regression equation for each factor's effect on the residual film pick-up rate $\mu_1$ was as follows:

$$\mu_1 = 88.06 + 1.42X_1 + 0.50X_3 + 0.77X_1X_2 + 2.18X_1X_3 - 4.39X_1^2 - 1.89X_2^2 - 2.04X_3^2 \quad (30)$$

(2) Establishment of the regression equation and significance analysis of the residual film impurity rate

**Table 4.** The analysis of variance of the regression equation.

| Source of Variation | *DOF* | Residua Film Pick-Up Rate $\mu_1$/% Sum of Squares | *F* | Significant Level *p* | Residua Film Impurity Rate $\mu_2$/% Sum of Squares | *F* | Significant Level *p* |
|---|---|---|---|---|---|---|---|
| Models | 9 | 165.94 | 4.46 | <0.0001 ** | 74.63 | 65.92 | <0.0001 ** |
| $X_1$ | 1 | 16.24 | 19.22 | 0.0001 ** | 7.41 | 58.92 | 0.0001 ** |
| $X_2$ | 1 | 1.44 | 3.01 | 0.0671 | 1.05 | 8.36 | 0.0233 * |
| $X_3$ | 1 | 2.00 | 1.61 | 0.0382 * | 3.13 | 24.84 | 0.0016 ** |
| $X_1 X_2$ | 1 | 2.4025 | 9.68 | 0.0268 * | 0.8100 | 6.44 | 0.0388 * |
| $X_1 X_3$ | 1 | 18.92 | 3.58 | <0.0001 ** | 8.70 | 69.19 | <0.0001 ** |
| $X_2 X_3$ | 1 | 0.4225 | 0.70 | 0.2799 | 0.0625 | 0.4969 | 0.5036 |
| $X_1^2$ | 1 | 81.24 | 0.23 | <0.0001 ** | 33.96 | 269.99 | <0.0001 ** |
| $X_2^2$ | 1 | 15.08 | 1.89 | 0.0002 ** | 7.56 | 60.11 | 0.0001 ** |
| $X_3^2$ | 1 | 17.57 | 0.25 | 0.0001 ** | 7.28 | 57.88 | 0.0001 ** |
| Residual | 7 | 2.16 | | | 0.8805 | | |
| Lack of fit | 3 | 1.70 | 1.96 | 0.0763 | 0.6125 | 3.05 | 0.1549 |
| Pure error | 4 | 0.4520 | | | 0.2680 | | |
| Total | 16 | 168.10 | | | 75.51 | | |

Note: ** means highly significant ($p < 0.01$), and * means significant ($0.01 \le p < 0.05$).

According to the model ANOVA, the order of significance of the test factors affecting the residual film impurity rate is $X_1 > X_3 > X_2$. The *p*-value of 0.1549, for the lack of fit of the residual film impurity rate, was not significant, indicating that the regression model was valid and accurate, and that it could be used to analyze and predict the residual film impurity rate. After removing the non-significant term, the regression equation for each factor's effect on the residual film impurity rate $\mu_2$ was as follows:

$$\mu_2 = 10.22 - 0.96X_1 + 0.36X_2 - 0.62X_3 - 0.45X_1X_2 - 1.47X_1X_3 + 2.84X_1^2 + 1.34X_2^2 + 1.32X_3^2 \quad (31)$$

*3.1. Response Surface Analysis*

Field trials were conducted in October 2021 in Beiquan Town, Shihezi City, Xinjiang, to test the machine's reliability and operational effectiveness, and to determine the optimum operating parameters. Figure 12 depicts the prototype field tests. At the same time, we used Design-Expert 12.0 software to create a response surface diagram. Figure 13 depicts an analysis of the interaction of the factors affecting pick-up rate and impurity rate.

(a) (b) (c)

**Figure 12.** A field test of the experimental device. (**a**) Before the test; (**b**) The test process; (**c**) After the test.

(1) Analysis of the influence of the residual film pick-up rate

Figure 13a depicts the effect of the interaction between $X_1$ and $X_2$ on the residual film pick-up rate when $X_3$ is at the central level (234 rpm). As shown in the graph, the residual film pick-up rate in $X_1$ and $X_2$ of interaction increases first and then decreases, with $X_1$ increasing first and then decreasing, and $X_2$ increasing first and then decreasing. On the residual film pick-up rate response surface, Figure 13b for $X_2$ is located in the center level

(40 mm) of the $X_1$ and $X_3$ interaction. As shown in the graph, the residual film pick-up rate increases and then decreases due to the interaction between $X_1$ and $X_3$, and increases and then decreases as $X_1$ increases, and increases and then decreases as $X_3$ increases; Figure 13c depicts the response of $X_1$ at the center level (1.5 m/s), as well as the interaction between $X_2$ and $X_3$. The graph depicts the effect of $X_2$ and $X_3$ interaction on the film pick-up rate. As shown in the graph, the residual film pick-up rate in $X_2$ and $X_3$ interactions increases first and then decreases, with $X_2$ increasing first and then decreasing, and $X_3$ increasing first and then decreasing [20,21].

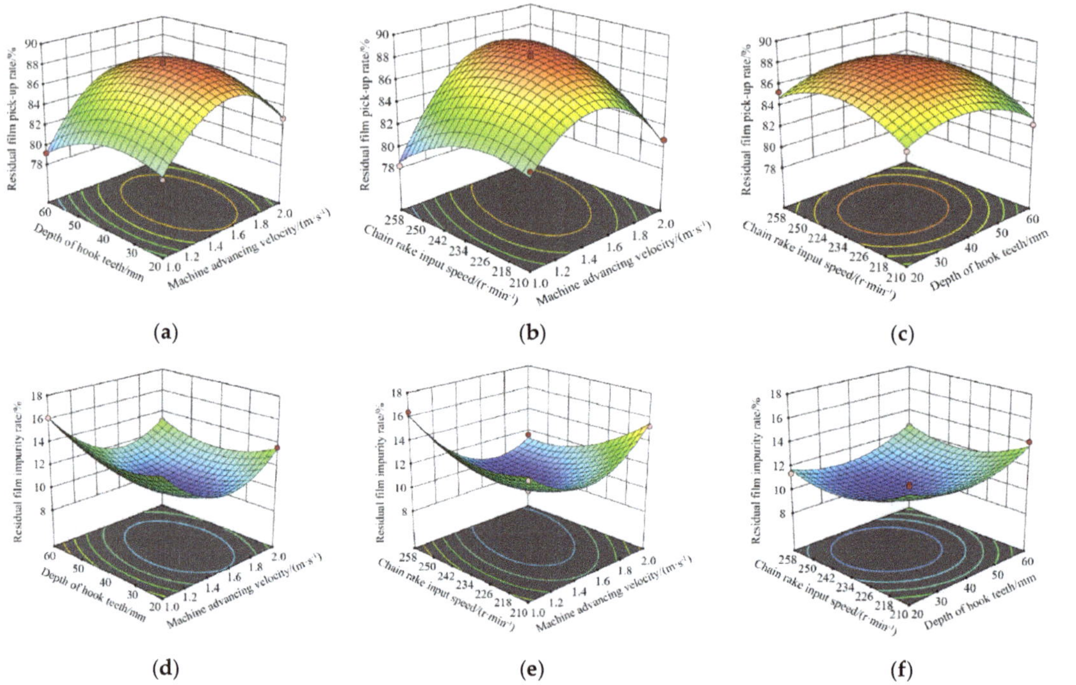


**Figure 13.** The effects of the interaction of various factors on the picking rate and trash content of cotton fallen on the ground. (**a**). $\mu_1$ = ($X_1$, $X_2$, 234); (**b**). $\mu_1$ = ($X_1$,40, $X_3$); (**c**). $\mu_1$ = (1.5, $X_2$, $X_3$); (**d**). $\mu_2$ = ($X_1$, $X_2$, 234); (**e**). $\mu_2$ = ($X_1$, 40, $X_3$); (**f**). $\mu_2$ = (1.5, $X_2$, $X_3$).

The main reasons for this are: the slower $X_1$, the poorer the ability of the hook teeth to pick up film continuously per unit length, and the subsequent hook teeth cannot pick up residual film sufficiently, resulting in a lower rate of picking up residual film. The faster $X_1$, the easier it is for the hook teeth to tear the residual film per unit length, which will also cause the residual film pick-up rate to decrease; $X_2$ has no significant effect on the residual film pick-up rate, which is due to the fact that the residual film in the field is mainly concentrated in the surface layer of the soil, so it is difficult to increase $X_2$ to improve the residual film pick-up rate; when $X_3$ is slower, the hook teeth are subject to soil resistance for a longer period of time and are easily de-formed and damaged, resulting in the pick-up residual film performance decline. When $X_3$ is faster, the hook teeth will cause a large impact on the residual film, which is not conducive to pick-up of the residual film.

(2) Effect of factor interaction of the residual film impurity rate

Figure 13d depicts the response surface plot of the interaction between $X_1$ and $X_2$ on the residual film rate when $X_3$ is at the central level (234 rpm). As shown in the figure, the residual film impurity rate in $X_1$ and $X_2$ of interaction under the influence of the first decrease and then increase, with $X_1$ increasing first decrease and then increasing, and $X_2$ increasing first decrease and then increasing; Figure 13e shows an interaction of $X_1$ and $X_3$ on the residual film impurity rate of response surface diagram with $X_2$ located in the center of the level (40 mm). The residual film impurity rate in $X_1$ and $X_3$ interactions can be seen in the figure, with $X_1$ increases first decreasing and then increasing, and $X_3$ increases first decreasing and then increasing. In Figure 13f, $X_1$ is located in the center of the level (1.5 m/s), and $X_2$ and $X_3$ interact on the residual film impurity rate response surface plot. The graph depicts the effect of the interaction between $X_2$ and $X_3$ on the residual film impurity rate. As shown in the graph, the residual film impurity rate in $X_2$ and $X_3$ interactions is first reduced and then increased, with $X_2$ increases first reduced and then increased, and $X_3$ increases first reduced and then increased.

The main reasons for the above are: the slower $X_1$, the faster the hook teeth disturb the soil per unit length, resulting in an increase in the proportion of impurities picked up. The faster $X_1$, the lower the residual film picking capacity of the hook teeth, but the picking of impurities on the film surface remains at a high level, resulting in a higher rate of residual film impurity. If $X_2$ is too shallow, the amount of residual film picked up decreases, but the rate of picking up impurities on the film surface increases. If $X_2$ is too deep, the impurities in the tillage layer will be picked up, resulting in a higher residual film impurity rate. When $X_3$ is too slow, the hook teeth are subject to greater resistance from the soil and are easily deformed and damaged, which is not conducive to the picking up of the residual film, but the effect of picking up the impurities on the residual film surface is not significant and can also cause the residual film impurity rate to be higher. When $X_3$ is too fast, the cleaning device's duration of time for cleaning impurities is shorter, and the impurities cannot be separated in time, resulting in a higher residual film impurity rate.

### 3.2. Parameter Optimization and Test Validation

To optimize the prototype's performance, a multi-objective optimization of the residual film pick-up rate and residual film impurity rate was performed using the Design-Experimentation module [22–24]. The optimization intervals were set to be the upper and lower limits of the test factors, with the goals of increasing the residual film pick-up rate and decreasing the residual film impurity rate. The optimization conditions in the software are set to dual-objective equal-weight optimization, with the following constraints:

$$\begin{cases} \mu_{1max} = F(X_1, X_2, X_3) \\ \mu_{2min} = F(X_1, X_2, X_3) \\ s.t \begin{cases} X_1 \in [1.0, 2.0] \\ X_2 \in [20.0, 60.0] \\ X_3 \in [210.0, 258.0] \end{cases} \end{cases} \quad (32)$$

The optimization and solution of the objective function for the optimal parameter combination was as follows: machine advancing velocity was 1.62 m/s, depth of hook tooth was 38.51 mm, and chain rake input speed was 241.42 rpm, yielding a predicted residual film pick-up rate of 88.27% and a residual film impurity rate of 9.96%. The optimized operating parameters for rounding were 1.6 m/s machine advancing velocity, a 39 mm depth of hook tooth, and a 241 rpm chain rake input speed based on the actual test conditions in the field. The above parameters were used for three replicate tests to obtain more accurate experimental results, and the average value was used as the final test result. The actual residual film pick-up rate was 87.52%, and the residual film impurity rate was 10.12%, as shown in Table 5. The relative error between actual and theoretical values was less than 5%, indicating that the optimisation results obtained with Design-Expert 12.0 software were reliable.

### 3.3. Discrete Element Modeling of Hook Tooth Motion Process

#### 3.3.1. Modeling and Parameter Setting of the Simulation Model

To further investigate the force on the hook teeth during the operation of the pickup device, simulation tests were performed using EDEM. To accurately simulate the interaction between the hook teeth and the soil during the actual operation, the radius of the soil particles, as shown in Figure 14a, was set to 3 mm, and the contact model of the soil particles was created in EDEM 2018 (DEM Solutions Ltd. Edinburgh, Scotland, UK) using the Hertz–Mindlin with Bonding model. Then, create a 3D model structure of the simplified hook tooth using SolidWorks 2018 software (Dassault Systèmes S.E., Massachusetts, Concord, MA, USA), save it as a ".x_t" format file, and import it into EDEM 2018.

**Table 5.** A comparison between the optimum theoretical and test results.

| Parameter | Residua Film Pick-Up Rate $\mu_1$/% | Residua Film Impurity Rate $\mu_2$/% |
|---|---|---|
| Theoretical optimization value | 88.27 | 9.96 |
| Test average | 87.52 | 10.12 |
| Relative error | 0.85 | 2.51 |

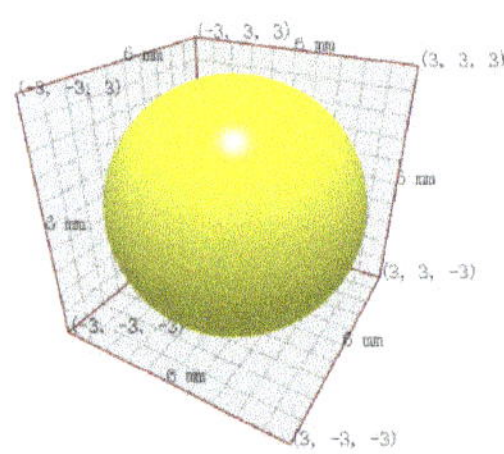

(**a**)

(**b**)

**Figure 14.** A particle and geometric simulation model. (**a**) Particle model of the soil; (**b**) simulation model of the EDEM.

The soil trough is modeled in EDEM software with dimensions L × W × H = (1250 × 1000 × 250) mm. In order to simulate the actual motion law of the hook tooth, according to the results of the previous analysis, set the front end of the hook tooth just touching the soil as the starting position, set the horizontal forward speed to 1.62 m/s, the chain rake input speed to 241 rpm, and the hook tooth simulation model operation process as shown in Figure 14b.

The simulation time step is set to $1.5 \times 10^{-6}$, the simulation time is one second, and the grid cell size is 3 times the minimum soil particle size. The contact model is chosen from soil particles and a hook tooth, and the main parameters are contact parameters and intrinsic parameters, with the contact parameters being soil recovery coefficient, static friction coefficient, and dynamic friction coefficient, and the intrinsic parameters being density, Poisson's ratio, and shear modulus. The data for the main parameters of the discrete element method test model were obtained using the method of calibration and optimization of the stacking test discrete element parameters [25–27], in which the soil parameters were obtained from the gray desert soil, which itself is the most widely distributed soil type in Shihezi, Xinjiang. The relevant parameters are shown in Table 6.

#### 3.3.2. Analysis of Simulation Results

When the simulation process is finished, the post-processing module of EDEM software is entered, and the maximum force on the hook tooth is obtained as 60.7 N. ANSYS software analyzes the stress and deformation of the hook tooth under maximum force, and the loading area is set to the hook tooth's working surface with a depth of 38.51 mm [28].

Under the condition of complete constraint of the hook tooth installation center, the load on the surface of the hook tooth is 60.7 N, the loading direction of the surface of the hook tooth is set to be the same as the direction of the instantaneous rotation of the tooth tip, and the load on the surface of the hook tooth is configured to be evenly distributed in order to simplify the calculation and improve the calculation speed.

**Table 6.** The simulation parameter settings of soil particles and geometry.

| Item | Parameter | Value |
|---|---|---|
| Soil particles | Poisson's ratio | 0.30 |
| | Shear modulus/Pa | $5 \times 10^7$ |
| | Density/(kg·m$^{-3}$) | 2600.00 |
| Hook teeth | Poisson's ratio | 0.35 |
| | Shear modulus/Pa | $7.27 \times 10^{10}$ |
| | Density/(kg·m$^{-3}$) | 7890.00 |
| Soil particles-Soil particles | Recovery coefficient | 0.21 |
| | Static friction coefficient | 0.68 |
| | Dynamic friction coefficient | 0.27 |
| Soil particles-Hook teeth | Recovery coefficient | 0.32 |
| | Static friction coefficient | 0.54 |
| | Dynamic friction coefficient | 0.13 |

Figure 15a depicts the hook tooth deformation cloud. The maximum deformation of the hook tooth occurs at the end, with a deformation of 31.42 mm, which is relatively small in comparison to the entire hook tooth. The hook teeth are made of spring steel, which has the ability to deform elastically and can withstand a certain load without permanent deformation after the load is removed. Figure 15b depicts the stress cloud of the hook tooth. The maximum strain occurs at the connection between the hook tooth and the threaded fastener, and the stress value is 215.33 MPa. The yield strength of the hook tooth is $[\sigma_b]$ = 1230 MPa, and while this time the safety factor under the established working conditions is 5.71, the stress value is much less than the allowable stress, so the hook tooth design meets the requirements.

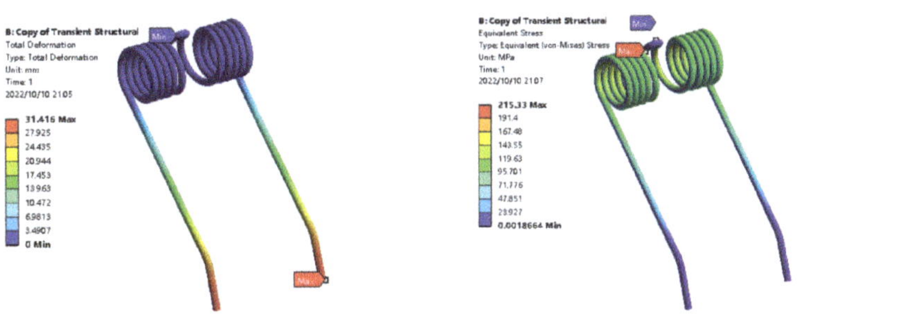


**Figure 15.** The analysis results. (**a**) Hook tooth deformation cloud map; (**b**) Hook tooth stress cloud.

## 4. Conclusions

1. In response to the problems of high residual film pick-up rate and poor reliability of the existing film recovery machine, a Hook-and-Tooth Chain Rail Type Residual Film Picking Device was designed, introducing the structure and working principle of the main components. By analyzing the main working components of the picking device, the main design parameters of its components were determined.

2. Field tests were carried out with the machine advancing velocity, depth of hook tooth, and chain rake input speed as influencing factors and the residual film pick-up rate and residual film impurity rate as test indicators. Additionally, the response surface data were analyzed using Design Expert software, and multiple fittings obtained the regression equation of the residual film pick-up rate and the residual film impurity rate. The influence of the interaction of various factors on the residual film pick-up rate and the residual film impurity rate was determined.
3. Experimental tests on the device proved that when the machine advancing velocity was 1.6 m/s, the depth of the hook tooth was 38 mm, and the chain rake input speed was 241 rpm. With these working parameters, field trials yielded a residual film pick-up rate of 87.52% and a residual film impurity rate of 10.21%. The optimized operating parameters were verified experimentally. Relative errors between the experimental results and optimized theoretical values of the residual film pick-up rate and residual film impurity rate were 0.85% and 2.51%, respectively, which is relatively low. Thus, the model was highly reliable.
4. EDEM simulated the motion process of the hook tooth and obtained the maximum force on the hook tooth during the working process. Then, ANSYS software was used to analyze the stress and deformation of the hook tooth under the state of maximum force, and the structural strength of the hook tooth was verified to meet the design requirements.

**Author Contributions:** Conceptualization, methodology, writing—original draft, J.X. and S.C.; writing—review and editing, H.W. and Y.Y.; software, H.W. and Y.Z.; investigation, H.W. and Y.Z.; data curation, J.Z. and S.W.; funding acquisition, J.X. and S.C.; validation, S.C.; All authors have read and agreed to the published version of the manuscript.

**Funding:** This research was funded by the National Natural Science Foundation of China funded project (51965058), Corps Young and Middle-Aged Leading Talents Program Project (2018CB011), and research and development of high quality and efficient separation machinery and equipment for cotton straw and residual film leaving the field (2022B02033-2).

**Institutional Review Board Statement:** Not applicable.

**Informed Consent Statement:** Not applicable.

**Data Availability Statement:** All relevant data presented in the article are stored according to institutional requirements and, as such, are not available online. However, all data used in this Manuscript can be made available upon request to the authors.

**Conflicts of Interest:** The authors declare no conflict of interest.

## References

1. Zhao, Y.; Chen, X.; Wen, H.; Zheng, X.; Niu, Q.; Kang, J. Research Status and Prospect of Control Technology for Residual Plastic Film Pollution in Farmland. *Trans. Chin. Soc. Mach.* **2017**, *48*, 1–14. [CrossRef]
2. Yan, C.; Mei, X.; He, W.; Zheng, S. Present situation of residue pollution of mulching plastic film and controlling measures. *Trans. Chin. Soc. Agric. Eng.* **2006**, *22*, 269–272.
3. Wang, Z.; Chen, X.; Yan, L.; Jiang, D.; Wang, M. Design and Experiment on Collecting and Removing Device for Profile Modeling Residual Plastic Film Collector. *Trans. Chin. Soc. Mach.* **2021**, *52*, 80–90. [CrossRef]
4. Hu, C.; Wang, X.F.; Chen, X.G.; Tang, X.Y.; Zhao, Y.; Yan, C. Current situation and control strategies of residual film pollution in Xinjiang. *Trans. Chin. Soc. Agric. Eng.* **2019**, *35*, 223–234. [CrossRef]
5. Lavo, G. Machine for Removing Wide Strips Laid Out on the Ground. 1995. Available online: https://patents.google.com/patent/US5386876A (accessed on 8 August 2022).
6. Parish, R.L. An automated machine for removal of plastic mulch. *Trans. ASAE* **1999**, *45*, 49. [CrossRef]
7. Rocca, A.R. Plastic Mulch Retrieve. 2012. Available online: https://patents.google.com/patent/US8302699B2 (accessed on 8 August 2022).
8. Shi, Z.L.; Tang, X.P.; Zhen, J.; Yan, J.S.; Zhang, X.J.; Jin, W. Performance test and motion simulation analysis of nail tooth type mechanism for collecting plastic residue. *Trans. Chin. Soc. Agric. Eng.* **2019**, *35*, 64–71. [CrossRef]
9. Liu, X.F.; Shi, X.; Guo, Z.F.; Wang, C.Y.; Wang, X.N. Performance test on roller type residual film recycling machine. *Trans. Chin. Soc. Agric. Eng.* **2017**, *33*, 26–31. [CrossRef]

10. Tang, Y.; Zhao, Y.; Wang, J.; Wang, Z. Design and experiment of film removing device for clamping finger-chain type residual film collector. *Trans. Chin. Soc. Agric. Eng.* **2020**, *36*, 11–19. [CrossRef]
11. Zhang, X.J.; Liu, J.Q.; Shi, Z.L.; Jin, W.; Yan, J.S.; Yu, M.J. Design and parameter optimization of reverse membrane and soil separation device for residual film recovery machine. *Trans. Chin. Soc. Agric. Eng.* **2019**, *35*, 46–55. [CrossRef]
12. Guo, W.; Jian, J.; San, Y.; Li, G.; Gao, M.; Hou, S. Design and Experimental Optimization of 4CML-1000 Type Chain Rake Film Recycling Machine. *Trans. Chin. Soc. Mach.* **2018**, *49*, 66–73. [CrossRef]
13. Xie, J.; Yang, Y.; Cao, S.; Zhang, Y.; Zhou, Y.; Ma, W. Design and experiments of rake type surface residual film recycling machine with guide chain. *Trans. Chin. Soc. Agric. Eng.* **2020**, *36*, 76–86. [CrossRef]
14. Yin, J.; Liu, D.; Li, Y. Design and Parameters Analyses of Automatic Height Profiling Device of Quadrate-Bale Baler Pickup. *Trans. Chin. Soc. Mach.* **2014**, *45*, 86–92. [CrossRef]
15. Kang, J.; Wang, S.; Yan, L. Design and Experiment of Loosen Shovel Installed on Plastic Film Collecting Machine. *Trans. Chin. Soc. Mach.* **2016**, *47*, 143–148. [CrossRef]
16. Cheng, D.X. *Handbook of Mechanical Design*; Chemical Industry Press: Beijing, China, 2008.
17. Zheng, Z.Q.; He, J.; Wang, Q.J.; Li, H.W.; Li, W.Y.; Chen, W.Z. Design and Experiment on Straw Pickup-chopping and Ditch-burying Integrated Machine. *Trans. Chin. Soc. Mach.* **2017**, *48*, 87–96. [CrossRef]
18. Yang, S.; Yan, L.; Mo, Y.; Chen, X.; Zhang, H.; Jiang, D. Design and Experiment on Collecting Device for Profile Modeling Residual Plastic Film Collector. *Trans. Chin. Soc. Mach.* **2018**, *49*, 109–115+164. [CrossRef]
19. Luo, K.; Yuan, P.; Jin, W.; Yan, J.S.; Bai, S.; Zhang, C.; Zhang, X.J. Design of chain-sieve type residual film recovery machine in plough layer and optimization of its working parameters. *Trans. Chin. Soc. Agric. Eng.* **2018**, *34*, 19–27. [CrossRef]
20. Yan, W.; Hu, Z.; Wu, N.; Xu, H.; You, Z.; Zhou, X. Parameter optimization and experiment for plastic film transport mechanism of shovel screen type plastic film residue collector. *Trans. Chin. Soc. Agric. Eng.* **2017**, *33*, 17–24. [CrossRef]
21. Xie, J.; Zhang, F.; Chen, X.; Han, Y.J.; Tang, W. Design and parameter optimization of arc tooth and rolling bundle type plastic film residue collector. *Trans. Chin. Soc. Agric. Eng.* **2019**, *35*, 26–37. [CrossRef]
22. You, J.; Chen, X.; Zhang, B.; Wu, J. Design and experiment of 4JSM-2000 type combined operation machine for cotton stalk chopping and residual plastic film collecting. *Trans. Chin. Soc. Agric. Eng.* **2017**, *33*, 10–16. [CrossRef]
23. You, Z.; Hu, Z.; Wu, H.; Zhang, Y.; Yan, J.; Yan, W.; Zhou, X. Design and experiment of 1MCDS-100A typed shovel-sieve residual film recovery machine. *Trans. Chin. Soc. Agric. Eng.* **2017**, *33*, 10–18.
24. Zhang, Z.; Li, J.; Wang, X.; Zhao, Y.; Xue, S.; Su, Z. Parameters Optimization and Test of an Arc-Shaped Nail-Tooth Roller-Type Recovery Machine for Sowing Layer Residual Film. *Agriculture* **2022**, *12*, 660. [CrossRef]
25. Zhong, W.Q.; Yu, A.B.; Liu, X.J.; Tong, Z.B.; Zhang, H. DEM/CFD-DEM Modelling of Non-spherical Particulate Systems: Theoretical Developments and Applications. *Powder Technol.* **2016**, *302*, 108–152. [CrossRef]
26. Wu, T.; Huang, W.; Chen, X.; Ma, X.; Han, Z.; Pan, T. Calibration of discrete element model parameters for cohesive soil considering the cohesion between particles. *J. South China Agric. Univ.* **2017**, *38*, 93–98. [CrossRef]
27. Wang, X.; Hu, H.; Wang, Q.; Li, H.; He, J.; Chen, W. Calibration method of soil contact characteristic parameters based on DEM theory. *Trans. Chin. Soc. Mach.* **2017**, *48*, 78–85. [CrossRef]
28. Wang, H.; Cao, S.; Liu, Y.; Yang, Y.; Meng, X.; Ji, P. Design of Cotton Recovery Device and Operation Parameters Optimization. *Agriculture* **2022**, *12*, 1296. [CrossRef]

*agriculture*

MDPI

Article

# A Seedbed Clearing and Shaping Device for Dry Direct-Seeded Rice

**Hui Li [1], Longyu Fang [2], Pingping Yuan [1], Wei Lu [1] and Wenwu Yang [2,3,4,*]**

1 Hunan Academy of Agricultural Sciences, Changsha 410125, China
2 Key Laboratory of Key Technology on Agricultural Machine and Equipment, Ministry of Education, South China Agricultural University, Guangzhou 510642, China
3 Huangpu Innovation Research Institute, South China Agricultural University, Guangzhou 510725, China
4 Maoming Branch, Guangdong Laboratory for Modern Agriculture, Maoming 525000, China
* Correspondence: yangwenwu@scau.edu.cn

**Abstract:** The soil in some areas of northern China is heavy owing to the presence of clay and stones, which significantly affects the normal operation of a planter as well as the growth of rice. In this regard, this study proposes a seedbed clearing and shaping device for dry direct-seeded rice, which can be used to remove stones in the seeding area, break soil blocks, for soil leveling, and groove forming. The overall structure and roller of the proposed device was developed based on theoretical calculations, discrete element modeling (DEM) simulations, and field tests. The soil-mixing tooth was distributed on the roller based on the double-helix rule, and the two sides of the helix were configured according to the right-hand and left-hand. Subsequently, DEM was used to develop a $3^3$ box-bench design. According to the agronomic requirements and operating speed ratio, the forward speed was set to 0.5 m/s. Furthermore, the optimization parameters combination of the device obtained by simulation experiments was: forward speed 0.5 m/s, soil depth 61 mm, and rotation speed 110 r/min, which obtained a stone removal rate of 85.65%, stone removal efficiency of 35.47 pieces/m, operating resistance of 719.23 N, and torque of 174.89 Nm. The field verification test results indicated that the stone removal rate was 77.23% under the optimization parameters combination, and the mean relative error of the simulated experiments value was 8.42%, which showed that the performance of the proposed device functioned stably and reliably, thereby providing a high-quality seedbed for sowing and rice growth. The developed device represents a useful solution for the seedbed clearing and shaping.

**Keywords:** seedbed clearing and shaping; stone removal rate; seeding furrow; dry direct-seeded rice; discrete element modeling

**Citation:** Li, H.; Fang, L.; Yuan, P.; Lu, W.; Yang, W. A Seedbed Clearing and Shaping Device for Dry Direct-Seeded Rice. *Agriculture* **2022**, *12*, 1740. https://doi.org/10.3390/agriculture12101740

Academic Editors: Mustafa Ucgul and Chung-Liang Chang

Received: 10 September 2022
Accepted: 20 October 2022
Published: 21 October 2022



## 1. Introduction

Rice is considered the main food crop, with over 50% of the world's population consuming rice [1]. Rice transplanting is a traditional rice cultivation method used in Asia with a long history, which can shorten the growth period of rice and ensure the number of seedlings in the field is conducive to a high and stable yield [2]. Rice transplanting is a highly labor-intensive and water-intensive method of cultivation [3–5]. However, there is shortage of labor resources owing to the development of the economy and society, and labor costs have gradually increased. As a result, agriculture has become less profitable [6,7]. Dry direct-seeded rice is an important choice for solving the shortage of labor and water resources [8,9]. Singh et al. [10] reported that compared to rice transplanting, direct-seeded rice could save 56% labor, 73% mechanical work, and 25% irrigation water. Therefore, direct-seeded rice technology could increase farmers' income and save natural resources. Furthermore, because this technology directly sows seeds into the field without seedling raising or transplanting, the process is simple and easy to operate [11,12]. Previous studies show that the yield of dry direct-seeded rice can be compared to that of the rice transplanting method [13–15]. However, because dry direct-seeded rice is more susceptible to climate and weeds, field management is an important factor that affects the yield [16–18]. To solve the problems of direct-seeded rice, foreign and domestic experts have carried out research on dry direct-seeded rice with film mulching [19,20].

The quality of the seedbed affects the sowing quality and growth of the rice [21,22], and dry direct-seeded rice with film mulching requires soft, fine, and smooth seedbed soil. However, in some areas of northern China, the soil is heavy with clay and stones, which significantly affects the normal operation of the planter as well as the growth of the rice. Moreover, owing to the hardness of the stone, when the planter, plastic film, and stone interact, the plastic film is scraped and broken, which significantly affects the performance of the plastic film.

In view of the problems of dry direct-seeded rice with film mulching in northern China, most research adopts high-power tractors that support stone-picking machines and tools for deep scavenging, or uses a sieve device to wash the soil. However, the above devices consume a large amount of power during operation, resulting in soil layer chaos and soil structure destruction. These devices focus on the number of stones picked up, and seldom care about how many stones are present in the topsoil, because disturbances in the deep soil can turn the bottom stones over to the surface of the soil, which does not solve the problems of stones in the seedbed. Although several studies have been conducted on the stone-removal equipment of the minimum soil layer, namely, the seeding area, in seedbeds in northern China, there are no relevant reports on the effect of the machines' working parameters on the operation effect and performance.

Therefore, this study developed a seedbed clearing and shaping device for dry direct-seeded rice aiming at the complex structure, high power, and large soil disturbance of the existing equipment. The proposed device can be used to remove stones in the seeding area, break soil blocks, and for soil leveling and groove forming, which works on the top soil without disturbing the deep soil to clean the sowing layer of the soil. Furthermore, we carried out the optimal design of the key structural components, and simulation experiments and field tests were conducted to analyze its operation performance and working effect to obtain the optimal combination of parameters of the device. The proposed device provides new technology and equipment for the development of precision agriculture.

## 2. Structure and Working Principles of the Proposed Device

### 2.1. Machine Structure

Figure 1 shows the overall structure of the proposed seedbed clearing and shaping device for dry direct-seeded rice. The device comprises a frame, belt-driving system, gear reducer, drip irrigation belt, guiding tube of the drip irrigation belt, protective board, soil-mixing tooth, roller, groove-pressing roller, plastic film, and a depth-limiting wheel, along with other components.

(a) (b)

**Figure 1.** Structural diagram of the proposed seedbed clearing and shaping device for dry direct-seeded rice: (**a**) mechanical structure of the device: 1. frame; 2. belt-driving system; 3. gear reducer; 4. drip irrigation belt; 5. guiding tube of the drip irrigation belt; 6. protective board; 7. soil-mixing tooth; 8. roller; 9. groove-pressing roller; 10. plastic film; and 11. depth-limiting wheel. (**b**) Roller with soil-mixing teeth.

### 2.2. Working Principle

The proposed device for dry direct-seeded rice is connected to a tractor using the trifilar suspension system, and parallel four-bar linkages are connected to sowing devices; each set of sowing devices can float freely depending on the terrain. Figure 2 shows the working principle of the proposed device. According to the agronomic requirements of rice cultivation, the parameters of the corresponding work module, such as the height of the groove-pressing roller and depth-limiting wheel, should be adjusted prior to field operation.

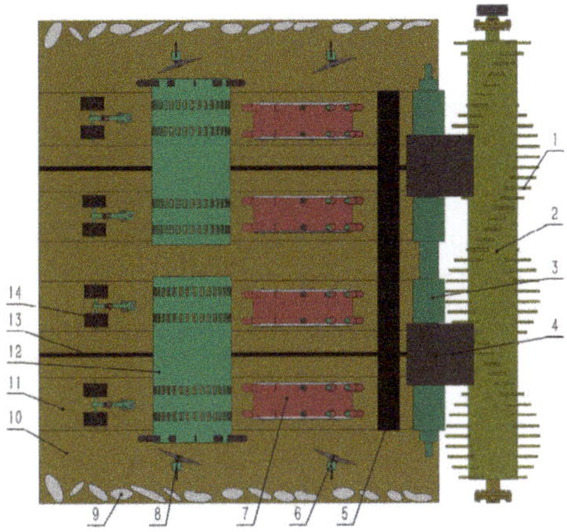


**Figure 2.** Working principle of the seedbed clearing and shaping device for dry direct-seeded rice. It comprises: 1. soil-mixing tooth; 2. roller; 3. groove-pressing roller; 4. drip irrigation belt; 5. plastic film; 6. soil-covering disk-1; 7. sowing device; 8. soil-covering disk-2; 9. stone and caking soil; 10. ridge; 11. seeding furrow; 12. soil-covering roller; 13. drip irrigation belt; and 14. soil-suppression device.

The power output shaft of the tractor provides an adequate running drive for the seedbed clearing and shaping device. The stones and soil blocks in the topsoil are thrown upward and forward using the rotary roller, and a part of the soil block is broken by collision with the soil-mixing tooth. Stones and soil blocks are conveyed to both sides of the device by the soil-mixing tooth with a spiral pattern distribution on the roller; the loose earth is not transported. The clearing operations are directed only at the sowing soil layer to reduce the disturbance to the deep soil and stones. When the groove-pressing roller rotates forward, it fills the shallow gullies and presses out the seeding furrow with a depth of 30–40 mm, thereby forming four regular seeding furrows and five ridges. Then, the seeds are sown in the corresponding seeding furrow, so the device provides a high-quality seedbed for the seeding operations and growth of rice.

The parameters of the proposed device have a direct effect on the quality of the seedbeds, and hence, are a key component of the planter for achieving the process of seedbed stone and soil block clearing and seeding furrow shaping. Therefore, according to the agronomic requirements of rice cultivation in the area of northern China, the sowing depth should be between 15 and 25 mm; that is, the stones and soil blocks should be removed from the 50 mm soil layer to form a clean seeding area, and the working depth of the soil-mixing tooth can be adjusted between 0 and 100 mm by controlling the height of the depth-limiting wheel. The width of the seeding furrow is determined to 200 mm based on the agronomy requirements (large and small row spacings are 250 mm and 120 mm, respectively) and combining the actual size of the sowing device to improve the utilization rate of natural rainfall, the depth of the seeding furrow is selected as 35 mm depending on the results of the preliminary experiments.

### 2.3. *Structural Design of Critical Component*

The soil-mixing tooth and roller are the key components of the device, since they directly affect the effects of seedbed clearing and shaping. The device adopts a harrow-type mechanism to pick up stones. The soil-mixing teeth are distributed on the roller according to the double-helix rule, and both sides of the helix are configured according to the right and left hands, respectively. The stones and soil blocks move to the two sides for the clearing of the seedbed.

#### 2.3.1. Kinematic Equation of the Endpoint of the Soil-Mixing Tooth

A Cartesian coordinate system is established, and the center of the roller is considered the original point O, whereas the forward direction and the direction perpendicular to the ground of the tractor is the positive direction of the X- and Y-axes, respectively. After time $t$, the tractor drives the roller from the original point O to $O_1$. The endpoint of the soil-mixing tooth was rotated by $\omega t$ degrees from point A to point $A_1$, whose coordinates are $A_1(x,y)$. Figure 3 shows the motion process. The parametric equation of the endpoint $A_1(x,Y)$ of the soil-mixing tooth is the equation with time $t$ as the variable, as shown in Equation (1). The trajectory of the endpoint of the soil-mixing tooth is an inverted trochoid formed by combining horizontal linear and circular motions.

$$\begin{cases} x = V_m t + R_m \cos(\omega t) \\ y = R_m \sin(\omega t) \end{cases} \quad (1)$$

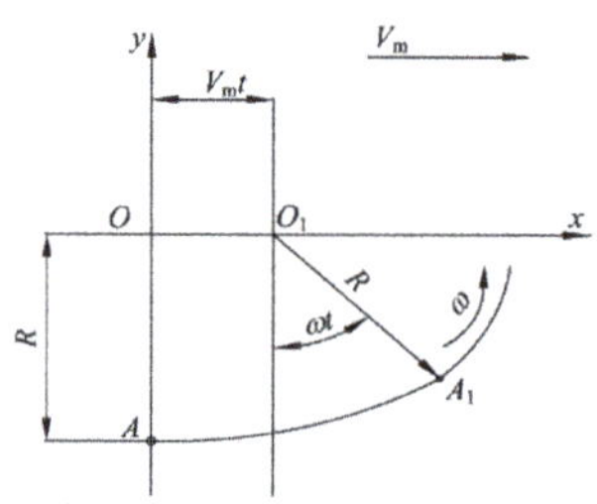


**Figure 3.** Motion model of the endpoint of the soil-mixing tooth.

$V_m$—speed of tractor, m/s;
$R_m$—radius of turn at the endpoint of the soil-mixing tooth, m;
ω—rotational angular velocity of the soil-mixing tooth, rad/s;
t—time, s.

#### 2.3.2. Operating Speed Ratio

Operating speed ratio $\lambda$ is the ratio of the line speed of the endpoint of the soil-mixing tooth and the speed of the tractor, which is crucial for the normal operation of the device:

$$\lambda = \frac{R_m \omega}{V_m} \quad (2)$$

Substituting Equation (2) into Equation (1), we get:

$$\begin{cases} x = R_m(\frac{\omega t}{\lambda} + \cos(\omega t)) \\ y = R_m \sin(\omega t) \end{cases} \quad (3)$$

According to Equation (3), if $R_m$, $\omega$ and $\lambda$ are known, the coordinates of the endpoint of the soil-mixing tooth can be calculated to obtain the motion trajectory. Figure 4 shows $R_m = 210$mm, ω = 3.2π, and the motion trajectory of the endpoint of the soil-mixing tooth when the operation speed ratio λ > 1, λ = 1 and λ > 1. As seen from the figure, when $\lambda \leq 1$, the motion trajectory is a short trochoid without overlap, that is, there is a soil pushing and leakage area, and the device cannot work normally. When λ > 1, the motion trajectory is a trochoid. Therefore, λ > 1 is an important condition for the normal operation of the device.

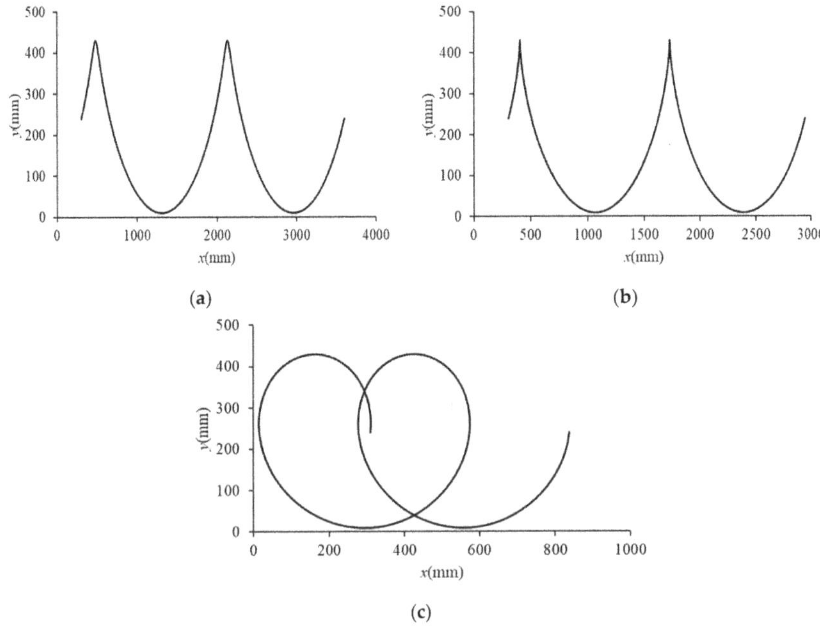


**Figure 4.** Motion trajectory of the endpoint of the soil-mixing tooth: (**a**) λ < 1; (**b**) λ = 1; (**c**)λ > 1.

2.3.3. Speed of the Endpoint of the Soil-Mixing Tooth

Equation (3) expresses the position of the endpoint of soil-mixing tooth at any time. The motion trajectory changes with the speed of tractor $V_m$, the radius of turn $R_m$, and the rotational angular velocity $\omega$. We take the derivative of Equation (3) with respect to time to obtain the partial velocities of the endpoint of the soil-mixing tooth in the $x$ and $y$ directions:

$$\begin{cases} V_x = \frac{dx}{dt} = R_m\left(\frac{\omega}{\lambda} - \omega \sin \omega t\right) \\ V_y = \frac{dy}{dt} = \omega \cos \omega t \end{cases} \quad (4)$$

Then, the instantaneous velocity $V$ of the endpoint of the soil-mixing tooth is:

$$V = \sqrt{V_x^2 + V_y^2} = V_m\sqrt{1 - 2\lambda \sin \omega t + \lambda^2} \quad (5)$$

## 3. Materials and Methods

Discrete element modeling (DEM) was adopted to carry out the simulation experiments of the seedbed clearing and shaping to improve the stone removal rate of the 50 mm soil layer and reduce the operating resistance to decrease the power consumption of the tractor, as well as to avoid manual repeatability experiments and reduce the intensity of labor. Accordingly, scientific and reliable working parameters of the device can be obtained.

### *3.1. Simulation Model Creation and Parameter Setting*

The interaction between the dispersed particles and the soil operation machine involves complex mechanical problems, making the study of soil-touching parts difficult.

Furthermore, there is currently no precise mathematical model for predicting the interactions of different operating environments. At present, the design and optimization of the soil-touching device mostly adopts the method of virtual simulation, among which DEM is an advanced and mature numerical simulation method for the discontinuous medium problem. In recent years, DEM has been widely applied in agricultural engineering, which mainly focuses on agricultural materials science and the application and research of the interaction between the soil and soil-touching parts. Therefore, this study adopted the EDEM 2018 software to simulate the seedbed clearing and shaping process and conduct the simulation experiments. The purpose of the DEM simulation experiments is to establish the regression equations of stone removal rate ($y_1$), stone removal efficiency ($y_2$), operating resistance ($y_3$), and torque ($y_4$), to understand the operation law and mechanism of the device, and to obtain the optimal combination of operation parameters to achieve high efficiency and energy-saving operations.

Figure 5a shows the simplified simulation model of the seedbed clearing and shaping device imported into the software preprocessing module.

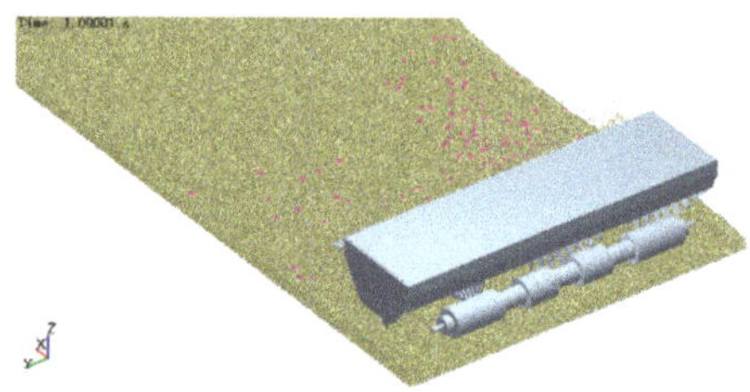

(**a**)

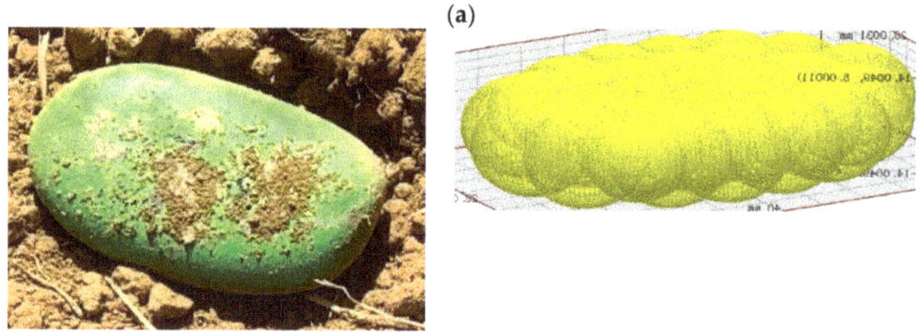

(**b**)

**Figure 5.** Simulation model: (**a**) simplified model of the device; (**b**) stone and stone particle model.

Multi-spherical polymerization was selected based on the characteristics of the stone shapes. A total of 35 sphere models were connected to construct the stone particle models (as shown in Figure 5b). The distribution of soil particle size was statistically analyzed using a size-grading sieve. The soil particles were divided into four grades, and nearly 50% of the soil particles were in the range of 1–2.36 mm. Combined with the computer performance and the actual situation of the simulation, the soil particle model considered spheres with a diameter of 2 mm, and the diameter was generated by the normal random distribution method to accord with the realistic irregularity of actual soil. The cutting-ring method was used to measure the soil density, which was repeated five times, and the average soil density was 1914 $kg \cdot m^{-3}$. Other soil parameters were determined based on the available parameters and relevant literature [23–25]. The soil particle model was calibrated with the angle of repose as the response value, and the relative error between the actual and the model was 2.89%. The Hertz–Mindlin (no-slip) contact model included in the EDEM 2018 software was used for the simulation analysis, and with reference to the relevant literature [26–28]. Table 1 summarizes the simulation experiment parameters, which were obtained through experimental measurement and references [29,30], and calibrated by EDEM. Based on the actual number of stones in the 50 mm soil layer, 240 stone particles were generated, the size of the soil tank was 4500 × 2400 × 150 mm (length × width × height), and the fixed time step was set to 20% of the Rayleigh time step ($8.86 \times 10^{-7}$ s).

**Table 1.** Simulation experiment parameters.

| Materials | Parameters | Value |
|---|---|---|
| Stone | Density (kg·m$^{-3}$) | 2470 |
| | Poisson ratio | 0.22 |
| | Shear modulus (Mpa) | 4500 |
| Soil | Density (kg·m$^{-3}$) | 1914 |
| | Poisson ratio | 0.38 |
| | Shear modulus (Mpa) | 1 |
| Steel | Density (kg·m$^{-3}$) | 7850 |
| | Poisson ratio | 0.3 |
| | Shear modulus (Mpa) | 79,000 |
| Stone–Stone | Elastic restitution coefficient | 0.58 |
| | Static friction coefficient | 0.55 |
| | Rolling friction coefficient | 0.05 |
| Soil–Soil | Elastic restitution coefficient | 0.11 |
| | Static friction coefficient | 0.6 |
| | Rolling friction coefficient | 0.4 |
| Stone–Steel | Elastic restitution coefficient | 0.54 |
| | Static friction coefficient | 0.5 |
| | Rolling friction coefficient | 0.05 |
| Soil–Steel | Elastic restitution coefficient | 0.12 |
| | Static friction coefficient | 0.3 |
| | Rolling friction coefficient | 0.05 |
| Stone–Soil | Elastic restitution coefficient | 0.21 |
| | Static friction coefficient | 0.4 |
| | Rolling friction coefficient | 0.15 |

### *3.2. Scheme and Method of Simulation Experiment*

According to the previous analysis and preliminary experimental results, the main operating parameters that affect the rice seedbed clearing and shaping, and the device working performance, namely, the forward speed $V$, soil depth $H$, and rotation speed $n$, were selected as the experimental factors. Furthermore, the stone removal rate ($y_1$), stone removal efficiency ($y_2$), operating resistance ($y_3$), and torque ($y_4$) were analyzed.

The simulation experiment using the box–bench design with three factors and three levels was conducted to evaluate the operating performance of the proposed device. Furthermore, preliminary combined factor experiments were conducted to determine the range of the experimental factors; the forward speed of the tractor is 0.4–0.8 m/s, soil depth is 40–80 mm, and the rotation speed is 70–130 r/min. Table 2 summarizes the code value of the simulation experimental factors, and Table 3 summarizes the 20 groups of simulation experiment schemes.

**Table 2.** Code of simulation experiment factors.

| Code Value | Forward Speed $V$ (m·s$^{-1}$) | Soil Depth $H$ (mm) | Rotation Speed $n$ (r/min) |
|---|---|---|---|
| 1.682 | 0.9364 | 93.636 | 150.45 |
| 1 | 0.8 | 80 | 130 |
| 0 | 0.6 | 60 | 100 |
| −1 | 0.4 | 40 | 70 |
| −1.682 | 0.2636 | 26.364 | 49.55 |

**Table 3.** Simulation experiment scheme and results.

| No | Forward Speed $V$ ($m \cdot s^{-1}$) | Soil Depth $H$ (mm) | Rotation Speed $n$ (r/min) | Stone Removal Rate $y_1$ (%) | Stone Removal Efficiency $y_2$ (pieces/m) | Operating Resistance $y_3$ (n) | Torque $y_4$ (Nm) |
|---|---|---|---|---|---|---|---|
| 1 | 0 | 0 | 0 | 91.833 | 34 | 763.582 | 173.014 |
| 2 | 1 | −1 | −1 | 71.250 | 18 | 483.361 | 92.290 |
| 3 | 0 | 0 | −1.682 | 85.896 | 44 | 802.878 | 137.276 |
| 4 | −1 | 1 | 1 | 87.083 | 40 | 655.782 | 183.634 |
| 5 | 0 | 0 | 0 | 88.667 | 34 | 773.245 | 177.796 |
| 6 | 0 | −1.682 | 0 | 72.042 | 8 | 308.313 | 63.606 |
| 7 | −1 | 1 | −1 | 92.229 | 48 | 872.667 | 197.637 |
| 8 | 0 | 0 | 0 | 91.042 | 35 | 783.080 | 181.173 |
| 9 | 0 | 0 | 0 | 90.646 | 31 | 790.377 | 184.534 |
| 10 | 1 | 1 | 1 | 94.604 | 35 | 1163.770 | 284.721 |
| 11 | 1 | 1 | −1 | 91.833 | 41 | 1354.200 | 260.389 |
| 12 | 1 | −1 | 1 | 86.688 | 9 | 587.393 | 123.762 |
| 13 | 0 | 1.682 | 0 | 94.208 | 34 | 1171.120 | 291.075 |
| 14 | −1 | −1 | −1 | 83.521 | 32 | 363.198 | 74.709 |
| 15 | 0 | 0 | 0 | 88.667 | 35 | 794.315 | 182.652 |
| 16 | 0 | 0 | 1.682 | 87.875 | 28 | 734.938 | 189.452 |
| 17 | −1 | −1 | 1 | 83.125 | 12 | 408.024 | 94.846 |
| 18 | 1.682 | 0 | 0 | 90.646 | 33 | 985.338 | 202.627 |
| 19 | 0 | 0 | 0 | 88.667 | 31 | 780.920 | 179.834 |
| 20 | −1.682 | 0 | 0 | 83.521 | 43 | 432.149 | 114.310 |

During the experiments, when the work of the device was stable, the post-processing module of the EDEM 2018 software was used to establish the upper and lower two-layer grids with soil depths between 0–50 mm and 50–100 mm, respectively, in the region of determination. The index $y_1$ is the ratio between the number of stones removed to both sides of the device and the total amount of stones during the simulation process. The index $y_2$ is the number of stones removed to both sides of the device among the unit length during the normal operation section. The index $y_3$ and $y_4$ can be obtained directly by using the post-processing module of the EDEM 2018 software.

### *3.3. Data Analysis*

The mechanical properties of the proposed seedbed clearing device was evaluated through the simulation experiment and field tests. The data processing and analysis were performed using Microsoft Excel and Design Expert software.

## 4. Results and Discussion

### *4.1. Variance Analysis and Discussion*

Table 3 summarizes the results of the simulation experiments. The simulation data were then used to establish the regression equations, and variance analysis was carried out to verify the validity of the regression equations and significance of the items of the regression equations. Simultaneously, the secondary optimization of the regression equation was carried out to eliminate insignificant items. The optimized regression equations are shown in Equations (6)–(9), which were subjected to ANOVA again, as shown in Tables 4–7, respectively.

$$y_1 = 72.311 - 75.359V + 0.974H - 0.04n + 0.495VH + 0.495VN - 0.004Hn - 0.005H^2 \quad (6)$$

$$y_2 = 3.135 - 1.53V + 0.1732H - 0.0459n + 0.0005Hn - 0.0014H^2 \quad (7)$$

$$y_3 = -696.2 + 274.7925V + 16.3075H + 6.0424n + 21.5622VH - 0.1159VN - 629.9684V^2 - 0.0356H^2 \quad (8)$$

$$y_4 = -133.7567 + 40.8698V + 2.0403H + 1.6016n + 3.667VH + 1.0348VN - 0.0086Hn - 194.1504V^2 - 0.0067n^2 \quad (9)$$

**Table 4.** ANOVA of the regression equation of stone removal rate.

| Source | Sum Of Squares | df | Mean Square | *F*-Value | *p*-Value |
|---|---|---|---|---|---|
| Model | 679.78 | 7 | 97.11 | 17.23 | <0.0001 * |
| *V* | 7.92 | 1 | 7.92 | 1.4 | 0.2588 |
| *H* | 450.6 | 1 | 450.6 | 79.94 | <0.0001 * |
| *n* | 18.73 | 1 | 18.73 | 3.32 | 0.0933 |
| *VH* | 31.34 | 1 | 31.34 | 5.56 | 0.0362 * |
| *Vn* | 70.51 | 1 | 70.51 | 12.51 | 0.0041 * |
| *Hn* | 37.92 | 1 | 37.92 | 6.73 | 0.0235 * |
| $H^2$ | 62.76 | 1 | 62.76 | 11.13 | 0.0059 * |
| Residual | 67.64 | 12 | 5.64 | | |
| Lack of Fit | 57.48 | 7 | 8.21 | 4.04 | 0.0717 |
| Pure Error | 10.16 | 5 | 2.03 | | |
| Cor Total | 747.41 | 19 | | | |

$R^2$ = 0.9095; $R^2_{adj}$ = 0.8567; $CV$ = 2.72%; adequate precision = 15.9099. Note: * shows the term is significant ($p < 0.05$).

**Table 5.** ANOVA of the regression equation of stone removal efficiency.

| Source | Sum of Squares | df | Mean Square | F-Value | *p*-Value |
|---|---|---|---|---|---|
| Model | 24.93 | 5 | 4.99 | 63.75 | <0.0001 * |
| *V* | 1.28 | 1 | 1.28 | 16.35 | 0.0012 * |
| *H* | 14.89 | 1 | 14.89 | 190.35 | <0.0001 * |
| *n* | 3.37 | 1 | 3.37 | 43.07 | <0.0001 * |
| *Hn* | 0.6872 | 1 | 0.6872 | 8.79 | 0.0103 * |
| $H^2$ | 4.71 | 1 | 4.71 | 60.18 | <0.0001 * |
| Residual | 1.09 | 14 | 0.0782 | | |
| Lack of fit | 0.9629 | 9 | 0.107 | 4.06 | 0.0687 |
| Pure error | 0.1319 | 5 | 0.0264 | | |
| Cor total | 26.02 | 19 | | | |

$R^2$ = 0.9579; $R^2_{adj}$ = 0.9429; $CV$ = 5.11%; adequate precision = 28.3661. Note: * shows the term is significant ($p < 0.05$).

**Table 6.** ANOVA of the regression equation of the operating resistance.

| Source | Sum Of Squares | df | Mean Square | F-Value | *P*-Value |
|---|---|---|---|---|---|
| Model | 1459000 | 7 | 208400 | 678.96 | <0.0001 * |
| *V* | 360700 | 1 | 360700 | 1175.06 | <0.0001 * |
| *H* | 978500 | 1 | 978500 | 3187.73 | <0.0001 * |
| *n* | 10172.1 | 1 | 10172.1 | 33.14 | <0.0001 * |
| *VH* | 59510.6 | 1 | 59510.6 | 193.88 | <0.0001 * |
| *Hn* | 38666.1 | 1 | 38666.1 | 125.97 | <0.0001 * |
| $V^2$ | 9241.95 | 1 | 9241.95 | 30.11 | <0.0001 * |
| $H^2$ | 2954.75 | 1 | 2954.75 | 9.63 | 0.0091 * |
| Residual | 3683.35 | 12 | 306.95 | | |
| Lack of fit | 3050.32 | 7 | 435.76 | 3.44 | 0.0962 |
| Pure error | 633.04 | 5 | 126.61 | | |
| Cor total | 1463000 | 19 | | | |

$R^2$ = 0.9948; $R^2_{adj}$ = 0.9910; $CV$ = 3.57%; adequate precision = 56.0155. Note: * shows the term is significant ($p < 0.05$).

**Table 7.** ANOVA of the regression equation of torque.

| Source | Sum of Squares | df | Mean Square | F-Value | *P*-Value |
|---|---|---|---|---|---|
| Model | 77034.4 | 8 | 9629.29 | 262.91 | < 0.0001 * |
| $V$ | 9430.16 | 1 | 9430.16 | 257.47 | < 0.0001 * |
| $H$ | 62425.7 | 1 | 62425.7 | 1704.41 | < 0.0001 * |
| $n$ | 1640.63 | 1 | 1640.63 | 44.79 | < 0.0001 * |
| $VH$ | 1721.18 | 1 | 1721.18 | 46.99 | < 0.0001 * |
| $Vn$ | 308.39 | 1 | 308.39 | 8.42 | 0.0144 * |
| $Hn$ | 213 | 1 | 213 | 5.82 | 0.0345 * |
| $V^2$ | 877.81 | 1 | 877.81 | 23.97 | 0.0005 * |
| $n^2$ | 530.14 | 1 | 530.14 | 14.47 | 0.0029 * |
| Residual | 402.89 | 11 | 36.63 | | |
| Lack of fit | 320.39 | 6 | 53.4 | 3.24 | 0.1092 |
| Pure error | 82.49 | 5 | 16.5 | | |
| Cor total | 77437.2 | 19 | | | |

$R^2$ = 0.9948; $R^2_{adj}$ = 0.9910; $CV$ = 3.57%; adequate precision = 56.0155. Note: * shows the term is significant ($p < 0.05$).

The variance analysis of the regression equations of the stone removal rate ($y_1$), stone removal efficiency ($y_2$), operating resistance ($y_3$), and torque ($y_4$) showed that the regression equations were significant without the lack of fit, thereby indicating that the regression equations were effective and can be used to predict the response value. The significance ordering of the forward speed $V$, soil depth $H$, and rotation speed $n$ on the stone removal rate ($y_1$), stone removal efficiency ($y_2$), operating resistance ($y_3$), and torque ($y_4$) were soil depth $H$ > forward speed $V$ > rotation speed $n$, soil depth $H$ > rotation speed $n$ > forward speed $V$, soil depth $H$ > forward speed $V$ > rotation speed $n$, and soil depth $H$ > forward speed $V$ > rotation speed $n$, respectively.

### 4.2. *Interactive Analysis and Discussion*

The stone removal rate ($y_1$), stone removal efficiency ($y_2$), operating resistance ($y_3$), and torque ($y_4$) were not only affected by the forward speed $V$, soil depth $H$, rotation speed $n$, but also by the interactions between the factors, considering that partial interactions significantly affect the response value. Based on the analysis of variance of the regression equations, the response surfaces were analyzed and discussed.

#### 4.2.1. Stone Removal Rate

The soil depth has a significant effect on the stone removal rate. Although the forward speed and rotation speed affect the stone removal rate, it is not significant. However, there were significant interaction effects among the three factors, and the quadratic term of the soil depth significantly affects the stone removal rate. As shown in Figure 6a, the effect of soil depth on the stone removal rate is significantly stronger than the forward speed; when the soil depth is small, the stone removal rate changes slightly or decreases as the forward speed increases. When the soil depth is approximately 60 mm, the stone removal rate increases as the forward speed increases, and the increasing trend increases. The increasing soil depth increases the stone removal rate, whereas the increasing trend of the stone removal rate increases with the increasing forward speed. As shown in Figure 6b, when the forward speed is small, the stone removal rate changes slightly as the rotation speed increases. When the forward speed is large, the stone removal rate gradually increases as the rotation speed increases; the effect of the forward speed on the change of the stone removal rate is almost the same as that of the rotation speed. As shown in Figure 6c, when the soil depth is less than 70 mm, the stone removal rate increases rapidly as the rotation speed increases. When the soil depth is more than 70 mm, the stone removal rate changes slightly as the rotation speed increases, and exhibits a downward trend. When the rotation speed is small, the stone removal rate increases rapidly as the soil depth increases; however, when the rotation speed increases, the changes in the stone removal rate decrease as the soil depth increases. Finally, the tone removal rate reaches the maximum and does not change even if the soil depth increases.

#### 4.2.2. Stone Removal Efficiency

The forward speed, soil depth, rotation speed, and the interaction between the soil depth and rotation speed have significant influence on the stone removal efficiency. As shown in Figure 6d, with the increase in the soil depth, the stone removal efficiency gradually increases and finally does not change. With the increase in the rotation speed, the stone removal efficiency gradually decreases, and the amount of the reduction is small.

#### 4.2.3. Operating Resistance

The forward speed, soil depth, and rotation speed, as well as the interactions between forward speed and soil depth, have significant influence on the operating resistance. As shown in Figure 6e, the operating resistance increases as the forward speed increases; however, the operating resistance increases more strongly as the soil depth increases. As shown in Figure 6f, the operating resistance increases as the soil depth increases. When the soil depth is less than 70 mm, the operating resistance gradually increases as the rotation speed increases; however, the increase is relatively gentle. When the soil depth is greater than 70 mm, the operating resistance gradually decreases as the rotation speed increases.

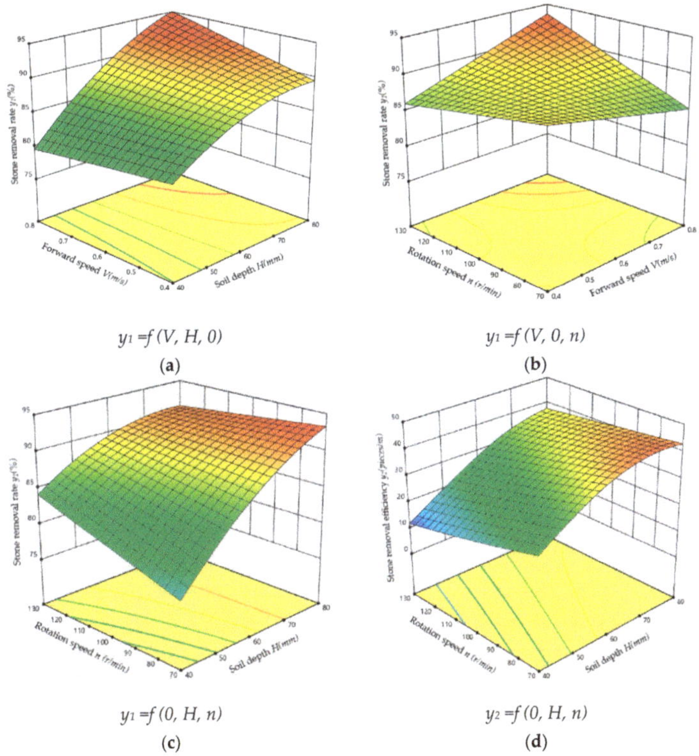


**Figure 6.** *Cont.*

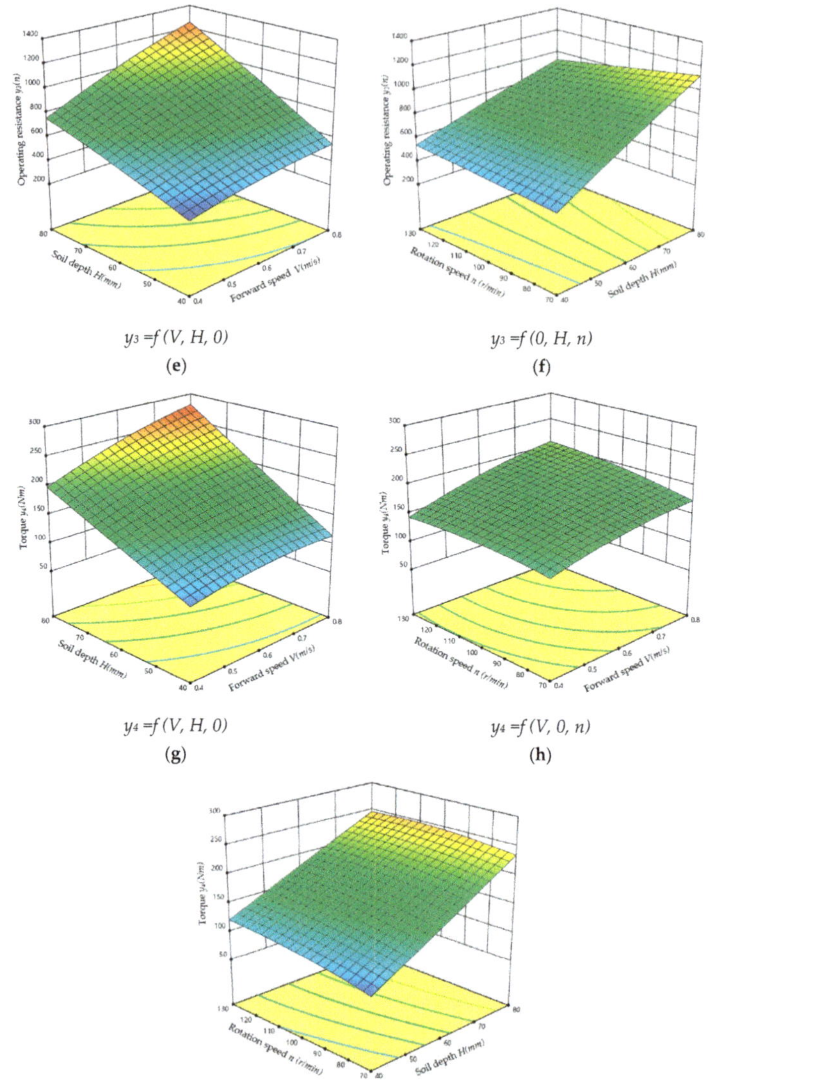


**Figure 6.** Response surfaces for interaction between factors: (**a**) effect of forward speed and soil depth on the stone removal rate; (**b**) effect of forward speed and rotation speed on the stone removal rate; (**c**) effect of soil depth and rotation speed on the stone removal rate; (**d**) effect of soil depth and rotation speed on the stone removal efficiency; (**e**) effect of forward speed and soil depth on the operating resistance; (**f**) effect of soil depth and rotation speed on the operating resistance; (**g**) effect of forward speed and soil depth on the torque; (**h**) effect of forward speed and rotation speed on the torque; and (**i**) effect of soil depth and rotation speed on the torque.

#### 4.2.4. Torque

The forward speed, soil depth, and rotation speed, as well as the interactions among the three factors, have a significant influence on the torque. As shown in Figure 6g, the torque increases as the forward speed and soil depth increase. As shown in Figure 6h, the increase in the forward and rotation speeds leads to a gentle increase in torque, and the two factors basically have the same trend of influence on the torque. As shown in Figure 6i, the rotation speed has little influence on the torque, and the torque gently increases as the rotation speed increases. When the rotation speed is approximately 100 r/min, the torque does not increase as the rotation speed increases. The soil depth has a great influence on the torque, and the torque increases rapidly as the soil depth increases.

### *4.3. Comprehensive Optimal Design*

To rank the importance of the response values depending on the operation principle of "good, fast, and effortless", and according to Equation (10), the optimal solution set was determined. To meet the requirements of the forward speed and operating speed ratio, the optimal solution is selected as forward speed V= 0.5 m/s, soil depth H = 61.28 mm, and rotation speed N = 110.85 r/min. The theoretical response values under the combination of parameters are stone removal rate $y_1$ = 88.37%, stone removal efficiency $y_2$ = 34.86 pieces/m, operating resistance $y_3$ = 692.66 N, and torque $y_4$ = 169.6 Nm. The expected value that the optimal solution satisfies the optimization condition is 0.746. The optimal operating parameters were rounded to forward speed V = 0.5 m/s, soil depth H = 61 mm, and rotation speed N = 110 r/min, and the optimal operating parameters were used to perform the simulation experiment to verify the validity and usability of the equations and results. The simulation experiment results were the stone removal rate $y_1$ = 85.65%, stone removal efficiency $y_2$ = 35.47 pieces/m, operating resistance $y_3$ = 719.23 N, and torque $y_4$ = 174.89 Nm, which were consistent with the theoretical results. Therefore, the equations and results were credible and could provide support for the field test. Figures 7 and 8 show the simulation experiment effect of the optimal operating parameters and motion trajectory of the stones.

$$\left\{\begin{array}{l}\left\{\begin{array}{l} max\ y_1(V,H,n) \\ max\ y_2(V,H,n) \\ min\ \ y_3(V,H,n) \\ min\ y_4(V,H,n) \end{array}\right. \\ \left\{\begin{array}{l} v = 0.5\ m/s \\ 40 \leq H \leq 80 \\ 70 \leq n \leq 130 \end{array}\right.\end{array}\right. \quad (10)$$

**Figure 7.** Simulation experiment effect of the device: (**a**) primary distribution of stones and (**b**) stone distribution after processing.

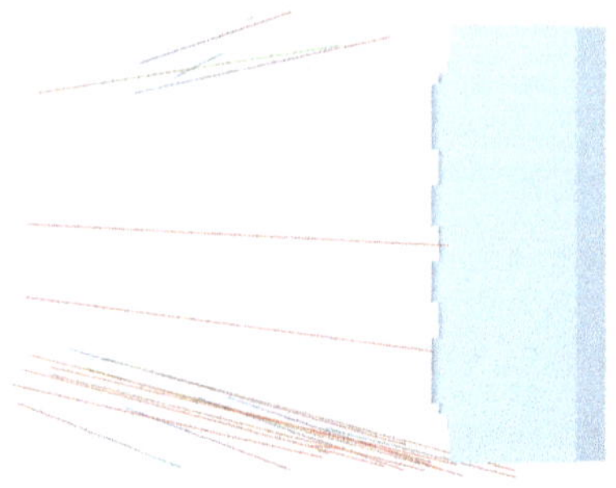

**Figure 8.** Motion trajectory of stone.

## *4.4. Comparisons with Field Measurements*

### 4.4.1. Field Test Conditions and Method

Owing to the COVID-19 pandemic and the rainy weather, the field test using the proposed device was conducted at the research and teaching base of South China Agricultural University, located in Yuezhou Road, Tianhe District, Guangzhou City, Guangdong Province (23°10′13.8″ N, 113°22′28.2″ E) on 31 December 2020. The experimental area was approximately 2000 $m^2$, and the average soil moisture content and compaction were 20.81% (0–150 mm depth) and 198.5 kPa (0–100 mm depth), respectively. Two squares with the same area (800 × 800 mm) were arranged 2000 mm apart in the operation area, and the number of stones in the two squares was 10 (sparse square) and 30 (dense square), respectively, which were dyed red and green, respectively, to facilitate stone picking up and data statistics, as shown in Figure 9a. The size of the stones in the two squares was randomly distributed in the range of 40–80 mm. The operating parameters of the device were set according to the optimal combination obtained from the simulation experiment. The field test was repeated six times, and the operating parameters of the device were calibrated before each operation to ensure the accuracy of the parameters, as shown in Figure 9b. The device was connected to the tractor (LOVOL M804-B wheeled tractor, Weifang, China) using a trifilar suspension system. Owing to the limitations of the experimental conditions, only the stone removal rate was counted, and the groove-forming effect was observed.

**(a)** **(b)**

**Figure 9.** Preparation for the field test: (**a**) distribution of stone and (**b**) calibration of operation parameters.

### 4.4.2. Field Test Results

The field test results are shown in Table 8. The coefficient of variation of the stone removal rate of six repeated field tests was found to be 8.98%; therefore, the operating performance of the device was stable. The average stone removal rate of six repeated field tests was 77.23%, and the difference compared with the simulation experiment results (85.65%) was less than 10%, which confirms the credibility of the simulation experiment results. In the six groups of field tests, the average stone removal rate of the two kinds of stone with different distribution densities were 82.78% and 71.67%, respectively. The stone removal rate in densely distributed areas were higher than those in less distributed areas. However, the difference between them was not significant, that is, the distribution density of the stones had no significant effect on the stone removal rate of the device. Figure 10 shows the field test site. The observation method was used to evaluate the groove-forming effect, and the comprehensive analysis showed that the effect reached the design requirements. Figure 11 shows the effect of groove forming and seedling growth.

**Table 8.** Field test results.

| No | 1 | | 2 | | 3 | | 4 | | 5 | | 6 | |
|---|---|---|---|---|---|---|---|---|---|---|---|---|
| Total stone | Dense square | Sparse square | Dense square | Sparse square | Dense square | Sparse square | Dense square | Sparse square | Dense square | Sparse square | Dense square | Sparse square |
| Number of stone removed | 26 | 8 | 27 | 6 | 23 | 7 | 27 | 6 | 20 | 7 | 26 | 9 |
| Stone removal rate (%) | 86.67 | 80 | 90 | 60 | 76.67 | 70 | 90 | 60 | 66.67 | 70 | 86.67 | 90 |
| Single stone removal rate (%) | 83.34 | | 75 | | 73.33 | | 75 | | 68.34 | | 88.34 | |
| Average stone removal rate (%) | 77.23 | | | | | | | | | | | |

(a) (b)

**Figure 10.** Distribution and statistics of stones: (**a**) stone distribution after processing and (**b**) collection of stones.

**Figure 11.** Effect of groove formation and seedling growth.

## 5. Conclusions

In order to improve the quality of the seedbed, we proposed a seedbed clearing and shaping device for dry direct-seeded rice, which was developed using theoretical calculations, discrete element modelling simulations, and field tests. The device can be used to remove stones in the seeding area, break soil blocks, and form grooves without disturbing the deep soil. The research results showed that:

(1) The primary and secondary factors that affect the stone removal rate ($y_1$), stone removal efficiency ($y_2$), operating resistance ($y_3$), and torque ($y_4$) of the device were: soil depth H > forward speed V > rotation speed n and soil depth H > rotation speed n > forward speed V. Furthermore, the optimization parameters combination of the device was: forward speed V = 0.5 m/s, soil depth H = 61 mm, and rotation speed N = 110 r/min, which obtained a stone removal rate of $y_1$ = 85.65%, stone removal efficiency of $y_2$ = 35.47 pieces/m, operating resistance of $y_3$ = 719.23 N, and torque of $y_4$ = 174.89 Nm.

(2) The field verification tests results indicated that the stone removal rate $y_1$ was 77.23% under the optimization parameters combination. The mean relative error of the simulated experiments value was 8.42%, thereby indicating that the accuracy and effectiveness of the simulation model and regression equation and the good effect of the groove formation and seedling growth were obtained.

The field test of the proposed device successfully performed the removal of stones in the seedbed and the formation of the seeding furrow, thereby providing a high-quality seedbed for sowing and rice growth. In the future, subsequent trials of the device in other regions will be carried out to explore the adaptability of the machine to different working conditions and potential problems to optimize and improve the device.

**Author Contributions:** Methodology, H.L. and W.Y.; software, H.L. and L.F.; validation, W.Y.; formal analysis, H.L.; investigation, H.L. and L.F.; data curation, H.L.; writing—original draft preparation, H.L.; writing—review and editing, P.Y. and W.L.; visualization, H.L.; supervision, W.Y.; project administration, W.Y.; funding acquisition, W.Y. and H.L. All authors have read and agreed to the published version of the manuscript.

**Funding:** This research was funded by the National Key R&D Program (2021YFD2000405-4), Special Projects in Key Area of "Artificial Intelligence" of General Colleges and Universities in Guangdong Province (2019KZDZX1003), and Independent Research and Development Projects of Maoming Laboratory (2021ZZ001).

**Institutional Review Board Statement:** Not applicable.

**Data Availability Statement:** The data presented in this study are available on demand from the first author at lihui2021@hunaas.cn.

**Acknowledgments:** The authors would like to thank the teacher and supervisor for their technical support. Moreover, we would like to thank Hebei Zhengrong Agricultural Machinery Manufacturing Company for manufacturing the device. Additionally, we sincerely appreciate the work of the editor and the reviewers of the present paper.

**Conflicts of Interest:** The authors declare no conflict of interest.

## References

1. Khush, G.S. Strategies for increasing the yield potential of cereals: Case of rice as an example. *Plant Breed.* **2013**, *132*, 433–436. [CrossRef]
2. Chen, S.; Cai, S.G.; Chen, X.; Zhang, G.P. Genotypic differences in growth and physiological responses of transplanting and direct seeding cultivation in rice. *Rice Sci.* **2009**, *16*, 143–150. [CrossRef]
3. Tuong, T.; Bouman, B. Rice production in water scarce environments. In *Water Productivity in Agriculture: Limits and Opportunities for Improvement*; Kijne, J., Barker, R., Molden, D., Eds.; CABI Publishing: Wallingford, UK, 2003; pp. 53–67.
4. Kakumanu, K.R.; Kaluvai, Y.R.; Nagothu, U.S.; Lati, N.R.; Kotapati, G.R.; Karanam, S. Building farm-level capacities in irrigation water management to adopt to climate change. *Irrig. Drain.* **2018**, *67*, 43–54. [CrossRef]
5. Tuong, T.P.; Castillo, E.G.; Cabangon, R.C.; Boling, A.; Singh, U. The drought response of lowland rice to crop establishment practices and N-fertilizer sources. *Field Crop. Res.* **2002**, *74*, 243–257. [CrossRef]
6. Liu, H.Y.; Hussain, S.; Peng, S.B.; Huang, J.L.; Cui, K.H.; Nie, L.X. Potentially toxic elements concentration in milled rice differ among various planting patterns. *Field Crop. Res.* **2014**, *168*, 19–26. [CrossRef]
7. Peng, S.B.; Tang, Q.Y.; Zou, Y.B. Current status and challenges of rice production in China. *Plant Prod. Sci.* **2009**, *12*, 3–8. [CrossRef]
8. Kumar, V.; Ladha, J.K. Direct Seeding of Rice: Recent Developments and Future Research Needs. *Adv. Agron.* **2011**, *111*, 297–413.
9. Farooq, M.; Siddique, K.H.M.; Rehman, H.; Aziz, T.; Lee, D.J.; Wahid, A. Rice direct seeding: Experiences, challenges and opportunities. *Soil Till. Res.* **2011**, *111*, 87–98. [CrossRef]
10. Singh, S.; Elamathi, S.; Ghosh, G.; Anandhi, P.; Masih, L.P. Performance of Direct-Seeded Rice as Influenced by Dual Cropping with Nitrogen Levels and Weed Management Practices in Prayagraj Region of Eastern Uttar Pradesh. *Natl. Acad. Sci. Lett.* **2020**, *43*, 399–402. [CrossRef]
11. Sandhu, N.; Subedi, S.R.; Singh, V.K.; Sinha, P.; Kumar, S.; Singh, S.P.; Ghimire, S.K.; Pandey, M.; Yadaw, R.B.; Varshney, R.K.; et al. Deciphering the genetic basis of root morphology, nutrient uptake, yield, and yield-related traits in rice under dry direct-seeded cultivation systems. *Sci. Rep.* **2019**, *9*, 9334. [CrossRef]
12. Kakumanu, K.R.; Kotapati, G.R.; Nagothu, U.S.; Kuppanan, P.; Kallam, S.R. Adaptation to climate change and variability: A case of direct seeded rice in Andhra Pradesh, India. *J. Water Clim. Chang.* **2019**, *10*, 419–430. [CrossRef]
13. Huang, M.; Zou, Y.B.; Jiang, P.; Xia, B.; Feng, Y.H.; Cheng, Z.W.; Mo, Y.L. Yield component differences between direct-seeded and transplanted super hybrid rice. *Plant Prod. Sci.* **2011**, *14*, 331–338. [CrossRef]
14. Liu, H.Y.; Hussain, S.; Zheng, M.M.; Peng, S.B.; Huang, J.L.; Cui, K.H.; Nie, L.X. Dry direct-seeded rice as an alternative to transplanted-flooded rice in Central China. *Agron. Sustain. Dev.* **2015**, *35*, 285–294. [CrossRef]
15. Xu, L.; Li, X.X.; Wang, X.Y.; Xiong, D.L.; Wang, F. Comparing the Grain Yields of Direct-Seeded and Transplanted Rice: A Meta-Analysis. *Agronomy* **2019**, *9*, 767. [CrossRef]
16. Anjum, S.A.; Tanveer, M.; Akbar, N.; Ashraf, U.; Ali, I.; Zohaib, A.; Manzoor, N. Comparative Efficacy of Various Weed Control Measures in Weed Dynamics, Yield and Profitability of Direct Seeded Fine Rice. *Pak. J. Agri. Sci.* **2017**, *54*, 129–134.
17. Singh, M.; Bhullar, M.S.; Chauhan, B.S. Influence of tillage, cover cropping, and herbicides on weeds and productivity of dry direct-seeded rice. *Soil Till. Res.* **2015**, *147*, 39–49. [CrossRef]
18. Muhammad, S.; Muhammad, I.; Sajid, A.; Muhammad, L.; Maqshoof, A.; Nadeem, A. The Effect of Different Weed Management Strategies on the Growth and Yield of Direct-Seeded Dry Rice (*Oryza sativa*). *Planta Daninha* **2016**, *34*, 57–64. [CrossRef]
19. Yao, Z.S.; Zheng, X.H.; Liu, C.Y.; Lin, S.; Zuo, Q.; Butterbach-Bahl, K. Improving rice production sustainability by reducing water demand and greenhouse gas emissions with biodegradable films. *Sci. Rep.* **2017**, *7*, 39855. [CrossRef] [PubMed]
20. Jabran, K.; Ullah, E.; Hussain, M.; Farooq, M.; Zaman, U.; Yaseen, M.; Chauhan, B.S. Mulching Improves Water Productivity, Yield and Quality of Fine Rice under Water-saving Rice Production Systems. *J. Agron. Crop Sci.* **2015**, *201*, 389–400. [CrossRef]
21. Hakansson, I.; Arvidsson, J.; Etana, A.; Rydberg, T.; Keller, T. Effects of seedbed properties on crop emergence. 6. Requirements of crops with small seeds. *Acta Agric. Scand. Sect. B—Soil Plant Sci.* **2013**, *63*, 554–563. [CrossRef]
22. Constantin, J.; Dürr, C.; Tribouillois, H.; Justes, E. Catch crop emergence success depends on weather and soil seedbed conditions in interaction with sowing date: A simulation study using the simple emergence model. *Field Crops Res.* **2015**, *176*, 22–33. [CrossRef]
23. Shi, L.R.; Zhao, W.Y.; Sun, W. Parameter calibration of soil particles contact model of farmland soil in northwest arid region based on discrete element method. *Trans. Chin. Soc. Agric. Eng.* **2017**, *33*, 181–187.
24. Dai, F.; Song, X.F.; Zhao, W.Y.; Zhang, F.W.; Ma, H.J.; Ma, M.Y. Simulative Calibration on Contact Parameters of Discrete Elements for Covering Soil on Whole Plastic Film Mulching on Double Ridges. *Trans. Chin. Soc. Agric. Eng.* **2020**, *50*, 49–56, 77.
25. Shi, Y.Y.; Sun, X.; Wang, X.C.; Hu, Z.C.; Newman, D.; Ding, W.M. Numerical simulation and field tests of minimum-tillage planter with straw smashing and strip laying based on EDEM software. *Comput. Electron. Agric.* **2019**, *166*, 105021. [CrossRef]

26. Hao, J.J.; Long, S.F.; Li, J.C.; Ma, Z.K.; Zhao, X.S.; Zhao, J.G.; Li, H. Effect of granular ruler in discrete element model of sandy loam fluidity in Ma yam planting field. *Trans. Chin. Soc. Agric. Eng.* **2020**, *36*, 56–64.
27. Barr, J.; Desbiolles, J.; Ucgul, M.; Fielke, J.M. Bentleg furrow opener performance analysis using the discrete element method. *Biosyst. Eng.* **2020**, *189*, 99–115. [CrossRef]
28. Yuan, J.; Li, J.G.; Zou, L.L.; Liu, X.M. Optimal Design of Spinach Root-cutting Shovel Based on Discrete Element Method. *Trans. Chin. Soc. Agric. Eng.* **2020**, *51*, 85–98.
29. Chen, Y.; Qin, Z.Y.; Rong, X.F. Study on Material's Breakage Behavior in a Jaw Crusher Based on EDEM. *Mach. Des. Manuf.* **2017**, *2*, 46–49.
30. Xing, J.J.; Zhang, R.; Wu, P.; Zhang, X.R.; Dong, X.H.; Chen, Y.; Ru, S.F. Parameter calibration of discrete element simulation model for latosol particles in hot areas of Hainan Province. *Trans. Chin. Soc. Agric. Eng.* **2020**, *5*, 158–166.

*agriculture*

MDPI

*Article*

# Tyre Configuration and Axle Load of Front-Wheel Assist and Four-Wheel Drive Tractors Effects on Soil Compaction and Rolling Resistance under No-Tillage

**David Rivero [1], Guido F. Botta [2,*], Diogenes L. Antille [3], Alejandra Ezquerra-Canalejo [4], Fernando Bienvenido [5] and Mustafa Ucgul [6]**

1 Facultad de Agronomia, Universidad Nacional de La Pampa, Santa Rosa 6300, Argentina
2 Departamento de Tecnologia, Universidad Nacional de Lujan, Lujan 6700, Argentina
3 CSIRO Agriculture and Food, Black Mountain Science and Innovation Precinct, Canberra 2601, Australia
4 Escuela Tecnica Superior de Ingenieria de Montes, Forestal y del Medio Natural, Universidad Politecnica de Madrid, 28040 Madrid, Spain
5 Centro de Investigacion Mediterraneo de Economia y Desarrollo Sostenible (CIMEDES), Universidad de Almeria, 04120 Almeria, Spain
6 Faculty of Science and Engineering, Southern Cross University, Lismore 2480, Australia
* Correspondence: gfbotta@agro.uba.ar; Tel.: +549-2323-555-247

**Abstract:** Selecting the appropriate tyre configuration and settings for heavy farm vehicles is important to ensure that soil compaction and power loss in rolling resistance are minimised and traction is optimised. This study investigated the effect of front-wheel assist (FWA, ≈75 kN) and four-wheel drive (4 WD, ≈100 kN) tractors fitted with different tyre configurations (single, dual), tyre sizes and inflation pressures on soil strength (a proxy for soil compaction), and rolling resistance. Single-pass tests were performed on a *Typic Argiudoll* (≈23% clay, bulk density: 1305 kg $m^{-3}$) managed under permanent no-tillage. Results showed that average power losses in rolling resistance were 7.5 kN and 5 kN for the 4 WD and FWA tractors, respectively. The average rut depth increased by approximately 1.4 times after a pass of the 4 WD compared with the FWA tractor. The soil cone index (0–600 mm depth) increased from 2023 kPa (before traffic) to 2188 and 2435 kPa after single passes of the FWA and 4WD tractors, respectively ($p < 0.05$). At the centreline of the tyre rut, dual tyres reduced the soil cone index a little compared with single tyres, but they significantly increased the volume of soil over which soil strength, and therefore soil compaction, was increased. For both tractors (regardless of tyre configuration or settings), soil strength increased to the full measured depth (600 mm), but relative changes before vs. after traffic became progressively smaller with increased soil depth. The power loss in rolling resistance was consistently greater with the heavier tractor, and rut depth was directly related to tyre inflation pressure.

**Keywords:** dual vs. single tyres; rut depth; soil bearing capacity; soil displacement; tractive efficiency; tyre size and inflation pressure

**Citation:** Rivero, D.; Botta, G.F.; Antille, D.L.; Ezquerra-Canalejo, A.; Bienvenido, F.; Ucgul, M. Tyre Configuration and Axle Load of Front-Wheel Assist and Four-Wheel Drive Tractors Effects on Soil Compaction and Rolling Resistance under No-Tillage. *Agriculture* **2022**, *12*, 1961. https://doi.org/10.3390/agriculture12111961

Academic Editor: Tao Cui

Received: 25 September 2022
Accepted: 17 November 2022
Published: 20 November 2022



## 1. Introduction

The drawbar performance of tractors depends primarily on the engine power and transmission system, the weight distribution on drive wheels, and the ground conditions over which the tractor is driven [1]. This performance is also affected by the tyre configuration (whether single or multiple tyres), tyre settings (size and inflation pressure) and tyre construction (radial, bias-ply, high-flexion), and factors related to soil–tyre interactions [2,3]. Mechanical power is obtained from internal combustion engines, but at less than 100% efficiency [4]. Power losses consist of gear and bearing friction (which represent the power loss in ground drive transmission) and drive wheel slip and rolling resistance (which represent the power loss in the ground drive) [5]. Rolling resistance includes the force required to deform the soil and deflect the tyres and the wheel bearing friction [6,7]. The coefficient of rolling resistance ($C_{rr}$) is defined as the ratio of the force required to roll a wheel over an untracked surface to the load supported by the wheel [8–10]. Numerically, $C_{rr}$ may be estimated by computing the cone index of the soil, tyre parameters that relate to its (unloaded) dimensions, and dynamic wheel loading, as shown in Equation (1) (after Hunt [11]; Crossley et al. [12]):

$$C_{rr} = \frac{1.2}{C_n} + 0.04 \tag{1}$$

and

$$C_n = \frac{CI \times b \times d}{W} \tag{2}$$

where: $C_n$ is a decimal dimensionless ratio, $CI$ is soil cone index expressed in N m$^{-2}$ [13], $b$ is the unloaded tyre section width (m), $d$ is the overall diameter of the unloaded tyre (m), and $W$ is the dynamic wheel load (force units normal to the soil surface, N).

Therefore, wide variations in gross tractive efficiency may occur as operating conditions vary; for example, changes in soil strength (soil cone index) influence the load-carrying capacity of the soil. For pneumatic tyres operated at inflation pressures that allow for tyre deflections up to about 20%, typical values of $C_n$ can vary between 15 for soft and light-textured soils to 50 for hard soils [10]. The overall tractive performance is also influenced by the efficiency of the ground drive, which is primarily determined by tyre performance [14]. Agricultural tyres are deformable and so are soils used for cropping, including soils that have been under long-term (e.g., >10 years) no-tillage [15–18]. The vertical force exerted by a loaded tyre is met by the reaction of the ground [19]. This loaded tyre will compact the soil until both opposing forces balance out. More significant soil deformation (sinkage) will increase rolling resistance and therefore the amount of energy wasted in compacting the soil, which could, otherwise, be used for developing traction [20–22]. The net result is a concurrent reduction in fuel-use efficiency [23]. When traffic is performed on soft soil conditions, increased tyre inflation pressure above the manufacturer's designated pressure for load, speed, and slope will tend to increase sinkage into the ground (rut depth) and rolling resistance [24,25], and a rule of thumb suggests that 10 mm of additional depth can increase fuel consumption by approximately 10% [26]. Similarly, overinflation of the tyre increases the slip rate due to the reduced soil–tyre contact area [27,28], which at 30% can increase fuel consumption by 20% or greater and cause significant compaction [23,26]. Options for improving traction and minimising rolling resistance and soil compaction include an increase in the tyre–soil contact area, which can be achieved by fitting larger tyres (preferably, of larger diameter) [29], reducing wheel load and tyre inflation pressure, fitting low/ultra-low ground pressure tyres such as IF (increased flexion) and VF (very-high flexion) marked tyres, or converting the system to controlled traffic farming [1,30–35]. The length of the contact area is a function of the tyre diameter; hence, an increase in tyre diameter will result in improved slip performance.

In Argentina, approximately 23 million hectares of arable land are managed under permanent (>20 years) no-tillage [36]. Over the past 20 years, there has been a trend towards the adoption of articulated four-wheel drive (4WD) and front-wheel assist tractors (FWA) motivated by the need to improve field efficiency and timeliness of farm operations [37]. A drawback of this trend has been an increased risk of soil damage due to compaction, which has been attributed to the combined effects of high traffic footprint (e.g., 40–60% of field-cropped area each time a crop is produced) and improper tyre selection or tyre settings for the type of farm vehicle, field operation, and soil condition [37–39]. Brand new tractors in Argentina are released from the factory or imported into the country untyred, and so tractors are offered by local dealers without tyres. Newly acquired tractors are fitted with tyres selected by customers upon sealing the deal on the equipment purchase or resolved after-market privately. The tyres selected for brand-new tractors are often not fit for purpose, with tyre dimensions considered to be suboptimal for the intended equipment and equipment use. Surveys conducted by Botta [40] and Botta et al. [37] showed that mismatches between the overall mass and power of the equipment and the tyres' configuration, dimensions, or settings were common in FWA and 4WD tractors. While buyers may choose to fit the correct set of tyres for a given equipment, their preferred option is often slightly smaller tyres as this reduces the upfront cost of the equipment. A hidden problem associated with this preference is the potential impact that underdimensioned tyres can have on energy-use efficiency (due to increased rolling resistance and wheel slip), soil compaction (due to increased tyre inflation pressure), and premature tyre wear.

Whilst the principles governing the complex soil–tyre–vehicle interactions are well-understood and well-documented (e.g., [41–44]), these relationships have not been quantified for local conditions in Argentina. Such data are needed to assist farmers and machinery operators make informed decisions about tyre selection for specific applications. The work reported in this article was conducted to address this data gap by undertaking a series of field-based experiments aimed at quantifying the impact of tractor–tyre combinations on rolling resistance and soil strength used, respectively, as indicators of energy-use efficiency and soil compaction. The specific objectives of this study were to determine the effect of front-wheel assist (FWA) and four-wheel drive (4WD) tractors on (1) rolling resistance and (2) soil compaction. For this, the two tractors were fitted with either single or dual tyres. They included the original set of tyres (those selected by the customer upon acquisition of the equipment or supplied by the dealer) and the correct set of tyres for the equipment. The FWA and 4WD tractors chosen for this study are both popular models in Argentina, and the soil type (*Typic Argiudoll*) at the experimental site had been under no-tillage for 22 years, which was, therefore, representative of the local farming conditions in the country. It was hypothesised that (1) given the axle load of the tractors used in the study, subsoil compaction would occur independently of the tyres fitted to the equipment, and (2) increased rolling resistance would be observed with the wider and dual tyre configurations, consistent with well-established theory and despite the soil being fairly consolidated (mean $\rho_b \approx 1305$ kg m$^{-3}$, 0–600 mm) after such a long-term no-tillage management and high ($\approx$45%) traffic footprint [39]. It was also postulated that the expected increase in rolling resistance when inappropriate tyres were used would be due to the bulldozer effect at the soil–tyre interface (soil pushed in front of the tyre) and less so to tyre sinkage into the ground. It was envisaged that this work would go some way to alert advisers and farm machinery operators in Argentina about hidden costs associated with improper tyre selection that may result in reduced energy-use efficiency and increased soil damage due to compaction.

## 2. Materials and Methods

### 2.1. Experimental Site

Field tests were conducted on a Typic Argiudoll from a farm located near Lujan (Buenos Aires, Argentina) 34°32′ S, 59°07′ W [45]. The soil at the site had been continuously managed under no-tillage winter cereal and soybean cropping for 22 years. A summary of measured soil physical and mechanical properties is provided in Table 1. The soil water contents derived from the Proctor test were used as a reference for comparison with field measurements of soil water content to denote the soil's susceptibility to compaction when the tractors were driven over.

**Table 1.** Characterisation of the soil (Typic Argiudoll) from the experimental site near Lujan (Buenos Aires, Argentina). $DUL_{100}$ is drained upper limit measured at 100 cm suction and is considered to be the laboratory determination of field capacity; SOC is soil organic carbon. Values are means of five determinations ($n$ = 5).

| Property Measured | Analytical Method | Analytical Value | | | |
|---|---|---|---|---|---|
| Depth interval, mm | - | 0–150 | 150–300 | 300–450 | 450–600 |
| Particle size analysis | [46] | - | - | - | - |
| Clay (<2 μm), % | - | 20.1 | 24.8 | 27.9 | 34.2 |
| Silt (2–50 μm), % | - | 75.6 | 70.8 | 67.2 | 61.3 |
| Sand (>50 μm), % | - | 4.3 | 4.4 | 4.9 | 4.5 |
| Textural class | [47] | Silt loam | Silt loam | Silty clay loam | Silty clay loam |
| Soil bulk density, kg $m^{-3}$ | [48] | 1240 | 1270 | 1330 | 1370 |
| Proctor density, kg $m^{-3}$ | [49] | 1490 | 1530 | 1680 | 1710 |
| Proctor soil water content, % (*w/w*) | [49] | 22.3 | 23.1 | 24.4 | 25.2 |
| $DUL_{100}$, % (*w/w*) | [50] | 26.6 | 28.5 | 26.8 | 28.7 |
| SOC, % (*w/w*) | [51] | 1.85 | 1.44 | 0.95 | 0.61 |
| $pH_{1:2.5}$ (soil:water ratio) | [52] | 5.8 | 5.8 | 6.0 | 6.2 |

*2.2. Description of Farm Equipment and Experimental Treatments*

The front-wheel assist JD7515 (henceforth FWA) and four-wheel drive Zanello-500 (henceforth 4WD) tractors used in these experiments are described in Table 2a,b. Both tractors are popular models in Argentina. The engine power of the tractors was 92 CV (68 kW) and 194 CV (143 kW) for the FWA and 4WD, respectively. The tyres fitted to the FWA tractor ensured the spoke ratios remained close to constant at approximately 1.25. In all tractor–tyre combinations, the tyre inflation pressures were adjusted to match the manufacturer's specifications for load and speed. Treatments marked with the symbol (†) indicate that the tractor was fitted with tyres supplied by the local dealer (or selected by the customer when the equipment was first acquired). As discussed earlier, while these particular tyres can be used with the two tractors available for the study, they were considered to be suboptimal and likely to underperform compared with the correct tyres based on the guidelines given by The European Tyre and Rim Technical Organization (https://www.etrto.org/, accessed on 30 October 2022). For the FWA tractor, the front and rear tyres had different dimensions for all three sets of tyres used in the tests. For the 4WD tractor, the front and rear tyres were identical and only differed between tests. The tractors were ballasted to perform field operations that typically demand for high pull (e.g., subsoiling, chisel ploughing, heavy-disc harrowing), and ballasting remained unchanged during the tests. Before the field tests were conducted, the two tractors were weighed using a weighbridge. The overall weight of the tractor–tyre combination was first recorded, followed by the individual weights of the front and rear axles. Experimental plots were laid out in a completely randomised block design with three replications ($n$ = 3) based on the approach used by Botta et al. [37]. The dimensions of each plot were 60 m wide by 100 m long, and the plots were separated by a 15 m buffer zone for turning and manoeuvring that ensured edge effects due to field traffic were avoided. Each plot was subdivided into 10 m to be able to perform rolling resistance measurements.

**Table 2.** (Top): Description of tractors, specification of front tyres and settings, and front axle loads used in the experiments. Notation: 'FWA' (front-wheel assist), 4WD (four-wheel drive), 'D' (dual tyres), and 'S' (single tyres). The symbol (†) denotes the tractor fitted with tyres supplied by the local dealer (or selected by customers when the equipment was first acquired), and (Bottom): Description of tractors, specification of rear tyres and settings, and rear axle loads used in the experiments. Notation: 'FWA' (front-wheel assist), 4WD (four-wheel drive), 'D' (dual tyres), and 'S' (single tyres). The symbol (†) denotes the tractor fitted with tyres supplied by the local dealer (or selected by customers when the equipment was first acquired).

| (a) Front Axle | | | | | | | |
|---|---|---|---|---|---|---|---|
| **Tractor** | **Treatment** | **Front Tyres** | **Specification** | **Inflation Pressure** | **Mean Ground Pressure** | **Front Axle Load** | **Load/Tyre** |
| **Units** | - | - | - | kPa | | kN | |
| FWA | FWA_D | Single | 520/70 R26 | 80 | 60.35 | 30.30 | 15.15 |
| FWA | FWA_S | Single | 18.4 R26 | 60 | 60.01 | 23.30 | 11.65 |
| FWA | (†) FWA_S | Single | 480/70 R26 | 70 | 65.10 | 28.84 | 14.42 |
| 4WD | 4WD_D | Single | 18.4 R38 | 190 | 48.80 | 64.96 | 32.48 |
| 4WD | 4WD_S | Single | 24.5 R32 | 110 | 52.81 | 67.20 | 33.60 |
| 4WD | (†) 4WD_S | Single | 23.1 R30 | 180 | 61.25 | 63.70 | 31.85 |
| **(b) Rear Axle** | | | | | | | |
| **Tractor** | **Treatment** | **Rear Tyre** | **Specification** | **Inflation Pressure** | **Mean Ground Pressure** | **Rear Axle Load** | **Load/Tyre** |
| **Units** | - | - | - | kPa | | kN | |
| FWA | FWA_D | Dual | 18.4 R38 | 140 | 45.25 | 45.43 | 22.71 |
| FWA | FWA_S | Single | 24.5 R32 | 60 | 55.40 | 46.60 | 23.30 |
| FWA | (†) FWA_S | Single | 23.1 R30 | 100 | 68.90 | 43.56 | 21.78 |
| 4WD | 4WD_D | Dual | 18.4 R38 | 80 | 31.80 | 35.00 | 17.50 |
| 4WD | 4WD_S | Single | 24.5 R32 | 40 | 39.20 | 36.20 | 18.10 |
| 4WD | (†) 4WD_S | Single | 23.1 R30 | 70 | 56.20 | 34.30 | 17.15 |

### 2.3. Experimental Variables Measured

Replicated measurements ($n$ = 30) of soil cone index (CI) [13], gravimetric soil water content [53], rolling resistance [54], and rut depth were performed as described here. The number of replications ensured that any spatial variability along the traffic path could be captured [55].

*Soil cone index and soil water content*: CI was determined with an SC-900 FieldScout™ digital cone penetrometer (Spectrum Technologies Inc., Aurora, IL, USA, https://www.specmeters.com/, accessed 7 May 2022). Measurements were taken at the centreline of the tyre rut to a depth of 600 mm at regular increments of 25 mm, and the data were digitally recorded. Samples for gravimetric determination of soil water content were collected from locations near those used for measurements of CI. Penetration resistance values were adjusted by covariance analysis using gravimetric soil water content measurements simultaneously determined [56]. For the 4WD tractor, CI was measured immediately after the pass of the front tyres and subsequently after the pass of the rear tyres, allowing the effect of an individual set of tyres to be quantified. Mean tyre–ground contact pressures were measured with a Tekscan® pressure sensor (https://www.tekscan.com/, accessed 7 May 2022). For all tests, the soil water content at the time traffic was near both the soil's $DUL_{100}$ (laboratory determination of drained upper limit at 100 cm suction) and the Proctor water contents (Table 1), and the susceptibility of the soil to compaction was considered to be high.

*Rolling resistance*: measurements were performed through single-pass tests, with the front and rear tyres of the pulled tractor running over the same path [54]. The FWA and 4WD tractors were towed at a speed of 5 km $h^{-1}$, offset from the towing vehicle, and the motion resistance force was recorded [37]. Subsequently, these data were corrected to account for the travelling angle between the two tractors [54]. Measurements were conducted using an electronic dynamometer (strain gauge type). For this, a cab-mounted unit collected the motion resistance force data with a sampling frequency of 200 Hz. These data were stored in a datalogger and downloaded as Excel files for subsequent analysis and reporting.

*Rut depth*: measurements were conducted using a profile meter that had 500 mm long, 5 mm diameter sliding rods spaced 25 mm apart on a 1 m aluminium H-frame. The profile meter was placed over the tyre rut perpendicular to the direction of travel, and the rods were allowed to drop until they made contact with the ground to replicate the shape of the rut [57]. The rods were then locked and the frame was lifted off the ground, placed on a gridded paper, and photographed. The maximum rut depth was then read off the image [58]. Relationships between tyre inflation pressure and rut depth were derived for each tractor by pulling together the data from all tests.

### 2.4. Statistical Analyses

The statistical package Statgraf 7.1 [59] was used to analyse the soil cone index, soil water content, rolling resistance, and rut depth data and involved ANOVA. The post hoc Duncan's multiple range test was used to determine differences between pairs of means.

## 3. Results and Discussion

### 3.1. Soil Cone Index and Soil Water Content

Measurements of soil water content showed no significant differences between treatments or depth intervals ($p > 0.05$) and reported an overall mean value ($\pm$ SD) for the full measured depth of 19.53 $\pm$ 1.63 % (*w/w*). The soil water content at the $DUL_{100}$, averaged across all depth intervals, was 27.65 $\pm$ 1.10 % (*w/w*). The mean soil water content derived from the Proctor test (23.75 $\pm$ 1.29%, *w/w*) suggested that this soil condition provided moderately high susceptibility to deformation and compaction [58,60,61]. While penetration resistance values were adjusted by covariance analysis, the lack of statistical differences in soil water content at the time field tests were performed provided confidence that CI, rolling resistance, and rut depth measurements reflected treatment as opposed to site effects (the latter being associated, for example, with variability in soil strength both spatially and at depth).

Soil CI data are summarised in Table 3a,b. Overall, there were significant differences in CI (0–600 mm) before (1991 $\pm$ 385 kPa) and after (2304 $\pm$ 376 kPa) the tests were performed ($p < 0.05$). Except for FWA_D and FWA_S, treatment effects on CI were significant to the full measured depth of 600 mm ($p < 0.05$). At the 450–600 mm depth interval, the FWA_D and FWA_S tractor–tyre combinations did not induce significant changes in CI, and therefore, values remained close to those recorded prior to traffic ($p > 0.05$). Table 3a also shows that values of CI were highest with the heavier tractor–tyre system combinations; however, subsoil compaction (depth range: 150 to 450 mm) was evident across all treatments and despite axle loads being lower than the limits suggested in earlier studies [35,62]. This observation confirmed the first hypothesis formulated prior to this study that subsoil compaction would occur with both tractors regardless of the tyres fitted to the equipment. However, these effects may be mitigated or avoided if advanced tyre technology, such as VF/IF marked tyres, were to be used as shown by other studies with much heavier (34 Mg) harvesting equipment [63]. Table 3b shows the individual effect of the front tyres and the additive effect of the rear tyres on soil CI after a single pass of the 4WD tractor. Soil CI (0–600 mm) values recorded after the pass of the front tyres represented between 69% and 74% of the CI measured after the pass of the whole vehicle (front and rear tyres), which suggested that the rear tyre caused additional compaction to that initially induced by the leading front tyre. However, the compactive effect of the front tyre appeared to be greater than the follower rear tyre, consistent with the knowledge that up to 80–90% of the maximum compaction may occur when the soil is first trafficked, provided soil conditions were conducive to such effects [64].

**Table 3.** (Top): Summary of soil cone index (CI) data recorded after single-pass tests with the FWA and 4WD tractors used in this study. Different letters horizontally indicate that mean values are significantly different ($p < 0.05$). Description of tractors and treatments' notation are as shown in the caption of Table 2. (Bottom): Summary of soil cone index (CI) data recorded after driving the front tyres and the whole vehicle (front and rear tyres) over the soil, and the CI of front tyres expressed as a percentage of the CI recorded after the pass of the whole vehicle. Single-pass tests performed with the 4WD tractor. Description of the 4WD tractor and treatments' notation are as shown in the caption of Table 2.

| (a) Tractor Type: FWA and 4WD | | | | | | |
|---|---|---|---|---|---|---|
| **Treatment** | **FWA_D** | **FWA_S** | **(†) FWA_S** | **4WD_D** | **4WD_S** | **(†) 4WD_S** |
| Depth interval, mm | CI, kPa | | | | | |
| 0–150 | 1699 b | 1778 b | 1800 b | 1789 b | 1749 b | 2020 c |
| 150–300 | 2010 b | 2170 c | 1930 b | 2380 c | 2545 d | 2290 c |
| 300–450 | 2322 b | 2410 c | 2290 b | 2688 c | 2790 d | 2610 c |
| 450–600 | 2565 a | 2605 b | 2508 a | 2762 c | 2890 c | 2701 c |

| (b) Tractor Type: 4WD | | | | | | | | | |
|---|---|---|---|---|---|---|---|---|---|
| **Measurement** | **CI after Pass of front Tyre** | | | **CI after Pass of Whole Vehicle** | | | **CI of front Tyre Relative to CI of Whole Vehicle** | | |
| **Treatment** | **4WD_D** | **4WD_S** | **(†) 4WD_S** | **4WD_D** | **4WD_S** | **(†) 4WD_S** | **4WD_D** | **4WD_S** | **(†) 4WD_S** |
| Depth interval, mm | CI, kPa | | | | | | % | | |
| 0–150 | 1319 | 1270 | 1450 | 1789 | 1749 | 2020 | 73.7 | 72.6 | 71.7 |
| 150–300 | 1590 | 1930 | 1578 | 2380 | 2545 | 2290 | 71.0 | 62.4 | 68.9 |
| 300–450 | 2000 | 1928 | 1845 | 2688 | 2790 | 2610 | 74.4 | 69.1 | 70.6 |
| 450–600 | 1900 | 2140 | 1950 | 2762 | 2890 | 2701 | 68.7 | 74.0 | 72.1 |

Consolidation of the soil under long-term no-tillage did not prevent compaction from occurring under the prevailing experiment conditions; thus, suggesting controlled traffic farming (CTF) may be an effective alternative to the widespread 'random' and high-footprint (40–60%) traffic systems that are common in Argentina [39,65]. By confining traffic compaction to the least possible area of permanent traffic lanes (such as in fully matched CTF systems), significant improvements in year-round trafficability may be realised with associated benefits in terms of reduced rolling resistance and overall improvements in energy-use efficiency [38,66].

### *3.2. Rolling Resistance and Rut Depth*

Rolling resistance and rut depth data are presented in Table 4 and Figure 1a,b, respectively. A strong relationship between tyre inflation pressure and rut depth was observed with both tractors ($p$-values < 0.001), consistent with earlier studies (e.g., [17,18,67,68]). Rut depths were the shallowest in the FWA_S and 4WD_S tractor–tyre combinations, respectively (Table 4).

**Table 4.** Rolling resistance, rut depth, and power losses recorded for the FWA and 4WD tractors after single-pass tests. Different letters (horizontally) indicate that mean values are significantly different ($p < 0.05$). Description of the tractors and treatments' notation are as shown in the caption of Table 2a,b.

| Measurement/Treatment | FWA_D | FWA_S | (†) FWA_S | 4WD_D | 4WD_S | (†) 4WD_S |
|---|---|---|---|---|---|---|
| Overall load, kN | 75.73 | 77.70 | 72.06 | 99.96 | 104.40 | 98.00 |
| Rolling resistance, kN | 6.2 c | 3.6 a | 5.2 b | 8.9 e | 6.5 c | 7.1 d |
| Power loss, kW | 8.6 c | 5.0 a | 7.2 b | 12.3 d | 9.0 c | 9.8 c |
| Rut depth, mm | 73 b | 62 a | 65 a | 100 d | 84 c | 97 d |

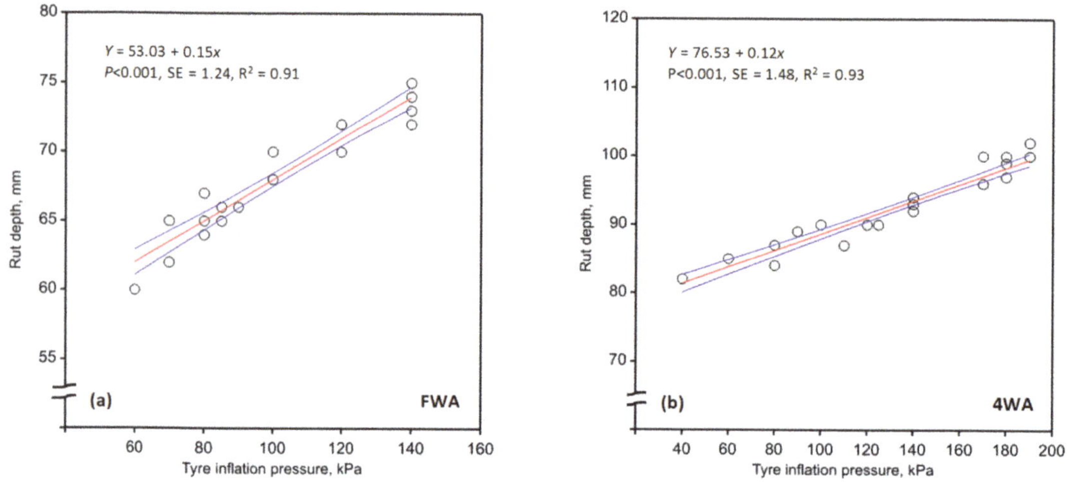


**Figure 1.** Observed relationships ($n$ = 30) between tyre inflation pressure and rut depth for (**a**) front-wheel assist (FWA) and (**b**) four-wheel drive (4WD) tractors used in this study (bulked data for all treatments). A description of treatments is shown in the caption of Table 2a,b.

There were significant differences in rolling resistance depending upon the tractor–tyre combinations, but overall values were higher with the 4WD tractor, which was about 25% heavier than the FWA tractor. Differences in rolling resistance between tractors were also explained by differences in rut depth, which were deeper for all 4WD–tyre combinations. For both tractors, rolling resistance was significantly lower with single compared with dual tyres. The single configuration system equipped with the tyres supplied by the dealer (marked †) reported higher rolling resistance than the same configuration equipped with the correct set of tyres ($p$-values < 0.05). The 'bulldozer' effect expected when using suboptimal tyres/tyre settings was less clear than the relationship between tyre inflation pressure and rut depth and the associated effect on rolling resistance. Hence, the second hypothesis formulated prior to this study could not be confirmed. Tyre inflation pressure was, therefore, a key factor influencing rolling resistance and power loss in ground drive [69].

Power loss estimates and rut depth measurements were consistent with rolling resistance data, with more significant penalties observed when the dual tyre configuration and the single tyres supplied by the dealer were used ($p$-values < 0.05). The single configuration system fitted with the 23.1 R30 tyre to the rear axle offered the least desirable combination of wheel load, tyre inflation pressure, and section width, as the mean ground contact pressures were the highest (Table 2b). Ground contact pressures higher than about 50 kPa were reported to be the threshold above which crop yield penalties may occur due to compaction [70]. The 24.5 R32 tyre in the single configuration system reported acceptable results considering that mean ground contact pressures beneath this tyre remained below or close to 50 kPa (Raper [70]'s suggested threshold value) and wheel loadings were the highest. The relatively low mean ground contact pressures recorded for the 18.4 R38 tyre were mainly due to the dual configuration, which allowed for increased contact area, and not to the inflation pressures used in this tyre. As highlighted earlier, CI (0–600 mm) measurements were also lower with this tyre, but the compaction footprint underground would be much wider than the soil–tyre contact area of the dual tyre system because of the distribution of compressive stresses beneath the tyres [63,71]. Other studies (e.g., [20,67,72]) have also shown that using wider tyres may reduce ground contact pressures and therefore sinkage and compaction but can potentially have adverse effects on traction and rolling resistance. A reduction in the tyre inflation pressure will not only increase the soil–tyre contact area but also the deformability of the tyre relative to the ground [5]. Consequently, the depth of sinkage along the length of the contact patch becomes more uniform and the rut depth shallower, both of which reduce rolling resistance. Since rolling resistance diminishes the thrust produced at the driving wheels, care should be exercised to minimize it so that production of drawbar pull is not significantly affected [29].

## 4. Conclusions

The work reported in this article investigated the effects of tyre configuration (namely, single and dual tyres, tyre dimensions, and inflation pressure) on rolling resistance and soil strength (which was used as a proxy for assessment of soil compaction). A series of single-pass tests with front-wheel assist (FWA, JD7515, ≈75 kN) and four-wheel drive (4WD, Zanello-500, ≈100 kN) tractors were conducted to compare tractor × tyre performance. The main conclusions derived from this work are summarised below:

(1) Average rolling resistance values across all tyre combinations were 5 kN for the FWA and 7.5 kN for the 4WD tractors, with average power losses of 6.9 and 10.4 kW, respectively. These observations were consistent with rut depth measurements, which increased 1.4 times on average after a single pass of the 4WD compared with the FWA tractor. Rolling resistance increased significantly with dual compared with single tyres, and while soil cone index at the centreline of the rut was lower after the dual tyre tests, the final volume of soil over which soil strength (and therefore soil compaction) increased was significantly greater than that of the single tyre configuration.

(2) Linear relationships between rut depth and tyre inflation pressure were established, which confirmed that greater soil deformation (sinkage) induced by wheeling is evidence of increased rolling resistance and soil compaction. If the same contact area was to be achieved with different tyres or tyre configurations, the depth of sinkage would be similar, but the resultant width of the rut would be increased with wider section tyres. This would result in increased rolling resistance. The rut depth–tyre inflation pressure-rolling resistance relationships documented in this study reinforced the knowledge that if the tyres could be operated at the lowest permissible inflation pressure, not only would the contact area be increased, but also (and importantly) their deformability relative to the soil. The net result would be shallower sinkage, reduced rolling resistance, and improved slip performance.

(3) Soil cone index data (depth range: 150 to 450 mm) provided evidence that subsoil compaction occurred across all treatments and despite axle loads being lower than thresholds suggested in earlier studies. This was observed with both the FWA and 4WD tractors regardless of the tyres fitted to the equipment. These adverse effects may be mitigated, or possibly avoided, if advanced tyre technology (e.g., VF/IF marked tyres) were to be used, which should be considered in future studies. There is also a need to assess the cost-effectiveness of low-ground pressure tyre technology for heavy farm equipment relative to the adoption of controlled traffic farming (CTF). This research is required for improving the overall performance (timeliness, trafficability, energy-use efficiency) of mechanisation systems in Argentina and reducing impacts on soil associated with traffic compaction.

**Author Contributions:** Conceptualisation, D.R., G.F.B., D.L.A. and M.U.; methodology, investigation, and data validation, D.R., A.E.-C. and G.F.B.; electronic calibration devices and formal analysis; D.R., A.E.-C. and F.B.; writing—original draft preparation, D.L.A., D.R., M.U. and G.F.B.; writing—review and editing. All authors have read and agreed to the published version of the manuscript.

**Funding:** The work reported in this article was supported by grants (I-153/19 FA Research Group) provided by Universidad Nacional de La Pampa, Argentina (https://www.agro.unlpam.edu.ar, accessed on 28 January 2022).

**Institutional Review Board Statement:** Not applicable.

**Data Availability Statement:** The data reported in this article are available from Guido F. Botta (gfbotta@agro.uba.ar).

**Acknowledgments:** Operational support provided by the owners of the farm where this study was conducted is gratefully acknowledged. The authors are also grateful to CSIRO Agriculture and Food (Canberra, Australia).

**Conflicts of Interest:** The authors declare no conflict of interest. The funders had no role in the design of the study; in the collection, analyses, or interpretation of data; in the writing of the manuscript; or in the decision to publish the results.

## References

1. Goering, C.E.; Stone, M.L.; Smith, D.W.; Turnquist, P.K. *Off-Road Vehicle Engineering Principles*; ASAE Publication No.: 801M0103; American Society of Agricultural Engineers: St. Joseph, MI, USA, 2006; ISBN 1-892769-26-3.
2. Soane, B.D.; Dickson, J.W.; Campbell, D.J. Compaction by agricultural vehicles: A review. III. Incidence and control of compaction in crop production. *Soil Tillage Res.* **1982**, *2*, 3–36. [CrossRef]
3. Misiewicz, P.A.; Blackburn, K.; Richards, T.E.; Brighton, J.L.; Godwin, R.J. The evaluation and calibration of pressure mapping system for the measurement of the pressure distribution of agricultural tyres. *Biosyst. Eng.* **2015**, *130*, 81–90. [CrossRef]
4. Srivastava, A.K.; Goering, C.E.; Rohrbach, R.P.; Buckmaster, D.R. *Engineering Principles of Agricultural Machines*, 2nd ed.; ASAE Pub. No.: 801M0206; American Society of Agricultural and Biological Engineers: St. Joseph, MI, USA, 2006; ISBN 1-892769-50-6.
5. Inns, F.M.; Kilgour, J. *Agricultural Tyres*; Dunlop Ltd.: London, UK, 1978.
6. Perdok, U.D.; Tijink, F.G.J. Developments in IMAG research on mechanization in soil tillage and field traffic. *Soil Tillage Res.* **1990**, *16*, 121–141. [CrossRef]
7. Abeels, P.F.J. Tire deflection and contact studies. *J. Terramech.* **1976**, *13*, 183–196. [CrossRef]
8. Wismer, R.D.; Luth, H.J. Off-road traction prediction for wheeled vehicles. *Trans. ASAE* **1974**, *17*, 8–10. [CrossRef]
9. Kurjenluomar, J.; Alakukku, L.; Ahokas, J. Rolling resistance and rut formation by implement tires on tilled clay soil. *J. Terramech.* **2009**, *46*, 267–275. [CrossRef]
10. ASABE. *ASAE-Standard ASAE-D497.7 MAR2011(R2020): Agricultural Machinery Management Data*; American Society of Agricultural and Biological Engineers: St. Joseph, MI, USA, 2020; p. 15.
11. Hunt, D.R. *Engineering Models for Agricultural Production*; The Avi Publishing Co., Inc.: Westport, CT, USA, 1986; ISBN 0-87055-494-8.
12. Crossley, C.P.; Kibiwot, V.N.; Reynolds, A.J.; Rickson, R.J. Rut formation and rolling resistance on earth roads. *J. Agric. Eng. Res.* **2001**, *78*, 99–107. [CrossRef]
13. ASABE. *ASAE-Standard EP542.1-NOV2019: Procedures for Using and Reporting data Obtained with the Soil Cone Penetrometer*; American Society of Agricultural and Biological Engineers: St. Joseph, MI, USA, 2019; p. 6.
14. Domier, K.W.; Friesen, D.H.; Townsend, J.S. Traction characteristics of two-wheel drive, four-wheel drive and crawler tractors. *Trans. ASAE* **1971**, *14*, 0520–0522. [CrossRef]
15. Botta, G.F.; Tolón-Becerra, A.; Lastra-Bravo, X.; Tourn, M.; Balbuena, R.; Rivero, D. Continuous application of direct sowing: Traffic effect on subsoil compaction and maize (*Zea mays* L.) yields in Argentinean Pampas. *Soil Tillage Res.* **2013**, *134*, 111–120. [CrossRef]
16. Burt, E.C.; Wood, R.K. Soil-tire interface stress measurements. *Trans. ASAE* **1987**, *30*, 1254–1258. [CrossRef]
17. Antille, D.L.; Ansorge, D.; Dresser, M.L.; Godwin, R.J. *The Effects of Tyre Size on Soil Deformation and Soil Bulk Density Changes*; ASABE Paper No.: 083879; American Society of Agricultural and Biological Engineers: St. Joseph, MI, USA, 2008; Volume 4, pp. 2294–2304. [CrossRef]
18. Antille, D.L.; Ansorge, D.; Dresser, M.L.; Godwin, R.J. Soil displacement and soil bulk density changes as affected by tire size. *Trans. ASABE* **2013**, *56*, 1683–1693. [CrossRef]
19. Koolen, A.J.; Kuipers, H. *Agricultural Soil Mechanics*; Advanced Series in Agricultural Sciences; Springer: Berlin/Heidelberg, Germany, 1983; Volume 13, pp. 1–252. [CrossRef]
20. Antille, D.L.; Peets, S.; Galambošová, J.; Botta, G.F.; Rataj, V.; Macák, M.; Tullberg, J.N.; Chamen, W.C.T.; White, D.R.; Misiewicz, P.A.; et al. Review: Soil compaction and controlled traffic farming in arable and grass cropping systems. *Agronomy Res.* **2019**, *17*, 653–682. [CrossRef]
21. Plackett, C.W. The ground pressure of some agricultural tyres at low load and with zero sinkage. *J. Agric. Eng. Res.* **1984**, *29*, 159–166. [CrossRef]
22. Liljedahl, J.B.; Turnquist, P.K.; Smith, D.W.; Hoki, M. Chapter 10: Traction. In *Tractors and Their Power Units*, 4th ed.; Springer: Boston, MA, USA, 1989; pp. 240–265. [CrossRef]
23. Jensen, T.A.; Tullberg, J.N.; Antille, D.L. Chapter 3: Improving farm machinery operation and maintenance to optimise fuel use efficiency. In *Energy-Smart Farming: Efficiency, Renewable Energy, and Sustainability. Part 1: Efficient Use of Electricity, Heat, and Fuel*; Sims, R.E.H., Ed.; Burleigh Dodds Series in Agricultural Science No.: 115; Burleigh Dodds Science Publishing Limited: Cambridge, UK, 2022; pp. 71–102. [CrossRef]

24. Chico-Santamarta, L.; Richards, T.; Godwin, R.J. *A Laboratory Study into the Mobility of Travelling Irrigators in Air Dry, Field Capacity and Saturated Sandy Soils*; ASABE Paper No.: 096106; American Society of Agricultural and Biological Engineers: St. Joseph, MI, USA, 2009; Volume 4, pp. 2629–2646. [CrossRef]
25. Sitkei, G. Sinkage and rolling resistance of wheels: Some new results on an old problem. *Prog. Agric. Eng. Sci.* **2015**, *11*, 85–94. [CrossRef]
26. Desbiolles, J.; Saunders, C.; Barr, J.; Riethmuller, G.; Northover, G.; Tullberg, J.; Antille, D. Chapter 6: Machinery evolution for conservation agriculture. In *Australian Agriculture in 2020: From Conservation to Automation*; Pratley, J., Kirkegaard, J., Eds.; Agronomy Australia and Charles Sturt University: Wagga Wagga, NSW, Australia, 2019; pp. 81–105. ISBN 13: 978-0-6485819-0-1.
27. Davis, D.B.; Finney, J.B.; Richardson, R.J. Relative effects of tractor weight and wheel-slip in causing soil compaction. *J. Soil Sci.* **1973**, *24*, 399–409. [CrossRef]
28. Raghavan, G.S.V.; McKyes, E.; Chassé, M. Effect of wheel slip on soil compaction. *J. Agric. Eng. Res.* **1977**, *22*, 79–83. [CrossRef]
29. McKibben, E.G.; Davidson, J.B. Transport wheels for agricultural machines IV. Effect of outside and cross-section diameters on the rolling resistance of pneumatic implement tires. *Agric. Eng.* **1940**, *21*, 57–58.
30. Tijink, F.G.J.; Döll, H.; Vermeulen, G.D. Technical and economic feasibility of low ground pressure running gear. *Soil Tillage Res.* **1995**, *35*, 99–110. [CrossRef]
31. Soane, B.D.; Dickson, J.W.; Blackwell, P.S. Some options for reducing compaction under wheels on loose soil. In Proceedings of the 8th Conference of the International Soil and Tillage Research Organization, Stuttgart, Germany, 10–14 September 1979; ISTRO: Stuttgart, Germany, 1979; Volume 2, pp. 347–352.
32. Antille, D.L.; Bennett, J.M.c.L.; Jensen, T.A. Soil compaction and controlled traffic considerations in Australian cotton-farming systems. *Crop Pasture Sci.* **2016**, *67*, 1–28. [CrossRef]
33. Harris, B.J. Increased Deflection Agricultural Radial Tires Following the Tire and Rim Association IF, VF, and IF/CFO Load and Inflation Standards. In Proceedings of the Agricultural Equipment Technology Conference, ASABE Distinguished Lecture Series Tractor Design No.: 39. Louisville, Kentucky, 13–15 February 2017; ASABE: St. Joseph, MI, USA, 2017. [CrossRef]
34. Godwin, R.J.; Misiewicz, P.A.; Smith, E.K.; Millington, W.A.J.; White, D.R.; Dickin, E.T.; Chaney, K. Summary of the effects of three tillage and three traffic systems on cereal yields over a four-year rotation. *Asp. Appl. Biol.* **2017**, *134*, 233–241.
35. McPhee, J.E.; Antille, D.L.; Tullberg, J.N.; Doyle, R.B.; Boersma, M. Managing soil compaction—A choice of low-mass autonomous vehicles or controlled traffic? *Biosyst. Eng.* **2020**, *195*, 227–241. [CrossRef]
36. Gauthier, G.F.; Carruthers, R.; Millán-Placci, F. *Logística de la Soja: Argentina-Paraguay-Uruguay. Report No.: 105151. Serie de Informes Técnicos del Banco Mundial en Argentina, Paraguay y Uruguay. N°4, 2016*; World Bank Group: Washington, DC, USA, 2016. Available online: http://documents.worldbank.org/curated/en/923401468272770160/Logistica-de-la-soja-Argentina-Paraguay-Uruguay (accessed on 27 October 2022).
37. Botta, G.F.; Tolón-Becerra, A.; Tourn, M.; Lastra-Bravo, X.; Rivero, D. Agricultural traffic: Motion resistance and soil compaction in relation to tractor design and different soil conditions. *Soil Tillage Res.* **2012**, *120*, 92–98. [CrossRef]
38. Antille, D.L.; Imhoff, S.C.; Alesso, C.A.; Chamen, W.C.T.; Tullberg, J.N. Potential to increase productivity and sustainability in Argentinean agriculture with controlled traffic farming: A short discussion. *Acta Technol. Agric.* **2015**, *18*, 83–87. [CrossRef]
39. Botta, G.F.; Antille, D.L.; Nardon, G.F.; Rivero, D.; Bienvenido, F.; Contessotto, E.E.; Ezquerra-Canalejo, A.; Ressia, J.M. Zero and controlled traffic improved soil physical conditions and soybean yield under no-tillage. *Soil Tillage Res.* **2022**, *215*, 105235. [CrossRef]
40. Botta, G.F. Armonización del Peso y Rodado del Tractor para Reducir la Compactación del Suelo. Master's Thesis, Universidad Nacional de La Plata, Facultad de Ciencias Agrarias y Forestales, La Plata/Buenos Aires, Argentina, 1997.
41. Bekker, M.G. Latest developments in off-the-road locomotion. *J. Frankl. Inst.* **1957**, *263*, 411–423. [CrossRef]
42. Bekker, M.G. *Off the Road Locomotion: Research and Development in Terramechanics*; University of Michigan Press: Ann Arbor, MI, USA, 1960; p. 220.
43. Wong, J.Y. On the study of wheel-soil interaction. *J. Terramech.* **1984**, *21*, 117–131. [CrossRef]
44. Ageikin, I.S. *Off-the-Road Wheeled and Combined Tractor Devices*; Amerind Publishing Co.: New Delhi, India, 1987; p. 202.
45. INTA. *Carta de suelos de la Provincia de Buenos Aires*; Instituto Nacional de Tecnología Agropecuaria: Buenos Aires, Argentina, 2015; Available online: https://inta.gob.ar/documentos/carta-de-suelos-de-la-provincia-de-buenos-aires (accessed on 20 April 2022).
46. Bouyoucos, G.J. Hydrometer method improved for making particle size analyses of soils. *Agronomy J.* **1962**, *54*, 464–465. [CrossRef]
47. Soil Survey Staff. *Soil Taxonomy: A Basic System of Soil Classification for Making and Interpreting Soil Surveys*, 2nd ed.; Agricultural Handbook No.: 436; US Department of Agriculture, National Resources Conservation Service: Washington, DC, USA, 1999; p. 886.
48. Blake, G.R.; Hartge, K.H. Bulk density. In *Methods of Soil Analysis, Part 1. Physical and Mineralogical Methods*, 2nd ed.; Klute, A., Ed.; Agronomy Monograph; ASA-SSSA: Madison, WI, USA, 1986; Volume 9, pp. 363–375. [CrossRef]
49. Ray, P.N.; Chapman, T.G. The British Standard compaction test for soils: A study of some factors affecting the test results. *Geotechnique* **1954**, *4*, 169–177. [CrossRef]
50. Marshall, T.J. *Relations between Water and Soil*; Technical Communication No.: 50 of The Commonwealth Bureau of Soils; Commonwealth Agricultural Bureaux: Harpenden, UK, 1959; p. 91.
51. Walkley, A.; Black, I.A. An examination of the Degtjareff method for determining soil organic matter, and a proposed modification of the chromic acid titration method. *Soil Sci.* **1934**, *37*, 29–38. [CrossRef]

52. Davis, L.E. Measurements of pH with the glass electrode as affected by soil moisture. *Soil Sci.* **1943**, *56*, 405–422. [CrossRef]
53. MAFF. *The Analysis of Agricultural Materials*, 3rd ed.; Reference Book 427; Ministry of Agriculture, Fisheries, and Food, The Stationery Office: London, UK, 1986.
54. Hassan, A.E.; Sirois, D.L. Traction and rolling resistance of a dual-tired skidder on wetland. *Trans. ASAE* **1985**, *28*, 1038–1042. [CrossRef]
55. Cassel, D.K. Tillage effects on soil bulk density and mechanical impedance. In *Predicting Tillage Effects on Soil Physical Properties and Processes*; Unger, P.W., Van Doren, D.M., Jr., Whisler, F.D., Skidmore, E.L., Eds.; ASA-SSSA Inc.: Madison, WI, USA, 1982; Volume 44, Chapter 4; pp. 45–67. [CrossRef]
56. Yasin, M.; Grisso, R.D.; Bashford, L.L.; Jones, A.J.; Mielke, L.N. Normalizing cone resistance values by covariance analysis. *Trans. ASAE* **1993**, *36*, 1267–1270. [CrossRef]
57. Aikins, K.A.; Jensen, T.A.; Antille, D.L. Three-dimensional scanning of soil surface and furrow profiles using a portable and affordable unit. *Biosyst. Eng.* **2020**, *193*, 279–289. [CrossRef]
58. Vero, S.E.; Antille, D.L.; Lalor, S.T.J.; Holden, N.M. Field evaluation of soil moisture deficit thresholds for limits to trafficability with slurry spreading equipment on grassland. *Soil Use Manag.* **2014**, *30*, 69–77. [CrossRef]
59. Bretschneider, S. Software Reviews: Statgraf 1.1. (1985). *Soc. Sci. Microcomput. Rev.* **1983**, *3*, 163–164. [CrossRef]
60. Earl, R. Prediction of trafficability and workability from soil moisture deficit. *Soil Tillage Res.* **1997**, *40*, 155–168. [CrossRef]
61. Spoor, G.; Tijink, F.G.J.; Weisskopf, P. Subsoil compaction: Risk, avoidance, identification and alleviation. *Soil Tillage Res.* **2003**, *73*, 175–182. [CrossRef]
62. Håkansson, I.; Reeder, R.C. Subsoil compaction by vehicles with axle load extend persistence and crop response. *Soil Tillage Res.* **1994**, *29*, 277–304. [CrossRef]
63. Tekeste, M.Z.; Antille, D.L.; Baillie, C.P. Modeling the effect of low ground pressure (LGP) tires on soil stress distribution beneath a cotton picker wheel loading. In Proceedings of the ISTVS 20th International and 9th Americas Conference, Online, 27–29 September 2021; International Society for Terrain-Vehicle Systems. [Paper No.: 119]. 2021. Available online: https://docs.google.com/document/d/12fQuuDEjR2pUK8WZeeYbiGzMfaZZ1nlkP2_GBTtU8J0/edit (accessed on 16 November 2022).
64. Hovanesian, J.D.; Buchele, W.F. Development of a recording volumetric transducer for studying effects of soil parameters on compaction. *Trans. ASAE* **1959**, *2*, 0078–0081. [CrossRef]
65. Botta, G.F.; Pozzolo, O.; Bomben, M.; Rosatto, H.; Rivero, D.; Ressia, M.; Tourn, M.; Soza, E.; Vazquez, J. Traffic alternatives in harvest of soybean (*Glycine max* L.): Effect on yields and soil under direct sowing system. *Soil Tillage Res.* **2007**, *96*, 145–154. [CrossRef]
66. Tullberg, J.N.; Yule, D.F.; McGarry, D. Controlled traffic farming-From research to adoption in Australia. *Soil Tillage Res.* **2007**, *97*, 272–281. [CrossRef]
67. Botta, G.F.; Rivero, D.; Tourn, M.; Bellora Melcon, F.; Pozzolo, O.; Nardon, G.F.; Balbuena, R.; Tolón-Becerra, A.; Rosatto, H.G.; Stadler, S. Soil compaction produced by tractor with radial and cross-ply tyres in two tillage regimes. *Soil Tillage Res.* **2008**, *101*, 44–51. [CrossRef]
68. Ansorge, D.; Godwin, R.J. The effects of tyres and a rubber track at high axle loads on soil compaction: Part 1. Single-axle studies. *Biosyst. Eng.* **2007**, *98*, 115–126. [CrossRef]
69. Pytka, J.D.; Dąbrowski, J.; Zając, M.; Tarkowski, P. Effects of reduced inflation pressure and vehicle loading on off-road traction and soil stress and deformation state. *J. Terramech.* **2006**, *43*, 469–485. [CrossRef]
70. Raper, R.L. Agricultural traffic impacts on soil. *J. Terramech.* **2005**, *42*, 259–280. [CrossRef]
71. Vanden Berg, G.E.; Gill, W.R. Pressure distribution between a smooth tire and the soil. *Trans. ASAE* **1962**, *5*, 0105–0107. [CrossRef]
72. Hidalgo, R.J.; Pozzolo, O.R.; Domínguez, J.F.; Giménez, L.; Botta, G.F. Soil compaction caused by the impact of machinery traffic during corn (*Zea mays*) harvest. *Rev. Fac. Cienc. Agrar.* **2020**, *52*, 161–174.

agriculture

MDPI

Article

# Tillage-Depth Verification Based on Machine Learning Algorithms

**Jing Pang [1,*], Xuwen Zhang [1], Xiaojun Lin [1], Jianghui Liu [2], Xinwu Du [1] and Jiangang Han [2]**

[1] College of Agricultural Equipment Engineering, Henan University of Science and Technology, Luoyang 471003, China
[2] Luoyang Xiyuan Vehicle and Power Inspection Institute Co., Ltd., Luoyang 471000, China
* Correspondence: pangjing@haust.edu.cn

**Abstract:** In an analysis of the penetration resistance and tillage depth of post-tillage soil, four surface-layer discrimination methods, specifically, three machine learning algorithms—Kmeans, DBSCAN, and GMM—and a curve-fitting method, were used to analyze data collected from the cultivated and uncultivated layers. Among them, the three machine learning algorithms found the boundary between the tilled and untilled layers by analyzing which data points belonged to which layer to determine the depth of the soil in the tilled layer. The curve-fitting method interpreted the intersection among data from the fitted curves of the ploughed layer and the un-ploughed layer as the tillage depth. The three machine learning algorithms were used to process a standard data set for model evaluation. DBSCAN's discrimination accuracy of this data set reached 0.9890 and its F1 score reached 0.9934, which were superior to those of the other two algorithms. Under standard experimental conditions, the ability of DBSCAN clustering to determine the soil depth was the best among the four discrimination methods, and the discrimination accuracy reached 90.63% when the error was 15 mm. During field-test verification, the discriminative effect of DBSCAN clustering was still the best among the four methods. However, the soil blocks encountered in the field test affected the test data, resulting in large errors in the processing results. Therefore, the combined RANSCA robust regression and DBSCAN clustering algorithm, which can eliminate interference from soil blocks in the cultivated layer and can solve the problem of large depth errors caused by soil blocks in the field, was used to process the data. After testing, when the RANSCA and DBSCAN combined method was used to process all samples in the field and the error was less than 20mm, the accuracy rate reached 82.69%. This combined method improves the applicability of discrimination methods and provides a new method of determining soil depth.

**Keywords:** Kmeans; DBSCAN; GMM; tilling depth

**Citation:** Pang, J.; Zhang, X.; Lin, X.; Liu, J.; Du, X.; Han, J. Tillage-Depth Verification Based on Machine Learning Algorithms. *Agriculture* **2023**, *13*, 130. https://doi.org/10.3390/agriculture13010130

Academic Editors: Mustafa Ucgul and Chung-Liang Chang

Received: 29 November 2022
Revised: 24 December 2022
Accepted: 29 December 2022
Published: 4 January 2023



## 1. Introduction

Soft soil is beneficial to the growth of crops, and fields that have been cultivated for many years will form a solid plough bottom, which hinders the penetration of precipitation into deep soil and is not conducive to the absorption of water by the crop roots. The depth of the tillage layer has a significant impact on agroecosystems and on crop yield and quality [1–4]. After the soil is loosened, the plow bottom can be broken, the plow-layer thickness can be increased, the soil structure can be improved, and the quality of cultivated land can be improved. In recent years, China has attached great importance to the quality of subsoiling operations, has encouraged farmers to prepare land for subsoiling operations, and has issued subsidies for subsoiling operations. At present, no unified standard has been accepted for automatically measuring the tillage depth in subsoiling operations. Manual sampling methods are often adopted but require manual cleaning of the ditch bottom, and human factors and soil conditions affect the measurement accuracy [5]. Moreover, manual determination of the tillage depth has a low efficiency, is labor-intensive, takes a long time, and provides a limited description of the tillage depth. Due to the importance of subsoiling land preparation and the need for supervision by the Ministry of Agriculture of China on, subsidies for subsoiling operations must be based on accurate information about the front-line operations of subsoiling machinery; thus, the demand for informatization identification and testing is quite high. Tillage depth is an important evaluation index of subsoiling quality, and accurate tillage-depth detection and control are very important. The determination of tillage depth by agricultural machinery equipment is mostly realized by one or more angle sensors, inclination sensors, attitude sensors, ultrasonic sensors, and infrared sensors.

The latest research on tillage-depth detection by domestic and foreign scholars can be divided into two types according to the detection method: non-contact and contact. Non-contact measurement methods are commonly used in the installation of ultrasonic sensors, particularly during the implementation of optical rangefinders on a replica wheel of the plow. Kim et al. [6] installed inclinometers and optical distance sensors on the left and right axes of linear potentiometers and used linear potentiometers and optical distance sensors to measure the depth of soil penetration. In addition, a single-axis inclinometer was used to measure the inclination angle during the tillage process, and the tillage depth was calculated based on the depth of the soil penetration and the pitch angle of the attached equipment. The conservation tillage popularized and applied in recent decades usually breaks up crop residues and mixes them with soil as nutrients for the next crop. However, neither ultrasonic sensors nor optical rangefinders can accurately distinguish crop residues from the ground. This affects the tillage-depth measurement, which greatly reduces the precision of the non-contact measurement method, and research on non-contact measurement methods has gradually decreased in recent years. In contrast, the contact measurement method has been increasingly studied, specifically, contact between the sensor and the actual soil surface. Mouazen et al. [7] connected a linear variable displacement sensor with a stroke length of 0–2 m to the frame of a metal wheel, with the height sensor located at the axle, and measured the swing-arm frame height. The distance between the soil surface and the frame changes, and an analytical–statistical hybrid model was built to calculate the change in the height of the frame. Xie et al. [8] developed a contact tillage-depth measurement method based on measuring the lift-arm inclination and the geometric relationship between the unit and the tractor. When measuring the tillage depth with this method, it is necessary to ensure that the plane of the plough body or the beam of the plough frame remains horizontal when the implemented frame reaches the maximum tillage depth. An irregular surface or a change in the slope affects the applicability and accuracy of the model. Afterwards, Jia Honglei et al. [9] designed an adaptive tillage-depth monitoring system that uses a photoelectric encoder to measure the angle between the adjustable swing arm. Based on different cases and corresponding mathematical models, they developed a LabVIEW program adapted to the measurement, which was slightly lacking in terms of real-time processing and visualization of the tillage depth. Overall, most non-contact technologies use ultrasonic rangefinders and optical rangefinders to indirectly measure the tillage depth, and most contact technologies use inclination sensors and angle-measuring instruments to indirectly measure the tillage depth. Although these methods provide an effective way to detect tillage depth and tillage quality online, each of these methods has certain problems. For example, when an inclination sensor is used to calculate the tillage depth, the sensor needs to be re-calibrated, which is inconvenient. Additionally, the detection accuracies of the angle-measuring instrument and the ultrasonic sensor are affected by surface debris.

To detect tillage depth, two methods are used: one is real-time monitoring of the tillage depth, and the other is identification detection. The testing equipment is usually placed on the machine for testing a rotary tillage operation in order to obtain real-time data of the operation. The experiments conducted by Kim et al. and Xie Bin et al., as described above, aimed to test the tillage depth in real time, as well as to verify and test the effect of their special testing equipment on rotary tillage. The goals of the abovementioned subsoiling preparation subsidies issued by the Ministry of Agriculture of China are to verify and test the subsoiling-preparation quality by measuring the tillage depth. However, some problems are found in the currently accepted methods of inspection, such as having inconsistent standards, producing large manual-measurement errors, and being labor-intensive. With the state strongly advocating for deep loose-land preparation, the Ministry of Agriculture and the Ministry of Finance attaches great importance to the subsoiling operations of the land, and subsidies for subsoiling operations must be issued based on accurate information about the front-line operations of subsoiling machinery. At present, the demand for information acceptance is quite strong. With the rapid development of autonomous navigation technology and information sensing technology in recent years, cultivated-land precision-monitoring technology has begun to become more refined and smarter. Additionally, verification and detection have become development trends in providing a quantitative basis for evaluating the quality of subsoiling operations by using sensors to determine tillage depth.

Machine learning algorithms often solve practical problems more efficiently due to the regularity information inherently found in data. In order to improve the automation and applicability of discriminating tillage depth, data on penetration resistance and tillage depth were analyzed. Finally, we decided to use four discrimination methods, specifically, three machine learning algorithms—Kmeans, DBSCAN, and GMM—as well as the curve fitting method, to analyze and compare the data from the cultivated layer and the uncultivated layer obtained in a laboratory experiment. Among them, the Kmeans algorithm clusters according to the distance similarity between the data points and the data points. The DBSCAN algorithm clusters according to the density of the data points, while the GMM algorithm clusters according to the assumption that the data points of each cluster conform to a Gaussian distribution of the corresponding cluster. The DBSCAN algorithm has the best effect in terms of determining tillage depth without any soil disturbance. In this paper, the Affinity propagation, Mean-shift, and OPTICS algorithms were also considered for use in distinguishing the surface layer. However, as Affinity propagation clustering [10] is carried out by sending messages between sample pairs until they converge, the number of clusters is determined according to the data provided, so distinguishing the binary classification problem between the cultivated layer and the uncultivated layer is ineffective. Mean-shift [11] is a density-based nonparametric clustering algorithm that needs to specify the candidate centroid and automatically sets the number of clusters. As the purpose of this experiment is to classify the number of specified clusters, this algorithm cannot be applied to distinguish the surface layer. The OPTICS algorithm [12] is a generalization of the DBSCAN algorithm. Clustering is performed according to density, but the number of clusters is automatically matched. Compared with DBSCAN, it is too sensitive and will divide the data into several clusters, meaning that it cannot solve the problem of plough layer discrimination. Furthermore, a field experiment was carried out to verify the ability of the methods to determine soil depth without any soil disturbance. The RANSCA robust regression algorithm is a kind of regression model that can be fitted into a regression model even if there are outliers or errors in the model. The use of a combination of this algorithm and the DBSCAN algorithm can solve the problem of soil disturbance discrimination and can accurately determine the tillage depth. The performance of the hybrid algorithm in field discrimination is not bad. The hybrid machine learning algorithm can greatly improve the production efficiency, reduce the labor intensity, and reduce the production cost. Thus, it provides a new method of determining topsoil depth.

## 2. Principle of Tillage-Depth Identification Based on Soil-Penetration Resistance

### *2.1. Test Instruments and Data Collection*

In this paper, soil samples from Luoyang, Henan (34°39′47″ N, 112°26′4″ E), were used as the soil samples. First, the soil water content and density were measured by sampling in the field. The samples were measured with the sieving method and the hydrometer method, and obtained. The soil moisture content in the field ranged from 10% to 20%, and the soil density ranged from 1100 kg·m$^{-3}$ to 1300 kg·m$^{-3}$. Before conducting the soil-sample test to obtain the experimental data of the soil in the tilled and un-ploughed layers, the soil moisture content and density were divided into three levels for the orthogonal tests; the moisture content was taken as 10%, 15%, and 20%; and the density was taken as $1.1 \times 10^3$ kg/m$^3$, $1.2 \times 10^3$ kg/m$^3$, and $1.3 \times 10^3$ kg/m$^3$.

The equipment used to measure the soil moisture content and density range during field sampling in the indoor test includes a probe (material 65 Mn steel, length 530 mm, and maximum diameter of tip 14 mm), a universal testing machine (DNS02-1KW), and a barrel (inner diameter 124 mm and height 400 mm).

The specific operation steps in the test process are as follows:

(1) Put about 5 kg of soil into the dryer for 48 h, and fully dry the soil.
(2) Ensure that the height of the soil in the bucket is 300 mm. Calculate the required dry soil weight and water weight according to the inner diameter of the bucket, soil height, soil density, and moisture content.
(3) Spray the soil until wet, and stir well to distribute the water evenly in the soil. Put the well-stirred soil into the bucket, turn the bucket upside down to decrease the distance between the soil surface and the top of the bucket to 100 mm, and ensure that the soil density reaches the required value.
(4) Put the bucket filled with soil onto the frame of the universal testing machine, and install the probe. Open the analysis software, and adjust the probe height to ensure that the bottom of the probe is close to the soil surface. Adjust the penetration speed of the probe to 8 mm/s and the penetration depth to 300 mm, and run the software. For data acquisition, use the software installed on the computer. The software then exports the relationship between penetration resistance and penetration depth in an Excel file format. The simulated artificial soil-sample test diagram is shown in Figure 1.
(5) The curve of penetration resistance and penetration depth can be obtained in real time during the test, and the test data can be saved after the test. The relationship between penetration resistance and penetration depth is shown in Figure 2 below.
(6) Taking the moisture content of the tilled layer, the depth and density of the tilled layer, and the moisture content and density of the untilled layer as factors, the regression equation between the resistance of the stratification point and the maximum value of the untilled layer on the significant factors of tilled layer, moisture content, and density of untilled layer was established. If the surface layer depth is a significant factor, the surface-layer depth can be directly obtained from the maximum resistance value of the un-tilled layer. If the surface depth was not a significant factor, the resistance value of the stratified point was obtained from the regression equation and the surface depth was obtained from the experimental data. Thus, 32 groups of soil samples were prepared, each group was tested twice repeatedly, and a total of 64 sample data were obtained.

### 2.2. The Principle of Machine Learning Algorithm in Discriminating the Depth of Cultivation

Soil-penetration resistance refers to the resistance of a conical or plunger-shaped penetration needle when penetrating into soil at a constant rate in an up–down direction due to the comprehensive effects of soil particle friction and extrusion. At the same sampling frequency, the soil in the cultivated layer is soft, so the penetration resistance of the cultivated layer increases with increasing penetration depth of the probe, and the collected data points are closely distributed. After years of cultivation in the field, a solid plough bottom layer is formed under the plough layer. When the probe enters the plough bottom layer, due to the soil firmness, the penetration resistance of the soil increases steeply, so the data points collected in the plough bottom layer are loosely distributed. Using a scatter plot and its corresponding mathematical analysis, the value of the cultivated depth of the measured point can be determined, and then, the method of determining the cultivated depth is verified as being feasible.

**Figure 1.** Probe-penetration laboratory test.

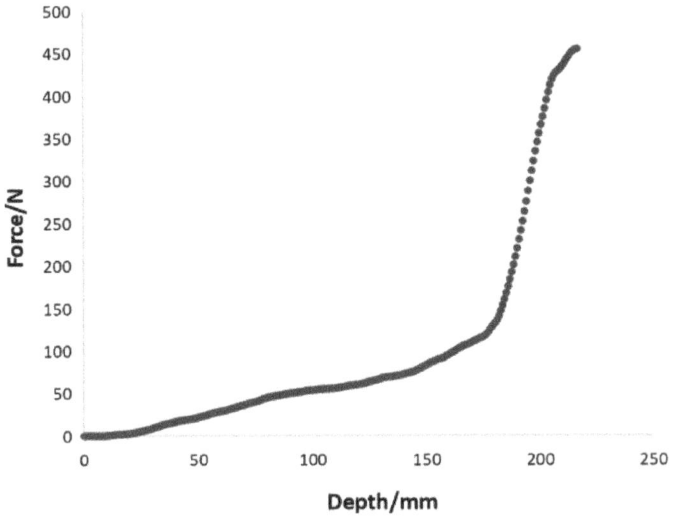


**Figure 2.** The relationship between penetration resistance and penetration depth.

## 3. Basic Discriminant Methods and Their Comparisons

### *3.1. Kmeans Clustering Algorithm to Discriminate Cultivated Layer*

The Kmeans clustering algorithm is one of the most commonly used methods in clustering. According to the similarity of the distance between points, the samples are separated into n groups with equal variance for clustering [13–18]. The Kmeans algorithm first randomly selects two data points as the centroids $\mu_1^t$ and $\mu_2^t$ from all the data points of the collected penetration resistance and penetration depth, with t as the number of iterations. The centroids chosen before iteration are marked as $\mu_1^{(0)}$ and $\mu_2^{(0)}$, and these two centroids serve as the cluster centers before the tilled and un-tilled layers [13]. Based on the relationships between the data points, the objective of the optimization is defined before the clustering starts:

$$J(c,\mu) = \min\sum_{i=1}^{M}\left|\left|x_i - \mu_{c_j}\right|\right|^2 \tag{1}$$

where $c$ is the division of the data into two categories to distinguish between the cultivated layer and the uncultivated layer, $c_1$ and $c_2$, and $M$ represents all data points, i.e., the total sample size.

The loop iteratively calculates the distance between each data point and the two cluster centers $\mu_1^t$ and $\mu_2^t$, and the cluster center with the shortest distance $x_i$ when obtaining category $c_j$ is selected by dividing data point a into category $c_j^t$ at the t-th iteration.

$$c_j^t < -\arg\min_k ||x_i - \mu_k^t||^2 \tag{2}$$

where $k$ is $c_j^t (j = 1,2)$ and the two clusters are obtained by processing all the data. The average distance of all sample points that are assigned to the same cluster is then calculated. The position of the average distance is used as the position of the new centroid of this iteration, that is, the new cluster centers $\mu_1^{t+1}$ and $\mu_2^{t+1}$ of this iteration, which must meet the following:

$$\mu_k^{t+1} < -\arg\min_u \sum_{i:c_j^t=k}^{b} ||x_i - \mu||^2 \tag{3}$$

The iterative process is repeated until J converges, that is, the centroids of all clusters no longer change and the Kmeans discrimination of the tillage layer is completed.

Figure 3 shows the effect of data discrimination using Kmeans clustering. The position of the centroid is at "+" in the figure. When the centroid position remains unchanged, Kmeans iteratively processes all data and determines the soil depth of the plough layer.

All data points $x_i$ of the same cluster $c_j$ are marked with the same cluster number, and the boundary data of different clusters are found as the result of the discrimination of the soil depth of the tillage layer. After processing 64 groups of simulated surface-soil samples, the results show that four groups obtained absolute errors of less than 5 mm, twelve groups obtained absolute errors of less than 10 mm, twenty-six groups obtained absolute errors of less than 15 mm, and forty-seven groups obtained absolute errors of less than 20 mm.

### *3.2. DBSCAN Density Clustering Algorithm to Discriminate Tillage Layer*

DBSCAN (density-based spatial clustering of applications with noise) density clustering treats clusters as low-density regions and high-density regions [19–22] and identifies a cluster class by classifying closely connected samples into one class [23,24]. The DBSCAN clustering algorithm is used to discriminate the soil depth of the cultivated layer. The radius eps (Eps-neighborhood of a point) of the cultivated layer data is set to 3.5, and the threshold min_samples is 4, that is, the data point of the cultivated layer is located in the radius eps of 3.5 and there are at least four samples in the field.

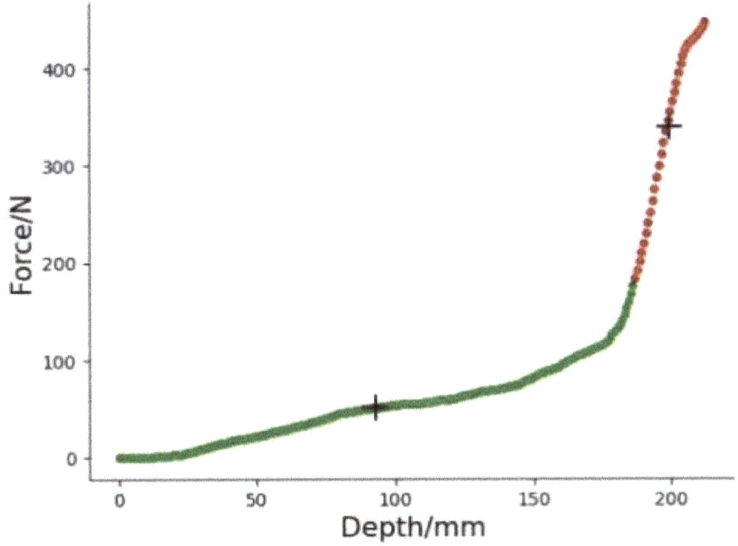


**Figure 3.** Kmeans processing data-discrimination effect diagram.

Since the penetration resistance of the un-ploughed layer increases due to the increased penetration resistance of the plow layer, the Euclidean distance between the collected sample points a($x_1, y_1$) and b($x_2, y_2$) increases, and the Euclidean distance $d$ is as follows:

$$d = \sqrt{(x_1 - x_2)^2 + (y_1 - y_2)^2} \quad (4)$$

According to the data-distribution law of the cultivated layer and the uncultivated layer, a data point is randomly selected as the core point $x_1$, and the Euclidean distance between other data points and the core point is used to determine whether other data points are densely connected. The basic algorithm principle model of DBSCAN is shown in Figure 4 below.

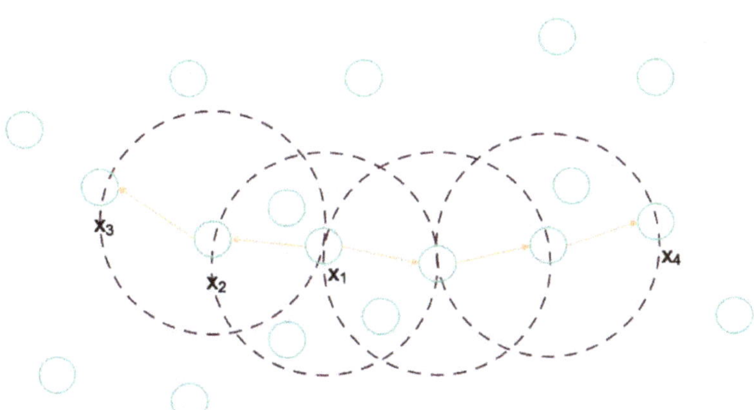


**Figure 4.** Principle model diagram of the basic algorithm of DBSCAN.

The core point $x_1$ is randomly selected; a circle with eps as the radius is drawn; and the point in the center of the circle is the density direct access, that is, $x_2$ is the density direct access of $x_1$; $x_3$, $x_4$, and $x_1$ are density accessible, $x_3$ and $x_4$ are density-connected, and all densities are connected. Cluster points are marked as data points of the same cluster. The density of the core point is compared with the threshold min_samples. If it is greater than or equal to the threshold min_samples, it is considered as a point in the same cluster. If it is less than the threshold min_samples, it is marked as a noise point and the noise point is marked as $-1$.

Due to the increase in penetration resistance, the data points collected in the un-ploughed layer cannot be densely connected with the data points of the cultivated layer and are marked as $-1$, which is an outlier. The first data point marked as an outlier, as discussed above, is the demarcation point between the tilled and uncultivated layers. The effect diagram for discriminating the DBSCAN's soil tillage layer is shown in Figure 5 below.

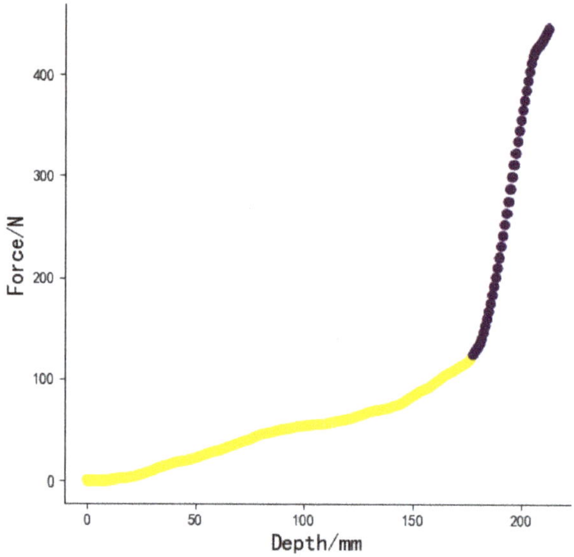


**Figure 5.** Discrimination effect diagram of a soil tillage layer by DBSCAN.

The results obtained after clustering 64 groups of soil samples in the simulated tillage layer using the DBSCAN algorithm show that thirty-two groups had an absolute error of less than 5 mm, fifty-two groups had an absolute error of less than 10 mm, fifty-eight groups had an absolute error of less than 15 mm, and sixty groups had an absolute error of less than 20 mm.

### 3.3. *Gaussian Mixture Model Clustering (GMM)*

The Gaussian mixture model is usually referred to as GMM. It first assumes that there are *k* Gaussian distributions and then assesses the probability that each sample conforms to each distribution [25–27]. To determine the tillage depth, the samples are divided into two clusters: the tilled layer and the un-tilled layer, that is, *k* is 2. The sample data are then divided into the distribution cluster with the highest probability, maximum likelihood estimation is performed, the probability of conforming to the distribution of the cultivated layer and the uncultivated layer is calculated based on the new distribution, and the Gaussian distribution parameters are iteratively updated until the model uses the EM algorithm [28,29]. The convergence reaches the local optimal solution, the Gaussian function has good computational performance, and the probability density is often recorded as follows:

$$\mathrm{f}(x) = p(x|\mu_i, \Sigma) \tag{5}$$

The Gaussian mixture distribution function required to define clustering is as follows:

$$p(X) = \sum_{i=1}^{k} \alpha_i p(X|\mu_i, \sum_i) \quad (6)$$

Among them, $\alpha$ represents the probability of each cluster, and the probability sum is 1, that is, $\Sigma_{i=1}^{k} \alpha_i = 1$. The data points are divided into corresponding clusters according to the probability $p(z_i = i) = \alpha_i$. Finally, the conditional probability formula of whether a data point conforms to the cultivated layer or the uncultivated layer is as follows:

$$p(z_j = i|x_j) = \frac{P(z_j = i)p(x_j|z_j = i)}{p(x_j)} = \frac{\alpha_i p(x_i|\mu_i, \Sigma)}{\Sigma_{l=1}^{k} \alpha_l p(x_j|\mu_l, \Sigma)} \quad (7)$$

Each datum j then is found to conform to the corresponding cluster distribution probability $\gamma_{ji}$ for comparison, and the data points are divided into the corresponding clusters.

When the probability is calculated according to the formula, the probability density of a certain distribution in both the numerator and the denominator conforms to the Gaussian distribution, and the maximum likelihood method is used, that is, the maximum likelihood is performed according to the probability product of the corresponding distribution of the data:

$$\mathrm{LL}(D) = \ln(\prod_{j=1}^{m} p(x_j)) = \sum_{j=1}^{m} \ln(\sum_{i=1}^{k} \alpha_i p(x_j|u_i, \Sigma)) \quad (8)$$

Then, take the derivative of the above function, set the derivative to 0, and find the corresponding $\mu$ and $\Sigma$:

$$\mu_i = \frac{\Sigma_{j=1}^{m} \gamma_{ji} x_j}{\Sigma_{j=1}^{m} \gamma_{ji}} \quad (9)$$

$$\sum_i = \frac{\Sigma_{j=1}^{m} \gamma_{ji} (x_j - \mu_i)(x_j - \mu_i)^T}{\Sigma_{j=1}^{m} \gamma_{ji}} \quad (10)$$

Additionally, when calculating the corresponding Gaussian mixture coefficient $\alpha$, the Lagrange multiplier method needs to be added as a constraint, and the equation is obtained as follows:

$$\alpha_i = \frac{1}{m} \sum_{j=1}^{m} \gamma_{ji} \quad (11)$$

The algorithm is as follows:

1. Determine the two clusters for the cultivated layer and the un-ploughed layer, that is, $k$ is 2;
2. Initialize the Gaussian distribution parameters $\mu$, $\Sigma$, and $\alpha$ of each cluster, and randomly assign them to construct the Gaussian mixture probability density;
3. Traverse all sample data, and calculate the conditional probability that each sample conforms to each distribution, that is $\gamma_{ji}$;
4. Then, use the calculated $\gamma_{ji}$ to calculate new values for $\mu$, $\Sigma$, and $\alpha$; update the distribution model with the new parameters; and obtain a new probability density;
5. Use the new distribution model to calculate the conditional probability of each sample, repeat steps (2)–(4), continuously update the model distribution parameters until the model converges, and then stop the iteration;
6. Obtain the conditional probability of each sample based on the new model parameters, and then, obtain the maximum value of $\lambda_j = \arg\max_{i \in \{1,2\}} \gamma_{ji}$, as the clustering result of these data.

The schematic diagram of the GMM clustering density estimation is shown in Figure 6. The darker the color is, the larger the density estimation is, and the lighter the color is, the smaller the density estimation is. The GMM discriminant effect is shown in Figure 7.

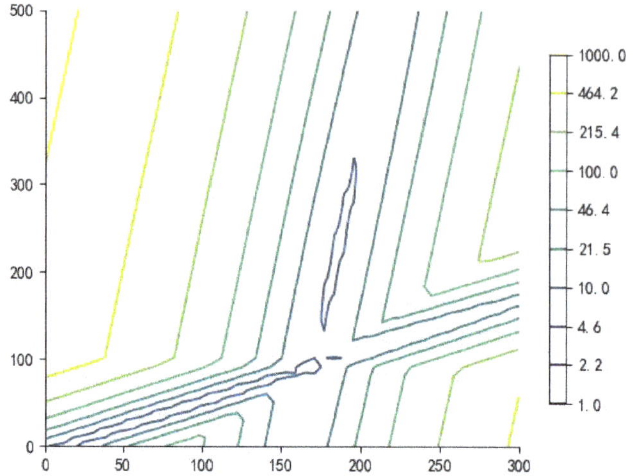


**Figure 6.** Schematic diagram of the density estimation of the GMM.

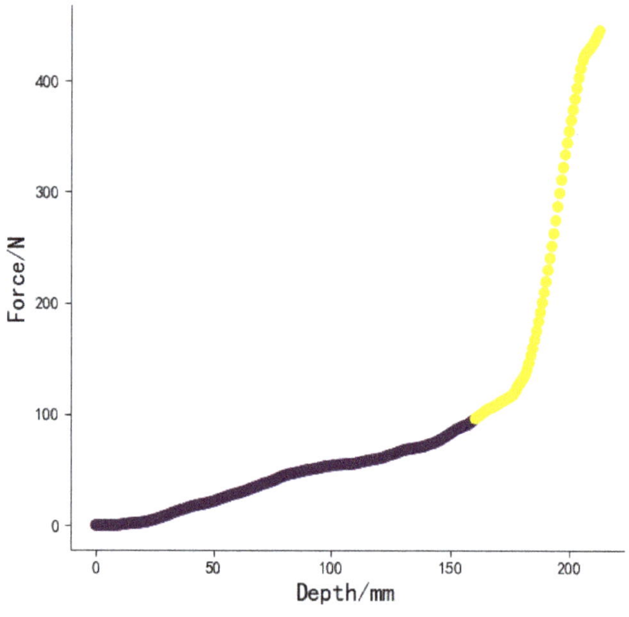


**Figure 7.** GMM clustering discrimination effect diagram.

The clustering result of the GMM algorithm marks the cultivated layer as a cluster 1, which is the yellow data point in Figure 7; the un-ploughed layer data are a cluster marked as 0, which is the black and purple data point in Figure 7. The results obtained after processing 64 groups of soil samples in the simulated tillage layer show that the number of samples with an absolute error less than 5 mm is 31, the number of samples with an absolute error less than 10 mm is 48, the number of samples with an error less than 15 mm is 52, and the number of samples with an error less than 20 mm is 59 groups.

### *3.4. Data Fitting to Determine Tillage Depth*

After the soil is compacted, physical properties such as bulk density, porosity, and temperature of the soil change compared with the soil before compaction, and the root penetration resistance also increases. The working principle of the rotary tiller is that the blade of the rotary tiller rotates at a certain speed and moves forward in random groups to make up the milling process: soil cutting, crushing, and throwing back. Therefore, physical properties such as bulk density, porosity, and temperature of the soil in the plough layer that have been rotary tilled tend to be the same. Since the soil in the plough bottom layer has not been ploughed for many years, it forms a solid plough bottom layer under the plough layer. Thus, a plow bottom layer 40–50 cm below the plough layer tends to maintain the same physical properties.

According to Yan Ben [29], depth and penetration resistance share a quadratic polynomial relationship, and the soil-sample data were collected at a uniform speed. Therefore, data fitting was performed according to the theory that depth and penetration resistance share a quadratic polynomial relationship.

A portion of the data for the tilled layer and a portion of the data for the uncultivated layer were intercepted, and quadratic curves were fitted to the depth and penetration resistance. The penetration resistance of the soil in the plough layer increases with the penetration depth, so it is inferred that this point is the boundary where the plough layer meets the un-ploughed layer. The original data before data fitting are shown in Figure 8, and the fitted quadratic curve is shown in Figure 9:

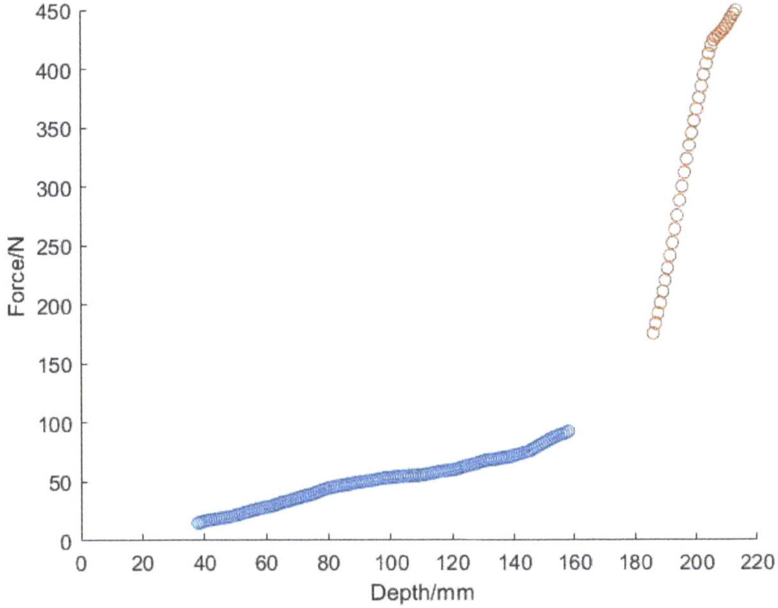


**Figure 8.** Schematic diagram of the original data points in the cultivated layer and the un-ploughed layer.

Sixty-four groups of soil samples were processed by fitting quadratic curves to find the intersection points. The number of soil samples with an error of less than 5 mm was 24, the number of samples with errors less than 10 mm was 47, the number of samples with errors less than 15 mm was 55, and the number of samples with errors less than 20 mm was 60.

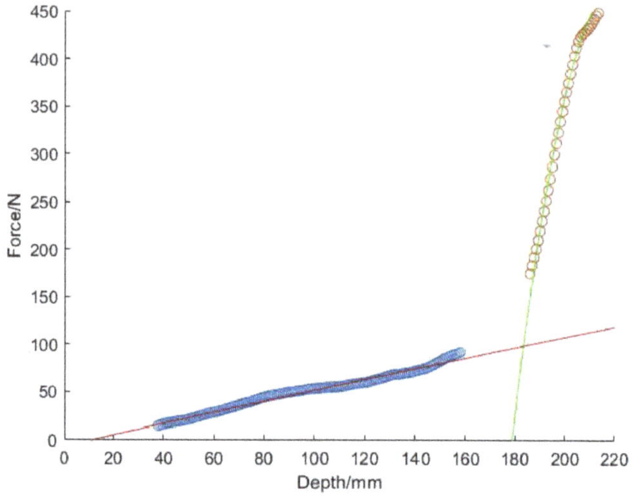


**Figure 9.** Schematic diagram of MATLAB fitting curve.

### 3.5. *Comparison of Evaluation Results of Three Kinds of Machine Learning*

This paper mainly deals with the problem of binary classification and takes the above curve data as a standard data set of the simulated soil samples. The accuracy rate, accuracy rate, recall rate, and F1 function of the clustering results from the three machine learning algorithms were evaluated [30]:

TP (True Positive) = True positive: a positive sample is correctly predicted to be positive

FP (False Positive) = False positive: a negative sample is incorrectly predicted to be positive

FN (False Negative) = False negative: a positive sample is incorrectly predicted to be negative

TN (True Negative) = True negative: a negative sample is correctly predicted to be negative

Accuracy refers to the percentage of cases in which predictions are correct:

$$\text{Accuracy} = \frac{\text{TP} + \text{TN}}{\text{TP} + \text{FP} + \text{FN} + \text{TN}} \tag{12}$$

Precision refers to the percentage of positive cases in which the prediction is correct:

$$\text{Precision} = \frac{\text{TP}}{\text{TP} + \text{FP}} \tag{13}$$

Recall refers to the percentage of cases that are actually positive and are predicted correctly:

$$\text{Recall} = \frac{\text{TP}}{\text{TP} + \text{FN}} \tag{14}$$

F1 score takes the precision rate and recall rate into consideration. When both are high, the values are balanced:

$$\text{F}_1 = \frac{2 * \text{Precision} * \text{Recall}}{\text{Precision} + \text{Recall}} \tag{15}$$

The above three machine learning algorithms are evaluated against the results of the standard data set, and the results are shown in Table 1 below.

**Table 1.** Results from the comparison of the three machine learning algorithms.

| | Accuracy | Precision | Recall | F1 Score |
|---|---|---|---|---|
| DBSCAN | 0.9890 | 1 | 0.9868 | 0.9934 |
| Kmeans | 0.9706 | 0.966 | 1 | 0.9827 |
| GMM | 0.9080 | 1 | 0.889 | 0.9412 |

It can be seen from the above table that the DBSCAN algorithm has the best performance for the classification of this standard data set, with an accuracy of 0.9890 and an $F_1$ score of 0.9934, which are superior to those of the other two algorithms.

### *3.6. Comparison of the Effects of Several Methods in Discriminating the Plough Layer*

The 64 groups of simulated soil samples are summarized in Table 2 according to the statistical data on the clustering effect.

**Table 2.** Statistical data on the clustering effect of the simulated soil samples.

| | Kmeans Clustering | | DBSCAN Clustering | | GMM Clustering | | Data Fitting | |
|---|---|---|---|---|---|---|---|---|
| | Sample Size | Accuracy | Sample Size | Accuracy | Sample Size | Accuracy | Sample Size | Accuracy |
| Error < 5 mm | 4 | 6.25% | 32 | 50% | 31 | 48.44% | 24 | 37.5% |
| Error < 10 mm | 12 | 18.75% | 52 | 81.25% | 48 | 75% | 47 | 73.44% |
| Error < 15 mm | 26 | 40.63% | 58 | 90.63% | 52 | 81.25% | 55 | 85.94% |
| Error < 20 mm | 47 | 73.44% | 60 | 93.75% | 59 | 92.19% | 60 | 93.75% |

According to the above table, the best clustering effect is clearly DBSCAN. When the error is within 15 mm, the accuracy of the DBSCAN clustering algorithm in discriminating the tillage layer is 90.63%; when the error is less than 20 mm, the accuracy rate reaches 93.75%. In terms of the discriminant effect, DBSCAN clustering has the best discriminative effect, data fitting has the second-best discriminant effect, GMM clustering has the third best discriminant effect, and Kmeans clustering has the worst discriminant effect.

## 4. Field-Test-Specific Situation and Hybrid Algorithm

### *4.1. Field-Data Collection*

The equipment used in the field experiment was a multi-point probe-type tillage section detection vehicle designed by our team (Figure 10). The inspection vehicle has six probes, with a side-by-side spacing of 20 cm. Each probe has a force sensor and a displacement sensor to collect data at the same time. The data from the six probes were collected and transmitted to the computer for subsequent data analysis. After collecting the data for a section, the inspection vehicle moves forward by 20 cm. The location of the field experiment was Mengjin County, Luoyang City (34.819° N, 112.3901° E).

Because the firmness of the soil in the field is much larger than that of the soil samples in the laboratory, it was found in the actual field experiment that the data collected by the probe have a disturbance area; the disturbance area is found in the ploughed layer and is close to the plow bottom. When the probe enters the disturbance area of the soil, the penetration resistance increases significantly before it even touches the plow bottom. In the simulation of soil penetration resistance in the field, our team found that the shear modulus of the soil at the bottom of the plough is significantly larger than that of the tilled layer, such that the soil firmness is greater than that of the tilled layer. When the disturbance area is close to the bottom of the plow, the probe continues to penetrate, and the depth of the disturbance area gradually decreases. However, the existence of a disturbance area causes the machine learning algorithm to make a larger error when discriminating the soil depth of the ploughed layer. Our team simulated a large amount of data on the soil ploughed layer in the field to determine that the disturbed area is located 20–30 mm above the ploughed and un-ploughed layers. In order to determine the applicability of the algorithm, an average value of 25 mm was selected as the systematic error for discriminating the cultivated layer in the field test.

**Figure 10.** Image of multi-point probe-type tillage section inspection vehicle used in the field test.

### *4.2. Discrimination Effect of the Four Methods for Discriminating the Plough Layer in the Field Test*

According to the agronomic requirements, the rotary tillage operation allows for the existence of clods within a certain range. When collecting data in the field experiment, since the maximum diameter of the probe is 14 mm, it is inevitable that the probe collects data vertically in a downwards direction and hits the clods. Therefore, the field experiment collects two types of data: one is data that do not encounter the clods, and the other is data that encounter the clods. The data collected without encountering clods and the Kmeans clustering, DBSCAN clustering, and GMM clustering discrimination effect are shown in Figure 11.

Among the methods, Kmeans clustering sets the parameter k as 2 to obtain two clusters, and the data point output of the first cluster is used as the boundary at which Kmeans clustering discriminates the cultivated layer. DBSCAN sets the parameter eps to 7 and min_samples to 5, that is, when the radius is 7, there are 5 sample points, and the data point is output when the first outlier is used is the boundary at which DBSCAN clustering discriminates the cultivated layer. Parameter k from GMM clustering is set to 2, and the data point output from the first different cluster is used as the boundart at which GMM clustering discriminates the cultivated layer. Additionally, MATLAB data fitting takes the intersection as the cut-off point of the tillage layer. A total of 52 groups of field samples were collected, and 41 groups of field samples were obtained in the case of no soil blocks. The clustering effect statistics for the cultivated layer data are shown in Table 3 below.

**Table 3.** Statistical data on clustering effect of field soil samples.

| | Kmeans Clustering | | DBSCAN Clustering | | GMM Clustering | | Data Fitting | |
|---|---|---|---|---|---|---|---|---|
| | Sample Size | Accuracy | Sample Size | Accuracy | Sample Size | Accuracy | Sample Size | Accuracy |
| Error < 5 mm | 5 | 12.2% | 14 | 34.15% | 10 | 24.4% | 9 | 21.95% |
| Error < 10 mm | 7 | 17.07% | 30 | 73.17% | 21 | 51.22% | 20 | 48.78% |
| Error < 15 mm | 11 | 26.83% | 36 | 87.8% | 26 | 63.41% | 26 | 63.41% |
| Error < 20 mm | 25 | 60.98% | 37 | 90.24% | 32 | 78.05% | 31 | 75.61% |

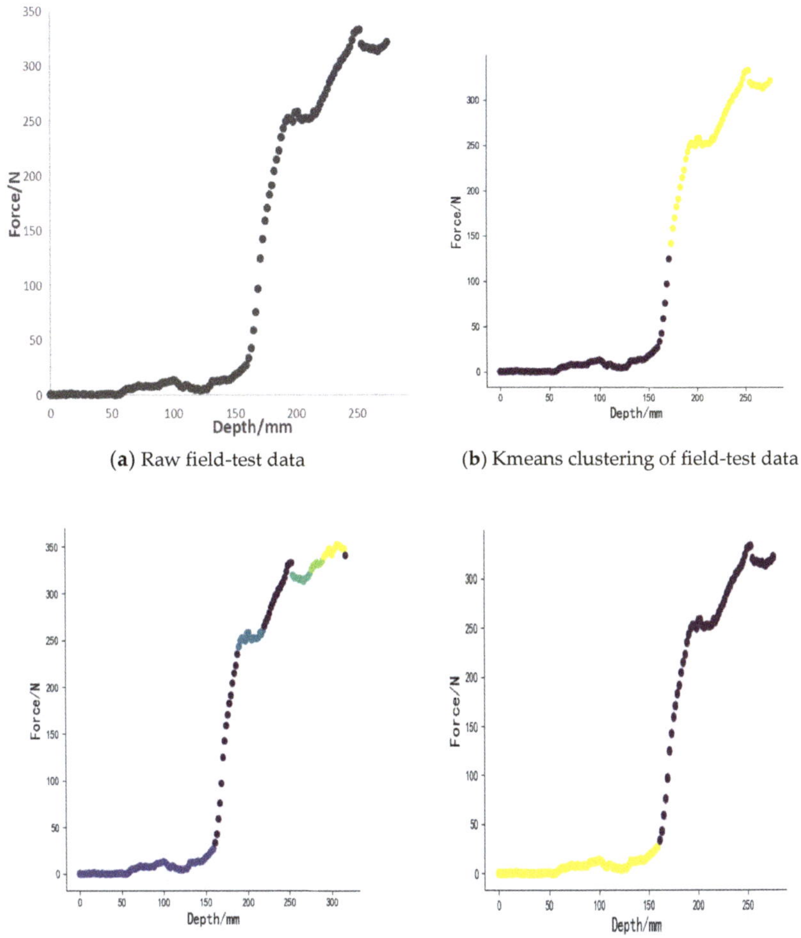


(**a**) Raw field-test data (**b**) Kmeans clustering of field-test data

(**c**) DBSCAN clustering of field-test data (**d**) GMM clustering of field-test data

**Figure 11.** Discrimination-effect of the clustering methods in the field test.

It can be seen from the above table that when the error is less than 20 mm, the accuracy rate of DBSCAN in discriminating the cultivated layer is 90.24%. The effect of DBCSAN clustering when discriminating the cultivated layer is still the best in the field test; followed by data fitting; and then, GMM clustering when discriminating the cultivated layer, and Kmeans still has the worst results.

*4.3. RANSCA Robust Regression Algorithm to Deal with the Error of the Soil Block*

When collecting the data, 11 groups of sample data were found to have encountered soil blocks. The sample data collected for the soil blocks are shown in Figure 12.

When encountering clods, several clustering methods determined the collected soil block data as the boundary point of the cultivated layer, resulting in a particularly large discrimination error. Thus, a new method is needed to solve the problem of encountering clods. In this paper, a hybrid RANSCA (random sample consensus) robust regression algorithm and DBSCAN clustering algorithm is proposed to discriminate the tillage layer.

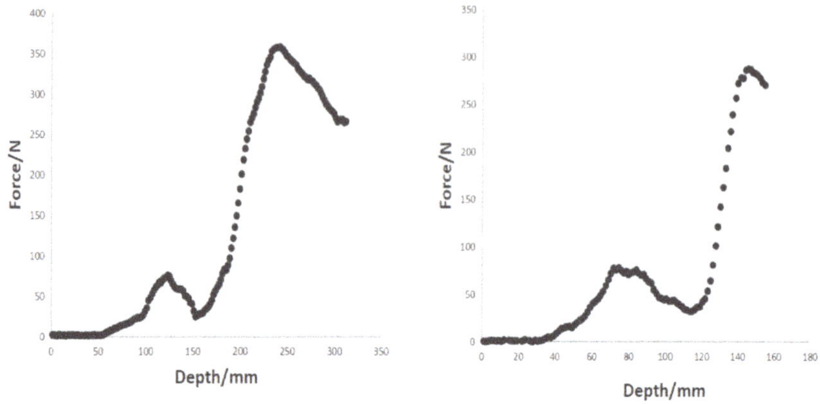


(**a**) Data set 1 encountered clods (**b**) Data set 2 encountered clods

**Figure 12.** The original data collected when encountering clods in the field.

The RANSCA algorithm fits a regression model in the presence of bad data (i.e., there are outliers or errors in the model) [31]. RANSAC is a non-deterministic algorithm that produces a reasonable result with a certain probability [32]. The basic assumption of the RANSCA algorithm is that the sample contains correct data (data that can be described by the model), called "inliers". Data that that cannot fit the mathematical model are called "outliers", i.e., the data set contains noise. The basic assumptions of the RANSCA algorithm are as follows:

(1) The data consists of correct data, namely "inliers", and the distribution of "inliers" can be used to calculate a model;
(2) "Outliers" are parameters that cannot be adapted to the model;
(3) Other data are noise.

Through an analysis of the curve diagram of penetration resistance and penetration depth, it can be found that, when no soil clods are encountered, the model calculated by fitting all the data of the surface layer is identified as the "local point". When dirt clods are encountered, the data points collected do not fit the calculated model and are identified as "external points". Data points that are determined to be "intra-office" after RANSCA robust regression are marked as true, and data points that are determined to be external are marked as false. It can be seen from the above that the data points before those considered true by the RANSAC robust regression algorithm are all points collected in the cultivated layer. RANSCA fits the regression model for the cultivated layer data and allows for noise. Even under discrimination by DBSCAN density clustering, the soil clod data encountered cannot be divided into a cluster with other cultivated layer data [33], but because the data are identified as a cultivated layer by the RANSCA algorithm, the data encountering soil clods are not regarded as the result of DBSCAN discriminating a plough layer. The RANSCA algorithm solves the problem of the inability of the DBSCAN algorithm to cluster soil clod data and other cultivated layer data.

### *4.4. The Hybrid RANSCA and DBSCAN Algorithm Identifies the Soil Topsoil Layer*

The cultivable layer is simultaneously identified by each part of the hybrid algorithm. That is, the first data point considered an outlier by DBSCAN after RANSCA discriminates the last true data point is used as the boundary between the cultivable layer and the uncultivable layer. The steps in RANSCA robust regression and DBSCAN clustering hybrid algorithm are as follows:

(1) First, the data from the cultivated layer are used as the data set for RANSAC training, and a set of random subsets in the data set are selected as "inliers" to train the model. The unknown parameters of the model are calculated from hypothetical "inliers".
(2) The obtained model is then used to test other data. If other data are also applicable to this model, they are considered "inliers". If enough points are considered "inliers", the model is proven to be reasonable enough.
(3) The model is then re-estimated with all of the "inliers". If the generated model has too few "inliers", it is discared and replaced with a better model until the final model is determined.
(4) After the model is determined, data points that encounter a soil block and are not suitable for the model are considered "outliers" and marked as false, and all data before the last data point considered "inliers" are regarded as the cultivated layer data.
(5) All data are clustered by DBSCAN clustering, and the first "outliers" after all of the data points for the cultivated layer are used as the basis for the discrimination of the cultivated layer.

The discriminant flow chart and a diagram of the effect are shown in Figure 13. The rendering diagram of the data set for discriminating the encounter of soil clods is shown in Figure 14.

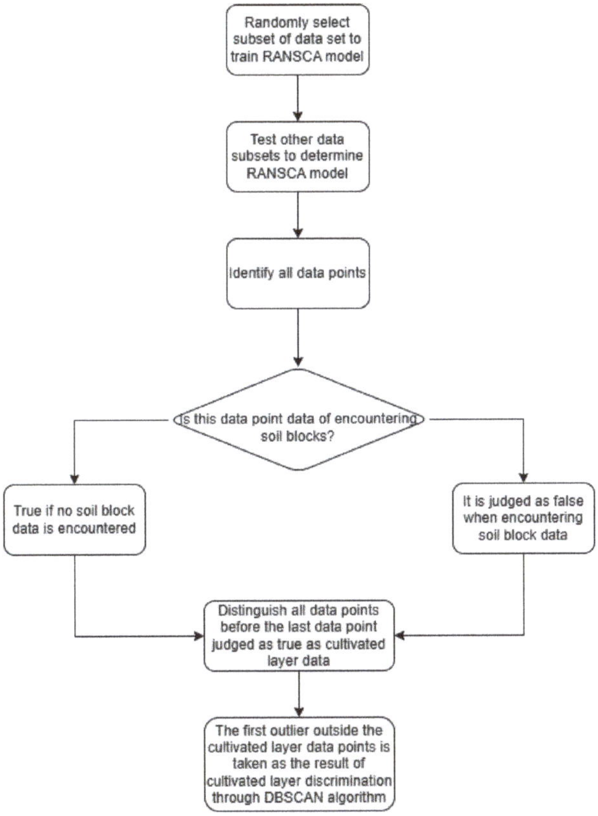


**Figure 13.** Data flow diagram of RANSCA and DBSCAN hybrid algorithm processing.

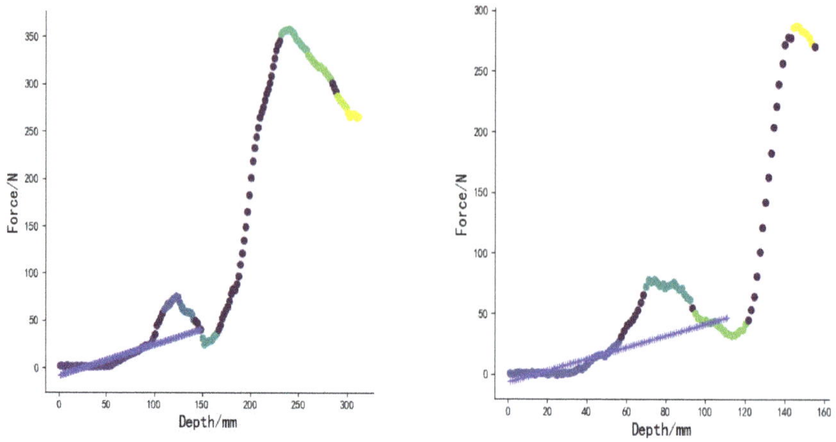


(a) (b)

**Figure 14.** Schematic diagram of the RANSCA and DBSCAN hybrid algorithm processing the data encountered in the clod. (**a**) The hybrid algorithm processes the encountered clod data set 1. (**b**) The hybrid algorithm processes the encountered clod data set 2.

The hybrid algorithm can discriminate the tillage layer, and the discriminant effect basically is ineffective when no soil block is encountered. The discriminant effect diagram is as follows. The discriminant effect diagram is shown in Figure 15 below.

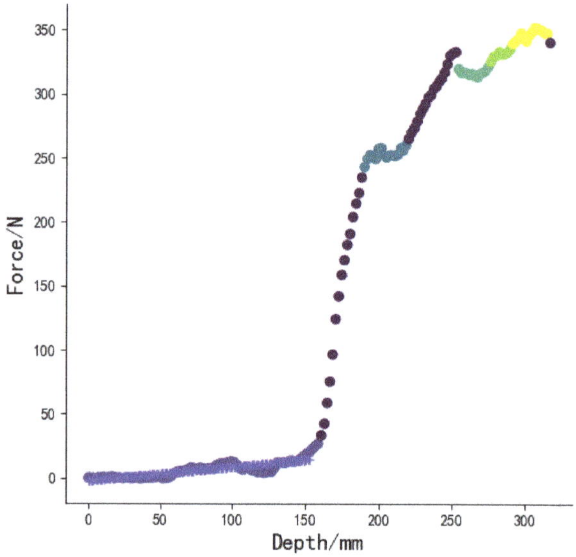


**Figure 15.** Schematic diagram of the RANSCA and DBSCAN hybrid algorithm processing data that did not encounter clods.

The 41 groups of field samples that did not encounter soil blocks were processed with DBSCAN and the combined RANSCA and DBSCAN algorithm, as well as the combined RANSCA and DBSCAN clustering to process all of the field samples of the 51 groups. The results are shown in Table 4 below.

**Table 4.** Comparison of DBSCAN clustering and hybrid clustering effects.

| | DBSCAN Clustering Results | | RANSCA and DBSCAN Did Not Encounter Clod Results | | RANSCA and DBSCAN Encountered Clod Results | |
|---|---|---|---|---|---|---|
| | Sample Size | Accuracy | Sample Size | Accuracy | Sample Size | Accuracy |
| Error < 5 mm | 14 | 34.15% | 14 | 34.15% | 16 | 30.77% |
| Error < 10 mm | 30 | 73.17% | 28 | 68.29% | 32 | 53.85% |
| Error < 15 mm | 36 | 87.8% | 32 | 78.05% | 38 | 69.23% |
| Error < 20 mm | 37 | 90.24% | 38 | 92.68% | 45 | 82.69% |

By comparing the results of DBSCAN clustering and the hybrid algorithm when dealing with unencountered soil blocks, it can be found that the hybrid algorithm has almost no influence on the discrimination of the surface layer when no soil blocks are encountered.

Moreover, the hybrid algorithm can solve the problem of a soil block encountered by the probe in the field, avoids interference from the soil block in the surface discrimination, and makes the discrimination method more applicable. When the error is less than 20 mm, the accuracy of the RANSCA and DBSCAN hybrid algorithm is 82.69%, and the performance of the hybrid algorithm in the field experiment is fair.

## 5. Conclusions

In this paper, three machine learning algorithms were used to process a standard data set for model evaluation in an indoor manual soil-sample test. Among them, the discrimination accuracy of DBSCAN for the data set reached 0.9890 and the F1 score reached 0.9934, which was better than those of the other two algorithms. At the same time, 64 groups of simulated soil-sample data were processed with DBSCAN and the discrimination accuracy was 90.63% when the error was less than 15 mm and 93.75% when the error was less than 20 mm, which verified that DBSCAN had a better discrimination effect than Kmeans clustering, GMM clustering, and data fitting.

In the field experiment, since the maximum diameter of the tip of the probe is 14 mm, it is almost impossible to rotate until all of the soil in the tilled layer becomes loose during mechanical operation. Whether rotating the till or the deep till, there can be soil clods within the range allowed by agronomy. That is to say, it is inevitable that there will be soil clods with a diameter of more than 14mm in the soil during a rotary tillage operation. When the probe goes straight down to assess penetration resistance, it inevitably hits small clods that are not completely broken. Therefore, data processing in the field can be divided into two conditions: no clods encountered and clods encountered. When no soil clods were encountered and the error was less than 20 mm, the accuracy of DBSCAN in identifying a plough layer was 90.24%. When comparing the four discriminant methods, DBSCAN still had the best discriminant effect on the soil layer. However, the soil clods encountered in the field experiment will affect the test data and cause significant errors in the processing results. Therefore, the combined RANSCA robust regression and DBSCAN clustering algorithm was used in this paper to process the collected data. When the mixed algorithm processed the data without encountering soil clods, the accuracy was not affected. When the error was less than 20 mm, the accuracy of the mixed algorithm reached 92.68%. At the same time, the hybrid algorithm eliminates the disturbance due to soil clods and determines the ploughing depth, which solves the problem of the large ploughing depth error caused by soil clods in the field. It has been verified that the combined RANSCA and DBSCAN method of discriminating the soil depth of topsoil in the field with an error of less than 20 mm can reach an accuracy of 82.69%. The error in distinguishing soil depth caused by soil clods in the field is thus solved. The combined RANSCA robust regression and DBSCAN clustering algorithm improved the applicability of the discrimination method and provided a new method for verifying and discriminating topsoil depth. It can be widely used to evaluate the quality of subsoiling operations and in assessing whether subsoiling land preparation subsidies should be issued by the Ministry of Agriculture of China. The tillage depth can be accurately determined by measuring only the penetration resistance and depth, making the information acceptance tests refined and smart.

**Author Contributions:** Conceptualization, J.P.; methodology, X.Z. and X.L.; software, X.Z.; formal analysis, J.L., X.D. and J.H.; writing—original draft preparation, J.P. and X.Z. All authors have read and agreed to the published version of the manuscript.

**Funding:** This work was supported by the National Key R&D Program of China (Grant No. 2017YFD0700300).

**Institutional Review Board Statement:** Not applicable.

**Informed Consent Statement:** Informed consent was obtained from all subjects involved in the study.

**Data Availability Statement:** All data sets in this article were collected by the team.

**Conflicts of Interest:** The authors declare no conflict of interest.

## References


1. Jin, C.; Pang, D.; Min, J.I.N.; Luo, Y.L.; Li, H.Y.; Yong, L.I.; Wang, Z.L. Improved soil characteristics in the deeper plough layer can increase grain yield of winter wheat. *J. Integr. Agric.* **2020**, *19*, 1215–1226. [CrossRef]
2. Arvidsson, J.; Håkansson, I. A model for estimating crop yield losses caused by soil compaction. *Soil Tillage Res.* **1991**, *20*, 319–332. [CrossRef]
3. Olesen, J.E.; Munkholm, L.J. Subsoil loosening in a crop rotation for organic farming eliminated plough pan with mixed effects on crop yield. *Soil Tillage Res.* **2007**, *94*, 376–385. [CrossRef]
4. Yamazaki, H.; Yoshida, T. Various scarification treatments produce different regeneration potentials for trees and forbs through changing soil properties. *J. For. Res.* **2020**, *25*, 41–50. [CrossRef]
5. Zikeli, S.; Gruber, S. Reduced tillage and no-till in organic farming systems, Germany—Status quo, potentials and challenges. *Agriculture* **2017**, *7*, 35. [CrossRef]
6. Kim, Y.S.; Kim, T.J.; Kim, Y.J.; Lee, S.D.; Park, S.U.; Kim, W.S. Development of a real-time tillage depth measurement system for agricultural tractors: Application to the effect analysis of tillage depth on draft force during plow tillage. *Sensors* **2020**, *20*, 912. [CrossRef]
7. Mouazen, A.M.; Anthonis, J.; Saeys, W.; Ramon, H. An automatic depth control system for online measurement of spatial variation in soil compaction, Part 1: Sensor design for measurement of frame height variation from soil surface. *Biosyst. Eng.* **2004**, *89*, 139–150. [CrossRef]
8. Xie, B.; Li, H.; Zhu, Z.X. Automatic measurement method of tillage depth of tractor suspension unit based on inclination sensor. *Trans. Chin. Soc. Agric. Eng.* **2013**, *29*, 15–21. [CrossRef]
9. Jia, H.; Guo, M.; Yu, H.; Li, Y.; Feng, X.; Zhao, J.; Qi, J. An adaptable tillage depth monitoring system for tillage machine. *Biosyst. Eng.* **2016**, *151*, 187–199. [CrossRef]
10. Bodenhofer, U.; Kothmeier, A.; Hochreiter, S. APCluster: An R package for affinity propagation clustering. *Bioinformatics* **2011**, *27*, 2463–2464. [CrossRef]
11. Peng, K.; Leung, V.C.M.; Huang, Q. Clustering approach based on mini batch kmeans for intrusion detection system over big data. *IEEE Access* **2018**, *6*, 11897–11906. [CrossRef]
12. Ankerst, M.; Breunig, M.M.; Kriegel, H.P.; Sander, J. OPTICS: Ordering points to identify the clustering structure. *ACM Sigmod Rec.* **1999**, *28*, 49–60. [CrossRef]
13. Ahmed, M.; Seraj, R.; Islam, S.M.S. The k-means algorithm: A comprehensive survey and performance evaluation. *Electronics* **2020**, *9*, 1295. [CrossRef]
14. Abubaker, M.; Ashour, W.M. Efficient data clustering algorithms: Improvements over Kmeans. *Int. J. Intell. Syst. Appl.* **2013**, *5*, 37–49. [CrossRef]
15. Sinaga, K.P.; Yang, M.S. Unsupervised K-means clustering algorithm. *IEEE Access* **2020**, *8*, 80716–80727. [CrossRef]
16. Likas, A.; Vlassis, N.; Verbeek, J.J. The global k-means clustering algorithm. *Pattern Recognit.* **2003**, *36*, 451–461. [CrossRef]
17. Adnan, R.M.; Khosravinia, P.; Karimi, B.; Kisi, O. Prediction of hydraulics performance in drain envelopes using Kmeans based multivariate adaptive regression spline. *Appl. Soft Comput.* **2021**, *100*, 107008. [CrossRef]
18. Zahra, S.; Ghazanfar, M.A.; Khalid, A.; Azam, M.A.; Naeem, U.; Prugel-Bennett, A. Novel centroid selection approaches for KMeans-clustering based recommender systems. *Inf. Sci.* **2015**, *320*, 156–189. [CrossRef]
19. Hahsler, M.; Piekenbrock, M.; Doran, D. dbscan: Fast density-based clustering with R. *J. Stat. Softw.* **2019**, *91*, 1–30. [CrossRef]
20. Schubert, E.; Sander, J.; Ester, M. DBSCAN revisited, revisited: Why and how you should (still) use DBSCAN. *ACM Trans. Database Syst. (TODS)* **2017**, *42*, 1–21. [CrossRef]
21. Khan, K.; Rehman, S.U.; Aziz, K.; Fong, S.; Sarasvady, S. DBSCAN: Past, present and future. In Proceedings of the Fifth International Conference on the Applications of Digital Information and Web Technologies (ICADIWT 2014), Chennai, India, 17–19 February 2014; IEEE: Toulouse, France, 2014; pp. 232–238. [CrossRef]

22. Weng, S.; Gou, J.; Fan, Z. $ h $-DBSCAN: A simple fast DBSCAN algorithm for big data. *Asian Conf. Mach. Learn. PMLR* **2021**, *157*, 81–96.
23. Chen, Y.; Zhou, L.; Bouguila, N.; Wang, C.; Chen, Y.; Du, J. BLOCK-DBSCAN: Fast clustering for large scale data. *Pattern Recognit.* **2021**, *109*, 107624. [CrossRef]
24. Luchi, D.; Rodrigues, A.L.; Varejão, F.M. Sampling approaches for applying DBSCAN to large datasets. *Pattern Recognit. Lett.* **2019**, *117*, 90–96. [CrossRef]
25. Windmeijer, F. A finite sample correction for the variance of linear efficient two-step GMM estimators. *J. Econom.* **2005**, *126*, 25–51. [CrossRef]
26. Roodman, D. How to do xtabond2: An introduction to difference and system GMM in Stata. *Stata J.* **2009**, *9*, 86–136. [CrossRef]
27. Campbell, W.M.; Sturim, D.E.; Reynolds, D.A. Support vector machines using GMM supervectors for speaker verification. *IEEE Signal Process. Lett.* **2006**, *13*, 308–311. [CrossRef]
28. McLachlan, G.J.; Krishnan, T. *The EM Algorithm and Extensions*; John Wiley & Sons: Hoboken, NJ, USA, 2007.
29. Yan, B. *Research and Measurement Device Design of the Penetration Resistance of Granular Materials*; Northeast Agricultural University: Harbin, China, 2018.
30. Nabipour, M.; Nayyeri, P.; Jabani, H.; Shahab, S.; Mosavi, A. Predicting stock market trends using machine learning and deep learning algorithms via continuous and binary data; a comparative analysis. *IEEE Access* **2020**, *8*, 150199–150212. [CrossRef]
31. Guo, Q.; Hu, X. Power line icing monitoring method using binocular stereo vision. In Proceedings of the 2017 12th IEEE Conference on Industrial Electronics and Applications (ICIEA), Siem Reap, Cambodia, 18–20 June 2017; IEEE: Toulouse, France, 2017; pp. 1905–1908. [CrossRef]
32. Ruiz, L.A.; Fdez-Sarría, A.; Recio, J.A. Texture feature extraction for classification of remote sensing data using wavelet decomposition: A comparative study. In Proceedings of the 20th ISPRS Congress, Istanbul, Turkey, 12–23 July 2004; Volume 35, pp. 1109–1114.
33. Kumar, K.M.; Reddy, A.R.M. A fast DBSCAN clustering algorithm by accelerating neighbor searching using Groups method. *Pattern Recognit.* **2016**, *58*, 39–48. [CrossRef]

agriculture

MDPI

Article

# Optimal Design and Analysis of Cavitating Law for Well-Cellar Cavitating Mechanism Based on MBD-DEM Bidirectional Coupling Model

**Gaowei Xu [1,*], Huimin Fang [2], Yumin Song [1] and Wensheng Du [3]**

1 Department of Automotive Engineering, Shandong Jiaotong University, Jinan 250357, China
2 School of Agricultural Engineering, Jiangsu University, Zhenjiang 212013, China
3 College of Mechanical and Electronic Engineering, Shandong Agricultural University, Taian 827101, China
* Correspondence: 202107@sdjtu.edu.cn

**Abstract:** A variable velocity parallel four-bar cavitating mechanism for well-cellar can form the well-cellar cavitation which suits for well-cellar transplanting under a continuous operation. In order to improve the cavitating quality, this paper analyzed the structural composition and working principle of the cavitating mechanism and established the bidirectional coupling model of multi-body dynamics and the discrete element between the cavitating mechanism and soil through Recurdyn and EDEM software. Based on the model, a three-factor, five-level quadratic orthogonal rotational combination design test was conducted with the parameters of the cavitating mechanism as the experimental factors and the parameters of the cavitation as the response index to obtain the optimal parameter combination, and a virtual simulation test was conducted for the optimal parameter combination in order to study the cavitating law of the cavitating mechanism and soil. The test results showed that the depth of the cavitation was 188.6 mm, the vertical angle of the cavitation was 90.4°, the maximum diameter of the cavitation was 76.1 mm, the minimum diameter of the cavitation was 68.5 mm, and the variance in the diameters for the cavitation was 5.42 mm$^2$. The cavitating mechanism with optimal parameters based on the Recurdyn–EDEM bidirectional coupling mode could further improve the cavitating quality.

**Keywords:** well-cellar cavitating mechanism; MBD-DEM bidirectional coupling model; optimal design; cavitating law

**Citation:** Xu, G.; Fang, H.; Song, Y.; Du, W. Optimal Design and Analysis of Cavitating Law for Well-Cellar Cavitating Mechanism Based on MBD-DEM Bidirectional Coupling Model. *Agriculture* **2023**, *13*, 142. https://doi.org/10.3390/agriculture13010142

Academic Editors: Mustafa Ucgul and Chung-Liang Chang

Received: 12 December 2022
Revised: 2 January 2023
Accepted: 2 January 2023
Published: 5 January 2023



## 1. Introduction

Well-cellar transplanting is a kind of agronomic technology which transplants the seedlings of some of the crops used for the planting of the hilly and mountainous areas, such as of tobacco [1] and pepper [2], into the well-cellar cavitation. The well-cellar cavitation is a cavity with a certain depth and height, as well as a consistent diameter. According to Darcy's law, the hydraulic gradient is generated in the well-cellar cavitation with the change in the external temperature and humidity, which forces the soil moisture in the well cellar to evaporate or condense, so as to keep the relative stability of the temperature and humidity inside the well-cellar cavitation [3]. Therefore, by taking the advantage of the relatively stable of the humidity and temperature in the well-cellar cavitation, watering, applying pesticides, and covering with a plastic film is immediately carried out after transplanting the crop seedlings into the well-cellar cavitation, which can realize the moderate early planting and deep planting inside the high ridge for the crop transplanting in hilly and mountainous areas, so as to increase the drought resistance and reduce the pests and diseases of transplanting seedlings [4]. At present, it has been partially promoted in the hilly and mountainous areas of Guizhou, Yunnan, Shandong, and Guangxi Province [5,6].

Nowadays, the transplanting equipment in the hilly and mountainous areas is mainly based on the duckbill transplanting machine according to the traditional planting agronomy [7–13]. It cannot form the stable well-cellar cavitation and it is difficult to adapt to the agronomic requirements of the well-cellar transplanting for the crops in the hilly and mountainous area. There are currently three types of cavitating devices for hilly and mountainous transplanters. First, the man-powered knapsack cavitating device [14] is driven by a gasoline engine and its cavitating position and depth is manually controlled, which results in a low cavitating efficiency and high labor intensity. Second, the shape of the cavitation formed by the multi-link cavitating mechanism is a "horn shape", which cannot meet the agronomic requirements of the well-cellar cavitation in the cavitating process [15,16] because it cannot match the forward velocity of the transplanter. Finally, the intermittent well-cellar cavitating device works with the intermittent powered chassis. As the powered chassis reaches the specified position, the device forms the cavitation along the direction which is perpendicular to the ridge surface [17,18]. The device avoids the matching problems between the cavitating mechanism and powered chassis, which formed the cavitation with a better quality. However, due to the intermittent movement of the chassis, the machines and tools have a low operating efficiency and reliability.

In the view of the above situation, the project team designed a kind of well-cellar-type cavitating mechanism for the hilly and mountainous transplanting machine [19], which is based on the noncircular gear parallel four-bar mechanism. It used the characteristics of a nearly uniform linear motion of the non-circular gear linkage to offset the forward working velocity of the mechanism. Under the continuous operation of the tool, the well-cellar cavitation which suited for transplanting the large depth and high specifications of the hilly and mountainous crops was formed, so as to improve the working efficiency and operation quality of the mechanized well-cellar cavitation and reduce the labor intensity of the artificial well-cellar cavitation.

At present, the cavitating mechanism was designed from the perspective of a mechanical design, and the interaction between the machine and soil was not studied. Accordingly, this paper used MBD-DEM bidirectional coupling technology to establish the interactive coupling model between the cavitating mechanism and soil in order to study the law of cavitation between the cavitating mechanism and the soil and optimize the parameters of the cavitating mechanism.

## 2. Materials and Methods

### *2.1. Composition and Working Principle of the Cavitating Mechanism*

The structural diagram for the well-cellar cavitating mechanism is shown in Figure 1. The mechanism was mainly composed of the non-circular gear mechanism, the spur gear mechanism, the parallel four-bar mechanism, the cavitator, and the frame. One and two were the non-circular gear which constituted the non-circular gear mechanism. Three, four, and five were the straight tooth cylindrical gear which composed the spur gear mechanism. Six, seven, and eight were the double crank, frame, and linkage which composed the parallel four-bar mechanism. Nine and ten were the powershuttle mechanism and cavitator which composed the cavitating device.

The non-circular gear mechanism and the spur gear mechanism were installed on the frame, and the non-circular gear 2 was fixedly connected with the spur gear 3. The double crank 6 of the parallel four-bar mechanism was fixedly connected with the gears on both sides of the cylindrical gear mechanism 3 and 5, separately. The cavitator was fixedly connected with the connecting rod of the parallel four-bar mechanism.

**Figure 1.** Structural diagram for the well-cellar cavitating mechanism: 1. non-circular gear 1; 2. non-circular gear 2; 3. spur gear 3; 4. spur gear 4; 5. spur gear 5; 6. crank; 7. linkage; 8. frame; 9. cavitating device; 10. cavitator.

The working principle of the cavitating mechanism is shown in Figure 2. During the operation, the power of uniform rotation was transmitted to the non-circular gear 1. The non-circular gear mechanism converted the uniform rotation into the variable-velocity rotation and transmitted it to the spur gear mechanism. The spur gear mechanism transmitted the variable velocity rotation to the double crank of the parallel four-bar mechanism and drove its connecting rod to oscillate at a variable velocity, and the cavitating device oscillated with the connecting rod. In the process of invading and extracting from the soil, the cavitator in the horizontal direction formed the velocity, which was similar to the forward velocity and opposite to the direction of the mechanism, so as to offset the forward velocity of the well-cellar cavitating mechanism and form the well-cellar cavitation with a uniform diameter. The cavitating device oscillated with the connecting rod and the power of the uniform rotation was transmitted to the cavitator through the flexible shaft. It drove the cavitating device to rotate at a constant velocity so as to compact the inner wall of the cavitation during the cavitating process. The shape of the well-cellar cavitation formed by the cavitating mechanism was approximately cylindrical, the depth of the well-cellar cavitation was 180~200 mm, and the diameter was 60~80 mm. The overall diameter of the well-cellar cavitation should be uniform, and its inner wall should be solid without the collapse.

### 2.2. Establishment of Bidirectional Coupling Model for Cavitating Mechanism Based on DEM and MBD

The bidirectional coupling of the discrete element method and multi-body dynamics could realize the interaction between complex mechanisms and discrete particles and has been widely used in the field of agricultural machinery [20,21]. The cavitating process of the cavitation under the action of the cavitating mechanism was the bidirectional interaction between the mechanism and the soil. Therefore, the method of the bidirectional coupling of DEM and MBD was used to analyze the cavitating process of the mechanism [22].

2.2.1. Multibody Dynamics Modeling

The multi-body dynamics model of the cavitating mechanism was established in the Recurdyn software, and the kinematic pairs between the components are set as shown in Table 1.

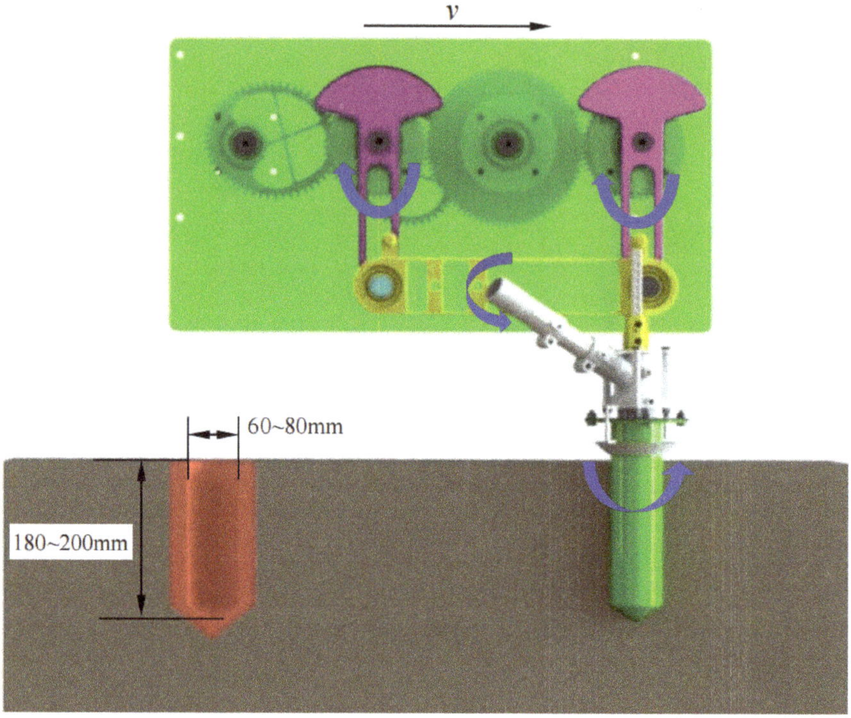


**Figure 2.** Working principle of the cavitating mechanism. Notes: $v$ represents the operating velocity of the mechanism.

**Table 1.** Kinematic pair configuration table of multi-body dynamics model.

| No. | Part 1 | Part 2 | Kinematic Pair |
|---|---|---|---|
| 1 | Frame | Ground | Moving pair |
| 2 | Crank | Frame | Rotary pair |
| 3 | Crank | Linkage | Rotary pair |
| 4 | Non-circular gear | Crank | Fixed pair |
| 5 | Non-circular gear 1 | Non-circular gear 2 | Contact pair |
| 6 | Spur gear 1 | Spur gear 2 | Contact pair |
| 7 | Spur gear 2 | Spur gear 3 | Contact pair |
| 8 | Non-circular gear 2 | Spur gear 1 | Fixed coupling pair |
| 9 | Cavitator | Powershuttle mechanism | Rotary pair |
| 10 | Powershuttle mechanism | Linkage | Fixed pair |

In the model, the forward velocity of the mechanism was set as 0.25 m/s, the rotary velocity of the non-circular 1 gear was set as 30 r/min, and the rotary velocity of the cavitating device was set as 500 r/min.

2.2.2. Discrete Element Modeling

Soil Particle Model

The basic structure of the soil particles for the plough layer mainly included block particles, core particles, flake particles, and columnar particles after tillage. In order to improve the authenticity of the model, spherical, linear, triangular, and tetrahedral particles were established in the discrete element software to approximate the massive, nuclear, flaky, and columnar soil particles, respectively [23,24].

The filling radius of the soil particles was set to 5 mm, and the model parameters of the soil are seen in Table 2 [25–27]. The soil particle parameters were obtained by referring to the relevant literature [28,29]. The soil moisture content and soil accumulation angle of the selected test site were measured. The basic parameters of the selected discrete element model are shown in Table 3.

**Table 2.** Model parameters of soil.

| Particle Model | | Radius | Contact Radius | Coordinate Position | | |
|---|---|---|---|---|---|---|
| | | | | X | Y | Z |
| Spherical | | 5 | 5 | 0 | 0 | 0 |
| Linear | Particle 1 | 5 | 5 | −4 | 0 | 0 |
| | Particle 2 | 5 | 5 | 0 | 0 | 0 |
| | Particle 3 | 5 | 5 | 4 | 0 | 0 |
| Triangular | Particle 1 | 5 | 5 | 0 | 2.2 | 0 |
| | Particle 2 | 5 | 5 | −2.2 | −1.5 | 0 |
| | Particle 3 | 5 | 5 | 2.2 | −1.5 | 0 |
| Tetrahedron | Particle 1 | 5 | 5 | 0 | 0 | 2.45 |
| | Particle 2 | 5 | 5 | 0 | 2.6 | −2.45 |
| | Particle 3 | 5 | 5 | −3 | −2.6 | −2.45 |
| | Particle 4 | 5 | 5 | 3 | −2.6 | −2.45 |

**Table 3.** Basic parameters of discrete element model.

| Parameters | Value |
|---|---|
| Poisson's ratio of soil | 0.35 |
| Soil particle density/(kg·m$^{-3}$) | 2550 |
| Shear modulus of soil particles/MPa | 1 |
| Coefficient of restitution between soil particles | 0.6 |
| Static friction factor between soil particles | 0.541 |
| Dynamic friction factor between soil particles | 0.31 |
| Poisson's ratio of steel | 0.3 |
| Density of steel/(kg·m$^{-3}$) | 7850 |
| Shear modulus of steel | $7.9 \times 10^{10}$ |
| Coefficient of restitution between soil and implements | 0.6 |
| Coefficient of static friction between soil and implements | 0.6 |
| Coefficient of dynamic friction between soil and implements | 0.6 |

Soil Particle Contact Model

The EEPA (Edinburgh Elasto-Plastic Adhesion) contact model includes the plasticity and viscosity of the particles and is suitable for simulating the farmland soils with a strong plasticity [30]. The soil particle contact model was set as the EEPA contact model, and the contact model between the cavitator and soil was set as the Hertz Mindlin with the JKR contact model with reference to the soil simulation routine of theEDEM 2020 (DEM Solutions Ltd., Edinburgh, Scotland, UK). According to the related literatures [31–33] on the discrete element parameters calibration of the physical characteristics of the soil particles, the configuration of the soil particles' simulation parameters is shown in Table 4.

**Table 4.** Contact model parameters of soil.

| Contact Model | Parameters | Value |
|---|---|---|
| Edinburgh Elasto-Plastic Adhesion contact model | Constant pull-off force/(N) | 0 |
| | Surface energy/(J/m$^2$) | 15.6 |
| | Contact plasticity ratio | 0.36 |
| | Slope exp | 1.5 |
| | Tensile exp | 4.24 |
| | Tangential stiff multiplier | 0.52 |

#### 2.2.3. Establishment of Multi-Body Dynamics-Discrete Element Bidirectional Coupling Model of Cavitating Mechanism

The bidirectional coupling interface between Recurdyn and EDEM was set up. On the basis of the multi-body dynamics model for the cavitating mechanism and the discrete element model of the cavitating soil, the bidirectional coupling model of Recurdyn–EDEM for the cavitating mechanism was constructed; the simplified model is shown in Figure 3.

**Figure 3.** Bidirectional coupling model of Recurdyn–EDEM for the cavitating mechanism: (**a**) front view of model in Recurdyn; (**b**) axonometric drawing of bidirectional coupling model; (**c**) bidirectional coupling model in EDEM.

### 2.3. *Orthogonal Simulation and Experimental Optimization*

In this section, through the multi-body dynamics-discrete element bidirectional coupling model between the cavitating mechanism and the soil, the regression equation between the optimization variables and optimization objectives of the cavitating mechanism was obtained in combination with the experimental design optimization method. On the basis of the regression equation, the multi-objective function was optimized to obtain the optimized combination of the parameters for the cavitating mechanism.

#### 2.3.1. Determination of Test Factors and Indicators

According to the kinematic equation, the crank's length $l_2$ of the parallel four-bar, the eccentricity $e$ of the non-circular gear, and the deformation coefficient $m_1$ of the non-circular gear in the kinematic equation were the significant factors which affected the quality of the cavitation. The standard of the cavitating quality was based on the results of the agronomic analysis for the cavitation, the depth of cavitation $y_1$, the vertical angle between the line connecting the top midpoint to the bottom midpoint of cavitation and the line of the ridge surface $y_2$, the maximum diameter of cavitation $y_3$, the minimum diameter of the cavitation $y_4$, the difference between the maximum and minimum value diameter of the cavitation $y_5$, and the variance in the diameters of the cavitation $y_6$ were used as the evaluation indicators [19]. In the above indicators, the perpendicularity for the cavitation was measured by the vertical angle of cavitation, the maximum diameter of cavitation, the minimum diameter of cavitation, the difference between the maximum and minimum diameter of cavitation, the variance in the diameters which measured the overall range of the diameter, and the consistency of the contour diameter for cavitation.

### 2.3.2. Test Contents and Methods

According to the single factor pre-test combined with the test index, the length range of the crank for the cavitating mechanism was 175~195 mm, the eccentricity of non-circular gear was 0.35~0.55, the deformation coefficient of the non-circular gear was 1.2~1.4, and the level of the factors was determined as five levels. On this basis, the optimized parameter combination of the cavitating mechanism was determined by using the bidirectional coupling model of Recurdyn–EDEM for the cavitating mechanism which was combined with the combination test method of the quadratic orthogonal rotation center. The diameter data of the 11 groups for the cavitation were measured through the bidirectional coupling model and the data were calculated to obtain the evaluation index. The test factor codes are shown in Table 5.

**Table 5.** Experimental factors and codes.

| Codes | Factors | | |
|---|---|---|---|
| | $x_1$ | $x_2$ | $x_3$ |
| | Length of Crank $l_2$/mm | Eccentricity of Non-Circular Gear $e$ | Deformation Coefficient of Non-Circular Gear $m_1$ |
| 1.682 | 195 | 0.55 | 1.40 |
| 1 | 191 | 0.51 | 1.36 |
| 0 | 185 | 0.45 | 1.30 |
| −1 | 179 | 0.39 | 1.24 |
| −1.682 | 175 | 0.35 | 1.20 |

### 2.3.3. Test Results and Analysis

The test scheme and results are shown in Table 6, the analysis of variance for each response index is shown in Table 7, and $x_1$, $x_2$, and $x_3$ were the factor coding values.

**Table 6.** Experimental plan and results.

| No. | Factors | | | Indexes | | | | | |
|---|---|---|---|---|---|---|---|---|---|
| | $x_1$ | $x_2$ | $x_3$ | $y_1$/mm | $y_2$/(°) | $y_3$/mm | $y_4$/mm | $y_5$/mm | $y_6$/mm$^2$ |
| 1 | −1 | −1 | −1 | 181.3 | 87.7 | 81.2 | 62.1 | 19.2 | 42.2 |
| 2 | 1 | −1 | −1 | 190.2 | 87.4 | 94.9 | 65.3 | 31.1 | 98.4 |
| 3 | −1 | 1 | −1 | 186.4 | 88.1 | 96.5 | 81.0 | 16.3 | 30.3 |
| 4 | 1 | 1 | −1 | 187.9 | 88.7 | 87.8 | 63.2 | 25.1 | 81.2 |
| 5 | −1 | −1 | 1 | 179.2 | 88.7 | 79.2 | 64.3 | 14.9 | 28.3 |
| 6 | 1 | −1 | 1 | 189.1 | 91.4 | 97.1 | 69.1 | 27.9 | 86.2 |
| 7 | −1 | 1 | 1 | 178.5 | 96.6 | 95.5 | 81.2 | 14.5 | 25.4 |
| 8 | 1 | 1 | 1 | 193.1 | 96.9 | 80.2 | 64.7 | 16.3 | 39.3 |
| 9 | −1.628 | 0 | 0 | 174.2 | 92.3 | 91.1 | 73.4 | 17.8 | 27.3 |
| 10 | 1.628 | 0 | 0 | 196.7 | 91.6 | 89.7 | 62.1 | 27.9 | 108.3 |
| 11 | 0 | −1.628 | 0 | 184.8 | 90.6 | 104.6 | 73.5 | 30.4 | 106.9 |
| 12 | 0 | 1.628 | 0 | 184.1 | 93.9 | 112.5 | 89.6 | 23.0 | 73.1 |
| 13 | 0 | 0 | −1.628 | 186.3 | 91.8 | 75.1 | 63.3 | 12.2 | 14.2 |
| 14 | 0 | 0 | 1.628 | 187.8 | 96.5 | 73.5 | 60.3 | 13.1 | 13.5 |
| 15 | 0 | 0 | 0 | 183.6 | 92.1 | 73.5 | 64.8 | 9.2 | 8.6 |

**Table 7.** Variance analysis of regression model.

| Testing Indicators | Sources of Variation | Sum of Squares | Freedom | F | $p$ | Testing Indicators | Sources of Variation | Sum of Squares | Freedom | F | $p$ |
|---|---|---|---|---|---|---|---|---|---|---|---|
| $y_1$ | Model | 446.55 | 9 | 14.65 | <0.0001 ** | $y_4$ | Model | 1139.46 | 9 | 62.22 | <0.0001 ** |
| | $x_1$ | 387.44 | 1 | 114.43 | <0.0001 ** | | $x_1$ | 150.29 | 1 | 73.86 | <0.0001 ** |
| | $x_2$ | 1.77 | 1 | 0.52 | 0.4819 | | $x_2$ | 232.73 | 1 | 114.38 | <0.0001 ** |
| | $x_3$ | 0.84 | 1 | 0.25 | 0.6277 | | $x_3$ | 0.52 | 1 | 0.25 | 0.6230 |
| | $x_1x_2$ | 0.91 | 1 | 0.27 | 0.6126 | | $x_1x_2$ | 223.66 | 1 | 109.92 | <0.0001 ** |
| | $x_1x_3$ | 24.85 | 1 | 7.34 | 0.0179* | | $x_1x_3$ | 1.05 | 1 | 0.52 | 0.4850 |
| | $x_2x_3$ | 0.031 | 1 | 0.01 | 0.9249 | | $x_2x_3$ | 2.31 | 1 | 1.14 | 0.3059 |
| | $x_1^2$ | 6.53 | 1 | 1.93 | 0.1881 | | $x_1^2$ | 6.94 | 1 | 3.41 | 0.0876 |
| | $x_2^2$ | 1.32 | 1 | 0.39 | 0.5439 | | $x_2^2$ | 487.69 | 1 | 239.68 | <0.0001 ** |
| | $x_3^2$ | 23.15 | 1 | 6.84 | 0.0214 * | | $x_3^2$ | 33.07 | 1 | 16.25 | 0.0014 ** |
| | Residual | 44.02 | 13 | | | | Residual | 26.45 | 13 | | |
| | Sum | 490.57 | 22 | | | | Sum | 1165.91 | 22 | | |
| $y_2$ | Model | 121.64 | 9 | 6.84 | 0.0011 ** | $y_5$ | Model | 1266.95 | 9 | 47.29 | <0.0001 ** |
| | $x_1$ | 0.33 | 1 | 0.17 | 0.6895 | | $x_1$ | 201.71 | 1 | 67.77 | <0.0001 ** |
| | $x_2$ | 31.22 | 1 | 15.80 | 0.0016 ** | | $x_2$ | 81.42 | 1 | 27.35 | 0.0002 ** |
| | $x_3$ | 64.17 | 1 | 32.47 | <0.0001 ** | | $x_3$ | 20.14 | 1 | 6.77 | 0.0219 * |
| | $x_1x_2$ | 0.28 | 1 | 0.14 | 0.7121 | | $x_1x_2$ | 25.56 | 1 | 8.59 | 0.0117 * |
| | $x_1x_3$ | 0.91 | 1 | 0.46 | 0.5090 | | $x_1x_3$ | 4.35 | 1 | 1.46 | 0.2482 |
| | $x_2x_3$ | 17.11 | 1 | 8.66 | 0.0114 * | | $x_2x_3$ | 1.20 | 1 | 0.4 | 0.5363 |
| | $x_1^2$ | 3.91 | 1 | 1.98 | 0.1892 | | $x_1^2$ | 345.97 | 1 | 116.23 | <0.0001 ** |
| | $x_2^2$ | 2.42 | 1 | 1.22 | 0.2887 | | $x_2^2$ | 577.26 | 1 | 193.94 | <0.0001 ** |
| | $x_3^2$ | 1.26 | 1 | 0.64 | 0.4389 | | $x_3^2$ | 17.85 | 1 | 6.00 | 0.0293 * |
| | Residual | 25.69 | 13 | | | | Residual | 38.70 | 13 | | |
| | Sum | 147.33 | 22 | | | | Sum | 1305.65 | 22 | | |
| $y_3$ | Model | 3075.00 | 9 | 75.22 | <0.0001 ** | $y_6$ | Model | 27,544.97 | 9 | 68.03 | <0.0001 ** |
| | $x_1$ | 2.01 | 1 | 0.44 | 0.5171 | | $x_1$ | 7271.36 | 1 | 161.63 | <0.0001 ** |
| | $x_2$ | 31.94 | 1 | 7.03 | 0.0199 * | | $x_2$ | 1349.26 | 1 | 29.99 | <0.0001 ** |
| | $x_3$ | 9.01 | 1 | 1.98 | 0.1826 | | $x_3$ | 401.81 | 1 | 8.93 | 0.0105 ** |
| | $x_1x_2$ | 386.42 | 1 | 85.07 | <0.0001 ** | | $x_1x_2$ | 303.81 | 1 | 6.75 | 0.0221 * |
| | $x_1x_3$ | 0.72 | 1 | 0.16 | 0.6970 | | $x_1x_3$ | 155.76 | 1 | 3.46 | 0.0855 |
| | $x_2x_3$ | 9.68 | 1 | 2.12 | 0.1681 | | $x_2x_3$ | 53.56 | 1 | 1.19 | 0.2950 |
| | $x_1^2$ | 451.78 | 1 | 99.46 | <0.0001 ** | | $x_1^2$ | 6125.08 | 1 | 136.15 | <0.0001 ** |
| | $x_2^2$ | 2193.45 | 1 | 482.88 | <0.0001 ** | | $x_2^2$ | 12,001.14 | 1 | 266.77 | <0.0001 ** |
| | $x_3^2$ | 2.06 | 1 | 0.45 | 0.5126 | | $x_3^2$ | 4.97 | 1 | 0.11 | 0.7448 |
| | Residual | 59.05 | 13 | | | | Residual | 584.82 | 13 | | |
| | Sum | 3134.05 | 22 | | | | Sum | 28,129.80 | 22 | | |

Notes, ** indicates that the difference is extremely significant ($p < 0.01$),* indicates that the difference is significant ($0.01 \leq p \leq 0.05$).

Regression analysis and factor variance analysis were carried out on the experimental data by Design-Expert 8.0 software (Stat-Ease Ltd., Minneapolis, MN, USA). After eliminating the insignificant items in the interactive and quadratic term, the regression equation between the index and the factor coding value was obtained:

$$y_1 = 183.60 + 5.33x_1 + 0.36x_2 - 0.25x_3 + 1.76x_1x_3 + 1.21x_3^2$$

$$y_2 = 92.15 + 0.16x_1 + 1.51x_2 + 2.17x_3 + 1.46x_2x_3$$

$$y_3 = 73.57 + 0.38x_1 + 1.53x_2 - 0.81x_3 - 6.95x_1x_2 + 5.33x_1^2 + 11.75x_2^2$$

$$y_4 = 64.84 - 3.32x_1 + 4.13x_2 + 0.19x_3 - 5.29x_1x_2 + 5.54x_2^2 - 1.44x_3^2$$

$$y_5 = 9.22 + 3.84x_1 - 2.44x_2 - 1.21x_3 - 1.79x_1x_2 + 4.67x_1^2 + 6.03x_2^2 + 1.06x_3^2$$

$$y_6 = 8.73 + 23.07x_1 - 9.94x_2 - 5.42x_3 - 6.16x_1x_2 + 19.63x_1^2 + 27.48x_2^2$$

In order to analyze the influence of each factor on each response index, the response surface was obtained by using Design-Expert 8.0 software, as shown in Figure 4.

Figure 4a was the influence of the crank's length and non-circular gear deformation coefficient on the depth of the cavitation at the zero level of other factors. The deformation coefficient for the non-circular gear was below the zero level; the depth of cavitation showed an upward trend with the increase in the crank's length. The deformation coefficient $m_1$ for the non-circular gear was above the zero level, the depth of cavitation showed an upward trend with the increase in the crank's length, and the upward trend was significantly larger than that of the former. The reason was that the longer the crank's length, the lower the position of the cavitator in the vertical direction, and the greater the depth of the cavitation. The deformation coefficient for the non-circular gear changed the velocity of the cavitator in the horizontal direction, resulted in the change in the angle for the cavitation, and then affected the vertical depth of the cavitation. The larger the deformation coefficient for the non-circular gear, the smaller the influence on the verticality of the cavitation, and the greater the vertical depth of the cavitation. The crank's length was below the zero level; the depth of the cavitation decreased with the rise of the non-circular gear deformation coefficient. The reason was that the crank's length changed below the zero level, the interaction between the non-circular gear deformation coefficient and the crank's length reduced the velocity of the cavitator in the horizontal direction, and then increased the inclined angle for the cavitation, which resulted in the decrease in the vertical depth for the cavitation. The crank's length was above the zero level; the length of the crank increased with the rise of the non-circular gear deformation coefficient. The reason was that the crank's length changed above the zero level, the interaction with the deformation coefficient of the non-circular gear continuously increased the velocity of the cavitator in the horizontal direction and gradually approached the forward velocity of the mechanism so that the inclined angle of the cavitation was continuously reduced and the vertical depth of the cavitation was increased.

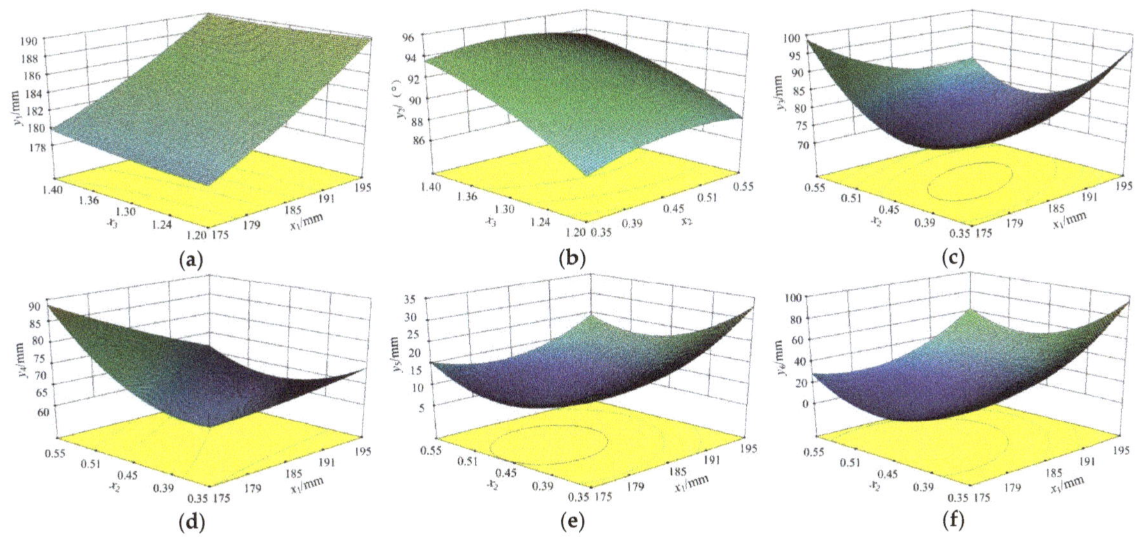


**Figure 4.** Response surface of factor to test index: (**a**) influence of crank's length and non-circular gear deformation coefficient on the depth of cavitation at the zero level of other factors; (**b**) influence of the deformation coefficient and the eccentricity for the non-circular gear on the vertical angle of cavitation at the zero level of other factors; (**c**) influence of crank's length and non-circular gear eccentricity on the maximum diameter of cavitation at the zero level of other factors; (**d**) influence of crank's length and non-circular gear eccentricity on the minimum diameter of cavitation at the zero level of other factors; (**e**) influence of crank's length and non-circular gear eccentricity on the difference between the maximum and minimum diameter of cavitation at the zero level of the other factors; (**f**) influence of the length for the crank and the eccentricity for the non-circular gear on the variance in diameters for cavitation at the zero level of other factors.

Figure 4b was the influence of the deformation coefficient and the eccentricity for the non-circular gear on the vertical angle between the line connecting the top midpoints to the bottom midpoint of cavitation and the line of the ridge surface at the zero level of the other factors. The deformation coefficient for the non-circular gear was below the zero level and the vertical angle of the cavitation and the line of the ridge surface showed a significant upward trend with the increase in the eccentricity for the non-circular gear. The deformation coefficient for the non-circular gear was above the zero level and the vertical angle of cavitation and the line of the ridge surface showed an upward trend significantly with the increase in the eccentricity of the crank's non-circular gear. The eccentricity of the non-circular gear was below the zero level and the vertical angle of cavitation and the line of the ridge surface showed an upward trend with the increase in the deformation coefficient for the non-circular gear. The eccentricity of the non-circular gear was above the zero level and the vertical angle of cavitation showed the upward trend with the increase in the deformation coefficient of the non-circular gear. The reason was that the increase in the deformation coefficient and the eccentricity for the non-circular gear constantly changed the velocity of the cavitator in the horizontal direction, which resulted in the continuous increase in the vertical angle of cavitation.

Figure 4c,d were the influence of the crank's length and non-circular gear eccentricity on the maximum and minimum diameter of cavitation at the zero level of the other factors. The eccentricity for the non-circular gear was above the zero level, the maximum and minimum diameters of cavitation both showed the downward trend with the increase in the crank's length. The non-circular gear coefficient was below the zero level and the maximum and minimum diameters of cavitation showed a rising trend with the increase in the crank's length. The reason was that the increase in the crank's length led to the increase in the difference between the velocity of the crank in the horizontal direction and the forward velocity of the mechanism, which made the maximum and minimum value of the cavitation diameter show a rising trend. With the increasing eccentricity for the non-circular gear (above the zero level), the influence on the velocity of the crank's length in the horizontal direction was increasing, which was reducing the velocity of the crank's length in the horizontal direction constantly, so that the maximum and minimum value of the cavitation diameter showed a rising trend. The length of the crank was above the zero level and the maximum and minimum diameter of cavitation showed a downward trend first and then an upward trend with the increase in the coefficient for the non-circular gear. The crank's length was below the zero level and the maximum diameter of the cavitation diameter showed a rising trend with the increase in the crank's length. The reason was that the increase in the non-circular gear eccentricity led to the increase in the difference between the horizontal velocity of the crank and the forward velocity of the mechanism, which made the maximum and minimum diameter of the cavitation diameter show a rising trend. With the increasing length of the crank (above the zero level), the increase in the eccentricity for the non-circular gear led to the difference between the velocity of the crank in the horizontal direction and the forward velocity of the mechanism, which decreased first and then increased, so that the maximum and minimum diameter of cavitation showed a downward trend first and then an upward trend.

The eccentricity of the non-circular gear was below the zero level, which resulted in there being less influence on the velocity of the crank's length in the horizontal direction, and the diameter of cavitation showed an upward trend. As the eccentricity of the non-circular gear increased continuously (above the zero level), the influence on the velocity of the crank's length in the horizontal direction also increased continuously, which reduced the velocity of the crank's length in the horizontal direction, and the diameter of cavitation showed a downward trend. The eccentricity of the non-circular gear was above the zero level and the maximum and minimum diameter of cavitation showed a downward trend with the increase in the crank's length. The non-circular gear coefficient was below the zero level and the maximum and minimum diameter of cavitation showed an upward trend with the increase in the crank's length. This was because the velocity of the crank's length $l_2$ in the horizontal direction increased and the difference between the velocity of the crank's length and the velocity of the forward direction of the mechanism also increased, which resulted in the increase in the diameter for the formed cavitation. The eccentricity of the non-circular gear was below the zero level, the effect on the velocity of the crank's length in the horizontal direction was less, and the diameter of cavitation showed an upward trend. As the eccentricity of the non-circular gear increased continuously (above the zero level), the influence on the velocity of the crank's length in the horizontal direction increased continuously, which reduced the velocity of the crank in the horizontal direction, and the diameter of cavitation showed a downward trend.

Figure 4e was the influence of the crank's length and non-circular gear eccentricity on the difference between the maximum and minimum diameter of cavitation at the zero level of the other factors. The eccentricity for the non-circular gear was constant and the difference between the maximum and minimum diameter of cavitation showed a downward trend first and then an upward trend with the increase in the length of the crank. The length of the crank was constant and the difference between the maximum and minimum diameter of cavitation showed a downward trend first and then an upward trend with the increase in the eccentricity of the non-circular gear. It meant that the crank's length and the eccentricity of the non-circular gear were near the zero level and the velocity of the cavitating mechanism in the horizontal direction was close to the forward velocity of the mechanism, so as to minimize the difference between the maximum and minimum diameter of cavitation. The eccentricity of the non-circular gear was below the zero level and the maximum and minimum diameter of cavitation showed an upward trend with the increase in the crank's length. The crank's length and the eccentricity of the non-circular gear was near the zero level, the velocity of the cavitating mechanism in the horizontal direction was closed to the forward velocity of the mechanism, and the difference between the maximum and minimum value of the hole diameter was the least.

Figure 4f was the influence of the length for the crank and the eccentricity for the non-circular gear on the variance in the diameters for cavitation at the zero level of the other factors. The length of the crank was constant and the variance in the diameters for cavitation showed a downward trend first and then an upward trend with the increase in the eccentricity $e$ of the non-circular gear. The eccentricity of the non-circular gear was constant and the variance in the diameters for cavitation showed a downward trend first and then an upward trend with the increase in the crank's length. It meant that the change in the crank's length and non-circular gear eccentricity led to the fluctuation of the difference between the velocity of the cavitator in the horizontal direction and the forward velocity of the mechanism, which resulted in a large change in the diameter of each part of the cavitation and increased the variance in the diameters for cavitation.

#### 2.3.4. Parameters Optimization

In order to obtain the best level combination of the experimental factors, the experimental factors of the optimal design were carried out, and the mathematical model of the parameter's optimization was established. The optimal target of the evaluation index was determined according to the agronomic requirements for the well-cellar cavitation with the depth of 180~200 mm, diameters of 60~80 mm, uniform overall diameter, and a high verticality. For the combination with the boundary conditions of the experimental factors, the regression equation of the response index for the cavitating mechanism was analyzed to obtain the optimization model of the nonlinear programming as follows:

$$\begin{cases} 180\text{ mm} \le y_1 \le 185\text{ mm} \\ 85^\circ \le y_2 \le 95^\circ \\ 60\text{ mm} \le y_3 \le 80\text{ mm} \\ 60\text{ mm} \le y_4 \le 80\text{ mm} \\ \min y_5 \\ \min y_6 \\ \text{s.t.} \begin{cases} 175\text{ mm} \le l_1 \le 195\text{ mm} \\ 0.35 \le e \le 0.55 \\ 1.2 \le m_1 \le 1.4 \end{cases} \end{cases}$$

This section used the Design-Expert 8.0 software optimization module to optimize the multi-objective parameters of the regression equation. The optimized parameters of the cavitating mechanism were obtained, the length of the crank $l_2$ was 185.71, the eccentricity $e$ of the non-circular gear was 0.456, and the deformation coefficient $m_1$ was 1.2857.

## 3. Results and Analysis

The bidirectional coupling model of Recurdyn and EDEM was used to simulate the cavitating mechanism under the optimal parameters. The cavitating process of the cavitating mechanism is shown in Figure 5. The cavitator was driven by the cavitating mechanism to invade the soil and the cavitator rotated under the drive of the driving device to interact with the soil simultaneously.

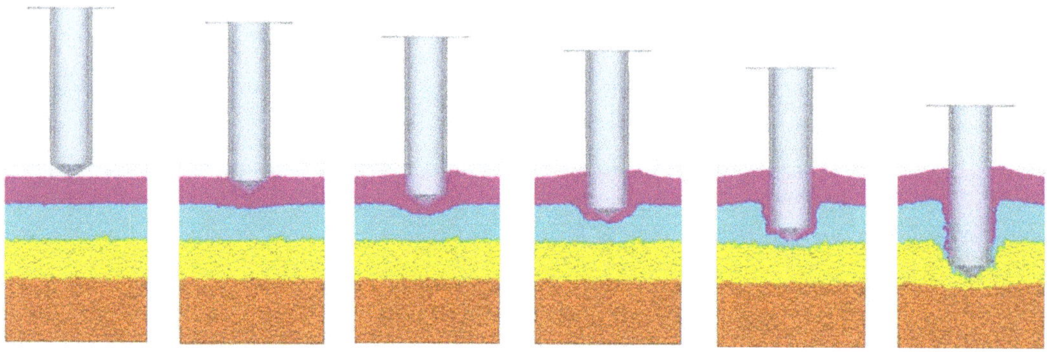

(**a**)

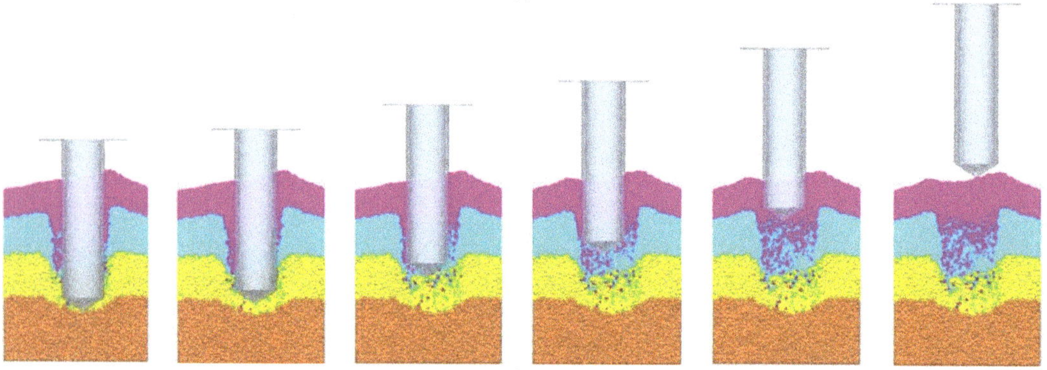

(**b**)

**Figure 5.** Cavitating process of the cavitating mechanism: (**a**) process of the cavitator starting to interact with the soil driven by the cavitating mechanism until reaching the maximum depth of the cavitation; (**b**) process of the cavitator extracting from the cavitation driven by the cavitating mechanism after the cavitator reached the maximum cavitating depth.

Figure 5a showed the process of the cavitator starting to interact with the soil until it reached the maximum depth of cavitation. As there was lack of the compaction and restriction of the other soil layers on the top soil layer, the gap between the soil particles was large, but the force between the soil particles was opposite. After the cavitator invaded the soil, the soil of the top layer near the cavitator rose, and the rotation of the cavitator would rotate and squeeze the soil particles, which made the soil particles generate the centrifugal force. Under the action of the centrifugal force, the soil particles on the top layer were thrown out and compacted to leave the cavitator, which resulted in a larger cavitating diameter for the soil of the top layer. With the increase in the cavitating depth, the force between the soil particles gradually increased, and the diameters of cavitation gradually decreased until it was the same as the diameter of the cavitator.

Figure 5b shows the process of the cavitator extracting from the cavitation driven by the cavitating mechanism after the cavitator reached the maximum cavitating depth. As the cavitator extracted from the cavitation, the velocity of the endpoint for the cavitator in the horizontal direction could not offset the forward velocity of the cavitating mechanism completely, which resulted in the small forward movement of the cavitator. At this time, the rotating cavitator acted on the front side of the cavitation, which changed the shape of the hole for a well-cellar cavitation from the circular to the approximate elliptical, as shown in Figure 6a. As the cavitator only acted on the front side of the well-cellar cavitation, the soil of the top layer on the right side for the well-cellar cavitation was higher than that of the left side. The sectional shape of the well-cellar cavitation is shown in Figure 6b.

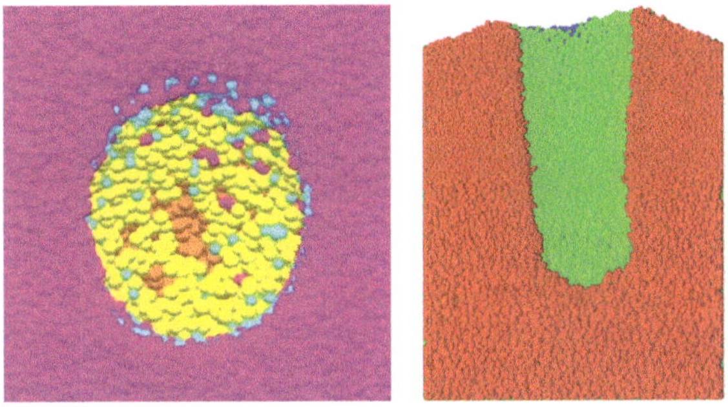

(a) (b)

**Figure 6.** Three-dimensional shape of the well-cellar cavitation: (**a**) shape of the hole for the well-cellar cavitation; (**b**) sectional shape of the well-cellar cavitation.

According to the optimal parameters, the discrete element model of cavitation was obtained through the bidirectional coupling model of the cavitating mechanism and the soil (see Figure 6). The measuring tools in the EDEM software were used to measure the parameters of cavitation, as shown in Table 8.

**Table 8.** Parameter comparison before and after optimization of mechanism parameters.

| | $y_1$/mm | $y_2$/(°) | $y_3$/mm | $y_4$/mm | $y_5$/mm | $y_6$/mm$^2$ |
|---|---|---|---|---|---|---|
| Before optimization | 180.8 | 90.3 | 71.8 | 64.5 | 7.3 | 6.05 |
| After optimization | 188.6 | 90.4 | 76.1 | 68.6 | 7.5 | 5.42 |

The depth of the cavitation $y_1$ after the optimization increased by 4.3% compared with that obtained before the optimization; the vertical angle of the cavitation $y_2$ before and after the optimization was basically unchanged; the maximum diameter of the cavitation $y_3$ after the optimization increased by 5.9% compared with that obtained before the optimization; the minimum diameter of the cavitation $y_4$ after the optimization increased by 6.3% compared with that obtained before the optimization; the difference between the maximum and minimum diameter of the cavitation $y_5$ before and after the optimization was basically unchanged; and the variance in the diameter for the cavitation $y_6$ decreased by 10.4% compared with that obtained before the optimization. The cavitating mechanism with optimal parameters based on the Recurdyn–EDEM bidirectional coupling mode could further improve the quality of the cavitation.

## 4. Conclusions

The work reported in this article established the multi-body dynamics and discrete element bidirectional coupling model of the cavitating mechanism and soil through the software of Recurdyn and EDEM. With the aid of the model for the cavitating mechanism and soil, combined with the quadratic orthogonal rotation center combination test method, the regression equation between the parameters of the cavitating mechanism and cavitation was established through the Design Expert 8.0 software, and the influence trend and interaction relationship between the parameters of the cavitating mechanism and the cavitation were obtained through the response surface. On the basis of the regression equation, the optimal parameters combination of the cavitating mechanism was obtained by using multi-objective function optimization. According to the optimal parameters combination of the cavitating mechanism, the multi-body dynamics and discrete element bidirectional coupling model of the cavitating mechanism and soil was used for virtual simulation tests, and the cavitating law of the mechanism was analyzed. The test results showed that the cavitating mechanism with optimal parameters based on the Recurdyn–EDEM bidirectional coupling mode could further improve the cavitating quality.

**Author Contributions:** Conceptualization, G.X. and H.F.; data curation, G.X. and Y.S.; formal analysis, W.D.; investigation, G.X. and Y.S.; methodology, H.F. and Y.S.; project administration, H.F.; resources, G.X. and H.F.; software, G.X. and W.D.; supervision, G.X. and Y.S.; validation, G.X., H.F. and W.D.; visualization, G.X. and W.D.; writing—original draft, G.X. and H.F.; writing—review and editing, Y.S. and W.D. All authors have read and agreed to the published version of the manuscript.

**Funding:** This research was supported by the Optimal design and experiment of parallel Combination-planting mechanism (202211510017), which is the National College Students Innovation and Entrepreneurship Training.

**Institutional Review Board Statement:** Not applicable.

**Data Availability Statement:** The data presented in this study are available on demand from the first author at (202107@sdjtu.edu.cn).

**Conflicts of Interest:** The authors declare no conflict of interest.

## References

1. Qiu, X.; Gao, W.; Chen, W.; Liang, G.; Tu, Y. Effects of Soil Temperature and Humidity on Early Stage Growth and Development of Tobacco Plants by Well-Cellar Transplanting. *Tobacco Sci. Technol.* **2015**, *48*, 20–24+35.
2. Iqbal, M.; Islam, M.; Ali, M.; Kabir, M.; Park, T.; Kang, T.; Park, K.; Chung, S. Kinematic analysis of a hopper-type dibbling mechanism for a 2.6 kW two-row pepper transplanter. *J. Mech. Sci. Technol.* **2021**, *35*, 2605–2614. [CrossRef]
3. Cheng, Y.; Tan, X.; Liu, Q.; Yang, C.; Shi, P.; Jia, M.; Zhang, H.; Wang, J. Effects of Different Seeding Height on Growth, Yield and Quality of Tobacco under Well-Cellar Transplanting. *Shandong Agric. Sci.* **2019**, *51*, 44–48.
4. Yu, L.; Xu, Z.; Han, Z.; Pan, D.; Zhang, F. Insertion trajectory analysis and experiment of rape combine harvester reel. *J. Agric. Mech. Res.* **2020**, *42*, 184–188.
5. Liu, L.; Zhang, Y.; Cao, D.; Gu, S.; Jiao, S.; Yan, Y.; Leng, P. Current situation and development suggestions of facility pepper industry in linyi city. *China Fruit Veg.* **2019**, *39*, 49–51.
6. Wen, Y.; Zhang, L.; Huang, X.; Yuan, T.; Zhang, J.; Tan, Y.; Feng, Z. Design of and Experiment with Seedling Selection System for Automatic Transplanter for Vegetable Plug Seedlings. *Agronomy* **2021**, *11*, 2031. [CrossRef]
7. Yu, G.; Liao, Z.; Xu, L.; Zhao, P.; Wu, C. Optimization design and test of large spacing planetary gear train for vegetable pot-seedling planting mechanisms. *Trans. Chin. Soc. Agric. Mach.* **2015**, *46*, 38–44.
8. Hu, J.; Pan, J.; Zhang, C.; Zhang, S.; Fei, W.; Pan, H. Optimization design and experiment on planetary gears planting mechanism of self-propelled transplanting machine. *Trans. Chin. Soc. Agric. Mach.* **2018**, *49*, 78–86.
9. Sun, L.; Shen, J.; Zhou, Y.; Ye, Z.; Yu, G.; Wu, C. Design of non-circular gear linkage combination driving type vegetable pot seedling transplanting mechanism. *Trans. CSAE* **2019**, *35*, 26–33.
10. Zhou, M.; Xue, X.; Qian, M.; Yin, D. Optimal design and experiment of opotted vegetable seedling transplanting mechanism with punching hole. *Trans. CSAE* **2020**, *51*, 77–83.
11. Jin, X.; Du, X.; Yang, C.; Ji, J.; Wang, S.; Yan, H. Design and experiment on crank-chute planting mechanism of transplanting machine. *Trans. Chin. Soc. Agric. Mach.* **2016**, *47*, 83–90.

12. Zhou, M.; SHAN, Y.; Xue, X.; Yin, D. Theoretical analysis and development of a mechanism with punching device for transplanting potted vegetable seedlings. *Int. J. Agric. Biol. Eng.* **2020**, *13*, 85–92. [CrossRef]
13. Stubbs, S.; Colton, J. The Design of a Mechanized Onion Transplanter for Bangladesh with Functional Testing. *Agriculture* **2022**, *12*, 1790. [CrossRef]
14. Yu, L.; Liu, G.; Zhang, F.; Wu, X.; Hu, C. Test and analysis for performance on the knapsack well-cellar making machine for seedling transplanting. *J. Agric. Mech. Res.* **2018**, *40*, 141–145.
15. Li, M.; Yan, Y. Design to 2YZS self-propelled tobacco transplanting machine. *J. Agric. Mech. Res.* **2012**, *34*, 154–156+160.
16. Dong, X.; Yang, B.; Xu, Y.; Shao, F.; Na, M. Kinematic analysis and simulation of cavitation mechanism of tobacco seedling transplanter. *J. Agric. Mech. Res.* **2017**, *39*, 37–41.
17. Fan, Y.; Zhang, X.; Li, J.; Wang, D. Design and test of tobacco cellar-type transplanter. *J. Chin. Agric. Mech.* **2020**, *41*, 40–46.
18. Zhang, D.; Zhang, Y.; Zhang, Y.; Li, Z. Design and locomotion characteristics analysis of cavitation mechanism of intermittent mobile tobacco seedlings transplanter. *J. Agric. Mech. Res.* **2021**, *43*, 106–112.
19. Xu, G.; Jian, S.; Song, Y.; Fang, H.; Qiu, X.; Ming, X. Design and experiment of cellar cavitation mechanism for crops of hilly mountains transplanter. *Trans. Chin. Soc. Agric. Mach.* **2022**, *53*, 105–113+125.
20. Jin, C.; Liu, G.; Ni, Y.; Yang, T.; Wang, T.; Qi, Y. Design and experiment of header profiling mechanism for combine harvester based on MBD-DEM coupling. *Trans. Chin. Soc. Agric. Eng.* **2022**, *38*, 1–10.
21. Niu, C.; Xu, L.; Ma, S.; Yan, C.; Tan, H. Design and experiment of the horizontal brush roll soil clearing device for covering soil above the cold-proof cloth of furrow planting grape. *Trans. Chin. Soc. Agric. Eng.* **2021**, *37*, 301–306.
22. Ma, S.; Xu, L.; Yuan, Q.; Niu, C.; Wang, S.; Yuan, T. Development of automatic obstacle-avoiding grapevine cold-proof soil cleaners. *Trans. Chin. Soc. Agric. Eng.* **2020**, *36*, 1–10.
23. Wang, C.; Liu, C.; Li, H.; Wang, Q.; He, J.; Lu, C. Design and experiment of asymmetric large-small double discs ditching device. *Trans. Chin. Soc. Agric. Eng.* **2018**, *34*, 28–36.
24. Fang, H.; Ji, C.; Ahmed, A.; Zhang, Q.; Guo, J. Simulation analysis of straw movement straw-soil-rotary blade system. *Trans. Chin. Soc. Agric. Mach.* **2016**, *47*, 60–67.
25. Shi, L.; Zhao, W.; Sun, W. Parameter calibration of soil particles contact model of farmland soil in northwest arid region based on discrete element method. *Trans. Chin. Soc. Agric. Eng.* **2017**, *33*, 181–187.
26. Fang, H.; Ji, C.; Farman, A.; Guo, J.; Zhang, Q.; Chaudhry, A. Analysis of soil dynamic behavior during rotary tillage based on discrete element method. *Trans. Chin. Soc. Agric. Mach.* **2016**, *47*, 22–28.
27. Janda, A.; Ooi, J. DEM modeling of cone penetration and unconfined compression in cohesive solids. *Powder Technol.* **2016**, *293*, 60–68. [CrossRef]
28. Thakur, S.; Morrissey, J.; Sun, J.; Chen, J.; Ooi, J. Micromechanical analysis of cohesive granular materials using the discrete element method with an adhesive elasto-plastic contact model. *Granul. Matter* **2014**, *16*, 383–400. [CrossRef]
29. Pue, J.; Emidio, G.; Verastegui-Flores, R.; Bezuijen, A.; Cornelis, W. Calibration of DEM material parameters to simulate stress-strain behavior of unsaturated soils during uniaxial compression. *Soil Tillage Res.* **2019**, *194*, 104303. [CrossRef]
30. Wang, X.; Zhong, X.; Geng, Y.; Wei, Z.; Hu, H.; Geng, D.; Zhang, X. Construction and parameter calibration of the nonlinear elastoplastic discrete element model for no-tillage soil compaction. *Trans. Chin. Soc. Agric. Eng.* **2021**, *37*, 100–107.
31. Han, C.; Xu, Y.; You, J.; Zhang, J.; Yuan, P. Parameter optimization of opener of semi-automatic transplanter for watermelon seedlings raised on compression substrate. *Trans. Chin. Soc. Agric. Eng.* **2019**, *35*, 48–56.
32. Wang, X.; Hu, H.; Wang, Q.; Li, H.; He, J.; Chen, W. Calibration method of soil contact characteristic parameters based on DEM theory. *Trans. Chin. Soc. Agric. Mach.* **2017**, *48*, 78–85.
33. Dong, X.; Su, C.; Zheng, H.; Han, R.; Li, Y.; Wan, L.; Song, J.; Wang, J. Analysis of soil disturbance process by vibrating subsoiling based on DEM-MBD coupling algorithm. *Trans. Chin. Soc. Agric. Eng.* **2022**, *38*, 34–43.

agriculture

*Article*

# Prediction of Single Disc Seeding System Forces, Using a Semi-Analytical and Discrete Element Method (DEM) Considering Rotation Effects

**Ali Khosravani [1], Jacky M. A. Desbiolles [1], Chris Saunders [1], Mustafa Ucgul [1,2,*] and John M. Fielke [1]**

1 STEM, University of South Australia, Adelaide, SA 5000, Australia
2 Faculty of Science and Engineering, Southern Cross University, Lismore, NSW 2084, Australia
* Correspondence: mustafa.ucgul@scu.edu.au

**Abstract:** Disc seeders are commonly used in no-till farming systems, and their performance evaluation generally rely on expensive and time-consuming field experiments. Mathematical models can help speed up force-related evaluations and improve the understanding of soil-disc interactions, to assist the performance optimisation processes. Previous analytical force prediction models of disc blades have not accounted for the free rotation aspect of the disc blade. This paper develops an analytical force prediction model from the wide blade failure theory adapted to suit rotating flat disc blades operating at different sweep and tilt angles and compares predictions with Discrete Element Method (DEM) simulations. To validate the two models, experiments were performed on a remoulded sandy soil condition using a rotating flat disc set at two tilt angles of 0° and 20°, and four sweep angles of 6, 26, 45 and 90° the 3-dimensional force components of draught, vertical and side forces were measured. Results showed a higher coefficient of determination ($R^2$ = 0.95) was obtained with analytical model predictions compared to DEM predictions ($R^2$ = 0.85) for their agreement with the test results. It was found that both the developed analytical approach and the DEM model can be used to predict tillage forces at different sweep and tilt angles acting on a rotating flat disc blade.

**Keywords:** flat disc; analytical force prediction model; discrete element method (DEM); soil-tool interaction

**Citation:** Khosravani, A.; Desbiolles, J.M.A.; Saunders, C.; Ucgul, M.; Fielke, J.M. Prediction of Single Disc Seeding System Forces, Using a Semi-Analytical and Discrete Element Method (DEM) Considering Rotation Effects. *Agriculture* **2023**, *13*, 202. https://doi.org/10.3390/agriculture13010202

Academic Editor: Tao Cui

Received: 15 December 2022
Revised: 6 January 2023
Accepted: 10 January 2023
Published: 13 January 2023



## 1. Introduction

During the last two decades, Australian no-till farmers have increasingly adopted disc seeders due to their characteristics of low soil disturbance and low soil throw, which facilitates working at higher speeds and operating at narrow rows spacings with high amounts of residue [1,2]. However, the disc seeder soil-reaction forces, particularly the draught force, must be optimised to minimise power requirements and fuel consumption [1] at higher speeds. Designing and evaluating tillage tools are generally based on time and resource-intensive field tests, which can only be undertaken at certain times of the year. If the interaction between soil and disc can be modelled, a large majority of the field tests can be avoided.

Currently, there is no published analytical force prediction model for a flat rotating disc blade with a low sweep angle (3–8°) and tilt angle (0–20°), as used in Australian no-till disc seeders. The analytical force prediction models developed by [3–7] modelled spherical/conical disc blades with large sweep angles (15–80°) and tilt angles (up to 30°), as normally fitted on implements used for soil inversion and residue incorporation. Limited work has been conducted using DEM for modelling soil-disc interaction by [8,9].

In past work by the authors [10], the interaction between soil and disc was modelled using an analytical method and a discrete element method (DEM) model. For the analytical method, tests were first conducted in an outdoor soil bin which revealed that only the leading part of the disc (active part) was involved in generating soil failure. Based on this information and using the wide blade passive failure theory [11,12], a fixed/non-rotating circular blade, operating at 90° to the direction of travel at two tilt angles of 0 and 20°, was modelled to prove the suitability of the wide blade passive failure theory for modelling a flat circular disc [13,14]. In a second step, the authors extended the fixed blade model to account for the effect of different sweep (0 to 8°) and tilt angles (0 to 20°) relevant to the configurations of zero-tillage disc seeders as used on single disc openers. Although the model provided some helpful information at that stage, it was limited as it assumed that the disc did not rotate.

As an alternative approach, the interaction between soil and a fixed disc blade was also modelled using DEM, which was proven as an effective method when its parameters were accurately calibrated [15–19]. In this study, the modelling work of [10] is extended to predict the draught, vertical and side forces reactions applied onto a flat rotating disc blade. Both analytical and DEM predictions are compared to the experimental results.

## 2. Materials and Methods

### 2.1. Experimental Work

The tillage forces (draught, vertical and side forces) acting on a circular disc were measured in Tillage Test Track (TTT) at the University of South Australia. The TTT is a continuous outdoor soil bin. It has two straights of 50 m length joining together by two curves by 50 m diameter. The test soil was placed between rails (2.5 m wide by 0.3 m deep). In the TTT, a tractor tows two trolleys, each capable of tillage testing [20]. A hydraulic cylinder on the test frame was used to adjust the frame height to obtain the desired operating depth (Figure 1). The TTT had sandy loam soil (85% sand, 3% silt and 12% clay with no residue), and during testing, it had a 7% moisture content (dry basis). The soil was reconstituted by tilling the full width, levelling and consolidating it using a 2 t troller.

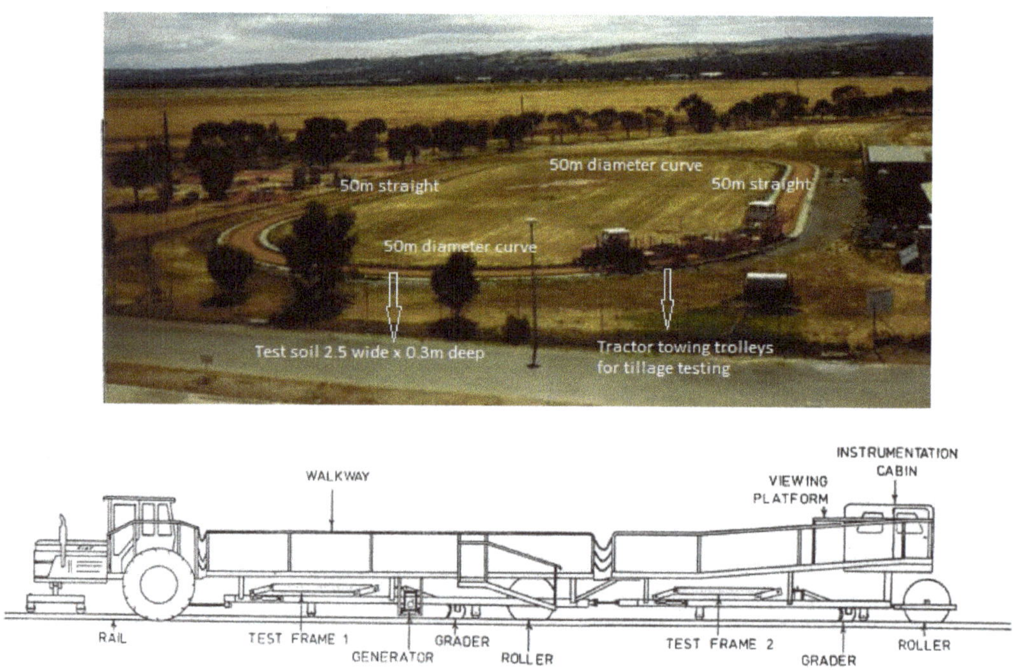


**Figure 1.** Tillage test track (TTT) rig adapted from [20].

To perform the tests, a flat disc blade of 460 mm diameter, 6 mm thickness and wedge angle of 16° was used in the tests. Experiments were performed at 67 mm operating depth (a typical seeding depth in Australia) and 2 km h$^{-1}$ forward speed. The operating speed of the disc seeders widely used in Australia is usually in the range of 12 to 15 km h$^{-1}$. However, this was undertaken to establish an analytical model based on the wide blade theory developed for quasi-static conditions. The lowest practical speed for testing of 2 km h$^{-1}$ was therefore selected to approximate quasi-static conditions and matched in the development of the analytical model. A test frame was developed that allowed setting different sweep (0 to 45°) and tilt (−20 to 40°) angles (Figure 2). The tillage forces were measured using a 3D dynamometer frame with 5 kN capacity S-type load cells. The data was logged at a frequency of 100 Hz using LabView TM V 7.1 software. Tests were performed for four sweep angles (6, 26, 45 and 90°) and two tilt angles (0 and 20°) using a randomised complete block design with three replications. Measurements were conducted over 40 m within the straight sections of the TTT (where a steady state condition was reached).

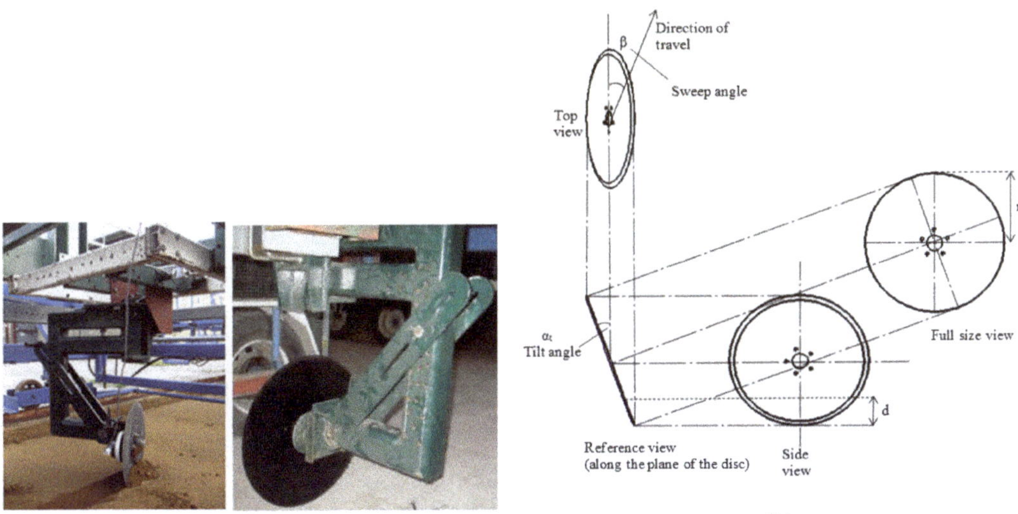


**(a)** **(b)**

**Figure 2.** (**a**) Force measurement frame fitted with adjustable disc mounting frame and (**b**) the definition of sweep and rake angles.

## 2.2. *Analytical Model Development*

The rotation of the disc was observed during testing to affect the direction of soil flow. Thus, the analytical model accounted for the change in soil flow by changing the direction of action of friction and resultant force. This affected the magnitude and direction of its components of draught, vertical and side forces.

### 2.2.1. To Determine the Active Part of the Disc

The active portion of the disc involved in soil failure was calculated to calculate the soil reaction forces onto the disc, as explained in [13]. Due to the difference in the *m* ratio (the ratio of forward rupture distance f to operating depth d, both measured in situ) between a fixed and free rotating disc at different sweep angles, the value of the active part of the disc, $L_a$ needed to be determined.

The active part of the disc ($L_a$) and ($\frac{L_a}{L}$) ratio was determined and is shown in Figure 3. As shown in Figure 3, the calculated ($\frac{L_a}{L}$) ratios at different sweep angles were similar to the measured values.

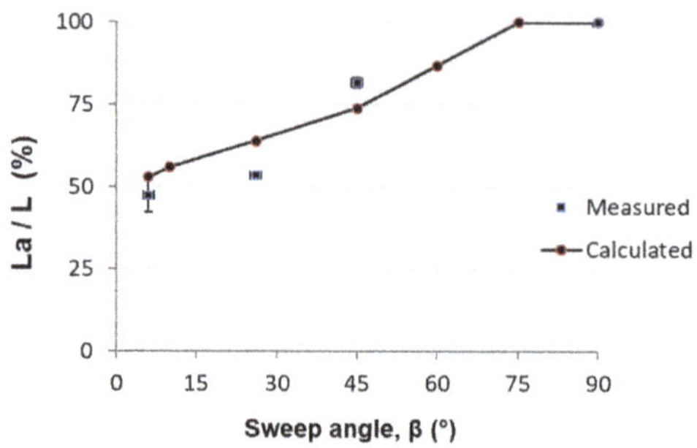


**Figure 3.** Comparison of calculated and measured ($\frac{L_a}{L}$) ratio at $\alpha = 90°$ ($\alpha$ is the rake angle).

### 2.2.2. Calculation of the Force

Once the active part of the disc, $L_a$, was determined, the method explained in [13,14] was used to calculate the forces at different sweep and tilt angles. The calculated forces at rake angles of $\alpha = 90°$ and 70° at 2 km h$^{-1}$ forward speed were compared with the measured forces with the difference shown in Table 1.

**Table 1.** Analytical force predictions compared to measured results.

| Sweep Angle β (°) | Force Difference with Rake Angle α = 70° | Force Difference with Rake Angle α = 90° |
|---|---|---|
| 6 | 41.2% | 28.3% |
| 26 | 2.4% | 14.6% |
| 45 | 6.3% | 5.9% |
| 90 | 3.9% | 1.0% |

The large difference between the predicted and measured force at a sweep angle of 6° can be explained by the rear side scrubbing reaction force, which was not accounted for by the model used in this work. Thus, the wide blade theory predicts well the force of a rotating disc blade only at the larger sweep angles.

Since the rotation of the disc affects the direction of soil flow and the direction of action of friction, the friction components of $\delta_x$ and $\delta_z$ were calculated and used to determine the force components of the rotating disc's draught, vertical and side forces.

### 2.2.3. Effect of Disc Rotation on the Direction of Soil Reaction Forces

With a free-rotating disc, the rotation of the disc changes the soil flow direction and thus affects the direction of friction at different sweep angles. The direction of friction alters the direction of the resultant force, P. To determine the effect of rotation on the direction of force, P, the direction of soil flow must be determined. The direction of overall soil movement over the disc face depends on the result of pressures applied on the disc surface due to sweep angle, disc rotation and the resistance of soil to compressive failure. It was assumed soil moves over the disc face as a rigid block. The direction of friction can be assumed as opposed to the direction of soil movement. Thus, the elemental friction forces act in the opposite direction to the velocity vectors [21]. To determine the overall soil movement direction, the relative movement of the soil to the disc, which is a function of both the forward and rotational speed of the disc, was determined. In order to determine the average soil-disc relative velocity, the soil–disc relative velocity at any point on the disc surface was calculated. A kinematic analysis of the disc [21] showed a considerable variation in the relative velocity over the soil-disc contact area. Therefore, the point of action of the resultant force on the soil-disc contact area was assumed to be the point at which the relative velocity is equal to the average relative velocity. That means the friction vector at this point is representative of the average friction vectors that apply to the soil-disc contact area.

#### 2.2.4. Method to Determine the Point of Action of force P on the Soil-Disc Contact Zone

In order to determine the point of action of the resultant force P on the soil-disc interface plane, the point of action of force P on the Z (vertical plane) and X (horizontal plane) axis was determined. The joining of these two points is the point of action of force P on the disc-soil contact area.

##### Method to Determine the Point of Action of Force P on a Vertical Plane

To determine the point of action of force P on a vertical plane, the value and point of action of any individual force acting on this plane were determined. For this, the percentages of the required force for soil failure in each element (e.g., the ratio of required force to overcome soil cohesion, surcharge, etc., to the total force, P) at different sweep angles were required. All applied forces on the active side of the disc–soil contact area are shown in Figure 4. It was assumed that the cohesion force, $P_c$, adhesion force, $P_{ca}$ and surcharge, $P_q$, act at half of the operating depth and the force related to soil weight and speed act at 2/3 of the depth [11].

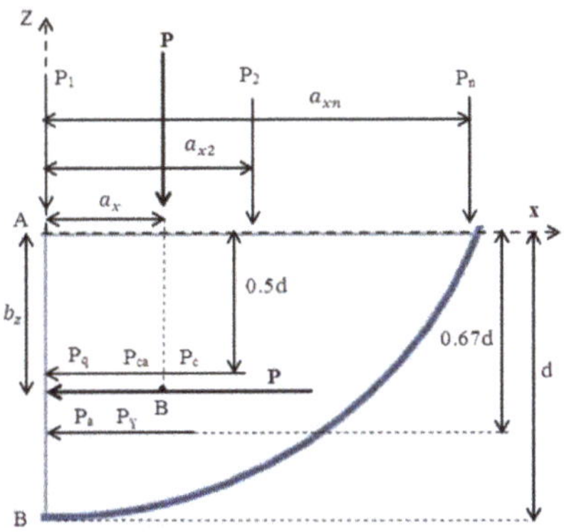


**Figure 4.** Horizontal and vertical applied force on the disc-soil contact area of the active side of the disc to determine the point of action.

The percentages of required force for each element were calculated from the force calculation equation [13]. Therefore, by knowing the value and point of action of any individual force, the point of action of a resultant elementary force ($P_i$) on a vertical plane was determined, as shown in Equation (1).

$$\frac{\left(P_c+P_{ca}+P_q\right)P_i}{2}*d_i+\frac{2(P_\gamma+P_a)P_i}{3}*d_i=P_i*b_{zi}$$
$$b_{zi}=\frac{3(P_c+P_{ca+}P_{q)+}4(P_\gamma+P_{a)}}{6}d_i \tag{1}$$

Since depth ($d_i$) is variable over the disc chord, the point of action of the resultant force P on a vertical plane ($b_z$), was calculated using Equations (2) and (3) and the integration equation described in [6] by replacing the r, d and L in the equation.

$$b_z=\frac{3(P_c+P_{ca+}P_{q)+}4(P_\gamma+P_{a)}}{6}*\frac{P_1d_1+P_2d_2+\cdots..P_nd_n}{P} \tag{2}$$

$$b_z=\frac{3(P_c+P_{ca+}P_{q)+}4\left(P_\gamma+P_{a)}A_1\int_0^{L_a}\left[\left(\sqrt{a^2-{L_a}^2}-(b-d)\right)^2\cdot d_i\right]dl+(A_{2+}A_3+A_{4+}A_5\right)\int_0^{L_a}\left[\left(\sqrt{a^2-{L_a}^2}-(b-d)\right)\cdot d_i\right]dl}{6P} \tag{3}$$

where $A_1 = \gamma_i \cdot N_\gamma,. A_2 = c \cdot N_c, A_3 = c_a \cdot N_{ca}\ , A_4 = q \cdot N_q, A_5 = \frac{\gamma_i}{g} \cdot N_a \sin \beta v^2$.

As $L_a$ > L/2, the above calculation was performed first by substituting $L_a$ with L/2 and then with ($L_a$ − L/2) as per Equation (4).

$$b_z = \frac{3\left(P_c + P_{ca+} P_q +\right) 4 \left(P_\gamma + P_a\right) \left[ A_1 \int_0^{\frac{L}{2}} \left[ \left( \sqrt{a^2 - L_a{}^2} - (b-d) \right)^2 \cdot d_i + \int_0^{\frac{L}{2} - L_a} \left( \sqrt{a^2 - L_a{}^2} - (b-d) \right)^2 \cdot d_i \right] dl + \right.}{6P}$$
$$\left(A_{2+} A_3 + A_{4+} A_5 \int_0^{\frac{L}{2}} \left[ \left( \sqrt{a^2 - L_a{}^2} - (b-d) \right) \cdot d_i + \int_0^{\frac{L}{2} - L_a} \left( \sqrt{a^2 - L_a{}^2} - (b-d) \right) d_i \right] \right] dl \tag{4}$$

To Determine the Point of Action of Force P on a Horizontal Plane

It was assumed that $P_1, P_2, \ldots . P_n$ are the forces acting at a distance $a_{x1}, a_{x2}, \ldots . a_{xn}$ from the Z axis. The Z axis was chosen as a reference axis. The point of action of the different applied forces ($P_1, P_2, \ldots . P_n$ ) on a horizontal plane (on X-axis direction, distance $a_x$) were determined by the summation of force times distance from the reference axis (Z) and dividing it by the total force P applied on the assumed plane, as shown in Equations (5)–(8).

$$P.a_x = \sum_0^{L_a} (P_i a_{xi}) \tag{5}$$

$$a_x = \frac{P_1 a_1 + P_2 a_2 + \cdots ..P_n a_n}{P} \tag{6}$$

$$a_x = \frac{A_1 \int_0^{L_a} \left[ \left( \sqrt{a^2 - L_a{}^2} - (b-d) \right)^2 . L_i \right] dl + (A_{2+} A_3 + A'_4 + A_5) \int_0^{L_a} \left[ \left( \sqrt{a^2 - L_a{}^2} - (b-d) \right) . L_i \right] dl}{P} \tag{7}$$

As $L_a$ > L/2, the above calculation was performed first by substituting $L_a$ by L/2 and then by ($L_a$ − L/2) as per Equation (8).

$$a_x = \frac{A_1 \int_0^{\frac{L}{2}} \left[ \left( \sqrt{a^2 - L_a{}^2} - (b-d) \right)^2 \cdot d_i + \int_0^{\frac{L}{2} - L_a} \left( \sqrt{a^2 - L_a{}^2} - (b-d) \right)^2 \cdot d_i \right] dl +}{P}$$
$$\left(A_{2+} A_3 + A_{4+} A_5 \int_0^{\frac{L}{2}} \left[ \left( \sqrt{a^2 - L_a{}^2} - (b-d) \right) \cdot d_i + \int_0^{\frac{L}{2} - L_a} \left( \sqrt{a^2 - L_a{}^2} - (b-d) \right) d_i \right] dl \tag{8}$$

Thus, point B, the joining point of action of forces P in X and Z directions (Figure 4), is the point of action of force P.

2.2.5. Method to Determine the Direction of Friction

Once the point of action of the resultant force P was determined (it was assumed an average resultant velocity act at this point), to determine the direction of soil-disc friction, it was required to calculate the resultant velocity at point B, as it was assumed that the elemental friction forces act in the opposite direction to the velocity vectors [21]. Therefore, a moving system of coordinates OXZ was used to study the kinematics of the disc [21]. By rotating the axis OX, through a sweep angle, β on a horizontal plane and rotating the axis OZ, through a tilt angle, $\alpha_t$ on a vertical plane, the new coordinate system of $OX_1Z_1$ will be formed (Figure 5). The axis OX is along with the direction of travel, and axis $OX_1$ and $OZ_1$ are in the plane of the disc cutting edge. The component of the resultant speed ($V_r$) in the X direction at point B is:

$$V_{rx} = V - V_t \cos(90^\circ - \theta) \tag{9}$$

$$V_t = V\lambda\ (r - (d - b_z))/rV_t \text{remains constant, but } V \text{ at OX1 become } V \cos\beta \tag{10}$$

$$V_{rx1} = V \cos\beta - (V_t \sin\theta) \tag{11}$$

$$V_{rz1} = V_t \cos\theta \tag{12}$$

$$\theta = \tan^{-1}\frac{AO}{AB} \tag{13}$$

$$V_r = \sqrt{V_{rx1}^{\ 2} + V_{rz1}^{\ 2}} \tag{14}$$

where $V$ = forward speed at the direction of travel, $V_t$ = tangential speed at point B, $V_r$ = resultant speed (absolute velocity), $\lambda$ = speed ratio from the experimental data.

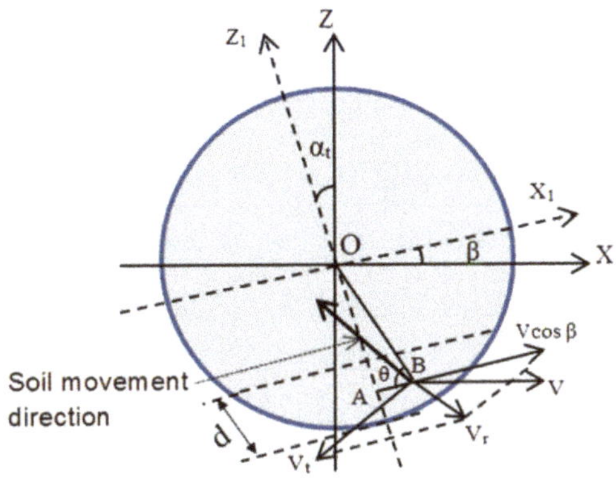


**Figure 5.** The coordinate system used to demonstrate velocity vectors and soil movement direction on the disc surface.

By substituting the value for $V$, β and θ, the resultant velocity was calculated. The measured data for a rotating disc at different sweep angles for λ (speed ratio) was used to calculate the tangential speed (Equation (10)).

By calculating the direction of average soil movement relative to the disc at different sweep angles, the angles $\lambda_x$ and $\lambda_z$ were determined (Figure 6) as follows:

$$\cos\lambda_x = \frac{V_{rx1}}{v_r} \tag{15}$$

$$\cos\lambda_z = \frac{V_{rz1}}{v_r} \tag{16}$$

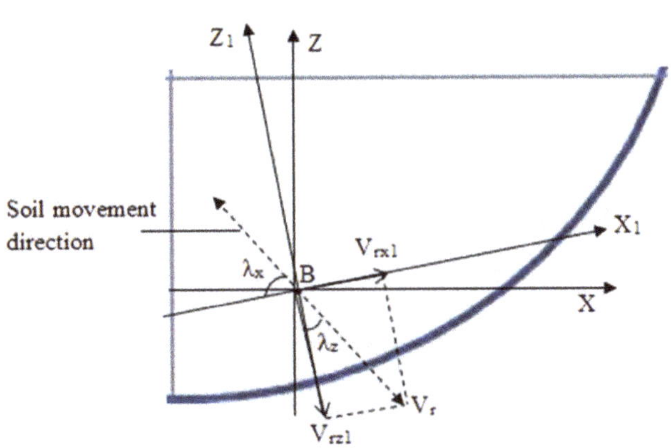


**Figure 6.** The direction of resultant soil movement friction.

The direction of force, P, that was assumed perpendicular to the disc face will be changed by the action of friction. In the wide blade theory, it is assumed that soil moves only upward on the disc face; therefore, friction only changes the direction of force P in a vertical plane, but for a rotating flat disc with a sweep angle, the soil moves with an angle relative to the horizontal and vertical planes (Figures 5 and 6). Friction has components in both vertical and horizontal directions (Figure 7). Therefore, it changes the direction of force, P, on both vertical (Z-axis) and horizontal planes (X-axis). The effect of friction on the orientation of force, P relative to a vertical plane, was determined by calculating the angle that force P makes with a vertical plane perpendicular to the direction of travel ($\delta_z$) (Figure 7a). The effect of friction on the orientation of force P relative to the horizontal plane was determined by calculating the angle that force P makes with a vertical plane parallel to the direction of travel ($\delta_x$) (Figure 7b). When the disc sweep angle is 90° the situation is the same as a wide blade and soil moves vertically without a horizontal movement, so $\delta_z = \delta$ and $\delta_x = \lambda_z = 0$ (Figure 6). When the disc sweep angle is zero, soil moves only on a horizontal plane, and no vertical movement occurs, so $\delta_x = \delta$, $\lambda_z = 90°$ and $\lambda_x = 0$. Hence, the sine function was used to show the $\lambda_z$ and $\lambda_x$ variation as follows:

$$\delta_x = \delta * \sin \lambda_z \text{ Or } \delta_x = \delta * \cos \lambda_x \tag{17}$$

$$\delta_z = \delta * \cos \lambda_z \tag{18}$$

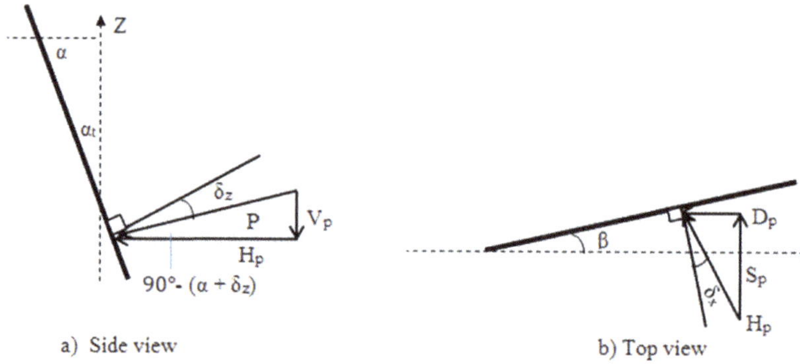


**Figure 7.** Partitioning of the draught $D_p$, vertical $V_p$ and side $S_p$ passive soil reactions acting on a disc at sweep angle β and tilt angle of $\alpha_t$.

By calculating the friction angle components on a vertical and horizontal plane, the direction of force P and their components (draught, vertical and side forces) at different sweep angles were determined.

2.2.6. Method to Determine the 3D Force Component of the Free-Rolling Disc

Figure 7 shows how the force P was partitioned into its draught, vertical and side force components at different sweep angles.

### *2.3. DEM Simulations*

In this study, the DEM simulations were carried out using a DELL Precision T5810 Intel® Xeon CPU E5-2687W v4 @ 3.00 GHz computer with the software EDEM 2020™. A linear cohesion integrated hysteretic spring contact model was suggested by [16] to model the soil–circular disc opener interactions. The hysteretic spring contact model accounts for the plastic deformation behaviour in the contact mechanics equations. Compressible materials such as soil can be modelled using the hysteretic spring contact model resulting in particles behaving in a linear elastic manner up to a predefined stress (Figure 8). When the total stress on the contact area exceeds the predefined compressive stress in the model, the particles behave as though they are undergoing plastic deformation [22]. On the other hand, the linear cohesion model allows users to add a cohesive force to the normal force direction with the ability to model soil cohesion.

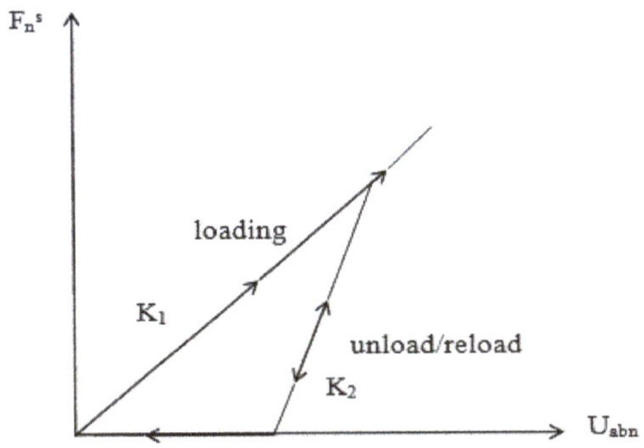


**Figure 8.** Hysteretic spring force-displacement relationship [14].

The hysteretic spring contact model calculates the total normal ($F_n$) and tangential ($F_t$) forces as;

$$F_n = F_n{}^s + F_n{}^d \tag{19}$$

$$F_t = F_t{}^s + F_t{}^d \tag{20}$$

$F_n{}^s$ was found as per [22];

$$F_n^s = -\begin{cases} K_1 \cdot U_{abn} & \text{loading} \\ K_2 \cdot (U_{abn} - U_0) & \text{unloading/reloading} \\ 0 & \text{unloading} \end{cases} \tag{21}$$

According to [23,24], $K_1$ *and* $K_2$ were calculated, as;

$$K_1 = 5\, r_{eq}\, min(Y_a, Y_b) \tag{22}$$

$$K_2 = K_1/e^2 \tag{23}$$

where $r_{eq}$ is [22],

$$1/r_{eq} = 1/r^*{}_a + 1/r^*{}_b \tag{24}$$

The residual overlap was updated in each time step as;

$$U_0 = \begin{cases} U_{abn} \cdot \left(1 - \frac{K_1}{K_2}\right) & \text{loading} \\ U_0 & \text{unloading/reloading} \\ U_{abn} & \text{unloading} \end{cases} \tag{25}$$

$F_t{}^s$, $F_n{}^d$, and $F_t{}^d$ were calculated as per [22];

$$F_t{}^s = -n_k\, K_1\, U_{abt} \tag{26}$$

$$F_n{}^d = -n_c\, (((4\, m_{eq}\, K_1)/(1+(\pi/\ln e)^2))\, \mathring{U}_{abn})^{-1/2} \tag{27}$$

$$F_t{}^d = -(((4\, m_{eq}\, n_k\, K_1)/(1+(\pi/\ln e)^2))\, \mathring{U}_{abt})^{-1/2} \tag{28}$$

where $m_{eq}$ is the equivalent mass and was defined in [22] as;

$$1/m_{eq} = 1/m^*_a + 1/m^*_b \tag{29}$$

$F_t$ was calculated using the following equation;

$$F_t = -\min\,(n_k\, K_1\, U_{abt} + F_t{}^d,\ \mu\, F_n{}^s) \tag{30}$$

$M$ and $M_r$ were computed as suggested by [25]:

$$M = r_{con}\, F_t \tag{31}$$

$$M_r = -\mu_r\, F_n{}^s\, r_{con}\, \lambda_\theta \tag{32}$$

The new position of a particle was updated by integrating the following equations;

$$\ddot{U} = (F_n + F_t)/m^* \tag{33}$$

$$R = (M + M_r)/I \tag{34}$$

The cohesion force was calculated as [22];

$$F_c = \xi\, A_c \tag{35}$$

As a result, Equation (18) becomes;

$$F_n = F_n{}^s + F_n{}^d + F_c \tag{36}$$

Using actual particulate sizes and shapes in 3D DEM simulations increases computation time (due to the high number of contacts) and computation costs. Therefore, particle sizes used in 3D DEM are generally selected larger than the actual particle sizes. In modelling soil, larger than actual spherical particle sizes are used to reduce the computation time. Hence, appropriate mechanical properties must be defined via calibration for the bulk behaviour of these larger particles. The DEM parameters used in this study are shown in Table 2. The DEM parameters of the coefficient of rolling friction and coefficient of restitution between soil and soil (a value less than 0.3 for compressible soil) were calibrated by matching simulation results to the actual measurement result of an angle of repose test.

**Table 2.** DEM parameters used for simulation.

| Property | Value | Source |
|---|---|---|
| Density of sand particles (kg $m^{-3}$) | 2600 | [26] |
| Density of steel (kg $m^{-3}$) | 7861 | [27] |
| Shear modulus of soil (Pa) | $5 \times 10^7$ | [28] |
| Shear modulus of steel (Pa) | $7.9 \times 10^{10}$ | [27] |
| Poisson's ratio of soil | 0.3 | [29] |
| Poisson's ratio of steel | 0.3 | [30] |
| Coefficient of restitution of soil-soil | 0.1 | Calibrated |
| Coefficient of friction of soil-soil | 0.7 | [31] |
| Coefficient of friction of soil-steel | 0.5 | [31] |
| Coefficient of rolling friction of soil-soil | 0.28 | Calibrated |
| Coefficient of rolling friction of soil-steel | 0.05 | [31] |
| Cohesive energy density between soil-soil (N $m^{-2}$) | 9000 | Measured using DST * |
| Adhesive energy density between soil-tool (N $m^{-2}$) | 3700 | Measured using DST * |

* DST = direct shear test.

In order to measure the angle of repose, a soil sample was placed in a pipe (100 mm diameter and 300 mm long). The pipe was then lifted upward (at around 500 mm $s^{-1}$). Soil flowed into a cylindrical tray (200 mm diameter and 22.5 mm high) until the soil overflowed and formed a pile. When at rest, the soil's angle of repose was measured using a digital level. The same test procedure was simulated in DEM. By using the calibrated parameters (Table 2), a static angle of repose of 29° (measured) was achieved in the simulation using a trial and error process (with an error margin of ±1° with 28.2° measured in DEM). The results of the trial and error process are shown in Table 3.

**Table 3.** Calibration results angle of repose simulation.

| Coefficient of Restitution | Coefficient of Rolling Friction | Angle of Repose (deg) |
|---|---|---|
| 0.3 | 0.3 | 30.1 |
| 0.1 | 0.3 | 29.9 |
| 0.1 | 0.32 | 31.4 |
| 0.1 | 0.28 | 28.2 * |
| 0.1 | 0.26 | 27.5 |
| 0.1 | 0.24 | 25.2 |

* selected values.

DEM simulations were undertaken using a virtual soil bin with 4000 mm length, 1500 mm width and 300 mm depth. A nominal 40 mm particle size (spherical) with the same particle size distribution of the test soil was used (359,188 particles) (Figure 9). The DEM bulk density was set to match the bulk density used in the experiments of 1860 kg $m^{-3}$ (dry basis). This was achieved by compressing the DEM particles using an upper physical plane until the bulk density of the virtual soil media reached the bulk density used in the field tests. The disc models were created using SolidWorks$^{TM}$ software and imported as a STEP file into the EDEM software. An operating depth of 67 mm and speed of 2 km $h^{-1}$ were used. As the circular disc is a passive-driven tool (as the soil's force drives its rotation on the disc), the rotational velocities measured during the TTT experiments (shown in Table 4) were used in the DEM simulations. In the future, the DEM simulation could be coupled with multibody dynamic (MBD) software to find the disc's rotational speed. From the simulations, the tillage forces were measured over the steady state section of the force vs. displacement graphs.

**Figure 9.** DEM simulation of the soil-disc interaction.

**Table 4.** Rotational speed measured in the TTT and used in the DEM simulations.

| Sweep Angle (°) | Rotational Speed with Tilt Angle = 0° (rpm) | Rotational Speed with Tilt Angle = 20° (rpm) |
|---|---|---|
| 6 | 24.22 | 21.98 |
| 26 | 22.6 | 19.38 |
| 45 | 18.45 | 17.07 |
| 90 | 0 | 0 |

## 3. Results

The relationships between the measured and predicted (both analytically and DEM) draught, vertical and side forces and sweep angles of the rotating disc are presented in Figures 10 and 11 for 0 and 20° tilt angles.

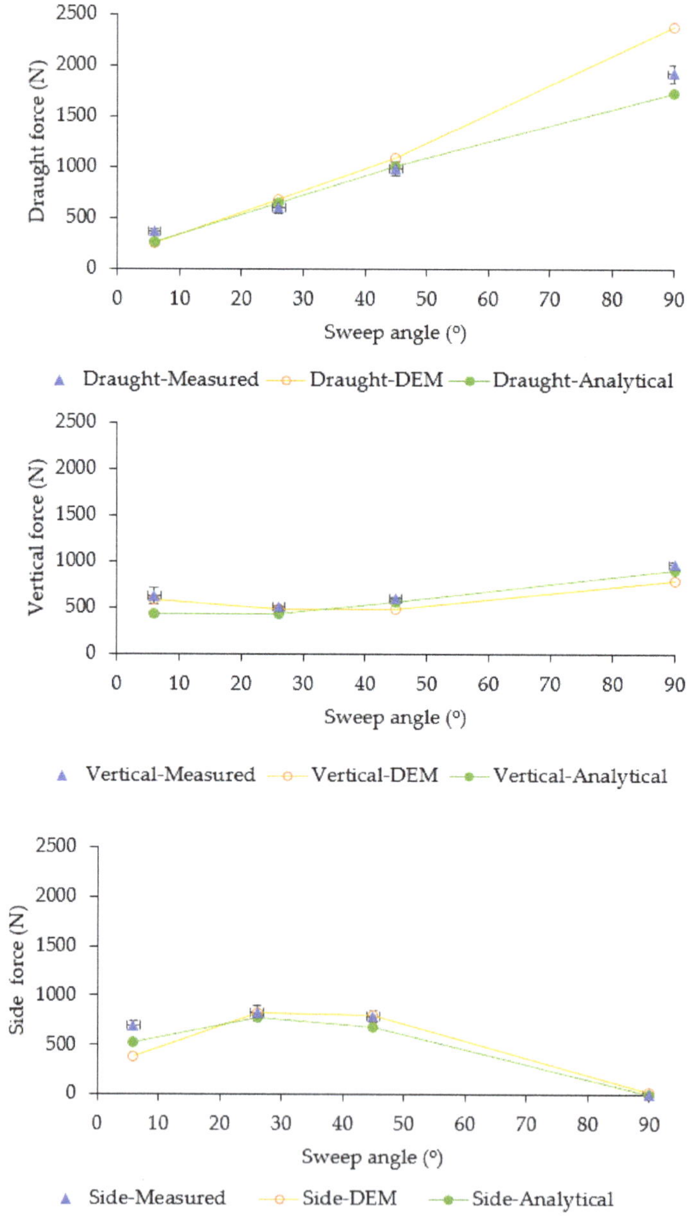


**Figure 10.** Comparison of measured analytically predicted and DEM predicted forces for a 0° tilt angle. The measured values include error bars for 1 standard deviation.

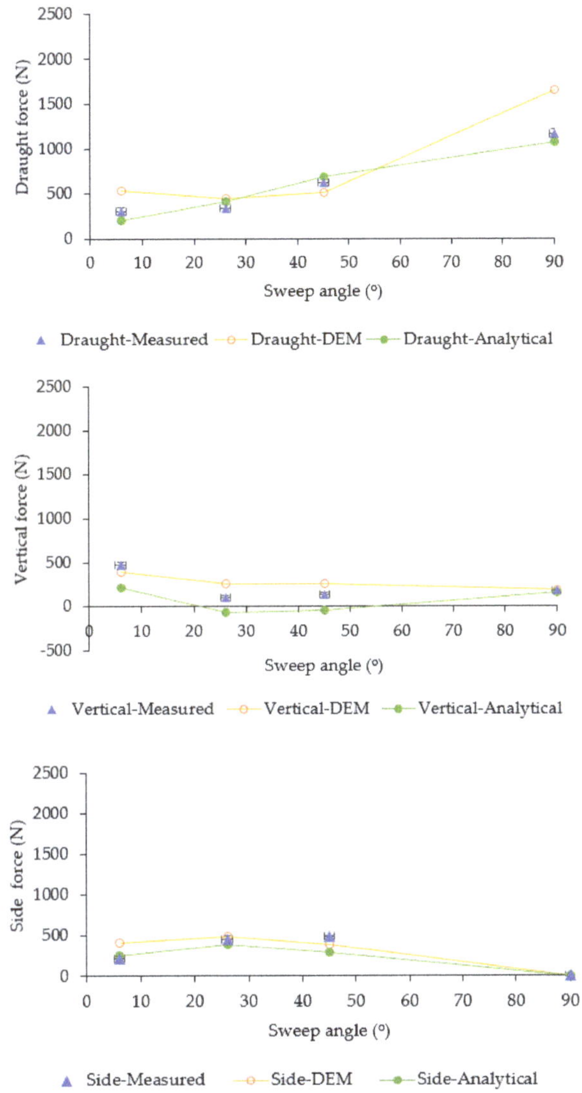


**Figure 11.** Comparison of measured analytically predicted and DEM predicted forces for a 20° tilt angle. The measured values include error bars for 1 standard deviation.

The test results showed that the accumulated soil in front of the disc at a higher sweep angle adds an additional force to the draught force while the side force first increased with the increase of sweep angle (up until 26°) and then reduced steeply as the disc pushes the soil forward rather than to the side. Also, the cosine function is reduced by increasing the angle. The vertical force decreased until the 26° sweep angle and then significantly increased with the increase of the sweep angle. Tilt angle has a major effect on soil reaction force, especially on the vertical force, as the tilted disc soil reaction force has downward components that help the disc penetrate the soil. Due to the close relationship between the rake angle and tilt angle in the disc, the tilt angle affects force partitioning in 3 dimensions. Overall, by tilting the disc 20°, the applied draught force was reduced by 19% at 6° and

40% at 90° sweep angle while the vertical force was decreased by 25% at 6° and 82% at a 90° sweep angle. Increasing the tilt angle to 20° also reduced the side forces by 70% at 6° and 38% at 90° sweep angles.

The analytically predicted draught, vertical and side force results showed a similar trend to the test results over the different sweep angles, but a large difference between the draught force component at $\beta = 90°$ and the side force component at $\beta = 6°$ was observed. The large difference between predicted and measured side force data at $\beta = 6°$ can be explained by scrubbing reactions which were not considered in this work. In comparison, the difference between the measured and predicted draught force $\beta = 90°$ can be attributed to the high amount of soil accumulated in front of the disc, which could not be accurately accounted for in the analytical model.

The DEM predicted draught, vertical and side forces of the vertical disc generally showed a good correlation with the measured data over the range of different sweep angles; however, DEM could not predict the draught forces at a 20° tilt angle well. When compared to the analytical results, a better coefficient of determination was obtained using analytical results simulations ($R^2 = 0.95$) than the DEM results ($R^2 = 0.85$) (Figure 12). The lower coefficient of determination for the DEM simulation results may be able to be improved by further refining the DEM calibrated parameters to better suit the larger than actual particles and using DEM-multibody dynamic coupling.

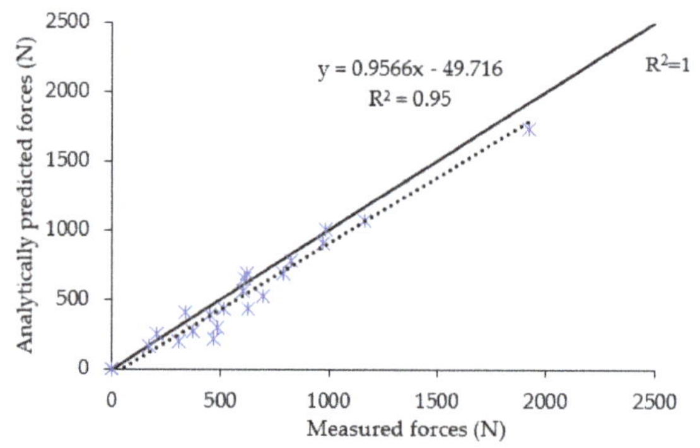


(**a**)

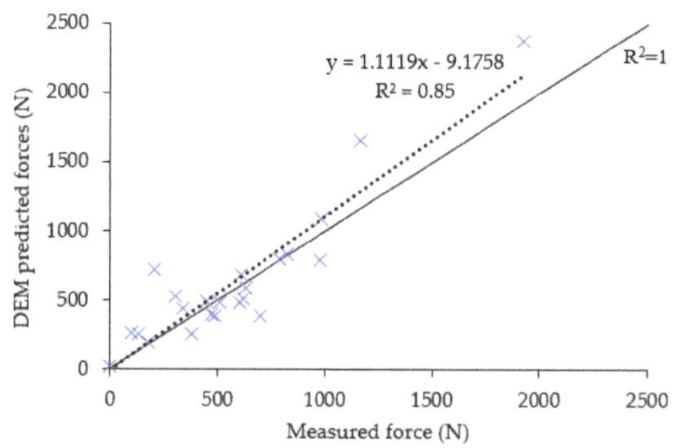


(**b**)

**Figure 12.** The whole of the data set correlations between the measured reaction forces and (**a**) analytical predictions and (**b**) DEM predictions (n = 24).

## 4. Conclusions

In this study, an analytical soil-force model initially developed for a fixed flat disc blade in [10] was extended to predict the forces for a rotating flat disc blade over a range of sweep and tilt angles. Its predictions were compared with experimental measurements, and results showed that this extended analytical model could accurately predict the soil reaction forces ($R^2 = 0.95$). The rotating flat disc blade interactions with a similar soil condition were carried out in simulated DEM space by assigning the disc blade rotational speed as a constant value. At this stage, the DEM could not predict the draught forces as accurately as the analytical model at the higher tilt angles. However, the benefits of DEM simulation include soil movement predictions and therefore can deliver more holistic results for optimising disc blade performance. Using multibody coupling to allow reactive simulations rather than assigning a fixed rotational velocity may improve accuracy and would also allow the DEM simulations to be extended beyond the quasi-static work conducted to date. Therefore, future research should focus on the multibody dynamic-DEM coupling to improve and extend the analysis. Beyond the current work conducted in low-moisture sandy-loam soil, it would be beneficial to undertake a further study into sticky soil conditions, a common challenge for most disc seeders.

**Author Contributions:** Conceptualisation, A.K., J.M.A.D. and J.M.F.; methodology, A.K. and J.M.A.D.; Software (DEM) assistance: M.U., C.S. and A.K.; validation, A.K.; formal analysis, A.K. and J.M.A.D.; writing—original draft preparation, A.K.; writing—review and editing, M.U., C.S., J.M.A.D. and J.M.F.; supervision, J.M.A.D. and J.M.F. All authors have read and agreed to the published version of the manuscript.

**Funding:** This research was funded by Grains Research and Development Corporation (GRDC) (Project Number: USA00005) and the University of South Australia (UniSA).

**Institutional Review Board Statement:** Not Applicable.

**Data Availability Statement:** The data presented in this study are available upon request from the first author A. K.

**Acknowledgments:** The authors gratefully acknowledge the Grains Research and Development Corporation (GRDC) and the University of South Australia (UniSA) in co-funding for this PhD research.

**Conflicts of Interest:** The authors declare no conflict of interest.

## Nomenclature

| | |
|---|---|
| $a$ | Indices for sphere or implement |
| $A_c$ | Contact area, (m$^2$) |
| $b$ | Indices for sphere or implement |
| $c$ | Cohesion (kPa) |
| $c_a$ | Adhesion (kPa) |
| $d$ | Blade working depth (m) |
| $D_p$ | Draught component of passive soil failure force (kN) |
| $e$ | Coefficient of restitution |
| $f$ | Forward rupture distance |
| $F_c$ | Cohesion force, (N) |
| $F^d{}_n$ | Normal damping force, (N) |
| $F^d{}_t$ | Tangential damping force, (N) |
| $F_n$ | Normal total contact force, (N) |
| $F^s{}_n$ | Normal contact force, (N) |
| $F^s{}_t$ | Tangential contact force, (N) |
| $F_t$ | Tangential total contact force, (N) |

| | |
|---|---|
| $I$ | Moment of inertia, (kg m$^2$) |
| $K_1$ | Stiffness for loading, (N m$^{-1}$) |
| $K_2$ | Stiffness for unloading/reloading, (N m$^{-1}$) |
| $L$ | Disc chord length measured at the soil surface (m) |
| $L_a$ | The active part of the disc chord that contributes to generating soil failure (m) |
| $M$ | Moment, (Nm) |
| $M_r$ | Moment due to rolling friction, (Nm) |
| $m$ | The ratio of forward rupture distance to the operating depth |
| $m^*$ | Mass, (kg) |
| $m_{eq}$ | Equivalent mass (kg) |
| $n$ | Number of replications |
| $n_c$ | Damping factor |
| $n_k$ | Stiffness factor |
| $p_i$ | Elemental passive force (kN) |
| $N_\gamma$, $N_c$, $N_{ca}$, $N_q$ | Dimensionless numbers associated with gravitational, cohesion, adhesion and surcharge effects (as per Hettiaratchi et al. (1966)) |
| $P$ | Passive cutting reaction force (kN) |
| $q$ | Total surcharge pressure acting on the soil failure area (kPa) |
| $r$ | Disc radius (m) |
| $r^*$ | Radius (m) |
| $r_{eq}$ | Equivalent radius, (m) |
| $r_{con}$ | Perpendicular distance of contact point from the centre of mass, (m) |
| $R$ | Rotational acceleration, (rad s$^{-2}$) |
| $S_p$ | Side force component of passive force (kN) |
| $U_{abn}$ | Normal component of the relative displacement, (m) |
| $U_{abt}$ | Tangential component of the relative displacement, (m) |
| $\mathring{U}_{abn}$ | Normal component of the relative velocity, (m s$^{-1}$) |
| $\mathring{U}_{abt}$ | Tangential component of the relative velocity, (m s$^{-1}$) |
| $U_0$ | Residual overlap, (m) |
| $\ddot{U}$ | Translational acceleration, (m s$^{-2}$) |
| $v$ | Forward speed (ms$^{-1}$) |
| $v_p$ | Passive soil failure velocity (m s$^{-1}$) |
| $V_p$ | Vertical component of passive force (kN) |
| $W_f$ | Furrow width(m) |
| $w$ | Tool width (m) |
| $Y$ | Yield strength (MPa) |
| Greek letters | |
| $\alpha_t$ | Tilt angle (deg) |
| $\alpha$ | Rake angle (deg) |
| $\beta$ | Sweep angle (deg) |
| $\beta_{cr}$ | Critical sweep angle (deg) |
| $\gamma_i$ | Initial soil bulk density (kNm$^{-3}$) |
| $\mu$ | Coefficient of friction |
| $\mu_r$ | Coefficient of rolling friction |
| $\lambda_\theta$ | Unit vector of angular velocity |
| $\delta$ | Angle of soil-metal friction (deg) |
| $\delta_x$ | Angle of soil-metal friction component projected on a horizontal plane (deg) |
| $\delta_z$ | Angle of soil-metal friction component projected on a vertical plane (deg) |
| $\xi$ | Cohesion energy density (J m$^{-3}$) |

## References


1. Ashworth, M.; Desbiolles, J.; Tola, E. *Disc Seeding in Zero-Till Farming Systems—A Review of Technology and Paddock Issues*; Western Australia No-Tillage Farmers Association: Northam, WA, Australia, 2010.
2. Desbiolles, J.; Saunders, C.; Barr, B.; Riethmuller, G.; Northover, G.; Tullberg, J.; Antille, D. Machinery evolution for conservation agriculture—In (Eds J Pratley and J Kirkegaard) "Australian Agriculture in 2020: From Conservation to Automation", 2020, pp. 81–105. Available online: www.csu.edu.au/research/grahamcentre/publications/e-books/australian-agriculture-in-2020 (accessed on 14 December 2022).

3. Godwin, R.J.; Seig, D.A.; Allott, M. Soil failure and force prediction for soil engaging discs. *Soil Use Manag.* **1987**, *3*, 106–114. [CrossRef]
4. Godwin, R.J.; O'Dogherty, M.J. Integrated soil tillage force prediction models. *Soil React.* **2007**, *44*, 3–14. [CrossRef]
5. Hettiaratchi, D.R.P. Prediction of soil forces acting on concave agricultural discs. *J. Agric. Eng. Res.* **1997**, *68*, 51–62. [CrossRef]
6. Alam, M.M. Soil Reaction Forces on Agricultural Disc Implements. Ph.D. Thesis, Newcastle University, Newcastle upon Tyne, UK, 1989; 129p.
7. Al-Ghazal, A.A. An Investigation into the Mechanics of Agricultural Discs. Ph.D. Thesis, Silsoe College, Cranfield Institute of Technology, Bedfordshire, UK, 1989; 299p.
8. Sadek, M.A.; Chen, Y.; Zeng, Z. Draft force prediction for a high-speed disc implement using discrete element modelling. *Biosyst. Eng.* **2021**, *202*, 133–141. [CrossRef]
9. Murray, S.E.; Chen, Y. Soil Bin Tests and Discrete Element Modeling of a Disc Opener. *Can. Biosyst. Eng.* **2018**, *60*, 1–10. [CrossRef]
10. Khosravani, A.; Desbiolles, J.; Fielke, J.M.; Ucgul, M.; Saunders, C. Prediction of single disc seeding system forces, using a semi-analytical and discrete element method (DEM). *Agriculture*, 2022; *Accepted for publication.*
11. Hettiaratchi, D.R.P.; Witney, B.D.; Reece, A.R. The calculation of passive pressure in two-dimensional soil failure. *J. Agric. Eng. Res.* **1966**, *11*, 89–107. [CrossRef]
12. Hettiaratchi, D.R.P.; Reece, A.R. The calculation of passive soil resistance. *Geotechnique* **1974**, *24*, 289–310. [CrossRef]
13. Khosravani, A. A Semi-Analytical Force Prediction Model for a Rotating Flat Disc Blade Opener. Ph.D. Thesis, University of South Australia, Magil, Australia, 2014; 246p.
14. Khosravani, A.; Desbiolles, J.; Fielke, J.M. Circular disc blade considerations in soil force prediction modelling. *J. Agric. Sci. Technol. A* **2014**, *4*, 371–383.
15. Tamás, K.; Jóri, I.J.; Mouazen, A.M. Modelling soil–sweep interaction with discrete element method. *Soil Till. Res.* **2013**, *134*, 223–231. [CrossRef]
16. Fielke, J.M.; Ucgul, M. Saunders. In *Discrete Element Modelling of Soil-Implement Interaction Considering Soil Plasticity, Cohesion and Adhesion. In 2013 Kansas City, Missouri, 21–24 July*; American Society of Agricultural and Biological Engineers: St. Joeseph, MI, USA, 2013; p. 1.
17. Tekeste, M.Z.; Balvanz, L.R.; Hatfield, J.L.; Ghorbani, S. Discrete element modelling of cultivator sweep-to-soil interaction: Worn and hardened edges effects on soil-tool forces and soil flow. *J. Terramechanics* **2019**, *82*, 1–11. [CrossRef]
18. Aikins, K.A.; Barr, J.B.; Antille, D.L.; Ucgul, M.; Jensen, T.A.; Desbiolles, J.M. Analysis of effect of bentleg opener geometry on performance in cohesive soil using the discrete element method. *Biosyst. Eng.* **2021**, *209*, 106–124. [CrossRef]
19. Aikins, K.A.; Ucgul, M.; Barr, J.B.; Jensen, T.A.; Antille, D.L.; Desbiolles, J.M. Determination of discrete element model parameters for a cohesive soil and validation through narrow point opener performance analysis. *Soil Till. Res.* **2021**, *213*, 105123. [CrossRef]
20. Fielke, J.M. Interactions of the Cutting Edge of Tillage Implements with Soil. Ph.D. Thesis, The University of Adelaide, Adelaide, Australia, 1994.
21. Nartov, P.S. *Disc, Soil-Working Implements*; Amerind Publishing: New Delhi, India, 1984; 148p.
22. EDEM. *EDEM Theory Reference Guide*; DEM Solutions: Edinburgh, UK, 2011.
23. Walton, O. *Elastic-Plastic Contact Model. Company Report*; DEM Solutions: Edinburgh, UK, 2006.
24. Walton, O.R.; Braun, R.L. Stress calculations for assemblies of inelastic spheres in uniform shear. *Acta Mech.* **1986**, *63*, 73–86. [CrossRef]
25. Raji, A.O. Discrete Element Modelling of the Deformation of Bulk Agricultural Particles. Ph.D. Thesis, Newcastle University, Newcastle upon Tyne, UK, 1999.
26. Huser, A.; Kvernvold, O. Prediction of sand erosion in process and pipe components. In Proceedings of the BHR Group Conference Series Publication, Banff, AB, Canada, 1 July 1998; Brill, J.P., Gregory, G.A., Eds.; Mechanical Engineering Publications: London, UK, 1998; Volume 31, pp. 217–227.
27. Hudson Tool Steel. P20 Mold steel. 2016. Available online: http://www.hudsontoolsteel.com/technical-data/steelP0 (accessed on 10 October 2020).
28. Academia. Some Useful Numbers for Rocks and Soils. Available online: http://www.academia.edu/4056287/Some_Useful_Numbers_for_rocks_and_soils2015 (accessed on 10 October 2020).
29. Asaf, Z.; Rubinstein, D.; Shmulevich, I. Determination of discrete element model parameters required for soil tillage. *Soil Till. Res.* **2007**, *92*, 227–242. [CrossRef]
30. Budynas, R.G.; Nisbett, K.J. *Shigley's Mechanical Engineering Design*; McGraw-Hill Education: New York, NY, USA, 2012.
31. Ucgul, M.; Saunders, C.; Fielke, J.M. Discrete element modelling of top soil burial using a full scale mouldboard plough under field conditions. *Biosyst. Eng.* **2017**, *160*, 140–153. [CrossRef]

*agriculture*

MDPI

*Article*

# Prediction of Single Disc Seeding System Forces, Using a Semi-Analytical and Discrete Element Method (DEM)

**Ali Khosravani [1], Jacky M. A. Desbiolles [1], John M. Fielke [1], Mustafa Ucgul [2,*] and Chris Saunders [1]**

1 UniSA STEM, University of South Australia, Mawson Lakes, SA 5095, Australia
2 Faculty of Science and Engineering, Southern Cross University, Lismore, NSW 2084, Australia
* Correspondence: mustafa.ucgul@scu.edu.au

**Abstract:** There is a rising interest amongst Australian farmers to use disc seeders due to their ability to operate in high residue conditions and at higher speeds, commonly in the range of 12 to 15 km h$^{-1}$. This paper reports on developing an analytical and discrete element method (DEM) force prediction model suited to a rotating flat disc blade operating at different sweep and tilt angles. To validate the models, field experiments were carried out with a flat disc blade at two tilt angles of 0 and 20° and four sweep angles of 6, 26, 45 and 90° in sandy soil. An analytical approach was developed following an experimental investigation that showed that only the forward portion of the disc blade is actively involved in generating soil failure, while the magnitude of this active portion of the soil-disc interface varied with sweep angle. The predicted active proportions correlated well with the experimental observations. As applying different sweep angles affects the direction of soil movement relative to the disc face, the directions of the friction and resultant forces at different sweep and tilt angles were determined. The equation of soil acceleration force was adapted to account for different sweep angles. Results showed that the predicted force fits relatively well with the measured data at 90, 45 and 26° sweep angle, while the low correlation between predicted and measured force at 6° sweep angle was due to the scrubbing reaction force not accounted for in the model. Results also showed that a better coefficient of determination ($R^2$ = 0.93) was obtained between DEM vs. test results compared to the analytical model predictions ($R^2$ = 0.86), particularly for predicting side forces. It was found from the study that both the developed analytical approach and DEM model enabled the prediction of soil forces at different sweep and tilt angles acting on a flat disc blade, which can assist in optimising disc design to lower the specific resistance.

**Keywords:** disc seeder; disc blade; discrete element method (DEM); force prediction; semi-analytical model; sandy soil

**Citation:** Khosravani, A.; Desbiolles, J.M.A.; Fielke, J.M.; Ucgul, M.; Saunders, C. Prediction of Single Disc Seeding System Forces, Using a Semi-Analytical and Discrete Element Method (DEM). *Agriculture* **2023**, *13*, 206. https://doi.org/10.3390/agriculture13010206

Academic Editor: Massimo Cecchini

Received: 14 December 2022
Revised: 7 January 2023
Accepted: 11 January 2023
Published: 13 January 2023



## 1. Introduction

No-tillage (zero tillage) is widely adopted in Australia, where farms are large, and labour is expensive. Zero-tillage disc seeders can be operated at higher operation speeds and in heavier stubble residue conditions [1]. Accordingly, there has been increasing interest amongst Australian farmers in using disc seeders. However, the optimum disc angle settings of disc seeders and their impacts on soil forces are not well documented. This is particularly important to reduce the draught force requirements and hence the fuel consumption [2]. As a complement to field experiments which are time and resource-intensive and with seasonal limitations, suitable modelling methods can help speed up evaluation and improve the understanding of soil–disc interactions, to assist optimisation processes.

Analytical and numerical models are the two most common methods used to model soil–tool interactions. The interaction between soil and a concave disc was previously studied using analytical methods [3–8] based on wide blade theories [9–11]. Although these models provide some helpful information, they are limited as they assume that all of the disc–soil interfaces are involved in soil failure. This study developed a novel semi-analytical model considering that only the active part of the disc contributes to soil failure to predict the tillage forces under typical disc settings. Quantifying tillage forces accurately under different operation and geometry settings will contribute to improved disc seeder performance by determining optimum settings, which create less force without sacrificing the quality of the disc operation. The analytical model will also help designers and manufacturers who cannot afford to use expensive industry software to predict the tillage forces acting upon the disc seeder [11]. The basics of soil–disc blade interactions were experimentally reviewed under controlled soil conditions to help guide an improved basis of soil/disc blade modelling.

Many recent studies [12–20] have shown that the discrete element method (DEM) modelling can provide accurate soil force predictions when its parameters are accurately calibrated and applied. DEM was developed in the field of rock mechanics [21], and the technique is based on interacting particles that are able to simulate realistic soil bulk behaviour. Interactions between all particles are governed by contact models replicating physical laws and adjusted via a calibration process to suit specific soil conditions. Both tillage tool forces and soil movement can be predicted using DEM. The linear cohesion integrated hysteretic spring contact model can accurately simulate soil–tool interactions for tined sweep tillage tools [22], whereby contact forces between the particles are calculated using a hysteretic spring model, and cohesion is defined by adding a cohesive force between the particles. Although DEM is a valuable method, no comparison has been conducted to evaluate its ability to predict the tillage forces against an analytical model in the case of soil–disc interaction. Therefore, this study will provide valuable knowledge in comparing two different methods.

*Logic of Approach*

Initial observations and measurements of the soil loosening features created by a vertical, flat disc blade set at different sweep angles (The angle measured on a horizontal plane between the active face of the disc blade and the direction of travel) were conducted by the authors in an outdoor laboratory facility of the University of South Australia. The results (See Section 2.1) clearly showed that only the leading part of the disc blade was involved in generating the full extent of soil failure, while the trailing part contributed only to extra soil movement. This leading part of the disc blade could thus be described as the 'active' part of the blade, defined by its length at the soil surface.

From this initial work, the authors developed the following staged approach to model soil-to-flat disc blade interactions:

A. The wide blade passive failure theory was first used to model a fixed circular blade operated at 90° to the direction of travel in a case study most closely related to wide blade theory, considering a wide circular blade with varying depth of cut across its width. The blade tilt angle (The angle measured between the active face of the disc blade and the vertical plane) was then the complementary angle to the blade rake angle. This basic model demonstrated the suitability of the wide blade theory for the case of a fixed circular blade operating in the range of 70 to 90° rake angle and validated a method of accounting for the variable depth of cut across a circular disc blade shape, as well as for the speed and bulldozing effects. The draught and vertical forces acting on the vertical blade were predicted within 6–19% and 1.5–14% of the measured data, respectively. This work was reported in [11,23].
B. The above-fixed blade prediction model was extended to account for the additional effect of sweep angle with a focus on the 0 to 8° range combined with the effect of tilt angle in the 0 to 20° range. This combined range is directly relevant to the configurations of zero-tillage disc seeders using single disc openers. This model development stage introduced the concept of an active part of the blade involved in generating furrow soil

failure and accounted for the soil surcharge accumulation developing at higher sweep angle values. These details are being reported in this paper.

C. The impact of disc blade free-rotation was added to the stage 2 model with an analysis of the direction of soil flow impacting the direction of soil-tool frictional reactions.
D. Finally, the effect of having a single-sided bevel on the rear face of the cutting edge was included to account for additional soil scrubbing reaction when operating at smaller sweep angles, which is often witnessed in field operations.

This paper describes the Stage 2 modelling in detail and validates the analytical force predictions with data obtained in a remoulded sandy-loam soil. It also compares the findings with DEM simulations of the same soil conditions, following the method reported by [22]. Comparing the two methods will help researchers and machinery manufacturers choose the suitable modelling approach.

## 2. Materials and Methods

### 2.1. Force Measurement Experiment

Experiments were performed in a continuous outdoor soil bin (tillage test track) located at the University of South Australia. During testing, the tillage test track had sandy loam soil (with no residue) at 7% moisture content (dry basis). The porosity of the soil was 39.5%. The particle size distribution of the soil is given in Table 1.

**Table 1.** Particle size distribution of the test soil.

| Sieve Size (mm) | Percentage Retained (%) |
|---|---|
| 2.36 | 0.5 |
| 1.18 | 2.6 |
| 0.6 | 15.1 |
| 0.425 | 11.7 |
| 0.3 | 14.0 |
| 0.15 | 36.7 |
| 0.075 | 10.8 |
| <0.075 | 8.6 |

Prior to testing, water was uniformly added and the soil was reconditioned via a loosening, levelling and multi-pass rolling process. A flat disc blade of 460 mm diameter, 6 mm thickness and wedge angle of 16° was used in the tests. All experiments were performed at a 67 mm disc operating depth within the range of typical seeding depth settings and at 2 km $h^{-1}$ forward speed, the lowest speed possible with the equipment, with the aim to approximate quasi-static conditions.

An experimental testing rig (Figure 1a) was used, which allowed the adjustment of the sweep angle (0 to 45° range) and tilt angle (−20 to +40° range). The draught, vertical, and lateral forces were simultaneously measured using a 3D force measuring frame with calibrated S-type load cells of 5 kN capacity (Figure 1b). The force measuring frame consisted of sets of 3 roller guided trays isolating each orthogonal force component, and its cross-sensitivities were calibrated under known static forces applied near the expected resultant force point of application on the disc blade. The forward speed was measured using a free-rolling side wheel (490 mm in diameter) equipped with a rotary encoder of 256 pulses/rev. The measured data were logged at a frequency of 100 Hz using LabView TM V 7.1 software.

Experiments were performed using four sweep angles (6, 26, 45 and 90°) and two tilt angles (0 and 20°) in a randomised complete block design with three (temporal) replications. Measurements were made over a distance of 40 m within the straight sections of the continuous, stadium-shaped track.

(**a**) (**b**)

**Figure 1.** Equipment used for disc blade experiments (**a**) disc blade adjustment rig and (**b**) three-dimensional force measuring frame.

## *2.2. Active Disc Section Evaluation*

To support the development of the analytical model, a series of tests were performed to determine the active section of a circular disc blade, which is involved in generating the full extent of soil failure. These exploratory experiments were performed in similar remoulded soil conditions and forward speed of 2 km $h^{-1}$. The experimental factors included three operating depths (40, 60 and 80 mm), two sweep angles (5 and 10°) and two tilt angles (0 and 20°), while contrasting the performance of a whole (Figure 2a) vs. a partial (Figure 2b) disc blade. The partial disc blade only had the forward 50% section of the disc blade engaged in the soil, while the rear section was cut out. The results of the active side evaluation are provided in the Supplementary Materials.

(**a**) (**b**)

**Figure 2.** Example soil failure from full (**a**) and partial (50%) (**b**) disc blade soil engagement, while fixed in rotation.

### 2.3. *Analytical Model Development*

#### 2.3.1. Predicting the Active Chord Length, La

To quantify the active portion of the soil–disc interface, the theoretical approach outlined in [11,23] was used. The furrow width data measured by laser profile meter (e.g., Figures 3 and 4) was used to determine the values of the variables *La* and *m* (the ratio of forward rupture distance f to operating depth d, both measured in situ) according to the geometrical relationship between the furrow boundary trace and the disc blade, at different sweep angles.

The chord length $L$ was defined as the horizontal line intersecting the disc blade face at the undisturbed soil surface. In the modelling, the active chord length $L_a$ was expected to extend beyond $L/2$ into the rear section of the disc blade. On the front section, the integration process was conducted over the variable $Li$ ($0 < Li < L/2$) referenced from the disc blade centre ($Li = 0$ at full depth, $d_{max}$) and extending to a maximum $L/2$ when coinciding with zero operating depth. The integration process over the rear section was similarly conducted with the variable $Li$ referenced from the disc blade centre to a maximum value ($=L_a - L/2$).

A geometrical analysis (Figure 3) defining $L$, $L_i$, $L_a$ shows the geometrical relationship between furrow size (from disc blade point of entry B to full furrow width $W_f$. In the rear section of the disc (from $L/2$ toward $L_a$), soil forward rupture starts to reduce due to reducing of the depth of the disc. At the particular point E along the disc chord length located at a distance $L_a$ from the disc blade point of entry B, the forward rupture distance f and the furrow width $W_f$ reach their maximum value:

$$W_f = L_a \sin\beta + f\ \cos\beta \tag{1}$$

where $L_a$ is the active proportion of the disc chord ($L$) at the soil surface, and which is measured from the disc blade point of entry into the soil.

The reducing forward rupture distance *f*, further along the chord length, indicates that section of the disc blade does not contribute to soil failure.

The $L_a$ relationship with sweep angle [11] depicts a significant increase of $L_a$ over the range, from 0.5–0.55 $L$ at the low sweep angle values encountered in zero-tillage disc seeders, and reaching 1.0 $L$ at larger sweep angles (>70°). The value of $L_a$ can be used to predict the force at different sweep angles. For a flat disc opener set at a low sweep angle (e.g., 6°) as used on zero-till single disc seeders, the predicted relationship shows that the forward 55% portion of the soil–disc interface actively generates soil failure.

The integration process of passive soil failure force P is initiated from the disc centre at (d max) and thus is only valid for $0 < L_i < L/2$. As shown in Figure 3, $L_a$ is expected to lie in the range of $L/2$ to $L$; therefore, force equation is further modified conceptually as the sum of the segment two integrations, namely:

$$P = \int_0^{L/2} P_i d_l + \int_0^{(L_a - L/2)} P_i d_l \tag{2}$$

Letting $k = L_a/L$, Equation (2) can be re-written, as follows:

$$P = \int_0^{L/2} P_i d_l + \int_0^{(k-\frac{1}{2})L} P_i d_l \tag{3}$$

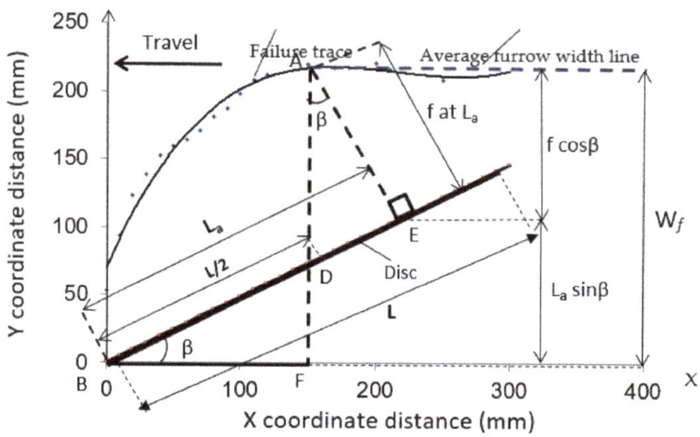


**Figure 3.** Plan view outlining the geometrical relationships from the record of furrow boundary trace relative to the disc blade at a sweep angle β and represented by its chord length L at the soil surface.

For the passive soil failure reaction, P applied to the disc when the sweep angle $\beta \leq 90°$ [23] is outlined below:

$$P = A_1\left[\int_0^{L/2}\left(\frac{b}{a}\sqrt{a^2-L_i{}^2}-(b-d)\right)^2 dl+\int_0^{(k-\frac{1}{2})L}\left(\frac{b}{a}\sqrt{a^2-L_i{}^2}-(b-d)\right)^2 dl\right](A_2+A_3+A_4 +A_5)\left[\int_0^{L/2}\left(\frac{b}{a}\sqrt{a^2-L_i{}^2}-(b-d)\right)dl+\int_0^{(k-\frac{1}{2})L}\left(\frac{b}{a}\sqrt{a^2-L_i{}^2}-(b-d)\right)dl\right] \quad (4)$$

where:

$$A_1 = \gamma_i \cdot N_\gamma; \quad (5)$$

$$A_2 = c \cdot N_c; \quad (6)$$

$$A_3 = c_a \cdot N_{ca}; \quad (7)$$

$$A_4 = q \cdot N_q; \quad (8)$$

$$A_5 = \frac{\gamma_i}{g} \cdot N_a v^2 \sin^2 \beta. \quad (9)$$

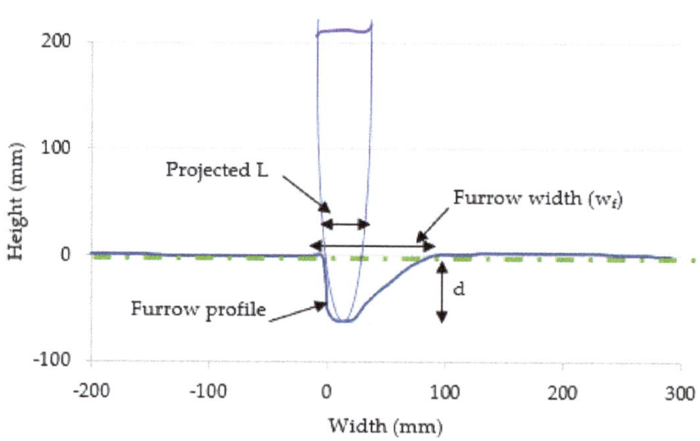


**Figure 4.** Example average soil furrow profile measured at 6° sweep angle of a full disc blade. Blue line indicates the furrow area and green line indicates the x-coordinate for drawing.

2.3.2. Validation of the Active Chord Length, La

To validate the $L_a$ concept, an experiment was conducted in Tillage Test Track conditions at a range of operating depths, sweep and tilt angles, comparing a fixed whole disc blade with its assumed partial equivalent with the rear 50% of soil/tool interface removed (Figure 2). The three-dimensional soil reaction forces were measured in the Tillage Test Track environment using a replicated and randomised experiment. The results of measurements of draught, vertical and side force data for the whole and partial disc blades are shown in Table 2.

**Table 2.** The measured forces of whole and partial disc blades at different angle settings.

| Sweep Angle (°) | Tilt Angle (°) | Depth (mm) | Draught (N) Whole | Draught (N) Partial | Vertical (N) Whole | Vertical (N) Partial | Side (N) Whole | Side (N) Partial |
|---|---|---|---|---|---|---|---|---|
| 5 | 0 | 40 | 210 | 173 | 78 | 57 | 75 | 60 |
| 5 | 0 | 80 | 477 | 367 | 120 | 88 | 292 | 237 |
| 5 | 0 | 60 | 332 | 251 | 114 | 78 | 174 | 131 |
| 10 | 0 | 60 | 325 | 241 | 75 | 71 | 256 | 234 |
| 5 | 20 | 60 | 313 | 245 | 118 | 122 | −19 | −22 |
| 10 | 20 | 60 | 231 | 177 | 26 | 31 | 130 | 109 |

Table 2 shows that the measured forces were typically higher for the whole blade, namely 18 to 26% higher draught on average, 3 to 32% larger upward vertical force, and 4 to 9% greater side force. As the operating depth became shallower, draught and side force differences tended to become less significant. The results indicate that, while the forward half of the soil–disc interface generates most of the soil failure work, in practice, a slightly greater proportion is involved in creating the total soil/tool forces. Some differences in soil forces can be attributed to secondary soil/tool interaction involving the loosened soil being displaced sideways.

2.3.3. Accounting for the Tangential Frictional Reaction due to Sweep Angle

The wide blade passive failure theory as reported by [11,23] was used to model a fixed (non-rotating) circular blade set at 90° to the direction of travel and at two tilt angles of 0 and 20°. When the sweep angle $\beta = 90°$, the disc blade is normal to the direction of travel and thus the direction of soil–tool friction applies in the vertical plane only, combining with the effect of rake angle, as modelled in the wide blade soil passive failure theory, and influences the direction of the passive soil resultant force $P$. With a sweep angle $\beta < 90°$, the asymmetry of soil movement in the horizontal plane creates an additional tangential reaction that influences the direction of the resultant horizontal force component $H_p$ (Figure 5). The forward movement of an angled disc blade was divided into orthogonal vector components that are perpendicular and tangential to the disc plane. The tangential component of blade movement generates a frictional reaction ($\delta_X$) at the soil/tool interface in the horizontal plane that is tangential to the disc face.

This component would be complementary to the traditional frictional reaction considered in the vertical plane ($\delta_Z$, when $\beta = 90°$), which is mobilised by the perpendicular vector component of movement. The modified soil force modelling approach for a fixed blade at a sweep angle $\beta < 90°$, therefore, considers two elements of friction, namely $\delta_z$ in the vertical plane (Figure 5a) and $\delta_x$ in the horizontal plane (Figure 5b).

Observations with a fixed circular disc blade in the Tillage Test Track environment showed that the soil movement relative to the blade shifts from (i) a situation of relative movement occurring solely in the vertical plane at $\beta = 90°$ to (ii) a situation where relative soil movement in the vertical plane reduces significantly with decreasing sweep angle, and (iii) a situation where movement in the horizontal plane predominates solely as the sweep angle approaches very low values. When $\beta = 90°$, the following assumptions were therefore adopted $\delta_z = \delta$ (friction fully mobilised in the vertical plane), and $\delta_x = 0°$ (friction not mobilised in the horizontal plane). When β is near 0°, the soil moves relative to the disc blade only in the horizontal direction and $\delta_x = \delta$ (friction fully mobilised in the horizontal plane) and $\delta_z = 0°$ (friction not mobilised in the vertical plane). Similarly to [3], the transition in the $\delta_x$ and $\delta_z$ values from 90° sweep angle down to 0° was assumed to follow a reducing sine function for $\delta_z$ and an increasing cosine function for $\delta_x$, as follows:

$$\delta_x = \delta \cos \beta \quad (10)$$

$$\delta_z = \delta \sin \beta \quad (11)$$

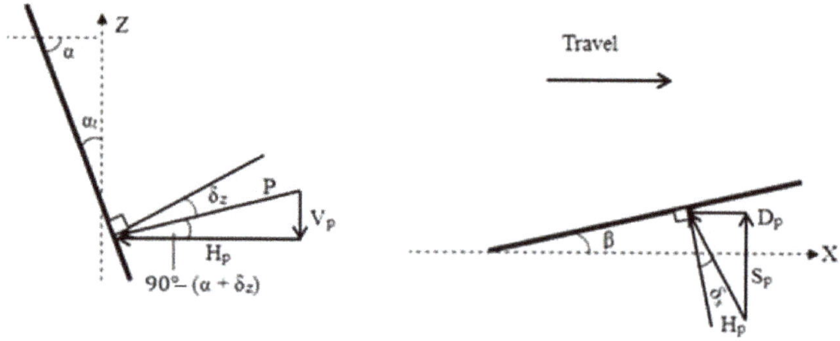


**Figure 5.** Partitioning of the passive soil reaction force P on a tilted disc blade at a sweep angle β; (**a**) side view (along the plane of the disc) and (**b**) top view.

### 2.3.4. Accounting for the Effect of Velocity Differential

To account for the effect of velocity, the additional component ($\frac{\gamma_i}{g} \cdot N_a d v^2$), as developed by [24], was added to the force equation. The inertial effect of forward velocity ($v$) on the disc forces assumed when $\beta = 90°$ was reported in [23]. Additionally, it is necessary to account for the velocity differential occurring in the direction of the soil failure plane when inclined with a sweep angle. The following relationship was used for the passive soil failure velocity ($v_p$) of the disc:

$$v_P = v \sin \beta \quad (12)$$

At working speeds such as 20 km h$^{-1}$, these inertial effects have to be taken into account above a certain critical speed. This was suggested by [8] to be equal to $\sqrt{5gw}$, where $g$ is the gravitational acceleration and $w$ is the width of the tool.

In this study, the total force for the fixed blade was calculated over the sweep and tilt angle combinations using Equation (4) for the soil input parameters, including soil cohesion, adhesion, bulk density and soil surcharge. The effect of sweep angle on soil frictional reaction force was included in the partitioning of the soil passive soil failure force into its three force components, as illustrated in Figure 5 and calculated as follows:

$$H_p = P \sin(\alpha + \delta_z) \quad (13)$$

$$V_p = P\cos\,(\alpha + \delta_z) \quad (14)$$

$$S_p = H_p\cos\,(\beta + \delta x) \quad (15)$$

$$D_p = \mathrm{H_p} \sin(\beta + \delta x) \quad (16)$$

The total draught force, $D_T$, in a bulldozing situation ($\beta > 45°$) was calculated by adding a proportion ($\sin \beta$) of the additional drag force $F$ and the horizontal component of the adhesion force $R_a$ to the component $D_p$, as follows:

$$D_T = D_p + (F + R_a \cos\alpha) \sin \beta \quad (17)$$

The total vertical force, $V_T$, in a bulldozing situation was estimated by combining the soil failure vertical force $Vp$, the soil body self-weight $W_3$, the interface force reaction $H$ and the vertical component of the adhesion force $R_a$ similarly to the methods expressed in [23].

The side force $S_T$ was calculated as follows:

$$S_T = S_p + (F + R_a \cos\alpha) \cos \beta \quad (18)$$

*2.4. DEM Simulations*

The DEM simulations were carried out using a DELL Precision T5810 Intel ® Xeon CPU E5-2687W v4 @ 3.00 GHz computer with the software EDEM 2020™. The linear cohesion integrated hysteretic spring contact model allowed the particles to behave in a linear elastic manner up to predefined stress, and when the total stress on the contact area exceeded the predefined stress (=yield strength) in the model, the particles behaved as though undergoing plastic deformation. The cohesion between the particles was defined by adding a cohesion force to the normal contact forces. In the hysteretic spring contact model, the total normal ($F_n$) and tangential ($F_t$) forces were computed as

$$F_n = F_n{}^s + F_n{}^d \quad (19)$$

$$F_t = F_t{}^s + F_t{}^d \quad (20)$$

where $F_n{}^s$, $F_n{}^d$, $F_t{}^s$ and $F_t{}^d$ are the normal contact force, normal damping force, tangential contact force and tangential damping force, respectively. $F_n{}^s$ was determined as per [25]:

$$F_n^s = -\begin{cases} K_1 \cdot U_{abn} & \text{loading} \\ K_2 \cdot (U_{abn} - U_0) & \text{unloading/reloading} \\ 0 & \text{unloading} \end{cases} \quad (21)$$

where $U_{abn}$ is the normal component of the relative displacement, $U_o$ is the residual overlap. $K_1$ and $K_2$ are the loading and unloading stiffnesses, respectively. As per [26,27], $K_1$ was computed as

$$K_1 = 5\, r_{eq}\, min(Y_a, Y_b) \quad (22)$$

where $Y$ is the yield strength, and $r_{eq}$ is the equivalent radius and defined as where req is defined as [24]

$$1/r_{eq} = 1/r^*{}_a + 1/r^*{}_b \quad (23)$$

where $r^*$ is the radius for the individual particles $a$ and $b$. Following [26,27], $K_2$ was computed as

$$K_2 = K_1/e^2 \quad (24)$$

where $e$ is the coefficient of restitution. The residual overlap was updated in each time step as

$$U_0 = \begin{cases} U_{abn} \cdot \left(1 - \frac{K_1}{K_2}\right) & \text{loading} \\ U_0 & \text{unloading/reloading} \\ U_{abn} & \text{unloading} \end{cases} \quad (25)$$

The tangential contact force, $F_t{}^s$, was calculated as per [25] as

$$F_t{}^s = -n_k\, K_1\, U_{abt} \quad (26)$$

where $U_{abt}$ is the tangential component of the relative displacement. $n_k$ is the stiffness factor defined as the ratio of tangential stiffness to normal loading stiffness. The normal and the tangential damping forces ($F_n{}^d$ and $F_t{}^d$) were calculated using

$$F_n{}^d = -n_c\ (((4\ m_{eq}\ K_1)/(1+(\pi/\ln e)^2))\ \mathring{U}_{abn})^{-1/2} \tag{27}$$

$$F_t{}^d = -(((4\ m_{eq}\ n_k\ K_1)/(1+(\pi/\ln e)^2))\ \mathring{U}_{abt})^{-1/2} \tag{28}$$

where $\mathring{U}_{abn}$ and $\mathring{U}_{abt}$ are the normal and tangential components of the relative velocity, respectively. $n_c$ is the damping factor that controls the amount of velocity-dependent damping. $m_{eq}$ is the equivalent mass and is defined in [25] as

$$1/m_{eq} = 1/m^*{}_a + 1/m^*{}_b \tag{29}$$

where $m^*$ is the mass for the individual particles $a$ and $b$. The total tangential force ($F_t$) was limited to the lesser of either the calculated tangential force or the sliding friction force. $F_t$ was determined as

$$F_t = -min\ (n_k\ K_1\ U_{abt} + F_t{}^d,\ \mu\ F_n{}^s) \tag{30}$$

The magnitude of the moments caused by total tangential force ($M$) and the rolling resistance ($M_r$) were calculated following [28]:

$$M = r_{con}\ F_t \tag{31}$$

$$M_r = -\mu_r\ F_n{}^s\ r_{con}\ \lambda_\theta \tag{32}$$

where $r_{con}$ is the perpendicular distance of the contact point from the centre of mass, $\mu_r$ is the coefficient of rolling friction, and $\lambda_\theta$, is the unit vector of angular velocity at the contact point. The new position of a particle was computed by integrating Equations (33) and (34):

$$\ddot{U} = (F_n + F_t)/m^* \tag{33}$$

$$R = (M + M_r)/I \tag{34}$$

The cohesion force was calculated as [25]:

$$F_c = \xi\ A_c \tag{35}$$

where $\xi$ is the cohesion energy density which is defined as the energy needed to remove a particle from its nearest neighbours divided by the total volume of the removed particle, and $A_c$ is the contact area. Hence, Equation (19) becomes

$$F_n = F_n{}^s + F_n{}^d + F_c \tag{36}$$

DEM is based on modelling the contacts between particles where many particles are used to represent a soil bulk. Hence, increasing the number of contacts increases the calculation time. In modelling soil, larger than actual spherical particle sizes are used to reduce the computation time (i.e., coarse graining). The DEM parameters were calibrated so that the larger than actual size DEM particles act as the actual soil. The DEM parameters used in the study are shown in Table 3, comprising a combination of measured and literature-based parameters. The DEM parameters of the coefficient of rolling friction and coefficient of restitution between soil and soil (a value less than 0.3 for a compressible soil) were calibrated by matching simulation results to the actual measurement result of an angle of repose test.

In order to measure the angle of repose, a soil sample was placed in a pipe (100 mm diameter and 300 mm long). After that, the pipe was lifted upward (at around 500 ms$^{-1}$). Soil flowed into a cylindrical tray (200 mm diameter and 22.5 mm long) until the soil overflowed and formed a pile. When at rest, the soil's angle of repose flowed on the surface and was measured using a digital level. The test procedure was also simulated in DEM.

By using the calibrated parameters (given in Table 3), an angle of repose of 29° (measured in the test) was achieved in the simulation using a trial and error process (with an error margin of ±1° i.e., −28.2° measured in DEM)

**Table 3.** EDEM parameters used for simulation.

| Property | Value | Source |
|---|---|---|
| Density of sand particles (kg $m^{-3}$) | 2600 | [29] |
| Density of steel (kg $m^{-3}$) | 7861 | [30] |
| Shear modulus of soil (Pa) | $5 \times 10^7$ | [31] |
| Shear modulus of steel (Pa) | $7.9 \times 10^{10}$ | [30] |
| Poisson's ratio of soil | 0.3 | [32] |
| Poisson's ratio of steel | 0.3 | [33] |
| Coefficient of restitution of soil-soil | 0.1 | Calibrated |
| Coefficient of friction of soil-soil | 0.7 | [34] |
| Coefficient of friction of soil-steel | 0.5 | [34] |
| Coefficient of rolling friction of soil-soil | 0.28 | Calibrated |
| Coefficient of rolling friction of soil-steel | 0.05 | [34] |
| Cohesive energy density between soil-soil (N $m^{-2}$) | 9000 | Measured using DST * |
| Adhesive energy density between soil-tool (N $m^{-2}$) | 3700 | Measured using DST * |

* DST = Direct shear test.

In order to run the simulations, a virtual soil bin of 4000 mm long × 1500 mm wide × 300 mm deep was generated. A nominal 40 mm particle size was used with the particles generated based on the measured particle size distribution of the sandy loam soil. A total of 359,188 particles were used in the simulations.

The DEM bulk density was set to match the bulk density used in the experiments. To do so, particles were compressed using an upper physical plane until the desired bulk density was achieved.

The disc models were created using SolidWorks$^{TM}$ software and then imported into the EDEM environment. After the depth and the speed of the disc models were set, simulations were carried out (Figure 6). The tillage forces were averaged over the stable section of the force vs. displacement graphs.

**Figure 6.** DEM simulation of the soil–disc interaction in a cohesionless medium.

## 3. Results

The relationships between force components and sweep angles of a fixed disc are presented in Figures 7 and 8 for the blade at two tilt angles and operating at a low speed

(2 km $h^{-1}$). The analytically predicted draught and vertical forces correlate well with the measured data over the different sweep angles (Figure 9a). Draught force expectedly increased with a greater sweep angle due to increasing the active part of the disc and associated furrow disturbance (Figures 7a and 8a). The rate of increase was lower at the 20° tilt angle (Figure 8a), but, overall, was well below that shown in the literature for concave discs (e.g., [3]). This can be explained by the lower amount of soil displacement induced by a flat disc blade compared to concave discs of harrows and ploughs used in previous literature, which inverts a large amount of soil. In line with experimental observations, the modelled effect of bulldozing (=accumulation of soil body ahead of the blade) becomes increasingly significant as the sweep angle increases from 45 to 90°.

The upward vertical force increases with a greater sweep angle, explained by the proportionality of the vertical reaction to increasing soil passive failure reaction force with an increasing sweep angle (Figures 7b and 8b). This strongly contrasts with the decreasing relationships observed by [3,35], who found a decreasing upward vertical reaction by increasing the sweep angle with concave discs. Their findings can be attributed to the effect of disc concavity, inducing (i) a much stronger push upward at low sweep angles from the compacting scrubbing reaction on the convex face of the blade and (ii) the beneficial effect of lower apparent rake angle on the concave face as the sweep angle increases. The modelled effect of bulldozing on the vertical force also becomes increasingly significant as the sweep angle increases from 45 to 90° while being slightly lower than that for the draught force. It was also found that the analytical model yields downward vertical forces (instead of the upward vertical forces measured) for a 20° tilt angle (Figure 8b).

The results for the side force show an expected rise to a maximum at approximately 30 to 40° sweep angle and a decrease beyond that (Figures 7c and 8c). Contrary to the literature data for concave discs, the side force does not reverse in direction at low sweep angles due to the limited scrubbing reaction occurring at the bevel edge angle, but overall, was well below that shown in the literature; e.g., [3] with concave discs. This can be explained by the lower extent of soil displacement induced by a flat disc blade compared to concave discs of harrows and ploughs used in previous literature, which invert a large amount of soil.

In line with experimental observations, the modelled effect of bulldozing becomes increasingly significant as the sweep angle increases from 45 to 90° (Figures 7c and 8c). The side force becomes nearly equal to the draught force when $\beta + \delta_X = 45°$. The analytical model significantly overpredicts the side force at the lower sweep angles, as the effect of scrubbing on the rear of the blade is not taken into account at this stage of the model. Overall, the draught and vertical forces acting on the vertical blade at 2 km $h^{-1}$ speed were predicted for the four experimental sweep angles within 17% to 27% of the measured values. As detailed above, the side force was significantly over-predicted at sweep angles below 45° due to scrubbing. The model accurately reflected the beneficial effect of a 20° tilt on reducing the draught force and on the side force at the larger sweep angles (Figure 7 vs. Figure 8). The vertical force reaction was significantly under-predicted at lower sweep angles as the scrubbing reaction at the bevel edge was not accounted for in the model. This scrubbing would increase the vertical reaction force. Prediction for the side force fell within 6.7% of the measured values at sweep angles of 45° and above. There was a significant overprediction of side force at lower sweep angles.

The DEM predicted draught, vertical and side forces of the vertical disc showed a good correlation with the measured data over the range of different sweep angles even though the analytical solution performs better for some angles. Compared to the analytical results, a better coefficient of determination was obtained using DEM simulations ($R^2$ = 0.93 vs. $R^2$ = 0.86) (Figure 9). A better side force prediction was obtained using DEM simulations, while vertical forces were slightly over-predicted for 6 and 25° sweep angles. The vertical force over-prediction at sweep angles below 45° can be attributed to the larger than actual size particles used in the simulations. In the draught and side force predictions, the DEM particles contact the disc's large surface, while only a few are in contact with the disc's surface in the vertical force predictions. An extensive scaling effect study, which was not considered in this study, might also help to determine contact parameters more precisely when the particle sizes are scaled up. Additionally, using clump particles (a combination of multiple spheres to obtain a more realistic particle shape) might also increase the accuracy of the simulations. In addition, in this study, the calibration procedure was based on an angle of repose test; performing a second test for the calibration procedure, such as a penetration test, might help to improve the accuracy of the DEM parameters and hence predicted vertical force prediction.

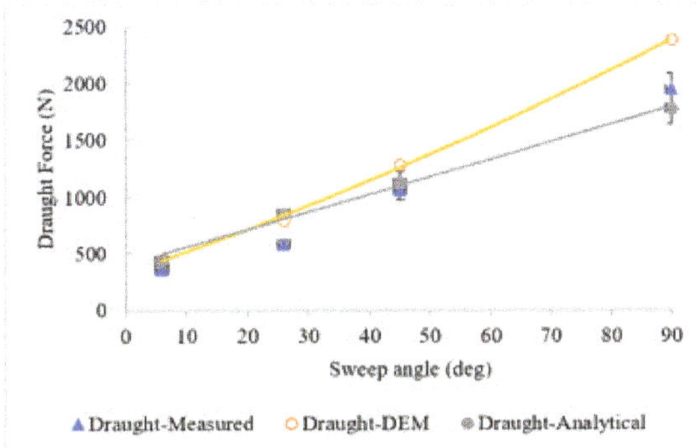


(**a**)- draught forces at 0° tilt angle

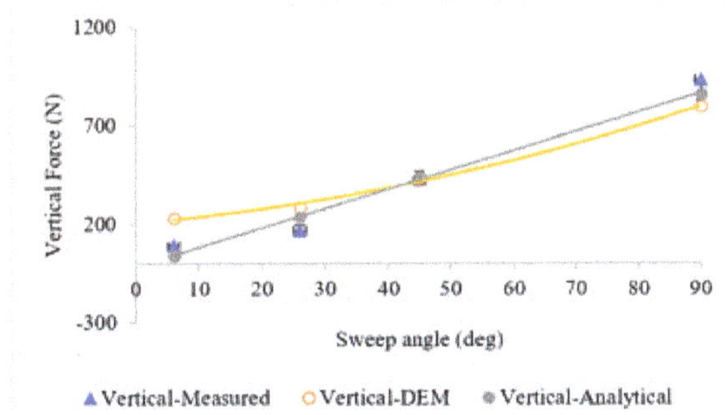


(**b**) vertical forces at 0° tilt angle

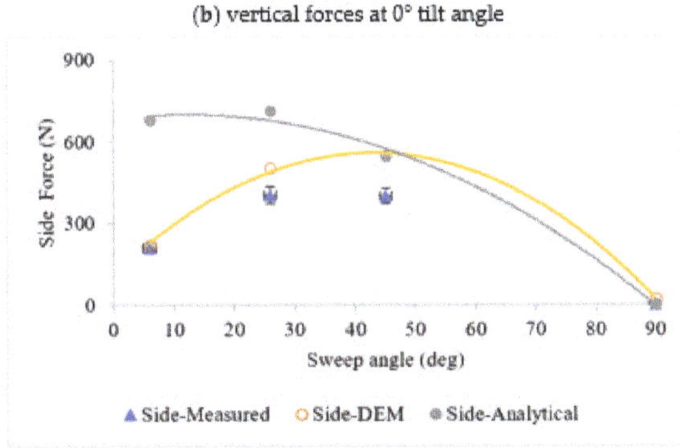


(**c**) side forces at 0° tilt angle

**Figure 7.** Comparison of measured, analytically predicted, and DEM predicted forces for 0° tilt angle (**a**) draught forces, (**b**) vertical forces, and (**c**) side forces. Error bars for measured results are plus and minus standard error.

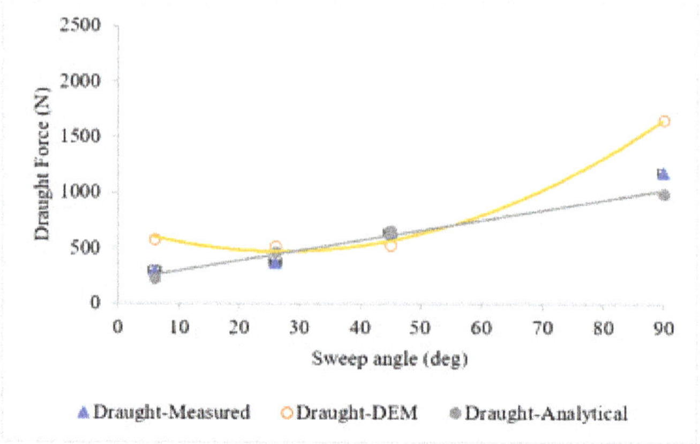


**(a)** draught forces at 20° tilt angle

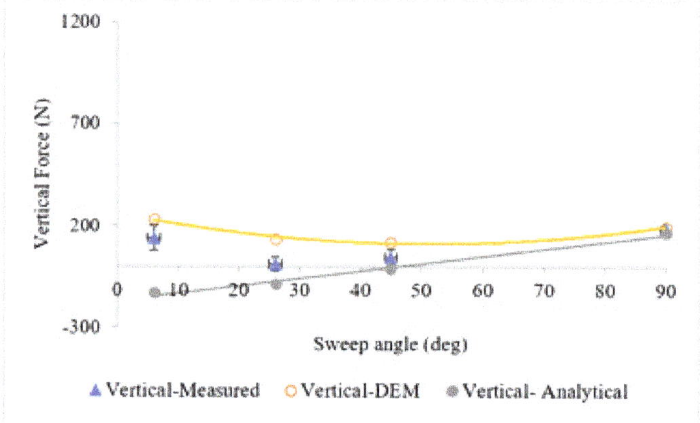


**(b)** vertical forces at 20° tilt angle

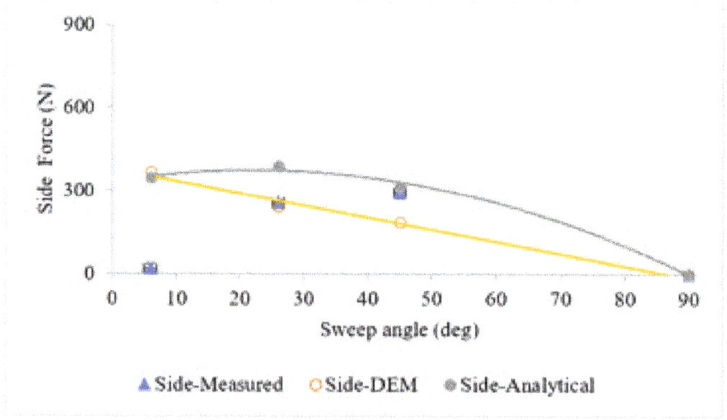


**(c)** side forces at 20° tilt angle

**Figure 8.** Comparison of measured, analytically predicted and DEM predicted forces for 20° tilt angle. (**a**) draught forces, (**b**) vertical forces, and (**c**) side forces.

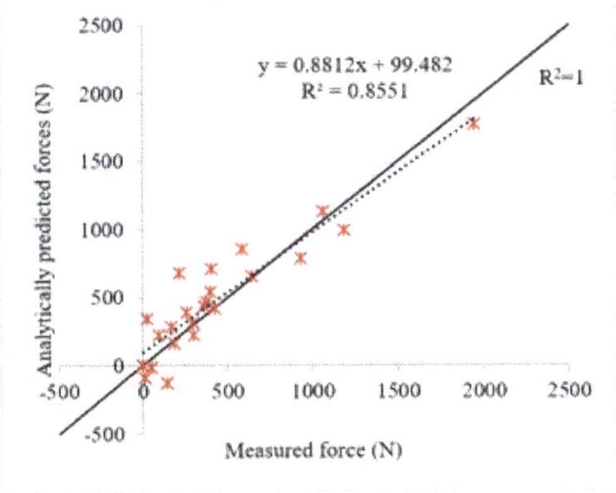


(a)

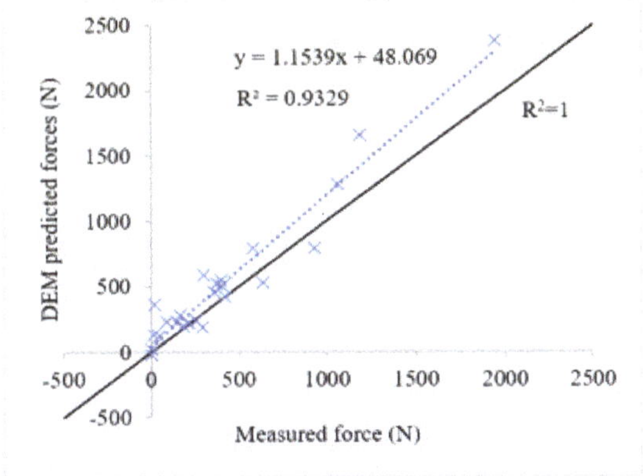


(b)

**Figure 9.** Whole data set of coefficients of determination between measured reaction forces and (**a**) analytical predictions or (**b**) DEM predictions ($n$ = 24). The red dots represent the measured force vs. analytically predicted force (as (x,y)), and the blue dots represent the measured force vs. DEM predicted force (as (x,y)). On the other hand, the black line represents the perfect prediction ($R^2 = 1$) which was added to visually show how closely the models predicted the measured data.

## 4. Conclusions

In this study, a fixed circular blade was tested in remoulded sandy soil at a range of sweep and tilt angles. An analytical soil reaction force model previously reported [11,23] was modified to suit the case of a flat circular disc blade (fixed in rotation) operating at various sweep and tilt angles suited to disc seeder configurations. It was found that the analytical model developed can predict draught forces very well, whereas it does not accurately predict the vertical and side forces for sweep angles of 6°. To improve the results of the analytical model future work will need to account for the scrubbing reaction at the rear bevel edge and to account for disc blade-free rotation.

Such model can then be used in sensitivity analyses to identify optimum points of operation including specific draught resistance, tilt and sweep angles and bevel edge geometry under contrasting soil conditions. Opportunities for further work are also identified, including modelling the association of disc blade and depth gauge wheel, often used on many zero-till disc seeders.

DEM simulation of the soil and fixed flat disc blade interaction was also carried out. The DEM results over-predicted the draught force at high sweep angles for a 0° tilt angle and overpredicted the draught force for both low and high sweep angles at a tilt angle of 20°. The DEM gave reasonable predictions of upward vertical force. However, the DEM predictions gave much higher side forces than that measured at sweep angles of 6 and 25° for both of the tilt angles of 0 and 20°. This was despite the DEM model accounting for the scrubbing reaction that occurs on the rear disc face at low sweep angles. Further work is needed to improve the model and adapt it to a freely rotating flat disc blade.

**Supplementary Materials:** The following supporting information can be downloaded at: https://www.mdpi.com/article/10.3390/agriculture13010206/s1, Table S1: Measured force and furrow area of whole/partial disc blades at different settings. Figure S1: Measured draught force of (Full) whole/partial disc blades at different settings. Figure S2: Measured vertical force of whole/partial disc at different settings. Figure S3: Measured side force of whole/partial (cut) disc at different settings. Figure S4: Comparison of furrow area of full and partial disc.

**Author Contributions:** Conceptualisation, A.K., J.M.A.D. and J.M.F.; methodology, A.K. and J.M.A.D.; Software (DEM) assistance: M.U. and A.K.; validation, A.K.; formal analysis, A.K. and J.M.A.D.; writing—original draft preparation, A.K.; writing—review and editing, M.U., C.S., J.M.A.D. and J.M.F.; supervision, J.M.A.D. and J.M.F. All authors have read and agreed to the published version of the manuscript.

**Funding:** This research was funded by Grains Research and Development Corporation (GRDC) (Project Number: USA00005) and the University of South Australia (UniSA).

**Institutional Review Board Statement:** Not applicable.

**Data Availability Statement:** The data presented in this study are available upon request from the first author A.K.

**Acknowledgments:** The authors gratefully acknowledge the Grains Research and Development Corporation (GRDC) and the University of South Australia (UniSA) in co-funding for this PhD research.

**Conflicts of Interest:** The authors declare no conflict of interest.

## Nomenclature

| | |
|---|---|
| a | Indices for sphere or implement |
| $A_c$ | Contact area, ($m^2$) |
| b | Indices for sphere or implement |
| c | Cohesion (kPa) |
| $c_a$ | Adhesion (kPa) |
| d | Blade working depth (m) |
| $D_p$ | Draft component of passive soil failure force (kN) |
| e | Coefficient of restitution |
| f | Forward rupture distance |
| $F_c$ | Cohesion force, (N) |
| $F^d{}_n$ | Normal damping force, (N) |
| $F^d{}_t$ | Tangential damping force, (N) |
| $F_n$ | Normal total contact force, (N) |
| $F^s{}_n$ | Normal contact force, (N) |
| $F^s{}_t$ | Tangential contact force, (N) |
| $F_t$ | Tangential total contact force, (N) |
| I | Moment of inertia, (kg $m^2$) |
| $K_1$ | Stiffness for loading, (N $m^{-1}$) |
| $K_2$ | Stiffness for unloading/reloading, (N $m^{-1}$) |
| L | Disc chord length measured at soil surface (m) |
| $L_a$ | The active part of disc chord that contributes in generating soil failure (m) |

| | |
|---|---|
| M | Moment, (N m) |
| $M_r$ | Moment due to rolling friction, (N m) |
| m | The ratio of forward rupture distance to the operating depth |
| m* | Mass, (kg) |
| n | Number of replications |
| $n_c$ | Damping factor |
| $n_k$ | Stiffness factor |
| $p_i$ | Elemental passive force (kN) |
| $N_\gamma$, $N_c$, $N_{ca}$, $N_q$ | Dimensionless numbers associated with gravitational, cohesion, adhesion and surcharge effects (as per Hettiaratchi et al. (1966)) |
| P | Passive cutting reaction force (kN) |
| q | Total surcharge pressure acting on the soil failure area (kPa) |
| r | Disc radius (m) |
| r* | Radius (m) |
| $r_{eq}$ | Equivalent radius, (m) |
| $r_{con}$ | Perpendicular distance of contact point from the centre of mass, (m) |
| R | Rotational acceleration, (rad $s^{-2}$) |
| $S_p$ | Side force component of passive force (kN) |
| $U_{abn}$ | Normal component of the relative displacement, (m) |
| $U_{abt}$ | Tangential component of the relative displacement, (m) |
| $\mathring{U}_{abn}$ | Normal component of the relative velocity, (m $s^{-1}$) |
| $\mathring{U}_{abt}$ | Tangential component of the relative velocity, (m $s^{-1}$) |
| $U_0$ | Residual overlap, (m) |
| $\ddot{U}$ | Translational acceleration, (m $s^{-2}$) |
| v | Forward speed ($ms^{-1}$) |
| $v_p$ | Passive soil failure velocity ($ms^{-1}$) |
| $V_p$ | Vertical component of passive force (kN) |
| $W_f$ | Furrow width (m) |
| w | Tool width (m) |
| Y | Yield strength (MPa) |
| Greek letters | |
| $\alpha_t$ | Tilt angle (deg) |
| $\alpha$ | Rake angle (deg) |
| $\beta$ | Sweep angle (deg) |
| $\beta_{cr}$ | Critical sweep angle (deg) |
| $\gamma_i$ | Initial soil bulk density ($kNm^{-3}$) |
| $\mu$ | Coefficient of friction |
| $\mu_r$ | Coefficient of rolling friction |
| $\lambda_\theta$ | Unit vector of angular velocity |
| $\delta$ | Angle of soil–metal friction (deg) |
| $\delta_x$ | Angle of soil–metal friction component projected on a horizontal plane (deg) |
| $\delta_z$ | Angle of soil–metal friction component projected on a vertical plane (deg) |
| $\xi$ | Cohesion energy density (J $m^{-3}$) |

## References


1. Ashworth, M.B.; Desbiolles, J.M.A.; Tola, E.K.H. *Disc Seeding in Zero-till Farming Systems—A Review of Technology and Paddock Issues*; Western Australia No-Tillage Farmers Association: Northam, Australia, 2010.
2. Aybek, A.; Başer, E.; Arslan, S.; Üçgül, M. Determination of the effect of biodiesel use on power take-off performance characteristics of an agricultural tractor in a test laboratory. *Turk. J. Agric. For.* **2011**, *35*, 103–113. [CrossRef]
3. Godwin, R.; Seig, D.; Allott, M. Soil failure and force prediction for soil engaging discs. *Soil Use Manag.* **1987**, *3*, 106–114. [CrossRef]
4. Al-Ghazal, A.A. An Investigation into the Mechanics of Agricultural Discs. Unpublished Ph.D. Thesis, Silsoe College, Cranfield Institute of Technology, Bedford, UK, 1989; 299p.
5. Alam, M.M. Soil Reaction Forces on Agricultural Disc Implements. Unpublished Ph.D. Thesis, University of Newcastle upon Tyne, Newcastle upon Tyne, UK, 1989; 129p.
6. Hettiaratchi, D. The Soil Contact Zones of Concave Agricultural Discs: Part 1, Theoretical Analysis. *J. Agric. Eng. Res.* **1997**, *67*, 113–125. [CrossRef]

7. Hettiaratchi, D. Prediction of Soil Forces Acting on Concave Agricultural Discs. *J. Agric. Eng. Res.* **1997**, *68*, 51–62. [CrossRef]
8. Godwin, R.J.; Dogherty, M.J.O. Integrated soil tillage force prediction models. *J. Terramechanics* **2007**, *44*, 3–14. [CrossRef]
9. Hettiaratchi, D.R.P.; Witney, B.D.; Reece, A.R. The calculation of passive pressure in two-dimensional soil failure. *J. Agric. Eng. Res.* **1966**, *11*, 89–107. [CrossRef]
10. Hettiaratchi, D.R.P.; Reece, A.R. The calculation of passive soil resistance. *Geotechnique* **1974**, *24*, 289–310. [CrossRef]
11. Khosravani, A. A semi-Analytical Force Prediction Model for a Rotating Flat Disc Blade Opener. Unpublished Ph.D. Thesis, University of South Australia, Adelaide, Australia, 2014; 246p.
12. Tamás, K.; Jóri, I.J.; Mouazen, A.M. Modelling soil–sweep interaction with discrete element method. *Soil Tillage Res.* **2013**, *134*, 223–231. [CrossRef]
13. Fielke, J.M.; Ucgul, M. Saunders. Discrete element modelling of soil-implement interaction considering soil plasticity, cohesion and adhesion. In *2013 Kansas City, Missouri, 21–24 July 2013*; American Society of Agricultural and Biological Engineers: St. Joseph, MI, USA, 2013; p. 1.
14. Li, P.; Ucgul, M.; Lee, S.-H.; Saunders, C. A new approach for the automatic measurement of the angle of repose of granular materials with maximal least square using digital image processing. *Comput. Electron. Agric.* **2020**, *172*, 105356. [CrossRef]
15. Barr, J.B.; Desbiolles, J.M.; Fielke, J.M.; Ucgul, M. Development and field evaluation of a high-speed no–till seeding system. *Soil Tillage Res.* **2019**, *194*, 104337. [CrossRef]
16. Tekeste, M.Z.; Balvanz, L.R.; Hatfield, J.L.; Ghorbani, S. Discrete element modeling of cultivator sweep-to-soil interaction: Worn and hardened edges effects on soil-tool forces and soil flow. *J. Terramechanics* **2019**, *82*, 1–11. [CrossRef]
17. Aikins, K.A.; Barr, J.B.; Antille, D.L.; Ucgul, M.; Jensen, T.A.; Desbiolles, J.M. Analysis of effect of bentleg opener geometry on performance in cohesive soil using the discrete element method. *Biosyst. Eng.* **2021**, *209*, 106–124. [CrossRef]
18. Zeng, Z.; Ma, X.; Chen, Y.; Qi, L. Modelling residue incorporation of selected chisel ploughing tools using the discrete element method (DEM). *Soil Tillage Res.* **2020**, *197*, 104505. [CrossRef]
19. Saunders, C.; Ucgul, M.; Godwin, R.J. Discrete element method (DEM) simulation to improve performance of a mouldboard skimmer. *Soil Tillage Res.* **2021**, *205*, 104764. [CrossRef]
20. Aikins, K.A.; Ucgul, M.; Barr, J.B.; Jensen, T.A.; Antille, D.L.; Desbiolles, J.M. Determination of discrete element model parameters for a cohesive soil and validation through narrow point opener performance analysis. *Soil Tillage Res.* **2021**, *213*, 105123. [CrossRef]
21. Cundall, P.A.; Strack, O.D.L. A discrete numerical model for granular assemblies. *Géotechnique* **1971**, *29*, 47–65. [CrossRef]
22. Ucgul, M.; Fielke, J.M.; Saunders, C. Three-dimensional discrete element modelling (DEM) of tillage: Accounting for soil cohesion and adhesion. *Biosyst. Eng.* **2015**, *129*, 298–306. [CrossRef]
23. Khosravani, A.; Desbiolles, J.; Fielke, J.M. Circular disc blade considerations in soil force prediction modelling. *J. Agric. Sci. Technol A.* **2014**, *4*, 371–383.
24. Mckeys, E. *Soil cutting and Tillage*; Elsevier, Science Publishers: Amsterdam, The Netherlands, 1985.
25. EDEM. *EDEM Theory Reference Guide*; DEM Solutions Ltd.: Edinburgh, UK, 2011.
26. Walton, O. *Elastic-Plastic Contact Model*; Company Report; DEM Solutions: Edinburgh, UK, 2006.
27. Walton, O.R.; Braun, R.L. Stress calculations for assemblies of inelastic spheres in uniform shear. *Acta Mech.* **1986**, *63*, 73–86. [CrossRef]
28. Raji, A.O. Discrete element modelling of the deformation of bulk agricultural particles. Ph.D. Thesis, University of Newcastle upon Tyne, Newcastle upon Tyne, UK, 1999.
29. Huser, A.; Vernvold, O.K. Prediction of sand erosion in process and pipe components. In *BHR Group Conference Series Publication*; Mechanical Engineering Publications Limited: St. Joseph, MI, USA, 1998; Volume 31, pp. 217–228.
30. Hudson Tool Steel. P20 Mold Steel. 2016. Available online: http://www.hudsontoolsteel.com/technical-data/steelP0 (accessed on 10 October 2020).
31. Academia. Some Useful Numbers for Rocks and Soils. 2015. Available online: www.academia.edu/4056287/Some_Useful_Numbers_for_rocks_and_soils (accessed on 10 October 2020).
32. Asaf, Z.; Rubinstein, D.; Shmulevich, I. Determination of discrete element model parameters required for soil tillage. *Soil Tillage Res.* **2007**, *92*, 227–242. [CrossRef]
33. Budynas, R.G.; Nisbett, K.J. *Shigley's Mechanical Engineering Design*; McGraw-Hill Education: New York, NY, USA, 2012.
34. Ucgul, M.; Saunders, C.; Fielke, J.M. Discrete element modelling of top soil burial using a full scale mouldboard plough under field conditions. *Biosyst. Eng.* **2017**, *160*, 140–153. [CrossRef]
35. Nalavade, P.P.; Salokhe, V.M.; Niyamapa, T.; Soni, P. Performance of Free Rolling and Powered Tillage Discs. *Soil Tillage Res.* **2010**, *109*, 87–93. [CrossRef]

*agriculture*

MDPI

*Article*

# Determination of Performance of No-Till Seeder and Stubble Cutting Prototype

**Mehmet Emin Bilgili [1], Yasemin Vurarak [1,*] and Ali Aybek [2]**

[1] Eastern Mediterranean Agricultural Research Institute Directorate, 01375 Adana, Turkey
[2] Biosystems Engineering Department, Faculty of Agriculture, Kahramanmaraş Sütçü İmam University, 46000 Kahramanmaraş, Turkey
* Correspondence: yasemin.vurarak@tarimorman.gov.tr

**Abstract:** One of the most common problems in maize production is the management of plant residues. Small agricultural enterprises, which cannot allocate capital for acquiring stalk cutting machines for their operation, face many technical problems in preparing the sowing bed for the products that will be planted after maize. Stalks of maize that cannot be shredded adequately and on time cause machinery to clog, prevent the preparation of a proper sowing bed, increase fuel consumption and increase costs. The aim of this study is to compare the no-till sowing machine prototype and stalk cutting machine prototype with the classical stalk cutter in terms of some management values. The prototype stubble cutting machine used in the study was manufactured with a cylindrical structure and equipped with 24 cutting blades 1 cm thick and 8 cm wide. İn addition, the prototype stubble cutting machine used in the study was manufactured with a cylindrical structure and equipped with 24 cutting blades 1 cm thick and 8 cm wide. İn addition, the no-till seeder prototype was manufactured as a bucket-type seed hopper equipped with granular fertilizer capable of sowing four rows. It was concluded that the stubble cutting machine prototypes resulted in less fuel consumption with lower penetration resistance when compared with the classical stalk shredder.

**Keywords:** stubble management; no-till sowing; stalk cutting; post-harvest; prototype

**Citation:** Bilgili, M.E.; Vurarak, Y.; Aybek, A. Determination of Performance of No-Till Seeder and Stubble Cutting Prototype. *Agriculture* **2023**, *13*, 289. https://doi.org/10.3390/agriculture13020289

Academic Editors: Mustafa Ucgul and Chung-Liang Chang

Received: 19 December 2022
Revised: 16 January 2023
Accepted: 23 January 2023
Published: 25 January 2023



## 1. Introduction

The Çukurova region in the south of Turkey is comprised of plains with one of the country's most fertile agricultural lands and good soil structure suitable for irrigated agriculture and production potential. Irrigated agriculture with high production potential is common in the region, and it is possible to grow 2–3 crops per year due to the Mediterranean climate. The superior advantages provided by the ecology have allowed the cultivation of many field crops in the Çukurova region, and one of these plants is maize (*Zea mays* L.). However, as reported in the survey study examining the harvesting and threshing traditions of Çukurova farmers, the farmers burn all the remaining maize stalks in the field (stubble burning) after harvest, which is a well-known problem waiting for a solution [1]. In addition to the low organic matter content of the soil in the region [2], the traditional high number of tillage operations and illegal burning of stubble residues [3] are among the leading problems that limit the sustainability of soil productivity. Designing and manufacturing agricultural machinery suitable for these activities that help sustainable management of agricultural lands are extremely important for the agricultural future of the region. Soil resources in Turkey as well as in the world are limited and cannot be increased. Reckless use of soil resources is one of the most important factors that increases the severity of the problem [4]. In addition, from an ecological point of view, it can be said that millions of tons of plant waste such as stubble, straw and paper raw materials are being burned [5]. A study carried out in the Çukurova region revealed that stubble is an important source of organic matter, and that 93–98% of stubble buried at a depth of 12 cm is transformed into humus within 18 months. The decomposition rate of the stubble left on the surface, however, was approximately 1/3 of the buried amount [6].

In order to cope with an increase in precipitation in the autumn grain sowing season in recent years, producers have burned plant residues left after the harvest of maize to avoid delaying the sowing time, causing the deterioration of soil, air and water resources over time. The existing technology should be changed and no-till seeders and stalk cutters should be included in the machine parks to ensure that stubble is collected without burning and some of it is mixed with the soil. With the help of the pressure of local governments and non-governmental organizations, and the subsidies granted for the machinery by the government to prevent stubble burning, a decrease in burning maize stubble has ocurred in the region, albeit partially. However, the amount of capital allocated to these machines and the tractor power requirement for their use limit the purchase demand of small producers. The no-till seeders and stalk cutters on the market mainly consist of heavy tonnage machines that require high tractor power. This situation has created a driving force for researchers to design prototype machine sets with low costs, which can be an alternative to existing machine sets and can perform no-till seeding and stalk cutting work as required. The region is located in the Çukurova region, which is Turkey's largest plain, and the airport is frequently closed to flights due to stubble burning. There is an urgent need for machines that farmers can easily buy to prevent stubble formation. A conventional tillage seeder and stubble mower costs around $15,000 on average and is very expensive for our farmers. The prototypes cost about $2000.

Using no-till machines in the sowing system is now also a necessity for maintaining soil fertility in terms of global warming and climate change. However, it is known that access to machines used in this context is limited for small farm operations. The aim of this study is to present no-till seeder and stubble cutter prototypes to maize producers that are affordable for small enterprises. Another aim is to increase the farmers' options for preparing more economical seedbeds without burning stubble.

## 2. Materials and Methods

### 2.1. The Trial Area and Trial Materials

In the study area, the Çukurova region, there is no problem in terms of soil depth, except for the coastal part. This region is slightly different from the typical Mediterranean climate. An evaluation of climatic data for many reveals that the long annual precipitation average is 647.5 mm and the annual average temperature is 18.74 °C [7]. The study was carried out for 2 years (2018, 2019 years) in the "36° 51′ 18″ N and 35° 20′ 51″ E" coordinated trial fields of the Eastern Mediterranean Agricultural Research Institute Directorate. The general soil properties at the station where the trial was carried out were determined as follows. The lime status of the soil is between 16 and 21%, organic matter amount is between 1.27% and 2.33%, cation exchange capacity ratios are between 21.24 and 38.02 cmol $kg^{-1}$, soil pH value is between 7.50 and 7.99 and $P_2O_5$ values are between 16 and 179 kg $ha^{-1}$. In addition, the silt ratio of the soil has been determined as between 25.3 and 53.7%, the sand ratio is between 9.7 and 51.6% and the clay ratio is between 23.1 and 41.8% [8]. The soil surface in the trial areas is stone-free and the slope is in the range of 0–1°. It is in the alluvial soil series in terms of soil structure. Soil texture is homogeneous.

Some data of the trial area are presented in Table 1, and some climatic parameters recorded during the trial are presented in Table 2.

**Table 1.** Soil properties of the trial area.

| Year | Depth (cm) | Saturation (%) | E.C. (dS m$^{-1}$) | pH | Lime (%) | Organic Matter (%) | Plant Availability (kg ha$^{-1}$) | | Physical Structure (%) | | |
|---|---|---|---|---|---|---|---|---|---|---|---|
| | | | | | | | $P_2O_5$ | $K_2O$ | Sand | Silt | Clay |
| 2018 | 0–30 | 55.50 | 1.05 | 7.60 | 13.43 | 1.28 | 56.0 | 901.5 | 32.4 | 35.5 | 31.5 |
| 2019 | 0–30 | 62.7 | 0.295 | 7.75 | 16.42 | 1.63 | 121.0 | 577.0 | 31.4 | 37.7 | 30.9 |

**Table 2.** Some of the climatic data recorded by years during the trial.

| Years | Months | | | | | | | | | | | | |
|---|---|---|---|---|---|---|---|---|---|---|---|---|---|
| | Parameters | I | II | III | IV | V | VI | VII | VIII | IX | X | XI | XII |
| 2018 | Temperature (°C) | 8.7 | 13.9 | 15.7 | 20.5 | 21.7 | 27.3 | 29.5 | 29.9 | 26.1 | 23.1 | 15.5 | 9.0 |
| 2019 | | 8.7 | 10.9 | 15.2 | 18.6 | 21.7 | 26.3 | 30.4 | 29.9 | 27.7 | 22.1 | 15.8 | 12.6 |
| 2018 | Relative Humidity (%) | 61.3 | 67.7 | 60.7 | 57.8 | 68.1 | 63.8 | 67.5 | 67.4 | 59.9 | 56.5 | 52.4 | 64.8 |
| 2019 | | 61.9 | 50.7 | 63.5 | 60.7 | 69.1 | 69.2 | 64.4 | 67.3 | 66.1 | 53.5 | 66.3 | 74.4 |
| 2018 | Precipitation (mm) | 24.5 | 9.7 | 12 | 1.8 | 12.7 | 18.4 | 0.2 | 8.2 | 24.3 | 3.2 | 22 | 194.8 |
| 2019 | | 49.6 | 0.6 | 65.6 | 62.4 | 51 | 10.6 | 0 | 0 | 11.2 | 40.6 | 119 | 49 |
| Sowing, harvest and stubble cutting dates during trial | | | | | | | | | | | | | |
| 2018 | Sowing | | | | | 03.05.2018 | | | | | | | |
| | Harvest | | | | | | | | | 01.09.2018 | | | |
| | Stubble cutting | | | | | | | | | 09.09.2018 | | | |
| 2019 | Sowing | | | | | 12.03.2019 | | | | | | | |
| | Harvest | | | | | | | | | 05.09.2019 | | | |
| | Stubble cutting | | | | | | | | | 12.09.2019 | | | |

The technical specifications of all tools and machines used during the trial are given in Table 3.

**Table 3.** Technical specifications of agricultural tools and machines used in the trial.

| Equipment | Technical Specifications | | |
|---|---|---|---|
| | Working Width (cm) | No. of Units (pcs) | Working Depth (cm) |
| Goble | 280 | 18 | 15 |
| Chisel | 260 | 9 | 25 |
| Disk harrow | 220 | 22 | 15 |
| Back double blade plough | 210 | 5 | 20 |
| Back plug roller | 280 | 4 | 20 |
| Hoeing machine | 280 | 3 | 15 |
| Band fertilizer machine | 280 | 5 | 15 |
| Spraying machine | 1200 | 1 | - |
| Round baler | 140 | 1 | - |
| No-till seeder (pneumatic) | 280 | 4 | 5–8 |
| Stubble shredder | 400 | 36 | - |

The weights of the no-till seeder and stubble shredder machines, the technical specifications of which are given in Table 3, are 1100 kg and 1740 kg, respectively. The stubble shredder takes action from the PTO with 540 rpm (Figure 1).

**Figure 1.** No-till seeder and stubble shredder used in the trial (traditional).

The technical specifications of the no-till seeder prototype and stubble cutter prototype are given in Table 4.

**Table 4.** The technical specifications of the no-till seeder prototype and stubble cutter prototype used in the trial.

| Technical Specifications | Stubble Cutter Machine Prototype | No-Till Seeder Prototype (Mechanic) |
|---|---|---|
| Working width road condition (cm) | 280 | 300 |
| Working width (cm) | 240 | 280 |
| No. of blades/units | 24 blades | 4 units |
| Working depth (cm) | 5 | 5–8 |
| Weight (kg) | Empty weight: 1200 Full weight: 1450 | Empty weight: 775 |

Working principles of the prototype machines:

Stubble cutter machine prototype: The machine consists of assembling a cylinder with a working width of 240 cm and a diameter of 80 cm with the chassis. The top of the cylinder is equipped with a 240 cm blade section, parallel to the axis, with 24 blades which are 1 cm thick and 8 cm wide. The material of the cutter blades is steel and the distance between the two blades is 11 cm. They are joined to the cylinder by longitudinal welding. The cylinder of the stalk shredder, which has four pressure pallets on the front, can be used full or empty. The empty weight is approximately 1200 kg, and the full weight (with waste engine oil) is 1450 kg. Field and road positions are adjustable (Figures 2 and 3).

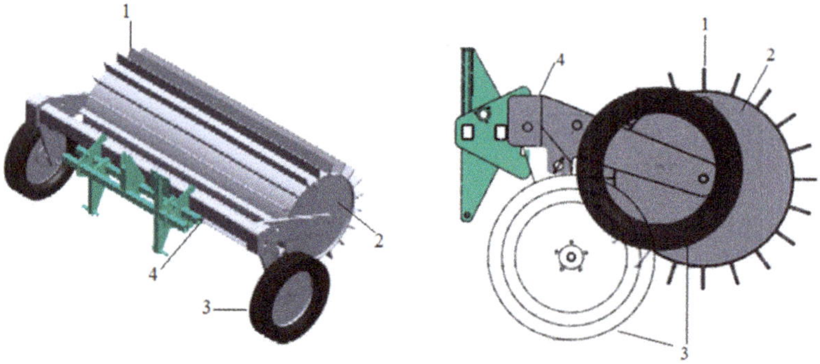


**Figure 2.** Design images of the stubble cutter prototype ((**1**) Stubble cutter blades; (**2**) Cylinder chassis; (**3**) Road condition wheel; (**4**) Main carrier chassis).

**Figure 3.** Recorded in experiments: (**a**) traditional stubble shredder; (**b**) efficiency of traditional stubble shredder; (**c**) prototype stubble cutter; (**d**) images of the effect of the prototype stubble cutter.

No-till seeder prototype: There is a spring-loaded stalk cutter disc in front of the machine and a 3 cm wide and 10 cm long soil cultivator rung for seedbed pre-preparation on the same axis behind the disc. These rungs are fitted with stubble cleaning flaps. After the seeds are released by the seeder, they are covered (pressure wheel). The elements that make up the whole machine are combined with a chassis and sowing can be completed in four rows. In addition, the seed hopper is designed with buckets, and there are two fertilizer hoppers in total, one for each sowing bucket, to fertilize with sowing (Figures 4 and 5).

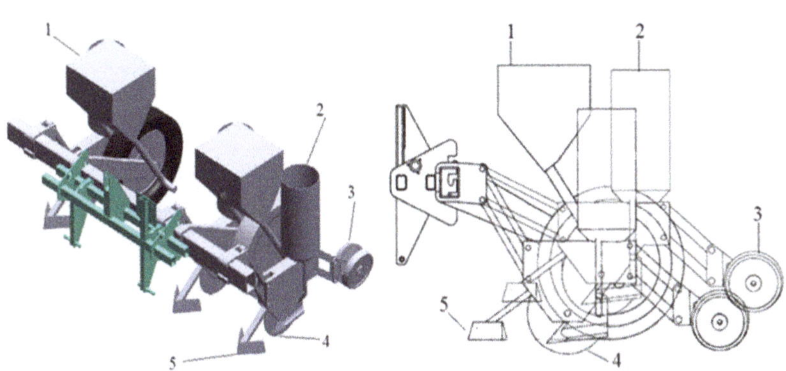


**Figure 4.** Design images of the no-till seeder prototype ((**1**) Fertilizer hopper; (**2**) Seed hopper; (**3**) Seed press roller; (**4**) Sowing coulter with cutting disc; (**5**) Row cleaning attachment).

Figure 5. Construction and usage stages of the prototype sowing machine used in the research.

### 2.2. Agricultural Activities Carried out During the Trial

A registered maize variety planted in the plain was used as plant material in the trial. This variety is grain, P1429 corn. The FOA maturity group of the variety is GDD 1577 °C. The number of physiological maturation days is 115–117 days and its purity is 100%. Sowing was carried out at a depth of 4–6 cm with 70 cm between rows and intrarow of 18–20 cm, according to the subject (70 000–75 000 seeds ha$^{-1}$). Taking into account the deficiencies revealed by the soil analysis, 80 kg ha$^{-1}$ of $P_2O_5$ fertilizer and half of the nitrogen fertilizer (140 kg ha$^{-1}$) were applied to the strips during sowing, and the other half was applied before the first irrigation [9]. Trial plots comprised of 4 rows with a length of 55 m and a distance of 2.8 m between plots were prepared. Weed control was carried out and measures were taken to deal with maize stem worms, cob worms and aphids, as well as agricultural control for soybean in case of any disease or pest according to the relevant procedure [10]. In maize production, the first irrigation was performed approximately one month after the sowing, the second irrigation was completed when the tassel emergence was completed, the third irrigation took place when the cob emerged, the fourth irrigation was applied at grain formation and the fifth irrigation was carried out during the milk formation phases [11,12]. Maize leaves were harvested with a combine harvester when the husks dried, the cobs separated from the stem and the grains hardened (moisture 15%) [13].

### 2.3. Trial Subjects

Two no-till seeder and stalk shredder systems were examined comparatively within the scope of the study. Soil tillage and maintenance works in both subjects were conducted in the same way, at the same time, with the same machines. Accordingly, the mechanical systems that constitute the trial subjects are as follows:

System 1: No-till seeding method by stalk shredding (subject to control): The previous crop was shredded with a stalk shredder, collected and then the stubble was directly ridge planted with a no-till planter (traditional).

Machines used in System 1: Goble disc + ridge planter + no-till seeder + hoe + furrow plow with fertilizer + pulverizer + combine harvester + stalk shredder + cylindrical straw baling machine.

System 2: No-till sowing method by cutting the stubble (prototype): The plant stubble from the previous crop was shredded with the stubble cutter prototype, collected, then followed by ridge sowing with the no-till seeder prototype (alternative prototype).

Machines used in System 2: Goble disc + ridge planter + no-till seeder prototype + hoe + furrow plow with fertilizer + pulverizer + combine harvester + stubble cutter prototype + cylindrical straw baling machine.

The goble disc was used before sowing in both systems. This is because the Çukurova region is a region that receives a lot of precipitation during the winter months, and the weed problem is quite abundant and the farmers cannot cope with it. Soil tillage is performed

5 months before sowing the main crop of corn, which is planted on pre-prepared ridges with a no-till seeder. For our region, the sowing beds must be prepared in advance in order to eliminate the weed problem before sowing the main crops.

### 2.4. Measurement, Analysis and Calculation Methods

References and methodologies of the measurement, analysis and calculation methods are given in Table 5. All measurements and calculations were made according to the methods reported in these references.

**Table 5.** Measurement method references.

| Measurement | References | Methodology |
|---|---|---|
| Measurements related to machine performance | | |
| Stalk cutting stress, MPa | [14] | The maximum pressure applied by the blade to the plant stem has been calculated as an average, according to the method in the reference. |
| Machine labor requirement, h $ha^{-1}$ | [15,16] | The main usage times per unit of the machines in the system are measured with 3 replications from the starts to the end of the work and calculated to determine an average value as "ha" unit. The base time was taken into account in the measurements. |
| Human labor requirement, h $ha^{-1}$ | [15,16] | The main usage times per unit of the human labor in the system are measured with 3 replications from the starts to the end of the work and calculated to determine an average value as "ha" unit. The base time was taken into account in the measurements. |
| Tractor power requirement, kW | [17] | The tractor power required by the prototype machines was quoted from the reference guide according to operating width and weight. |
| Fuel consumption, L $ha^{-1}$ | [16–18] | Fuel consumption was measured in 3 repetitions per hectare for all machines in the system and calculated according to the tank filling method, and the average value was taken. |
| Measurements for plant growth | | |
| Number of plants, Unit per $m^{-2}$ | [13,19] | The number of plants per 1 $m^2$ on a row after emergence was measured with 3 replications, and their average was taken. |
| Plant height, cm | [13,19] | 10 randomly selected plants from each study plot were measured with a wooden meter in 3 replicates. |
| Stem thickness, cm | [20] | Measurements were made with a digital caliper for the first 5 cm of the stem level from the soil surface of the plant before harvesting 5 randomly selected plants from each plot in 5 repetitions, and the average value was taken. |
| Height of first cob from the ground, cm | [13,19] | Measurements were made in each system using a tape measure from the soil surface to the point where the first cob was tied in 5 randomly selected plants before harvest, and the averages were taken. |
| Stem moisture, % | [13,19] | 25 cm samples were taken from the stem of 5 randomly selected plants in each system before harvesting, and the average stem moisture was calculated by taking the difference between their weights before and after they were kept in an oven set at 105 °C for 24 h according to the dry base. |
| Stubble (Biomass) density and surface coverage rate, Kg $m^{-2}$ | [21–24] | A meter was placed on a 1 m × 1 m piece of wood material at an angle of 45° and markings were made at 1 cm intervals, and the points corresponding to the shredded stems were marked, counted, the total of the points in contact with the parts were divided by 100 cm and the surface coverage ratio was calculated by using the cross line method (%). Calculations were made with 3 replications for the systems and the averages were taken. This process was conducted separately for 2 different situations, before and after baling. |

**Table 5.** *Cont.*

| Measurement | References | Methodology |
|---|---|---|
| Biomass moisture, % | [13–19] | Samples weighing 500 g were taken from the stubble remaining on the soil surface after the harvest, with 3 repetitions, and the average stubble moisture was determined by taking the difference between their weights before and after they were kept in an oven set at 105 °C for 24 h on a dry basis. |
| Yield, t $ha^{-1}$ | [13,19] | For each system, the average grain moisture was expected to decrease to 15%, and cobs in 5 $m^2$ were harvested with 3 replications before harvesting and the averages were taken. |
| Measurements and analyzes related to soil structure | | |
| Bulk density, g $m^{-3}$ | [25] | Undisturbed soil samples at different depths with 3 replications were processed using 100 $cm^3$ core sets according to the cylindrical method, and the averages were taken to determine the soil volumetric mass. |
| Soil humidity, % | [26] | The average soil humidity was calculated by determining the difference in weight before and after the undisturbed soil samples were kept in an oven set at 105 °C for 24 h on a dry basis with 3 replications. |
| Soil penetration resistance, MPa | [27,28] | The averages were achieved by using a penetrelogger with memory, LSD display and GPS source that can measure in the range of 0–80 cm soil depth and 0–5 MPa to measure rows with 3 replications at different depths after. |
| Cost calculations | | |
| Input/output ratio | [15] | The input/output ratio was calculated by considering the annual records per unit area. Calculations were made as cost–benefit. |

At the end of the study that lasted for two years, the results of the measurements of the two systems were compared using the average, ratio and proportion methods, and the operating data of the machines in the alternative system were determined.

## 3. Results

The germination rate of the seed used was 90%, and the average field shoot emergence rate for 2 years according to the systems was 95% for system 1 and 86% for system 2, respectively. Some vegetative development data of the plants in the main crop maize production are given in Table 6 according to the years and systems. When an evaluation is made according to the average data, it is possible to say that the efficiency in System 1 is less than System 2, albeit at a low rate. If an evaluation is made in general, there is no considerable difference between the parameters for the whole development.

**Table 6.** Data obtained from the maize plant according to years in the trial.

| Years | Subject | Plant Height (cm) | Stem Thickness (mm) | Number of Plants (per $m^{-2}$) | Height of First Cob from the Ground (cm) | Stem Humidity (%) | Yield (t $ha^{-1}$) |
|---|---|---|---|---|---|---|---|
| 2018 | System 1 | 233 | 23.5 | 5.5 | 102 | 17 | 14.81 |
| | System 2 | 232 | 24.1 | 5.4 | 109 | 20 | 16.02 |
| 2019 | System 1 | 243 | 23.9 | 4.5 | 103 | 18 | 15.17 |
| | System 2 | 240 | 24.5 | 4.0 | 104 | 21 | 15.35 |
| System 1 ave. | | 238 | 23.7 | 5.0 | 102.5 | 17.5 | 15.01 |
| System 2 ave. | | 236 | 24.3 | 4.7 | 106.5 | 20.5 | 15.68 |

The stubble coverage rates on the land after harvest are given as before and after baling in Table 7. The stubble coverage rate, which is over 95% after the maize harvest every year, has been collected by baling in such a way that 24–25% stubble remains. On average, stubble moisture was determined to be between 20 and 25% during both years.

**Table 7.** Biomass status of maize stubble on the field surface.

| Subject | Biomass Status | Years | |
|---|---|---|---|
| | | 2018 | 2019 |
| System 1 | Coverage rate after harvest (%) | 96 | 95 |
| | Coverage rate after baling (%) | 25 | 27 |
| | Biomass moisture (%) | 22 | 23 |
| System 2 | Coverage rate after harvest (%) | 97 | 96 |
| | Coverage rate after baling (%) | 24 | 25 |
| | Biomass moisture (%) | 20 | 21 |

The volume weight and soil moisture measurements determined in the plots where the systems were tested are given in Table 8. Considering the average soil volume weight and soil moisture values of both subjects, it is seen that there is no significant effect in terms of these parameters.

**Table 8.** The volume weights and moisture of the soils taken from the trial area of the systems.

| Year | Subject | Bulk Density (g $m^{-3}$) | | Soil Moisture (%) | |
|---|---|---|---|---|---|
| | | 0–15 cm | 15–30 cm | 0–15 cm | 15–30 cm |
| 2018 | System 1 | 1.68 | 1.69 | 15.94 | 16.14 |
| | System 2 | 1.63 | 1.71 | 15.34 | 15.73 |
| 2019 | System 1 | 1.69 | 1.70 | 15.97 | 16.06 |
| | System 2 | 1.64 | 1.72 | 15.35 | 15.72 |
| System 1 ave. | | 1.68 | 1.69 | 15.95 | 16.10 |
| System 2 ave. | | 1.63 | 1.71 | 15.34 | 15.72 |

Soil penetration resistance between 0 and 30 cm layers is given in Table 9. The soil penetration resistance values obtained in the current study were lower than the threshold of penetration resistance for field crops.

**Table 9.** Penetration resistances of the soils taken from the trial area of the systems.

| Year | Subject | Penetration Resistance (MPa) | | | | | |
|---|---|---|---|---|---|---|---|
| | | 0–5 cm | 5–10 cm | 10–15 cm | 15–20 cm | 20–25 cm | 25–30 cm |
| 2018 | System 1 | 0.65 | 1.25 | 1.37 | 1.40 | 1.39 | 1.53 |
| | System 2 | 0.52 | 0.98 | 1.07 | 1.26 | 1.41 | 1.63 |
| 2019 | System 1 | 0.58 | 1.23 | 1.35 | 1.43 | 1.43 | 1.50 |
| | System 2 | 0.48 | 0.95 | 1.05 | 1.20 | 1.40 | 1.50 |
| System 1 ave. | | 0.56 | 1.24 | 1.36 | 1.41 | 1.41 | 1.51 |
| System 2 ave. | | 0.50 | 0.96 | 1.07 | 1.23 | 1.40 | 1.56 |

The fuel consumption of the systems, the machine labor requirements and the income and expense ratios of the system are given in Table 10.

**Table 10.** Some operational data and cost status of the systems.

| Year | Subject | Machine Labor Requirement (h ha$^{-1}$) | Human Labor Requirement (h ha$^{-1}$) | Fuel Consumption (L ha$^{-1}$) | Input/Output Ratio |
|---|---|---|---|---|---|
| 2018 | System 1 | 1.33 | 1.49 | 104.2 | 34.70 |
| | System 2 | 1.51 | 1.69 | 95.4 | 40.85 |
| 2019 | System 1 | 1.26 | 1.41 | 103.8 | 27.25 |
| | System 2 | 1.43 | 1.60 | 96.0 | 30.35 |
| System 1 ave. | | 1.29 | 1.45 | 104.0 | 30.97 |
| System 2 ave. | | 1.47 | 1.65 | 95.7 | 35.60 |

The total machine labor requirement values on the basis of systems was taken into consideration; the averages for 2 years, an average of 1.29 h ha$^{-1}$ for System 1 and System 2, are required, while this rate is 1.47 h ha$^{-1}$ for system 2 where a prototype machine is used. In addition, when the human labor requirements of the systems are examined, it has been determined that the labor requirement of system 1 is 1.45 h ha$^{-1}$ on average, and System 2 is 1.65 h ha$^{-1}$. In terms of fuel consumption, it has been determined that the average for 2 years is 104 L ha$^{-1}$ for system 1 and 95.7 L ha$^{-1}$ for system 2 where prototype machines are used. In terms of income/expense ratio, System 2 (35.60) was determined to be a more profitable investment than System 1 (30.97).

The fuel consumption and forward speeds of prototype machines and conventional machines on the market are given in Table 11. While the average fuel consumption of the prototype stubble cutter is 3.25 L ha$^{-1}$, the average fuel consumption of a conventional stalk shredder is 16.80 L ha$^{-1}$. It has been determined that the classical stalk shredder consumes 5.2 times more fuel than the prototype stubble cutter. In this respect, it is possible to say that the prototype stubble cutter will be effective in reducing costs. At the same time, it was determined that the speed of the prototype stubble cutter was slightly slower because it sunk into the soil while operating. During operation, the maximum pressure applied to the stubble cutting point was calculated as 0.81 MPa. The operating values of the prototype no-till seeder, on the other hand, consumed 9.5 L ha$^{-1}$, while the fuel consumption of the traditional no-till seeder was approximately 1/3 less. It has been determined that they are similar (4.4 km h$^{-1}$) in terms of forward speeds.

**Table 11.** Some operational parameters of prototype stalk cutters and classical stalk shredders.

| Equipment | Operating Width (cm) | Fuel Consumption (L ha$^{-1}$) | Forward Speed (km h$^{-1}$) | Necessary Tractor Power (kW) |
|---|---|---|---|---|
| Classic stalk shredder | 400 | 16.80 | 8.5 | 80 |
| Stubble cutter machine prototype | 280 | 3.25 | 7.8 | 37 |
| No-till seeder | 280 | 3.5 | 4.5 | 60 |
| No-till seeder prototype | 280 | 9.5 | 4.4 | 50 |

Possible malfunctions were also observed during the trials. During the two-year studies, no malfunctions occurred in the prototypes. However, in the no-till seeder prototype, there is a possibility of malfunctions in mechanical parts, chain and gear system over time. No malfunction is expected in the prototype stubble-cutting machine, except for the rolling bearing.

## 4. Discussion

Soil compaction is a factor that affects the germination rate of the crop to be planted. Whether in conventional or direct sowing, it is not desirable to have soil with a compaction level that affects germination. However, it is known that soil penetration resistance increases as soil depth increases [28]. Looking at the 2-year averages of the systems, it has been determined that the penetrometer values in the range of 0–5 cm, 5–10 cm, 10–15 cm and 15–20 cm increase as the depth increases in System 1. That is, the mean penetration resistance measured for System 1 and System 2 at 0–5 cm depth was 0.56 MPa and 0.50 MPa, respectively, while this value was 1.51 MPa and 1.56 MPa at 25–30 cm depth, respectively. However, for both systems, there is no compression in the range that will restrict plant growth. It is thought that this can be associated with the total weights of the machines used in the systems. It is stated that the limit penetration resistance value, which prevents plant root growth, is 3 MPa [29]. It has been reported that this value is 3.6 MPa in soils where a conventional tillage system is applied and 5 MPa in soils without tillage [30]. If there are no permanent root canals and cracks in the soil, the critical value is considered to be 2 MPa [31]. In another study, it is reported that the 2 MPa limit value cannot be accepted as a limiting value for root growth for different tillage systems [32]. In addition, it is reported that in soils with high clay content, the critical limit value should be 2 MPa when a conventional tillage system is applied, 3 MPa when chisel cultivation is applied and 3.5 MPa when a direct sowing system is applied [33]. The soil penetration resistance values obtained in the current study were lower than the threshold limit of penetration resistance for field crops. It was suggested that the threshold of penetration resistance for field crops should be in the range of 2 to 3.5 MPa [34,35].

In terms of volume weight value, 1.68 g $cm^{-3}$ and 1.63 g $cm^{-3}$ are found for the traditional System 1 and alternative System 2, respectively, at a depth of 0–15 cm and 1.69 $cm^{-3}$ and 1.71 $cm^{-3}$, respectively, at a depth of 15–30 cm. As the depth increased, the volume weight increased slightly in both systems. However, there were no extreme differences between subjects. In a study, it was stated that the limit volume weight value for plant root development for clay loam soils above 1.40 g $cm^{-3}$ should be considered in terms of compaction potential [36,37]. It is stated that the ideal volume weight in clay loam soils is <1.40 g $cm^{-3}$, the volume weight that affects root growth is 1.63 g $cm^{-3}$ and the volume weight that inhibits root growth is >1.80 g $cm^{-3}$ [38]. In another study, it is stated that plants develop well when the soil volume weight is in the range of 1.15–1.45 g $cm^{-3}$, while a volume weight value greater than 1.55 g $cm^{-3}$ is not suitable for plant growth [39]. In a study conducted in Mexico, the effects of direct sowing and traditional tillage methods on the physical properties of a clay loam soil were investigated. The results showed that there was little difference between the tillage systems and the volume weight values were 1.21 g $cm^{-3}$–1.39 g $cm^{-3}$, respectively [40]. In a study investigating the effects of direct sowing and conventional tillage on soil properties and corn yield in southwest China, it was determined that the direct sowing method decreased soil volume weight by 7% in the upper 30 cm and increased soil moisture content by 3% compared to the conventional tillage method. In addition, they stated that corn yield increased by 11% in the direct sowing method compared to the traditional tillage method [41]. In our study, it was determined that the prototypes used in System 2 increased the efficiency by 4.2% compared to the other system. In parallel with our study, it has been stated that, at least in the short term, the direct sowing method may be a suitable alternative to reduce soil mobilization and its negative effects without any significant loss in crop yield [42].

The stubble coverage on the fields after harvest was measured twice, before and after baling. The stubble coverage rate, which was over 95% after the maize harvest during both years, is the main cause of stubble burning in the region. Maize stubble is mostly burned by the producers due to the lack of suitable equipment. A study revealed that a stubble yield of 7 t $ha^{-1}$ was reported, even for no-till plots where maize was planted directly as a second crop with an average yield was 8.2 t $ha^{-1}$, and this forced the producer in dealing with the most important problem in terms of stubble management [43]. In the same study, it was emphasized that it was necessary to work on different alternatives that would be easily accessible to producers to manage maize stubble after harvest.

The moisture of the soil and the humidity of the plant stubble or stems must be suitable to enable machines that are equipped to cut and process stubble to be able to do the cutting work at a sufficient level. In many studies, it has been reported that if the soil surface is hard enough, the discs cut the stubble, and if it is soft, the stubble is buried into the soil. It has also been reported in the study that large diameter discs cut the stubble more easily [44,45]. Therefore, it is desirable that the surface soil moisture is slightly less to increase the performance of the prototype stubble cutter, as in System 2. In terms of trial results, it was an advantage that the humidity in the trial plots in System 2 was less than in System 1. It is thought that this may have had an impact on the prototype no-till seeder used in the system.

At the end of the study, the average total machine labor requirement on the basis of systems was 1.29 h ha$^{-1}$ and 1.47 h ha$^{-1}$ for System 1 and System 2, respectively. In addition, when the human labor requirements of the systems are examined, it has been determined that the labor requirement of System 1 is 1.45 h ha$^{-1}$ on average, and System 2 is 1.65 h ha$^{-1}$. In the system with prototype machines (System 1), the use of machine labor was found to be 12.3% higher. In a study conducted in Turkey, different sets were compared in terms of machine labor use. It was determined that 4.27 h ha$^{-1}$ machine labor is required for chisel + disc harrow + pneumatic sowing, 3.64 h ha$^{-1}$ for milling + pneumatic sowing set, and direct sowing set 2.44 h ha$^{-1}$ [46]. In this respect, it is possible to say that System 2, where our prototypes are located, has a very low machine labor requirement.

Fuel consumption was determined as 104 L ha$^{-1}$ and 95.7 L ha$^{-1}$, respectively. In terms of income/expenditure ratio, it was determined that System 2 (35.6) would be a more profitable investment than System 1 (30.97). In addition, some operating costs have been determined for prototype machines. It has been determined that the fuel consumption of the prototype stubble cutter is 3.25 L ha$^{-1}$, while a conventional stalk shredder consumes 16.80 L ha$^{-1}$ on average. Therefore, it has been determined that the conventional stalk shredder consumes 5.2 times more fuel than the prototype stubble cutter. The operating values of the prototype no-till seeder, on the other hand, consumed 9.5 L ha$^{-1}$, while the fuel consumption of the traditional no-till seeder was approximately 1/3 less. It is thought that the reason for this is that the no-till seeder prototype is mechanical, and it tends to dive deep from time to time and is exposed to excessive load. Researchers in different regions continue to work on different types of prototype stalk shredders according to the products grown in the regions and the soil structure. For example, in a study conducted in the Thrace Region of Turkey, a prototype stalk shredder was used together with a disc harrow and milling type shredders used by the producers to shred sunflower stalks. It has been determined that the most important indicator of economy in shredders is fuel consumption, which has been determined as 11.03 L ha$^{-1}$ for disc harrows, 14.47 L ha$^{-1}$ for milling type shredders, 6.74 L ha$^{-1}$ for prototype stalk shredders (suspended type) and 5.76 L ha$^{-1}$ for prototype stalk shredders (towed type) [47].

When the stubble coverage rate was examined, it was determined that the average data of 2 years in System 1 and System 2 subjects were 26% and 24.5%, respectively. In the study, it was determined that the corn stubble surface cover was 7%, 35% and 27%, respectively, after ear plough + two times disc harrow + sowing, chisel + disc harrow + sowing, rotary tiller + sowing and direct mulch sowing. In addition, it was reported as 39% in direct stubble sowing [48].

In general, the stalks do not lose contact with the soil during the shredding process performed with conventional type stalk shredders, which has an adverse effect on the performance of the sowing machines for subsequent plantings. While long stubble causes a decrease in the sowing depth and sowing uniformity, and seeds are more dispersed into the environment, at the same time, they are bent and buried deeper into the soil by the furrow opened compared to short stubble [49]. This was not encountered in the performance of the prototype stubble cutter, and the root part remaining in the soil was cut and fragmented by the vertical penetration of the machine's cutter blade.

Researchers suggest that we cannot think to save our soil if we keep treating it as a factory, sticking to a "fuel in–food out" model. Soil is not a machine; it is a living being of astonishing, and still largely unexplored, complexity. To seriously address the soil issue, we should agree that it is necessary to rethink the structure and functioning of the whole farming system to start with [50]. Stubble management is the most important part of sustainable soil management. Alternative prototype machines will continue to increase in importance day by day, without burning the stubble.

Looking at international studies, it is seen that there are different designs on the stalk-cutting prototype machines. Some of these are prototypes that are mounted on the combine, while others are designed to be used by connecting to the tractor. However, it has been reported that stalk-cutters mounted on the combine reduce the combine's performance [51]. This also shows that stalk-cutters should be able to deal with different alternatives without being dependent on the combine. In many countries, such as Turkey, studies on no-tillage sowing have increased in recent years. In order to eliminate the straw coverage problem, there are many different stubble-sowing machine prototype studies. One of them is an essential prototype of no-tillage sowing made in southwest China. In this study, a no-tillage sowing machine with a bidirectional stubble-cutting apparatus was designed, and working parameters were determined in field conditions [52]. It is thought that increasing the amount of stubble cultivation area will be a success for every country by conducting more studies like this.

At the end of the study, it is thought that the stubble cutter recommended to the producers is likely to be used by the producers over time. However, it should be known that change will not be rapid, and habits cannot be abandoned quickly.

## 5. Conclusions

Only the traces of the cutter blades are visible on the field surface because of the rolling operation of the machine on the field surface. According to the 2-year data of the study, it was determined that the classical shredder used in the study consumed 5.2 times more fuel than the prototype stubble cutter. It can be said that this has a direct impact on the income/expense ratios and reduces costs. While the tractor power required for prototype machines is 37–60 kW, a more powerful (>60 kW) tractor is needed for the other machine. Some of the advantages of the prototype stubble cutter include simple production technology for small industrial enterprises, easy production and short maintenance and repair time in case of malfunction, and the availability of spare parts. This will also allow small businesses to expand their machine parks by reducing the acquisition capital of the machines. Furthermore, it can be said that especially the stubble cutter can be used for other purposes such as weed crushing/cutting and silage compaction. However, using the prototype no-till seeder can have a negative impact on costs since the fuel consumption is three times higher than the classic stalk shredder.

**Author Contributions:** Conceptualization, M.E.B., Y.V. and A.A.; methodology, M.E.B. and Y.V.; formal analysis, M.E.B., Y.V. and A.A.; writing—original draft preparation, Y.V. and M.E.B.; writing—review and editing, Y.V. and A.A.; supervision, M.E.B., Y.V. and A.A.; funding acquisition, M.E.B., Y.V. and A.A. All authors have read and agreed to the published version of the manuscript.

**Funding:** This project: number TAGEM/TSKAD/14/A13/P08/07, was funded by the Ministry of Agriculture and Forestry Eastern Mediterranean Agricultural Research Institute of Turkey.

**Institutional Review Board Statement:** Not applicable.

**Informed Consent Statement:** Informed consent was obtained from all subjects involved in the study.

**Data Availability Statement:** The data presented in this study are available on request from the authors.

**Conflicts of Interest:** The authors declare no conflict of interest.

## References


1. Şenesen, İ. Threshing and harvesting traditions in the folk economy related to agriculture in Çukurova. *Kültür Araştırmaları Derg.* **2021**, *9*, 277–297. (In Turkish)
2. Dinç, U.; Sarı, M.; Şenol, S.; Kapur, S.; Sayın, M.; Derici, R.; Ağca, N. *Çukurova Region Soils*; Supplementary Textbook; Çukurova University, Faculty of Agriculture: Adana, Türkiye, 1995; Volume 26, p. 172. (In Turkish)
3. Celik, I.; Barut Bereket, Z.; Ortas, I.; Gok, M.; Demirbas, A.; Tulun, Y.; Akpinar, C. Impacts of different tillage practices on some soil microbiological properties and crop yield under semi-arid Mediterranean conditions. *Int. J. Plant Prod.* **2011**, *3*, 237–254.
4. Koca, Y.K.; Acar, M.; Turgut, Y.Ş. Evaluation of quality of agricultural soils with geostatistical modeling. *Harran Tarım Ve Gıda Bilim. Derg.* **2019**, *23*, 489–499. (In Turkish) [CrossRef]
5. Bulut, İ. *A New Human-Environmental Disaster That Is Ignored in Turkey, a Geographical Approach to Stubble Burning and Its Consequences*; Seminarnotes; Ataturk University Faculty of Arts and Sciences, Department of Geography: Erzurum, Türkiye, 2000; *unpublished*. (In Turkish)
6. Özbek, H.; Dinç, U.; Güzel, N.; Kapur, S. *The Effect of Burning Stubbles on the Physical and Chemical Properties of the Soil in the Çukurova Region*; Project no: TOAG-182; TUBITAK: Ankara, Türkiye, 1976.
7. MGM. *Climatedata*; General Directorate of Meteorology: Ankara, Türkiye, 2019.
8. Irmak, S.; Semercioğlu, T. The relationship between soil-plant selenium content in some wheat (*Tritkum* spp.) varieties cultivated in Çukurova Region. *J. Agric. Sci. Res.* **2012**, *5*, 19–23. (In Turkish)
9. Öktem, M.O.; Biçer, Y.; Altındişli, G. Fertilizer demand of maize for nitrogen in Çukurova. In *Soil and Water Resources Research Yearbook*; Publication no: 108; KHGM: Ankara, Türkiye, 1999. (In Turkish)
10. Öztemiz, S.; Güllü, M.; Tatlı, F.; Üremiş, İ. Evaluation of integrated research, implementation and training project in maize in the Eastern Mediterranean Region. *Harran J. Agric. Food Sci.* **2007**, *1*, 33–42. (In Turkish)
11. Eylen, M.; Kanber, R.; Yazar, A. The water-yield relationship of the second crop maize grown after wheat in Çukurova conditions. In *Soil and Water Resources Research Yearbook*; General Publication No: 173, Series No: 108; KHGM: Tarsus, Türkiye, 1990. (In Turkish)
12. Biber, Ç.; Tekin, K. Plant water consumption and limited irrigation applications of the maize plant. *Anatol. J. Agric. Sci.* **2006**, *21*, 140–146. (In Turkish)
13. Cerit, İ. The Effect of Some Tillage Methods, Which Can be an Alternative to Burning Wheat Stubble in Second Crop Cultivation, on Grain Yield and Agricultural Properties in Maize Plant. Master's Thesis, Çukurova University Graduate School of Natural and Applied Sciences, Adana, Turkey, 2001. (In Turkish)
14. İnce, A.; Uğurluay, S.; Özcan, M.T. Bending and shearing characteristics of sunflower stalk residue. *Biosyst. Eng.* **2005**, *92*, 175–181. [CrossRef]
15. Zeren, Y.; Özgüven, F.; Işık, A.; Busse, N.; Fichtel, H.; Delusehıte, R. Comparison of different mechanization systems in field farming in Çukurova. In Proceedings of the 4th International Agricultural Mechanization and Energy Congress, Adana, Türkiye, 1–4 October 1990. (In Turkish)
16. Sümer, S.K.; Sabancı, A. Effects of different tire configurations on tractor performance. *Turk. J. Agric. For.* **2005**, *29*, 461–468. (In Turkish)
17. Anonymous. *Turkey Agricultural Tools and Machinery Business Values Guide*; Prime Ministry General Directorate of Village Services, A.P.K, Department of Soil and Water Resources Research Branch Directorate: Ankara, Türkiy, 1996. (In Turkish)
18. Özden, D.M.; Soğancı, A. *Turkey Agricultural Machinery Operation Values Guide*; Village Services General Management, Soil and Water Resources Research Branch Directorate: Ankara, Türkiye, 1996. (In Turkish)
19. Yalçın, H. A research on the Determination of Suitable Tillage Methods in Second Crop Maize Production for Silage. Ph.D. Thesis, Ege University, Graduate School, Bornova, Türkiye, 1998. (In Turkish)
20. Şahin, G. Research on the Removal of Sunflower Stalks from the Field. Master's Thesis, Tekirdağ University Graduate School of Sciences Agricultural Machinery Department, Tekirdağ, Türkiye, 1998. (In Turkish)
21. Dickey, E.C.; Jasa, P.J.; Shelton, D.P.; Estimating residue cover. Biol. *Syst. Eng. Pap. Sand Publ.* **1986**. Available online: https://digitalcommons.unl.edu/biosysengfacpub/255 (accessed on 29 November 2022).
22. Shelton, D.P.; Kachman, S.D.; Dickey, E.C.; Fairbanks, K.T.; Jasa, P.J. Tillage and planting system, stalk chopper, and knife applicator influencies on corn residue cover. *Appl. Eng. Agric.* **1994**, *10*, 255–261. [CrossRef]
23. Shelton, D.; Jasa, P.J. Estimating percent residue cover using the line-transect method. G1931. In *Nebraska Extension, Neb Guide, Conservation and Management*; University of Nebraska: Lincoln, NE, USA, 2009.
24. Karabacak, H. Determination of Plant Surface Cover Rate by Image Processing Technique. Master's Thesis, Ankara University Graduate School of Sciences, Department of Agricultural Machinery, Ankara, Türkiye, 2007. (In Turkish)
25. Tüzüner, A. *Soil and Water Analysis Laboratories Handbook*; KHGM: Ankara, Türkiye, 1990. (In Turkish)

26. Kirişçi, V.; Say, S.M.; Işık, A.; Akıncı, İ. Effective soil properties in working with agricultural machinery. In Proceedings of the 16th National Congress of Agricultural Mechanization, Bursa, Türkiye, 5–7 September 1995; pp. 490–501. (In Turkish)
27. Say, M.S. Determination of Soil Penetration Resistance with Soil Conditions and Development of Mathematical Models. Master's Thesis, Çukurova University, Institute of Science, Department of Agriculture Machinery, Adana, Türkiye, 1995. (In Turkish)
28. Bereket Barut, Z.; Kara, O.; Bilgili, M.E.; Baran, M.F. The possibilities of direct sowing of second crop maize on the stubby ridges for in Çukurova Region. In Proceedings of the 27th Agricultural Mechanization National Congress, Samsun, Türkiye, 3–5 October 2012; congress book. pp. 5–7. (In Turkish)
29. Hakansson, I.; Lipiec, J. A review of the usefulness of relative bulk density values in studies of structure and compaction. *Soil Tillage Res.* **2000**, *53*, 71–85. [CrossRef]
30. Ehlers, W.; Kopke, V.; Hesse, F.; Bohm, W. Penetration Resistance and Growth of Oats in Tilled and Untilled Loess. *Soil Tillage Res.* **1983**, *3*, 261–275. [CrossRef]
31. Bengough, A.G.; Bransby, M.F.; Hans, J.; McKenna, S.J.; Roberts, T.J.; Valentine, T.A. Root responses to soil physical conditions; growth dynamics from field to cell. *J. Exp. Bot.* **2005**, *57*, 437–447. [CrossRef] [PubMed]
32. Sa, J.C.M.; Tivet, F.; Lal, R.; Briedis, C.; Hartman, D.C.; dos Santos, J.Z.; dos Santos, J.B. Long-term tillage systems impacts on soil C dynamics, soil resilience and agronomic productivity of a Brazilian Oxisol. *Soil Tillage Res.* **2014**, *136*, 38–50.
33. Çelik, İ.; Ortaş, İ.; Bereket Barut, Z.; Gök, M.; Sarıyev, A.; Demirbaş, A.; Akpınar, Ç.; Tülün, Y. *The Effects of Different Tillage Systems on Soil Quality Parameters and Crop Yields. Final Report of the TUBITAK Project (also available in Turkish)*; Grant No: 106O023; The Scientific and Technological Research Council of Turkey: Ankara, Turkey, 2009; p. 163. (In Turkish)
34. Marshall, S.E.; Tokunaga, A. Soil compaction and strength: Measurement methods and influences on perennial grass growth. In *CAL-PAC Society for Range Management Symposium–Grazing for Biological Conservation*; Barry, S., Risberg, D., Eds.; California-Pacific Section, Society for Range Management: San Diego, CA, USA, 2006.
35. Bengough, A.G.; McKenzie, B.M.; Hallett, P.D.; Valentine, T.A. Root elongation, water stress, and mechanical im-pedance: A review of limiting stresses and beneficial root tip traits. *J. Exp. Bot.* **2011**, *62*, 59–68. [CrossRef]
36. Lhotsky, J.; Vachal, J.; Ehrlich, J. Soustava Opatrenik Zurodnovani Zhutnelych Pud. Metodika Pro Zavadeni Vysledku Vyzkumu Do Zemedelske Praxec 14-38 s. in: Badalikova, B. 2010. Chapter 2. Influence of Soil Tillage on Soil Compaction. In *Soil Engineering*; Deduosis, A.P., Bartzanas, T., Eds.; Springer: Berlin/Heidelberg, Germany, 1984; pp. 19–30.
37. Badalikova, B. Chapter 2. Influence of soil tillage on soil compaction. In *Soil Engineering*; Deduosis, A.P., Bartzanas, T., Eds.; Springer: Berlin/Heidelberg, Germany, 2010; pp. 19–30.
38. Pierce, F.J.; Larson, W.E.; Dowdy, R.H.; Graham, W.A.P. Productivity of Soils: Assessing long-term changes due to erosion. *J. Soil Water Cons.* **1983**, *38*, 39–44.
39. Birkás, M.; Jug, D.; Kisic, I. Reasonable tillage to alleviate climate damage and cropping losses. In *Books of Soils*; Birkás, M., Ed.; Szent Istvan University Press: Budapest, Hungary, 2014; ISBN 978-963-269-447-4.
40. Ordoñez-Morales, K.D.; Cadena-Zapata, M.; Zermeño-González, A.; Campos-Magaña, S. Effect of tillage systems on physical properties of a clay loam soil under oats. *Agriculture* **2019**, *9*, 62. [CrossRef]
41. Bai, L.; Kong, X.; Li, H.; Zhu, H.; Wang, C.; Ma, S. Effects of conservation tillage on soil properties and maize yield in karst regions, Southwest China. *Agriculture* **2022**, *12*, 1449. [CrossRef]
42. Martins, R.N.; Portes, M.F.; Moraes, H.M.F.; Furtado Junior, M.R.; Rosas, J.T.F.; Orlando Junior, W.A. Influence of tillage systems on soil physical properties, spectral response and yield of the bean crop. *Remote Sens. Appl. Soc. Environ.* **2021**, *22*, 100517. [CrossRef]
43. Sessiz, A.; Akıncı, C.; Turgut, M.M. The effects of reduced tillage and direct seeding application on yield of second crop maize in GAP Region. *J. Agric. Mach. Sci.* **2010**, *6*, 203–208.
44. Morrison, J.E.; Allen, R.R. Planter and drill requirements for soils with surface residues. Conservation Tillage: Today and Tomorrow Southern Region No-till Conference. In Proceedings of the Southern Region No-Tillage Conference Proceedings, College Station, TX, USA, July 1987; pp. 44–58.
45. Price, T. What Should My No-Till Planter Look Like? 1999. Available online: https://industry.nt.gov.au/__data/assets/pdf_file/0011/233102/311.pdf (accessed on 29 November 2022).
46. Bayram, E. Comparison of Different Tillage Methods in the Second Crop Silage Corn Farming in Terms of Management. Master's Thesis, Gaziosmanpaşa University Institute of Science and Technology, Department of Agricultural Machinery, Tokat, Türkiye, 2010.
47. Kocabıyık, H.; Kayişoğlu, B. Evaluation of performances and energy cost of stalk choppers. *J. Agric. Mach. Sci.* **2005**, *1*, 183–188.
48. Shelton, D.P.; Dickey, E.C.; Jasa, P.J.; Hirschi, M.C.; Brown, L.C. Water Erosion (Chapter 3). In *Conservation Tillage Systems and Management*, 1st ed.; MWPS-45; Iowa State University: Ames, IA, USA, 1992; pp. 8–11.
49. Doan, V.; Chen, Y.; Irvine, B. Effect of oat stubble height on the performance of no-till seeder openers. *Can. Bio-Syst. Eng.* **2005**, *47*, 237–244.
50. Gomiero, T. Alternative land management strategies and their impact on soil conservation. *Agriculture* **2013**, *3*, 464–483. [CrossRef]
51. Yore, M.W.; Summers, M.D.; Jenkins, B.M. Development of a stubble cutting system for a combine harvester. *ASAE Annu. Meet.* **2001**, 011087. [CrossRef]

52. Zhu, H.; Wu, X.; Qian, C.; Bai, L.; Ma, S.; Zhao, H.; Li, H. Design and experimental study of a bi-directional rotating stubble-cutting no-tillage planter. *Agriculture* **2022**, *12*, 1637. Available online: https://www.mdpi.com/2077-0472/12/10/1637 (accessed on 27 November 2022). [CrossRef]

*agriculture*

MDPI

*Article*

# Simulating Soil–Disc Plough Interaction Using Discrete Element Method–Multi-Body Dynamic Coupling

**Mustafa Ucgul** [1,2]

[1] Faculty of Science and Engineering, Southern Cross University, Lismore, NSW 2084, Australia; mustafa.ucgul@scu.edu.au

[2] UniSA STEM, University of South Australia, Mawson Lakes, SA 5095, Australia

**Abstract:** Due to their (a) lower draught force requirements and (b) ability to work at deeper operation depths and faster operation speeds, disc ploughs have gained interest in Australia. A modified version of the disc plough that involves removing every second disc and fitting larger and often more concave discs has become popular. However, the development of the one-way modified disc plough is in its infancy, and a detailed analysis is required, particularly on soil movement. Historically, the soil movement analysis of the soil–tool interactions is conducted using empirical methods. However, the experimental tests are resource and labour intensive. When the soil and tool interaction can be accurately modelled, more efficient tools can be designed without performing expensive field tests, which may only be undertaken at certain times of the year. This study modelled the interaction between soil and a one-way modified disc plough using the discrete element method (DEM). As the disc plough is a passive-driven tool, the rotational speed of the disc plough was modelled using DEM-MBD (multi-body dynamic) coupling. The results of the study show that DEM-MBD coupling can predict the rotational speed of the disc plough with a maximum relative error of 6.9%, and a good correlation was obtained between the DEM-predicted and actual soil movement ($R^2$ = 0.68).

**Keywords:** discrete element method (DEM); multi-body dynamics (MBD); DEM-MBD coupling; topsoil burial

**Citation:** Ucgul, M. Simulating Soil–Disc Plough Interaction Using Discrete Element Method–Multi-Body Dynamic Coupling. *Agriculture* **2023**, *13*, 305. https://doi.org/10.3390/agriculture13020305

Academic Editor: Francisco J. Castillo Ruiz

Received: 15 December 2022
Revised: 13 January 2023
Accepted: 23 January 2023
Published: 27 January 2023



## 1. Introduction

A disc plough is a primary tillage tool used when a mouldboard plough is unsuitable, such as in stony, stiff, and dry soils [1]. Although a mouldboard plough provides complete soil inversion, disc ploughs are cheaper and can be used in broader soil conditions. Additionally, the draught force requirement of the disc ploughs is lower than that of the mouldboard plough, which also reduces fuel consumption [2]. In addition, disc ploughs can be used at faster operating speeds. In Australia, disc ploughs are used to bury (a) non-wetting topsoil, (b) weed seeds, or (c) surface-applied amendments (organic matter, nutrients, and lime) from the surface to 250–350 mm depth. A modified version of the disc plough that involves removing every second disc and fitting larger and often more concave discs has also become popular. Using larger discs creates a larger break-out pressure and allows the disc plough to be used deeper, whereas removing every second disc provides more space for soil to turn over and a greater degree of inversion [3]. However, to improve the effectiveness of the burial performance of the modified disc ploughing, understanding the soil movement process is essential. Soil movement due to tillage operations are generally investigated using empirical methods. Empirical methods create relationships from specific test results and therefore have no extrapolative capabilities outside this context [4,5]. Although the finite element method (FEM) and computational fluid dynamics (CFD) can be used to model soil flow around a tool, these methods are limited, as they cannot model mixing between soil layers or crack propagation [6]. Modelling the soil–tool interaction is a highly complex process due to the non-linear behaviour of the soil. The discrete element method (DEM) can be used to simulate the soil–disc plough interaction. Recent studies have shown that DEM can simulate the soil–implement interaction (considering tillage forces and soil movement) [7–12].

To model the performance of the soil–disc plough interaction, the determination of the disc's rotational speed is essential. A disc plough is a passive-driven tool whose rotation is driven by the soil reaction forces. Therefore, using DEM simulation is not sufficient to simulate soil–disc plough interaction. However, DEM simulation can be coupled with multi-body dynamic (MBD) software to predict the rotational speed of the disc plough. Nevertheless, using DEM-MBD coupling in real-scale simulations is computationally expensive due to the longer simulation time.

This study aims to model a disc plough's topsoil burial performance using DEM-MBD coupling, which was not investigated in prior studies. The DEM parameters of the field soil were determined using the angle of repose test. Then, the rotational speed of the disc plough was predicted using DEM-MBD coupling. After that, the topsoil burial performance of the disc plough was simulated using DEM and then validated with field tests. The colour difference between the deep- and topsoil was used along with a digital image processing algorithm to validate the DEM simulation results. The proposed simulation approach will help designers and researchers to evaluate the performance of any passive-driven soil-engaging tool without performing cost- and time-intensive field tests.

## 2. Materials and Methods

The Material and Methods section is two-fold: simulating the soil–disc plough interaction using DEM and DEM-MBD coupling and performing field tests to validate the results.

### 2.1. DEM and DEM-MBD Coupling Simulations

This study conducted the DEM simulations using commercial DEM software, EDEM 2019™ (developed by DEM Solutions Ltd. In Edinburgh (United Kingdom). A DELL Precision T7610 Dual Xeon E5-2680 v3 @ 2.5 GHz and 128 GB RAM—workstation computer was used to run the EDEM 2019™ software. A hysteretic spring contact model (HSCM) integrated with a linear cohesion model suggested by [13] was used as the mathematical model for contact calculations.

Compressible materials such as soil can be modelled using the HSCM. In the HSCM, the particles behave as a linear elastic material until their stress level reaches a predefined stress (yield) (Figure 1). Once the total stress on the contact area exceeds the yield stress, the particles are deformed plastically [14]. On the other hand, the linear cohesion model adds a cohesive force through normal contact direction (for modelling soil cohesion).

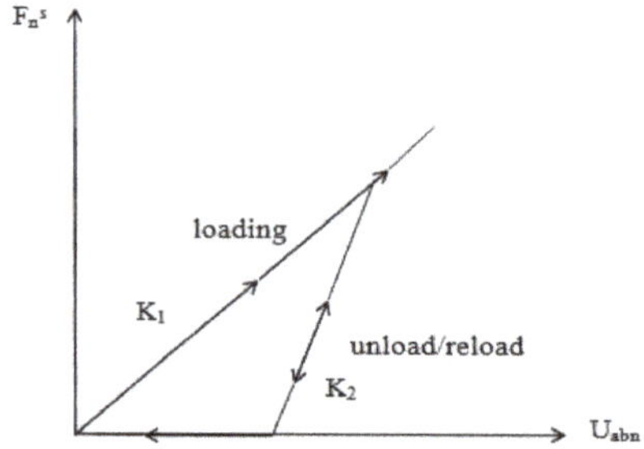


**Figure 1.** Hysteretic spring force-displacement relationship (adapted from EDEM Theory Reference Guide, 2011 [14]).

The HSCM uses the following equations to calculate the total normal ($F_n$) and tangential ($F_t$) forces [14].

$$F_n = F_n^s + F_n^d \tag{1}$$

$$F_n = F_n^s + F_n^d \tag{2}$$

$$F_n^s = -\begin{cases} K_1 \cdot U_{abn} & \text{loading} \\ K_2 \cdot (U_{abn} - U_0) & \text{unloading/reloading} \\ 0 & \text{unloading} \end{cases} \tag{3}$$

where $F_n^s$, $F_n^d$, $F_t^s$, and $F_t^d$ are the normal contact force, normal damping force, tangential contact force, and tangential damping force, respectively. $U_{abn}$ is the normal component of the relative displacement, and $U_0$ is the residual overlap. $K_1$ and $K_2$ are the loading and unloading stiffnesses, respectively, and [15,16] suggest the following equation for calculating $K_1$ and $K_2$.

$$K_1 = 5 \times r_{eq} \times \min(Y_a, Y_b) \tag{4}$$

$$K_2 = \frac{K_1}{e^2} \tag{5}$$

where Y is the yield strength and e is the coefficient of restitution; $r_{eq}$ is the equivalent radius, and [14] calculates $r_{eq}$ as

$$\frac{1}{r_{eq}} = \frac{1}{r_a} + \frac{1}{r_b} \tag{6}$$

For each time step, the $U_0$ is updated as

$$U_0 = \begin{cases} U_{abn} \cdot \left(1 - \frac{K_1}{K_2}\right) & \text{loading} \\ U_0 & \text{unloading/reloading} \\ U_{abn} & \text{unloading} \end{cases} \tag{7}$$

HSCM calculates the $F_t^s$, $F_n^d$, and $F_t^d$ using the following formulas [14]:

$$F_t^s = n_k \cdot K_1 \cdot U_{abt} \tag{8}$$

$$F_n^d = -n_c \times \sqrt{\frac{4 \times m_{eq} \times K_1}{1 + \left(\frac{\pi}{\ln e}\right)^2}} \cdot \dot{U}_{abn} \tag{9}$$

$$F_t^d = -\sqrt{\frac{4 \times m_{eq} \times n_k \times K_1}{1 + \left(\frac{\pi}{\ln e}\right)^2}} \cdot \dot{U}_{abt} \tag{10}$$

where $U_{abt}$ is the tangential component of the relative displacement, $n_k$ is the stiffness factor, defined as the ratio of tangential stiffness to normal loading stiffness, $\mathring{U}_{abn}$ and $\mathring{U}_{abt}$ are the normal and tangential components of the relative velocity, respectively, $n_c$ is the damping factor that controls the amount of velocity-dependent damping, and $m_{eq}$ is the equivalent mass and calculated as suggested by [14]:

$$\frac{1}{m_{eq}} = \frac{1}{m_a} + \frac{1}{m_b} \tag{11}$$

The following should also be considered when calculating $F_t$:

$$F_t = -\min\left(n_k \cdot K_1 \cdot U_{abt} + F_t^d,\ \mu \cdot F_n^s\right) \tag{12}$$

The magnitude of the moments caused by total tangential force (M) and the rolling resistance ($M_r$) were computed as per [17]:

$$M = r_{con} \times F_t \tag{13}$$

$$M_r = -r_{con} \times \mu_r \cdot F_n^s \cdot \lambda_\theta . \tag{14}$$

where $r_{con}$ is the perpendicular distance of the contact point from the centre of mass, $\mu_r$ is the coefficient of rolling friction, and $\lambda_\theta$ is the unit vector of angular velocity at the contact point. The position of the particle is calculated by integrating Equations (15) and (16).

$$\ddot{U} = \frac{(F_n + F_t)}{m} \tag{15}$$

$$R = \frac{(M + M_r)}{I} \tag{16}$$

Cohesion is calculated and added to Equation (1) as [14]:

$$F_c = \xi \times A_c \tag{17}$$

where $\xi$ is the cohesion energy density and $A_c$ is the contact area. Hence, Equation (1) becomes

$$F_n = F_n^s + F_n^d + F_c \tag{18}$$

Using actual particulate sizes and shapes in 3D DEM simulations increases computation time (due to the increased number of contacts) and computation costs. Therefore, larger particle sizes are used in 3D DEM simulations. In addition, spherical particles are generally preferred due to their computational simplicities. However, when larger particle sizes are used, the parameters of these larger particles should be determined via a calibration process, so that these larger particles act like the simulated granular material.

In this study, randomly generated 7.5 mm nominal radii spherical particles (between nominal radii x (0.5 to 2)) were used. This particle size distribution was based on the actual particle size distribution (Table 1). A 0.3 mm sieve size was considered as the base value (7.5 mm nominal radii), particle sizes less than 0.3 mm sieve size were considered 0.5 × 7.5 mm nominal radii, and particle sizes greater than 0.3 mm sieve size were considered 2 × 7.5 mm nominal radii. In other words, very small and large particle sizes were grouped in ×0.5 and ×2 nominal radii.

**Table 1.** Particle size distribution of the test soil.

| Sieve Size (mm) | Percentage Retained (%) |
|---|---|
| 2.36 | 7.2 |
| 1.18 | 10.6 |
| 0.6 | 20.2 |
| 0.3 | 39.0 |
| 0.15 | 19.6 |
| 0.075 | 3.2 |
| <0.075 | 0.2 |

Note that considering the number of particles created in the DEM simulations and available computational software licenses, 7.5 mm nominal radii was the smallest particle size that could be selected.

The DEM parameters were determined from the literature or some physical tests (Table 2). The friction coefficient of soil–soil and soil–steel (using a piece of polished steel placed in the upper portion of the shear box that was sheared over the soil) were determined using direct shear box tests. The coefficient of rolling friction between soil and steel was measured by performing inclined plane tests using a steel sphere (9.5 mm radius). Firstly, a container was filled with test soil, and the steel ball was placed. After that, the tilt angle of the container gradually increased. The tilt angle was measured when the ball began to roll over the soil. The coefficient of rolling friction was then calculated from the measured tilt angle.

However, the coefficient of rolling friction and cohesive energy density of soil particles were calibrated using an angle of repose test. This calibration process is based on matching simulation results to actual measured results using a soil sample taken from the field [18]. The soil sample from the field was collected using a sealed bag (to keep the soil's moisture), and the angle of repose test was conducted on the same day in the soil laboratory. Note that the soil samples collected from the field were used for the abovementioned soil tests.

A soil sample was placed in a pipe (100 mm diameter and 300 mm long) to measure the angle of repose. After that, the pipe was lifted upward at a constant speed of 500 mm $s^{-1}$ using a Hounsfield Tensiometer. Soil flowed into a cylindrical tray (200 mm diameter and 22.5 mm long) until the soil overflowed and formed a pile. When at rest, images of the angle of repose were captured and processed as per [18]. The flow chart of the process for the measurement of the angle of repose is given in Figure 2. The method to measure the angle of repose was based on an automatic search of the best-fitted line formula along the boundary of the material pile using linear regression analysis. After the images of the soil pile were taken, digital image processing techniques were used to segment the conical shape of the pile from the original image. For this process, a global batch clustering algorithm developed by [19] was employed: "Morphological operations were then applied to segment the boundary pixel points of the pile. In the final step, the angle of repose was determined by searching the maximum correlation coefficients using linear regression, using a series of gradually increasing sizes of windows along the entire boundary of the material pile" [18].

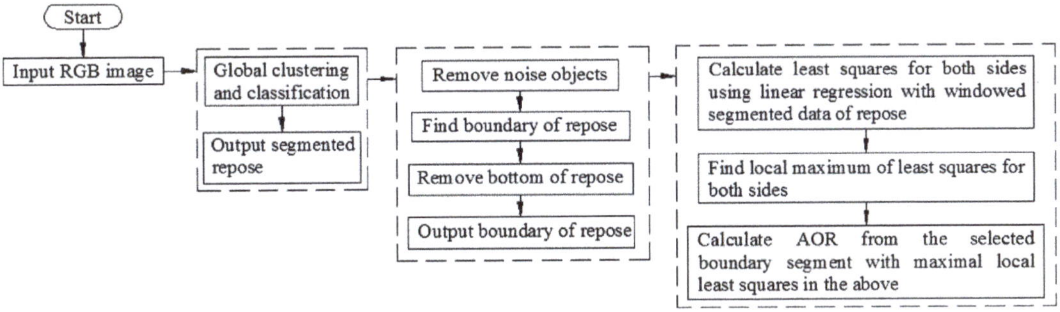


**Figure 2.** Image processing scheme for automatic measurement of the angle of repose.

**Table 2.** EDEM parameters used for simulation.

| Property | Value | Source |
|---|---|---|
| The density of sand particles (kg $m^{-3}$) | 2600 | [20] |
| The density of steel (kg $m^{-3}$) | 7865 | [21] |
| Shear modulus of soil (Pa) | $1 \times 10^7$ | [22] |
| Shear modulus of steel (Pa) | $7.9 \times 10^{10}$ | [21] |
| Poisson's ratio of soil | 0.3 | [23] |
| Poisson's ratio of steel | 0.3 | [24] |
| Yield strength of the soil (Pa) | $1.16 \times 10^6$ | The default value in EDEM |
| Coefficient of friction (soil–soil) | 0.5 | Direct shear test |
| Coefficient of friction (soil–steel) | 0.5 | Direct shear test |
| Coefficient of rolling friction (soil–soil) | 0.15 | Calibrated by the angle of repose test |
| Coefficient of rolling friction (soil–steel) | 0.05 | Inclined surface test |
| Cohesive energy density between soil–soil (N $m^{-2}$) | 2000 | Calibrated by the angle of the repose test |
| Particle size distribution | 0.5–2 | Selected based on the PSD of soil |

After measuring the angle of repose, by varying the coefficient of rolling friction of soil–soil and cohesive energy density, the measured angle of repose of 31.48° (±0.5°) was achieved by using a trial-and-error process in the DEM simulation (Figure 3).

(a) (b)

**Figure 3.** (**a**) Angle of repose test and (**b**) angle of repose simulation.

Simulating the rotational speed of the disc plough cannot be carried out without using multi-body dynamic coupling, as the disc is rotated due to the force from soil particles (as the disc plough is a passive-driven tool). To model disc rotation, MSC ADAMS software was coupled with EDEM. However, using EDEM-ADAMS coupling for real-scale simulation is not viable due to the available computational power. Therefore, this study determined the rotational speed of the disc plough with one disc using a shorter soil bin (10,000 mm long × 3250 mm wide × 500 mm deep).

To simulate the actual field operation, an open furrow was produced at 300 mm depth (Figure 4). Into this open furrow, a furrow was ploughed using the disc plough at 300 mm depth.

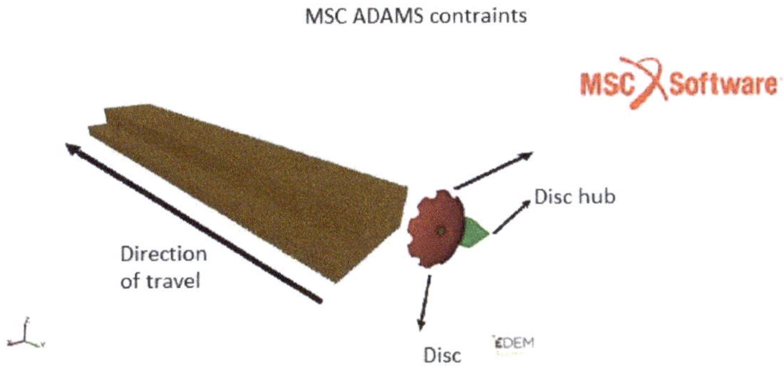


**Figure 4.** EDEM-ADAMS co-simulation to calculate the rotational speeds of the disc.

The CAD model of the disc plough assembly was created using SolidWorks$^{TM}$ software as an assembly (which consists of a hub and a disc). The designed assembly was then imported to MSC ADAMS$^{TM}$ software. In MSC ADAMS$^{TM}$ software, a cylindrical joint was added between the hub's centre and the disc plough's centre. A translational connection was also added between the ground and the disc hub, which was also restrained from all other motions except the direction of travel. After that, a translational motion was assigned to the disc hub through the direction of travel. Finally, the EDEM$^{TM}$ and MSC ADAMS $^{TM}$ were coupled, and the simulations were carried out. The rotational speeds of the soil–disc plough interactions were predicted at 300 mm depth and 2.5, 5, 7.5, and 10 km h$^{-1}$ forward speeds for a single disc.

In order to simulate the real-scale simulation of the soil–disc plough interaction, a virtual field soil with dimensions of 20,000 mm long × 3250 mm wide × 500 mm deep was generated to predict topsoil burial. The DEM soil bulk density was set to 1481 kg/m$^3$ (the bulk density used in the field test). To achieve the desired bulk density, particles (8,819,576 in total) were compressed using an upper physical plane until the desired bulk density was achieved. After that, the disc plough assembly was imported, the depth, speed, and rotational speed (predicted in the first set of simulations) were set, and simulations were carried out (Figure 5). The simulation was run at 300 mm operation depth and 5 and 7.5 km h$^{-1}$ forward speeds, the same operational depth and speed used in the field experiment.

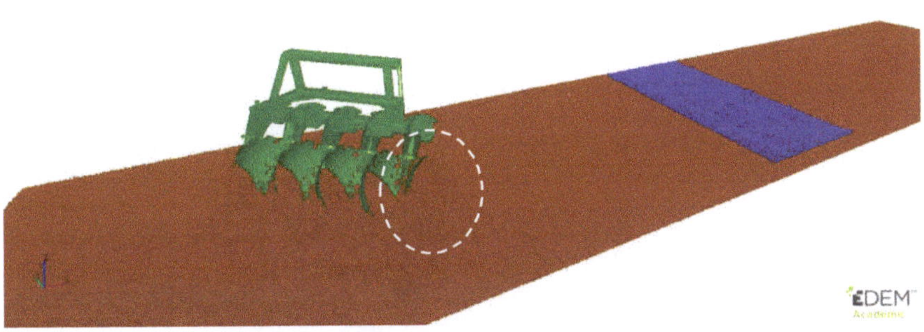


**Figure 5.** DEM simulation of soil–disc plough interaction to predict topsoil burial.

In order to compare the DEM-predicted soil movement with the field experiment, the colour of the DEM particles in a 2000 mm wide × 1000 mm long × 40 mm deep trench was changed from brown to blue (Figure 5). After the simulations were carried out, the brown-coloured sand particles were removed from the simulation, and the coordinates of the blue-coloured particles were exported to Microsoft Excel. Then, the percentage of the volumes of blue-coloured particles at each 50 mm increment was calculated.

A 5-furrow disc plough was designed and used to reduce the simulation size (in opposition to the 13-furrow disc plough used in the field tests). As the first disc creates an open furrow, the soil burial of this furrow is more excessive than that of the other discs. Therefore, the topsoil burial results of the first furrow were not considered in the analysis.

*2.2. Field Tests*

In order to validate the DEM simulation results of the modified one-way disc plough–soil interaction, a field experiment was performed at Malinong (35.5106° S, 139.5157° E), South Australia, in May 2019. The soil was sandy loam with a bulk density, and moisture contents were 1481 kg $m^{-3}$ and 1.6% (dry basis), respectively. The moisture content of the soil was measured by the oven drying method (ASABE Standards, 2008), and the bulk density of the soil was measured using a core sample (50 mm diameter and 77 mm long). A thirteen-furrow modified disc plough was used in the experiment (Figure 6).

**Figure 6.** One-way disc plough used in the field tests.

Field experiments were undertaken at 300 mm depth and 5 and 7.5 km $h^{-1}$ forward speeds. Due to the long-term no-till practice, the top 40 mm of the topsoil had a darker colour than deep soil. This colour difference was used in the tests to validate the DEM simulation results of topsoil burial. After ploughing, three vertical excavations were conducted (using a shovel) at 750 mm increments across the direction of travel. The digital photos of each slice were taken. A metal frame was designed to ensure the camera was always the same distance from the cut when taking the pictures to assist in the accuracy of the later image processing. Note that the topsoil burial profile of a plough (mouldboard or disc) is continuous and uniform, multiple slices were investigated in the field test, and the quantification method is not subjective; therefore, no repeated tests were performed.

The images were then digitally analysed to determine the pixel locations of the dark-coloured topsoil. For this process, the photos were cropped to remove undesired features such as excessive background and shadows to help the software detect the target features. A clustering algorithm developed by [25] was then run over the image to highlight the location of the dark-coloured topsoil and produce a binary plot of the pixels.

In the algorithm, after the region of interest (ROI) is selected, the original RGB (Red, Green, Blue) components are transformed into the perceptual colour components (i.e., a* and b* from Lab colour space and Hue component from CIE standard). After that, the PCA solution from the image data set or subset is used in a hierarchical manner along with the termination measure under an ellipsoidal property. Then, the initial value of centroids of the number of colour clusters is found in the first part of the algorithm (Figure 7). In the second part, the initial parameters are refined further by a modified global clustering algorithm. Then, the refined centroids are used to label the different colour clusters (Figure 7).

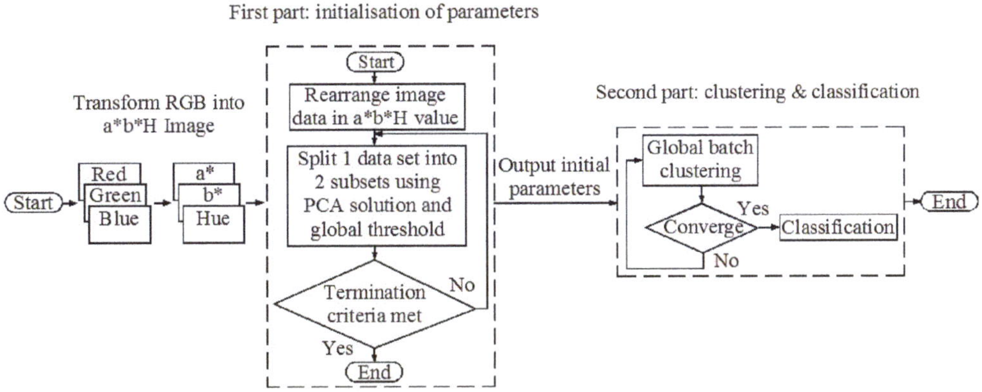


**Figure 7.** Digital image processing (DIP) method for acquiring the featured coloured soil cluster image.

The pixel coordinates were then processed in MS Excel, where all the pixel coordinates from the three pictures were combined for each test condition, and the percentages of the topsoil burial at each 50 mm interval were calculated.

## 3. Results

Measured and DEM-MBD coupled predicted rotational speeds are given in Figure 8. As shown in Figure 8, increasing the forward speed increases the rotational speed. Increasing the forward speed from 3 km $h^{-1}$ to 9 km $h^{-1}$ increases the rotational speed from 13.07 rpm to 42.73 rpm. The results show that the measured and DEM-MBD-predicted rotational speed values are in good agreement, with a maximum relative error of 6.9%. This finding indicates that the proposed modelling approach can be effectively used to predict the rotational speed of the disc plough.

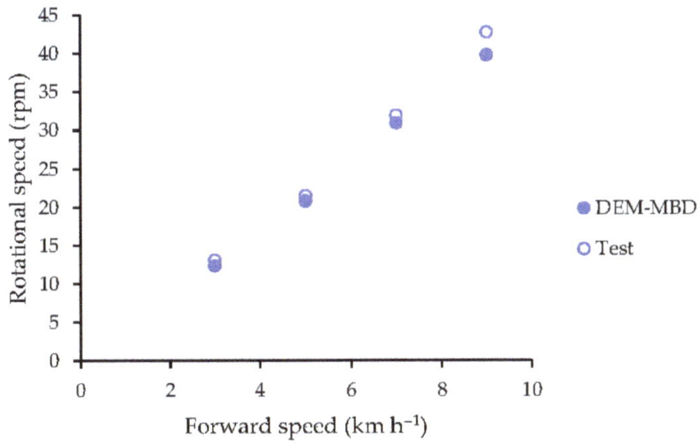


**Figure 8.** Measured (field test) and DEM-MBD predicted rotational speed vs. forward speed.

The predicted rotational speed results in Figure 8 were used in full-scale DEM simulations to evaluate the topsoil burial performance of the one-way modified disc plough.

Examples of images captured from the field experiments for each test are given in Figure 9. The combined cross-sectional locations of the buried topsoil observed in the excavated profiles (sum of three images) are represented in Figure 10. The results of the DEM simulations are also shown in Figure 10. The field test and DEM simulated results using a deep working modified one-way plough showed similar visual trends. At slower speeds, more distinct and vertical burial profiles were obtained, while burial profiles became horizontal at higher speeds [26].

300 mm depth and 5 km $h^{-1}$ speed

300 mm depth and 7.5 km $h^{-1}$ speed

**Figure 9.** An example of digital pictures taken from each excavation.

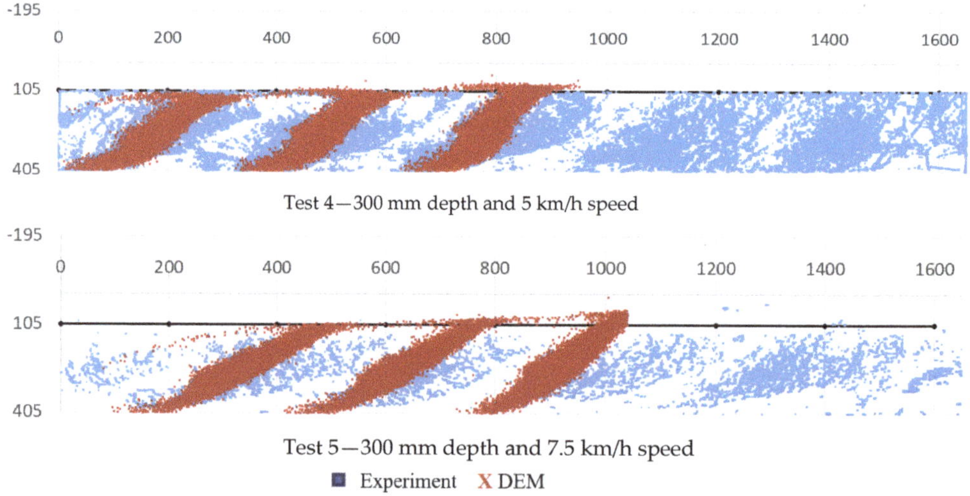


**Figure 10.** Comparison of the experiment and DEM results.

The percentage of particles in each layer computed for both field soil and DEM simulations is shown in Figure 11. Results show that increasing forward speed decreases the topsoil buried to 200–300 mm layers. Decreases of 16.1% and 27.9% were found in field tests and DEM simulations, respectively. It was also found that increasing speed creates a less uniform topsoil burial and creates a bulge in concentration between 100–200 mm depths. The concentration of the bulge between 100–200 mm increases with speed. It was found that the bulge concentration between 100–200 mm increases by 21.3% and 15.3% in the field test and DEM simulation results, respectively. It was also determined that only a small portion of the topsoil was buried in 250–300 mm layers at both speeds (12.4% and 7.4% at 5 and 7.5 km $h^{-1}$, respectively).

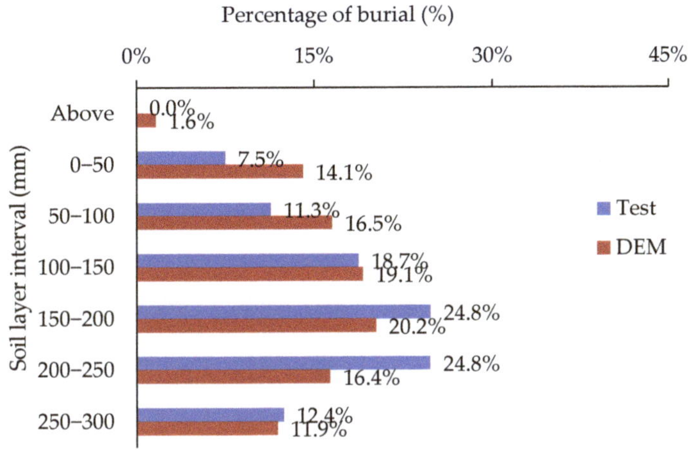


(**a**)

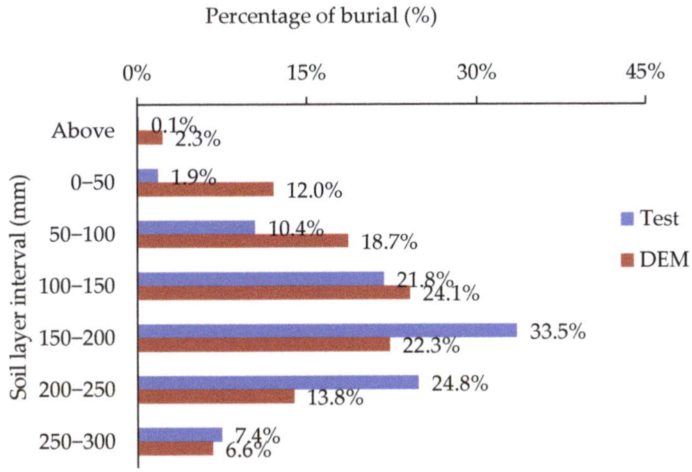


(**b**)

**Figure 11.** Percentage of particles in each layer for an experiment vs. DEM simulation for 300 mm depth and (**a**) 5 km $h^{-1}$ and (**b**) 7.5 km $h^{-1}$ forward speeds.

As shown in Figure 11, the percentage of the topsoil buried at 5 and 7.5 km $h^{-1}$ forward speeds in the field was comparable to topsoil buried at 5 and 7.5 km $h^{-1}$ forward speeds in the DEM. The good correlation between the test and DEM results ($R^2$ = 0.68) also validates this finding (Figure 12).

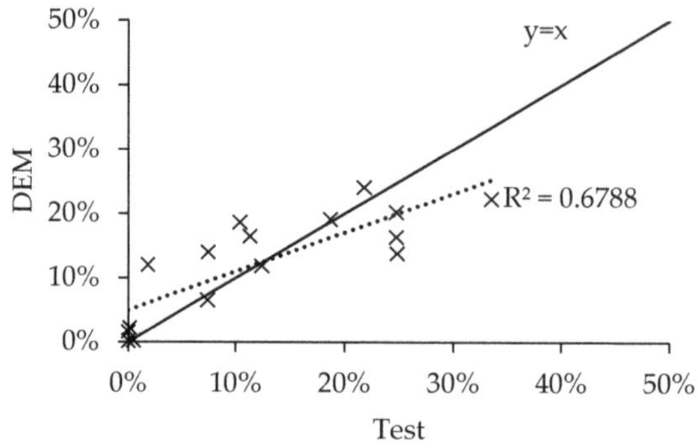


**Figure 12.** Correlation between experiment vs, DEM simulation. The cross dots represent the test vs. DEM predicted percentage of burial (as (x,y)), and, the black line represents the perfect prediction (y = x or $R^2 = 1$).

The results summarised above indicate that there is a need to improve the design of the disc plough for improved topsoil burial. The results also showed that DEM-MBD coupling could be used to simulate the soil–disc plough interaction, which also helps to optimise the operational and design characteristics of the one-way plough.

The differences between the DEM and field results can be attributed to the experimental error during the excavation process when taking the images (each time a slice was dug, there was always a chance that some soil particles could fall into the sliced section) and the larger DEM particles used in the simulations. Using a more powerful computer and more software licences might help to improve the accuracy of the results. The simulation results may also be able to be improved by further refining the DEM-calibrated parameters to better suit the larger than actual particles.

## 4. Conclusions

This study simulated the interaction between soil and a modified one-way disc plough using DEM to investigate the topsoil burial. The rotational speed of the disc plough (as a passively driven tool) was precited using DEM-MBD co-simulation. The results were validated using field experiments. The results showed that the rotational speed of the disc plough was predicted with a maximum relative error of 6.9%. In contrast, the coefficient of determination between the test and DEM-predicted topsoil burial was $R^2 = 0.68$. This shows that the developed DEM and DEM-MBD models can be used to model the soil–disc plough interaction.

It was also found from the test and DEM results that the used, modified one-way disc plough does not (1) effectively bury the topsoil into a deeper layer or (2) provide uniform topsoil burial throughout the surface. Therefore, future work is required to improve the design and geometry of the disc plough to achieve a more effective topsoil burial using DEM.

**Author Contributions:** Conceptualisation, M.U.; methodology, M.U.; validation, M.U.; formal analysis, M.U.; writing—original draft preparation, M.U.; writing—review and editing, M.U. All authors have read and agreed to the published version of the manuscript.

**Funding:** This research was funded by Australia's Grains Research and Development Corporation (GRDC) and the Department of Agriculture and Food Western Australia (DAWFA) (Project Number: DAW 00244).

**Institutional Review Board Statement:** Not applicable.

**Data Availability Statement:** The data presented in this study are available upon request from the first author, M.U.

**Acknowledgments:** The research was funded by Australia's Grains Research and Development Corporation (GRDC) and the Department of Agriculture and Food Western Australia (DAWFA) DAW 00244. The technical and field work assistance of Dr Chris Saunders is gratefully acknowledged.

**Conflicts of Interest:** The authors declare no conflict of interest.

## Nomenclature

| | |
|---|---|
| $a$ | Indices for sphere or implement |
| $A_c$ | Contact area, ($m^2$) |
| $b$ | Indices for sphere or implement |
| $e$ | Coefficient of restitution |
| $F_c$ | Cohesion force, (N) |
| $F^d{}_n$ | Normal damping force, (N) |
| $F^d{}_t$ | Tangential damping force, (N) |
| $F_n$ | Normal total contact force, (N) |
| $F^s{}_n$ | Normal contact force, (N) |
| $F^s{}_t$ | Tangential contact force, (N) |
| $F_t$ | Tangential total contact force, (N) |
| $I$ | Moment of inertia, (kg $m^2$) |
| $K_1$ | Stiffness for loading, (N $m^{-1}$) |
| $K_2$ | Stiffness for unloading/reloading, (N $m^{-1}$) |
| $M$ | Moment, (N m) |
| $M_r$ | Moment due to rolling friction, (N m) |
| $m$ | Mass, (kg) |
| $m_{eq}$ | Equivalent mass (kg) |
| $n_c$ | Damping factor |
| $n_k$ | Stiffness factor |
| $r$ | Radius (m) |
| $r_{eq}$ | Equivalent radius, (m) |
| $r_{con}$ | Perpendicular distance of contact point from the centre of mass, (m) |
| $R$ | Rotational acceleration, (rad $s^{-2}$) |
| $U_{abn}$ | Normal component of the relative displacement, (m) |
| $U_{abt}$ | Tangential component of the relative displacement, (m) |
| $\mathring{U}_{abn}$ | Normal component of the relative velocity, (m $s^{-1}$) |
| $\mathring{U}_{abt}$ | Tangential component of the relative velocity, (m $s^{-1}$) |
| $U_0$ | Residual overlap, (m) |
| $\ddot{U}$ | Translational acceleration, (m $s^{-2}$) |
| $Y$ | Yield strength (MPa) |
| *Greek letters* | |
| $\mu$ | Coefficient of friction |
| $\mu_r$ | Coefficient of rolling friction |
| $\lambda_\theta$ | Unit vector of angular velocity |
| $\xi$ | Cohesion energy density (J $m^{-3}$) |

## References

1. Nkakini, S.O. Draught force requirements of a disc plough at various tractor forward speeds in loamy sand soil during ploughing. *Int. J. Adv. Res. Eng. Tech.* **2015**, *6*, 52–68.
2. Aybek, A.; Baser, E.; Arslan, S.; Ucgul, M. Determination of the effect of biodiesel use on power take-off performance characteristics of an agricultural tractor in a test laboratory. *Turk. J. Agric. For.* **2011**, *35*, 103–113. [CrossRef]
3. Pratley, J.E.; Kirkegaard, J. (Eds.) *Australian Agriculture in 2020: From Conservation to Automation*; Agronomy Australia by Graham Centre for Agricultural Innovation; Charles Sturt University: Wagga Wagga NSW, Australia, 2019.
4. Van Muysen Van Oost, W.K.; Govers, G. Soil translocation resulting from multiple passes of tillage under normal field operating conditions. *Soil Tillage Res.* **2006**, *87*, 218–230. [CrossRef]
5. Li, S.; Lobb, D.; Lindstrom, M.J. Tillage translocation and tillage erosion in cereal-based production in Manitoba, Canada. *Soil Tillage Res.* **2007**, *94*, 164–182. [CrossRef]

6. Plouffe, C.; Lague, C.; Tessier, S.; Richard, M.J.; McLaughlin, N.B. Moldboard plow performance in a clay soil: Simultaneous and experiment. *Trans. ASAE* **1999**, *42*, 1531–1539. [CrossRef]
7. Aikins, K.A.; Antille, D.L.; Ucgul, M.; Barr, J.B.; Jensen, T.A.; Desbiolles, J.M. Analysis of effects of operating speed and depth on bentleg opener performance in cohesive soil using the discrete element method. *Comput. Electron. Agric.* **2021**, *187*, 106236. [CrossRef]
8. Aikins, K.A.; Barr, J.B.; Antille, D.L.; Ucgul, M.; Jensen, T.A.; Desbiolles, J.M. Analysis of effect of bentleg opener geometry on performance in cohesive soil using the discrete element method. *Biosyst. Eng.* **2021**, *209*, 106–124. [CrossRef]
9. Aikins, K.A.; Ucgul, M.; Barr, J.B.; Jensen, T.A.; Antille, D.L.; Desbiolles, J.M. Determination of discrete element model parameters for a cohesive soil and validation through narrow point opener performance analysis. *Soil Tillage Res.* **2021**, *213*, 105123. [CrossRef]
10. Ucgul, M.; Saunders, C.; Fielke, J.M. A Method of quantifying discrete element method simulations of top soil burial from a mouldboard plough. In Proceedings of the 2016 ASABE Annual International Meeting, Orlando, FL, USA, 17–20 July 2016.
11. Zeng, Z.; Ma, X.; Chen, Y.; Qi, L. Modelling residue incorporation of selected chisel ploughing tools using the discrete element method (DEM). *Soil Tillage Res.* **2020**, *197*, 104505. [CrossRef]
12. Sadek, M.A.; Chen, Y.; Zeng, Z. Draft force prediction for a high-speed disc implement using discrete element modelling. *Biosyst. Eng.* **2021**, *202*, 133–141. [CrossRef]
13. Fielke, J.M.; Ucgul, M.; Saunders, C. Discrete element modelling of soil-implement interaction considering soil plasticity, cohesion and adhesion. In *2013 Kansas City, Missouri, July 21–July 24, 2013*; American Society of Agricultural and Biological Engineers: St. Joseph, MI, USA, 2013.
14. EDEM. *EDEM Theory Reference Guide*; DEM Solutions Ltd.: Edinburgh, UK, 2011.
15. Walton, O. *Elastic-Plastic Contact Model*; Company Report, DEM Solutions: Edinburgh, UK, 2006.
16. Walton, O.R.; Braun, R.L. Stress calculations for assemblies of inelastic spheRes. in uniform shear. *Acta Mech.* **1986**, *63*, 73–86. [CrossRef]
17. Raji, A.O. Discrete Element Modelling of the Deformation of Bulk Agricultural Particles. Ph.D. Thesis, University of Newcastle, Newcastle upon Tyne, UK, 1999.
18. Li, P.; Ucgul, M.; Lee, S.H.; Saunders, C. A new approach for the automatic measurement of the angle of repose of granular materials with maximal least square using digital image processing. *Comput. Electron. Agric.* **2020**, *172*, 105356. [CrossRef]
19. Li, P.; Lee, S.-.H.; Park, J.-S. Development of a global batch clustering with gradient descent and initial parameters in colour image classification. *IET Image Process* **2019**, *13*, 161–174. [CrossRef]
20. Huser, A.; Kvernvold, O. Prediction of sand erosion in process and pipe components. In *BHR Group Conference Series Publication*; Mechanical Engineering Publications Limited: London, UK, 1998; Volume 31, pp. 217–228.
21. Hudson Tool Steel. P20 Mold Steel. 2016. Available online: http://www.hudsontoolsteel.com/technical-data/steelP0 (accessed on 10 October 2022).
22. Academia. Some Useful Numbers for Rocks and Soils. 2015. Available online: www.academia.edu/4056287/Some_Useful_Numbers_for_rocks_and_soils (accessed on 10 October 2022).
23. Asaf, Z.; Rubinstein, D.; Shmulevich, I. Determination of discrete element model parameters required for soil tillage. *Soil Tillage Res.* **2007**, *92*, 227–242. [CrossRef]
24. Budynas, R.G.; Nisbett, K.J. *Shigley's Mechanical Engineering Design*; McGraw-Hill Education: New York, NY, USA, 2012.
25. Li, P.; Ucgul, M.; Lee, S.H.; Saunders, C. A new method to analyse the soil movement during tillage operations using a novel digital image processing algorithm. *Comput. Electron. Agric.* **2019**, *156*, 43–50. [CrossRef]
26. Ucgul, M.; Saunders, C.; Davies, S. Insights into One-Way Disc Ploughs. Available online: https://groundcover.grdc.com.au/innovation/precision-agriculture-and-machinery/insights-into-one-way-disc-ploughs (accessed on 20 October 2022).

*agriculture*

MDPI

*Review*

# Application of Computational Intelligence Methods in Agricultural Soil–Machine Interaction: A Review

**Chetan Badgujar [1,*], Sanjoy Das [2], Dania Martinez Figueroa [2] and Daniel Flippo [1]**

[1] Biological and Agricultural Engineering, Kansas State University, Manhattan, KS 66502, USA
[2] Electrical & Computer Engineering, Kansas State University, Manhattan, KS 66502, USA
* Correspondence: chetan19@ksu.edu

**Abstract:** Rapid advancements in technology, particularly in soil tools and agricultural machinery, have led to the proliferation of mechanized agriculture. The interaction between such tools/machines and soil is a complex, dynamic process. The modeling of this interactive process is essential for reducing energy requirements, excessive soil pulverization, and soil compaction, thereby leading to sustainable crop production. Traditional methods that rely on simplistic physics-based models are not often the best approach. Computational intelligence-based approaches are an attractive alternative to traditional methods. These methods are highly versatile, can handle various forms of data, and are adaptive in nature. Recent years have witnessed a surge in adapting such methods in all domains of engineering, including agriculture. These applications leverage not only classical computational intelligence methods, but also emergent ones, such as deep learning. Although classical methods have routinely been applied to the soil–machine interaction studies, the field is yet to harness the more recent developments in computational intelligence. The purpose of this review article is twofold. Firstly, it provides an in-depth description of classical computational intelligence methods, including their underlying theoretical basis, along with a survey of their use in soil–machine interaction research. Hence, it serves as a concise and systematic reference for practicing engineers as well as researchers in this field. Next, this article provides an outline of various emergent methods in computational intelligence, with the aim of introducing state-of-the-art methods to the interested reader and motivating their application in soil–machine interaction research.

**Keywords:** tillage; traction; compaction; neural networks; support vector regression; fuzzy inference system; adaptive neuro-fuzzy inference system

**Citation:** Badgujar, C.; Das, S.; Figueroa, D.M.; Flippo, D. Application of Computational Intelligence Methods in Agricultural Soil–Machine Interaction: A Review. *Agriculture* **2023**, *13*, 357. https://doi.org/10.3390/agriculture13020357

Academic Editors: Mustafa Ucgul and Chung-Liang Chang

Received: 25 October 2022
Revised: 18 January 2023
Accepted: 19 January 2023
Published: 31 January 2023



## 1. Introduction

Soil-engaging tools or machines are an indispensable part of mechanized agriculture. A soil–machine interaction deals with a behavior of tools or machines with soil that results in either tillage, traction, or compaction. The soil–machine interaction is mainly categorized into tillage, traction, and compaction [1]. In traction, a powered traction element (wheel/track) of the vehicle operates on deformable soil, causing soil shear to generate the traction [2]. The soil-derived traction force overcomes the vehicle's resisting forces and maintains its constant motion with its slip and terrain damage [3]. The slip is a principal form of vehicle power loss and one of the prime reasons behind the off-road vehicle's worst traction efficiency. Tractors are the prime movers in agriculture and are mainly used for drawbar work. The drawbar is the most used but the least efficient power outlet, and approx. 20 to 55% of the tractor's available energy is wasted at the soil-tire interface, often resulting in soil compaction and tire wear [4]. Therefore, the traction performance of the vehicle is quantified in terms of traction, slip, and power efficiencies on certain terrain. In the off-road vehicle, it is critical to optimize or increase the working capacity, efficiency, and reduce slip and terrain damage. Multiple variables, such as traction element geometry, operating variables, and soil physical conditions, influence vehicle traction performance. Therefore, traction models are often proposed to optimize off-road vehicle performance (e.g., drawbar, slip, traction efficiency).

Tillage alters the soil mechanically to create favorable conditions for crops [2]. It employs either powered or unpowered mechanical devices (tool/implement) to apply forces to the soil, resulting in soil cutting, inversion, pulverization, displacement, mixing, or a collective action aiming to obtain desired conditions [5]. Most tillage devices are passive (unpowered), known as conventional tillage, where a drawbar is applied to the device, and its movement through soil results in tillage. In contrast, active tillage, also known as rotary tillage, employs a powered device to transmit power to the soil. The powered tool comparatively moves greater soil volume than required, and energy cost increases with a working width and depth. Tillage is the most energy-intensive agricultural operation and accounts for nearly half of the total crop production energy [6].

Tillage energy is influenced by multiple factors, including soil conditions (initial condition, texture, bulk density, moisture content, and crop residue cover), tool parameters (shape, size, and cutting-edge sharpness), and operating parameters (depth and speed) [5,6]. Therefore, extensive literature is available that aims to reduce tillage input energy by optimizing those factors. The research efforts mainly revolved around the soil failure pattern, soil movement, and force or energy prediction models [5,6]. The information on tillage force or energy is critical to select tillage types, tools, control variables, energy management, optimization, and reducing excessive soil pulverization. For example, knowing the tool draft in specific soil helps select the tractor size with a matching implement, reducing operation costs and negative soil impacts. Therefore, tillage force or energy prediction models are necessary from an energy optimization perspective.

Soil compaction is a leading factor in degrading productive farmland [7,8]. It has degraded an estimated 83 Mha of farmland [7,9] and affected around 45% of agrarian soil [10,11]. Natural and artificial activities are responsible for soil compaction. The artificial activities involved in crop production can severely affect soil compaction. These activities include heavyweight machinery, and its intense use, uncontrolled vehicle traffic, multiple passes, operating machines under unfavorable conditions (e.g., wet soil), repeated tillage, and bad crop rotation [7,12,13]. In addition to topsoil compaction, a subsoil or plow pan is caused by heavy vehicular movement, heavy plow weight, downward forces from a plow bottom/disk, and repeated tillage. Soil compaction resulting from the soil–machine interaction influences the soil structure, porosity, permeability, and density [7,14], which impacts the crop yield and may degrade the soil. The soil compaction evolved from soil-machine interaction is a complex process that involves multiple interrelated factors. Hence, optimizing the vehicle parameters (e.g., tire type, orientation, inflation pressure, axle weight, traffic), tillage parameters (tool shape, weight, depth, speed, and tillage intensity), and assessing the initial field conditions (soil moisture) can minimize or eliminate the soil compaction.

The soil–machine interaction is a dynamic and intricate process that includes multivariate. However, understanding and accurately describing (models) the soil–machine interaction may provide a solution to sustainable agricultural production. For example, a slight improvement in the tillage tool design or practice could significantly reduce the input energy and avoid excessive soil pulverization or compaction. Likewise, improving the vehicle traction efficiency may increase the working capacity, save energy, and avoid terrain damage or compaction.

In recent years, computational intelligence (CI) methods have succeeded in solving intricate problems in agriculture and life science. The literature shows that researchers, scholars, and engineers have implemented cutting-edge CI methods, including neural networks, fuzzy logic, neuro-fuzzy systems, support vector machines, and genetic algorithms, to solve a challenging problem in the soil tillage and traction domain. However, it lacks a comprehensive, curated source of reference and a detailed and well-organized discussion on the application of CI methods on the soil–machine interaction. Therefore, this study aimed to survey and analyze the recent research efforts in the soil–machine interaction and critically review the existing methods with a detailed discussion. The article provides brief information, progress, and future direction on CI methods in the soil tillage and traction domain. The proposed study would serve as a concise reference to the reader, engineers, researchers, and farm managers who are further interested in the soil–machine interaction. It is also a quick and systematic way to understand the applicable methods that allow crucial decision-making in farm management.

The review is organized into the following sections: Section 2 discusses the traditional approach used in soil–machine interaction modeling. It also discusses the strengths and limitations of the traditional approach. Section 3 presents a brief overview of popular CI methods, while Section 4 discusses the popular CI methods. These sections provide a piece of fundamental and in-depth information to the readers. Section 5 provides a brief literature survey on CI methods that are employed in soil–machine interaction studies. It also contains the followed literature survey methodology. In Section 6, the strengths and limitations of CI methods were identified. Section 7 talks about the other emergent CI models that may provide a better solution than popular CI methods. Section 8 discusses the significance, scope, and future direction.

## 2. Traditional Modeling Methods

In recent decades, numerous methods have been proposed to evaluate, analyze, model, and understand the soil–machine interaction, which aims to optimize energy, time, efficiency, and machine or tool service life with reduced wear. The methods are explained as follows:

### *2.1. Analytical Method*

Analytical methods are based on physical principles of soil/terrain, machine parameters, and simple assumptions. The traction force is often computed from a soil–tire contact–surface interface and stress distribution (shear and normal) [5,15]. However, both soil and tire deform during the process, making it challenging to describe in mathematical terms. Moreover, machine dynamics, varying soil conditions, its elastic-plastic nature, and inadequate information on boundary conditions make the soil–machine interaction a very complex problem to model accurately. These challenges raise questions on its adaptability [3,5,15,16].

Likewise, in tillage, soil resistance is computed with a logarithmic spiral method and passive earth pressure theory [6]. These are assumption-based methods that do not include the actual soil failure patterns that vary with tool parameters (shape, rake angle, speed) and soil parameters (moisture, density, and structure) [17–20]. Moreover, the analytical methods are suitable for simple shapes, but difficult for the complex shape tool [5,6]. Thus, it exhibits limited applicability for tool design and energy or force prediction.

### *2.2. Empirical Method*

Empirical methods are derived from a large amount of experimental data, where the best-fit regression curve explains a relationship among the selected variables. Empirical equations are simple and consist of a few variables with constants specific to the soil, track/wheel, tool type, machine configuration, and operating conditions. Thus, these equations cannot be extrapolated to the other problems, restricting their broad applicability [3,5,15]. Thus, precautions are necessary on a new tire, tillage tool, and test environment [3,5]. Moreover, it requires a large amount of experimental data, which is laborious and costly. Additionally, it is subjected to a multi-collinearity problem, arising from not truly dependent factors [21].

### 2.3. Semi-Empirical Method

Semi-empirical method combines experimental data, empirical formulations, and analytical methods. In traction studies, the stress (normal and shear) and soil deformation are computed by assuming stress under a flat plate, and bevameter is used [3,5,15]. The flat plate is non-flexible, but the tire or track is flexible, working on deformable soil. Thus, this method requires improvement. Similarly, a passive earth pressure theory explains the soil-failure pattern for a simple shaped tillage tool [5,6,19]. However, adopting the earth pressure theory to other complex shaped tools is challenging [5,17,22,23]. Semi-empirical is a hybrid, reliable, and the most common method, although equations derived from assumptions limit its accuracy in varying terrain.

### 2.4. Numerical Method

Numerical methods such as finite element and discrete element were extensively studied, and lately in soil tillage and traction domain [3,5,15,19,24–28]. The detailed examples can be found here [25,26]. These methods have successfully modeled the complex, dynamic, and non-linear soil–machine interaction problems with greater accuracy and fewer assumptions [3,25,26]. However, it is a highly computational method consisting of a virtual simulation with commercial software installed on a high-speed computer. Therefore, it is time-consuming and requires special and costly resources. Moreover, the simulation setup needs an accurate description of a soil medium that varies on a spatial-temporal basis, making it challenging.

In short, the traditional modeling methods have a few limitations and are very specific to machine or tool types and experimental conditions, which restricts their wide applicability.

## 3. Computational Intelligence: An Overview

Broadly speaking, the term "computational intelligence" refers to a wide class of approaches that rely on approximate information to solve complex problems [29–32]. There are a vast array of such problems (e.g., classification, regression, clustering, anomaly detection, function optimization), where CI models have been extensively used. In the available literature on soil–machine interactions, these models have been used for regression tasks. Accordingly, this article describes CI models from a regression standpoint. However, as some articles have used CI optimization approaches for training the models (i.e., model parameter optimization for best performance), CI-based optimization algorithms are also addressed here, albeit in the context of training.

Many CI models are derived from paradigms observed in the natural world. Artificial neural networks (NN), deep neural networks (DNN), and radial basis function networks (RBFN) are structures that loosely resemble the organization of neurons in higher animals. Fuzzy inference systems (FIS) perform computations in a manner analogous to verbal reasoning. Adaptive neuro-fuzzy systems (ANFIS) are designed to combine the attractive features of NN and FIS. These models are very well suited for regression tasks.

Other CI regression models, which are designed from purely mathematical considerations, do not have any natural counterparts. This class of machine learning methods includes support vector regression (SVR) and Bayesian methods.

Natural phenomena also provide the backdrop of CI optimization methods. Genetic algorithms (GA) are modeled after Darwinian evolution, while particle swarm optimization (PSO) simulates the foraging strategy of a swarm of organisms. These methods have been routinely used to train other CI models [33–36].

CI models for regression are data-driven approaches. In soil–machine interaction studies, the data are typically collected from field experiments. Each sample in the data is a pair $(\mathbf{x}(n), \mathbf{t}(n))$, where $n$ is a sample index. The quantity $\mathbf{x}(n) \in \mathbb{R}^M$ in a sample is an $M$-dimensional input, $\mathbf{t}(n) \in \mathbb{R}^N$ is its corresponding $M$-dimensional target (or desired output) vector. The symbol $\mathbf{\Theta}$ is used in this article to denote the set of all trainable parameters of any CI model. Wherever necessary, it may be treated as a vector. Note that throughout this article, italicized fonts are used to represent scalar quantities, and bold fonts for vectors (lowercase) and multi-dimensional arrays (uppercase).

### 3.1. Data Preprocessing

Preprocessing is often essential before using data to train a CI model. It renders the data more suitable for CI models.

(*i*) Data Normalization: This is the most rudimentary form of preprocessing. Each field of the data are normalized separately so that the entries lie in some desired range, usually $[0, 1]$ or $[-1, 1]$.

(*ii*) Data Cleaning: Experimental data may contain some missing entries. One option to deal with the issue is to remove every sample, which contains a missing (scalar) field. This practice may be wasteful, particularly when the data are limited. If so, missing fields may be filled with means, medians, or interpolated values. Corrupted entries can also be treated in this manner [37]. Noise reduction is another form of data cleaning. When the noise follows a non-skewed distribution around a zero mean, noise removal may not be necessary in regression tasks. Convolution with Gaussian or other filters is a common filtering tool for time series data [38].

(*iii*) Dimension Reduction: Dimension reduction is useful when the number of input dimensions, say $M'$ is too high. Principal component analysis (PCA) is widely used for this purpose. More advanced techniques for dimension reduction include nonlinear PCA [39] and independent component analysis (ICA) [40].

(*iv*) Spectral Transformation: This technique can be used with periodic data. The classical Fourier transform is regularly used to extract frequency components of such data; it does not preserve the time information of the input. Wavelet transforms can be used when the data must incorporate frequency and temporal components.

The samples are randomly divided into three disjoint sets—the training set $\mathcal{S}_t$, the test set $\mathcal{S}_s$, and the validation set $\mathcal{S}_v$. Training samples are used directly to adjust the model parameters in small increments. Unlike them, samples in $\mathcal{S}_s$, are not used explicitly to compute parameter increments. Instead, testing samples are used intermittently during training to monitor progress. Validation samples in $\mathcal{S}_v$ are used as surrogates for the real world. The performance of the CI model is evaluated with respect to $\mathcal{S}_v$ only after training is completely accomplished. Approximately 60%–80% of the samples are assigned to $\mathcal{S}_t$ and the remainder divided roughly equally between $\mathcal{S}_s$ and $\mathcal{S}_v$.

### 3.2. Loss Functions

The purpose of training any model is to minimize the differences between the targets and the true outputs, quantified in terms of its loss [41,42], which is the average of the penalties incurred by all samples. The symbol $\mathcal{L}$ is used to represent the loss. The model's loss with respect to samples in the dataset $\mathcal{S}$ is,

$$\mathcal{L}_{\mathbf{\Theta}}(\mathcal{S}) = \frac{1}{|\mathcal{S}|} \sum_{n \in \mathcal{S}} l(t(n) - y(n)) \tag{1}$$

The optional subscript $\mathbf{\Theta}$ in Equation (1) above is used to highlight the loss's dependence on the model parameters. Each term $l(\cdot)$ is a sample penalty or error. Using this convention, $\mathcal{L}_{\mathbf{\Theta}}(\mathcal{S}_t)$, $\mathcal{L}_{\mathbf{\Theta}}(\mathcal{S}_s)$, and $\mathcal{L}_{\mathbf{\Theta}}(\mathcal{S}_v)$ are the training and validation losses.

Several loss functions have been proposed. The following are the most commonly used.

(*i*) Mean squared ($\mathcal{L}_2$) loss: For scalars, this loss is the average of the squared differences between the network's outputs $y(n)$, for inputs $\mathbf{x}(n)$ and the corresponding targets, so that, $\mathcal{L}_2 = |\mathcal{S}|^{-1} \sum_n [y(n) - t(n)]^2$. For vector outputs, the Euclidean norm $||\mathbf{y}(n) - \mathbf{t}(n)||$ is used, where $\mathbf{y}(n)$ is the model's vector output. The $\mathcal{L}_2$ loss is the most commonly used function. Using quadratic penalty terms makes the function quite sensitive to statistical outliers.

(*ii*) Averaged absolute ($\mathcal{L}_1$) loss: This is the average of the absolute difference, $\mathcal{L}_1 = |\mathcal{S}|^{-1} \sum_n |y(n) - t(n)|$. The $\mathcal{L}_1$ loss is used to avoid assigning excessive penalties to noisier samples. On the other hand, its effectiveness is compromised for data with copious noise.

(*iii*) Hüber loss: The Hüber loss represents a trade-off between the $\mathcal{L}_2$ and $\mathcal{L}_1$ losses [43]. Samples where the absolute difference is less than a threshold $\delta$ incur a quadratic penalty, while the remaining ones have a linear penalty. It is obtained as the average $|\mathcal{S}|^{-1} \sum_n l_\delta(n)$, where $l_\delta(n)$ is the penalty of the $n^{\text{th}}$ sample,

$$l_\delta(n) = \begin{cases} \frac{1}{2}[y(n) - t(n)]^2, & \text{if } |y(n) - t(n)| < \delta; \\ \delta|y(n) - t(n)| - \frac{1}{2}\delta^2, & \text{otherwise.} \end{cases} \quad (2)$$

As the Hüber loss function is not twice differentiable at $\pm\delta$, the similarly shaped log-cosh function below can be used in its place,

$$l_{lc}(n) = \log_e \frac{1}{2}\left(e^{y(n)-t(n)} + e^{t(n)-y(n)}\right). \quad (3)$$

(*iv*) $\epsilon$-Loss: This loss does not apply a penalty when the difference $y(n) - t(n)$ lies within a tolerable range $[-\epsilon, +\epsilon]$, for some constant $\epsilon$. A linear penalty is incurred whenever the numerical difference lies outside this range. In other words, $\mathcal{L}_\epsilon = |\mathcal{S}|^{-1} \sum_n l_\epsilon(n)$, where,

$$l_\epsilon(n) = \begin{cases} 0, & \text{if } |y(n) - t(n)| < \epsilon; \\ |y(n) - t(n)| - \epsilon, & \text{otherwise.} \end{cases} \quad (4)$$

Since the loss function is not differentiable at $\pm\epsilon$, if needed, a subgradient in $[0, 1]$ can be used as a substitute for its derivative at $\pm\epsilon$.

The shapes of the above loss functions are illustrated in Figure 1. The log-cosh loss, which is similar to the Hüber loss, is not shown. There are several other loss functions, including those that are specific to the application, that have not been listed here.

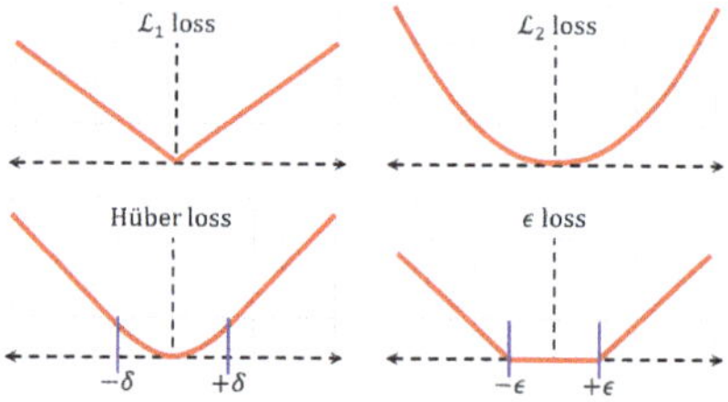


**Figure 1.** Loss Functions. Losses $\mathcal{L}$ as functions of the difference between the model output $y$, and the corresponding target (desired output) $t$.

*3.3. Model Selection*

Model complexity is a key concept in statistical learning theory, closely related to overfitting. The complexity of a model can be quantified as the number of independent scalar parameters used to compute its output and their ranges. The V-C (Vapnik–Chervonenkis) dimensionality of a model is one such measure of complexity [44] that has led to the development of support vector machines.

Model complexity is a critical factor that should be considered during model selection. Low-complexity models tend to exhibit a bias towards specific input-output maps. For instance, a linear model, which is the least complex regression model, cannot be used to capture nonlinear input–output relationships. Conversely, increasing a CI model's complexity endows it with more degrees of freedom to fit the training data. Due to its lower bias, training the model yields significantly lower training error $\mathcal{L}_{\boldsymbol{\Theta}}(\mathcal{S}_t)$. Unfortunately, a model with too large a complexity becomes too sensitive to extraneous artifacts present in its training dataset $\mathcal{S}_t$, such as random noise, sampling, or aliasing. These are extraneous artifacts that do not reflect any underlying input–output relationship. As stated in another manner, as the model's complexity increases, so do its variance. The model with higher variance performs poorly in the real world, with inputs outside $\mathcal{S}_t$. This is reflected in terms of its higher validation loss $\mathcal{L}_{\mathbf{\Theta}}(\mathcal{S}_v)$. In general, the model's effective loss can be decomposed into three components,

$$\mathcal{L}_{\mathbf{\Theta}} = \text{bias}^2_{\mathbf{\Theta}} + \text{var}_{\mathbf{\Theta}} + \text{noise}. \tag{5}$$

The square of the bias term is used in (5), as it can acquire positive and negative values. The noise component is an artifact introduced by the external environment, and is independent of the model $\mathbf{\Theta}$. Selecting a CI model with the optimal complexity is a well-known bias-variance dilemma in machine learning. This phenomenon is depicted in Figure 2.

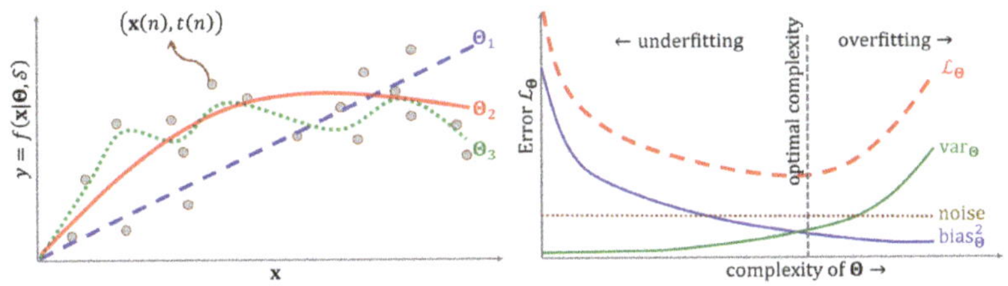


**Figure 2.** Bias, Variance, and Model Complexity. (**Left**) Performance of three models, $\mathbf{\Theta}_1$ (dashed blue), $\mathbf{\Theta}_2$ (solid red), and $\mathbf{\Theta}_3$ (dotted green) with low, optimum, and high model complexities. Small grey circles are training samples $(\mathbf{x}, \mathbf{t}) \in \mathcal{S}$. (**Right**) Squared bias (solid blue), variance (solid green), noise (dotted brown), and loss (dashed red) as functions of model complexity.

A widely used approach to keep the model's complexity at lower levels is by adding a regularization term $\mathcal{R}(\cdot)$ to the loss function. Regularizers are routinely devised in terms of the model parameters in $\mathbf{\Theta}$. If $\mathbf{\Theta}$ is treated as a vector of parameters, $\mathcal{R}(\mathbf{\Theta}) = ||\mathbf{\Theta}||_1$ and $\mathcal{R}(\mathbf{\Theta}) = ||\mathbf{\Theta}||_2^2$ are used as LASSO (least absolute shrinkage and selection operator) and ridge regularizers. The elastic net function, which is the convex combination of the LASSO and ridge terms, so that $\mathcal{R}(\mathbf{\Theta}) = r||\mathbf{\Theta}||_2^2 + (1-r)||\mathbf{\Theta}||_1$ (where $0 < r < 1$ is a constant), is another popular choice for regularization [45].

### 3.4. Training Algorithms

At present, almost all training algorithms rely on the basic gradient descent. If $\mathcal{L}_{\mathbf{\Theta}}(\cdot)$ is the loss function (which may include a regularization term), the parameters of the model are incremented using the training samples in $\mathcal{S}_t$, as shown in the following expression,

$$\mathbf{\Theta} \leftarrow \mathbf{\Theta} + \eta \nabla_{\mathbf{\Theta}} \mathcal{L}_{\mathbf{\Theta}}(\mathcal{S}_t) \tag{6}$$

The quantity $\eta$ in the above expression is the gradient descent step size, commonly referred to as the learning rate in CI terminology. The operator $\nabla_{\mathbf{\Theta}}$ is the gradient (vector derivative) w.r.t. $\mathbf{\Theta}$.

Since the loss $\mathcal{L}_{\mathbf{\Theta}}(\mathcal{S}_t)$ is the sum of all sample penalties $l(n)$, where $n \in \mathcal{S}_t$, a direct implementation of Equation (6) would require a pass over all samples in $\mathcal{S}_t$ before $\mathbf{\Theta}$ can be updated. As this is computationally burdensome (particularly for large datasets), training algorithms invariably use stochastic gradient descent (SGD). Before every training epoch of SGD, the samples in $\mathcal{S}_t$ are rearranged randomly. The vector parameter $\mathbf{\Theta}$ is incremented once for each sample $n$, using the gradient $\nabla_{\mathbf{\Theta}} l(n)$.

In theory, SGD can lead to a speed up the training algorithm by a factor $|\mathcal{S}_t|$. However, as the directions of the gradients $\nabla_{\mathbf{\Theta}} l(n)$ are not perfectly aligned with one another, the actual speed up is considerably less than $|\mathcal{S}_t|$. Adding a momentum term to the gradient step helps alleviate this situation. If in step $n-1$ the parameter $\mathbf{\Theta}$ is incremented by an amount $\Delta\mathbf{\Theta}(n-1)$, in the next step $n$, the increment would be $\Delta\mathbf{\Theta}(n) = \eta\nabla_{\mathbf{\Theta}} l(n) + \mu\Delta\mathbf{\Theta}(n-1)$. The quantity $\mu$ $(0 \leq \mu < \eta)$ is the momentum rate.

The convergence rate of the training algorithm can be significantly improved by Newton's algorithm, which requires the Hessian matrix $\nabla^2_{\mathbf{\Theta}}$. It can be readily established that the outer product $\nabla_{\mathbf{\Theta}}\nabla^{\mathrm{T}}_{\mathbf{\Theta}}$ is a close approximation of the Hessian. In the Levenberg-Marquardt algorithm [46], the diagonal elements of this matrix are incremented by an amount $\mu$ to improve the conditioning. Accordingly, incremental updates with the Levenberg-Marquardt algorithm are implemented as per the following rule,

$$\mathbf{\Theta} \leftarrow \mathbf{\Theta} + \left(\nabla_{\mathbf{\Theta}}\nabla^{\mathrm{T}}_{\mathbf{\Theta}} + \mu\mathbf{I}\right)^{-1}\nabla_{\mathbf{\Theta}}\mathcal{L}_{\Theta}(\mathcal{S}_t) \tag{7}$$

Overtraining—a problem that is closely related to overfitting, is frequently encountered during training. This is shown in Figure 3. Since samples in $\mathcal{S}_t$ are used to compute the gradient, as long as $\eta$ is small enough, the training loss decreases each time the parameters are incremented. Initially, the test loss $\mathcal{L}(\mathcal{S}_t)$ also drops with training. However, after the model has undergone a significant amount of training, $\mathcal{L}(\mathcal{S}_t)$ begins to rise. Applying (6) further will cause overtraining.

$K$-fold cross-validation [47] is an effective means for performance evaluation with sparse data. The samples in $\mathcal{S}_t$ are randomly shuffled and split into $K$ groups or folds of equal size. One of the folds is used as the test set $\mathcal{S}_s$, and the rest are used to increment $\mathbf{\Theta}$. This process is repeated $K$ times, with each fold acting as the test set. The loss averaged over all $K$ folds is a reliable estimate of its true (real-world) loss.

Premature convergence is another issue that is sometimes observed during training (see Figure 3). This occurs if the training algorithm encounters a local minimum of the loss function's landscape, where the gradient $\nabla_{\mathbf{\Theta}}$ is very close to zero. Applying (6) or (7) would have little effect on the parameter $\mathbf{\Theta}$. A simple method to rectify the situation is to restart the training process from some other (randomly generated) initial point.

The presence of narrow ridges with "V"-shaped cross sections is another reason why the loss may remain unchanged (see Figure 3), giving the appearance of a local minimum for several iterations. Although there is no perceptible drop in the loss, the amount of increment to $\mathbf{\Theta}$ is not negligible. Restarts are unnecessary in such situations, for the algorithm eventually leaves such a ridge after multiple updates.

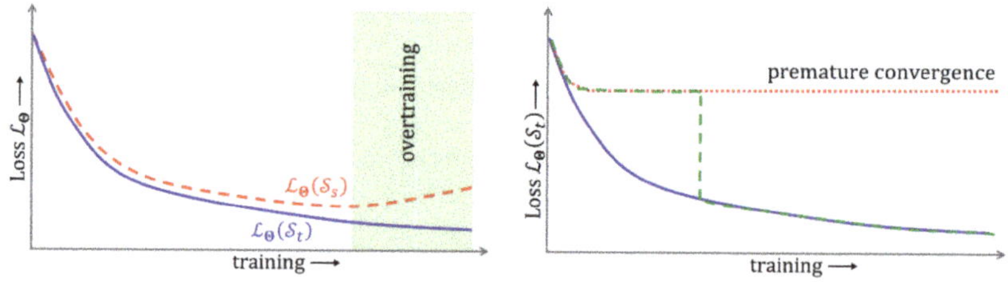


**Figure 3.** Overtraining and premature convergence. Overtraining is illustrated (**left**), showing how test loss (dashed red) begins to rise with overtraining (shaded green region) although training loss (solid blue) decreases. Premature convergence (**right**) of the training loss is shown (dotted red) in contrast to desired convergence (dashed green, solid blue). Due to "V" shaped narrow ridges in the loss function's landscape, there may not be any perceptible decrease in the loss for many training iterations (dashed green)

### 3.5. Optimization Metaheuristics

Existing training algorithms apply optimization metaheuristics, such as GA and PSO, to avoid getting trapped in local minima. These algorithms maintain a set of many candidate solutions, referred to as its population. In each step of the optimization algorithm, a new population is formed out of the existing one, using a variety of stochastic and heuristic search operators. Stochastic operators help the algorithm escape from local minima, while heuristics aid in its convergence towards the global maximum of the objective function.

GAs are useful in training CI regression models. Let the population of such a GA be the set $\{\mathbf{\Theta}^j | j = 1, 2, \cdots\}$, where each $\mathbf{\Theta}$ is a candidate model parameter. During each iteration, pairs of solutions are selected from the population in a random manner, but with better ones (in terms of the inverse loss function) being more likely to be picked. Using the crossover operator, a new pair of new solutions is generated from the old ones. For example, in a convex crossover, the existing pair $(\mathbf{\Theta}^i, \mathbf{\Theta}^j)$ can be used to generate a new pair, $(\mathbf{\Theta}^{i'}, \mathbf{\Theta}^{j'}) = \left(\mu\mathbf{\Theta}^i + (1-\mu)\mathbf{\Theta}^j, (1-\mu)\mathbf{\Theta}^i + \mu\mathbf{\Theta}^j\right)$.

In the mutation operator, a small amount of perturbation $\Delta\mathbf{\Theta}^i$ is added to each new candidate parameter so that it becomes equal to $\mathbf{\Theta}^i + \Delta\mathbf{\Theta}^i$.

In Gaussian mutation, the perturbation $\Delta\Theta^i$ follows a Gaussian distribution centered around the origin. This process is repeated many times until no further improvement can be found.

Although GA and PSO have found widespread use in many optimization applications, their use in machine–soil interaction studies is rather limited. GAs have been used during model training. In these cases, the GA is hybridized with (6) and (7), or any other related method. Gradient descent steps can be incorporated into the GA in different ways. For instance, $\mathbf{\Theta}^i$ can be mutated into $\mathbf{\Theta}^i + \Delta\mathbf{\Theta}^i + \eta\nabla_{\mathbf{\Theta}_i}\mathcal{L}$, where $\Delta\mathbf{\Theta}^i$ is the random perturbation and $\nabla_{\mathbf{\Theta}}\mathcal{L}$, the gradient of the training loss $\mathcal{L}$.

Similar hybrid techniques exist for PSO (cf. [48]). However, PSO has not been used in the existing literature on machine–soil interactions. On the other hand, a relatively unknown population-based stochastic algorithm has been used in [49,50].

## 4. Current Computational Intelligence Models

### 4.1. Neural Networks

Neural network (NN) models have been routinely used in various regression applications since the mid-eighties, wherever a significant amount of data are involved. Neural networks are layered structures consisting of an input layer, one or more intermediate layers, called hidden layers, and an output layer. Each layer comprises elementary processing units or neurons. In a fashion resembling the mammalian cortex, the neurons in each hidden and output layer receive the outputs of those of the preceding layer, as their inputs via weighted synaptic connections.

Figure 4 shows the layout of a neural network with $L$ layers. The indices of the input and output layers are 1 and $L$, where $M$ and $N$ are the number of neurons in the input and output layers. The vectors $\mathbf{x}$ ($\mathbf{x} \in \mathbb{R}^M$) and $\mathbf{y}$ ($\mathbf{y} \in \mathbb{R}^N$) denote the network's input and output.

The size of an NN can be written succinctly as $\prod_{l=1}^{L} N^{(l)}$ where $N^{(l)}$ is the number of neurons in layer $l$. For instance, a $3 \times 5 \times 6 \times 2$ NN has three input neurons, a hidden layer with five neurons, another hidden layer with six neurons, and two output neurons. Note that indices of layers (superscripts) are shown within parentheses so as not to confuse them with exponents.

Until recently, NNs were equipped with only one or two hidden layers (so that $L = 3$ or $L = 4$)—an approach used everywhere in the published literature on soil–machine interaction studies. To distinguish them from deep neural networks (DNNs), which have multiple hidden layers, models with only one hidden layer are referred to as shallow networks. However, for the purpose of this review, networks with two hidden layers are also included in this category. This section focuses on classical methods that are common to both shallow and deep networks. Advanced features that are relevant to DNNs are addressed separately in a subsequent section.

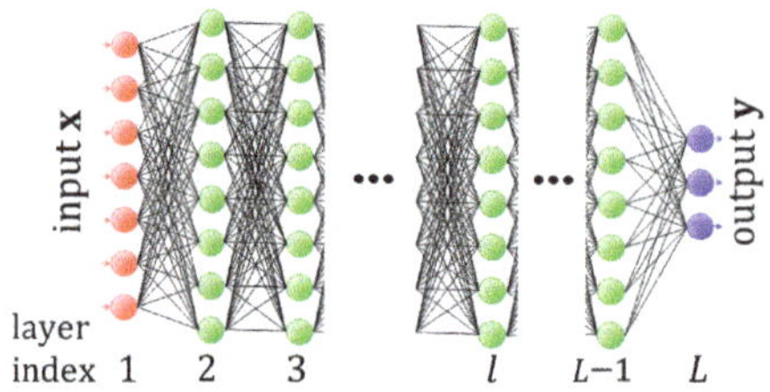


**Figure 4.** Neural network. Neurons are depicted as small circles and synaptic connections as straight lines. The network has an input layer (red), hidden layers (green), and an output layer (blue). Since the network shown has multiple hidden layers, it is a deep neural network.

The output of the $k^{\text{th}}$ neuron in a layer indexed $l$ ($l \in \{1 \cdots L\}$) is denoted as $y_j^{(l)}$. Thus, the $i^{\text{th}}$ element of $\mathbf{x}$ is $x_j = y_j^{(1)}$; similarly $y_j = y_j^{(L-1)}$. Figure 4 shows all quantities associated with the $k^{\text{th}}$ neuron in the layer $l$ ($l > 1$). The neuron's input is the weighted sum of the outputs of all neurons in the preceding layer ($l - 1$), as shown in the following expression,

$$s_k^{(l)} = w_{k,0}^{(l)} + \sum_j w_{k,j}^{(l)} y_j^{(l-1)}. \tag{8}$$

The summation in (8) is carried out over the outputs $y_j^{(l-1)}$ of all neurons (indexed $j, j \geq 1$) of the previous layer, and the associated weight is $w_{k,j}^{(l)}$. The quantity $w_{k,0}^{(l)}$ is the neuron's bias. Figure 5 shows a neuron in a hidden layer. The weights and biases are the trainable parameters of the neural network that are included in $\mathbf{\Theta}$.

Neurons in the input layer are linear elements; their role is merely to transmit the incoming vector to hidden neurons. However, those in the hidden layers, and optionally in the output layer as well, incorporate a monotonically increasing nonlinear function $f(\cdot)$, where either $f : \mathbb{R} \to (0,1)$ or $f : \mathbb{R} \to (-1,1)$, that is referred to as the activation function. The output of the neuron is,

$$y_k^{(l)} = f\left(s_k^{(l)}\right). \tag{9}$$

The logistic function $\sigma(s) = (1 + \exp(-s))^{-1}$ and the hyperbolic tangent function ($\tanh(\cdot)$) are the most commonly used activation functions used in shallow networks. Due to their characteristic 'S' shapes, such activation nonlinearities fall under the category of sigmoid functions.

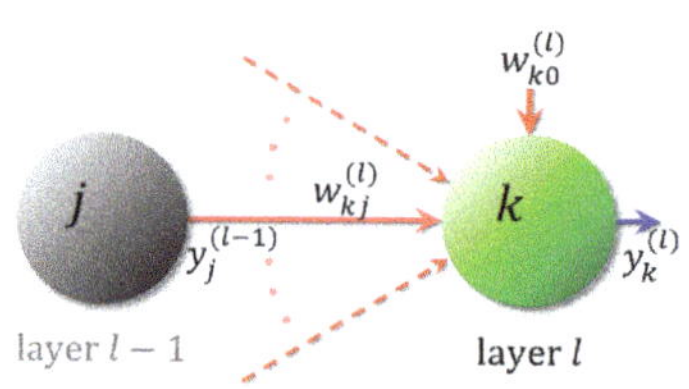


**Figure 5.** Neuron. Quantities associated with a neuron (green circle). Also shown is a neuron in the preceding layer (grey circle)

Historically, the popularity of NNs surged with the introduction of the back-propagation (BP) algorithm [51], which is a reformulation of SGD designed to train layered structures. The error $\delta_k^{(l)}$ of the $k^{\text{th}}$ neuron in the $l^{\text{th}}$ layer is defined as the derivative of the penalty term $l_{\mathbf{\Theta}}$ (in the loss $\mathcal{L}_{\mathbf{\Theta}}$) with respect to the neuron's input $s_k^{(l)}$ (see Equation (8)). Such penalties can be readily differentiated for neurons in the output layer ($l = L$). The back-propagation rule shows how $\delta_k^{(l)}$ can be computed for hidden neurons ($l < L$), using the errors of the next layer, $l+1$. The schematic in Figure 6 illustrates how errors back-propagate. The general expression to compute the errors is,

$$\delta_k^{(l)} = \begin{cases} \frac{\partial}{\partial s_k^{(L)}} l_{\boldsymbol{\Theta}}, & \text{if } l = L; \\ \sum_j w_{kj}^{(l+1)} \delta_j^{(l+1)}, & \text{otherwise.} \end{cases} \quad (10)$$

The weights in $\boldsymbol{\Theta}$ can be updated in the following manner,

$$w_{kj}^{(l)} = w_{kj}^{(l)} + \eta y_j^{(l-1)} \delta_k^{(l)}. \quad (11)$$

It is common practice to include a momentum term to BP. Additionally, BP can be extended to apply Levenberg-Marquardt updates. This is the Levenberg-Marquardt BP (LMBP) algorithm [52].

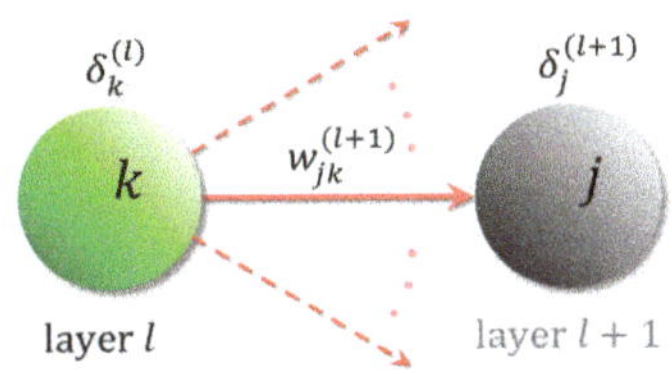


**Figure 6.** Back-propagation of errors. Shown here are the quantities relevant to the back-propagation of error from a neuron (solid circle) to another in the previous layer (green circle).

The VC-dimensionality of a neural network is typically specified in terms of the total number of weights and biases [53]. The number of training samples should be about ten times this quantity. The number of epochs to achieve training is independent of the data size.

### *4.2. Radial Basis Function Networks*

The radial basis function network (RBFN) [54,55] is another popular computational intelligence regression model that is topologically identical to an $M \times K \times 1$ neural network. In other words, an RBFN has $M$ input neurons, a single hidden layer of $K$ neurons, and only one output neuron. The sole purpose of the input layer, which contains $M$ linear neurons, is to pass on $M$ dimensional inputs to the hidden layer. The $K$ neurons in the hidden layer are incorporated with nonlinear activation functions. The output neuron computes the weighted summation of the outputs from the hidden layer. Due to its strong resemblance to a shallow neural network, the RBFN is sometimes treated as a specific kind of NN. RBFNs have been successfully used in agricultural applications [56–58].

Unlike in NNs, the hidden neurons of the RBFN are designed to produce localized responses. The activation function of any hidden neuron has an $M$ dimensional parameter called its center. The closer an input is to its center, the higher the neuron's output. In this manner, the network's hidden neurons simulate sensory cells of the peripheral nervous system, which have localized receptive fields.

Suppose $\mathbf{x}$ ($\mathbf{x} \in \mathbb{R}^M$) is the network's input. Each hidden neuron $k$ (where $k \in \{1 \cdots K\}$) receives $\mathbf{x}$ from the input layer, and produces an output $f(||\mathbf{x} - \mathbf{c}_k||)$, where $||\cdot||$ denotes a vector norm operator (e.g., length). Gaussian nonlinearities are the most widely used activation functions. In this case, the output of the $k^{\text{th}}$ hidden neuron, denoted as $\phi_k$, is obtained according to the following expression,

$$\phi_k = e^{\frac{1}{\sigma_k}||\mathbf{x}-\mathbf{c}_k||^2} \quad (12)$$

The quantity $\sigma_k$ in (12) is an optional width parameter of the $k^{\text{th}}$ hidden neuron. When dealing with training samples that are distributed evenly within the input space, all hidden neurons may be assigned the same widths $\sigma$.

With $w_k$ ($k \in 1 \cdots K$) being network weights, the RBFN's output $y$ is the weighted sum $\sum_k w_k \phi_k$. Using (12), $y$ can be expressed directly in terms of the input $\mathbf{x}$ as,

$$y = \sum_k w_k e^{\frac{1}{\sigma_k}||\mathbf{x}-\mathbf{c}_k||^2} \tag{13}$$

Figure 7 depicts the main quantities of an RBFN.

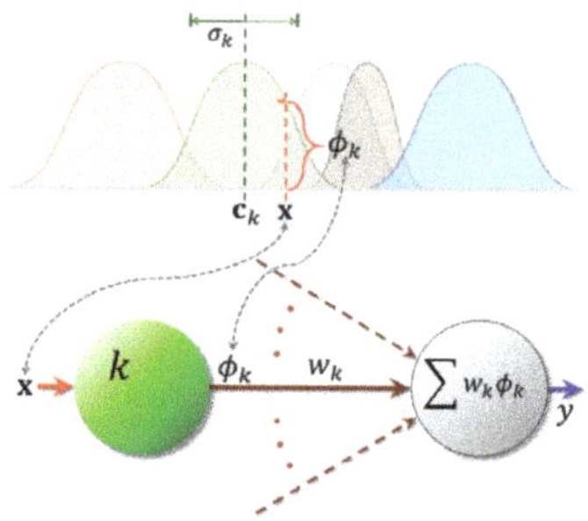


**Figure 7.** Radial Basis Function. Shown are a hidden neuron (green) and the output neuron (grey).

The RBFN's parameters in $\mathbf{\Theta}$ are all its weights $w_k$ and centers $\mathbf{c}_k$. If necessary to train the network's widths $\sigma_k$, they are also included in $\mathbf{\Theta}$. Due to the use of localized activation functions, the number of hidden neurons $K$ required by the RBFN increases exponentially with the input dimensionality $M$. Hence the effectiveness of RBFNs is limited to tasks involving low dimensional data (up to $M = 6$ or 7). Even in such tasks, RBFNs require significantly more hidden neurons than NNs. As the trade-off for this limitation, RBFNs offer faster training, often by a few orders of magnitude. This speedup over Equation (6) is achieved when the centers, widths, and weights are trained separately [59].

A popular method to train the centers of the hidden neurons is by using $K$-means clustering [60]. For each hidden neuron $k$, a subset $\mathcal{N}_k$ of samples in training set $\mathcal{S}_t$ is obtained. This subset consists of all samples that are closer to the neuron's center $\mathbf{c}_k$ than to $\mathbf{c}_{k'}$ of any other neuron $k'$, $k' \neq k$. The center of each hidden neuron is made equal to the average of all samples $\mathbf{x}(n)$ in $\mathcal{N}_k$. The two steps can be expressed, as shown below,

$$\begin{aligned} \mathbf{c}_k &\leftarrow \frac{1}{|\mathcal{N}_k|} \sum_{n \in \mathcal{N}_k} \mathbf{x}(n), \quad \text{where,} \\ \mathcal{N}_k &= \{n \in \mathcal{S}_t | k' \neq k, ||\mathbf{x}(n) - \mathbf{c}_k|| < ||\mathbf{x}(n) - \mathbf{c}_{k'}||\} \end{aligned} \tag{14}$$

A relatively small number of iterations of (14) is enough to train the centers of all hidden neurons. Their widths can be fixed at some constant value such that $\sigma_k = \sigma$, $k \in \{1 \cdots K\}$. Alternately, the nearest neighbor heuristic can be applied to determine each $\sigma_k$ separately, such as,

$$\sigma_k = c \operatorname*{argmin}_{k' \neq k} ||\mathbf{c}_k - \mathbf{c}_{k'}|| \tag{15}$$

The quantity $c$ in (15) is an algorithmic constant.

For the $\mathcal{L}_1$ or the Hüber loss functions, the weight parameters $\mathbf{w}_k$ must be trained in an iterative manner using (6). When the $\mathcal{L}_2$ loss is used, the Moore-Penrose pseudoinverse formula provides a simpler method to obtain the weights. Let $\mathbf{w} \in \mathbb{R}^K$ be the vector of all weights. Similarly, let $\boldsymbol{\phi}(n) \in \mathbb{R}^K$ ($n \in \mathcal{S}_t$) be the vector of outputs of the hidden neurons, determined using (12)) with input $\mathbf{x}(n)$. It can be observed that the RBFN's output is $y(n) = \boldsymbol{\phi}^{\mathrm{T}}(n)\,\mathbf{w}$ (where $\cdot^{\mathrm{T}}$ is the transpose operator).

To observe how the pseudoinverse formula works, let us construct an activation matrix $\boldsymbol{\Phi} \in \mathbb{R}_+^{|\mathcal{S}_t| \times K}$, whose $n^{\text{th}}$ row is $\boldsymbol{\phi}^{\mathrm{T}}(n)$. Whence $\mathbf{y} = \boldsymbol{\Phi}\mathbf{w}$ is the $|\mathcal{S}_t| \times 1$ vector of all outputs of the RBFN. If $\mathbf{t} \in \mathbb{R}^{|\mathcal{S}_t|}$ is the corresponding vector of all target values, the mean squared $\mathcal{L}_2$ loss is the expression $\mathcal{L}_2 = ||\mathbf{y} - \mathbf{t}||^2$. The weight vector that minimizes this loss is $\underset{\mathbf{w}}{\operatorname{argmin}}||\boldsymbol{\Phi}\mathbf{w} - \mathbf{t}||^2$.

If the number of training samples is more than the number of hidden neurons (i.e., $|\mathcal{S}_t| > K$), which is always the case in a real application, the matrix $\boldsymbol{\Phi}^{\mathrm{T}}\boldsymbol{\Phi}$ is non-singular. In this case, the expression for the loss minimizing weight vector $\mathbf{w}$ is determined as,

$$\mathbf{w} = \left(\boldsymbol{\Phi}^{\mathrm{T}}\boldsymbol{\Phi}\right)^{-1}\boldsymbol{\Phi}^{\mathrm{T}}\mathbf{t} \tag{16}$$

The factor $\left(\boldsymbol{\Phi}^{\mathrm{T}}\boldsymbol{\Phi}\right)^{-1}\boldsymbol{\Phi}^{\mathrm{T}}$ in (16) is a matrix of size $K \times |\mathcal{S}_t|$. It is referred to as the pseudoinverse of $\boldsymbol{\Phi}$ and denoted as $\boldsymbol{\Phi}^{+}$.

In theory, all RBFN parameters can be trained iteratively using gradient descent. Although training the RBFN parameter vectors $[\mu_k]$ and $[\sigma_k]$ in this manner is fairly uncommon, and gradient descent is often used to train the weight vector $\mathbf{w}$. This is carried out as in Equation (6), with $\boldsymbol{\Theta}$ replaced with $\mathbf{w}$. This method is applied to avoid numerical issues with matrix pseudoinversion and wherever the training algorithm is not based on the mean squared loss function.

Recent RBFN models use multivariate Gaussian distributions, where Equation (12) is replaced with,

$$\phi_k = e^{\frac{1}{2}(\mathbf{x}-\mathbf{c}_k)^{\mathrm{T}}\boldsymbol{\Sigma}_k(\mathbf{x}-\mathbf{c}_k)} \tag{17}$$

In the above expression, the quantity $\boldsymbol{\Sigma}_k \in \mathbb{R}^{M \times M}$ is a covariance matrix.

### *4.3. Support Vector Regression*

SVRs are another class of CI models [61,62] that are widely used in various engineering and other applications [63]. SVRs have been used for regression applications in agriculture [64–67]. Unlike the other CI models discussed earlier, SVRs do not have any strong parallels in nature. Instead, they are specifically aimed at addressing the issue of model complexity, which is addressed below.

The simplest formulation is the linear SVR with $\epsilon$-loss, as shown in Figure 8. Sample targets that lie within a margin of $\pm\epsilon$ from the regression line do not incur any penalty, while those outside the margin incur penalties. Hence, the error arising from a sample pair $(\mathbf{x}(n), t(n))$ is obtained as shown in (4). Denoting this error as $\xi(n)$, it can be readily established that the following constraints are satisfied,

$$\begin{cases} \xi(n) \geq 0 \\ t(n) - \mathbf{w}^{\mathrm{T}}\mathbf{x}(n) - b \leq \epsilon + \xi(n) \\ \mathbf{w}^{\mathrm{T}}\mathbf{x}(n) + b - t(n) \leq \epsilon + \xi(n) \end{cases} \tag{18}$$

When the above conditions are satisfied, the loss is simply the sum of all errors, $\sum_n \xi(n)$.

It has been demonstrated that the gap between the validation and training losses (i.e., $\mathcal{L}(\mathcal{S}_v) - \mathcal{L}(\mathcal{S}_t)$) can be lowered by increasing $\epsilon$, or alternately by decreasing $||\mathbf{w}||_2$ while $\pm\epsilon$ is a constant [68]. This term can be recognized as the LASSO regularizer.

With $C$ being an algorithmic constant, the optimal regression model can be obtained as the solution to the following constrained optimization problem,

$$\begin{cases} \min\limits_{\mathbf{w},b,\xi} \;\; \frac{1}{2}\mathbf{w}^{\mathrm{T}}\mathbf{w} + C\sum_{n \in \mathcal{S}_t}\xi(n) \\ \text{s.t.} \quad (18) \quad \text{is true} \end{cases} \tag{19}$$

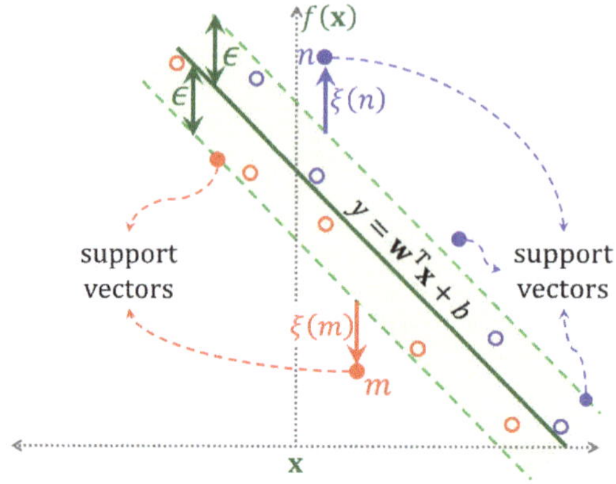


**Figure 8.** Linear support vector regression. The regression line $y = \mathbf{w}^{\mathrm{T}}\mathbf{x} + b$ (solid green), and the $\epsilon$ region (shaded green) of zero penalties around it, are shown. Also shown are samples (small circles) including the support vectors (filled circles) indexed $m$ and $n$.

The above problem (19) to obtain the SVR is in primal form. Classical optimization theory (cf. [69,70]) illustrates that for every primal problem, a dual problem can be constructed using the Lagrange multipliers of the primal constraints as its variables. The optimization theory establishes that under certain constraint qualifications, the optima of the primal and dual problems coincide at a saddle point. The dual form of (19) can be derived readily [68]. Ignoring the constraints $\xi(n) \geq 0$ and using the symbols $\lambda_{+} \in \mathbb{R}_{+}^{|\mathcal{S}_t|}$ and $\lambda_{-} \in \mathbb{R}_{+}^{|\mathcal{S}_t|}$ as the Lagrange multiplier vectors of the other constraints in (18), the dual problem can be formulated in the following manner,

$$\begin{cases} \min\limits_{\lambda_{+},\lambda_{-}} & \frac{1}{2}(\lambda_{+}-\lambda_{-})^{\mathrm{T}}\mathbf{K}(\lambda_{+}-\lambda_{-}) + \lambda_{+}^{\mathrm{T}}(\epsilon\mathbf{1}-\mathbf{t}) + \lambda_{-}^{\mathrm{T}}(\epsilon\mathbf{1}+\mathbf{t}) \\ \text{s.t.} & \mathbf{1}^{\mathrm{T}}(\lambda_{+}-\lambda_{-}) = 0 \\ & \mathbf{0} \leq \lambda_{+}, \lambda_{+} \leq C\mathbf{1} \end{cases} \tag{20}$$

The element in the $m^{\text{th}}$ row and $n^{\text{th}}$ column of the symmetric matrix $\mathbf{K} \in \mathbb{R}_{+}^{|\mathcal{S}_t|}$ in (20) is $\mathbf{x}(m)^{\mathrm{T}}\mathbf{x}(n)$. The bias $b$ and the normal vector $\mathbf{w}$ can be obtained from the dual solution, although $\mathbf{w}$ is not required.

In more generalized settings, input samples can lie in any arbitrary Hilbert space. The inner product of the $m^{\text{th}}$ and $n^{\text{th}}$ samples is represented as $\langle \mathbf{x}(m), \mathbf{x}(n) \rangle$. The matrix $\mathbf{K}$ will contain pairwise inner products of such samples.

Nonlinear SVRs implicitly apply a transformation $\phi(\cdot)$ from the input space $\mathcal{S}$ to an unknown Hilbert space [61]. Under these circumstances, the $(m,n)^{\text{th}}$ element of $\mathbf{K}$, which we now denote as $K(\mathbf{x}(m), \mathbf{x}(n))$, is obtained as provided below,

$$K(\mathbf{x}(m), \mathbf{x}(n)) = \langle \phi(\mathbf{x}(m)), \phi(\mathbf{x}(n)) \rangle \tag{21}$$

The function $K : \mathcal{S} \times \mathcal{S} \rightarrow \mathbb{R}_{+}$ is referred to as the kernel function. Mercer's theorem states that as long as the kernel satisfies a few conditions, there must exist some transformation $\phi : \mathcal{S} \rightarrow \mathbb{H}$ satisfying (21). As long as these conditions are met, the matrix $\mathbf{K}$ obtained from every possible sample set will be symmetric and positive definite. In other words, kernel functions can be devised without even considering the mapping $\phi(\cdot)$; this mapping, along with its range in Hilbert space $\mathbb{H}$ can remain unknown. This is a remarkable feature of SVR models. In engineering applications, any symmetric, non-negative measure of similarity between pairs of samples can be adopted as the kernel function. For instance, Gaussian kernels $e^{\frac{1}{\sigma}||\mathbf{x}(m)-\mathbf{x}(n)||^2}$, or $\mathcal{L}_p$-normed kernels $||\mathbf{x}(m) - \mathbf{x}(n)||_p$ can be adopted for inputs that lie in a Euclidean space $\mathbb{R}^M$. In bio-informatics, where samples may consist of

DNA strands that are sequences of the letters C, T, G, and A, the kernel may vary negatively with the minimum edit distance between every pair of samples.

For each sample, $\mathbf{x}(n)$ that is strictly within the $\pm\epsilon$ margin of the regression line (see Figure 8), the corresponding dual variables $\lambda_{\pm}(n)$ obtained from (20) will be zeros. It is only when the sample lies either on the margin's boundaries or outside it that yields either $\lambda_{+}(n) > 0$ or $\lambda_{-}(n) > 0$. These samples are the support vectors. The set of all support vectors is,

$$\mathcal{V} = \{n|\lambda_{+}(n) > 0 \,\text{or}\, \lambda_{-}(n) > 0\}. \quad (22)$$

Given an unknown sample $\mathbf{x}$, the estimated output $y$ can be obtained using the kernels of $\mathbf{x}$ and the support vectors in $\mathcal{V}$,

$$y = \sum_{n \in \mathcal{V}} (\lambda_{+}(n) - \lambda_{-}(n)) K(\mathbf{x}(n), \mathbf{x}) + b.$$

Although not provided in this article, the bias $b$ can be obtained readily from the dual form in (20).

As long as the training set $\mathcal{S}_t$ is small enough so that it is computationally feasible to compute the matrix $\mathbf{K}$ and store in memory, quadratic programming can be applied directly to solve (20). Otherwise, there are a plethora of iterative training algorithms [71–73], that are well-equipped to train SVRs with larger data sets. SVRs can be formulated using other losses and regularizers as well.

### 4.4. *Fuzzy Inference Systems*

FIS is a CI model that is inspired by decision-making processes in humans. It uses fuzzy sets to capture the inherent vagueness in human verbal reasoning. The fuzzy set theory extends the classical concept of a set (called a 'crisp' set in fuzzy terminology) by incorporating such imprecision. The manner in which it does so is described next.

Any element $x$ from the universe of discourse $\mathbb{U}$ can either be in a given crisp set $A$, where $A \subset \mathbb{U}$ (i.e., $x \in A$) or not in it (i.e., $x \notin A$). Accordingly, a binary membership function $\mu_A : \mathbb{U} \rightarrow \{0,1\}$ can be defined such that $\mu_A(x) = 1$ iff $x \notin A$, otherwise $\mu_A(x) = 0$ iff $x \notin A$. The membership function of a fuzzy set $A$ is allowed to have any real value within the interval $[0,1]$, i.e., $\mu_A : \mathbb{U} \rightarrow [0,1]$. The numerical value of $\mu_A(x)$ indicates the degree to which $x$ is included in $A$. For example, let $T$ be the set of tall students in a class. If $T$ is a crisp set, there must be a minimum cutoff for tallness. Let this cutoff be 5′10″. Hence, Jack and Jill, whose heights are 5′9″ and 6′1″, have memberships $\mu_T(\text{Jack}) = 0$, and $\mu_T(\text{Jill}) = 1$ in . On the other hand, if $T$ is a fuzzy set, then memberships such as $\mu_T(\text{Jack}) = 0.7$, and $\mu_T(\text{Jill}) = 0.99$ are possible, indicating that Jack is very close to being tall, whereas Jill is definitely tall.

When the universe of the discourse is a continuous variable, memberships can be defined in terms of functions of real arguments $\mu : \mathbb{R} \rightarrow [0,1]$. The Gaussian, trapezoidal, and triangular functions are commonly used for memberships. The Gaussian membership of a scalar input $x$ to the fuzzy set $A \subset \mathbb{U}$ is $e^{-(x-\mu)/\sigma}$. The trapezoidal membership can be defined using four parameters, $a$, $b$, $c$, and $d$ ($a \leq b < c \leq d$),

$$\mu_A(x) = \begin{cases} 0, & \text{if } x < a; \\ \frac{x-a}{b-a}, & \text{if } a \leq x < b; \\ 1, & \text{if } b \leq x < c; \\ \frac{d-x}{d-c}, & \text{if } c \leq x < d; \\ 0, & \text{if } d \leq x. \end{cases} \quad (23)$$

The triangle membership function requires only three parameters, $a$, $b$, and $c$ ($a \leq b \leq c$),

$$\mu_A(x) = \begin{cases} 0, & \text{if } x < a; \\ \frac{x-a}{b-a}, & \text{if } a \leq x < b; \\ \frac{c-x}{c-b}, & \text{if } b \leq x < c; \\ 0, & \text{if } c \leq x. \end{cases} \tag{24}$$

Gaussian memberships, as well as those in (23) and (24), have peak values of unity. Although this is common practice in real-world applications, fuzzy sets can also admit any other membership function as long as its maximum lies anywhere in $(0, 1]$. The complement $\bar{A}$ of the fuzzy set $A$ can be readily defined in terms of the membership function as $\mu_{\bar{A}} = 1 - \mu_A$. A fuzzy singleton—say $B$, is a fuzzy set that is fully parametrized by a constant $v_B$, where $v_B \in \mathbb{R}$, such that for the input $y \in \mathbb{R}$,

$$\mu_B(y) = \begin{cases} 1, & \text{if } y = v_B; \\ 0, & \text{otherwise.} \end{cases} \tag{25}$$

The operations of union ($\cup$) and intersection ($\cap$) in crisp sets correspond to conjunction (AND) and disjunction (OR) in Boolean algebra. In terms of membership functions, the union $A \cup B$ and intersection $A \cap B$ of the sets $A$ and $B$ are $\mu_{A \cup B} = \mu_A \,\text{OR}\, \mu_A$, and $\mu_{A \cap B} = \mu_A \,\text{AND}\, \mu_A$. Union and intersection of fuzzy sets can be realized in various ways [74], using t-conorms and t-norms. A popular choice is to use $\max(\cdots)$ as the t-conorm operator and $\min(\cdots)$ as the t-norm. In this case, $\mu_{A \cup B} = \max\{\mu_A, \mu_B\}$ and $\mu_{A \cap B} = \min\{\mu_A, \mu_B\}$. In our previous example, suppose $S$ is the fuzzy set of smart students and $\mu_S(\text{Jill}) = 0.75$, then $\mu_{S \cup T}(\text{Jill}) = \max\{0.75, 0.99\} = 0.99$ and $\mu_{S \cap T}(\text{Jill}) = \min\{0.75, 0.99\} = 0.75$.

A FIS encapsulates human knowledge through a fuzzy rule base. Each rule in the base consists of two parts, an antecedent and a consequent, and is written in the format, "**If** ANTECEDENT **then** CONSEQUENT". If the input to the model is an $M$-dimensional vector $\mathbf{x}$ and its output is an $M$-dimensional vector $\mathbf{y}$, the antecedents and consequents are made up of $M$ and $N$ fields. The generic format of a rule with index $k \in 1, 2, \cdots, K$ is as shown below,

$$\textbf{If } \underbrace{x_1 \text{ is } A_1^k \diamond x_2 \text{ is } A_2^k \cdots \diamond x_M \text{ is } A_M^k}_{\text{ANTECEDENT}} \textbf{ then } \underbrace{y_1 \text{ is } B_1^k \cdots \diamond y_N \text{ is } B_N^k}_{\text{CONSEQUENT}}. \tag{26}$$

Each diamond symbol ($\diamond$) in (26) represents an AND or an OR operator.

The order in which these operations are applied may either be in accordance with an established convention or, alternately, specified explicitly by inserting brackets at appropriate places. Mathematically speaking, the $j^{\text{th}}$ field in the antecedent of the fuzzy rule in Equation (26), "$x_j$ is $A_j$" is the membership, $\mu_{A_j}(x_j)$. In a similar fashion, the $i^{\text{th}}$ field in the consequent is $\mu_{B_i}(y_i)$. Figure 9 shows a simple rule base with $K = 6$ rules.

There are two kinds of FIS, differing only in the way the sets $B_i$ in the consequent's $i^{\text{th}}$ field ($i \in \{1, 2, \cdots, N\}$) are defined. In the Mamdani model [75], they are allowed to be fuzzy sets. As a result of this flexibility, a Mamdani FIS can easily apply verbal descriptions of the consequents. On the other hand, in the Takagi-Sugeno-Kang (TSK) model [76,77], each $B_i$ must be a singleton, as in Equation (25). A TSK model renders the FIS more amenable to mathematical treatment. Figure 9 shows examples of the Mamdani and TSK models.

The various steps involved in mapping an input to its output will be illustrated using the examples shown in Figure 10 (Mamdani model) and Figure 11 (TSK model). The steps are briefly described below.

(*i*) Fuzzification: This step is carried out separately in each antecedent field "$x_j$ is $A_j^k$" and for each rule $k$. It involves computing the values of the memberships $\mu_{A_j^k}(x_j)$ using the numerical values of the input element $x_j$.

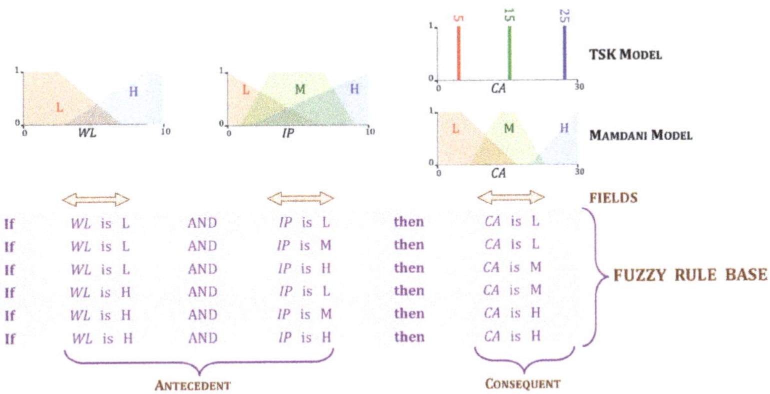


**Figure 9.** Fuzzy inference system. The figure shows membership functions (**top**), and fuzzy rule base (**bottom**). The input **x** has two elements, $WL \in [0, 10]$ (wheel load), that can be L (Low), M (Medium), or H (High), and $IP \in [0, 10]$ (inflation pressure) that can be L (Low) or H (High). The output $y \in [0, 30]$ is a scalar ($N = 1$). This is the quantity $CA$ (contact area), which can be L (Low), M (Medium), or H (High). The membership functions, $\mu_{\mathrm{L}}(CA)$, $\mu_{\mathrm{M}}(CA)$, and $\mu_{\mathrm{H}}(CA)$ are trapezoids/triangle (Mamdani) or singletons (TSK).

*(ii)* Aggregation: In this step, AND and OR operations are applied as appropriate to each rule in the FIS. The rules in the FIS shown in Figures 10 and 11 only involve conjunctions (AND) that are implemented through the min$(\cdot)$ t-norm. The aggregated membership is referred to as its rule strength. The strength of rule $k$ is,

$$\mu_A^k = \bigcup_j \mu_{A_j^k}(x_j). \tag{27}$$

*(iii)* Inference: The strength of each rule is applied to its consequent. Each rule $k$ in our example contains only one consequent field. Its membership function $\mu_{B^k}$ is limited to a maximum of $\mu_{A^k}$. For every rule, $k$ in $k$, a two-dimensional region $\mathcal{R}^k$ is identified in the Mamdani model. Since the TSK model involves only singletons at this step, only a two-dimensional point $\mathcal{R}^k$ is necessary. Accordingly,

$$\mathcal{R}^k = \begin{cases} \{(y^k, z^k) | y^k \in [0, y_{max}],\ z^k \in [0, \max(\mu_{B^k}(y), \mu_{A^k})]\}, & \text{Mamdani;} \\ (v_{B^k}, \mu_{A^k}), & \text{TSK.} \end{cases} \tag{28}$$

In the example shown in Figure 10, the upper limit $y_{max} = 30$.

*(iv)* Defuzzification: The value of the FIS's output is determined in the last step. The Mamdani FIS in Figure 10 uses the centroid defuzzification method. The regions $\mathcal{R}^k$ are unified into a single region $\mathcal{R}$. The final output is the x-coordinate of the centroid of $\mathcal{R}$. The TSK model in Figure 11 uses a weighted sum to obtain the output $y$ of the FIS. Mathematically,

$$y = \begin{cases} \left[\int_{\mathcal{R}} d\mathcal{R}\right]^{-1} \int_{\mathcal{R}} z^k \, d\mathcal{R}, & \text{Mamdani;} \\ \left[\sum_k \mu_{A^k}\right]^{-1} \sum_k v_{B^k} \mu_{A^k}, & \text{TSK.} \end{cases} \tag{29}$$

In the above expression, $\mathcal{R} = \bigcup_k \mathcal{R}$. It is evident from the above description, that the inference and defuzzification step in a Mamdani FIS is more computationally intensive in comparison to that in the TSK model. There are several other methods to obtain the output of a FIS. For details, the interested reader is referred to [78,79]. The

Mamdani model [80–82] as well as the TSK model [83–87] have been used frequently in agricultural research.

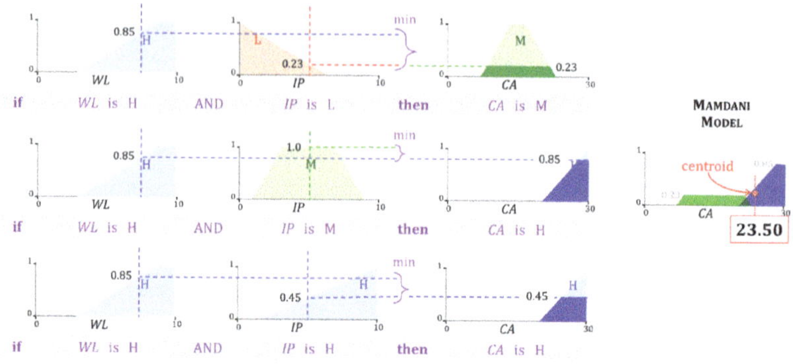


**Figure 10.** Mamdani FIS. The inputs to the FIS in Figure 9 are $WL = 7.5$ and $WL = 5.0$ (dotted vertical lines), and the output is $y = 23.50$). The first three rules in Figure 9 with $WL = \text{H}$ are ignored since $\mu_{\text{L}(7.5)=0)}$. The dark-shaded regions are $\mathcal{R}^k$ of the relevant rules.

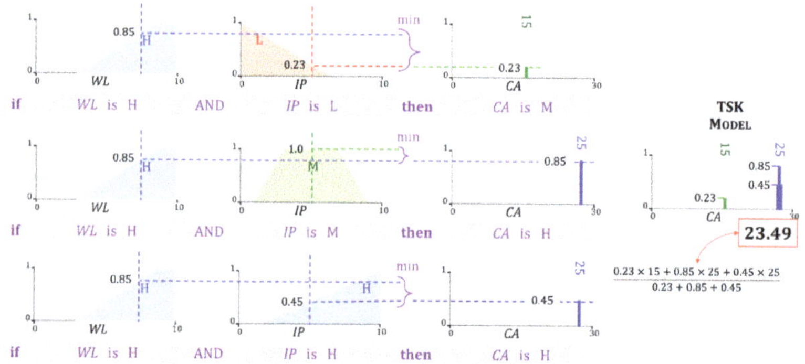


**Figure 11.** TSK FIS. The inputs to the FIS in Figure 9 are $WL = 7.5$ and $WL = 5.0$ (dotted vertical lines), and the output is $y = 23.49$. The first three rules in Figure 9 with $WL = \text{H}$ are ignored since $\mu_{\text{L}}(7.5) = 0)$. In the other rules, the values of $v_{B^k}$ are 15, 25, and 25.

### *4.5. Adaptive Neuro-Fuzzy Inference Systems*

A TSK model with fuzzy rules as in (26) is often referred to as a zero-order FIS. An ANFIS is based on a zero or higher order TSK model, that is arranged in a manner resembling an NN [88–90]. ANFIS models frequently use first-order TSK rule bases. Assuming a scalar output $y$, the format of the $k^{\text{th}}$ rule in such a first-order TSK model is,

$$\textbf{If } x_1 \text{ is } A_1^k \diamond \cdots \diamond x_M \text{ is } A_M^k \textbf{ then } y = b_0^k + b_1^k x_1 + \cdots + b_M^k x_M. \quad (30)$$

The consequent in Equation (30) is a linear expression for $y$ in terms of $\mathbf{x}$, with $M \times K$ coefficients, $b_j^k$ (where $j \in 1, \cdots, M$ and $k \in 1, \cdots, K$). To simplify its training, the membership functions in the ANFIS rules' antecedents are usually restricted to Gaussian [90]. Figure 12 shows an example of a first-order TSK model.

Figure 13 illustrates the ANFIS corresponding to the first-order TSK rule set shown in Figure 12. The parameters of the membership functions of each input variable are trainable quantities. For Gaussian memberships, they are $\sigma_j^k, \mu_j^k$, where $j \in \{1, \cdots, M\}$, and $k$ is the index of a rule). The coefficients in the consequent side of each such rule, which are $b_j^k$,

$j \in \{0, 1, \cdots, M\}$ are also trainable. All trainable quantities constitute the parameter vector $\boldsymbol{\Theta}$ of the ANFIS.

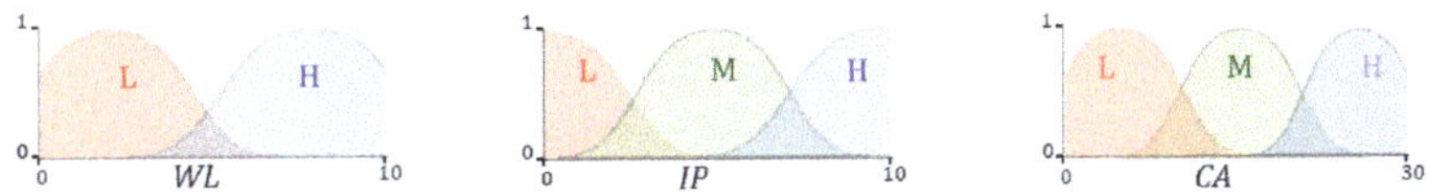


| | | | | | |
|---|---|---|---|---|---|
| If | $WL$ is L | AND | $IP$ is L | then | $CA = 5WL + 5IP$ |
| If | $WL$ is L | AND | $IP$ is M | then | $CA = 8WL + 7IP$ |
| If | $WL$ is L | AND | $IP$ is H | then | $CA = 10WL + 10IP$ |
| If | $WL$ is H | AND | $IP$ is L | then | $CA = 5WL + 5IP + 10$ |
| If | $WL$ is H | AND | $IP$ is M | then | $CA = 8WL + 7IP + 10$ |
| If | $WL$ is H | AND | $IP$ is H | then | $CA = 10WL + 10IP + 10$ |

**Figure 12.** Type 1 TSK FIS. Shown are the membership functions of the fields (top) and the fuzzy rule base (bottom). The antecedents of the rules are the same as those in Figure 9.

There are five layers in the ANFIS model, which are as follows.

(*i*) Fuzzifying layer: The role of the first layer is to fuzzify scalar elements $x_j, j \in \{1 \cdots M\}$ of the input $\mathbf{x}$. It involves computing the memberships $\mu_{A_j^k}(x_j)$ in (30).

(*ii*) Aggregating layer: This layer performs aggregation. When all $\diamond$ operators in (30) are conjunctions, the output of the $k^{\text{th}}$ unit in the second layer is obtained using the expression,

$$\mu_A^k = \prod_j \mu_{A_j^k}(x_j). \tag{31}$$

(*iii*) Normalizing layer: This is the third layer of the ANFIS, whose role is to normalize the incoming aggregated memberships, $\mu_A^k$ from the previous layer. The output of its $k^{\text{th}}$ unit is,

$$\hat{\mu}_A^k = \frac{\mu_A^k}{\sum_{k'} \mu_A^{k'}}. \tag{32}$$

(*iv*) Consequent layer: The output of the $k^{\text{th}}$ unit of the fourth layer is,

$$y^k = \hat{\mu}_A^k \left( b_0^k + \sum_j b_j^k x_j \right). \tag{33}$$

(*v*) Output layer: The final layer of the ANFIS performs a summation of the consequent outputs $y^k$,

$$y = \sum_k y^k. \tag{34}$$

The quantity $y$ is the output of the ANFIS.

Several methods have been proposed to train the parameters of an ANFIS model [91]. Much research has been directed towards gradient descent approaches ((6)) resembling BP [89,92]. Such approaches have been used in agriculture [93–95]. A Levenberg-Marquardt approach has been suggested recently [96]. Stochastic metaheuristics such as GAs [97] and PSO [98] have also been investigated. Hybrid approaches combining them are widely used to train ANFISs [99]. A comparison of three metaheuristics has been reported in [100] for an agriculture-related application.

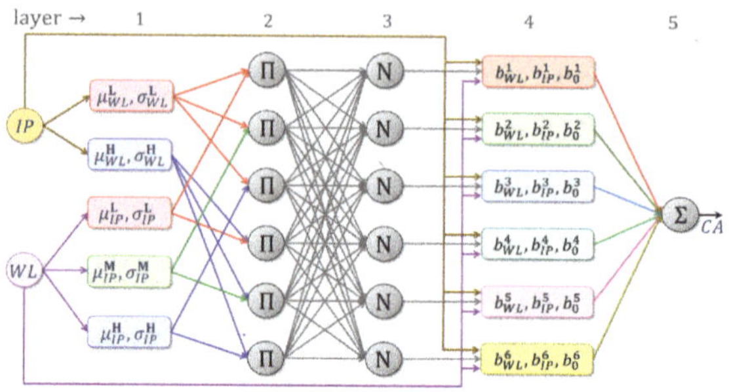


**Figure 13.** Adaptive Neuro-Fuzzy Inference System. Shown are the inputs and the five layers of an ANFIS.

## 5. Soil–Machine Interaction Studies: A Brief Survey

### *5.1. Literature Survey Methodology*

In recent decades, CI methods have been extensively studied in agriculture, particularly in crop management, insect–pest management, irrigation scheduling, precision agriculture, input application optimization, yield prediction, and so on [80]. Initially, we collected research articles for the period ranging from 1990 to 2022, from multiple online databases such as Web of Science, Scopus, Science Direct, Google Scholar, Wiley, and Springer Link. More than 150 research articles were collected in the preliminary screening stage. Out of the 150 articles, only 50 articles directly related to the CI application on traction, tillage, and compaction were selected. Figure 14 shows the year-wise and categorical distribution of the selected articles where CI methods were employed.

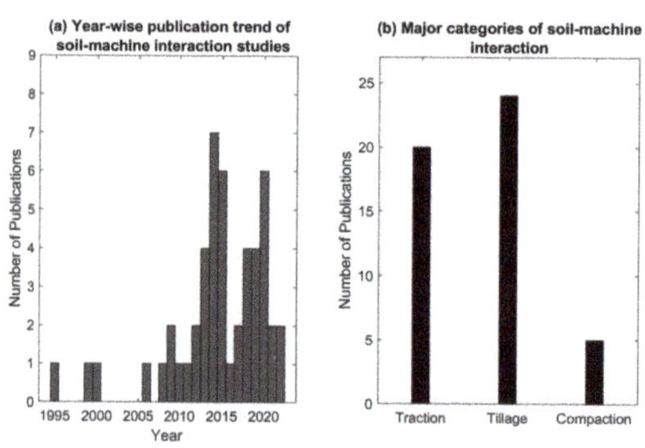


**Figure 14.** Soil–machine interaction studies that employed CI methods: (**a**) Year-wise publication trend; (**b**) Major categories.

### *5.2. Traction*

In traction studies, an individual traction element (wheel, track) or entire off-road vehicle is tested in a controlled laboratory setup or prepared or un-prepared fields for its performance optimization. The major performance parameters are drawbar pull, traction efficiency, slip, and fuel consumption, which are optimized as a function of numerous variables pertaining to the machine, operational setup, and soil properties. A summary of CI models developed in selected traction studies is shown in Table 1.

An off-road vehicle (tractor or skidder) used in agriculture and forestry is specially designed for drawbar work, i.e., pulling or pushing the implements. Drawbar power is a product of drawbar pull and vehicle velocity in the travel direction. The vehicle tire size and its inflation pressure increase the soil contact area, which improves the drawbar performance. The drawbar performance of the forestry skidder was studied [101] in soft soil to develop a multiple linear regression (MLR) and fully connected NN to predict the drawbar pull. For tractor energy management and optimization, the NN hybridized with GA, and the ANFIS was implemented to predict the drawbar pull energy [50], and net traction power [102]. The tractor's drawbar pull varies with vehicle configuration, weight, and operating mode (2WD and 4WD). Thus, a FIS was proposed to estimate a drawbar pull [103]. In addition to tire size, the drawbar pull is also influenced by the tire geometrical parameters, which can be defined with 3D footprints. Thus, NN was implemented to understand the complex relationship between 3D tire footprints and generated the drawbar pull [104].

The traction device develops a force, parallel to the travel direction and transfers to the vehicle. The traction efficiency is a ratio of output power to input power to the device [105]. It is one of the most critical factors in traction studies and relates to energy saving. Several studies were conducted in a laboratory setup with a single-wheel tester to study the influence of the traction device's operational parameters and soil properties on traction efficiency. Table 1 contains the various CI methods proposed to model the traction efficiency [106–112].

Motion resistance is an opposing force, that works against the traction device's forward motion and accounts for all energy loss unrelated to slip [105]. Motion resistance is the difference between gross traction and net traction. A series of experiments were conducted on a driven wheel in a soil bin (clay loam) [113–115] that aimed to study the motion resistance influenced by various operating parameters, and predict motion resistance with the CI methods (NN & FIS).

The tractor is a major power source in agriculture. Therefore, it is essential to understand how tractor power can be best utilized for varying field conditions for efficient operation. The tractor loses the most power at the soil–tire interface, and its performance is influenced by operational and soil/terrain parameters. Therefore, the field performance of a 75 HP tractor was evaluated [116], and NN was proposed for predicting the tractor performance as a function of soil and tractor-implement variables. Likewise, NN and ANFIS were proposed to study the performance of tractor-implement operational parameters on traction efficiency [94] and wheel slip [117]. Specific fuel consumption is the most used and common indicator of tractor performance. Thus, NN was proposed to predict the fuel consumption of a 60 HP 2WD tractor [118].

Mobile robots and autonomous ground vehicles (AGV) are becoming popular on smart farms. Thus, the traction behavior of the ground vehicle was studied on a sloped soil bin, and NN predicted the traction, mobility, and energy requirement of the AGV [119].

### 5.3. Tillage

Tillage is classified into two major categories: (1) primary and (2) secondary tillage, based on purpose, tillage depth, and energy requirement. Primary tillage is an initial major soil working operation, aiming to open up any cultivable land, reduce soil strength, cover plants/residues, and rearrange soil aggregates [2]. It manipulates soil at a greater depth (15 to 30 cm), and moldboard, disk, chisel plow, and subsoiler are commonly used primary tillage tools.

Moldboard (MB) plow shatters soil, inverts furrow slices, and covers crop residues or grasses. As the plow bottom advances, it cuts, fails the ground, and forms furrow slices, which shows a periodic variation in the draft force. Therefore, a time-lagged recurrent neural network (RNN) was proposed to predict the dynamic draft as a function of one step ahead prediction [120] for various shaped tillage tools (MB, Korean, model plow). The MB plow consumes the highest energy compared to other tillage tools for a given depth [121,122]. Therefore, the researchers studied the performance of various types of MB plow in varying soil conditions for energy optimization. The developed CI methods are listed in Table 2 and explained briefly as follows: The ANFIS models were proposed for predicting the draft and specific draft of three-bottom MB plow [123]. The NN predicted the specific draft and fuel consumption of a tractor-mounted MB plow under varying operating conditions [124]. Similarly, the NN was proposed for general-purpose MB plow's draft, and energy requirement [122].

**Table 1.** Traction studies which employed the CI methods.

| Author & Year | Traction Device | Method | Input | Output |
|---|---|---|---|---|
| Hassan and Tohmaz (1995) [101] | Rubber-tire skidder | NN | Tire size, tire pressure, normal load, line of pull angle | Drawbar pull |
| Çarman and Taner (2012) [106] | Driven wheel | NN | Travel reduction | Traction efficiency |
| Taghavifar et al. (2013) [113] | Driven wheel | NN | Velocity, tire pressure, normal load | Rolling resistance |
| Taghavifar and Mardani (2013) [114] | Driven wheel | FIS | Velocity, tire pressure, normal load | Motion resistance coeff. |
| Taghavifar and Mardani (2014) [107] | Driven wheel | ANFIS | Velocity, wheel load, slip | Energy efficiency indices (Traction coeff. and traction efficiency) |
| Taghavifar and Mardani (2014) [108] | Driven wheel | NN | Velocity, wheel load, slip | Energy efficiency indices (Traction coeff. and traction efficiency) |
| Taghavifar and Mardani (2014) [109] | Driven wheel | NN | Soil texture, tire type, wheel load, speed, slip, inflation pressure | Traction force |
| Taghavifar and Mardani (2014) [115] | Driven wheel | NN & SVR | Wheel load, inflation pressure, velocity | Energy wasted |
| Taghavifar and Mardani (2015) [50] | Driven wheel | ANFIS | Wheel load, inflation pressure, velocity | Drawbar pull energy |
| Taghavifar et al. (2015) [102] | Driven wheel | NN-GA | Wheel load, inflation pressure, velocity | Available power |
| Ekinci et al. (2015) [110] | Single wheel tester | NN & SVR | Lug height, axle load, inflation pressure, drawbar pull | Traction efficiency |
| Almaliki et al. (2016) [116] | Tractor | NN | Moisture content, cone index, tillage depth, inflation pressure, engine speed, forward speed | Traction efficiency, drawbar pull, rolling resistance, fuel consumption |
| Pentos et al. (2017) [111] | Micro tractor | NN | Vertical load, horizontal deformation, soil Coeff., compaction, moisture content | Traction force and traction efficiency |
| Shafaei et al. (2018) [94] | Tractor | ANFIS, NN | Forward speed, plowing depth, tractor mode | Traction efficiency |
| Shafaei et al. (2019) [117] | Tractor | ANFIS, NN | Forward speed, plowing depth, tractor mode | Wheel slip |
| Shafaei et al. (2020) [103] | Tractor | FIS | Tractor weight, wheel slip, tractor driving mode | Drawbar pull |
| Pentos et al. (2020) [112] | Micro tractor | NN, ANFIS | Vertical load, horizontal deformation, soil Coeff., compaction, moisture content | Traction force and traction efficiency |
| Hanifi et al. (2021) [118] | Tractor (60 HP) | NN | Inflation pressure, axle load, drawbar force | Specific fuel consumption |
| Badgujar et al. (2022) [119] | AGV | NN | Slope, speed, drawbar | Traction efficiency, slip and power number |
| Cutini et al. (2022) [104] | Tractor | NN | Tire geometric parameters (area, length, width, depth), slip | Drawbar pull |

The MB plow has a sliding plow bottom that slides through the soil. The sliding friction is one of the primary reasons for the MB plow's higher draft and energy requirement. On the contrary, a disk plow is equipped with concave rolling disks, i.e., a rolling plow bottom designed to reduce friction through rolling action. The energy requirement of the disk plow is significantly lower than the MB plow. Thus, the NN was proposed to predict the disk plow draft and energy requirement [125,126].

Deep tillage (depth < 30 cm) is designed to shatter soil, breaking up hardpans and compacted soil layers to ease water and plant root movement. A chisel plow and subsoiler are mainly used for deep tillage. The chisel plow has a series of shovels or teeth spaced on a frame. Its draft requirement is comparatively low and varies with soil type and depth of operation. Hence, the NN was proposed to model the chisel plow draft using various soil textural indices [127]. TSK-type ANFIS was proposed for chisel plow draft prediction [86]. Likewise, NN was presented for modeling the chisel plow performance parameters [128]. More details on the proposed model inputs and output can be found in Table 2.

A subsoiler has a narrow straight shank to break and fracture the deep compacted soil zone at a greater depth (60–90 cm). The subsoiling demands high horsepower, ranging from 30 to 50 hp per shank [129]. Thus, to predict the draft and energy requirement of the subsoiler, the NN was presented as a function of soil parameters and operational variables [130]. The subsoiler is a non-inversion tillage tool, available in various shaped shanks, and selecting the right shank could reduce the draft [131]. The conventional straight shank requires a significantly higher draft and is often replaced with parabolic, bent leg, or paraplow shanks [132]. Therefore, the CI-based models (ANFIS, MLR, RSM) were presented for predicting the draft of three types of subsoiler shank (subsoiler, paraplow, and bent leg) [130]. Similarly, the ANFIS was proposed to predict the forces acting on paraplow having three different design configurations (bent wing with forward, backward, and without wing) [133].

Secondary tillage is performed for seedbed preparation, crop production practices, and moisture conservation. Examples of secondary tillage tools include a harrow (disk, spring or spike tooth, chisel), cultivator, and clod crushing roller. The energy requirement of secondary tillage tools is comparatively less than that of primary. The cultivator and harrow are often operated at a higher ground speed to produce finer tilth, soil pulverization, and weed control. Thus, its operational parameters (tool type, speed, depth) are often investigated to achieve finer tilth, prevent soil degradation and optimize the tillage energy. The NN predicted the draft of a cultivator, disk harrow, and MB plow in a soil bin setup [21]. The FIS was proposed for predicting the soil fragmentation resulting from a combination of primary and secondary tillage implements during the seedbed preparation [134]. The draft efficiency and soil loosening of the duck foot cultivator were predicted with the FIS in soil bin [135]. Similarly, the NN was proposed to predict the draft force of a chisel cultivator [136]. An RBF neural network was presented to simulate the soil–machine interaction of five narrow blades in field conditions [137].

Reduced tillage offers several benefits, such as reduced energy and soil disturbance over traditional tillage. Winged share is a reduced tillage tool, and the CI models (NN and FIS) were proposed for predicting the draft force of two different types of winged share tillage tools in a soil bin (loam soil) [138,139]. Likewise, a combined tillage implement is equipped with multiple tillage tools on a single frame to reduce the tractor passes. The combined tillage saves time, fuel, and energy to obtain the desired soil conditions compared to the conventional method [140,141]. Therefore, the CI models (NN and ANFIS) were proposed to predict the energy indices of the tractor-implement system during a combined tillage operation [142].

The model tool is a miniature-scale replica of an actual tool and is often studied in a laboratory environment. The NN models were developed for predicting the energy requirement of the rectangular cross-sectional model tool in a soil bin [143]. Similarly, NN was proposed to understand the design and technical insight of the plowing process of a multi-flat plate (model tool) and resulting soil fineness [144].

**Table 2.** Tillage studies that employed the CI methods.

| Author and Year | Tillage Tool | CI Method | Input | Output |
|---|---|---|---|---|
| Zhang and Kushwaha (1999) [137] | Narrow blades (five) | RBF neural network | Forward speed, tool types, soil type | Draft |
| Choi et al. (2000) [120] | MB plow, Janggi plow, model tool | Time lagged RNN | One step ahead prediction | Dynamic draft |
| Aboukarima (2006) [127] | Chisel plow | NN | Soil parameters (textural index, moisture, bulk density), tractor power, plow parameters (depth, width, speed) | Draft |
| Alimardani et al. (2009) [130] | Subsoiler | NN | Travel speed, tillage depth, soil parameters (physical) | Draft and tillage energy |
| Roul et al. (2009) [21] | MB plow, cultivator, disk harrow | NN | Plow parameters (depth, width, speed), bulk density, moisture | Draft |
| Marakoğlu and Çarman(2010) [135] | Duckfoot cultivator share | FIS | Travel speed, working depth | Draft efficiency and soil loosening |
| Rahman et al. (2011) [143] | Rectangular tillage tool | NN | Plow depth, travel speed, moisture | Energy requirement |
| Mohammadi et al. (2012) [138] | Winged share tool | FIS | Share depth, width, speed | Draft requirement |
| Al-Hamed et al. (2013) [125] | Disk plow | NN | Soil parameters (texture, moisture, soil density), tool parameters (disk dia., tilt and disk angle), plow depth, plow speed | Draft, Unit draft and energy requirement |
| Saleh and Aly (2013) [144] | Multi-flat plowing tines | NN | Plow parameters (geometry, speed, lift angle, orientation, depth), soil conditions (moisture, density, strength) | Draft force, vertical force, side force, soil finess |
| Akbarnia et al. (2014) [139] | Winged share tool | NN | Working depth, speed, share width | Draft force |
| Abbaspour-Gilandeh and Sedghi (2015) [134] | Combine tillage | FIS | Moisture, speed, soil sampling depth | Median weight diameter |
| Shafaei et al (2017) [86] | Chisel plow | ANFIS | Plowing depth, speed | Draft force |
| Shafaei et al. (2018) [145] | MB plow | ANFIS | Plowing depth, speed | Draft (specific force and draft force) |
| Shafaei et al. (2018) [123] | Disk plow | NN, MLR | Plowing depth, speed | Draft |
| Shafaei et al. (2018) [126] | Disk plow | ANFIS, NN | Plowing depth, speed | Fuel efficiency |
| Shafaei et al. (2019) [117] | Conservation tillage | NN, ANFIS | Plowing depth, speed, tractor mode | Energy indices |
| Askari and Abbaspour-Gilandeh (2019) [132] | Subsoiler tines | MLR, ANFIS, RSM | Tine type, speed, working depth, width | Draft |
| Çarman et al. (2019) [124] | MB plow | NN | Tillage depth, speed | Draft, fuel consumption |
| Marey et al. (2020) [128] | Chisel plow | NN | Tractor power, soil texture, density, moisture, plow speed, depth | Draft, rate of soil volume plowed, fuel consumption |
| Al-Janobi et al. (2020) [122] | MB plow | NN | Soil texture, field working index | Draft, energy |
| Abbaspour-Gilandeh et al. (2020) [136] | MB plow, para-plow | ANFIS | Velocity, depth, type of implement | Draft, vertical and lateral force |
| Abbaspour-Gilandeh et al. (2020) [133] | Chisel cultivator | NN, MLR | Depth, moisture, cone index, speed | Draft |
| Shafaei et al. (2021) [146] | MB plow | FIS | Tillage depth, speed, tractor mode | Power consumption efficiency |

### 5.4. Compaction

Vehicular traffic is common during field operation, and it is estimated that the wheels traffic the soil more than five times a year. The vehicular traffic affects the soil structure, void ratio, and bulk density, which further influence crop yield. Therefore, the soil compaction resulting from vehicular traffic needs to be reduced or avoided. Hence, two agricultural tires were studied in a soil bin and the FIS-based models were developed to predict bulk density, penetration resistance, and soil pressure at a 20 cm depth [147].

Tire–soil contact area varies with tire parameters such as vertical load, inflation pressure, and thread type/pattern. The contact area determines the forces acting on soil and resulting stress–strain. Therefore, a series of experiments were conducted in a soil bin, and several CI models (i.e., NN, FIS, and Wavelet NN) were proposed to predict the wheel contact area, contact pressure, soil strength, and soil density based on tire parameter [148–150]. Multiple wheel passes cumulatively compact the soil. Hence, the NN was presented for predicting the penetration resistance and soil sinkage as a function of wheel pass and wheel operating parameters [151] mentioned in Table 3.

**Table 3.** Soil compaction studies that employed the CI methods.

| Author and Year | Traction Device | CI Method | Input | Output |
|---|---|---|---|---|
| Çarman (2008) [147] | Radial tire (2) | FIS | Tire contact pressure, velocity | Bulk density, penetration resistance, soil pressure at 20 cm depth |
| Taghavifar et al. (2013) [148] | Tire | NN | Wheel load, inflation pressure, wheel pass, velocity, slip | Penetration resistance, soil sinkage |
| Taghavifar and Mardani (2014) [149] | Tire | FIS | Wheel load, inflation pressure | Contact area, contact pressure |
| Taghavifar and Mardani (2014) [150] | Tire (size 220/65R21) | WNN, NN | Wheel load, velocity, slip | Contact pressure |
| Taghavifar (2015) [151] | Tire (size 220/65R21 and 9.5L-14) | NN | Soil texture, tire type, slip, wheel pass, load, velocity | Contact pressure, bulk density |

### 5.5. Implemented CI Methods

A summary of CI methods proposed in a selected article (50) is presented in Figure 15. The NNs were the most frequently employed, followed by multiple linear regression, ANFIS, and FIS. The NN-based models were proposed in 36 studies (50.7%), out of which 34 studies employed a fully connected feedforward (FF) NN type (Figure 15b). Other types of NNs, such as RNNs, wavelet NNs, and RBFNs, were reported once (Figure 16a). This indicates that shallow NN with only one or two layers was sufficient to model complex soil–machine interactions (Figure 16a). In most of these studies, NNs were trained with BP or LMBP (Figure 16b). A GA-based metaheuristic was used in one such research, and dimension reduction using ICA in another.

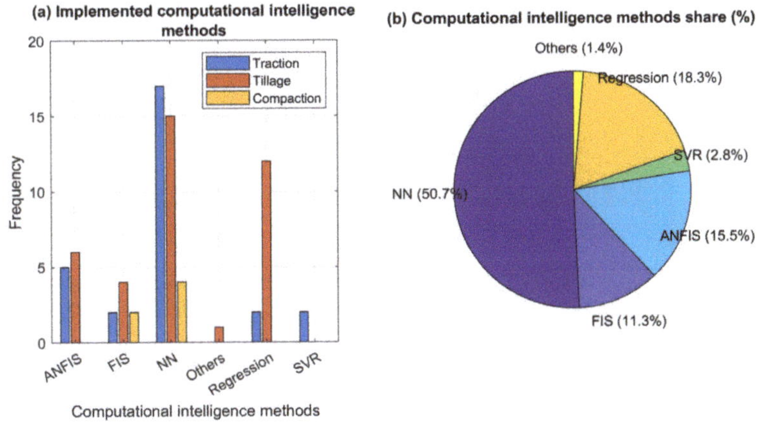


**Figure 15.** CI methods used in soil–machine interaction studies; (**a**) soft computing methods and their frequency; (**b**) percentage share of each method.

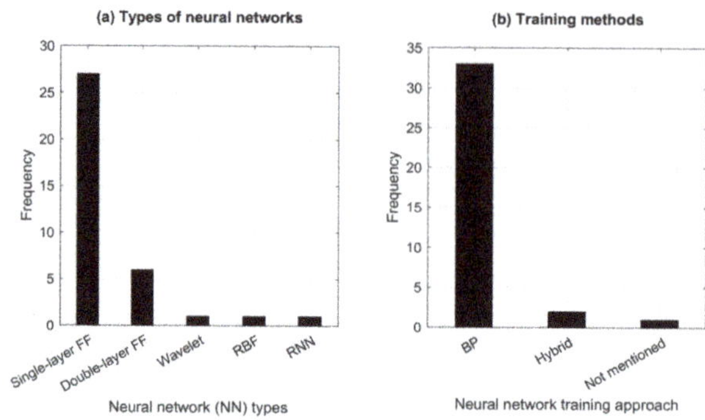


**Figure 16.** Neural network: (**a**) types of a neural network, (**b**) training methods.

Subsequently, the FIS was implemented in a total of eight studies (11.3%). The triangular, Gaussian, and linear were observed as the most popular membership functions. The ANFIS models were proposed in eleven studies (15.5%), with the first-order TSK fuzzy inference system being the preferred approach. ANFIS models were often trained using a combination of the least-squares method and BP. The SVR models were applied in two studies, which used various kernel functions.

Additionally, traditional regression methods were implemented in thirteen (11.3%) studies. These regression methods included MLR and the standard ASABE equations (tool draft equation). These methods were usually compared with CI methods in terms of prediction accuracy.

The performances of the models were evaluated with commonly used metrics. Figure 17a shows the frequencies of their usages. As is evident from Figure 17b, CI models consistently outperformed classical approaches. Although the performance of the traditional regression method was comparatively lower in terms of model accuracy, MATLAB was the most widely used platform to implement CI-based models (Figure 17c).

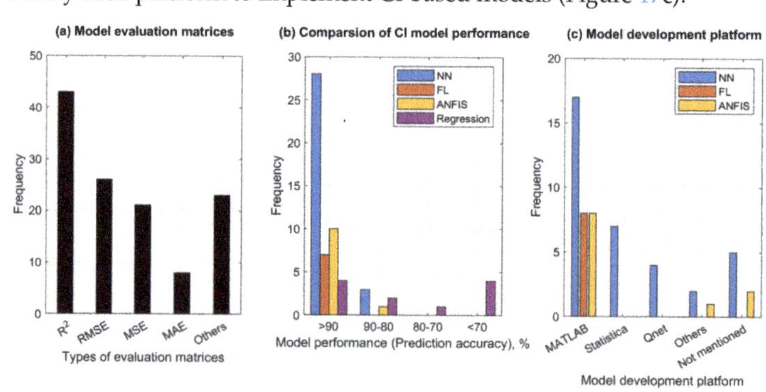


**Figure 17.** CI models: (**a**) evaluation metrics, (**b**) performance comparison, and (**c**) development platform.

## 6. Strengths and Limitations of CI Methods

CI models offer manifold advantages over traditional methods described earlier. The features that make these models so attractive are enumerated below.

(*i*) Data-driven models can handle copious amounts of data with relative ease [152]. With increasing data size, the corresponding growth in computational overheads is generally between linear and quadratic orders of magnitude. For instance, the number of iterations (called epochs) needed to train a neural network is fixed regardless of data size [53]. On the other hand, traditional methods regularly witness quadratic or higher growths.

(*ii*) To further enhance their performances after initial offline training, data-driven CI models (e.g., NNs and DNNs) can learn online during actual deployment [153]. In other words, they are capable of learning from experience.
(*iii*) FIS models can directly benefit from human domain experts; their expert knowledge can be incorporated into the model [154].
(*iv*) Conversely, FIS model outputs are amenable to direct human interpretation. NNs endowed with such capability have been recently proposed [155].
(*v*) CI-based algorithms can easily be hybridized with traditional algorithms as well as with one another, thereby offering the benefits of both (c.f. [48,156]). For instance, ANFIS is a combination of FIS and NN approaches.
(*vi*) These models offer the advantage of flexibility. A model developed for a specific task can be adapted to handle another similar task [157,158].
(*vii*) CI models are robust to various forms of imprecision, such as incomplete information, noise, and statistical outliers [152,159–161]. In some cases, they may even benefit from the presence of noise.

It is of little surprise that CI approaches have become very popular in agricultural soil-tillage, traction domains, and many other applications.

In spite of the several attractive features that CI models offer, they have a few shortcomings as well. These are outlined below.

(*i*) Interpretability: Several CI models such as NN & SVR are black box approaches. Unlike physics-based approaches, the nonlinear input-output relationships expressed by these models are not self-explanatory, i.e., do not render themselves to common sense interpretations. Although various schemes towards making these relationships more explainable are currently being explored, [162–165], this research is only at a preliminary stage.
(*ii*) Computational requirements: The development of CI models often requires specialized software (e.g., MATLAB). Moreover, training DNNs with reasonably sized data may prove to be too time-consuming unless using GPUs (graphics processing units), where processors can be run as a pipeline or in parallel [166].
(*iii*) Data requirements: In comparison to classical methods, CI models require relatively copious amounts of data for training. As such models are not equipped for extrapolation, data samples must adequately cover the entire input range of real-world inputs. In order to effectively train certain CI models such as RBFNs, the data should not be skewed in any direction. Unfortunately, experimentally generating such data can often be a resource-intensive and time-consuming process.
(*iv*) Output dimensionality: Unlike NNs, some other CI models are equipped to handle only scalar outputs. Although there are indirect methods to train SVRs with vector outputs [167,168], this is an inherent limitation of FIS models.

## 7. Emergent Computational Intelligence Models

The study critically analyzed the most popular CI methods found in the literature, particularly in the soil–machine interaction domain. Further, we suggest emergent CI methods that may provide better results and can be considered as alternatives to existing CI methods. Those methods are described in brief here.

### *7.1. Deep Neural Networks*

DNNs are NN models with multiple hidden layers [169–171]. In the past few years, this class of CI models has witnessed explosive growth in popularity. DDNs have emerged as a popular tool in a wide range of applications in agriculture [172–177], where they have been used for various image recognition tasks. Unfortunately, DNNs have yet to be explored in any soil–machine interaction application.

Figure 4 illustrates the layout of such a DNN with fully connected layers. State-of-the-art DNNs incorporate various other types of layers, including RBFN [178], SVR [179], and TSK fuzzy [180,181] layers. DNNs can be endowed with the ability to handle time series data by incorporating long short-term memory (LSTM) or gated recurrent unit (GRU) layers [176,182]. At each time step $t$, these layers can hold in memory essential features from earlier time steps (i.e., $t-1$, $t-2$, etc.) by means of time-delayed feedback. Such DNNs are called recurrent neural networks. An alternate to LSTM and GRU in DNNs is the attention mechanism [183], which has been applied in agriculture [184].

Although sigmoid functions are widely used as neuronal nonlinearities, the presence of a large number of layers in the DNN poses the problem of vanishing gradients [185]. This issue is addressed by using rectified linear (ReLu) units [186], which incorporate the ReLu function, $f(s) = \max(s, 0)$. Current training methods that are based on BP [187–189]. The Adam (adaptive momentum estimation) algorithm is currently the dominant approach to train DNNs [190]. In [191], Adam was used to successfully train DNNs for agriculture data. The use of metaheuristics in conjunction with gradient methods has been investigated [192,193].

Unlike FIS and ANFIS models, DNNs are black-box approaches, whose outputs are not readily amenable to human interpretation. However, recent studies are beginning to address this issue [164,165].

### 7.2. Regression Trees and Random Forests

Decision trees are CI methods that use graphical tree-based representations [194,195], with binary trees [196] being most frequently used. During training, each node in a binary tree is used to split sample pairs $(\mathbf{x}(n), t(n))$ $(n \in \mathcal{S}_t)$ into two subsets $\mathcal{S}_t^{\mathrm{L}}$ and $\mathcal{S}_t^{\mathrm{LR}}$. A threshold $\theta_j$ is applied to an element $x_j$. Hence,

$$y = \begin{cases} \mathcal{S}_t^{\mathrm{L}} = \{n \in \mathcal{S}_t | x_j(n) \leq \theta_j\}; \\ \mathcal{S}_t^{\mathrm{R}} = \{n \in \mathcal{S}_t | x_j(n) > \theta_j\}. \end{cases} \tag{35}$$

The threshold is computed so that at each node, the split is as evenly balanced as possible. Information theoretic and heuristic methods using values of the targets $t(n)$ in the training dataset. Regression trees have found agricultural applications in the past few years [197,198].

Random forests are CI methods that use multiple trees to obtain outputs [199,200]. There has been a steep rise in the use of this approach for various applications in agriculture [201–211]. An excellent survey of decision trees, random forests, and other CI models has been published in [212].

### 7.3. Extreme Learning Machines

Extreme learning machines (ELM) are CI models that are useful in regression problems [213–215]. Although in comparison to some other CI models (NNs, RBFNs, and SVRs), ELMs have not been as widely used in other engineering domains; surprisingly, they are very popular in various agricultural applications [216–223].

An ELM is structurally equivalent to an $M \times K \times N$ NN. The neurons in the hidden layer incorporate nonlinear activation functions in the same manner as in Equation (9). However, unlike in NNs, the hidden layer in an ELM is not fully connected to the input. The hidden weights of an ELM can be arranged as a $K \times M$ sparse matrix. These weights are assigned randomly and do not undergo any training. Only the output weights are trained using a matrix form of the pseudoinverse rule in Equation (16). This allows ELMs to be trained significantly faster than equivalent NNs. Hybrid training algorithms for ELMs have also been proposed in [224–226] for agricultural applications. DNN architectures that contain ELM layers are being investigated (cf. [227,228]).

### 7.4. Bayesian Methods

Bayesian methods are CI paradigms where the outcome renders itself to a probabilistic interpretation. Central to these methods is the Bayes rule. The rule can be applied to a parametric Bayesian model in the following manner,

$$\mathrm{p}(\boldsymbol{\Theta}|\mathcal{S}_t) = \frac{\mathrm{p}(\mathcal{S}_t|\boldsymbol{\Theta})\mathrm{p}(\boldsymbol{\Theta})}{\mathrm{p}(\mathcal{S}_t)}. \quad (36)$$

In this expression, $\mathrm{p}(\cdot)$ is the probability of the argument. The left-hand side of Equation (36) is the posterior probability. The factors $\mathrm{p}(\mathcal{S}_t|\boldsymbol{\Theta})$, and $\mathrm{p}(\mathcal{S}_t)$ in the right-hand side are the likelihood and the prior probability. It can be demonstrated that LASSO and ridge regularization discussed earlier in Section 3.3 are instances of Bayesian methods, where the prior probabilities follow Laplacian and Gaussian distributions.

Since the training data $\mathcal{S}_t$ is independent of the model, it can be dropped from the Bayes rule. The model parameter is obtained as the one that has the highest probability, $\mathrm{argmax}_{\boldsymbol{\Theta}}\mathrm{p}(\boldsymbol{\Theta}|\mathcal{S}_t)$. Given any unknown input $\mathbf{x}$, the output probability $\mathrm{p}(y|\mathbf{x},\boldsymbol{\Theta})$ can be obtained from $\boldsymbol{\Theta}$. Bayesian approaches have been used in several areas of agriculture [229–232].

A Bayesian network is a specific Bayesian modeling approach that uses a graphical structure that resembles an NN [233]. Inferring the output in this model relies heavily on statistical sampling techniques [234]. Bayesian networks have been used in [184,235,236].

A mixture of Gaussians [237,238] is a Bayesian model that uses hidden variables $z_i, i = 1, 2, \cdots$, which play an intermediate role between the inputs and outputs. Given any input $\mathbf{x}$, the output probability $\mathrm{p}(y)$ is determined as the following summation,

$$\mathrm{p}(y) = \sum_i \mathrm{p}(y|z_i)\mathrm{p}(z_i|\mathbf{x}). \quad (37)$$

The use of such methods has begun to be explored in agriculture [38,239,240].

Gaussian process regression [241–243] is a Bayesian approach that assumes the presence of Gaussian noise. As in SVR, kernel matrices are applied in this method. Gaussian process regression has been extensively used in various applications related to agriculture [244–248].

### 7.5. Ensemble Models

Ensemble models are approaches that combine multiple CI models for decision-making [249–251]. Bagging and boosting are two commonly used ensemble approaches. Random forests and Gaussian mixtures discussed earlier in this section are ensemble models.

There has been a surge in the use of these methods in the agriculture domain [252–259]. Recent research has been directed towards using bagging [260–263] and boosting [198]. Ensembles of NNs have been investigated in [264–270]. GA and PSO have also been studied in this context [255,264].

## 8. Future Direction and Scope

### 8.1. Online Traction Control

A sensing technology has reached its maturity, and ample research material is available, where numerous sensors were employed to sense, measure, and provide real-time information on the biological material (e.g., plant, soil, and field conditions). This review article taught us that CI methods can accurately and precisely model or predict complex soil–machine interactions. Therefore, future research efforts should target automatic and real-time traction-tillage control with the help of a sensing and prediction model. The online traction control system would optimize the machine parameters in real-time to increase traction efficiencies and reduce soil compaction. For example, traction control is a standard safety feature in today's automotive vehicles. The wheel sensor senses the road conditions (icy or slippery), and the control algorithm enables the traction control to adapt to road conditions in real-time. Moreover, the planetary rover developed by NASA is also equipped with a traction control algorithm that senses the terrain driving condition and predicts the chance of getting trapped in soil (immobility condition) [271,272].

### 8.2. Online Tillage Control

The agricultural soil and field conditions are dynamic and vary on a spatial and temporal scale. Hence, a single tillage tool or management system operating uniformly throughout the field would not be sufficing. Multiple factors, including soil type, texture, structure, moisture, field topography, slope, and crop rotation, play a vital role when deciding which implement is best for the field. The current tillage management approach involves employing a single tillage tool for the entire area. The soil moisture is the only parameter checked before performing the tillage operation. Therefore, future research should develop variable depth, variable-intensity, and adaptive tillage implements that can be controlled in real-time. This site-specific tillage management would collect real-time information on soil and operating terrain, and CI models would serve as decision-support tools, creating a fully automated tillage management system. Site-specific tillage has excellent potential since the intensity of the operation is adapted to the local needs, which can dramatically improve tillage. Recently, adaptive tillage has become a significant research focus, where the tillage tool adapts or changes its shape in real-time [273,274].

**Author Contributions:** Conceptualization, C.B., S.D. and D.F.; Methodology, C.B., S.D. and D.F.; Formal analysis, C.B. and S.D.; Resources, C.B., D.M.F., S.D. and D.F.; Data curation, C.B., D.M.F. and S.D.; Writing—original draft preparation, C.B. and S.D.; Writing—review and editing, C.B., S.D. and D.F.; Supervision, S.D. and D.F.; project administration, S.D. and D.F.; funding acquisition, S.D. and D.F. All authors have read and agreed to the published version of the manuscript.

**Funding:** This study was financially supported by the National Institute of Food and Agriculture (NIFA-USDA). The project is titled "National Robotics Initiative (NRI): Multi-Robot Farming on Marginal, Highly Sloped Lands" Project number- KS4513081.

**Institutional Review Board Statement:** Not applicable.

**Informed Consent Statement:** Not applicable.

**Data Availability Statement:** Not applicable.

**Conflicts of Interest:** The authors declare no conflict of interest.

## References

1. Ani, O.A.; Uzoejinwa, B.; Ezeama, A.; Onwualu, A.; Ugwu, S.; Ohagwu, C. Overview of soil-machine interaction studies in soil bins. *Soil Tillage Res.* **2018**, *175*, 13–27. [CrossRef]
2. ASABE. *Terminology and Definitions for Soil Tillage and Soil-Tool Relationships*; Technical Report ASAE EP291.3 Feb2005 (R2018); American Society of Agricultural and Biological Engineers: St. Joseph, MI, USA, 2018.
3. Sunusi, I.I.; Zhou, J.; Zhen Wang, Z.; Sun, C.; Eltayeb Ibrahim, I.; Opiyo, S.; korohou, T.; Ahmed Soomro, S.; Alhaji Sale, N.; Olanrewaju, T.O. Intelligent tractors: Review of online traction control process. *Comput. Electron. Agric.* **2020**, *170*, 105176. [CrossRef]
4. Zoz, F.; Grisso, R. Traction and Tractor Performance. American Society of Agricultural and Biological Engineers: St. Joseph, MI, USA, 2012.
5. Upadhyaya, S.K.; Way, T.R.; Upadhyaya, S.K.; Chancellor, W.J. Chapter 2. Traction Mechanics. Part V. Traction Prediction Equations. In *Advances in Soil Dynamics Volume 3*, 1st ed.; American Society of Agricultural and Biological Engineers: St. Joseph, MI, USA, 2009; pp. 161–186. [CrossRef]
6. Karmakar, S.; Kushwaha, R.L. Dynamic modeling of soil–tool interaction: An overview from a fluid flow perspective. *J. Terramech.* **2006**, *43*, 411–425. [CrossRef]
7. Johnson, C.E.; Bailey, A.C. Soil Compaction. In *Advances in Soil Dynamics Volume 2*, 1st ed.; American Society of Agricultural and Biological Engineers: St. Joseph, MI, USA, 2002; pp. 155–178. [CrossRef]
8. Acquah, K.; Chen, Y. Soil Compaction from Wheel Traffic under Three Tillage Systems. *Agriculture* **2022**, *12*, 219. [CrossRef]
9. Soane, B.; van Ouwerkerk, C. Soil Compaction Problems in World Agriculture. In *Developments in Agricultural Engineering*; Elsevier: Amsterdam, The Netherlands, 1994; Volume 11, pp. 1–21. [CrossRef]
10. Brus, D.J.; van den Akker, J.J.H. How serious a problem is subsoil compaction in the Netherlands? A survey based on probability sampling. *Soil* **2018**, *4*, 37–45. [CrossRef]

11. Zabrodskyi, A.; Šarauskis, E.; Kukharets, S.; Juostas, A.; Vasiliauskas, G.; Andriušis, A. Analysis of the Impact of Soil Compaction on the Environment and Agricultural Economic Losses in Lithuania and Ukraine. *Sustainability* **2021**, *13*, 7762. [CrossRef]
12. Keller, T. Soil Compaction and Soil Tillage—Studies in Agricultural Soil Mechanics. Ph.D. Thesis, Swedish University of Agricultural Sciences, Uppsala, Sweden, 2004.
13. DeJong-Hughes, J.; Moncrief, J.; Voorhees, W.; Swan, J. Soil Compaction: Causes, Effects and Control. The University of Minnesota Extension Service, St. Paul, MN, USA, 2001. Available online: https://hdl.handle.net/11299/55483 (accessed on 24 October 2022).
14. Badalíková, B. Influence of Soil Tillage on Soil Compaction. In *Soil Engineering*; Dedousis, A.P., Bartzanas, T., Eds.; Springer: Berlin/Heidelberg, Germany, 2010; Volume 20, pp. 19–30. [CrossRef]
15. Tiwari, V.; Pandey, K.; Pranav, P. A review on traction prediction equations. *J. Terramechan.* **2010**, *47*, 191–199. [CrossRef]
16. Wong, J.Y. *Theory of Ground Vehicles*, 3rd ed.; John Wiley: New York, NY, USA, 2001.
17. Godwin, R.; Spoor, G. Soil failure with narrow tines. *J. Agric. Eng. Res.* **1977**, *22*, 213–228. [CrossRef]
18. Makanga, J.; Salokhe, V.; Gee-Clough, D. Effect of tine rake angle and aspect ratio on soil failure patterns in dry loam soil. *J. Terramech.* **1996**, *33*, 233–252. [CrossRef]
19. Karmakar, S. Numerical Modeling of Soil Flow and Pressure Distribution on a Simple Tillage Tool Using Computational Fluid Dynamics. Ph.D. Thesis, University of Saskatchewan, Saskatoon, SK, Canada, 2005.
20. Tagar, A.; Ji, C.; Ding, Q.; Adamowski, J.; Chandio, F.; Mari, I. Soil failure patterns and draft as influenced by consistency limits: An evaluation of the remolded soil cutting test. *Soil Tillage Res.* **2014**, *137*, 58–66. [CrossRef]
21. Roul, A.; Raheman, H.; Pansare, M.; Machavaram, R. Predicting the draught requirement of tillage implements in sandy clay loam soil using an artificial neural network. *Biosyst. Eng.* **2009**, *104*, 476–485. [CrossRef]
22. Fielke, J.; Riley, T. The universal earthmoving equation applied to chisel plough wings. *J. Terramech.* **1991**, *28*, 11–19. [CrossRef]
23. Godwin, R.; Seig, D.; Allott, M. Soil failure and force prediction for soil engaging discs. *Soil Use Manag.* **1987**, *3*, 106–114. [CrossRef]
24. Kushwaha, R. L.; Shen, J. Finite Element Analysis of the Dynamic Interaction Between Soil and Tillage Tool. *Trans. ASAE* **1995**, *38*, 1315–1319. [CrossRef]
25. Upadhyaya, S.K.; Rosa, U.A.; Wulfsohn, D. Application of the Finite Element Method in Agricultural Soil Mechanics. In *Advances in Soil Dynamics Volume 2*, 1st ed.; American Society of Agricultural and Biological Engineers: St. Joseph, MI, USA, 2002; pp. 117–153. [CrossRef]
26. Shmulevich, I.; Rubinstein, D.; Asaf, Z. Chapter 5. Discrete Element Modeling of Soil-Machine Interactions. In *Advances in Soil Dynamics Volume 3*, 1st ed.; American Society of Agricultural and Biological Engineers: St. Joseph, MI, USA, 2009; pp. 399–433. [CrossRef]
27. Liu, J.; Kushwaha, R.L. Two-decade Achievements in Modeling of Soil—Tool Interactions. In Proceedings of the ASABE Annual International Meeting 2008, Providence, RI, USA, 29 June–2 July 2008; American Society of Agricultural and Biological Engineers: St. Joseph, MI, USA, 2008. [CrossRef]
28. Taheri, S.; Sandu, C.; Taheri, S.; Pinto, E.; Gorsich, D. A technical survey on Terramechanics models for tire–terrain interaction used in modeling and simulation of wheeled vehicles. *J. Terramech.* **2015**, *57*, 1–22. [CrossRef]
29. Ghosh, S.; Konar, A. An Overview of Computational Intelligence Algorithms. In *Call Admission Control in Mobile Cellular Networks*; Springer: Berlin/Heidelberg, Germany, 2013; pp. 63–94. [CrossRef]
30. Vasant, P. *Handbook of Research on Novel Soft Computing Intelligent Algorithms: Theory and Practical Applications*; IGI Global: Hershey, PA, USA, 2013. [CrossRef]
31. Xing, B.; Gao, W.J. *Innovative Computational Intelligence: A Rough Guide to 134 Clever Algorithms*; Springer: Manhattan, NY, USA, 2014; Volume 62.
32. Ibrahim, D. An overview of soft computing. *Procedia Comput. Sci.* **2016**, *102*, 34–38. [CrossRef]
33. Ding, S.; Li, H.; Su, C.; Yu, J.; Jin, F. Evolutionary artificial neural networks: A review. *Artif. Intell. Rev.* **2013**, *39*, 251–260. [CrossRef]
34. Stanley, K.O.; Clune, J.; Lehman, J.; Miikkulainen, R. Designing neural networks through neuroevolution. *Nat. Mach. Intell.* **2019**, *1*, 24–35. [CrossRef]
35. Elbes, M.; Alzubi, S.; Kanan, T.; Al-Fuqaha, A.; Hawashin, B. A survey on particle swarm optimization with emphasis on engineering and network applications. *Evol. Intell.* **2019**, *12*, 113–129. [CrossRef]
36. Karaboga, D.; Kaya, E. Adaptive network based fuzzy inference system (ANFIS) training approaches: A comprehensive survey. *Artif. Intell. Rev.* **2019**, *52*, 2263–2293. [CrossRef]
37. Ridzuan, F.; Zainon, W.M.N.W. A review on data cleansing methods for big data. *Procedia Comput. Sci.* **2019**, *161*, 731–738. [CrossRef]
38. Badgujar, C.; Das, S.; Flippo, D.; Welch, S.M.; Martinez-Figueroa, D. A Deep Neural Network-Based Approach to Predict the Traction, Mobility, and Energy Consumption of Autonomous Ground Vehicle on Sloping Terrain Field. *Comput. Electron. Agric.* **2022**, *196*, 106867. [CrossRef]
39. Scholz, M. Validation of nonlinear PCA. *Neural Process. Lett.* **2012**, *36*, 21–30. [CrossRef]
40. Stone, J.V. Independent component analysis: An introduction. *Trends Cogn. Sci.* **2002**, *6*, 59–64. [CrossRef]

41. Cherkassky, V.; Ma, Y. Comparison of loss functions for linear regression. In Proceedings of the 2004 IEEE International Joint Conference on Neural Networks, Budapest, Hungary, 25–29 July 2004; Volume 1, pp. 395–400. [CrossRef]
42. Čížek, P.; Sadıkoğlu, S. Robust nonparametric regression: A review. *Wiley Interdiscip. Rev. Comput. Stat.* **2020**, *12*, e1492. [CrossRef]
43. Huang, S.; Wu, Q. Robust pairwise learning with Huber loss. *J. Complex.* **2021**, *66*, 101570. [CrossRef]
44. Vapnik, V.; Levin, E.; Le Cun, Y. Measuring the VC-dimension of a learning machine. *Neural Comput.* **1994**, *6*, 851–876. [CrossRef]
45. Zou, H.; Hastie, T. Regularization and variable selection via the elastic net. *J. R. Stat. Soc. Ser. B (Stat. Methodol.)* **2005**, *67*, 301–320. [CrossRef]
46. Wilamowski, B.M.; Yu, H. Improved computation for Levenberg–Marquardt training. *IEEE Trans. Neural Netw.* **2010**, *21*, 930–937. [CrossRef]
47. Rodriguez, J.D.; Perez, A.; Lozano, J.A. Sensitivity analysis of k-fold cross validation in prediction error estimation. *IEEE Trans. Pattern Anal. Mach. Intell.* **2009**, *32*, 569–575. [CrossRef]
48. Das, S.; Koduru, P.; Gui, M.; Cochran, M.; Wareing, A.; Welch, S.M.; Babin, B.R. Adding local search to particle swarm optimization. In Proceedings of the 2006 IEEE International Conference on Evolutionary Computation, Vancouver, BC, Canada, 16–21 July 2006; pp. 428–433.
49. Taghavifar, H.; Mardani, A. Energy loss optimization of run-off-road wheels applying imperialist competitive algorithm. *Inf. Process. Agric.* **2014**, *1*, 57–65. [CrossRef]
50. Taghavifar, H.; Mardani, A. Evaluating the effect of tire parameters on required drawbar pull energy model using adaptive neuro-fuzzy inference system. *Energy* **2015**, *85*, 586–593. [CrossRef]
51. Rumelhart, D.E.; Hinton, G.E.; Williams, R.J. Learning representations by back-propagating errors. *Nature* **1986**, *323*, 533–536. [CrossRef]
52. Sapna, S.; Tamilarasi, A.; Kumar, M.P. Backpropagation learning algorithm based on Levenberg Marquardt Algorithm. *Comput. Sci. Inf. Technol. (CS IT)* **2012**, *2*, 393–398.
53. Abu-Mostafa, Y.S.; Magdon-Ismail, M.; Lin, H.T. *Learning from Data*; AMLBook: New York, NY, USA, 2012.
54. Ghosh, J.; Nag, A. An overview of radial basis function networks. In *Radial Basis Function Networks 2*; Springer: New York, NY, USA: 2001; pp. 1–36.
55. Ruß, G. Data mining of agricultural yield data: A comparison of regression models. In Proceedings of the Industrial Conference on Data Mining, Leipzig, Germany, 20–22 July 2009; Springer: New York, NY, USA, 2009; pp. 24–37.
56. da Silva, E.M., Jr.; Maia, R.D.; Cabacinha, C.D. Bee-inspired RBF network for volume estimation of individual trees. *Comput. Electron. Agric.* **2018**, *152*, 401–408. [CrossRef]
57. Zhang, D.; Zang, G.; Li, J.; Ma, K.; Liu, H. Prediction of soybean price in China using QR-RBF neural network model. *Comput. Electron. Agric.* **2018**, *154*, 10–17. [CrossRef]
58. Ashraf, T.; Khan, Y.N. Weed density classification in rice crop using computer vision. *Comput. Electron. Agric.* **2020**, *175*, 105590. [CrossRef]
59. Eide, Å.J.; Lindblad, T.; Paillet, G. Radial-basis-function networks. In *Intelligent Systems*; CRC Press: Boca Raton, FL, USA, 2018.
60. Bock, H.H. Clustering methods: A history of k-means algorithms. *Selected Contributions in Data Analysis and Classification*; Springer: New York, NY, USA, 2007; pp. 161–172.
61. Smola, A.J.; Schölkopf, B. A tutorial on support vector regression. *Stat. Comput.* **2004**, *14*, 199–222. [CrossRef]
62. Pisner, D.A.; Schnyer, D.M. Support vector machine. In *Machine learning*; Elsevier: Amsterdam, The Netherlands, 2020; pp. 101–121.
63. Cervantes, J.; Garcia-Lamont, F.; Rodríguez-Mazahua, L.; Lopez, A. A comprehensive survey on support vector machine classification: Applications, challenges and trends. *Neurocomputing* **2020**, *408*, 189–215. [CrossRef]
64. Mucherino, A.; Papajorgji, P.; Pardalos, P.M. A survey of data mining techniques applied to agriculture. *Oper. Res.* **2009**, *9*, 121–140. [CrossRef]
65. Mehdizadeh, S.; Behmanesh, J.; Khalili, K. Using MARS, SVM, GEP and empirical equations for estimation of monthly mean reference evapotranspiration. *Comput. Electron. Agric.* **2017**, *139*, 103–114. [CrossRef]
66. Patrício, D.I.; Rieder, R. Computer vision and artificial intelligence in precision agriculture for grain crops: A systematic review. *Comput. Electron. Agric.* **2018**, *153*, 69–81. [CrossRef]
67. Kok, Z.H.; Shariff, A.R.M.; Alfatni, M.S.M.; Khairunniza-Bejo, S. Support vector machine in precision agriculture: A review. *Comput. Electron. Agric.* **2021**, *191*, 106546. [CrossRef]
68. Burges, C.J. A tutorial on support vector machines for pattern recognition. *Data Min. Knowl. Discov.* **1998**, *2*, 121–167. [CrossRef]
69. Hindi, H. A tutorial on convex optimization. In Proceedings of the 2004 American Control Conference, Boston, MA, USA, 30 June–2 July 2004; Volume 4, pp. 3252–3265.
70. Hindi, H. A tutorial on convex optimization II: Duality and interior point methods. In Proceedings of the 2006 American Control Conference, Minneapolis, MN, USA, 14–16 June 2006; p. 11.
71. Chapelle, O. Training a support vector machine in the primal. *Neural Comput.* **2007**, *19*, 1155–1178. [CrossRef] [PubMed]
72. Liang, Z.; Li, Y. Incremental support vector machine learning in the primal and applications. *Neurocomputing* **2009**, *72*, 2249–2258. [CrossRef]

73. Wu, J.; Wang, Y.G. Iterative Learning in Support Vector Regression With Heterogeneous Variances. *IEEE Trans. Emerg. Top. Comput. Intell.* **2022**, 1–10. [CrossRef]
74. Zimmermann, H.J. Fuzzy set theory. *Wiley Interdiscip. Rev. Comput. Stat.* **2010**, *2*, 317–332. [CrossRef]
75. Iancu, I. A Mamdani type fuzzy logic controller. *Fuzzy Log. -Control. Concepts Theor. Appl.* **2012**, *15*, 325–350.
76. Guerra, T.M.; Kruszewski, A.; Lauber, J. Discrete Tagaki–Sugeno models for control: Where are we? *Annu. Rev. Control* **2009**, *33*, 37–47. [CrossRef]
77. Nguyen, A.T.; Taniguchi, T.; Eciolaza, L.; Campos, V.; Palhares, R.; Sugeno, M. Fuzzy control systems: Past, present and future. *IEEE Comput. Intell. Mag.* **2019**, *14*, 56–68. [CrossRef]
78. Nakanishi, H.; Turksen, I.; Sugeno, M. A review and comparison of six reasoning methods. *Fuzzy Sets Syst.* **1993**, *57*, 257–294. [CrossRef]
79. Ying, H.; Ding, Y.; Li, S.; Shao, S. Comparison of necessary conditions for typical Takagi-Sugeno and Mamdani fuzzy systems as universal approximators. *IEEE Trans. Syst. Man Cybern. -Part A Syst. Humans* **1999**, *29*, 508–514. [CrossRef]
80. Huang, Y.; Lan, Y.; Thomson, S.J.; Fang, A.; Hoffmann, W.C.; Lacey, R.E. Development of soft computing and applications in agricultural and biological engineering. *Comput. Electron. Agric.* **2010**, *71*, 107–127. [CrossRef]
81. Touati, F.; Al-Hitmi, M.; Benhmed, K.; Tabish, R. A fuzzy logic based irrigation system enhanced with wireless data logging applied to the state of Qatar. *Comput. Electron. Agric.* **2013**, *98*, 233–241. [CrossRef]
82. Zareiforoush, H.; Minaei, S.; Alizadeh, M.R.; Banakar, A.; Samani, B.H. Design, development and performance evaluation of an automatic control system for rice whitening machine based on computer vision and fuzzy logic. *Comput. Electron. Agric.* **2016**, *124*, 14–22. [CrossRef]
83. Kisi, O.; Sanikhani, H.; Zounemat-Kermani, M.; Niazi, F. Long-term monthly evapotranspiration modeling by several data-driven methods without climatic data. *Comput. Electron. Agric.* **2015**, *115*, 66–77. [CrossRef]
84. Valdés-Vela, M.; Abrisqueta, I.; Conejero, W.; Vera, J.; Ruiz-Sánchez, M.C. Soft computing applied to stem water potential estimation: A fuzzy rule based approach. *Comput. Electron. Agric.* **2015**, *115*, 150–160. [CrossRef]
85. Malik, A.; Kumar, A.; Piri, J. Daily suspended sediment concentration simulation using hydrological data of Pranhita River Basin, India. *Comput. Electron. Agric.* **2017**, *138*, 20–28. [CrossRef]
86. Shafaei, S.; Loghavi, M.; Kamgar, S. Appraisal of Takagi-Sugeno-Kang type of adaptive neuro-fuzzy inference system for draft force prediction of chisel plow implement. *Comput. Electron. Agric.* **2017**, *142*, 406–415. [CrossRef]
87. Shiri, J.; Keshavarzi, A.; Kisi, O.; Iturraran-Viveros, U.; Bagherzadeh, A.; Mousavi, R.; Karimi, S. Modeling soil cation exchange capacity using soil parameters: Assessing the heuristic models. *Comput. Electron. Agric.* **2017**, *135*, 242–251. [CrossRef]
88. Jang, J.S.; Sun, C.T. Neuro-fuzzy modeling and control. *Proc. IEEE* **1995**, *83*, 378–406. [CrossRef]
89. Babuška, R.; Verbruggen, H. Neuro-fuzzy methods for nonlinear system identification. *Annu. Rev. Control* **2003**, *27*, 73–85. [CrossRef]
90. Shihabudheen, K.; Pillai, G.N. Recent advances in neuro-fuzzy system: A survey. *Knowl.-Based Syst.* **2018**, *152*, 136–162. [CrossRef]
91. de Campos Souza, P.V. Fuzzy neural networks and neuro-fuzzy networks: A review the main techniques and applications used in the literature. *Appl. Soft Comput.* **2020**, *92*, 106275. [CrossRef]
92. Wu, W.; Li, L.; Yang, J.; Liu, Y. A modified gradient-based neuro-fuzzy learning algorithm and its convergence. *Inf. Sci.* **2010**, *180*, 1630–1642. [CrossRef]
93. Wang, L.; Zhang, H. An adaptive fuzzy hierarchical control for maintaining solar greenhouse temperature. *Comput. Electron. Agric.* **2018**, *155*, 251–256. [CrossRef]
94. Shafaei, S.; Loghavi, M.; Kamgar, S. An extensive validation of computer simulation frameworks for neural prognostication of tractor tractive efficiency. *Comput. Electron. Agric.* **2018**, *155*, 283–297. [CrossRef]
95. Petković, B.; Petković, D.; Kuzman, B.; Milovančević, M.; Wakil, K.; Ho, L.S.; Jermsittiparsert, K. Neuro-fuzzy estimation of reference crop evapotranspiration by neuro fuzzy logic based on weather conditions. *Comput. Electron. Agric.* **2020**, *173*, 105358. [CrossRef]
96. Wiktorowicz, K. RFIS: Regression-based fuzzy inference system. *Neural Comput. Appl.* **2022**, *34*, 12175–12196. [CrossRef]
97. Cheng, C.B.; Cheng, C.J.; Lee, E. Neuro-fuzzy and genetic algorithm in multiple response optimization. *Comput. Math. Appl.* **2002**, *44*, 1503–1514. [CrossRef]
98. Shihabudheen, K.; Mahesh, M.; Pillai, G.N. Particle swarm optimization based extreme learning neuro-fuzzy system for regression and classification. *Expert Syst. Appl.* **2018**, *92*, 474–484. [CrossRef]
99. Castellano, G.; Castiello, C.; Fanelli, A.M.; Jain, L. Evolutionary neuro-fuzzy systems and applications. In *Advances in Evolutionary Computing for System Design*; Springer: New York, NY, USA, 2007; pp. 11–45.
100. Aghelpour, P.; Bahrami-Pichaghchi, H.; Kisi, O. Comparison of three different bio-inspired algorithms to improve ability of neuro fuzzy approach in prediction of agricultural drought, based on three different indexes. *Comput. Electron. Agric.* **2020**, *170*, 105279. [CrossRef]
101. Hassan, A.; Tohmaz, A. Performance of Skidder Tires in Swamps—Comparison between Statistical and Neural Network Models. *Trans. ASAE* **1995**, *38*, 1545–1551. [CrossRef]
102. Taghavifar, H.; Mardani, A.; Hosseinloo, A.H. Appraisal of artificial neural network-genetic algorithm based model for prediction of the power provided by the agricultural tractors. *Energy* **2015**, *93*, 1704–1710. [CrossRef]

103. Shafaei, S.; Loghavi, M.; Kamgar, S. Benchmark of an intelligent fuzzy calculator for admissible estimation of drawbar pull supplied by mechanical front wheel drive tractor. *Artif. Intell. Agric.* **2020**, *4*, 209–218. [CrossRef]
104. Cutini, M.; Costa, C.; Brambilla, M.; Bisaglia, C. Relationship between the 3D Footprint of an Agricultural Tire and Drawbar Pull Using an Artificial Neural Network. *Appl. Eng. Agric.* **2022**, *38*, 293–301. [CrossRef]
105. *American National Standard ANSI/ASAE S296.5 DEC2003 (R2018)*; General Terminology for Traction of Agricultural Traction and Transport Devices and Vehicles. ASABE: St. Joseph, MI, USA, 2018.
106. Carman, K.; Taner, A. Prediction of Tire Tractive Performance by Using Artificial Neural Networks. *Math. Comput. Appl.* **2012**, *17*, 182–192. [CrossRef]
107. Taghavifar, H.; Mardani, A. On the modeling of energy efficiency indices of agricultural tractor driving wheels applying adaptive neuro-fuzzy inference system. *J. Terramech.* **2014**, *56*, 37–47. [CrossRef]
108. Taghavifar, H.; Mardani, A. Applying a supervised ANN (artificial neural network) approach to the prognostication of driven wheel energy efficiency indices. *Energy* **2014**, *68*, 651–657. [CrossRef]
109. Taghavifar, H.; Mardani, A. Use of artificial neural networks for estimation of agricultural wheel traction force in soil bin. *Neural Comput. Appl.* **2014**, *24*, 1249–1258. [CrossRef]
110. Ekinci, S.; Carman, K.; Kahramanlı, H. Investigation and modeling of the tractive performance of radial tires using off-road vehicles. *Energy* **2015**, *93*, 1953–1963. [CrossRef]
111. Pentoś, K.; Pieczarka, K. Applying an artificial neural network approach to the analysis of tractive properties in changing soil conditions. *Soil Tillage Res.* **2017**, *165*, 113–120. [CrossRef]
112. Pentoś, K.; Pieczarka, K.; Lejman, K. Application of Soft Computing Techniques for the Analysis of Tractive Properties of a Low-Power Agricultural Tractor under Various Soil Conditions. *Complexity* **2020**. [CrossRef]
113. Taghavifar, H.; Mardani, A.; Karim-Maslak, H.; Kalbkhani, H. Artificial Neural Network estimation of wheel rolling resistance in clay loam soil. *Appl. Soft Comput.* **2013**, *13*, 3544–3551. [CrossRef]
114. Taghavifar, H.; Mardani, A. A knowledge-based Mamdani fuzzy logic prediction of the motion resistance coefficient in a soil bin facility for clay loam soil. *Neural Comput. Appl.* **2013**, *23*, 293–302. [CrossRef]
115. Taghavifar, H.; Mardani, A. A comparative trend in forecasting ability of artificial neural networks and regressive support vector machine methodologies for energy dissipation modeling of off-road vehicles. *Energy* **2014**, *66*, 569–576. [CrossRef]
116. Almaliki, S.; Alimardani, R.; Omid, M. Artificial Neural Network Based Modeling of Tractor Performance at Different Field Conditions. *Agric. Eng. Int. CIGR J.* **2016**, *18*, 262–274.
117. Shafaei, S.; Loghavi, M.; Kamgar, S. Feasibility of implementation of intelligent simulation configurations based on data mining methodologies for prediction of tractor wheel slip. *Inf. Process. Agric.* **2019**, *6*, 183–199. [CrossRef]
118. Küçüksariyildiz, H.; Çarman, K.; Sabanci, K. Prediction of Specific Fuel Consumption of 60 HP 2WD Tractor Using Artificial Neural Networks. *Int. J. Automot. Sci. Technol.* **2021**, *5*, 436–444. [CrossRef]
119. Badgujar, C.; Flippo, D.; Welch, S. Artificial neural network to predict traction performance of autonomous ground vehicle on a sloped soil bin and uncertainty analysis. *Comput. Electron. Agric.* **2022**, *196*, 106867. [CrossRef]
120. Choi, Y.S.; Lee, K.S.; Park, W.Y. Application of a Neural Network to Dynamic Draft Model. *Agric. Biosyst. Eng.* **2000**, *1*, 67–72.
121. ASABE. *Agricultural Machinery Management Data*; Technical Report ASAE D497.4 MAR99; American Society of Agricultural and Biological Engineers: ASABE, St. Joseph, MI, USA, 2000.
122. Al-Janobi, A.; Al-Hamed, S.; Aboukarima, A.; Almajhadi, Y. Modeling of Draft and Energy Requirements of a Moldboard Plow Using Artificial Neural Networks Based on Two Novel Variables. *Eng. Agrícola* **2020**, *40*, 363–373. [CrossRef]
123. Shafaei, S.; Loghavi, M.; Kamgar, S.; Raoufat, M. Potential assessment of neuro-fuzzy strategy in prognostication of draft parameters of primary tillage implement. *Ann. Agrar. Sci.* **2018**, *16*, 257–266. [CrossRef]
124. Çarman, K.; Çıtıl, E.; Taner, A. Artificial Neural Network Model for Predicting Specific Draft Force and Fuel Consumption Requirement of a Mouldboard Plough. *Selcuk J. Agric. Food Sci.* **2019**, *33*, 241–247. [CrossRef]
125. Al-Hamed, S.A.; Wahby, M.F.; Al-Saqer, S.M.; Aboukarima, A.M.; Ahmed, A.S. Artificial neural network model for predicting draft and energy requirements of a disk plow. *J. Anim. Plant Sci.* **2013**, *23*, 1714–1724.
126. Shafaei, S.M.; Loghavi, M.; Kamgar, S. A comparative study between mathematical models and the ANN data mining technique in draft force prediction of disk plow implement in clay loam soil. *Agric. Eng. Int. CIGR J.* **2018**, *20*, 71–79.
127. Aboukarima, A.; Saad, A.F. Assessment of Different Indices Depicting Soil Texture for Predicting Chisel Plow Draft Using Neural Networks. *Alex. Sci. Exch. J.* **2006**, *27*, 170–180.
128. Marey, S.; Aboukarima, A.; Almajhadi, Y. Predicting the Performance Parameters of Chisel Plow Using Neural Network Model. *Eng. Agrícola* **2020**, *40*, 719–731. [CrossRef]
129. DeJong-Hughes, J. *Tillage Implements, 2021*; The University of Minnesota Extension Service: St. Paul, MN, USA, 2021.
130. Alimardani, R.; Abbaspour-Gilandeh, Y.; Khalilian, A.; Keyhani, A.; Sadati, S.H. Prediction of draft force and energy of subsoiling operation using ANN model. *J. Food, Agric. Environ.* **2009**, *7*, 537–542.
131. Bergtold, J.; Sailus, M.; Jackson, T. *Conservation Tillage Systems in the Southeast: Production, Profitability and Stewardship*; USDA: Washington, DC, USA; Sustainable Agriculture Research & Education: College Park, MD, USA, 2020.
132. Askari, M.; Abbaspour-Gilandeh, Y. Assessment of adaptive neuro-fuzzy inference system and response surface methodology approaches in draft force prediction of subsoiling tines. *Soil Tillage Res.* **2019**, *194*, 104338. [CrossRef]

133. Abbaspour-Gilandeh, M.; Shahgoli, G.; Abbaspour-Gilandeh, Y.; Herrera-Miranda, M.A.; Hernández-Hernández, J.L.; Herrera-Miranda, I. Measuring and Comparing Forces Acting on Moldboard Plow and Para-Plow with Wing to Replace Moldboard Plow with Para-Plow for Tillage and Modeling It Using Adaptive Neuro-Fuzzy Interface System (ANFIS). *Agriculture* **2020**, *10*, 633. [CrossRef]
134. Abbaspour-Gilandeh, Y.; Sedghi, R. Predicting soil fragmentation during tillage operation using fuzzy logic approach. *J. Terramech.* **2015**, *57*, 61–69. [CrossRef]
135. Marakoğlu, T.; Çarman, K. Fuzzy knowledge-based model for prediction of soil loosening and draft efficiency in tillage. *J. Terramech.* **2010**, *47*, 173–178. [CrossRef]
136. Abbaspour-Gilandeh, Y.; Fazeli, M.; Roshanianfard, A.; Hernández-Hernández, M.; Gallardo-Bernal, I.; Hernández-Hernández, J.L. Prediction of Draft Force of a Chisel Cultivator Using Artificial Neural Networks and Its Comparison with Regression Model. *Agronomy* **2020**, *10*, 451. [CrossRef]
137. Zhang, Z.X.; Kushwaha, R. Applications of neural networks to simulate soil-tool interaction and soil behavior. *Can. Agric. Eng.* **1999**, *41*, 119–125.
138. Mohammadi, A. Modeling of Draft Force Variation in a Winged Share Tillage Tool Using Fuzzy Table Look-Up Scheme. *Agric. Eng. Int. CIGR J.* **2012**, *14*, 262–268.
139. Akbarnia, A.; Mohammadi, A.; Alimardani, R.; Farhani, F. Simulation of draft force of winged share tillage tool using artificial neural network model. *Agric. Eng. Int. CIGR J.* **2014**, *16*, 57–65.
140. Usaborisut, P.; Prasertkan, K. Specific energy requirements and soil pulverization of a combined tillage implement. *Heliyon* **2019**, *5*, e02757. [CrossRef]
141. Upadhyay, G.; Raheman, H. Comparative assessment of energy requirement and tillage effectiveness of combined (active-passive) and conventional offset disc harrows. *Biosyst. Eng.* **2020**, *198*, 266–279. [CrossRef]
142. Shafaei, S.; Loghavi, M.; Kamgar, S. Prognostication of energy indices of tractor-implement utilizing soft computing techniques. *Inf. Process. Agric.* **2019**, *6*, 132–149. [CrossRef]
143. Rahman, A.; Kushwaha, R.L.; Ashrafizadeh, S.R.; Panigrahi, S. Prediction of Energy Requirement of a Tillage Tool in a Soil Bin using Artificial Neural Network. In Proceedings of the 2011 ASABE Annual International Meeting, Louisville, KY, USA, 7–10 August 2011; ASABE: St. Joseph, MI, USA, 2011. [CrossRef]
144. Saleh, B.; Aly, A. Artificial Neural Network Model for Evaluation of the Ploughing Process Performance. *Int. J. Control Autom. Syst.* **2013**, *2*, 1–11.
145. Shafaei, S.; Loghavi, M.; Kamgar, S. On the neurocomputing based intelligent simulation of tractor fuel efficiency parameters. *Inf. Process. Agric.* **2018**, *5*, 205–223. [CrossRef]
146. Shafaei, S.M.; Loghavi, M.; Kamgar, S. On the Reliability of Intelligent Fuzzy System for Multivariate Pattern Scrutinization of Power Consumption Efficiency of Mechanical Front Wheel Drive Tractor. *J. Biosyst. Eng.* **2021**, *46*, 1–15. [CrossRef]
147. Carman, K. Prediction of soil compaction under pneumatic tires a using fuzzy logic approach. *J. Terramech.* **2008**, *45*, 103–108. [CrossRef]
148. Taghavifar, H.; Mardani, A.; Taghavifar, L. A hybridized artificial neural network and imperialist competitive algorithm optimization approach for prediction of soil compaction in soil bin facility. *Measurement* **2013**, *46*, 2288–2299. [CrossRef]
149. Taghavifar, H.; Mardani, A. Fuzzy logic system based prediction effort: A case study on the effects of tire parameters on contact area and contact pressure. *Appl. Soft Comput.* **2014**, *14*, 390–396. [CrossRef]
150. Taghavifar, H.; Mardani, A. Wavelet neural network applied for prognostication of contact pressure between soil and driving wheel. *Inf. Process. Agric.* **2014**, *1*, 51–56. [CrossRef]
151. Taghavifar, H. A supervised artificial neural network representational model based prediction of contact pressure and bulk density. *J. Adv. Veh. Eng.* **2015**, *1*, 14–21.
152. Chen, X.W.; Lin, X. Big data deep learning: Challenges and perspectives. *IEEE Access* **2014**, *2*, 514–525. [CrossRef]
153. Hoi, S.C.; Sahoo, D.; Lu, J.; Zhao, P. Online learning: A comprehensive survey. *Neurocomputing* **2021**, *459*, 249–289. [CrossRef]
154. Wagner, C.; Smith, M.; Wallace, K.; Pourabdollah, A. Generating uncertain fuzzy logic rules from surveys: Capturing subjective relationships between variables from human experts. In Proceedings of the 2015 IEEE International Conference on Systems, Man, and Cybernetics, Hong Kong, China, 9–12 October 2015; pp. 2033–2038.
155. Evans, R.; Grefenstette, E. Learning explanatory rules from noisy data. *J. Artif. Intell. Res.* **2018**, *61*, 1–64. [CrossRef]
156. Mashwani, W.K. Comprehensive survey of the hybrid evolutionary algorithms. *Int. J. Appl. Evol. Comput.* **2013**, *4*, 1–19. [CrossRef]
157. Weiss, K.; Khoshgoftaar, T.M.; Wang, D. A survey of transfer learning. *J. Big Data* **2016**, *3*, 1–40. [CrossRef]
158. Zhuang, F.; Qi, Z.; Duan, K.; Xi, D.; Zhu, Y.; Zhu, H.; Xiong, H.; He, Q. A comprehensive survey on transfer learning. *Proc. IEEE* **2020**, *109*, 43–76. [CrossRef]
159. Abdella, M.; Marwala, T. The use of genetic algorithms and neural networks to approximate missing data in database. In Proceedings of the IEEE 3rd International Conference on Computational Cybernetics, Hotel Le Victoria, Mauritius, 13–16 April 2005; pp. 207–212.
160. Amiri, M.; Jensen, R. Missing data imputation using fuzzy-rough methods. *Neurocomputing* **2016**, *205*, 152–164. [CrossRef]

161. Capuano, N.; Chiclana, F.; Fujita, H.; Herrera-Viedma, E.; Loia, V. Fuzzy group decision making with incomplete information guided by social influence. *IEEE Trans. Fuzzy Syst.* **2017**, *26*, 1704–1718. [CrossRef]
162. Olden, J.D.; Jackson, D.A. Illuminating the "black box": A randomization approach for understanding variable contributions in artificial neural networks. *Ecol. Model.* **2002**, *154*, 135–150. [CrossRef]
163. Sheu, Y.h. Illuminating the Black Box: Interpreting Deep Neural Network Models for Psychiatric Research. *Front. Psychiatry* **2020**, *11*, 551299. [CrossRef] [PubMed]
164. Jeyakumar, J.V.; Noor, J.; Cheng, Y.H.; Garcia, L.; Srivastava, M. How can i explain this to you? an empirical study of deep neural network explanation methods. *Adv. Neural Inf. Process. Syst.* **2020**, *33*, 4211–4222.
165. Zhang, Y.; Tiňo, P.; Leonardis, A.; Tang, K. A survey on neural network interpretability. *IEEE Trans. Emerg. Top. Comput. Intell.* **2021**, *5*, 726–742. [CrossRef]
166. Awan, A.A.; Subramoni, H.; Panda, D.K. An in-depth performance characterization of CPU- and GPU-based DNN training on modern architectures. In Proceedings of the Machine Learning on HPC Environments, New. York, NY, USA, 12–17 November 2017; pp. 1–8.
167. Lázaro, M.; Santamaría, I.; Pérez-Cruz, F.; Artés-Rodríguez, A. Support vector regression for the simultaneous learning of a multivariate function and its derivatives. *Neurocomputing* **2005**, *69*, 42–61. [CrossRef]
168. Cheng, K.; Lu, Z.; Zhang, K. Multivariate output global sensitivity analysis using multi-output support vector regression. *Struct. Multidiscip. Optim.* **2019**, *59*, 2177–2187. [CrossRef]
169. LeCun, Y.; Bengio, Y.; Hinton, G. Deep learning. *Nature* **2015**, *521*, 436–444. [CrossRef] [PubMed]
170. Rusk, N. Deep learning. *Nat. Methods* **2016**, *13*, 35–35. [CrossRef]
171. Liu, W.; Wang, Z.; Liu, X.; Zeng, N.; Liu, Y.; Alsaadi, F.E. A survey of deep neural network architectures and their applications. *Neurocomputing* **2017**, *234*, 11–26. [CrossRef]
172. Kamilaris, A.; Prenafeta-Boldú, F.X. Deep learning in agriculture: A survey. *Comput. Electron. Agric.* **2018**, *147*, 70–90. [CrossRef]
173. Khan, S.; Tufail, M.; Khan, M.T.; Khan, Z.A.; Iqbal, J.; Wasim, A. Real-time recognition of spraying area for UAV sprayers using a deep learning approach. *PLoS ONE* **2021**, *16*, e0249436. [CrossRef]
174. Saleem, M.H.; Potgieter, J.; Arif, K.M. Automation in agriculture by machine and deep learning techniques: A review of recent developments. *Precis. Agric.* **2021**, *22*, 2053–2091. [CrossRef]
175. Hu, K.; Coleman, G.; Zeng, S.; Wang, Z.; Walsh, M. Graph weeds net: A graph-based deep learning method for weed recognition. *Comput. Electron. Agric.* **2020**, *174*, 105520. [CrossRef]
176. Godara, S.; Toshniwal, D. Deep Learning-based query-count forecasting using farmers' helpline data. *Comput. Electron. Agric.* **2022**, *196*, 106875. [CrossRef]
177. Altalak, M.; Alajmi, A.; Rizg, A. Smart Agriculture Applications Using Deep Learning Technologies: A Survey. *Appl. Sci.* **2022**, *12*, 5919. [CrossRef]
178. Hryniowski, A.; Wong, A. DeepLABNet: End-to-end learning of deep radial basis networks with fully learnable basis functions. *arXiv* **2019**. arXiv:1911.09257.
179. Li, Y.; Zhang, T. Deep neural mapping support vector machines. *Neural Netw.* **2017**, *93*, 185–194. [CrossRef] [PubMed]
180. Zhang, Y.; Ishibuchi, H.; Wang, S. Deep Takagi–Sugeno–Kang fuzzy classifier with shared linguistic fuzzy rules. *IEEE Trans. Fuzzy Syst.* **2017**, *26*, 1535–1549. [CrossRef]
181. Das, R.; Sen, S.; Maulik, U. A survey on fuzzy deep neural networks. *ACM Comput. Surv. (CSUR)* **2020**, *53*, 1–25. [CrossRef]
182. Yu, Y.; Si, X.; Hu, C.; Zhang, J. A review of recurrent neural networks: LSTM cells and network architectures. *Neural Comput.* **2019**, *31*, 1235–1270. [CrossRef] [PubMed]
183. Niu, Z.; Zhong, G.; Yu, H. A review on the attention mechanism of deep learning. *Neurocomputing* **2021**, *452*, 48–62. [CrossRef]
184. Wang, Y.; Wang, H.; Peng, Z. Rice diseases detection and classification using attention based neural network and bayesian optimization. *Expert Syst. Appl.* **2021**, *178*, 114770. [CrossRef]
185. Hanin, B. Which neural net architectures give rise to exploding and vanishing gradients? *Adv. Neural Inf. Process. Syst.* **2018**, *3*, 1–18. [CrossRef]
186. Talathi, S.S.; Vartak, A. Improving performance of recurrent neural network with relu nonlinearity. *arXiv* **2015**. arXiv:1511.03771.
187. Shrestha, A.; Mahmood, A. Review of deep learning algorithms and architectures. *IEEE Access* **2019**, *7*, 53040–53065. [CrossRef]
188. Lillicrap, T.P.; Santoro, A.; Marris, L.; Akerman, C.J.; Hinton, G. Backpropagation and the brain. *Nat. Rev. Neurosci.* **2020**, *21*, 335–346. [CrossRef]
189. Mathew, A.; Amudha, P.; Sivakumari, S. Deep learning techniques: An overview. In Proceedings of the International Conference on Advanced Machine Learning Technologies and Applications, Jaipur, India, 13–15 February 2020; Springer: New York, NY, USA, 2020; pp. 599–608.
190. Kingma, D.P.; Ba, J. Adam: A method for stochastic optimization. *arXiv* **2014**. arXiv:1412.6980.
191. Zhang, Z.; Liu, H.; Meng, Z.; Chen, J. Deep learning-based automatic recognition network of agricultural machinery images. *Comput. Electron. Agric.* **2019**, *166*, 104978. [CrossRef]
192. Jin, X.B.; Yang, N.X.; Wang, X.Y.; Bai, Y.T.; Su, T.L.; Kong, J.L. Hybrid deep learning predictor for smart agriculture sensing based on empirical mode decomposition and gated recurrent unit group model. *Sensors* **2020**, *20*, 1334. [CrossRef]

193. Ahn, S.; Kim, J.; Lee, H.; Shin, J. Guiding deep molecular optimization with genetic exploration. *Adv. Neural Inf. Process. Syst.* **2020**, *33*, 12008–12021.
194. Navada, A.; Ansari, A.N.; Patil, S.; Sonkamble, B.A. Overview of use of decision tree algorithms in machine learning. In Proceedings of the 2011 IEEE Control and System Graduate Research Colloquium, Shah Alam, Malaysia, 27–28 June 2011; pp. 37–42.
195. Loh, W.Y. Classification and regression trees. *Wiley Interdiscip. Rev. Data Min. Knowl. Discov.* **2011**, *1*, 14–23. [CrossRef]
196. Chen, X.; Wang, B.; Gao, Y. Symmetric Binary Tree Based Co-occurrence Texture Pattern Mining for Fine-grained Plant Leaf Image Retrieval. *Pattern Recognit.* **2022**, *129*, 108769. [CrossRef]
197. Saggi, M.K.; Jain, S. Reference evapotranspiration estimation and modeling of the Punjab Northern India using deep learning. *Comput. Electron. Agric.* **2019**, *156*, 387–398. [CrossRef]
198. Zhang, L.; Traore, S.; Ge, J.; Li, Y.; Wang, S.; Zhu, G.; Cui, Y.; Fipps, G. Using boosted tree regression and artificial neural networks to forecast upland rice yield under climate change in Sahel. *Comput. Electron. Agric.* **2019**, *166*, 105031. [CrossRef]
199. Segal, M.R. *Machine Learning Benchmarks and Random Forest Regression*; Kluwer Academic Publishers: Dordrecht, The Netherlands, 2004.
200. Cutler, A.; Cutler, D.R.; Stevens, J.R. Random forests. In *Ensemble Machine Learning*; Springer: New York, NY, USA, 2012; pp. 157–175.
201. Da Silva Júnior, J.C.; Medeiros, V.; Garrozi, C.; Montenegro, A.; Gonçalves, G.E. Random forest techniques for spatial interpolation of evapotranspiration data from Brazilian's Northeast. *Comput. Electron. Agric.* **2019**, *166*, 105017. [CrossRef]
202. Zhang, Y.; Sui, B.; Shen, H.; Ouyang, L. Mapping stocks of soil total nitrogen using remote sensing data: A comparison of random forest models with different predictors. *Comput. Electron. Agric.* **2019**, *160*, 23–30. [CrossRef]
203. Amirruddin, A.D.; Muharam, F.M.; Ismail, M.H.; Ismail, M.F.; Tan, N.P.; Karam, D.S. Hyperspectral remote sensing for assessment of chlorophyll sufficiency levels in mature oil palm (*Elaeis guineensis*) based on frond numbers: Analysis of decision tree and random forest. *Comput. Electron. Agric.* **2020**, *169*, 105221. [CrossRef]
204. Karimi, S.; Shiri, J.; Marti, P. Supplanting missing climatic inputs in classical and random forest models for estimating reference evapotranspiration in humid coastal areas of Iran. *Comput. Electron. Agric.* **2020**, *176*, 105633. [CrossRef]
205. Obsie, E.Y.; Qu, H.; Drummond, F. Wild blueberry yield prediction using a combination of computer simulation and machine learning algorithms. *Comput. Electron. Agric.* **2020**, *178*, 105778. [CrossRef]
206. Ramos, A.P.M.; Osco, L.P.; Furuya, D.E.G.; Gonçalves, W.N.; Santana, D.C.; Teodoro, L.P.R.; da Silva Junior, C.A.; Capristo-Silva, G.F.; Li, J.; Baio, F.H.R.; et al. A random forest ranking approach to predict yield in maize with uav-based vegetation spectral indices. *Comput. Electron. Agric.* **2020**, *178*, 105791. [CrossRef]
207. Rastgou, M.; Bayat, H.; Mansoorizadeh, M.; Gregory, A.S. Estimating the soil water retention curve: Comparison of multiple nonlinear regression approach and random forest data mining technique. *Comput. Electron. Agric.* **2020**, *174*, 105502. [CrossRef]
208. dos Santos Luciano, A.C.; Picoli, M.C.A.; Duft, D.G.; Rocha, J.V.; Leal, M.R.L.V.; Le Maire, G. Empirical model for forecasting sugarcane yield on a local scale in Brazil using Landsat imagery and random forest algorithm. *Comput. Electron. Agric.* **2021**, *184*, 106063. [CrossRef]
209. Mariano, C.; Monica, B. A random forest-based algorithm for data-intensive spatial interpolation in crop yield mapping. *Comput. Electron. Agric.* **2021**, *184*, 106094. [CrossRef]
210. Dhaliwal, J.K.; Panday, D.; Saha, D.; Lee, J.; Jagadamma, S.; Schaeffer, S.; Mengistu, A. Predicting and interpreting cotton yield and its determinants under long-term conservation management practices using machine learning. *Comput. Electron. Agric.* **2022**, *199*, 107107. [CrossRef]
211. Yoo, B.H.; Kim, K.S.; Park, J.Y.; Moon, K.H.; Ahn, J.J.; Fleisher, D.H. Spatial portability of random forest models to estimate site-specific air temperature for prediction of emergence dates of the Asian Corn Borer in North Korea. *Comput. Electron. Agric.* **2022**, *199*, 107113. [CrossRef]
212. Elavarasan, D.; Vincent, D.R.; Sharma, V.; Zomaya, A.Y.; Srinivasan, K. Forecasting yield by integrating agrarian factors and machine learning models: A survey. *Comput. Electron. Agric.* **2018**, *155*, 257–282. [CrossRef]
213. Huang, G.B.; Zhou, H.; Ding, X.; Zhang, R. Extreme Learning Machine for Regression and Multiclass Classification. *IEEE Trans. Syst. Man Cybern. Part B* **2012**, *42*, 513–529. [CrossRef] [PubMed]
214. Ding, S.; Zhao, H.; Zhang, Y.; Xu, X.; Nie, R. Extreme learning machine: Algorithm, theory and applications. *Artif. Intell. Rev.* **2015**, *44*, 103–115. [CrossRef]
215. Huang, G.; Huang, G.B.; Song, S.; You, K. Trends in extreme learning machines: A review. *Neural Netw.* **2015**, *61*, 32–48. [CrossRef]
216. Mohammadi, K.; Shamshirband, S.; Motamedi, S.; Petković, D.; Hashim, R.; Gocic, M. Extreme learning machine based prediction of daily dew point temperature. *Comput. Electron. Agric.* **2015**, *117*, 214–225. [CrossRef]
217. Gocic, M.; Petković, D.; Shamshirband, S.; Kamsin, A. Comparative analysis of reference evapotranspiration equations modelling by extreme learning machine. *Comput. Electron. Agric.* **2016**, *127*, 56–63. [CrossRef]
218. Patil, A.P.; Deka, P.C. An extreme learning machine approach for modeling evapotranspiration using extrinsic inputs. *Comput. Electron. Agric.* **2016**, *121*, 385–392. [CrossRef]
219. Feng, Y.; Peng, Y.; Cui, N.; Gong, D.; Zhang, K. Modeling reference evapotranspiration using extreme learning machine and generalized regression neural network only with temperature data. *Comput. Electron. Agric.* **2017**, *136*, 71–78. [CrossRef]

220. Sadgrove, E.J.; Falzon, G.; Miron, D.; Lamb, D. Fast object detection in pastoral landscapes using a colour feature extreme learning machine. *Comput. Electron. Agric.* **2017**, *139*, 204–212. [CrossRef]
221. Ali, M.; Deo, R.C.; Downs, N.J.; Maraseni, T. Multi-stage committee based extreme learning machine model incorporating the influence of climate parameters and seasonality on drought forecasting. *Comput. Electron. Agric.* **2018**, *152*, 149–165. [CrossRef]
222. Shi, P.; Li, G.; Yuan, Y.; Huang, G.; Kuang, L. Prediction of dissolved oxygen content in aquaculture using Clustering-based Softplus Extreme Learning Machine. *Comput. Electron. Agric.* **2019**, *157*, 329–338. [CrossRef]
223. Gong, D.; Hao, W.; Gao, L.; Feng, Y.; Cui, N. Extreme learning machine for reference crop evapotranspiration estimation: Model optimization and spatiotemporal assessment across different climates in China. *Comput. Electron. Agric.* **2021**, *187*, 106294. [CrossRef]
224. Nahvi, B.; Habibi, J.; Mohammadi, K.; Shamshirband, S.; Al Razgan, O.S. Using self-adaptive evolutionary algorithm to improve the performance of an extreme learning machine for estimating soil temperature. *Comput. Electron. Agric.* **2016**, *124*, 150–160. [CrossRef]
225. Wu, L.; Huang, G.; Fan, J.; Ma, X.; Zhou, H.; Zeng, W. Hybrid extreme learning machine with meta-heuristic algorithms for monthly pan evaporation prediction. *Comput. Electron. Agric.* **2020**, *168*, 105115. [CrossRef]
226. Zhu, B.; Feng, Y.; Gong, D.; Jiang, S.; Zhao, L.; Cui, N. Hybrid particle swarm optimization with extreme learning machine for daily reference evapotranspiration prediction from limited climatic data. *Comput. Electron. Agric.* **2020**, *173*, 105430. [CrossRef]
227. Yu, W.; Zhuang, F.; He, Q.; Shi, Z. Learning deep representations via extreme learning machines. *Neurocomputing* **2015**, *149*, 308–315. [CrossRef]
228. Tissera, M.D.; McDonnell, M.D. Deep extreme learning machines: Supervised autoencoding architecture for classification. *Neurocomputing* **2016**, *174*, 42–49. [CrossRef]
229. Abdelghafour, F.; Rosu, R.; Keresztes, B.; Germain, C.; Da Costa, J.P. A Bayesian framework for joint structure and colour based pixel-wise classification of grapevine proximal images. *Comput. Electron. Agric.* **2019**, *158*, 345–357. [CrossRef]
230. Khanal, A.R.; Mishra, A.K.; Lambert, D.M.; Paudel, K.P. Modeling post adoption decision in precision agriculture: A Bayesian approach. *Comput. Electron. Agric.* **2019**, *162*, 466–474. [CrossRef]
231. Tetteh, G.O.; Gocht, A.; Conrad, C. Optimal parameters for delineating agricultural parcels from satellite images based on supervised Bayesian optimization. *Comput. Electron. Agric.* **2020**, *178*, 105696. [CrossRef]
232. Fang, Y.; Xu, L.; Chen, Y.; Zhou, W.; Wong, A.; Clausi, D.A. A Bayesian Deep Image Prior Downscaling Approach for High-Resolution Soil Moisture Estimation. *IEEE J. Sel. Top. Appl. Earth Obs. Remote Sens.* **2022**, *15*, 4571–4582. [CrossRef]
233. Koller, D.; Friedman, N. *Probabilistic Graphical Models: Principles and Techniques*; MIT Press: Cambridge, MA, USA, 2009.
234. Hrycej, T. Gibbs sampling in Bayesian networks. *Artif. Intell.* **1990**, *46*, 351–363. [CrossRef]
235. Chapman, R.; Cook, S.; Donough, C.; Lim, Y.L.; Ho, P.V.V.; Lo, K.W.; Oberthür, T. Using Bayesian networks to predict future yield functions with data from commercial oil palm plantations: A proof of concept analysis. *Comput. Electron. Agric.* **2018**, *151*, 338–348. [CrossRef]
236. Kocian, A.; Massa, D.; Cannazzaro, S.; Incrocci, L.; Di Lonardo, S.; Milazzo, P.; Chessa, S. Dynamic Bayesian network for crop growth prediction in greenhouses. *Comput. Electron. Agric.* **2020**, *169*, 105167. [CrossRef]
237. Bilmes, J.A. A gentle tutorial of the EM algorithm and its application to parameter estimation for Gaussian mixture and hidden Markov models. *Int. Comput. Sci. Inst.* **1998**, *4*, 126.
238. Lu, J. A survey on Bayesian inference for Gaussian mixture model. *arXiv* **2021**. arXiv:2108.11753.
239. Mouret, F.; Albughdadi, M.; Duthoit, S.; Kouamé, D.; Rieu, G.; Tourneret, J.Y. Reconstruction of Sentinel-2 derived time series using robust Gaussian mixture models—Application to the detection of anomalous crop development. *Comput. Electron. Agric.* **2022**, *198*, 106983. [CrossRef]
240. Zhu, C.; Ding, J.; Zhang, Z.; Wang, J.; Wang, Z.; Chen, X.; Wang, J. SPAD monitoring of saline vegetation based on Gaussian mixture model and UAV hyperspectral image feature classification. *Comput. Electron. Agric.* **2022**, *200*, 107236. [CrossRef]
241. Quinonero-Candela, J.; Rasmussen, C.E. A unifying view of sparse approximate Gaussian process regression. *J. Mach. Learn. Res.* **2005**, *6*, 1939–1959.
242. Wilson, A.G.; Knowles, D.A.; Ghahramani, Z. Gaussian process regression networks. *arXiv* **2011**. arXiv:1110.4411.
243. Smola, A.; Bartlett, P. Sparse greedy Gaussian process regression. *Adv. Neural Inf. Process. Syst.* **2000**, *13*, 1–7.
244. Azadbakht, M.; Ashourloo, D.; Aghighi, H.; Radiom, S.; Alimohammadi, A. Wheat leaf rust detection at canopy scale under different LAI levels using machine learning techniques. *Comput. Electron. Agric.* **2019**, *156*, 119–128. [CrossRef]
245. Shabani, S.; Samadianfard, S.; Sattari, M.T.; Shamshirband, S.; Mosavi, A.; Kmet, T.; Várkonyi-Kóczy, A.R. Modeling daily pan evaporation in humid climates using gaussian process regression. *arXiv* **2019**. arXiv:1908.04267.
246. Nieto, P.G.; García-Gonzalo, E.; Puig-Bargués, J.; Solé-Torres, C.; Duran-Ros, M.; Arbat, G. A new predictive model for the outlet turbidity in micro-irrigation sand filters fed with effluents using Gaussian process regression. *Comput. Electron. Agric.* **2020**, *170*, 105292. [CrossRef]
247. Rastgou, M.; Bayat, H.; Mansoorizadeh, M.; Gregory, A.S. Prediction of soil hydraulic properties by Gaussian process regression algorithm in arid and semiarid zones in Iran. *Soil Tillage Res.* **2021**, *210*, 104980. [CrossRef]
248. Nguyen, L.; Nguyen, D.K.; Nghiem, T.X.; Nguyen, T. Least square and Gaussian process for image based microalgal density estimation. *Comput. Electron. Agric.* **2022**, *193*, 106678. [CrossRef]

249. Zhang, C.; Ma, Y. *Ensemble Machine Learning: Methods and Applications*; Springer: New York, NY, USA, 2012.
250. Sagi, O.; Rokach, L. Ensemble learning: A survey. *Wiley Interdiscip. Rev. Data Min. Knowl. Discov.* **2018**, *8*, e1249. [CrossRef]
251. Zhou, Z.H. Ensemble learning. In *Machine Learning*; Springer: New York, NY, USA, 2021; pp. 181–210.
252. Chaudhary, A.; Kolhe, S.; Kamal, R. A hybrid ensemble for classification in multiclass datasets: An application to oilseed disease dataset. *Comput. Electron. Agric.* **2016**, *124*, 65–72. [CrossRef]
253. Haagsma, M.; Page, G.F.; Johnson, J.S.; Still, C.; Waring, K.M.; Sniezko, R.A.; Selker, J.S. Model selection and timing of acquisition date impacts classification accuracy: A case study using hyperspectral imaging to detect white pine blister rust over time. *Comput. Electron. Agric.* **2021**, *191*, 106555. [CrossRef]
254. Kar, S.; Purbey, V.K.; Suradhaniwar, S.; Korbu, L.B.; Kholová, J.; Durbha, S.S.; Adinarayana, J.; Vadez, V. An ensemble machine learning approach for determination of the optimum sampling time for evapotranspiration assessment from high-throughput phenotyping data. *Comput. Electron. Agric.* **2021**, *182*, 105992. [CrossRef]
255. Chaudhary, A.; Thakur, R.; Kolhe, S.; Kamal, R. A particle swarm optimization based ensemble for vegetable crop disease recognition. *Comput. Electron. Agric.* **2020**, *178*, 105747. [CrossRef]
256. Chia, M.Y.; Huang, Y.F.; Koo, C.H. Improving reference evapotranspiration estimation using novel inter-model ensemble approaches. *Comput. Electron. Agric.* **2021**, *187*, 106227. [CrossRef]
257. Wu, T.; Zhang, W.; Jiao, X.; Guo, W.; Hamoud, Y.A. Evaluation of stacking and blending ensemble learning methods for estimating daily reference evapotranspiration. *Comput. Electron. Agric.* **2021**, *184*, 106039. [CrossRef]
258. Koyama, K.; Lyu, S. Soft-labeling approach along with an ensemble of models for predicting subjective freshness of spinach leaves. *Comput. Electron. Agric.* **2022**, *193*, 106633. [CrossRef]
259. Xu, C.; Ding, J.; Qiao, Y.; Zhang, L. Tomato disease and pest diagnosis method based on the Stacking of prescription data. *Comput. Electron. Agric.* **2022**, *197*, 106997. [CrossRef]
260. Aiken, V.C.F.; Dórea, J.R.R.; Acedo, J.S.; de Sousa, F.G.; Dias, F.G.; de Magalhães Rosa, G.J. Record linkage for farm-level data analytics: Comparison of deterministic, stochastic and machine learning methods. *Comput. Electron. Agric.* **2019**, *163*, 104857. [CrossRef]
261. Weber, V.A.M.; de Lima Weber, F.; da Silva Oliveira, A.; Astolfi, G.; Menezes, G.V.; de Andrade Porto, J.V.; Rezende, F.P.C.; de Moraes, P.H.; Matsubara, E.T.; Mateus, R.G.; et al. Cattle weight estimation using active contour models and regression trees Bagging. *Comput. Electron. Agric.* **2020**, *179*, 105804. [CrossRef]
262. Genedy, R.A.; Ogejo, J.A. Using machine learning techniques to predict liquid dairy manure temperature during storage. *Comput. Electron. Agric.* **2021**, *187*, 106234. [CrossRef]
263. Mohammed, S.; Elbeltagi, A.; Bashir, B.; Alsafadi, K.; Alsilibe, F.; Alsalman, A.; Zeraatpisheh, M.; Széles, A.; Harsányi, E. A comparative analysis of data mining techniques for agricultural and hydrological drought prediction in the eastern Mediterranean. *Comput. Electron. Agric.* **2022**, *197*, 106925. [CrossRef]
264. Ayan, E.; Erbay, H.; Varçın, F. Crop pest classification with a genetic algorithm-based weighted ensemble of deep convolutional neural networks. *Comput. Electron. Agric.* **2020**, *179*, 105809. [CrossRef]
265. Barbosa, A.; Hovakimyan, N.; Martin, N.F. Risk-averse optimization of crop inputs using a deep ensemble of convolutional neural networks. *Comput. Electron. Agric.* **2020**, *178*, 105785. [CrossRef]
266. e Lucas, P.d.O.; Alves, M.A.; e Silva, P.C.d.L.; Guimarães, F.G. Reference evapotranspiration time series forecasting with ensemble of convolutional neural networks. *Comput. Electron. Agric.* **2020**, *177*, 105700. [CrossRef]
267. Gonzalo-Martín, C.; García-Pedrero, A.; Lillo-Saavedra, M. Improving deep learning sorghum head detection through test time augmentation. *Comput. Electron. Agric.* **2021**, *186*, 106179. [CrossRef]
268. Gu, Z.; Zhu, T.; Jiao, X.; Xu, J.; Qi, Z. Neural network soil moisture model for irrigation scheduling. *Comput. Electron. Agric.* **2021**, *180*, 105801. [CrossRef]
269. Khanramaki, M.; Asli-Ardeh, E.A.; Kozegar, E. Citrus pests classification using an ensemble of deep learning models. *Comput. Electron. Agric.* **2021**, *186*, 106192. [CrossRef]
270. Li, Q.; Jia, W.; Sun, M.; Hou, S.; Zheng, Y. A novel green apple segmentation algorithm based on ensemble U-Net under complex orchard environment. *Comput. Electron. Agric.* **2021**, *180*, 105900. [CrossRef]
271. Gonzalez, R.; Iagnemma, K. Slippage estimation and compensation for planetary exploration rovers. State of the art and future challenges. *J. Field Robot.* **2018**, *35*, 564–577. [CrossRef]
272. Gonzalez, R.; Chandler, S.; Apostolopoulos, D. Characterization of machine learning algorithms for slippage estimation in planetary exploration rovers. *J. Terramech.* **2019**, *82*, 23–34. [CrossRef]
273. Jørgensen, M. Adaptive tillage systems. *Agron. Res.* **2014**, *12*, 95–100.
274. Jia, H.; Guo, M.; Yu, H.; Li, Y.; Feng, X.; Zhao, J.; Qi, J. An adaptable tillage depth monitoring system for tillage machine. *Biosyst. Eng.* **2016**, *151*, 187–199. [CrossRef]

*agriculture* MDPI

*Review*

# Review of Discrete Element Method Simulations of Soil Tillage and Furrow Opening

**Kojo Atta Aikins [1,2,*], Mustafa Ucgul [3,4], James B. Barr [5], Emmanuel Awuah [6], Diogenes L. Antille [2,7], Troy A. Jensen [2] and Jacky M. A. Desbiolles [4]**

1 Department of Agricultural and Biosystems Engineering, Kwame Nkrumah University of Science and Technology, PMB, Kumasi, Ghana
2 Centre for Agricultural Engineering, University of Southern Queensland, Toowoomba, QLD 4350, Australia
3 Faculty of Science and Engineering, Southern Cross University, Lismore, NSW 2480, Australia
4 Agricultural Machinery R&D Centre, STEM, University of South Australia, Adelaide, SA 5000, Australia
5 Seed Terminator, 23 Aldershot Rd., Lonsdale, SA 5160, Australia
6 College of Engineering, Nanjing Agricultural University, No. 40 Dianjiangtai, Pukou District, Nanjing 210095, China
7 CSIRO Agriculture and Food, Black Mountain Science and Innovation Precinct, Canberra, ACT 2601, Australia
* Correspondence: kaaikins.coe@knust.edu.gh

**Abstract:** In agricultural machinery design and optimization, the discrete element method (DEM) has played a major role due to its ability to speed up the design and manufacturing process by reducing multiple prototyping, testing, and evaluation under experimental conditions. In the field of soil dynamics, DEM has been mainly applied in the design and optimization of soil-engaging tools, especially tillage tools and furrow openers. This numerical method is able to capture the dynamic and bulk behaviour of soils and soil–tool interactions. This review focused on the various aspects of the application of DEM in the simulation of tillage and furrow opening for tool design optimization. Different contact models, particle sizes and shapes, and calibration techniques for determining input parameters for tillage and furrow opening research have been reviewed. Discrete element method predictions of furrow profiles, disturbed soil surface profiles, soil failure, loosening, disturbance parameters, reaction forces, and the various types of soils modelled with DEM have also been highlighted. This pool of information consolidates existing working approaches used in prior studies and helps to identify knowledge gaps which, if addressed, will advance the current soil dynamics modelling capability.

**Keywords:** calibration; DEM contact models; soil dynamics; soil failure; soil forces; cohesive and frictional soils

**Citation:** Aikins, K.A.; Ucgul, M.; Barr, J.B.; Awuah, E.; Antille, D.L.; Jensen, T.A.; Desbiolles, J.M.A. Review of Discrete Element Method Simulations of Soil Tillage and Furrow Opening. *Agriculture* **2023**, *13*, 541. https://doi.org/10.3390/agriculture13030541

Academic Editor: Tao Cui

Received: 7 February 2023
Accepted: 14 February 2023
Published: 23 February 2023



## 1. Introduction

In mechanized agriculture, the energy use for soil tillage operations can be as high as 50% of the total energy used in crop production [1,2]. The energy-use efficiency associated with tillage can be increased by improving the design of tillage implements and through their correct operation and settings [3–5]. The design optimization of tillage tools and furrow openers conventionally relies on repeated prototyping and evaluation through soil bins and field experimentation. This task is laborious, time-consuming, and expensive [6,7]. In order to reduce the resource intensity involved, various analytical and numerical models for predicting soil–tool interaction and soil forces have been developed. Analytical models are typically based on the universal earthmoving equation (Equation (1)) [8,9].

$$P = (\gamma d^2 N_\gamma + cdN_c + c_a dN_a + qdN_q)w \quad (1)$$

where $P$ is soil cutting force (N), $\gamma$ is the specific weight of soil (N $m^{-3}$), $d$ is working depth (m), $c$ is soil cohesion (Pa), $c_a$ is soil–metal adhesion (Pa), $q$ is surcharge stress (Pa), $w$ is tool width (m), and $N_\gamma$, $N_c$, $N_a$, and $N_q$ are dimensionless $N$ factors that are dependent on gravity, cohesion, adhesion, and surcharge, respectively.

Analytical models require a preliminary assumption of soil failure patterns to predict soil and implement forces [10,11]. These models use simple and approximate geometric profiles such as wedges and crescents that are easily expressed mathematically to model soil failure patterns [4,8,9,12,13]. They also aim to predict the maximum soil reaction force at incipient soil failure, rather than the average soil forces during the tillage process. However, in practice, soil failure occurs in more complex patterns. Analytical models also regard soil failure as bulk soil movement without accounting for interactions between individual soil particles, aggregates, and macro-organic matter evident in field conditions [7].

Numerical models, such as finite element modelling (FEM) and computational fluid dynamics (CFD), have also been employed to predict soil failure patterns and soil forces with some degree of accuracy [14,15]. However, their use is limited because they consider soil as a continuum body rather than being made up of discrete particles. Thus, continuum numerical models fail to account for variations in soil structure, physical conditions, flow, and the mixing and translocation of soil particles [6]. To overcome the shortcomings of analytical and continuum numerical methods, the discrete element method (DEM) can be employed. DEM is a discontinuum numerical method used for modelling the mechanical behaviour of granular materials. Initially developed by Cundall and Strack [16], DEM is used to model particle interactions within granular materials such as gravel, grain, soil, and powder. Unlike analytical and other numerical (continuum) approaches, DEM accounts for the discrete nature of granular particles and their interactions with neighbouring particles and interfacing objects such as contact walls and machine parts. The DEM approach involves the dynamic creation and breaking of bonds between contacting particles [7,16] and can simulate soil–tool reaction forces, as well as track particle movements and model mixing and translocating processes.

Several studies have shown that DEM can be successfully employed to model 2- and 3-dimensional space interactions between granular particles and machine parts. Operations such as grain flow in a hopper, soil movement in bulldozing, and soil deformation and displacement during field traffic, tillage and furrow opening processes have been modelled in DEM, achieving results that closely agree with experimental observations [17–21]. Figure 1 shows a comparison between a tine furrow opener interaction with soil (left) and its simulation with DEM (right) [21]. Other applications of DEM include modelling of cohesionless and cohesive/adhesive particles [10,18,20]. Discontinuum approaches such as DEM offer greater potential for more accurate prediction of soil–tool interactions in soil dynamics applications compared with continuum approaches. Thus, DEM is a powerful tool that can accurately guide and speed-up the design optimization process by researchers, developers, and manufacturers.

The objective of the work reported in this article was to review the various aspects of the application of DEM in the field of soil dynamics by focusing on soil tillage and furrow opening research for tool performance optimization. Contact models for soil particles; DEM particle size and shape considerations; calibration techniques for determining accurate input parameters; predictions of soil failure, particle movement and reaction forces; and the types of soils modelled with DEM have been reviewed. Knowledge gaps that need attention in future research, and that will go some way to advance soil dynamics modelling capability, have also been identified.

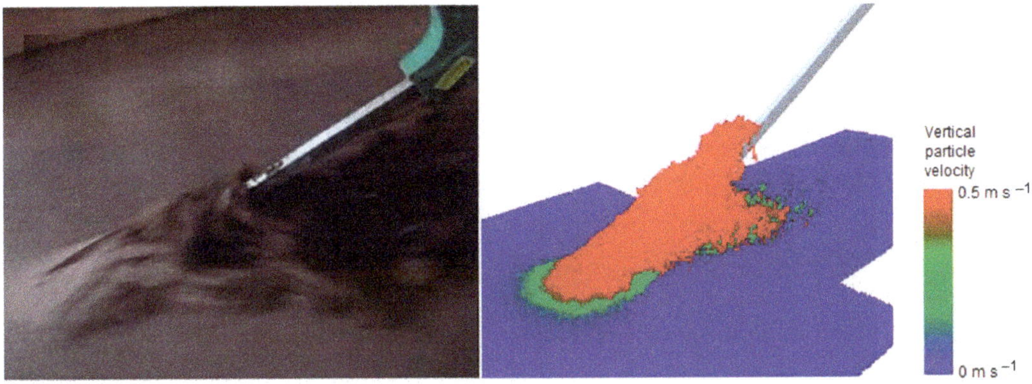


**Figure 1.** Discrete element method simulation of a narrow tine furrow opener operating in a moist sandy-loam soil. Modified from Barr et al. [21].

## 2. Modelling Agricultural Soils with DEM

The discrete element method is used to model soil as a collection of a finite number of individual spherical particles that interact with neighbouring particles and machine parts when subjected to external forces and forced displacement at the soil–tool interface, such as from a soil tillage tool [7]. This process induces the relative motion of particles within the bulk. Contact forces between these particles and their resultant motion are calculated using Newton's Second Law [22]. These calculations involve repeating the same algorithm at each time step of the simulation process, using results from previous calculations as input.

### *2.1. DEM Contact Models*

DEM contact models are developed to describe the mechanical and physical interactions of granular particles with neighbouring particles or external objects. The interactions are modelled using equations of motion and contact models expressed as linear, adhesive, and elastoplastic normal contact models, as well as viscosity, tangential force, and torque models. The tangential force and torque models account for friction, rolling, and torsion [23,24]. Physical interactions between particles are expressed via combining functional elements of springs, dampers, and tangential friction. Total contact forces are expressed as the sum of spring ($F^s$) and damping ($F^d$) forces. Some commonly used contact models in DEM simulations are listed in Table 1 and reviewed below. These contact models have been implemented in commercially available software such as Bulk Flow Analyst™, Chute Analyst™, Chute Maven®, DEMpack™, Altair® EDEM™, ELFEN, GROMOS-96, ITASCA PFC (2D & 3D), LiGGGHTS®, MIMES, PASSAGE/DEM, Rocky, SimPARTIX®, StarCCM+, UDEC, 3DEC, and YADE [25,26]. Table 1 also shows the various DEM software that have been used in tillage and furrow opening research and the types of soil the contact models have been used to model.

**Table 1.** Discrete element method contact models, their advantages and disadvantages, types of soil modelled, and software used.

| Contact Model | Advantages | Disadvantages | Types of Soil Modelled | References | Software Used by Researchers |
|---|---|---|---|---|---|
| Linear spring contact model | • Simple to use. | • Does not account for nonlinearity in loading and unloading cycles and plastic deformation of soil. | Sandy | Tanaka et al. [27], Asaf et al. [10], Shmulevich et al. [6], Ono et al. [28] | PFC2D, EDEM |
| Linear spring contact model with cohesion | • Allows users to consider cohesion in the linear spring contact model. | • Only considers the cohesion through the normal direction. | Vertosol | Bravo et al. [18] | DEMeter++ |
| Hertz–Mindlin contact model | • Simple to use.<br>• Although this model was designed for fine, dry particles, it can be used to model wet particles as well. | • Inaccurate prediction for vertical tillage force. | Sandy | Ucgul et al. [29] | EDEM |
| Parallel bond model (PBM) or Hertz–Mindlin contact model with cohesion | • Allows users to model cohesion. | • Excessive forces cause the bonds to be broken irrationally. | Coarse sand, loamy, sandy loam, loessal, clay, sandy clay loam, loamy clay | Tamas et al. [30], Chen et al. [7], Bo et al. [31], Hang et al. [32], Cheng et al. [33], Yang et al. [34], Hoseinian et al. [35] | EDEM PFC3D |
| Hertz–Mindlin contact model with Johnson–Kendall–Roberts (JKR) | • Enables the modelling of strongly adhesive bonds such as exist in dry powders or wet materials. | • - | Clay, silty clay loam | Cheng et al. [33], Du et al. [36], Zhai et al. [37] | EDEM |
| Hysteric spring contact model | • Accounts for plastic deformation during loading and unloading of soil.<br>• Suitable for both cohesive cohesionless soils. | • It requires a large number of input parameters, making its setup and calibration complex. | Sandy | Ucgul et al. [38], Ucgul et al. [29] | EDEM |
| Hysteric spring contact model with linear cohesion contact model | • Allow users to consider cohesion in the hysteretic spring contact model. | • Only considers the cohesion through the normal direction. | Sandy loam, clay (Vertosol) | Barr et al. [21], Barr et al. [39], Makange et al. [40], Aikins et al. [41], Awuah et al. [42], Wang et al. [43] | EDEM |
| Edinburgh elasto-plastic adhesion model | • It is versatile since it can be used as a linear or non-linear Hertzian spring model.<br>• It also allows tensile forces to develop and a non-linear force-displacement behaviour in compression. | • It requires a lot of input parameters. | Clay, clay loam, sandy loam, loam, sandy | Kim et al. [44], Wu et al. [45], Zhao et al. [46], Sun et al. [47] | EDEM, PFC3D, LiGGGHTS |

The movement of particles due to the contact forces are governed by Newton's equation of motion for linear and angular motion as expressed by Equations (2) and (3). By solving these equations, the motion of the particles can be determined. For two spherical particles ($i$ = 1,2) of masses $m_i$ and radii $r_i$, located at $x_i$ in contact, taking $F$ as contact force, $g$ as acceleration due to gravity, $I_i$ as the moment of inertia of a particle, $\omega_i$ as the angular velocity of a particle, and $T_i$ as torque due to the tangential component of the contact force:

$$m_i\ddot{x}_i = F_i + m_i g \tag{2}$$

$$I_i\dot{\omega}_i = T_i \tag{3}$$

2.1.1. Linear Spring Contact Model

This contact model is linearly elastic and is the simplest contact model often used to simulate soil–tool interactions (Table 1) [6,27]. A contact force is created between the two spherical particles in contact as described above. The contact force can be decomposed into normal and tangential force components. The overlap at the contact point generates a repulsive contact force and energy dissipation. When an overlap $\delta_n > 0$ is formed between the two particles at a relative velocity $\dot{\delta}_n$ in a direction normal to the contact surface, a normal contact force $F_n$ is created based on the spring and dashpot models (Figure 2) such that:

$$F_n = k_n\delta_n + d_n\dot{\delta}_n \tag{4}$$

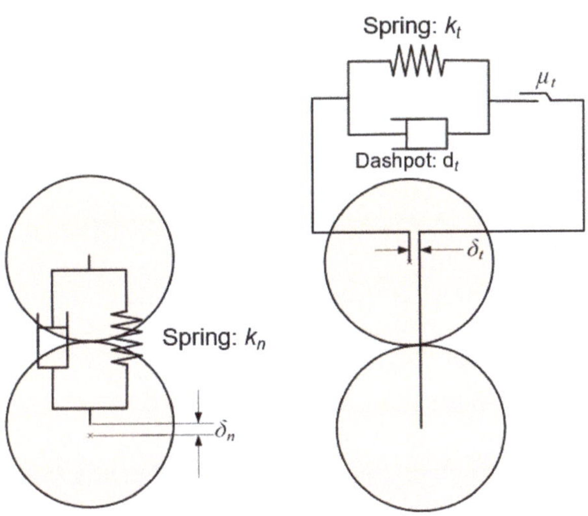


**Figure 2.** A schematic diagram of the normal (**left**) and tangential (**right**) components of the linear spring contact model.

The parameter $k_n$ is the normal stiffness, while $d_n$ is the damping coefficient. Considering an imaginary rod of radius $r = (r_1 + r_2)/2$ between the centres of the two particles and Young's Modulus $E$:

$$F_n = \pi E r/2 \tag{5}$$

When the tangential component of the contact force, $F_t > \mu_t F_n$, sliding friction occurs. The local friction coefficient $\mu_t = tan\varnothing$, where $\varnothing$ is the internal friction angle between the particles.

The tangential component of the contact force is also given by:

$$F_t = -k_t\delta_t - d_t\dot{\delta}_t \tag{6}$$

where $k_t$, $\delta_t$, $d_t$, and $\dot{\delta}_t$ are tangential components of stiffness, overlap, damping coefficient, and relative velocity.

2.1.2. Hertz–Mindlin Contact Model

The Hertz–Mindlin contact model (HMCM), especially when it is used with the parallel bond model (PBM, see Section 2.1.4), is the most popular contact model used by researchers [7,30,33–35,48] to simulate soil–tool interaction in tillage research. However, as an (non-linear) elastic contact model, it fails to predict vertical soil forces accurately (Table 1) [49]. In this model, the contact force consists of a non-linear Hertz component described by the hysteretic spring force-displacement relationship shown in Figure 3 and a damping component (second part of Equation (7)). It is also resolved into normal and tangential components. The HMCM and its parameters are described in Equation (7) to (16) [20,29].

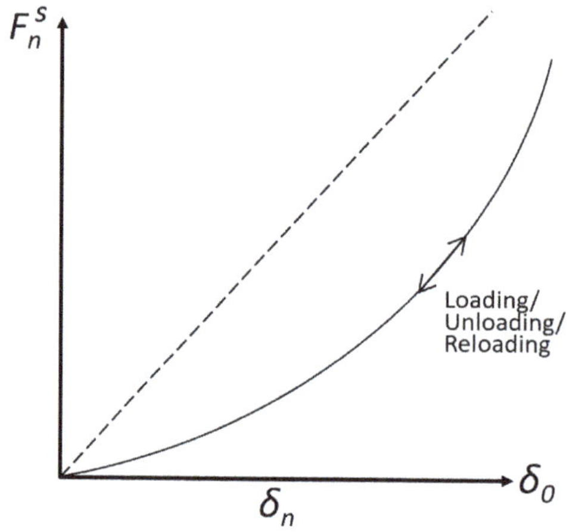


**Figure 3.** Hysteretic spring force –displacement relationship used in the Hertz–Mindlin contact model (HMCM). Redrawn from Ucgul et al. [29]. The dashed line contrasts the basic linear elastic model relationship.

Normal contact force, $F_n$:

$$F_n = -k_n \delta_n^{3/2} - d_n \delta_n^{1/4} \dot{\delta}_n \tag{7}$$

$$k_n = 2E_{eq}\sqrt{r_{eq}\delta_n} \tag{8}$$

$$d_n = \frac{\ln e}{\sqrt{\ln^2 e + \pi^2}}\sqrt{k_n m_{eq}} \tag{9}$$

where:

Equivalent radius,

$$r_{eq} = \left(\frac{1}{r_1} + \frac{1}{r_2}\right)^{-1} \tag{10}$$

Equivalent Young's modulus,

$$E_{eq} = \left(\frac{1-\nu_1^2}{E_1} + \frac{1-\nu_1^2}{E_2}\right)^{-1} \tag{11}$$

Equivalent particle mass,

$$m_{eq} = \left(\frac{1}{m_1} + \frac{1}{m_2}\right)^{-1} \quad (12)$$

Tangential contact force, $F_t$:

$$F_t = -k_t\delta_t - d_t\delta_t^{1/4}\dot{\delta}_t \quad (13)$$

$$k_t = 8G_{eq}\sqrt{r_{eq}\delta_n} \quad (14)$$

$$d_t = \frac{\ln e}{\sqrt{\ln^2 e + \pi^2}}\sqrt{k_t m_{eq}} \quad (15)$$

Equivalent shear modulus,

$$G_{eq} = \left(\frac{2-\nu_1}{G_1} + \frac{2-\nu_2}{G_2}\right)^{-1} \quad (16)$$

2.1.3. Hysteretic Spring Contract Model

The hysteretic spring contact model (HSCM) is an elastic–plastic contact model that accounts for the plastic deformation that occurs during the loading and unloading of soil. It makes the particles behave as though they undergo plastic deformation after the load reaches a yield point, as shown in Figure 4 [50]. The main disadvantage of this contact model is that it requires a large number of input parameters, making its setup and calibration of DEM material properties complex (Table 1) [49]. The HSCM comprises two parts: the spring characteristic illustrated in Figure 2 and damping. A comparative study by Ucgul et al. [29] revealed that the HSCM could model soil–tool interaction more accurately than the HMCM. The HSCM has been used to predict soil reaction forces as well as furrow profiles successfully, especially with the linear cohesion model [39–41,43]. The governing equations of the HSCM are described in Equation (17) to (20).

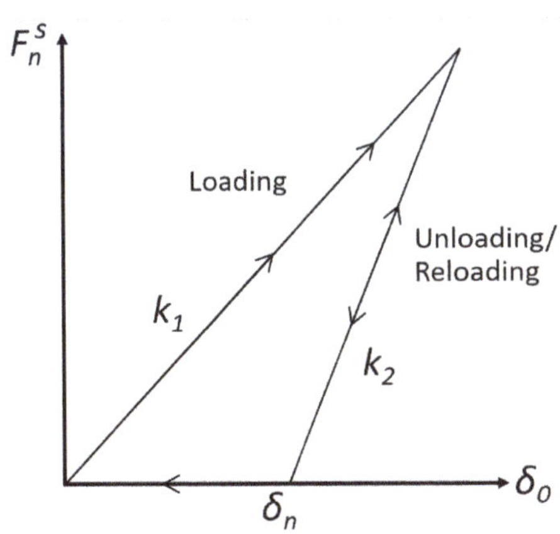


**Figure 4.** Hysteretic spring force-displacement relationship used in the hysteretic spring contact model (HSCM). Redrawn from Ucgul et al. [29].

Normal contact force, $F_n$:

During loading,

$$F_n = -k_1\delta_n - n_c\sqrt{\frac{4m_{eq}k_1}{1+\left(\frac{\pi}{\ln e}\right)^2}}\cdot\dot{\delta}_n \tag{17}$$

During unloading and reloading,

$$F_n = -k_2(\delta_n - \delta_0) - n_c\sqrt{\frac{4m_{eq}k_1}{1+\left(\frac{\pi}{\ln e}\right)^2}}\cdot\dot{\delta}_n \tag{18}$$

During unloading again,

$$F_n = 0 - n_c\sqrt{\frac{4m_{eq}k_1}{1+\left(\frac{\pi}{\ln e}\right)^2}}\cdot\dot{\delta}_n \tag{19}$$

where $k_1$ and $k_2$ are the loading and the unloading stiffnesses, respectively, and $e$ is the coefficient of restitution of the particles, and they are related as $e = \sqrt{k_1/k_2}$.

Tangential contact force, $F_t$:

$$F_t = -n_k k_1\delta_t - \sqrt{\frac{4m_{eq}n_k k_1}{1+\left(\frac{\pi}{\ln e}\right)^2}}\cdot\dot{\delta}_t \tag{20}$$

where $n_k$ is the stiffness factor equal to the tangential stiffness ratio to normal loading stiffness.

#### 2.1.4. Accounting for Cohesion with DEM Contact Models

In reality, agricultural soils exhibit varying levels of cohesion between particles and adhesion to walls and tools they come in contact with. Attractive pressure (that is, cohesive and adhesive forces) are induced due to the capillary effect and water bridge that exists between particles in unsaturated soils [25,51,52]. Thus, a more realistic contact model for agricultural soils should account for cohesion and adhesion. The linear cohesion and parallel bond models have been used in the DEM modelling of agricultural soils (Table 1).

##### Linear Cohesion Model

When this model is used, a cohesive or adhesive force is added to the normal force component of the contact model used for cohesionless soils. Even though the linear cohesion model itself does not include a tangential component, its addition increases the normal force, which consequently increases frictional force for greater resistance to slippage [41,50,52,53]. The linear cohesion model can be added to any of the three contact models discussed in Section 2.1.1 to Section 2.1.3 above [50]. If $F_{ca}$ (Equation (21)) is the cohesive or adhesive force, then the normal contact force is modified, as shown in Equation (22).

$$F_{ca} = r_c^2\pi\hat{c} \tag{21}$$

$$F_n = F_n^s + F_n^d + F_{ca} \tag{22}$$

$$r_c = \left(\frac{3r_{eq}F_n^s}{4E_{eq}}\right)^{\frac{1}{3}} \tag{23}$$

The parameter $r_c$ is the contact radius between particles and can be determined using Equation (23). Equation (21) is called the constant cohesion model because the cohesive stress $\hat{c}$ is a constant. The constant cohesion model makes the model particles too sticky [52]. A modification has therefore been proposed, depending on the degree of compression between two adjacent particles. If compression between two adjacent particles is given by Equation (24), then the cohesive stress increases with time $t$ according to Equation (25).

$$\sigma_n = \frac{F_n}{\pi r_c^2} \tag{24}$$

$$\hat{c}^{(t=n)} = k_c \max\left(\sigma_n^{(t=1)}, \sigma_n^{(t=2)}, \ldots ..\sigma_n^{(t=n-1)}\right) \tag{25}$$

Parallel Bond Model

An adaptation of the Hertz–Mindlin contact model (HMCM) for cohesive soils is the parallel bond model (PBM) developed by Potyondy and Cundall [54]. The PBM, based on beam theory, uses rectangular (2D) or cylindrical (3D) cement entities as parallel bonds at the point of contact between the two cohesive particles (Figure 5). This bond is modelled as an elastic beam whose length approaches zero and could be represented by a set of springs uniformly distributed over the contact plane and centred at the contact point [54]. After bond formation, normal and tangential bond forces and moment are calculated in addition to contact forces [50,55,56]. Thus, the bond can withstand or transmit both forces and moments between particles. The bond breaks when its predefined maximum normal or shear strengths are exceeded [57,58]. When no bond exists between particles, the PBM reverts to the HMCM [26]. The PBM is able to model clod formation and the brittle nature of agricultural soils in a more realistic manner [30,59]. It can be used only for particle-particle bonding, not particle-wall (tool) bonding [30,50]. The PBM is the most used model in cohesive soil tillage research [7,30,31,51,55–70]. Because the base contact model of the PBM is the HMCM, it also fails to predict vertical soil forces accurately as revealed in Table 2 [55,69]. Details of the PBM can be found in Potyondy and Cundall [54].

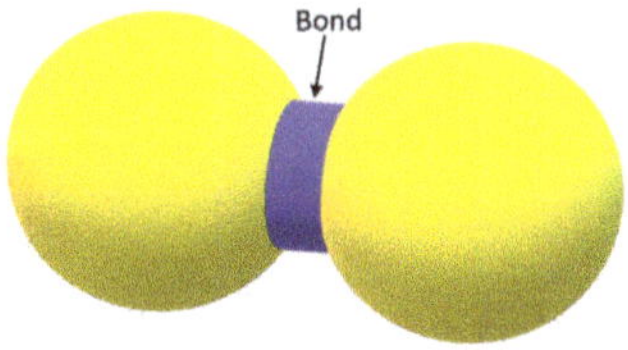


**Figure 5.** The parallel bond model depicted as cylindrical bond between two particles.

Johnson–Kendall–Roberts Cohesion

Another cohesion contact model combined with the HMCM is the Johnson–Kendall–Roberts (JKR) cohesion model [71]. It has been used by researchers such as Cheng et al. [33], Du et al. [36], and Zhai et al. [37] to model cohesive soils in soil tillage simulations with DEM (Table 1). Using this model, the tangential contact force and the normal and tangential damping forces of the HMCM are maintained, while the normal contact force is modified to include cohesion [36]. This modification enables the modelling of strongly adhesive bonds such as exist in dry powders or wet materials (e.g., wet soil). It captures the influence of Van der Waals forces due to contact between two surfaces [50]. A cohesion or adhesion parameter called surface energy is introduced. When this surface energy is zero, the model reverts to the HMCM. The normal contact force in HMCM-JKR is given by Equations (26) and (27), as follows:

$$F_n = \frac{4E_{eq}a^3}{3R_{eq}} - 4\sqrt{\pi\gamma E_{eq}a^3} \tag{26}$$

$$\delta_n = \frac{a^2}{R_{eq}} - 2\sqrt{\frac{\pi\gamma a}{E_{eq}}} \tag{27}$$

where $a$ is JKR contact radius and $\gamma$ is surface energy (J/m$^2$).

**Table 2.** Relative errors in DEM-predicted soil–tool reaction forces, travel speeds, and operating depths reported in the various literature across soil types.

| Reference | Relative Error in DEM Prediction (%) Relative to Measured Data | | Travel Speed (km $h^{-1}$) | Operating Depth (mm) | Tillage Tools | Soil Texture | Dry Bulk Density (kg $m^{-3}$) | Soil Water Content (%, *w/w*) | Cohesive Strength (kPa) | Contact Model |
|---|---|---|---|---|---|---|---|---|---|---|
| | Draught | Vertical Force | | | | | | | | |
| Sadek et al. [58] | n/a | n/a | n/a | n/a | n/a | Sandy soil | 990<br>1280<br>1360<br>1500 | 0.02<br>13<br>21.5 | 1.23-32.70 | PBM |
| Chen et al. [7] | 4 to 31 | n/a | 3.19 (average) | 100 | Sweep tine | Coarse sand<br>Loamy sand<br>Sandy loam | 1410<br>1330<br>1410 | 8.98<br>14.84<br>18.2 | 15.7<br>25.2<br>36 | PBM |
| Obermayr et al. [52] | n/a | n/a | 2.16–4.5 | 10–200 | Bulldozer blade | n/a | 1900 | n/a | 11.16 | LSCM + cohesion |
| Tamas et al. [30] | 4 to 12 | n/a | 1.8–8.64 | 200 | Sweep tine | Sandy soil | 1850 | 6.33 | 11.86 | PBM |
| Bravo et al. [18] | 9, 24 | n/a | - | 150–500 | Para-plough and mouldboard plough | Clay (Vertosol) | 1000<br>1200<br>1400 | 8<br>18<br>20<br>35 | 25–125 | LSCM + cohesion |
| Li et al. [56] | 3 to 15 | n/a | 3.6 | 180–260 | Subsoiler | n/a | n/a | 19 | n/a | PBM |
| Mak and Chen [61] | n/a | n/a | 2.2–6.59 | 50–200 | Sweep tine | Loamy sand | 1320 | 11.3 | 13.9 | PBM |
| Obermayr et al. [72] | n/a | n/a | | 100–200 | Straight-vertical blade and bulldozer blade | Sand | 1520<br>1980<br>1870 | 10<br>15 | 6–22.5 | LSCM + cohesion |
| Ucgul et al. [38] | ≤11.6 | ≤15.2 | 5–12.5 | 70 | Sweep tine | Sandy loam | 1750 | 8 | 6 | HSCM |
| Ucgul et al. [53] | n/a | n/a | 4–12 | 75 | Sweep tine | Sandy loam | 1320<br>1780<br>1880 | 1<br>15<br>13 | 3<br>15<br>22 | HSCM + LCM |
| Kotrocz et al. [60] | n/a | n/a | n/a | 50–150 | Cone penetrometer | Loamy sand | 1632 | 15.8 | 6.61–8.66 | PBM |
| Li et al. [70] | | | 2.99 | 3–18 | Claw | Sandy loam | 1300 | n/a | 17.5 | PBM |
| Murray [69] | 1.86 | 50.7 | 8 | 38 | Disc and hoe openers | Clayey lacustrine | 1560 | 19.6 | n/a | PBM |
| Hang et al. [32] | n/a | n/a | 3 | 300 | Subsoiler | Loamy clay | 1346 | 12.5 | 11.8 | PBM |
| Milkevych et al. [62] | n/a | n/a | 3.2 | 100 | Sweep tine | Coarse sand<br>Loamy sand | 1410<br>1330 | 9<br>14.8 | 15.8<br>25.1 | PBM |
| Tekeste et al. [55] | 9, 12 | -59, -49 | 0.79–9.65 | 102 | Sweep tine | Loam | 1307 | 8.99 | 33 | PBM |
| Tong et al. [73] | <10 | <10 | 7.2 | 300–450 | Subsoiler (straight shank-sweep tine, curved shank-chisel tine, curved shank-sweep tine, bentleg-chisel tine) | n/a | 1230–1420 | n/a | n/a | not stated |

**Table 2.** *Cont.*

| Reference | Relative Error in DEM Prediction (%) Relative to Measured Data | | Travel Speed (km $h^{-1}$) | Operating Depth (mm) | Tillage Tools | Soil Texture | Dry Bulk Density (kg $m^{-3}$) | Soil Water Content (%, *w*/*w*) | Cohesive Strength (kPa) | Contact Model |
|---|---|---|---|---|---|---|---|---|---|---|
| | Draught | Vertical Force | | | | | | | | |
| Kim et al. [44] | 5.16 to 9.9 | n/a | 7.64–7.9 | 5–200 | Mouldboard plough | Loam | 1496–1904 | 24.5–34.02 | n/a | EEPA |
| Aikins et al. [41] | 5 to 31 | 8, 20 and greater | 8 | 100 | Bentleg and narrow point openers | Clay (Vertosol) | 1504 | 23.7 | 46.4 | HSCM + LCM |
| Wang et al. [74] | 15.08 | n/a | 3 | 300 | Winged subsoiler | Sandy loam | 1404–1833 | n/a | n/a | PBM |
| Sadek et al. [75] | ≤20.2 | n/a | 4–16 | 127 | Disc | Sandy loam | 1700 | 16.32 | n/a | PBM |
| Saunders et al. [76] | n/a | n/a | 4.5–10 | 25–100 | Plough skimmers | Sandy loam | 1523.8 | 8.3 | n/a | HSCM + LCM |
| Ma et al. [77] | 2.88 to 5.97 | n/a | 1.08–2.16 | 120 | Scraper | Sandy loam | 1389 | 10 | n/a | not stated |
| Hoseinian et al. [35] | 2 | 2.5 | 0.9 | 150 | Dual sideway-share | Sandy clay loam | 1565 | 11.5 | 15.4 | PBM |

Edinburgh Elasto-Plastic Adhesion Model

Consolidation stress history is one of the main sources of cohesion in cohesive granular materials, and it must be accounted for to accurately model such materials in DEM [78]. The Edinburgh elasto-plastic adhesion model (EEPA) contact model uses a non-linear hysteretic spring model to account for the elastic–plastic contact deformation and an adhesive or cohesive force (pull-off strength) component—acting between dissimilar or similar materials, respectively based on the assumption that this force increases with increasing plastic contact area [50,79]. This model is versatile because, depending on its input parameters, it can be used as either a linear spring model (Figure 6a) or a non-linear Hertzian spring model (Figure 6b) [79]. Figure 6 shows "a schematic diagram of particle contact and normal force-overlap ($f_n$-$\delta$) curve" for the EEPA contact model. A full description of the EEPA contact model can be found in Morrissey et al. [78]. The EEPA contact model has been used recently for modelling the interaction between tillage tools and agricultural soils [44–46].

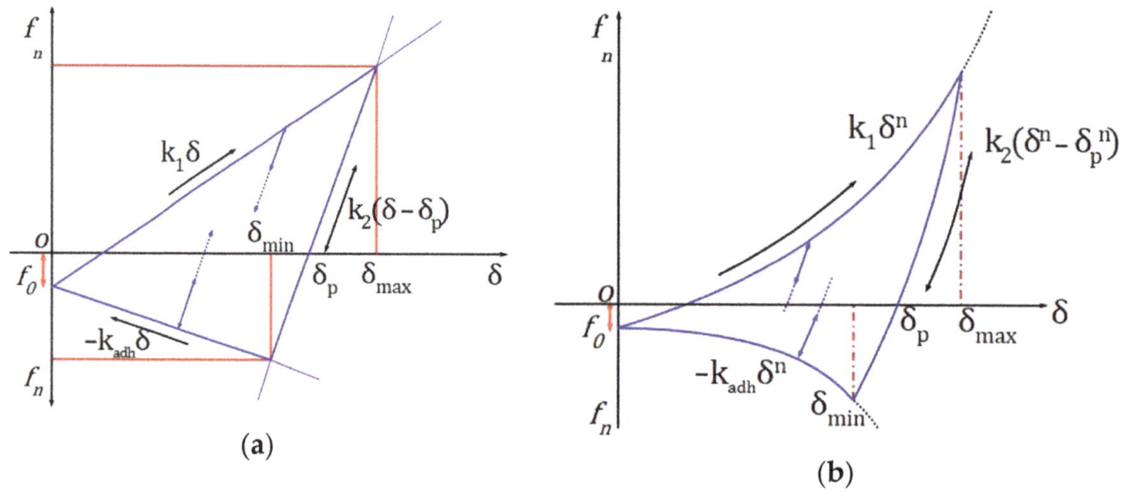


**Figure 6.** Edinburgh elasto-plastic adhesion model (EEPA) normal contact force-displacement relationship (**a**) linear and (**b**) non-linear [50].

### 2.2. Particle Size and Shape

Particle shape and size used in DEM significantly affect the necessary simulation time and the accuracy of simulation results [10,28,38]. They are input parameters that should be carefully chosen during calibration for DEM particle interactions to be as close to reality as possible [20,49]. In DEM simulations, it is ideal to use particles of similar sizes to the actual granular materials being modelled. For instance, the actual sizes of agricultural soil particles are relatively small, ranging from several nanometres to about 2 mm for very coarse sand [80]. To model actual particle sizes in DEM requires unrealistically long computation time and impractically high computer processing power [78]. The most time-consuming part of soil particle DEM simulations is contact detection, and is proportional to the number of particles [81]. For this reason, larger particle sizes than real soil particle sizes are generally employed in DEM [20,38,76]. The larger particles are sometimes implied to represent soil aggregates instead of individual soil particles and somewhat capture the bulk behaviour of a structured soil profile [78].

In reality, soil particles come in various irregular shapes. Thus, particles used in DEM should be not only of a similar size range but also of a similar shape range to actual soil particles to ensure simulations are more representative of realistic bulk behaviour. The basic shape of a DEM particle is a sphere (or circle in 2D modelling) under most DEM codes [20]. Spherical particles approximate and simplify simulations by improving contact detection efficiency and reducing computation time [82]. Usually, irregular (non-spherical) particles are created by clumping a number of spherical particles together, as shown in Figure 7 [28,32,46,83]. This enables the use of spherical particle contact detection algorithms that are simpler and require shorter computation time than those of irregular shapes [24,84].

**Figure 7.** Different associations of spherical DEM particles used to more realistically account for the effects of soil particle shape on bulk behaviour.

Nonetheless, clumped particles also require relatively higher computational time than when purely spherical particles are used. Therefore, most studies adopt spherical particles to represent soil particles or soil aggregates in DEM simulations. Spherical particle assemblies simulating the soil profile are characterized by considerably lower internal friction and shear strength than actual soil particles due to lower impact of rolling friction. However, this is usually overcome by introducing an arbitrary high rolling friction coefficient to simulate the interlocking tendencies that exist between the irregular shape soil particles [83,85].

## 3. Calibration Techniques for Determining DEM Input Parameters

Running a DEM model involves providing it with such input parameters aimed at simulating soil behaviour as close to real soils as possible. Accurate results can only be obtained with accurate input parameters [57,86]. Several approaches exist for calibrating DEM input parameters that accurately represent both soil to soil particle properties and soil particle to tool or machine interface properties. The most common calibration methods for the former include the angle of repose and hopper discharge, direct shear and triaxial tests, and corresponding in situ soil measurements. The most common calibration methods for the latter include the inclined plane test, the modified shear test, and corresponding in situ measurements. All these approaches are focused on bulk responses (i.e., natural stable state and force reactions) of soil under an applied load. After experimental runs in the laboratory or field, these experiments are then replicated numerically as closely as possible, optimizing parameters iteratively until bulk numerical responses agree with field or laboratory measurements [20]. Trial and error methods have traditionally been relied upon in the past while, more recently, the application of response surface methodology (RSM) is demonstrating benefits of significantly reducing the number of numerical simulations required for accurate calibration [34,45,87–91].

### 3.1. Angle of Repose Test

The angle of repose test is used to assess flowability and inter-particle friction of loose soil [92–94]. This test is also essential when there is a need for a qualitative assessment of soil surface and furrow profiles in tillage simulations [29]. Various researchers [21,42,84,88,95] used the angle of repose test to calibrate coefficients of static and rolling friction between soil particles.

In this test, the soil is allowed to flow by gravity onto a flat surface to form a cone pile. The angle of repose is measured as shown in Figure 8a [29]. Another approach for the angle of repose test is to confine the particles being modelled within the walls of a box, ensuring the top of the particles is levelled. By removing one of the sidewalls, the particles flow to form the angle of repose as shown in Figure 8b [83]. This test is usually used for cohesionless particles and particles with low cohesion with good flowability. However, the general principle is that repeatable observations can be made during key stages of the "angle of repose" experiment. For instance, Roessler and Katterfeld [96] reported successfully calibrating DEM parameters for cohesive soil using the angle of repose test. A cylinder was filled with the cohesive soil, and the cylinder was gradually lifted as shown in Figure 8c. Reproducible phases of soil flow were observed, namely "the build-up of a stable bulk material column, the convex bending of the column, and the beginning of collapse of the column." Aikins et al. [41] observed a reproducible dome-like pile of cohesive soil (Figure 9a,b) and used the results to calibrate soil–soil coefficients of static and rolling friction values.

(a) (b) (c)

**Figure 8.** Angle of repose test setup used by (**a**) Ucgul et al. [29] for a cohesionless soil, (**b**) Mousavi-raad et al. [83] for maize, and (**c**) Roessler and Katterfeld [96] for a cohesive soil.

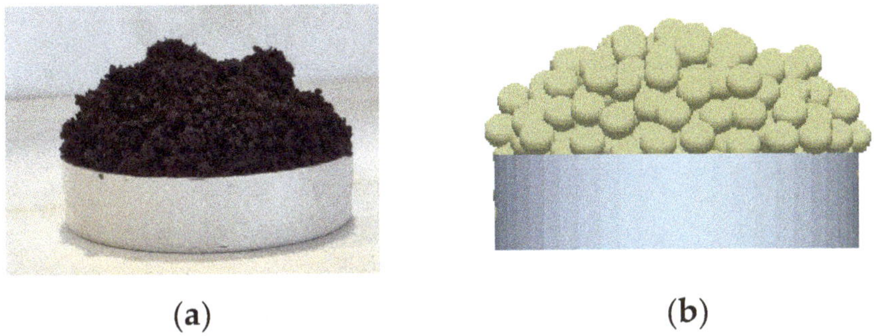


**Figure 9.** Repeatable dome-like soil pile obtained in the angle of repose test for a loose cohesive soil by Aikins et al. [41]: (**a**) laboratory experiment and (**b**) optimized DEM simulation.

### *3.2. Inclined Plane Test*

The inclined plane test has been used to determine soil–tool or soil–machine coefficients of static and rolling friction [29]. A schematic diagram of the setup for the inclined plane test is shown in Figure 10. A flat bed of the soil to be modelled is packed into a tray and held on a table with adjustable horizontal inclination. A block of tool material and ball bearings are separately placed on the flat bed, and the table is tilted to an angle $\Psi$ at which the block just starts sliding or the ball just starts rolling down the inclination. The block is used for the determination of the soil–tool coefficient of static friction ($\mu_s$), while the ball bearing is used in the determination of the soil–tool coefficient of rolling friction ($\mu_r$). If the mass of the block is $m_s$, the mass of the ball is $m_r$, and the angles at which sliding and rolling occur are $\Psi_s$ and $\Psi_r$, respectively, then the coefficients are calculated according to Equations (28) and (29) [29,97].

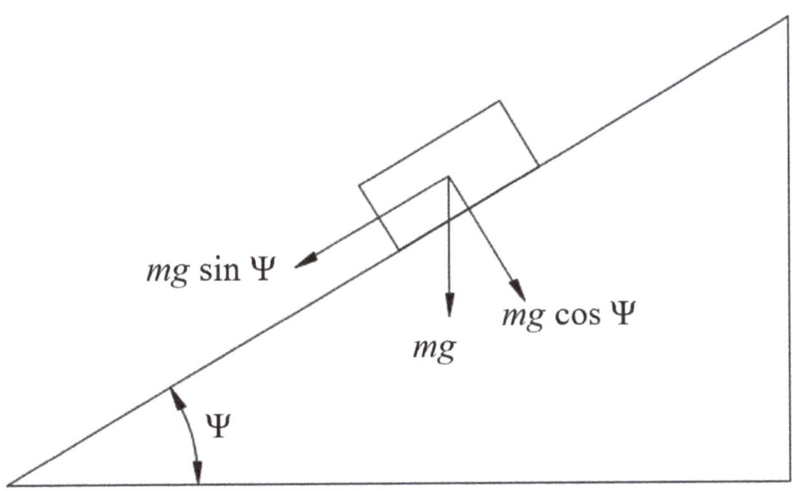


**Figure 10.** Schematic diagram of inclined plane test setup with a block [97] for static friction coefficient. A ball can be used as well for rolling friction coefficient as found in Ucgul et al. [29].

Soil–tool coefficient of static friction,

$$\mu_s = \frac{m_s g \sin \Psi_s}{m_s g \cos \Psi_s} = \tan \Psi_s \tag{28}$$

Soil–tool coefficient of rolling friction,

$$\mu_r = \frac{m_r g \sin \Psi_r}{m_r g \cos \Psi_r} = \tan \Psi_r \quad (29)$$

### *3.3. Direct Shear Test*

The direct shear test is used to determine internal soil parameters namely, cohesion and internal friction angle (for soil-to-soil particle interactions). The modified shear test is used to determine the adhesion and external friction angle (for soil to tool or machine interface properties). These are typically used as direct DEM input parameters to support the calibration of other parameters. It has been used for cohesive and adhesive soils, as well as cohesionless soils [41,44,53].

This approach uses the normal and shear stresses acting on a column of granular materials' cross-section. The experimental setup consists of two shear boxes, one placed on the other and filled with the granular material being modelled. One half is fixed while the other is made movable horizontally in one direction (Figure 11a). A specified normal force ($F_a$) is applied while an increasing horizontal (shearing) force ($F_b$) is applied to the movable half till a certain amount of displacement occurs [98]. At that point, the horizontal force would have reached a maximum value and remain constant or slightly increase or decrease afterward [30]. The experiment is repeated several times with different normal force values.

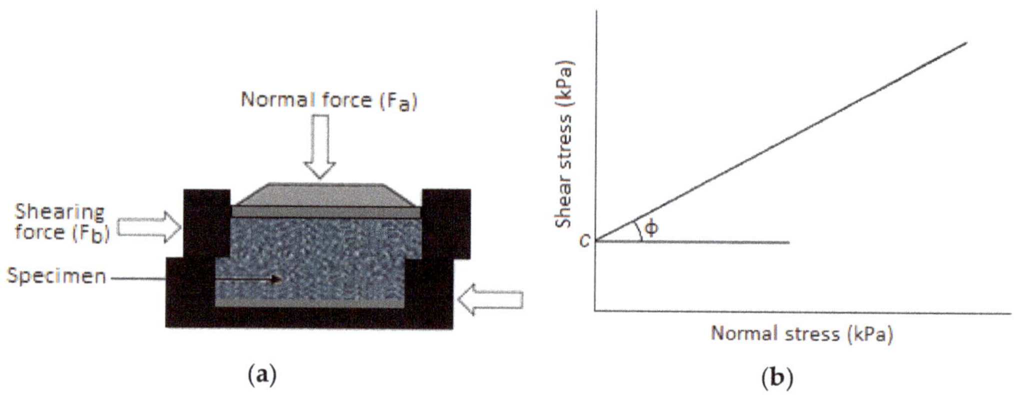


**Figure 11.** (**a**) Direct shear box schematic, and (**b**) normal and shear stress expected relationship.

In the modified shear box test, the bottom half of the shear box is replaced with a material matching that of the tool or machine part. Normal force and shearing forces are applied the same way as described above. Corresponding normal and shear stresses are plotted as shown in Figure 11b. Given that the cohesive strength of the soil is $c$, the cross-sectional area of the shear box is $A$, and the internal friction angle is $\phi$, the relationship between the normal force and the horizontal force is displayed in Equation (30).

$$F_b = cA + F_a \tan \phi \quad (30)$$

### *3.4. Triaxial Compression Test*

This test can also determine the soil cohesive strength and internal friction angle and used to calibrate DEM input parameters for cohesive-frictional soils [52,72]. A triaxial compression test consists of loading an undisturbed cylindrical soil specimen insulated with an impermeable membrane and subjected to an adjustable confining pressure within a water chamber [18,99,100]. The specimen is then subjected to combined axial ($\sigma_1$) and radial

($\sigma_3$) stresses as indicated in Figure 12 until soil failure is achieved [98]. The radial stress ($\sigma_3$) is first applied around the specimen to a set level via the confining water pressure. An axial strain is then mechanically applied at a controlled rate which generates a corresponding additional deviator stress ($q$) logged over time and combining with $\sigma_3$ to form a resultant axial stress $\sigma_1$. The above steps are repeated several times under increasing radial stress. The plots of the deviator stress ($q = \sigma_1 - \sigma_3$) against axial strain identify each deviator stress value at failure and a simple process—for instance using the Mohr circle method—is then used to quantify soil cohesion and internal friction angle [101].

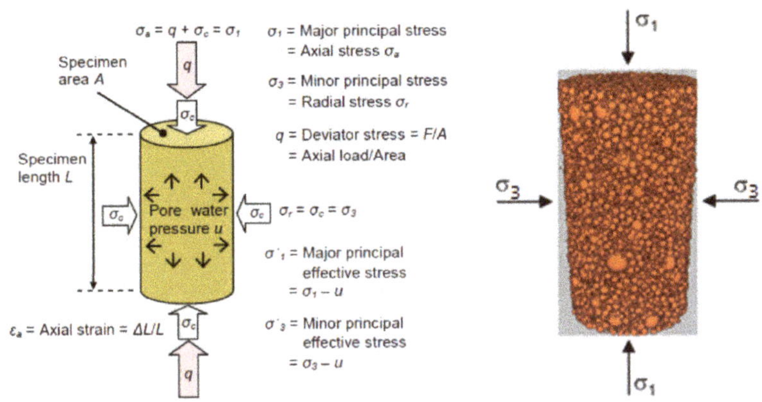


(a) (b)

**Figure 12.** Specimen stress state in a triaxial compression test: retrieved from (**a**) Rees [99] and (**b**) Ahlinhan et al. [102].

### 3.5. In Situ Approaches

Measurements of soil mechanical properties are most accurately done in the laboratory [103]. However, while laboratory methods can precisely measure soil properties, samples may not always be fully representative of field soil conditions due to sampling and handling limitations and time-related changes between sampling and testing. Hence, some researchers have used in situ approaches to measure soil properties for DEM calibration purposes. Kim et al. [44] used an on-site measurement system to determine soil mechanical properties such as shear modulus, Young's modulus, and soil–tool static and rolling friction (Figure 13). On the other hand, Aikins et al. [41] used an on-site cone penetration test (Figure 14) to calibrate Young's modulus of a Black Vertosol.

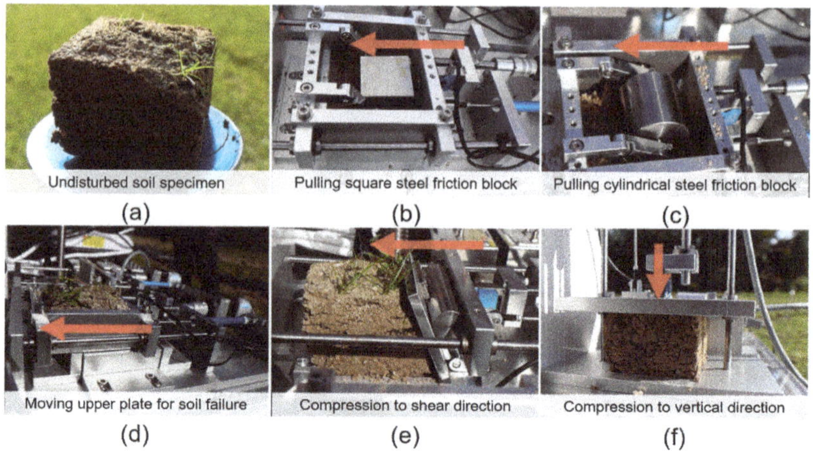


**Figure 13.** Instruments for in situ measurements of soil mechanical properties: (**a**) sampled undisturbed soil specimen, (**b**) soil–metal static friction, (**c**) soil–metal rolling friction, (**d**) direct shear test, (**e**) shear modulus, and (**f**) Young's modulus. Retrieved from Kim et al. [44].

**Figure 14.** (**a**) Motorised cone penetrometer for in situ measurements, and (**b**) DEM cone penetration simulation used by Aikins et al. [41].

Asaf et al. [10] proposed grouser shear and sinkage or penetration tests using wedges of different wedge angles and a plate for calibrating DEM contact parameters. Jang et al. [82] also used a rectangular plate, while Ucgul et al. [38] and Ucgul et al. [29] used circular disc and cone penetration tests to calibrate model parameters. Cheng et al. [33] used a soil adhesion mass test to determine DEM input parameters of wet clay soil by employing the Plackett–Burman test and response surface methodology (RSM) to optimise input parameters.

## 4. Prediction of Soil Failure, Loosening, and Disturbance Parameters

### 4.1. Soil Failure and Loosening

Tamas et al. [30] and Barr et al. [21] have revealed the ability of DEM to predict soil rupture and crack propagation, which is an advantage over FEM. Some researchers have used velocity profiles [7,32] or displacement profiles [41,69,104] as soil loosening indicators. Others [21,30] used porosity (in PFC3D Particle Flow Code) or voidage (in EDEM 2.7$^{TM}$), respectively, to measure the degree of particles loosening in DEM. In the work of Tamas et al. [30], for instance, it was found that the DEM modelled soil porosity and soil-break-up resulting from loosening by sweeps increased with both increasing speed and rake angle, which agrees with experimental results.

Identifying particle movement or loosening is mainly used in defining the boundary between disturbed and undisturbed particles to simulate soil failure boundary or furrow profile. Barr et al. [21] argued that using velocity and displacement profiles is based on the assumption that particle movement results in only soil loosening, ignoring the fact that particle movement also occurs during a soil compaction process. The validity of this assumption is therefore limited to tools operating above their critical depth. Additionally, these approaches are open to subjective decisions since a threshold has to be arbitrarily defined to differentiate between the "so-called" loosened and unloosened particles. For example, Murray [69] had to describe loosened particles as having a displacement magnitude above 5 mm. Barr et al. [21] instead proposed and used a voidage grid (Figure 15) to define failure boundaries. A voidage grid was applied to the DEM particles after the tillage process was completed and the particles had settled, which reflects experimental practice. Voidage is similar to soil porosity as it measures the proportion of volume not occupied by particles. An empty space will have a voidage of 100%, while a completely filled space will have a voidage of 0%. With $V_g$ being grid volume and $V_p$ the total volume of particles whose centroids are located within the grid, voidage can be calculated according to Equation (31).

$$Voidage = \frac{V_g - V_p}{V_g} \times 100\% \quad (31)$$

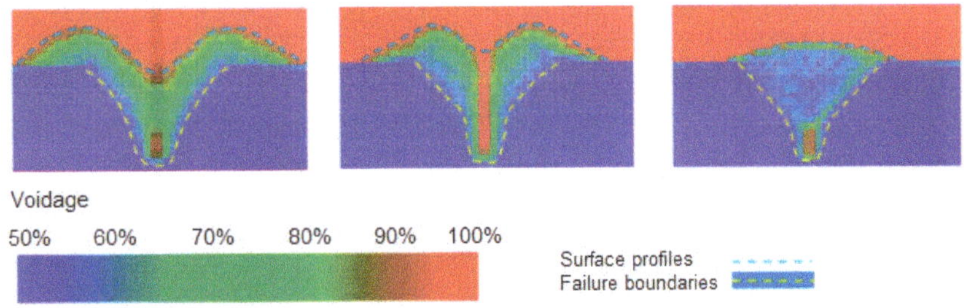


**Figure 15.** Furrow profiles created using voidage grid bins in EDEM 2.7$^{TM}$. Adapted from Barr [49].

Aikins et al. [41] used particle displacement (PD) analysis to determine the loosened furrow boundary in DEM. The displacement threshold was not set arbitrarily as was done by Murray [69]. Aikins et al. [41] defined the loosened furrow boundary based on two criteria:

1. Minimum particle displacement caused directly by an opener occurs with particles just adjacent to the bottom part of the opener (for wide tines) or particles aligning the walls of the slot below critical depth (for narrow tines).

2. To establish a sharp contrast between displaced and undisturbed particles, particle locations immediately after particle loosening (i.e., before the particle settle) has to be used.

Aikins et al. [41] traced the minimum particle displacement up the profile to produce a loosened furrow boundary as shown in Figure 16a. Figure 16a is a contour plot of the width and depth of the virtual soil bin profile against displacements (resultant) for each particle within the profile. In some studies [21,32], disturbed soil surface profile after tillage was determined using voidage grid binning and velocity profile. However, Aikins et al. [41] used the profile of the top surface of displaced DEM particles as the disturbed surface profile after the particles had settled because that gives more realistic results and is similar to what actually happens in field experiments. Wang et al. [74] employed another approach using the "clipping" module in EDEM simulation software to define the disturbed soil boundary (Figure 16b). The furrow profile was obtained by connecting the boundaries of the different layers of disturbed soil.

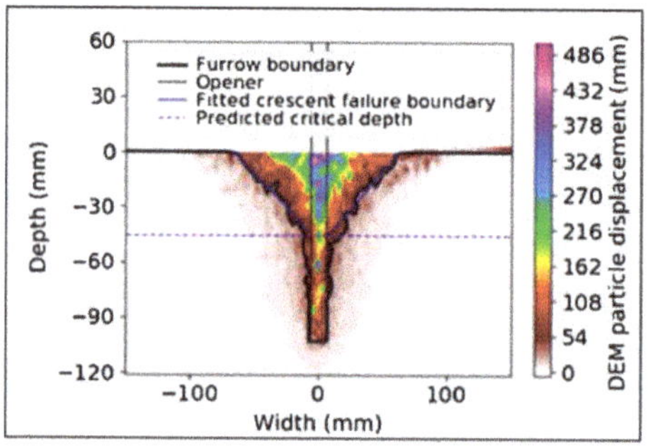


(a)

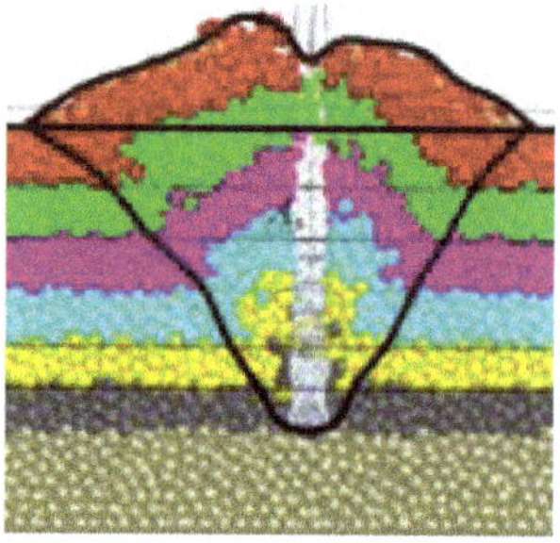

(b)

**Figure 16.** Definitions of furrow boundary used by (**a**) Aikins et al. [41] and (**b**) Wang et al. [74].

*4.2. Soil Movement and Disturbance Parameters*

From the loosened criteria described in the previous section, furrow profile, soil movement, and various soil disturbance parameters have been predicted or determined in DEM simulations with varying levels of relative error (RE) compared to experimental results. Such soil disturbance parameters include the lateral, forward, and upward movement of particles; furrow width at soil surface; loosened furrow cross-sectional area; furrow % backfill and dip area. Barr et al. [21] found an RE of 9% in loosened furrow cross-sectional area, 26% in furrow width, 14% in dip area, 0.8% in furrow backfill, 16% in ridge height, and 9% in lateral soil throw in a DEM prediction of soil disturbance parameters with narrow point openers operating in a sandy-loam soil. Barr and Fielke [105] closely predicted lateral soil throw and soil layer mixing using narrow tine openers with 35° and 90° rake angles and a bentleg opener (Figure 17). Using furrow openers with different rake angles and cutting edge cross-sections, Aikins et al. [41] closely predicted furrow profiles and similar patterns for surface profiles. The majority of DEM predictions of furrow cross-sectional area, furrow width, critical depth, and lateral soil throw had an RE of 1% to 20%. However, poor predictions were made for ridge height due to the use of large DEM particles (radius of 5 mm) [41].

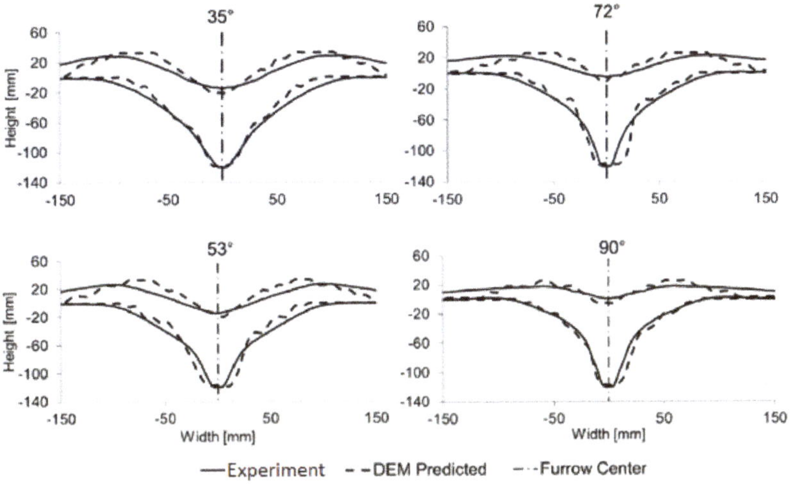


**Figure 17.** Comparison of furrow profiles for 16 mm wide narrow openers at different rake angles (35°, 53°, 72°, and 90°) obtained from soil bin experiment and DEM simulations. Retrieved and modified from Barr [49].

Wang et al. [74] determined soil looseness, furrow width, soil disturbance coefficient, and soil disturbance area ratio with an RE from 3.24% to 41.64% for a winged subsoiler and 0.24% to 28.74% for a non-winged subsoiler. There was also a satisfactory agreement in "shape and magnitude" of lateral, forward, and upward displacement of different soil layers resulting from a sweep cultivator between experimental and DEM simulation results [62]. Using a disc with tilt angles from 0° to 20°, Murray [69] estimated an average absolute RE of about 10.53% for lateral soil throw. The DEM simulation also revealed, in agreement with the experimental result, that lateral soil throw increased with increasing tilt angle. For a hoe furrow opener, a RE of 14.8% was recorded.

Reduction in the forward movement of soil particles at greater depth has been predicted through DEM simulations [21,32,41]. Other researchers have previously described this phenomenon [9,106–108] through experimental work and analytical models in the concept of critical depth. Below the critical depth, soil movement changes from forward, sideways, and upward directions (generating a loosened crescent-shaped soil failure) to mainly forward and sideways, generating a compaction type failure in a horizontal plane. Barr et al. [21] and Aikins et al. [41] closely predicted the narrowing of furrow down the profile and critical depth with a 90° rake angle (Figures 15 and 16a). Hang et al. [32] observed a reduction of the forward movement of particles in the inter-row zone between tines with increasing tine spacing in both DEM and experimental results. Greater inter-row zone soil movement with narrower tine spacing was attributed to more intensive interaction between tines. DEM can also simulate greater soil movement with wing attachment to tine cutting tools [31].

Greater soil upheaval observed with a low rake angle opener (35°) and increased lateral throw of soil (due to splashing effect) typically found with steeper (90°) rake angle openers have been successfully replicated with DEM [21]. However, Ucgul et al. [38] observed that the dynamic height of soil flow under sweep openers was under-predicted by 23% to 35% at speeds of 5 to 12.5 km/h and did not follow the observed shape using spherical particles of 10 mm radii. The prediction was improved by using smaller particles of radii 1.5 mm, which still underpredicted soil flow height by 15%. Using particles of smaller radii, however, considerably increased computation time. Lateral soil throw was also under predicted by about 32% and 9% with radii 10 and 1.5 mm, respectively.

Chen et al. [7] estimated a maximum of 3% RE in furrow cross-sectional area, up to 4% RE in disturbed width, from 14% to 26% more soil (by volume) heaped above the soil surface and 5% to 15% more emptied cross-section below the soil surface. Overall, a close agreement was observed between experimental and simulation results. Hang et al. [32] estimated less than 20% RE between DEM predicted and experimentally determined soil disturbance coefficient and soil looseness. Saunders et al. [76] reported significant underprediction and correlations between measured and predicted furrow area (r = 0.82) and maximum soil throw (r = 0.88) when optimizing the performance of a mouldboard skimmer in a sandy-loam soil. Several furrow parameters from bentleg openers operating in sandy-loam soil, including loosened cross-sectional area (RE = 14.9%), furrow dip area (RE = 14.4%), backfill (RE = 1.8%), ridge height (RE = 16.8%), and lateral soil throw (RE = 14.9%), were accurately predicted using the voidage grid bin approach [39]. Furthermore, Barr et al. [39] observed the same findings as those measured in soil bin investigations, with DEM simulations of bentleg openers also achieving 100% backfill and cancelling furrow spill over.

The DEM has also been used to simulate rotary tiller operations for design optimization. Zhang et al. [109] and Hirasawa et al. [110] closely predicted the height and pattern of soil surface undulations after rotary tillage in DEM. Soil movement pattern during rotary tillage and improving soil layer mixing with tillage depth and travel speed for a rotary tiller have also been closely predicted [36,109]. Cheng et al. [33] recorded an RE of 1.84% in mass of soil that adhered to rotary tiller blades.

## 5. Prediction of Tillage Forces

Most attention in the DEM simulation of soil cutting tools has been on predicting soil forces. Table 2 shows relative error (RE) values in draught and vertical force predictions with DEM, travel speed and operating depths for the stated tillage tools, soil types, and DEM contact models. Draught force cycles between peaks at incipient soil failure and troughs at the start of the reloading phase have been well captured in DEM simulations of sweeps by Tamas et al. [30]. In the same DEM study, draught was found to increase with greater speed and rake angle. Draught was predicted with 4% to 12% RE as speed was increased from 0.5 to 2.4 m $s^{-1}$ with a sweep tine. Bo et al. [31] observed a similar trend between draught force measured for four subsoilers in the soil bin and that obtained through DEM simulation. A winged subsoiler among the four had the highest draught force in both soil bin tests (up to 50% more) and simulations (up to 55% more). With all four subsoilers, DEM predicted draught force with relative errors below 4%.

Additionally, Wang et al. [74] showed that a winged subsoiler operating at a speed and depth of 3 km $h^{-1}$ and 300 mm, respectively, had an RE of 9.71%, whereas the non-winged subsoiler obtained an RE of 15.08% when the draught force was compared with the experimental result. Chen et al. [7] observed about 4–31% RE between the draught of experimental and DEM results. A good correlation was obtained between the measured and predicted draught forces (r = 0.95), whereas a more limited correlation was observed for vertical force (r = 0.71). With blunt (R90B) and chamfered (C2S) narrow openers with a 90° rake angle, a blunt opener with a 45° rake angle, and a bentleg opener, Aikins et al. [41] predicted draught force with REs of 20%, 22%, 31%, and 5%, respectively. Vertical force was also predicted with 8% and 20% relative error for R90B and C2S, respectively, but poorly for the two other openers.

DEM simulations with a mouldboard plough were also able to simulate the gradual entry of the mouldboard into the soil with a gradual draught increase. For the two soil conditions used in this study, 9% and 2.4% errors in cultivator tool draught were observed for a soft-wet soil and a hard-dry soil, respectively. DEM also closely predicted an increase in draught with increasing depth, with RE ranging from about 3% to 15% [56]. Kim et al. [44] observed that draught force increased with increasing depth with an overall average RE of 7.45%. Tong et al. [73] showed that, across four tillage depths, simulated draught and vertical forces were up to 10% smaller than those measured in the field.

In a DEM prediction of horizontal and vertical soil forces with DEM particles clumped to form different shapes (similar to that shown in Figure 7), Ono et al. [28] obtained the most accurate predictions with the three linearly overlapping spheres. The worst prediction was obtained with simple spherical particles. Ucgul et al. [38] observed a linear increase in draught force against sweep tine width measured experimentally and predicted using DEM with a maximum RE of 8%. Likewise, a non-linear increase in vertical force against width with a maximum RE of 13.7% was recorded. High correlations were recorded between measured and predicted draught forces (r = 0.978) and vertical forces (r = 0.971) with tool speeds from 5 to 12.5 km $h^{-1}$ and a depth of 70 mm. Prediction of the effect of rake angle on soil forces followed a similar trend (r values of 0.98 and 0.97) and had an RE of 11.6% and 15.2% for draught and vertical forces, respectively. Ucgul et al. [111] again obtained an accurate prediction of draught and vertical forces of a sweep tillage tool at varying speeds and geometry with r values ranging from 0.84 to 0.92. Murray [69] estimated an average RE of 1.86% for draught and 50.7% for vertical force with a flat single disc opener. For rotary tillers, Zhang et al. [109] reported a 12% RE in power consumption, while Du et al. [36] predicted increasing torque with tillage depth (150 to 180 mm) and travel speed (about 2 to 3 km $h^{-1}$).

## 6. Soils Modelled in DEM Simulations

Tables 1 and 2 list soil types used in various tillage and furrow opener DEM simulations and their bulk densities, soil water contents, and cohesive strengths. It can be seen that most of the soils modelled with DEM are of sandy to sandy-loam textures. DEM modelling of highly cohesive soils is still relatively scarce in the literature. Bravo et al. [18] used DEM to model highly cohesive clay soil (Vertosol) with cohesive strength of up to about 125 kPa when the soil was highly compacted (bulk density of about 1400 kg $m^{-3}$) and relatively dry (soil water content of about 18%). Aikins et al. [41] also modelled a Vertosol with cohesive strength of 46.4 kPa.

The properties and flow characteristics of sandy soils differ from those of clay soils, which show cohesive and adhesive properties in the presence of sufficient moisture. The successful modelling of a Vertosol and its interaction with tools in DEM by Bravo et al. [18] and Aikins et al. [41] revealed the ability of DEM to model cohesive soils. Although some authors [33,41,43,44,104,112,113] have recently modelled cohesive soils using DEM, more attention is needed in future research to cover the wide spectrum of agricultural soils.

## 7. Conclusions

Based on this review, the following conclusions can be drawn:

1. Even though the Hertz–Mindlin contact model (HMCM) has been used in most DEM studies of tillage and furrow opening, it consistently fails to predict vertical soil force accurately. The Hysteretic Spring contact model (HSCM) can more accurately predict soil forces and particle movement.
2. Angle of repose, inclined plane, direct shear, triaxial compression, and some in situ tests (grouser shear, plate sinkage, and cone penetration tests) have been used to measure and calibrate DEM input parameters. The angle of repose test has been used mainly for cohesionless soils due to the poor flowability of cohesive soils. However, using results from reproducible phases of the angle of repose experiment, successful calibrations for cohesive soils have been achieved.
3. Unlike other numerical models, DEM is able to closely predict not only soil forces, but it is also capable of modelling soil failure mechanisms, soil loosening, and soil particle movement. Soil rupture and crack propagation, critical depth, three-dimensional particle movement within the soil profile and lateral particle movement on top of the soil have all been predicted in DEM.

4. Using voidage or porosity grids to determine loosened furrow cross-sectional profiles has been found to be superior to using particle velocity and displacement profiles. However, some researchers have successfully used a particle displacement approach to determine accurate furrow profiles with a more objective criteria for defining loosened furrow boundary.
5. Close predictions of draught and vertical forces ($\leq$20%) have been obtained with DEM. These predictions can be improved by using smaller particles of a near-real shape. However, this must be balanced with computation time requirements.

Based on the review conducted, the following recommendations are made for future research:

1. The Edinburgh elasto-plastic adhesion model (EEPA) has been successfully used to model consolidated or cohesive powders. This contact model is recommended to be studied more extensively for cohesive soils, although some researchers have used it.
2. Due to pore water pressure within wet and soft soils, coupling DEM and CFD is likely to produce more accurate simulations. This idea can be explored in future research.
3. A comprehensive analysis of soil disturbance parameters has been successfully done using voidage grids in EDEM® DEM software. Replication of this approach in other DEM software is recommended.
4. The criteria introduced by Aikins et al. [41] for defining particle displacement threshold for DEM furrow profile identification need further investigation with particles of smaller radii than the 5 mm used in the study. This approach can provide greater details on the three-dimensional soil translocation process.

**Author Contributions:** Writing—original draft preparation, K.A.A.; writing—review and editing, K.A.A., M.U., J.B.B., E.A., D.L.A., T.A.J. and J.M.A.D.; visualization, K.A.A. and E.A.; supervision, M.U., J.B.B., D.L.A., T.A.J. and J.M.A.D.; project administration, D.L.A., T.A.J. and J.M.A.D.; funding acquisition, K.A.A. All authors have read and agreed to the published version of the manuscript.

**Funding:** This research was funded by CLAAS Stiftung (Harsewinkel, Germany) and the University of Southern Queensland (Toowoomba, QLD, Australia).

**Institutional Review Board Statement:** Not applicable.

**Informed Consent Statement:** Not applicable.

**Data Availability Statement:** Not applicable.

**Acknowledgments:** The authors are grateful to the Centre for Agricultural Engineering at the University of Southern Queensland (Toowoomba, Qld, Australia), the Agricultural Machinery Research and Design Centre at the University of South Australia (Adelaide, SA, Australia), CSIRO Agriculture and Food (Canberra, ACT, Australia), and the CLAAS Foundation (Harsewinkel, Germany, http://www.claas-stiftung.com/ accessed on 17 February 2023) for financial and operational support to conduct this review.

**Conflicts of Interest:** The authors declare no conflict of interest.

## Nomenclature

| | |
|---|---|
| $\gamma$ | Surface energy (J/m$^2$) |
| $ffi_n$ | Linear overlap |
| $\dot{ffi}_n$ | Normal component of relative velocity |
| $\dot{ffi}_t$ | Tangential component of relative velocity |
| $\hat{c}$ | Cohesive stress |
| $\dot{ffi}_n$ | Linear relative velocity |
| $ffi_t$ | Tangential component of overlap |
| $\bar{}_t$ | Friction coefficient |
| ∅ | Internal friction angle between the particles (Degree) |
| $\mu_r$ | Coefficient of rolling friction |
| $\mu_s$ | Coefficient of static friction |
| $A$ | Cross-sectional area of the shear box |
| a | JKR contact radius |
| $c$ | Soil cohesion (Pa) |
| $c_a$ | Soil–metal adhesion (Pa) |
| $d$ | Working depth (m) |
| $d_n$ | Damping coefficient |
| $d_t$ | Tangential component of damping coefficient |
| $e$ | Coefficient of restitution of the particles |
| E | Young's modulus |
| $E_{eq}$ | Equivalent Young's modulus |
| $F$ | Contact force |
| $F_a$ | Normal force in direct shear test |
| $F_b$ | Horizontal (shearing) force in direct shear test |
| $F_{ca}$ | Cohesive or adhesive force |
| $F^d$ | Damping force |
| $F_n$ | Normal contact force |
| $F^s$ | Spring force |
| $F_t$ | Tangential component of the contact force |
| $g$ | Acceleration due to gravity |
| $G_{eq}$ | Equivalent shear modulus |
| $I_i$ | Moment of inertia of a particle |
| $k_1$ | Loading stiffnesses |
| $k_2$ | Unloading stiffnesses |
| $k_n$ | Normal stiffness |
| $k_t$ | Tangential component of stiffness |
| $m_{eq}$ | Equivalent particle mass |
| $m_i$ | Mass of spherical particle |
| $m_r$ | Mass of the ball used in inclined plane test |
| $m_s$ | Mass of block used in inclined plane test |
| $N$ | N factor. Suffixes: $\gamma$ = gravitational, c = cohesive, a = adhesive, q = surcharge |
| $P$ | Soil cutting force (N) |
| $q$ | Surcharge stress (Pa) |
| $q$ | Deviator stress in triaxial compression test |
| $r_c$ | Contact radius |
| $r_{eq}$ | Equivalent particle radius |
| $r_i$, | Radius of spherical particle |
| $T_i$ | Torque due to the tangential component of the contact force |
| $V_g$ | Voidage grid volume |

| | |
|---|---|
| $V_p$ | Total volume of particles with centroids within voidage grid |
| $w$ | Tool width (m) |
| $x_i$ | Location of spherical particle |
| $\gamma$ | Specific weight of soil (N $m^{-3}$) |
| $\varepsilon_a$ | Axial strain in triaxial compression test |
| $\sigma_1$ | Axial stress in triaxial compression test |
| $\sigma_3$ | Radial stress in triaxial compression test |
| $\Psi$ | Inclined plane tilt angle. Subscripts: s = sliding, r = rolling |
| $\omega_i$ | Angular velocity of a particle |

## References


1. Aybek, A.; Baser, E.; Arslan, S.; Ucgul, M. Determination of the effect of biodiesel use on power take-off performance characteristics of an agricultural tractor in a test laboratory. *Turk. J. Agric. For.* **2011**, *35*, 103–113. [CrossRef]
2. Kushwaha, R.L.; Zhang, Z.X. Evaluation of factors and current approaches related to the computerized design of tillage tools: A review. *J. Terramechanics* **1998**, *35*, 69–86. [CrossRef]
3. Aikins, K.A.; Barr, J.B.; Ucgul, M.; Jensen, T.A.; Antille, D.L.; Desbiolles, J.M.A. No-tillage furrow opener performance: A review of tool geometry, settings and interactions with soil and crop residue. *Soil Res.* **2020**, *58*, 603–621. [CrossRef]
4. McKyes, E. *Soil Cutting and Tillage*; Development in Agricultural Engineering Volume 7; Elsevier: Amsterdam, The Netherlands, 1985; Volume 7.
5. Godwin, R.J. A review of the effect of implement geometry on soil failure and implement forces. *Soil Tillage Res.* **2007**, *97*, 331–340. [CrossRef]
6. Shmulevich, I.; Asaf, Z.; Rubinstein, D. Interaction between soil and a wide cutting blade using the discrete element method. *Soil Tillage Res.* **2007**, *97*, 37–50. [CrossRef]
7. Chen, Y.; Munkholm, L.J.; Nyord, T. A discrete element model for soil-sweep interaction in three different soils. *Soil Tillage Res.* **2013**, *126*, 34–41. [CrossRef]
8. Hettiaratchi, D.R.P.; Reece, A.R. Symmetrical three-dimensional soil failure. *J. Terramechanics* **1967**, *4*, 45–67. [CrossRef]
9. Godwin, R.J.; Spoor, G. Soil failure with narrow tines. *J. Agric. Eng. Res.* **1977**, *22*, 213–228. [CrossRef]
10. Asaf, Z.; Rubinstein, D.; Shmulevich, I. Determination of discrete element model parameters required for soil tillage. *Soil Tillage Res.* **2007**, *92*, 227–242. [CrossRef]
11. Shmulevich, I. State of the art modeling of soil–tillage interaction using discrete element method. *Soil Tillage Res.* **2010**, *111*, 41–53. [CrossRef]
12. McKyes, E.; Ali, O.S. The cutting of soil by narrow blades. *J. Terramechanics* **1977**, *14*, 43–58. [CrossRef]
13. Godwin, R.J.; O'Dogherty, M.J. Integrated soil tillage force prediction models. *J. Terramechanics* **2007**, *44*, 3–14. [CrossRef]
14. Fielke, J.M. Finite element modelling of the interaction of the cutting edge of tillage implements with soil. *J. Agric. Eng. Res.* **1999**, *74*, 91–101. [CrossRef]
15. Karmakar, S.; Ashrafizadeh, S.R.; Kushwaha, R.L. Experimental validation of computational fluid dynamics modeling for narrow tillage tool draft. *J. Terramechanics* **2009**, *46*, 277–283. [CrossRef]
16. Cundall, P.A.; Strack, O.D.L. A discrete numerical model for granular assemblies. *Geotechnique* **1979**, *29*, 47–65. [CrossRef]
17. Franco, Y.; Rubinstein, D.; Shmulevich, I. Prediction of soil-bulldozer blade interaction using discrete element method. *Trans. ASABE* **2007**, *50*, 345–353. [CrossRef]
18. Bravo, E.L.; Tijskens, E.; Suarez, M.H.; Cueto, O.G.; Ramon, H. Prediction model for non-inversion soil tillage implemented on discrete element method. *Comput. Electron. Agric.* **2014**, *106*, 120–127. [CrossRef]
19. Markauskas, D.; Ramírez-Gómez, Á.; Kačianauskas, R.; Zdancevičius, E. Maize grain shape approaches for DEM modelling. *Comput. Electron. Agric.* **2015**, *118*, 247–258. [CrossRef]
20. Coetzee, C.J. Review: Calibration of the discrete element method. *Powder Technol.* **2017**, *310*, 104–142. [CrossRef]
21. Barr, J.B.; Ucgul, M.; Desbiolles, J.M.A.; Fielke, J.M. Simulating the effect of rake angle on narrow opener performance with the discrete element method. *Biosyst. Eng.* **2018**, *171*, 1–15. [CrossRef]
22. Peng, B. Discrete Element Method (DEM) Contact Models Applied to Pavement Simulation. Unpublished. Master's Thesis, Virginia Polytechnic Institute and State University, Blacksburg, Virginia, 2014.
23. Luding, S. Introduction to discrete element methods: Basics of contact force models and how to perform the micro-macro transition to continuum theory. *Eur. J. Environ. Civ. Eng.* **2008**, *12*, 785–826. [CrossRef]
24. Horabik, J.; Molenda, M. Parameters and contact models for DEM simulations of agricultural granular materials. *Biosyst. Eng.* **2016**, *147*, 206–229. [CrossRef]
25. Tamás, K.; Bernon, L. Role of particle shape and plant roots in the discrete element model of soil–sweep interaction. *Biosyst. Eng.* **2021**, *211*, 77–96. [CrossRef]
26. Ucgul, M. Simulation of Sweep Tillage Using Discrete Element Modelling. Unpublished. Ph.D. Thesis, University of South Australia, Adelaide, Australia, 2014.

27. Tanaka, H.; Momozu, M.; Oida, A.; Yamazaki, M. Simulation of soil deformation and resistance at bar penetration by the Distinct Element Method. *J. Terramechanics* **2000**, *37*, 41–56. [CrossRef]
28. Ono, I.; Nakashima, H.; Shimizu, H.; Miyasaka, J.; Ohdoi, K. Investigation of elemental shape for 3D DEM modeling of interaction between soil and a narrow cutting tool. *J. Terramechanics* **2013**, *50*, 265–276. [CrossRef]
29. Ucgul, M.; Fielke, J.M.; Saunders, C. Three-dimensional discrete element modelling of tillage: Determination of a suitable contact model and parameters for a cohesionless soil. *Biosyst. Eng.* **2014**, *121*, 105–117. [CrossRef]
30. Tamas, K.; Jori, I.J.; Mouazen, A.M. Modelling soil-sweep interaction with discrete element method. *Soil Tillage Res.* **2013**, *134*, 223–231. [CrossRef]
31. Bo, L.; Rui, X.; Liu, F.Y.; Jun, C.; Han, W.T.; Bing, H. Determination of the draft force for different subsoiler points using discrete element method. *Int. J. Agr. Biol. Eng.* **2016**, *9*, 81–87. [CrossRef]
32. Hang, C.; Gao, X.; Yuan, M.; Huang, Y.; Zhu, R. Discrete element simulations and experiments of soil disturbance as affected by the tine spacing of subsoiler. *Biosyst. Eng.* **2018**, *168*, 73–82. [CrossRef]
33. Cheng, J.; Zheng, K.; Xia, J.; Liu, G.; Jiang, L.; Li, D. Analysis of adhesion between wet clay soil and rotary tillage part in paddy field based on discrete element method. *Processes* **2021**, *9*, 845. [CrossRef]
34. Yang, Y.; Wen, B.; Ding, L.; Li, L.; Chen, X.; Li, J. Soil Particle Modeling and Parameter Calibration for Use with Discrete Element Method. *Trans. ASABE* **2021**, *64*, 2011–2023. [CrossRef]
35. Hoseinian, S.H.; Hemmat, A.; Esehaghbeygi, A.; Shahgoli, G.; Baghbanan, A. Development of a dual sideway-share subsurface tillage implement: Part 1. Modeling tool interaction with soil using DEM. *Soil Tillage Res.* **2022**, *215*, 105201. [CrossRef]
36. Du, J.; Heng, Y.; Zheng, K.; Luo, C.; Zhu, Y.; Zhang, J.; Xia, J. Investigation of the burial and mixing performance of a rotary tiller using discrete element method. *Soil Tillage Res.* **2022**, *220*, 105349. [CrossRef]
37. Zhai, S.; Shi, Y.; Zhou, J.; Liu, J.; Huang, D.; Zou, A.; Jiang, P. Simulation optimization and experimental study of the working performance of a vertical rotary tiller based on the discrete element method. *Actuators* **2022**, *11*, 324. [CrossRef]
38. Ucgul, M.; Fielke, J.M.; Saunders, C. 3D DEM tillage simulation: Validation of a hysteretic spring (plastic) contact model for a sweep tool operating in a cohesionless soil. *Soil Tillage Res.* **2014**, *144*, 220–227. [CrossRef]
39. Barr, J.; Desbiolles, J.; Ucgul, M.; Fielke, J.M. Bentleg furrow opener performance analysis using the discrete element method. *Biosyst. Eng.* **2020**, *189*, 99–115. [CrossRef]
40. Makange, N.R.; Ji, C.; Torotwa, I. Prediction of cutting forces and soil behavior with discrete element simulation. *Comput. Electron. Agric.* **2020**, *179*, 105848. [CrossRef]
41. Aikins, K.A.; Ucgul, M.; Barr, J.B.; Jensen, T.A.; Antille, D.L.; Desbiolles, J.M.A. Determination of discrete element model parameters for a cohesive soil and validation through narrow point opener performance analysis. *Soil Tillage Res.* **2021**, *213*, 105123. [CrossRef]
42. Awuah, E.; Zhou, J.; Liang, Z.; Aikins, K.A.; Gbenontin, B.V.; Mecha, P.; Makange, N.R. Parametric analysis and numerical optimisation of Jerusalem artichoke vibrating digging shovel using discrete element method. *Soil Tillage Res.* **2022**, *219*, 105344. [CrossRef]
43. Wang, X.; Zhang, Q.; Huang, Y.; Ji, J. An efficient method for determining DEM parameters of a loose cohesive soil modelled using hysteretic spring and linear cohesion contact models. *Biosyst. Eng.* **2022**, *215*, 283–294. [CrossRef]
44. Kim, Y.-S.; Siddique, M.A.A.; Kim, W.-S.; Kim, Y.-J.; Lee, S.-D.; Lee, D.-K.; Hwang, S.-J.; Nam, J.-S.; Park, S.-U.; Lim, R.-G. DEM simulation for draft force prediction of moldboard plow according to the tillage depth in cohesive soil. *Comput. Electron. Agric.* **2021**, *189*, 106368. [CrossRef]
45. Wu, Z.; Wang, X.; Liu, D.; Xie, F.; Ashwehmbom, L.G.; Zhang, Z.; Tang, Q. Calibration of discrete element parameters and experimental verification for modelling subsurface soils. *Biosyst. Eng.* **2021**, *212*, 215–227. [CrossRef]
46. Zhao, Z.; Li, H.; Liu, J.; Yang, S.X. Control method of seedbed compactness based on fragment soil compaction dynamic characteristics. *Soil Tillage Res.* **2020**, *198*, 104551. [CrossRef]
47. Sun, J.; Chen, H.; Wang, Z.; Ou, Z.; Yang, Z.; Liu, Z.; Duan, J. Study on plowing performance of EDEM low-resistance animal bionic device based on red soil. *Soil Tillage Res.* **2020**, *196*, 104336. [CrossRef]
48. Dai, F.; Song, X.; Zhao, W.; Shi, R.; Zhang, F.; Zhang, X. Mechanism analysis and performance improvement of mechanized ridge forming of whole plastic film mulched double ridges. *Int. J. Agric. Biol. Eng.* **2020**, *13*, 107–116. [CrossRef]
49. Barr, J. Optimising bentleg opener geometry for higher speed no-till seeding. Unpublished Ph.D. Thesis, University of South Australia, Mawson Lakes, South Australia, Australia, 2018.
50. EDEM. *EDEM 2020 Documentation*; DEM Solutions: Edinburgh, UK, 2020.
51. Tamás, K.; Kovács, Á.; Jóri, I.J. The evaluation of the parallel bond's properties in DEM modeling of soils. *Period. Polytech. Mech. Eng.* **2016**, *60*, 21–31. [CrossRef]
52. Obermayr, M.; Vrettos, C.; Eberhard, P. A discrete element model for cohesive soil. In Proceedings of the III International Conference on Particle-Based Methods—Fundamentals and Applications, Stuttgart, Germany, 18–20 September 2013; pp. 783–794.
53. Ucgul, M.; Fielke, J.M.; Saunders, C. Three-dimensional discrete element modelling (DEM) of tillage: Accounting for soil cohesion and adhesion. *Biosyst. Eng.* **2015**, *129*, 298–306. [CrossRef]
54. Potyondy, D.O.; Cundall, P.A. A bonded-particle model for rock. *Int. J. Rock Mech. Min. Sci.* **2004**, *41*, 1329–1364. [CrossRef]

55. Tekeste, M.Z.; Balvanz, L.R.; Hatfield, J.L.; Ghorbani, S. Discrete element modeling of cultivator sweep-to-soil interaction: Worn and hardened edges effects on soil-tool forces and soil flow. *J. Terramechanics* **2019**, *82*, 1–11. [CrossRef]
56. Li, B.; Liu, F.Y.; Mu, J.Y.; Chen, J.; Han, W.T. Distinct element method analysis and field experiment of soil resistance applied on the subsoiler. *Int. J. Agric. Biol. Eng.* **2014**, *7*, 54–59. [CrossRef]
57. Mak, J.; Chen, Y.; Sadek, M.A. Determining parameters of a discrete element model for soil–tool interaction. *Soil Tillage Res.* **2012**, *118*, 117–122. [CrossRef]
58. Sadek, M.A.; Chen, Y.; Liu, J. Simulating shear behavior of a sandy soil under different soil conditions. *J. Terramechanics* **2011**, *48*, 451–458. [CrossRef]
59. Van der Linde, J. Discrete Element Modeling of a Vibratory Subsoiler. Unpublished MSc. Thesis, University of Stellenbosch, Matieland, South Africa, 2007.
60. Kotrocz, K.; Mouazen, A.M.; Kerenyi, G. Numerical simulation of soil-cone penetrometer interaction using discrete element method. *Comput. Electron. Agric.* **2016**, *125*, 63–73. [CrossRef]
61. Mak, J.; Chen, Y. Simulation of draft forces of a sweep in a loamy sand soil using the discrete element method. *Can.Biosyst.Eng.* **2014**, *56*, 2.1–2.7. [CrossRef]
62. Milkevych, V.; Munkholm, L.J.; Chen, Y.; Nyord, T. Modelling approach for soil displacement in tillage using discrete element method. *Soil Tillage Res.* **2018**, *183*, 60–71. [CrossRef]
63. Sadek, M.; Chen, Y. Feasibility of using PFC3D to simulate soil flow resulting from a simple soil-engaging tool. *Trans. ASABE* **2015**, *58*, 987–996. [CrossRef]
64. Zeng, Z.; Chen, Y. Simulation of soil-micropenetrometer interaction using the discrete element method (DEM). *Trans. ASABE* **2016**, *59*, 1157–1163. [CrossRef]
65. Zeng, Z.; Chen, Y.; Zhang, X. Modelling the interaction of a deep tillage tool with heterogeneous soil. *Comput. Electron. Agric.* **2017**, *143*, 130–138. [CrossRef]
66. Zeng, Z.W.; Chen, Y. Simulation of straw movement by discrete element modelling of straw-sweep-soil interaction. *Biosyst. Eng.* **2019**, *180*, 25–35. [CrossRef]
67. Zhang, R.; Li, J. Simulation on mechanical behavior of cohesive soil by Distinct Element Method. *J. Terramechanics* **2006**, *43*, 303–316. [CrossRef]
68. Gupta, V.; Sun, X.; Xu, W.; Sarv, H.; Farzan, H. A discrete element method-based approach to predict the breakage of coal. *Adv. Powder Technol.* **2017**, *28*, 2665–2677. [CrossRef]
69. Murray, S. Modelling of Soil-Tool Interactions Using the Discrete Element Method. Unpublished MSc. Thesis, University of Manitoba, Winnipeg, MB, Canada, 2016.
70. Li, B.; Chen, Y.; Chen, J. Modeling of soil-claw interaction using the discrete element method (DEM). *Soil Tillage Res.* **2016**, *158*, 177–185. [CrossRef]
71. Johnson, K.L.; Kendall, K.; Roberts, A.D. Surface energy and the contact of elastic solids. *Proc. R. Soc. Lond. Ser. A Math. Phys. Sci.* **1971**, *324*, 301–313.
72. Obermayr, M.; Vrettos, C.; Eberhard, P.; Dauwel, T. A discrete element model and its experimental validation for the prediction of draft forces in cohesive soil. *J. Terramechanics* **2014**, *53*, 93–104. [CrossRef]
73. Tong, J.; Jiang, X.-H.; Wang, Y.-M.; Ma, Y.-H.; Li, J.-W.; Sun, J.-Y. Tillage force and disturbance characteristics of different geometric-shaped subsoilers via DEM. *Adv. Manuf.* **2020**, *8*, 392–404. [CrossRef]
74. Wang, X.; Li, P.; He, J.; Wei, W.; Huang, Y. Discrete element simulations and experiments of soil-winged subsoiler interaction. *Int. J. Agric. Biol. Eng.* **2021**, *14*, 50–62. [CrossRef]
75. Sadek, M.A.; Chen, Y.; Zeng, Z. Draft force prediction for a high-speed disc implement using discrete element modelling. *Biosyst. Eng.* **2021**, *202*, 133–141. [CrossRef]
76. Saunders, C.; Ucgul, M.; Godwin, R.J. Discrete element method (DEM) simulation to improve performance of a mouldboard skimmer. *Soil Tillage Res.* **2021**, *205*, 104764. [CrossRef]
77. Ma, S.; Niu, C.; Yan, C.; Tan, H.; Xu, L. Discrete element method optimisation of a scraper to remove soil from ridges formed to cold-proof grapevines. *Biosyst. Eng.* **2021**, *210*, 156–170. [CrossRef]
78. Morrissey, J.P.; Thakur, S.C.; Ooi, J.Y. *EDEM Contact Model: Adhesive Elasto-Plastic Model*; School of Engineering, University of Edinburgh: Edinburgh, UK, 2014.
79. Thakur, S.C.; Morrissey, J.P.; Sun, J.; Chen, J.F.; Ooi, J.Y. Micromechanical analysis of cohesive granular materials using the discrete element method with an adhesive elasto-plastic contact model. *Granul Matter* **2014**, *16*, 383–400. [CrossRef]
80. Schoeneberger, P.J.; Wysocki, D.A.; Benham, E.C.; Soil Survey Staff. *Field Book for Describing and Sampling Soils*; version 3.0; Natural Resources Conservation Service, National Soil Survey Center: Lincoln, NE, USA, 2012.
81. Morrissey, J.P. Discrete Element Modelling of Iron Ore Pellets to Include the Effects of Moisture and Fines. Unpublished Ph.D. Thesis, University of Edinburgh, Edinburgh, UK, 2013.
82. Jang, G.; Lee, S.; Lee, K.J. Discrete element method for the characterization of soil properties in Plate-Sinkage tests. *J. Mech. Sci. Technol.* **2016**, *30*, 2743–2751. [CrossRef]

83. Mousaviraad, M.; Tekeste, M.; Rosentrater, K. Discrete element method (DEM) simulation of corn grain flow in commercial screw auger. In Proceedings of the 2016 American Society of Agricultural and Biological Engineers Annual International Meeting, Orlando, FL, USA, 17–20 July 2016.
84. Coetzee, C.J. Calibration of the discrete element method and the effect of particle shape. *Powder Technol.* **2016**, *297*, 50–70. [CrossRef]
85. Smith, W.; Melanz, D.; Senatore, C.; Iagnemma, K.; Peng, H.E. Comparison of discrete element method and traditional modeling methods for steady-state wheel-terrain interaction of small vehicles. *J. Terramechanics* **2014**, *56*, 61–75. [CrossRef]
86. Adajar, J.B.; Alfaro, M.; Chen, Y.; Zeng, Z. Calibration of discrete element parameters of crop residues and their interfaces with soil. *Comput. Electron. Agric.* **2021**, *188*, 106349. [CrossRef]
87. Hernández-Vielma, C.; Estay, D.; Cruchaga, M. Response surface methodology calibration for DEM study of the impact of a spherical bit on a rock. *Simul. Model. Pract. Theory* **2022**, *116*, 102466. [CrossRef]
88. Mudarisov, S.; Farkhutdinov, I.; Khamaletdinov, R.; Khasanov, E.; Mukhametdinov, A. Evaluation of the significance of the contact model particle parameters in the modelling of wet soils by the discrete element method. *Soil Tillage Res.* **2022**, *215*, 105228. [CrossRef]
89. Qiu, Y.; Guo, Z.; Jin, X.; Zhang, P.; Si, S.; Guo, F. Calibration and Verification Test of Cinnamon Soil Simulation Parameters Based on Discrete Element Method. *Agriculture* **2022**, *12*, 1082. [CrossRef]
90. Zeng, F.; Li, X.; Zhang, Y.; Zhao, Z.; Cheng, C. Using the discrete element method to analyze and calibrate a model for the interaction between a planting device and soil particles. *INMATEH Agric. Eng.* **2021**, 413–424. [CrossRef]
91. Zhu, J.; Zou, M.; Liu, Y.; Gao, K.; Su, B.; Qi, Y. Measurement and calibration of DEM parameters of lunar soil simulant. *Acta Astronaut.* **2022**, *191*, 169–177. [CrossRef]
92. Al-Hashemi, H.M.B.; Al-Amoudi, O.S.B. A review on the angle of repose of granular materials. *Powder Technol.* **2018**, *330*, 397–417. [CrossRef]
93. Qi, L.; Chen, Y.; Sadek, M. Simulations of soil flow properties using the discrete element method (DEM). *Comput. Electron. Agric.* **2019**, *157*, 254–260. [CrossRef]
94. Li, P.; Ucgul, M.; Lee, S.H.; Saunders, C. A new approach for the automatic measurement of the angle of repose of granular materials with maximal least square using digital image processing. *Comput. Electron. Agric.* **2020**, *172*, 105356. [CrossRef]
95. Rackl, M.; Hanley, K.J. A methodical calibration procedure for discrete element models. *Powder Technol.* **2017**, *307*, 73–83. [CrossRef]
96. Roessler, T.; Katterfeld, A. DEM parameter calibration of cohesive bulk materials using a simple angle of repose test. *Particuology* **2019**, *45*, 105–115. [CrossRef]
97. Aikins, K.A. Performance of Bentleg Opener for High Speed No-Tillage Sowing in Cohesive Soil. Unpublished. Ph.D. Thesis, University of Southern Queensland, Toowoomba, QLD, Australia, 2020.
98. Syed, Z.; Tekeste, M.; White, D. A coupled sliding and rolling friction model for DEM calibration. *J. Terramechanics* **2017**, *72*, 9–20. [CrossRef]
99. Rees, S. *Introduction to Triaxial Testing: Part 1*; GDS Instruments: Hampshire, UK, 2013.
100. Wang, Y.; Guo, P.; Dai, F.; Li, X.; Zhao, Y.; Liu, Y. Behavior and Modeling of Fiber-Reinforced Clay under Triaxial Compression by Combining the Superposition Method with the Energy-Based Homogenization Technique. *Int. J. Geomech.* **2018**, *18*, 04018172. [CrossRef]
101. Bardet, J.P. *Experimental Soil Mechanics*; Prentice-Hall: Upper Saddle River, NJ, USA, 1997.
102. Ahlinhan, M.F.; Houehanou, E.; Koube, M.B.; Doko, V.; Alaye, Q.; Sungura, N.; Adjovi, E. Experiments and 3D DEM of Triaxial Compression Tests under Special Consideration of Particle Stiffness. *Geomaterials* **2018**, *8*, 39–62. [CrossRef]
103. Das, B.M. *Advanced Soil Mechanics*, 5th ed.; CRC Press: Boca Raton, FL, USA, 2019. [CrossRef]
104. Aikins, K.A.; Barr, J.B.; Antille, D.L.; Ucgul, M.; Jensen, T.A.; Desbiolles, J.M.A. Analysis of effect of bentleg opener geometry on performance in cohesive soil using the discrete element method. *Biosyst. Eng.* **2021**, *209*, 106–124. [CrossRef]
105. Barr, J.; Fielke, J. Discrete element modelling of narrow point openers to improve soil disturbance characteristics of no-till seeding systems. In Proceedings of the ASABE Annual International Meeting, Orlando, FL, USA, 17–20 July 2016.
106. Spoor, G.; Godwin, R.J. An experimental investigation into the deep loosening of soil by rigid tines. *J. Agric. Eng. Res.* **1978**, *23*, 243–258. [CrossRef]
107. Spoor, G.; Fry, R.K. Soil disturbance generated by deep-working low rake angle narrow tines. *J. Agric. Eng. Res.* **1983**, *28*, 217–234. [CrossRef]
108. Conte, O.; Levien, R.; Debiasi, H.; Sturmer, S.L.K.; Mazurana, M.; Muller, J. Soil disturbance index as an indicator of seed drill efficiency in no-tillage agrosystems. *Soil Tillage Res.* **2011**, *114*, 37–42. [CrossRef]
109. Zhang, X.; Zhang, L.; Hu, X.; Wang, H.; Shi, X.; Ma, X. Simulation of soil cutting and power consumption optimization of a typical rotary tillage soil blade. *Appl. Sci.* **2022**, *12*, 8177. [CrossRef]
110. Hirasawa, K.; Kataoka, T.; Kubo, T. Prediction and evaluation for leveling performance in rotary tiller. In Proceedings of the 4th IFAC Conference on Modelling and Control in Agriculture, Espoo, FL, USA, 27–30 August 2013; pp. 315–320.
111. Ucgul, M.; Fielke, J.M.; Saunders, C. Defining the effect of sweep tillage tool cutting edge geometry on tillage forces using 3D discrete element modelling. *Inf. Process.* **2015**, *2*, 130–141. [CrossRef]

112. Aikins, K.A.; Antille, D.L.; Ucgul, M.; Jensen, T.A.; Barr, J.B.; Desbiolles, J.M.A. Analysis of effects of operating speed and depth on bentleg opener performance in cohesive soil using the discrete element method. *Comput. Electron. Agric.* **2021**, *187*, 106236. [CrossRef]
113. Karkala, S.; Davis, N.; Wassgren, C.; Shi, Y.; Liu, X.; Riemann, C.; Yacobian, G.; Ramachandran, R. Calibration of Discrete-Element-Method Parameters for Cohesive Materials Using Dynamic-Yield-Strength and Shear-Cell Experiments. *Processes* **2019**, *7*, 278. [CrossRef]

MDPI
St. Alban-Anlage 66
4052 Basel
Switzerland
Tel. +41 61 683 77 34
Fax +41 61 302 89 18
www.mdpi.com

*Agriculture* Editorial Office
E-mail: agriculture@mdpi.com
www.mdpi.com/journal/agriculture

www.ingramcontent.com/pod-product-compliance
Lightning Source LLC
Chambersburg PA
CBHW061456220926
51850CB00004B/101

*9783036572949*